Unsteady Aerodynamics, Aeroacoustics, and Aeroelasticity of Turbomachines and Propellers

Sixth International Symposium .

on

UNSTEADY AERODYNAMICS, AEROACOUSTICS AND AEROELASTICITY OF TURBOMACHINES AND PROPELLERS

presented at the

University of Notre Dame

September 15-19, 1991

under the sponsorship of

The National Aeronautics and Space Administration

Lewis Research Center

The General Electric Company

Aircraft Engine Division

United Technologies Research Center

The University of Notre Dame

Center for Applied Mathematics

College of Engineering

Department of Aerospace and Mechanical Engineering

Office of Advanced Studies and Research

Symposium Chairman

H. M. Atassi

The International Scientific Committee

H.M. Atassi (U. S. A.)
A. Bölcs (Switzerland)
H. Forsching (Germany)
T. H. Fransson (Sweden)
J.E. Ffowcs Williams (England)
H.E. Gallus (Germany)
M.E. Goldstein (U. S. A.)
M. Namba (Japan)
E. Széchényi (France)
Y. Tanida (Japan)
J.M. Verdon (U. S. A.)
D.S. Whitehead (England)

H.M. Atassi
Editor

Unsteady Aerodynamics, Aeroacoustics, and Aeroelasticity of Turbomachines and Propellers

With 450 Figures

Springer-Verlag
New York Berlin Heidelberg London Paris
Tokyo Hong Kong Barcelona Budapest

H.M. Atassi
Aerospace and Mechanical Engineering Department
Fitzpatrick Hall
University of Notre Dame
Notre Dame, IN 46556-5637 USA

Library of Congress Cataloging-in-Publication Data
Unsteady aerodynamics, aeroacoustics, and
aeroelasticity of turbomachines and propellers / H.M. Atassi, editor.
p. cm.
Includes bibliographical references.
ISBN 0-387-97977-8. — ISBN 3-540-97977-8
1. Turbomachines—Aerodynamics. 2. Unsteady flow (Fluid dynamics)
I. Atassi, H. M.
TJ267.A37 1993
629.132′3—dc20 92-43206

Printed on acid-free paper.

Production managed by Hal Henglein; manufacturing supervised by Jacqui Ashri.
Camera-ready copy prepared by the contributors.
Printed and bound by Edwards Brothers, Inc., Ann Arbor, MI.
Printed in the United States of America.

9 8 7 6 5 4 3 2 1

ISBN 0-387-97977-8 Springer-Verlag New York Berlin Heidelberg
ISBN 3-540-97977-8 Springer-Verlag Berlin Heidelberg New York

Preface

The first International Symposium on Unsteady Aerodynamics and Aeroelasticity of Turbomachines was held in Paris in 1976, and was followed by symposia at Lausanne in 1980, Cambridge in 1984, Aachen in 1987, Beijing in 1989, and Notre Dame in 1991. The proceedings published following these symposia have become recognized both as basic reference texts in the subject area and as useful guides to progress in the field. It is hoped that this volume, which represents the proceedings of the Sixth International Symposium on Unsteady Aerodynamics of Turbomachines, will continue that tradition.

Interest in the unsteady aerodynamics, aeroacoustics, and aeroelasticity of turbomachines has been growing rapidly since the Paris symposium. This expanded interest is reflected by a significant increase in the numbers of contributed papers and symposium participants. The timeliness of the topics has always been an essential objective of these symposia. Another important objective is to promote an international exchange between scientists and engineers from universities, government agencies, and industry on the fascinating phenomena of unsteady turbomachine flows and how they affect the aeroelastic stability of the blading system and cause the radiation of unwanted noise. This exchange acts as a catalyst for the development of new analytical and numerical models along with carefully designed experiments to help understand the behavior of such systems and to develop predictive tools for engineering applications.

Progress in unsteady aerodynamics of turbomachines and propellers has been hampered by inherent difficulties such as complex geometry, rotating blades, solid-fluid interaction, multistages, high frequency, transonic vortical flows, viscous effects, and acoustic interference. Some of these effects, such as blade vibration, upstream disturbances, or sound generation, are small by-products of the fluid motion. Yet they cause significant unwanted and deleterious effects and need to be carefully understood and properly accounted for both in theoretical modeling and experiment. In spite of these considerable difficulties, we have had numerous successes since the 1970s. Today, with massively powerful computers and certainly more elaborate experimental methods, we may be poised for significant progress in the field. However, turbomachine propulsion technology keeps evolving at a fast pace toward higher speed, smaller engine size, increased thrust per unit engine weight, and ever higher efficiency. This has constantly moved the frontier of our goals, making them more elusive than ever.

This series of symposia has been instrumental in fostering cooperation between individual researchers and institutions. The development of the cascade standard configurations is a significant example of this ongoing cooperation. These standard configurations were developed to compare and validate available computer codes with each other and with specific data for cascades in nonuniform flows, thus providing invaluable bench mark results for new theories, codes, and experiments.

The current volume contains four review articles and 39 contributed papers. They are divided into seven chapters: Computational Methods for Unsteady Flows, Unsteady Cascade Flows, Unsteady Transonic and Supersonic Flows, Unsteady Viscous Effects, Unsteady Flow Measurements, Aeroacoustics, and Aeroelastic Instabilities and Control. An eighth chapter contains the results of a special panel session on the cascade standard configurations. The four review articles correspond to the four invited lectures given at the symposium by Professor John E. Ffowcs Williams, Professor Heinz E. Gallus, Dr. Philip R. Gliebe, and Dr. Joseph M. Verdon. They define the state of knowledge at the time of the symposium and provide a background that should help in the reading of the individual papers.

This symposium would not have been possible without the generous financial support of the National Aeronautics and Space Administration, the General Electric Company Aircraft Engine Division, United Technologies Research Center, and the University of Notre Dame. Special thanks are due to Drs. Marvin E. Goldstein, Robert E. Kielb, and Joseph M. Verdon for their efforts in securing this financial support from their institutions. The editor gratefully acknowledges the help of the International Scientific Committee in selecting the papers to be presented at the symposium and the cooperation of the authors in preparation of the manuscripts. Special thanks are also due to the local organizing committee and, in particular, to Dr. Jisheng Fang for their help in organizing the symposium. Thanks are also due to the administration and staff of the Notre Dame Center for Continuing Education who provided the practical arrangements for the symposium. The staff of Springer-Verlag have been very helpful in guiding the editor through the difficulties posed by this publication. Finally, the patient and competent secretarial help of Mrs. Patricia A. Kipker through all the stages of symposium and proceedings preparation is warmly appreciated.

H.M. Atassi
Notre Dame, Indiana

Contents

Preface v

I. Computational Methods for Unsteady Flows 1

Unsteady Aerodynamic Methods 3
Joseph M. Verdon

An Innovative Algorithm for Periodic Flow Calculations Using a Parallel Architecture—Some Applications to Unsteady Aerodynamics 43
G. Carte, P. Fraunié, and P. Dussouillez

A Framework for Multi-Stage Unsteady Flow Calculations 57
Michael Giles

Long Time Behavior of Unsteady Flow Computations 73
S.I. Hariharan

Simulation of 3D-Unsteady Internal Flows 91
Najat Benbouta, Pascal Ferrand, and Francis Leboeuf

II. Unsteady Cascade Flows 107

Time Linearized Euler Calculations for Unsteady Quasi-3D Cascade Flows 109
Gerhard Kahl and Arno Klose

Unsteady Flow Field Predictions for Oscillating Cascades 127
Dennis L. Huff

Compressible Flows with Vortical Disturbances Around a Cascade of Loaded Airfoils 149
Jisheng Fang and Hafiz M. Atassi

Development of a Steady Potential Solver for Use with Linearized Unsteady Aerodynamic Analyses 177
Daniel Hoyniak and Joseph M. Verdon

Calculation of Unsteady Linearized Euler Flows in Cascades Using Harmonically Deforming Grids 195
Kenneth C. Hall and William S. Clark

2D Linearized Harmonic Euler Flow Analysis for Flutter and Forced Response 213
D.G. Holmes and H.A. Chuang

Prediction of Unsteady 2D-Flow in Turbomachinery Bladings 231
D. Peitsch, H.E. Gallus, and H.-P. Kau

Axial-Gap Effects on the Propagation of Unsteady Flow in Cascades 251
T. Korakianitis

III. Unsteady Transonic and Supersonic Flows 269

Unsteady Lifting Surface Theory for Supersonic Through-Flow Fan 271
Masanobu Namba and Toshiya Hanada

Time-Dependent Predictions and Analysis of Turbine Cascade Data in the Transonic Flow Region 289
A. Bölcs, A. Cargill, T.H. Fransson, A. Suddhoo, and K. Vogeler

Supersonic Stator-Rotor Interaction in a Turbine Stage 309
Gilles Billonnet

Oscillation of an Airfoil Behind a Detached Normal Shock 331
S. Kaji, M. Ogawa, and H. Doi

IV. Unsteady Viscous Effects 357

Experimental Determination of Nonlinear Transfer Functions Characterizing Shock Wave/Turbulent Boundary Layer Interactions 359
F.O. Thomas, H.C. Chu, and C.M. Putnam

Analysis of Blade Unsteady Boundary Layers and Wakes 377
M. Barnett and J.M. Verdon

Active Control of Centrifugal Compressor Rotating Stall 397
P.B. Lawless and S. Fleeter

Computational Investigation of Airfoil Stall Flutter 415
J.D. Clarkson, J.A. Ekaterinaris, and M.F. Platzer

Compressor Blade Gust Response to Attached and Separated Flow Forcing Functions 433
Kuk H. Kim and Sanford Fleeter

V. Unsteady Flow Measurements 453

Experimental Methods and Results in Unsteady Aerodynamics and Aeroelasticity 455
H.E. Gallus

Three-Dimensional Unsteady Flow in a Single Stage Axial-Flow Turbine and Compressor 487
H.E. Gallus, C.A. Poensgen, and J. Zeschky

Reduced Frequency Effects on Gust Response Unsteady Aerodynamics in Turbomachines 507
Steven R. Manwaring and Sanford Fleeter

The Response of a Low Solidity Symmetric Airfoil Cascade to Compressor Wake and Linear Theory Gusts 529
Gregory H. Henderson and Sanford Fleeter

Time-Dependent Pressure Fluctuations on an Oscillating Turbine Cascade at Transonic Off-Design Flow Conditions 547
A. Bölcs, T.H. Fransson, and H. Körbächer

Linear Oscillating Cascade Unsteady Aerodynamic Experiments 567
Daniel H. Buffum and Sanford Fleeter

Experimental Investigation of the Unsteady Pressure Field on a Vibrating Blade in Steady and Unsteady Flow 583
C.A. Poensgen and H.E. Gallus

Test Facility for Flutter Investigations with Variable Frequency of the Vibrating Blade 603
Th. Steer

VI. Aeroacoustics 617

Aeroacoustics in Turbomachines and Propellers—Future Research Needs 619
Philip R. Gliebe

Acoustic Control of Flow Instabilities 643
J.E. Ffowcs Williams

High-Speed Helicopter Rotor Noise—Shock Waves as a Potent Source of Sound 655
F. Farassat, Yung-Jang Lee, H. Tadghighi, and R. Holz

Unsteady Flowfield of a Propfan at Takeoff Conditions 669
M. Nallasamy and J.F. Groeneweg

Prediction of Noise Field of a Propfan at Angle of Attack 685
Edmane Envia

The Application of Experimental Data to Blade Wake Interaction Noise Prediction 705
Stewart Glegg and William J. Devenport

Aeroacoustics of Real Blade Cascades 725
Kenneth A. Kousen and Joseph M. Verdon

High Frequency Gust Interaction with Single Loaded Airfoils in Subsonic Flows 743
James R. Scott and Hafiz M. Atassi

Influence of Airfoil Shape and Incidence Angle on High-Frequency Gust Interaction Noise 765
E.J. Kerschen, C.T. Tsai, and M.R. Myers

VII. Aeroelastic Instabilities and Control 783

A Comparison of Mistuning Effects on Flutter Between Stalled and Unstalled Cascades 785
Ichiro Fujimoto, Hiroshi Nagai, and Hideo Tanaka

Splitter Blades for Passive Turbomachine Flutter Control 807
Hsiao-Wei Chiang and Sanford Fleeter

A Non-Synchronous Vibration of Moving Blades Coupled with the Bleed Chamber Resonance in an Axial Compressor 829
N. Yamaguchi, T. Sato, S. Umemura, and T. Ohwaki

Aeroelasticity of Fan Blades with Complex Modes 843
Edmond Széchényi

VIII. Standard Configurations for Unsteady Flows 857

Panel Discussion on "Standard Configurations for Unsteady Flow Through Vibrating Axial-Flow Turbomachine-Cascades" 859
T.H. Fransson and J.M. Verdon

I. Computational Methods for Unsteady Flows

Unsteady Aerodynamic Methods

Joseph M. Verdon[1]

ABSTRACT This paper reviews recent advances in the theoretical and computational modeling of turbomachinery unsteady flows. In particular, the unsteady aerodynamic analyses intended for aeroelastic and aeroacoustic predictions must be applicable over wide ranges of operating conditions and modes of unsteady excitation. Also, because of the large number of controlling parameters involved, there is a stringent requirement for computational efficiency. To date these requirements have been met only to a limited extent. As a result, aeroelastic and aeroacoustic design predictions are, for the most part, still based on the classical linearized unsteady aerodynamic analyses that were developed during the early 1970s. During the past decade, significant advances in unsteady aerodynamic prediction capabilities have been achieved. In particular, researchers have developed linearized analyses that account for the effects of realistic design features, such as real blade geometry, mean blade loading and operation at transonic Mach numbers, on the unsteady aerodynamic response of two-dimensional cascades. The improvements in physical modeling that such linearizations allow are motivating their current implementation into aeroelastic and aeroacoustic design prediction systems. Also, in recent years, considerable progress has been made on developing time-accurate numerical simulations, based on the Euler and Navier-Stokes equations, of nonlinear unsteady flows through blade rows. Although not yet suitable for design use, these analyses offer opportunities for an improved understanding of the unsteady aerodynamic processes associated with blade vibration and noise generation. In addition, they will allow an evaluation of the impact of flow unsteadiness on time-mean flow behavior, and therefore, on the aerodynamic performance and efficiency of blade rows.

1.1 Introduction

The development of theoretical analyses to predict unsteady flows in axial-flow turbomachines has been motivated primarily by the need to predict the aeroelastic and the aeroacoustic behaviors of the blading. For both

[1] United Technologies Research Center, 411 Silver Lane, East Hartford, CT 06108.

applications, the unsteady aerodynamic analysis is only one component of an overall design prediction system. However, because of the complexity of the unsteady aerodynamic problem, this component has generally been regarded as the one requiring the most research attention.

Blade vibration and noise generation are both undesirable consequences of the unsteady flow processes that occur within axial-flow turbomachines and are, therefore, important concerns to the turbomachinery designer. Of the two, vibration problems have received more attention because they can lead to structural failure of the blading and possibly, extensive damage to the engine. Although excessive noise is disturbing and, in the future, may limit access to certain airports, the associated pressure fluctuations rarely threaten structural integrity.

Aerodynamically induced blade vibrations are usually classified into two general categories — flutter and forced vibration. In the former, the aerodynamic forces that sustain the blade motion are regarded as being dependent solely on that motion; whereas in the latter, the aerodynamic forces that excite the blade motion are independent of that motion. In both circumstances the vibratory motion of the blading can lead to fatigue failure. In modern fans and compressors flutter can be encountered over a wide range of operating conditions. The common types have been designated as classical supersonic (unstalled) flutter, subsonic/transonic high-incidence (or stalled) flutter, supersonic positive-incidence flutter, and negative-incidence or choke flutter. Classical supersonic flutter can occur at design conditions; the others, at off-design with subsonic/transonic positive-incidence flutter having the highest frequency of occurrence [MASS75, Fle79, FJ87]. Because of performance trends toward higher flow velocities, turbine flutter has also become an increasingly important concern.

Destructive forced vibrations can occur in fan, compressor or turbine blading when a periodic aerodynamic excitation, with frequency close to a structural system natural frequency, acts on the blades in a given row. Such excitations are generated at multiples of the engine rotation frequency and arise from a variety of sources, including inlet and exit flow nonuniformities and the aerodynamic interactions that occur between a given blade row and its neighboring blade rows and/or structural supports which move at a different rotational speed. The two principal types of such interaction are usually referred to as potential-flow and wake interaction. The former is associated with pressure variations which impinge on a given array from upstream and/or downstream and is of serious concern when the axial spacings between neighboring blade rows are small or flow Mach numbers or excitation frequencies are high. Wake interaction is the effect upon the flow through a given row of the wakes shed by one or more upstream rows and can persist over considerable axial distances.

External aerodynamic excitations also contribute to the noise generated by blade rows. Most of the research on turbomachinery noise has focused on the fan-stage of modern high by-pass ratio engines because the fan controls fly over noise on landing approach and is a strong contributor along with jet noise at take off [GR87]. Typical spectra of the noise radiated from the inlet and exit of a fan operating at subsonic tip speed show tones at the rotor/stator blade passing frequency and its harmonics and broadband noise at a lower level elsewhere. At supersonic tip speeds multiple pure tones or combination tones, associated with rotor-locked leading-edge shocks, also appear at the inlet at multiples of the shaft rotation frequency. The primary mechanisms responsible for pure tone noise are the wake and potential-flow interactions between the various elements of the fan, i.e., the rotor, the exit guide vane (stator) and the supporting struts. Broadband noise is generated by the interaction of random disturbances, e.g., those due to inlet boundary-layer and rotor-wake turbulence, with the downstream blading.

1.1.1 Unsteady Aerodynamic Analyses

An unsteady aerodynamic analysis intended for turbomachinery aeroelastic and aeroacoustic applications must be applicable over wide ranges of structural geometries, mean-operating conditions, and unsteady excitation modes and frequencies. Such an analysis must be capable of predicting the unsteady aerodynamic response of a blade row to prescribed structural (i.e., blade motions) and external aerodynamic excitations. The latter include variations in total pressure, total temperature and static pressure at inlet and variations in static pressure at exit. For aeroelastic applications, the unsteady loads that act on the vibrating blades must be determined, whereas for aeroacoustic applications, the pressure waves that carry energy away from the blade row are the response quantities of interest. Because of the large number of controlling parameters involved, the analysis must be capable of providing the necessary aerodynamic response information efficiently and economically.

Until recently, the unsteady aerodynamic analyses used for aeroelastic and aeroacoustic design predictions were based on classical linearized inviscid flow theory (see [Whi87] for a review), which essentially applies to lightly loaded thin-airfoil cascades. Very efficient semi-analytic solution procedures have been developed for two-dimensional, attached, subsonic flows [Whi70, KO70, Smi71], high-frequency transonic flows [SA86], and supersonic flows with subsonic [Ver77, NW78, AG78, Ni79, BF81] or supersonic [Lan57] axial velocity component. Models based on variations or extensions of the two-dimensional classical linearization have also been developed to treat separated subsonic flows [Chi80] and supersonic flows with strong in-passage normal shocks [GBA77]. Finally, extensive efforts have also been made to develop three-dimensional classical linearized analyses, as reviewed by Namba [Nam87]. The two-dimensional subsonic and supersonic analyses, supplemented by a good deal of empiricism, have been used for most current designs. Those for supersonic flow with subsonic axial velocity component have been very important because they give conservative estimates for the onset of supersonic unstalled torsional flutter in fans. Classical analyses provide very efficient unsteady aerodynamic response predictions that are useful near design conditions, but they are not appropriate off-design, where blade loading effects are important, or for transonic flows with embedded shock discontinuities.

To overcome the limitations of the classical model, while retaining some measure of computational efficiency, general, two-dimensional, inviscid, unsteady aerodynamic linearizations have been developed [Ver89b, Ver90, Whi90]. These account for the effects of realistic design features, such as real blade geometry, finite mean or steady blade loading, and operation at transonic Mach numbers, on the unsteady aerodynamic response of the blading to imposed structural and aerodynamic excitations. Here, the unsteady flow is regarded as a small perturbation of a fully nonlinear steady flow. Thus, the steady flow is governed by a nonlinear equation set; the unsteady flow, by a system of linear equations with variable coefficients that depend upon the underlying steady flow. Unsteady linearizations relative to potential mean or steady background flows [AA80, Whi82, VC84, UV90, HV91, CD91] have received considerable attention over the past decade, and linearized Euler analyses [HC89, HC91] are currently being developed to address unsteady flows containing strong shocks. While solutions based on these analyses require significantly greater computing times than classical solutions, the improvements in physical modeling that they allow, coupled with advances in our ability to resolve numerically the resulting governing equation sets, are making this approach an increasingly attractive one for aeroelastic and aeroelastic design applications.

Although linearized inviscid analyses meet the needs of aeroelastic and aeroacoustic designers for efficient unsteady aerodynamic response predictions, they do not account for potentially important unsteady flow phenomena, e.g., phenomena associated with finite-amplitude excitation, viscous separation, large shock excursion, etc. Thus, work has been proceeding to develop time-accurate, nonlinear, inviscid (Euler) and viscid (Navier-Stokes) solution techniques for unsteady flows through isolated and aerodynamically coupled blade rows. Such analyses provide a comprehensive physical modeling capability, but they also require substantial computational resources.

Since the mid 1980's, a number of Euler and Navier-Stokes procedures have been developed to address flows through single blade rows in which the unsteadiness is caused by blade vibration [HR89, He90, GBQ90, HSR91, HD91, Sid91] or by aerodynamic disturbances at the inflow or outflow boundaries [Gil88], and flows through aerodynamically coupled blade rows in which the unsteadiness is caused by the relative motion between adjacent blade rows [KK85, Fou86, Rai87, Rai89, Gil90b, GH91] and by aerodynamic disturbances at inlet [RD90, KG90, DDEM90]. The research on isolated cascades of vibrating blades has focused on fan and compressor flutter applications. Here, the Euler or Navier-Stokes equations are solved on time-dependent grids that deform with the blade motion. The methods for unsteady flows excited by external aerodynamic excitations or by the aerodynamic coupling between adjacent stationary and rotating blade rows treat blades that are fixed in the stator or rotor frame of reference. For the most part, such analyses have been applied to study wake/turbine rotor, turbine stator/rotor or combustor hotstreak/turbine stator/rotor interactions.

The present paper reviews the unsteady aerodynamic models that are currently being developed for turbomachinery blade rows. As a convenience, these models will be described for two-dimensional unsteady flows through isolated or single blade rows. The extensions required to treat three-dimensional flows and flows through coupled blade rows are reasonably straightforward, at least conceptually. The reader will determine that although significant progress has been made, substantial future research will be required before fully acceptable unsteady aerodynamic analyses can be introduced into the turbomachinery aeroelastic and aeroacoustic design processes.

1.2 Nonlinear Unsteady Aerodynamic Analyses

We consider time-dependent flow, with negligible body forces, of a perfect gas with constant specific heats through a two-dimensional cascade, such as the one shown in Figure 1.1. The unsteady fluctuations in the flow arise from one or more of the following sources (Figure 1.2): blade motions, upstream total temperature and total pressure disturbances, and upstream and/or downstream static pressure disturbances which carry energy toward the blade row (entropy, vorticity and acoustic waves). In the absence of unsteady excitation, the mean flow far upstream from the blade row is assumed to be at most a small perturbation from a uniform free stream.

The mean or steady-state positions of the blade chord lines coincide with the line segments $\eta = \xi \tan\Theta + mG, 0 \leq \xi \leq \cos\Theta, m = 0,\ \pm 1,\ \pm 2,\ \ldots$, where ξ and η are coordinates in the axial flow and cascade tangential directions, respectively, m is a blade number index, Θ is the cascade stagger angle, and $\mathbf{G}$ is the cascade gap vector which is directed along the η-axis with magnitude equal to the blade spacing. The blade motions, i.e.,

$$\mathcal{R}(\mathbf{x} + m\mathbf{G}, t) = \mathrm{Re}\{\mathbf{r}(\mathbf{x})\exp[i(\omega t + m\sigma)]\}, \quad \mathbf{x} \in \mathrm{B}\ , \tag{1.1}$$

where $\mathcal{R}$ is the displacement of a point on a moving blade surface ($\mathcal{B}_m$) relative to its mean or steady-state position (B_m), are prescribed functions of position $\mathbf{x}$ and time t. In Eq. (1.1) $\mathbf{r}$ is a complex-amplitude vector, σ is the phase angle between adjacent blade motions, Re{ } denotes the real part of { } and B denotes the reference ($m = 0$) blade surface.

The entropic, $\tilde{s}_{-\infty}(\mathbf{x},t)$, vortical, $\tilde{\boldsymbol{\zeta}}_{-\infty}(\mathbf{x},t)$, and acoustic, $\tilde{p}_{I,\mp\infty}(\mathbf{x},t)$, excitations, where the subscripts $-\infty$ and $+\infty$ refer to regions far upstream

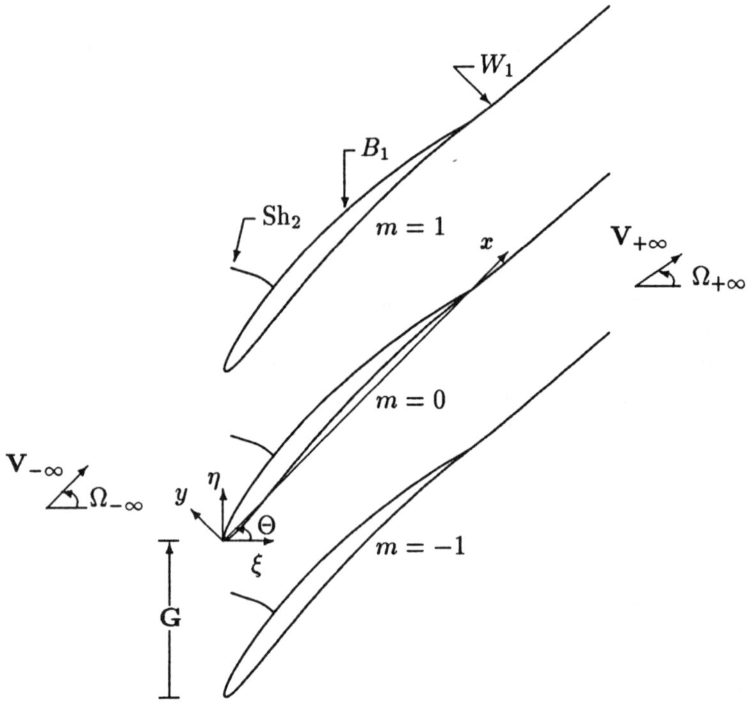

FIGURE 1.1. Two-dimensional transonic compressor cascade; $M_{+\infty} < M_{-\infty} < 1$.

$(\xi \leq \xi_-)$ and far downstream $(\xi \geq \xi_+)$ from the blade row, respectively, are also prescribed functions of $\mathbf{x}$ and t. The conditions at inlet and exit must be compatible with the fluid dynamic field equations, and the prescribed pressure disturbances must be of the type that carry energy toward the blade row. Although more general situations can be accommodated, here we assume that the external aerodynamic excitations are steady in a reference frame that moves at constant velocity V_{EXC} in the η-direction. Therefore, in the blade-row frame of reference, they are periodic in η with fundamental wave number $\kappa_\eta = \sigma^{-1}G$, and periodic in time with fundamental frequency $\omega = -\kappa_\eta V_{\text{EXC}}$.

In the present discussion all physical variables are dimensionless. Lengths have been scaled with respect to blade chord, time with respect to the ratio of blade chord to upstream freestream flow speed, density with respect to the upstream freestream density, velocity with respect to the upstream freestream flow speed, and pressure with respect to the product of the upstream freestream density and the square of the upstream freestream speed. The scalings for the remaining variables can be determined from the

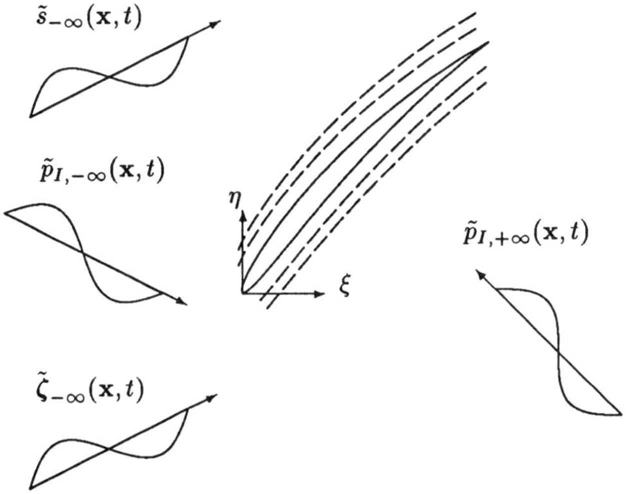

FIGURE 1.2. Unsteady excitations: blade motion; incident vortical and entropic disturbances from upstream; and incident acoustic disturbances from upstream and downstream.

equations given below, which are of the same form as their dimensional counterparts.

1.2.1 Conservation Laws and Thermodynamic Relations

Equations that govern the flow are determined from the conservation laws for mass, momentum and energy and the thermodynamic relations for a perfect gas. If we consider an arbitrary moving control volume $\mathcal{V}(t)$ bounded by a control surface $\mathcal{S}(\mathbf{x}, t)$, the conservation laws for the fluid within $\mathcal{V}$ at time t can be written as

$$\frac{d}{dt}\int_{\mathcal{V}} \tilde{\bar{\rho}} d\mathcal{V} + \int_{\mathcal{S}} \tilde{\bar{\rho}}(\tilde{\mathbf{V}} - \frac{\partial \mathcal{R}}{\partial t}) \cdot \mathbf{n} d\mathcal{S} = 0 \; , \tag{1.2}$$

$$\frac{d}{dt}\int_{\mathcal{V}} \tilde{\bar{\rho}}\tilde{\mathbf{V}} d\mathcal{V} + \int_{\mathcal{S}} \tilde{\bar{\rho}}[\tilde{\mathbf{V}} \otimes (\tilde{\mathbf{V}} - \frac{\partial \mathcal{R}}{\partial t})] \cdot \mathbf{n} d\mathcal{S} = \int_{\mathcal{S}} \tilde{\mathbf{T}} \cdot \mathbf{n} d\mathcal{S} \tag{1.3}$$

and

$$\begin{aligned} \frac{d}{dt}\int_{\mathcal{V}} \tilde{\bar{\rho}}(\tilde{E} + \tilde{V}^2/2) d\mathcal{V} + \int_{\mathcal{S}} \tilde{\bar{\rho}}(\tilde{E} + \tilde{V}^2/2)(\tilde{\mathbf{V}} - \frac{\partial \mathcal{R}}{\partial t}) \cdot \mathbf{n} d\mathcal{S} \\ = \int_{\mathcal{S}} \tilde{\mathbf{V}} \cdot \tilde{\mathbf{T}} \cdot \mathbf{n} d\mathcal{S} - \int_{\mathcal{S}} \tilde{\mathbf{q}} \cdot \mathbf{n} d\mathcal{S} \; . \end{aligned} \tag{1.4}$$

Here $\tilde{\bar{\rho}}$, $\tilde{\mathbf{V}}$ and $\tilde{E}$ are the fluid density, velocity and specific internal energy, respectively, $\mathcal{R}(\mathbf{x},t)$ defines the displacement of a point on the control surface, $\tilde{\mathbf{T}}$ is the stress tensor, $\tilde{\mathbf{q}}$ is the heat flux vector, $\mathbf{n}$ is a unit outward normal vector, and $\otimes$ denotes the tensor or dyadic product. For a Newtonian fluid having zero coefficient of bulk viscosity, the stress tensor (dyadic) is related to the fluid (thermostatic) pressure $\tilde{P}$ and velocity by

$$\tilde{\mathbf{T}} = -\tilde{P}\mathbf{I} - \tilde{\mu}Re^{-1}[2\nabla\cdot\tilde{\mathbf{V}}/3 + \nabla\otimes\tilde{\mathbf{V}} + (\nabla\otimes\tilde{\mathbf{V}})_c] \,, \tag{1.5}$$

where $\tilde{P}$ is the pressure, $\mathbf{I}$ is the unit tensor, $\tilde{\mu}$ is the coefficient of shear viscosity, $\mathrm{Re} = \rho^* V^* L^*/\mu^*$ is the Reynolds number, the superscript $*$ refers to a dimensional reference value of a flow variable, and the subscript c denotes the conjugate dyadic. The heat flux $\mathbf{q}$ is related to the gradient of temperature via Fourier's law, i.e., $\tilde{\mathbf{q}} = -\tilde{k}Pr^{-1}Re^{-1}\nabla\tilde{T}$, where $\tilde{k}$ is the coefficient of thermal conductivity, $Pr = \mu^* C_p^*/k^*$ is the Prandtl number of the flow, and C_p^* is the specific heat of the fluid at constant pressure. The coefficients $\tilde{\mu}$ and $\tilde{k}$ are assumed to be known functions of the temperature $\tilde{T}$.

In addition to the foregoing equations, we also require two relations from classical thermodynamics, in particular, the equation of state for a thermally perfect gas and the equation relating the internal energy and the temperature for a calorically perfect gas, i.e.,

$$\tilde{P} = \gamma^{-1}(\gamma - 1)\tilde{\bar{\rho}}\tilde{T} \quad \text{and} \quad \tilde{E} = \gamma^{-1}\tilde{T} = (\gamma-1)^{-1}\tilde{P}/\tilde{\bar{\rho}} \,, \tag{1.6}$$

respectively. Here, γ is the specific heat ratio of the fluid (constant pressure to constant volume). It is also useful to introduce the fundamental thermodynamic identity

$$d\tilde{S} = T^{-1}\left[d\tilde{E} + \tilde{P}d(\tilde{\bar{\rho}}^{-1})\right] = \gamma^{-1}d\tilde{P}/\tilde{P} - d\tilde{\bar{\rho}}/\tilde{\bar{\rho}} \,, \tag{1.7}$$

where $\tilde{S}$ is the specific entropy of the fluid.

The foregoing equations provide a sufficient set for determining the unknown fluid dynamic variables. For turbulent flows these equations describe the behavior of the ensemble- or statistical-mean values of the time-dependent flow variables. The effects of random turbulent fluctuations are represented by including turbulent velocity correlations in the definition of the stress tensor $\tilde{\mathbf{T}}$. To obtain a closure of the mean flow equations, these correlations are usually related to gradients in the ensemble-averaged or mean-flow velocity components via algebraic eddy viscosity models.

1.2.2 The Navier-Stokes Equations

Equations that describe the fluid motion at a (fixed) point in the field are obtained from the integral conservation laws by applying the Reynolds transport theorem to interchange the order of time differentiation and volume integration, Green's theorem to convert the surface integrals to volume integrals, and taking the limit of the resulting expressions as $\mathcal{V}(t) \to 0$. We find that

$$\frac{\partial \tilde{\tilde{\rho}}}{\partial t} + \nabla \cdot (\tilde{\tilde{\rho}} \tilde{\mathbf{V}}) = 0 \,, \tag{1.8}$$

$$\frac{\partial}{\partial t}(\tilde{\tilde{\rho}} \tilde{\mathbf{V}}) + \nabla \cdot (\tilde{\tilde{\rho}} \tilde{\mathbf{V}} \otimes \tilde{\mathbf{V}}) = \nabla \cdot \tilde{\mathbf{T}} \tag{1.9}$$

and

$$\frac{\partial}{\partial t}[\tilde{\tilde{\rho}}(\tilde{E} + \tilde{V}^2/2)] + \nabla \cdot [\tilde{\tilde{\rho}} \tilde{\mathbf{V}}(\tilde{E} + \tilde{V}^2/2)] = \nabla \cdot (\tilde{\mathbf{V}} \cdot \tilde{\mathbf{T}}) - \nabla \cdot \mathbf{q} \,. \tag{1.10}$$

These (conservation) forms of the Navier-Stokes equations correspond to the integral forms (1.2), (1.3) and (1.4) and are usually preferred in numerical simulations, but the convective forms, i.e.,

$$\frac{D\tilde{\tilde{\rho}}}{Dt} + \tilde{\tilde{\rho}} \nabla \cdot \tilde{\mathbf{V}} = 0 \,, \tag{1.11}$$

$$\tilde{\tilde{\rho}} \frac{D}{Dt} \tilde{\mathbf{V}} + \nabla \tilde{P} = 0 \,, \tag{1.12}$$

and

$$\tilde{\tilde{\rho}} \frac{D\tilde{S}}{Dt} = \tilde{T}^{-1}[\tilde{P} \nabla \cdot \tilde{\mathbf{V}} + \tilde{\mathbf{T}} : (\nabla \otimes \tilde{\mathbf{V}}) - \nabla \cdot \tilde{\mathbf{q}}] \,, \tag{1.13}$$

are also useful. Here, $D\ /Dt$ is the material or convective derivative operator; i.e., $D\ /Dt = \partial\ /\partial t + \tilde{\mathbf{V}} \cdot \nabla$, and the notation : indicates the scalar product of the two tensors. The entropy transport equation (1.13) is obtained by combining (1.10), (1.11), (1.12) and the fundamental thermodynamic identity (1.7), and performing the necessary algebra.

For application to turbomachinery unsteady flows the field equations must be supplemented by conditions at the vibrating blade surfaces and conditions at the inflow and outflow boundaries. Since transient unsteady aerodynamic behavior is usually not of interest, a precise knowledge of the initial state of the fluid is not required. The no-slip condition, i.e.,

$$\tilde{\mathbf{V}} = \frac{\partial \mathcal{R}}{\partial t}, \quad \mathbf{x} \in \mathcal{B}_m \,, \tag{1.14}$$

where the vector $\mathcal{R}$ is prescribed, applies at blade surfaces. In addition, either the heat flux $\tilde{\mathbf{q}} \cdot \mathbf{n}$ or the temperature $\tilde{T}$ must be prescribed at such surfaces. The total pressure and total temperature fluctuations at inlet, and the incident pressure disturbances at inlet and exit must also be specified. The total pressure and total temperature fluctuations at exit and the unsteady pressure response at inlet and exit are to be determined as part of the time-dependent solution.

The Navier-Stokes equations must be considered if convective nonlinearities and viscous effects are expected to have an important impact on the unsteady aerodynamic response of a blade row. However, for most flows of practical interest the Reynolds number (Re) is usually sufficiently high so that viscous effects are concentrated in relatively thin layers across which the flow properties vary rapidly but continuously. These layers generally lie adjacent to the blade surfaces (boundary layers), downstream of the blades (wakes) and in the vicinity of rapid compressions (shocks). In such cases separate sets of approximate equations, i.e., reduced forms of the Navier-Stokes equations, can be constructed and solved to describe the flows in the "outer" inviscid region and the "inner" viscid regions. Appropriate methods for coupling the inviscid and viscid solutions are also needed. If viscous effects are negligible, i.e., if Re $\to \infty$, it is sufficient to consider only the outer or inviscid flow.

1.2.3 Inviscid Flow — The Euler Equations

In the inviscid limit the thickness of the viscous layers becomes zero and the latter can be modeled as surfaces across which the flow variables are discontinuous. In particular, boundary layers and wakes become vortex sheets which support a discontinuity in tangential velocity, and shocks become surfaces which support a discontinuity in normal velocity. It is usually assumed that the boundary layers remain attached to the blade surfaces supporting jumps in velocity from zero at the "wall" to inviscid values at their "edges". Further, the vortex-sheet unsteady wakes emanate from the blade trailing edges and extend infinitely far downstream. The integral forms of the conservation laws are required to describe the flow over the entire domain of interest. These provide differential equations in regions where the flow variables are continuously differentiable and "jump" conditions at the surfaces at which the flow variables are discontinuous.

The field equations that govern continuous, inviscid, fluid motion (i.e., the Euler equations) are obtained from (1.8–1.10) or (1.11–1.13) by setting $\tilde{\mathbf{T}} = -\tilde{P}\mathbf{I}$ and $\tilde{\mathbf{q}} = 0$. These equations must be supplemented by jump conditions that apply at the moving boundary layers, $\mathcal{B}_m$, wakes, $\mathcal{W}_m$, and shocks, $\mathcal{S}h_{m,n}$, where the subscripts on $\mathcal{S}h$ refer to the nth shock associated with the mth blade. The far-field conditions used in the inviscid approximation are the same as those used in Navier-Stokes simulations.

The inviscid surface conditions are obtained from the integral conservation laws (1.1—1.3) by setting $\tilde{\mathbf{T}} = -\tilde{P}\mathbf{I}$ and $\tilde{\mathbf{q}} = 0$, considering a control volume that contains an element of a surface of discontinuity, and taking the limit as the length normal to this surface element approaches zero. The resulting conditions for conserving mass, momentum, and energy at a moving surface are

$$[\![\tilde{M}_f]\!] = 0 \ , \tag{1.15}$$

$$\tilde{M}_f[\![\tilde{\mathbf{V}}]\!] + [\![\tilde{P}]\!]\mathbf{n} = 0 \tag{1.16}$$

and

$$\tilde{M}_f[\![\tilde{E} + \tilde{V}^2/2]\!] + [\![\tilde{P}\tilde{\mathbf{V}}]\!]\mathbf{n} = 0 \ , \tag{1.17}$$

respectively, where $[\![\ ]\!]$ denotes the difference in a flow quantity across a surface of discontinuity, and

$$\tilde{M}_f = \tilde{\bar{\rho}}(\tilde{\mathbf{V}} - \frac{\partial \mathcal{R}}{\partial t}) \cdot \mathbf{n} \tag{1.18}$$

is the fluid mass flux across the moving surface.

Since vortex sheets support a jump in tangential velocity (i.e., $[\![\mathbf{V}]\!] \cdot \boldsymbol{\tau} \neq 0$), it follows from (1.16) that $\tilde{M}_f = 0$ at such surfaces. Hence, the conditions

$$\tilde{M}_f = 0, \quad [\![\tilde{P}]\!] = 0, \quad \text{and} \quad [\![\tilde{\mathbf{V}}]\!] \cdot \mathbf{n} = 0, \quad \mathbf{x} \in \mathcal{B}_m\ ,\ \mathcal{W}_m\ , \tag{1.19}$$

prevail at vortex sheet boundary-layer and wake surfaces. The first of these applies at the "outer edges" of the viscous layers, and therefore, also at the solid blade surfaces (the inviscid flow tangency condition). At shocks, $\tilde{M}_f \neq 0$; so it follows from (1.16) that the tangential component of the fluid velocity, $\tilde{\mathbf{V}}_\tau$, must be continuous. The remaining jump conditions, along with the thermodynamic equations of state, are then required to determine the shock velocity, $\partial\mathcal{R}/\partial t$, and the changes in the properties of the fluid as it passes through a shock. The surface displacement vector $\mathcal{R}(\mathbf{x}, t)$ is prescribed at blade surfaces, but at wake and shock surfaces it must be determined as part of the time-dependent solution.

Although the viscous displacement and curvature effects associated with actual boundary layers, wakes and shocks are neglected in a purely inviscid analysis, changes could be made in the foregoing jump conditions to accommodate them. If so, the inviscid "outer" region solution could be matched to viscid "inner" region solutions to provide a solution for the complete flow field. We refer the reader to the articles by Melnik [Mel80] and Lock and Firmin [LF82] for comprehensive reviews on the application of such viscid/inviscid interaction concepts in steady-state aerodynamics.

1.2.4 Numerical Simulations

Extensive research has been conducted in recent years to develop time-accurate Euler and Navier-Stokes procedures for unsteady flows through turbomachinery blade rows. These procedures build upon schemes developed for steady flows in which a sequence of iterative solutions are marched forward in "pseudo time" until a converged steady-state prediction is determined. Solutions for periodic unsteady flows are marched forward in "real" time from an initial steady state to a converged periodic unsteady state. Existing algorithms invariably treat the conservative forms of the nonlinear field equations in which the unknown variables are $\tilde{\bar{\rho}}$, $\tilde{\bar{\rho}}\mathbf{V}$ and $\tilde{\bar{\rho}}(E + V^2/2)$. The field equations are solved subject to boundary conditions at the blade surfaces, e.g., the flow tangency condition for inviscid flows or the no-slip condition for viscous flows. However, shock- and wake-jump conditions are usually not imposed in inviscid calculations. Instead, shock and wake phenomena are "captured" by approximating the conservative forms of the field equations across surfaces of discontinuity. This tends to smear shock and wake behavior, but considerably reduces the analytical effort.

The governing equations are approximated using finite-volume, finite-difference, or finite-element spatial discretizations, and the numerical solutions are advanced in time using single or multi-step explicit or implicit time-marching procedures. Explicit schemes are usually favored for inviscid flows; implicit schemes, for viscous flows, because the high mesh densities required impose stringent constraints on the stability of explicit schemes. Solutions are usually determined on sheared H-type grids which cover an extended blade-passage solution domain, but for viscous calculations "inner" O or C grids which wrap around the blades are used in conjunction with the "outer" H-grids. To simulate blade motion, the computational meshes deform so that the computational boundaries coincide with the moving blade boundaries at all instants of time. For viscous calculations the effects of turbulence are represented by the inclusion of a turbulence model, usually one of the algebraic eddy viscosity models [CS74, BL78] widely used in steady flow calculations. Transition is assumed to occur instantaneously at specified discrete blade-surface locations.

There are two weaknesses associated with algebraic turbulence models. First, they require extensive "fine tuning" to accurately predict eddy viscosities in regions of separation and in wakes. Second, algebraic models base the local turbulent viscosity on the instantaneous local velocity profile; hence, they do not account for the "history" effects associated with the temporal or spatial development of the unsteady flow. The use of one- or two-equation turbulence models has been suggested [Rai87, GH91] to overcome these weaknesses.

In addition to the limitations placed on the solutions by the use of simple turbulence and transition models, two other simplifications have been introduced which may limit the nonlinear capabilities of current time-marching analyses. These are associated with the conditions imposed at the inflow and outflow boundaries and the use of a time-lagged circumferential periodicity condition to relate the flows in adjacent blade passages. For steady flows the usual procedure is to specify the total pressure, the total enthalpy and the flow angle at inlet and the static pressure at exit. Compatibility relations, 1 upstream and 3 downstream, are then applied along with the prescribed inlet and exit information to compute the flow variables at the inflow and outflow boundaries. Also, steady flows, produced under uniform inflow conditions, will exhibit blade-to-blade periodicity, i.e., $F(\mathbf{x}+m\mathbf{G}) = F(\mathbf{x})$, $m = 0 \pm 1, \pm 2, \dots$, where F is any steady fluid property. Thus, the solution domain can be restricted to a single extended blade passage region (Fig. 1.3).

For unsteady flows the incoming disturbances must be prescribed as functions of $\mathbf{x}$ of t that are compatible with the nonlinear equations of motion.

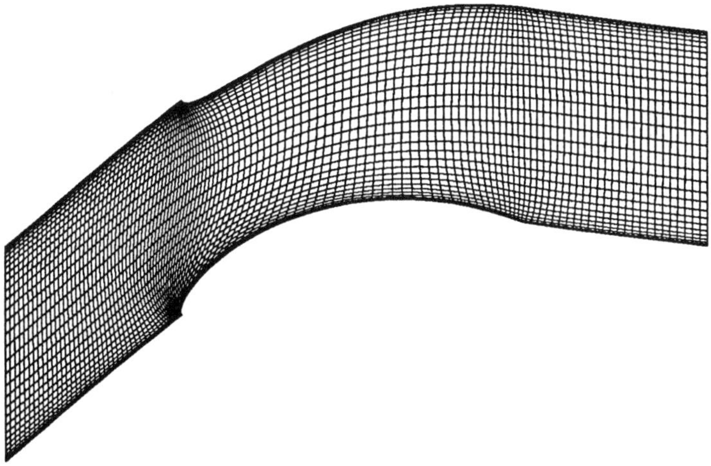

FIGURE 1.3. Extended blade-passage solution domain and typical H-grid used for unsteady flow calculations.

The inlet and exit conditions must also allow disturbance waves generated within the solution domain, i.e., pressure response waves, and entropy and vorticity waves at exit, to pass through the boundaries without producing artificial reflections. At present, linearized, one-dimensional, characteristic theory is usually applied to provide inlet and exit conditions for nonlinear unsteady flow calculations. The assumption is made that the unsteady perturbations are plane waves with minimal variation in the circumferential direction. Entropic, vortical and incoming pressure perturbations are specified at the inlet boundary, and the outgoing pressure characteristic is extrapolated from the interior of the domain. The incoming pressure perturbation is specified at the outlet boundary, and the entropic, vortical and outgoing pressure characteristics are extrapolated from the interior. Thus, linear unsteady inlet and exit conditions are imposed in nonlinear unsteady flow calculations. Since these are based on one-dimensional characteristic theory, they are not strictly non-reflecting. Approximate, linear, two-dimensional, non-reflecting inlet and exit conditions have been proposed [Gil90a], but these have not yet been widely implemented into time-marching codes.

In general, the unsteady flows produced by circumferentially and temporally periodic unsteady excitations must be computed over many, and in some cases all, blade passages. The number of passages needed depends upon the interblade phase angle $\sigma = \kappa_\eta G$ of the excitation. However, because multi-passage calculations involve enormous computing times, a restrictive assumption is usually introduced to limit the computational domain to a single passage. For example, most investigators impose a phase-shifted periodicity condition, i.e., $\tilde{F}(\mathbf{x} + m\mathbf{G}, t - m\sigma/\omega) = \tilde{F}(\mathbf{x}, t)$, $m = 0, \pm 1, \pm 2, \ldots$, where $\tilde{F}$ is any unsteady fluid property [EAM77]. This condition implies that an unsteady excitation at wave number, $\kappa_\eta = \sigma G^{-1}$, and frequency, ω, will produce a response at the same wave number and frequency, which is only true for linear systems. It allows the solution domain to be restricted to a single blade-passage region, but a large amount of computer memory is still needed to store the unsteady flow variables at two periodic boundaries over an entire period of excitation.

Another approach for reducing the computational domain employs a dependent variable transformation to rewrite the governing equations in "time-inclined" computational coordinates [Gil88]. This eliminates the storage requirements associated with phase-shifted periodicity, but introduces additional complexity into the solution procedure for the interior flow. Also, whereas the phase-shifted periodicity condition imposes restrictions on the circumferential and temporal behavior of the unsteady response, time-inclining places restrictions only on the circumferential behavior. Neither phase-shifted periodicity nor time inclining can be employed if excitations at different fundamental circumferential wave numbers occur simultaneously, say at the inlet and exit boundaries. The use of these procedures also places restrictions on the "nonlinear" response behavior that can be predicted.

Recent Applications

Based on a review of the recent literature, the present author estimates that 2500 grid points and 1000 time steps per cycle of excitation are used in a typical single-passage, time-marching, Euler calculation. Five (5) cycles are needed to converge the nonlinear inviscid solution to a periodic state, and the entire calculation requires about 10 CPU minutes on a CRAY YMP. The corresponding numbers for a Navier-Stokes calculation are 12,500 mesh points per passage, 6000 time steps per cycle, 5 cycles to converge, and 3 CPU hours of CRAY YMP time.

The number of grid points cited above are probably not sufficient to determine the first-harmonic or linear response, let alone the higher harmonic or nonlinear responses, for a number of important applications, including the prediction of transonic flutter and the response of the blading to excitations produced by an adjacent row. For example, Giles [Gil88] applied a time-marching Euler analysis to predict the response of a flat-plate cascade to a sinusoidal wake excitation. The cascade had a stagger angle Θ of 30 deg, a gap-to-chord ratio G of 0.5, and operated at an inlet Mach number of 0.7. The wake to blade count was approximately 1.1 resulting in an excitation frequency and interblade phase angle of 13.96 and −400 deg, respectively. Excellent agreement was achieved between the numerical predictions and predictions based on classical linearized theory for the first-harmonic, blade, pressure-difference distributions, but the numerical calculation required 18 cycles on a 200 × 25 H-grid, followed by 12 cycles on a 400 × 50 grid.

In spite of the current limitations, very sophisticated time-marching, nonlinear, unsteady, aerodynamic analyses and codes have been developed and applied to realistic single and multiple blade row configurations. Procedures for isolated cascades of vibrating blades have been developed for fan and compressor flutter applications. In these analyses the Euler or Navier-Stokes equations are solved on grids that deform with time to account for the blade motion. Inviscid unsteady solutions have been determined for subsonic inlet flows [He90, HSR91], and for supersonic inlet flows with subsonic [GBQ90] or supersonic [HR89] axial velocity component. Since strong inviscid/viscid interaction effects are important in fan and compressor applications, a coupled Euler/integral boundary-layer method [HD91] has been developed to efficiently account for such interactions. In addition, a time-marching, Navier-Stokes, solution procedure has been developed [Sid91] to more accurately account for the effects of viscous-layer separation on the aerodynamic response to compressor blade vibrations.

Several research groups have developed methods for unsteady flows excited by incident aerodynamic excitations or by the aerodynamic coupling between adjacent stationary and rotating blade rows. Here the blades are fixed in the stator or rotor frame of reference, and the incident excitations or the relative motions between the adjacent rows give rise to high-frequency unsteady flows that can excite forced blade vibration and discrete-tone noise. These analyses have been applied to study wake/turbine rotor, turbine stator/rotor or combustor hotstreak/turbine stator/rotor interactions. Euler calculations of wake/rotor interactions [Gil88] have been performed in which incoming wakes were specified at the inflow boundary of a single turbine blade row. The flow in the wakes was assumed to be parallel with uniform pressure and total enthalpy and a prescribed velocity defect distribution. Inviscid stator/rotor interactions have been calculated in three-dimensions [KK85], quasi-three dimensions [Fou86] and two-dimensions [Gil90b]. In particular, 2D calculations for a highly loaded transonic first turbine stage [Gil90b] have shown that the propagation and reflection of the shocks that originate at the trailing edge of the upstream stator can produce a forty percent variation in the lift on the downstream rotor.

Perhaps the best known solutions for coupled blade rows are those based on the Navier-Stokes calculations performed by Rai for stator/rotor interactions in two-dimensions [Rai87] and three-dimensions [Rai89]. Rai's method is actually a hybrid Navier-Stokes/Euler method in which the thin-layer Navier-Stokes equations are solved on an inner O-grid, and the Euler equations are solved on an outer H-grid. Recently, Giles and Haimes [GH91] have also provided such a hybrid analysis. Finally, Navier-Stokes solutions for the migration of a combustor hot streak through an isolated turbine stage (stator and rotor) have been calculated by several investigators [RD90, KG90, DDEM90].

The time-accurate, nonlinear, unsteady analyses have been applied to flat-plate cascades so that nonlinear predictions could be compared with those of classical linearized theory. In general, the time-accurate results for subsonic flows are in very good agreement with those of classical linearized theory, but problems exist for superresonant flows [HSR91], i.e., flows in which an acoustic response persists in the far field of the blade row. For supersonic flows the time-accurate surface-pressure predictions [GBQ90, HR89] are in qualitative agreement with those of classical theory, but the former indicate smeared oblique shocks and contain spurious blade-surface pressure variations in the regions between shock discontinuities. The time-accurate nonlinear analyses have also been applied to very challenging fan, compressor and turbine flows. Reasonable steady and unsteady flow predictions that are in qualitative agreement with experimental measurements have been determined. Reasons for the differences between the predictions and the measurements have not been firmly established.

The time-marching codes, although much too expensive, at present, for use in design, are receiving a great deal of attention. They are important because they give researchers a means to investigate and understand nonlinear unsteady aerodynamic phenomena. Therefore, they can serve a useful role in determining the impact of transonic, viscous, and finite-amplitude excitation effects on the aeroelastic and aeroacoustic response of blade rows.

1.3 Linearized Unsteady Aerodynamic Analyses

The computational resources required to simulate nonlinear inviscid and viscid unsteady fluid dynamic behavior will continue to prohibit the use of such simulations in the aeroelastic or aeroacoustic design process. Therefore, approximate analyses, based, for example, on small-disturbance and high Reynolds number flow concepts, are needed to provide efficient predictions of unsteady aerodynamic response phenomena. A significant advantage offered by the linearized, inviscid approximation is that the first-order unsteady fluid motions arising from various Fourier modes of unsteady excitation are not coupled. Hence, it is sufficient to develop solution procedures for a single harmonic (in t and η) component of a given excitation. Solutions for arbitrary periodic excitations and arbitrary combinations of such excitations can then be obtained by superposition.

For small-amplitude, i.e., of $\mathcal{O}(\epsilon)$, unsteady excitations a time-dependent inviscid flow can be regarded as a small perturbation of underlying nonlinear mean or steady background flow. Thus, for example, the nonlinear time-dependent fluid velocity $\tilde{\mathbf{V}}(\mathbf{x},t)$ can be expressed in the form

$$\tilde{\mathbf{V}}(\mathbf{x},t) = \mathbf{V}(\mathbf{x}) + \tilde{\mathbf{v}}(\mathbf{x},t) + \dots , \tag{1.20}$$

where $\mathbf{V}(\mathbf{x})$ is the local mean velocity and $\tilde{\mathbf{v}}(\mathbf{x},t)$ is the first-order (in ϵ)

unsteady velocity. In addition, Taylor series expansions, e.g.,

$$\tilde{\mathbf{V}}_{\mathcal{S}} = (\tilde{\mathbf{V}} + \mathcal{R} \cdot \nabla \tilde{\mathbf{V}} + \ldots)|_{\mathrm{S}} , \tag{1.21}$$

and surface-vector equations, i.e.,

$$\boldsymbol{\tau}_{\mathcal{S}} = \boldsymbol{\tau}_{\mathrm{S}} + \left(\mathbf{n}_{\mathrm{S}} \cdot \frac{\partial \mathcal{R}}{\partial \tau_{\mathrm{S}}}\right) \mathbf{n}_{\mathrm{S}} + \ldots \quad \text{and} \quad \mathbf{n}_{\mathcal{S}} = \mathbf{n}_{\mathrm{S}} - \left(\boldsymbol{\tau}_{\mathrm{S}} \cdot \frac{\partial \mathcal{R}}{\partial \tau_{\mathrm{S}}}\right) \cdot \boldsymbol{\tau}_{\mathrm{S}} + \ldots , \tag{1.22}$$

where $\mathbf{n} \times \boldsymbol{\tau}$ points out from the page, can be applied to refer information at a moving ($\mathcal{S}$) blade, wake or shock surface to the mean (S) position of this surface.

Equations for the steady and first-order unsteady flow properties are determined by substituting the foregoing expansions into the full time-dependent Euler equations, equating terms of like power in ϵ and neglecting terms of $\mathcal{O}(\epsilon^2)$ or higher. The first-order or linearized unsteady response to a harmonic unsteady excitation at temporal frequency ω and circumferential wave number $\kappa_\eta = \sigma G^{-1}$ will be harmonic in time, and it will satisfy a phase-lagged blade-to-blade periodicity condition, e.g.,

$$\tilde{\mathbf{v}}(\mathbf{x}, t) = \mathrm{Re}\{\mathbf{v}(\mathbf{x}) \exp(i\omega t)\} \quad \text{and} \quad \mathbf{v}(\mathbf{x} + m\mathbf{G}) = \mathbf{v}(\mathbf{x}) \exp(im\sigma) . \tag{1.23}$$

Thus, a solution to a time-independent, but complex, linearized unsteady flow problem is required only over a single extended blade-passage region of the cascade. In addition, since analytic far-field solutions can be determined [Ver89c], the numerical solution domain can be restricted to a single extended blade-passage region of finite extent in the axial flow direction.

The equations that govern the zeroth-order or steady background flow follow from those given previously by replacing the nonlinear time-dependent, flow variables, $\tilde{\mathbf{V}}$, $\tilde{P}$, $\tilde{\bar{\rho}}$, $\tilde{S}$, etc., by their zeroth-order counterparts, $\mathbf{V}$, P, $\bar{\rho}$, S, etc., and setting local temporal derivatives, $\partial\ /\partial t$, and surface displacements, $\mathcal{R}$, equal to zero. The steady flow can be described via the conservative or convective forms of the differential field equations. Here we select the conservative form of the mass equation, the convective form of the momentum equation, and the entropy transport equation, i.e.,

$$\nabla \cdot (\bar{\rho}\mathbf{V}) = 0 ; \tag{1.24}$$

$$\nabla(V^2/2) + \gamma(\gamma - 1)^{-1} P \bar{\rho}^{-\gamma} \nabla \bar{\rho}^{(\gamma-1)} = \mathbf{V} \times \bar{\boldsymbol{\zeta}} - \gamma P \bar{\rho}^{-1} \nabla S , \tag{1.25}$$

and

$$\mathbf{V} \cdot \nabla S = 0 , \tag{1.26}$$

where $\bar{\boldsymbol{\zeta}} = \nabla \times \mathbf{V}$ is the mean-flow vorticity vector.

Surface conditions for the zeroth-order or steady flow follow from (1.15) through (1.19). Since $\mathcal{R} = 0$, these conditions are imposed at the mean positions of the blade, wake and shock surfaces. Blade mean positions (B_m)

are prescribed, but the mean wake (W_m) and shock ($\mathrm{Sh}_{m,n}$) locations must be determined as part of the steady solution. Since the steady flow remains attached to the mean blade surfaces, $\mathbf{V}\cdot\mathbf{n} = 0$ for $\mathbf{x} \in B_m$. Also, the steady pressure and normal velocity component must be continuous across blade wakes, i.e., $[\![P]\!] = 0$ and $[\![\mathbf{V}]\!] \cdot \mathbf{n} = 0$ for $\mathbf{x} \in W_m$, and mass, momentum and energy must be conserved at the steady shock locations, $\mathrm{Sh}_{m,n}$. Note that, if a strong shock impinges on a blade, the entropy and the tangential component of the fluid velocity will be discontinuous across its wake.

In addition to the foregoing surface conditions, conditions must be imposed on the steady flow far from the blade row. If the departures from uniform flow conditions in the far field are due solely to the presence of the blade row, it is sufficient to specify the total pressure, total enthalpy and the direction of the uniform inlet flow and the static pressure of the exit flow. Small steady departures from these uniform flow conditions must be determined as part of the nonlinear mean-flow solution.

1.3.1 The Linearized Unsteady Equations

After introducing the first-order thermodynamic relation $s = \gamma^{-1}p/P - \rho/\bar{\rho}$, which follows from (1.7), to eliminate the density ρ, and performing the necessary algebra, we find that the first-order unsteady perturbation of a rotational and nonisentropic, steady flow is described by the following system of linearized Euler equations:

$$\frac{\bar{D}s}{Dt} + \mathbf{v}\cdot\nabla S = 0 . \tag{1.27}$$

$$\frac{\bar{D}}{Dt}(\mathbf{v}-s\mathbf{V}/2)+[(\mathbf{v}-s\mathbf{V}/2)\cdot\nabla]\mathbf{V}+\nabla(p/\bar{\rho})+\frac{\mathbf{V}}{2}\frac{\bar{D}s}{Dt}-\bar{\rho}^{-1}p\nabla S = 0\,, \tag{1.28}$$

and

$$\frac{\bar{D}}{Dt}[p/(\bar{\rho}A^2) - s] + \bar{\rho}^{-1}\nabla\cdot(\bar{\rho}\mathbf{v}) = 0 . \tag{1.29}$$

Here s, $\mathbf{v}$ and p are the complex amplitudes of the first-order entropy, velocity and pressure, respectively, and $\bar{D}/Dt = i\omega + \mathbf{V}\cdot\nabla$ is a convective derivative operator based on the mean flow velocity. The linearized unsteady flow can be determined by solving (1.27), (1.28) and (1.29), subject to the appropriate surface and far-field conditions.

The linearized flow tangency condition follows from (1.19) and is given by

$$\mathbf{v}\cdot\mathbf{n} = [i\omega\mathbf{r} + (\mathbf{V}\cdot\boldsymbol{\tau})(\boldsymbol{\tau}\cdot\nabla)\mathbf{r} - (\mathbf{r}\cdot\nabla)\mathbf{V}]\cdot\mathbf{n}, \quad \mathbf{x} \in B_m\,, \tag{1.30}$$

The linearized wake-jump conditions also follow from (1.19) and can be expressed in the form

$$[\![p]\!] = -\kappa\mathbf{r}\cdot\mathbf{n}[\![\bar{\rho}(\mathbf{V}\cdot\boldsymbol{\tau})^2]\!], \quad \mathbf{x} \in W_m \tag{1.31}$$

and

$$[\![\mathbf{v}]\!]\cdot\mathbf{n} = (\boldsymbol{\tau}\cdot\nabla)\{\mathbf{r}\cdot\mathbf{n}[\![\mathbf{V}\cdot\boldsymbol{\tau}]\!]\}+\mathbf{r}\cdot\mathbf{n}[\![\mathbf{V}\cdot\boldsymbol{\tau}(\boldsymbol{\tau}\cdot\nabla)\ln\bar{\rho}]\!] = 0, \quad \mathbf{x} \in W_m\,. \tag{1.32}$$

Here, $\mathbf{r}$ is the complex-amplitude of the wake displacement; the wake mean positions, W_m, are assumed to coincide with the mean-flow stagnation streamlines; and $\kappa = \boldsymbol{\tau}\cdot\partial\mathbf{n}/\partial\tau = -\mathbf{n}\cdot\partial\boldsymbol{\tau}/\partial\tau$ is the curvature of the mean wake surface. Equations (1.31) and (1.32) provide two independent relations for determining the jumps in the linearized unsteady pressure, $[\![p]\!]$, and normal velocity component, $[\![\mathbf{v}]\!]\cdot\mathbf{n}$, across each wake. However, since the wake normal displacement, $\mathbf{r}\cdot\mathbf{n}$, is also an unknown, these relations are not sufficient, unless the steady tangential velocity and density are continuous across the wakes.

Although not given explicitly here, the linearized equations that ensure that mass, momentum, and energy are conserved to within first order across a shock are obtained by expanding the nonlinear time-dependent conditions (1.15), (1.16), (1.17), according to (1.20) through (1.23), and performing the necessary algebra. The resulting expressions govern the jumps in the first-order fluid properties p, $\mathbf{v}$ and s across moving shock surfaces and the normal component of the shock displacement $\mathbf{r}\cdot\mathbf{n}$. Again, however, there is one more unknown associated with the unsteady jump conditions than independent equations. Since the foregoing conditions are not sufficient for fitting wakes and shocks into an unsteady solution based on the linearized Euler equations, additional information will be required.

We have assumed that the mean or steady flow at inlet ($\xi < \xi_-$) is at most a small perturbation from a uniform stream. By Kelvin's theorem, a similar condition will apply far downstream ($\xi > \xi_+$) if no strong shocks are present in the flow. Small steady or unsteady perturbations of a uniform freestream are governed by constant-coefficient field equations for which analytical solutions can be determined. In particular, it can be shown that a small-amplitude unsteady perturbation of a uniform mean flow can be decomposed into distinct entropic-, vortical-, and acoustic-type modes. The entropic and vortical disturbances are convected without distortion by the mean flow, and the vortical disturbance has a divergence-free velocity field. The acoustic disturbance is directly related to an irrotational velocity fluctuation, i.e., $p = -\bar{\rho}\bar{D}\phi/Dt$, and its kinematical behavior is quite different from that of a vortical or entropic disturbance.

Each of the foregoing modes of unsteady motion, i.e.,

$$s(\mathbf{x}) = s_{-\infty}\exp(i\boldsymbol{\kappa}_{-\infty}\cdot\mathbf{x}), \quad \xi < \xi_-\,, \tag{1.33}$$

$$\boldsymbol{\zeta}(\mathbf{x}) = \boldsymbol{\zeta}_{-\infty}\exp(i\boldsymbol{\kappa}_{-\infty}\cdot\mathbf{x}), \quad \xi < \xi_-\,, \tag{1.34}$$

and

$$p_{I,\mp\infty}(\mathbf{x}) = p_{I,\mp\infty}\exp(-\beta_{\mp\infty}\xi + i\boldsymbol{\kappa}_{\mp\infty}\cdot\mathbf{x}), \quad \xi \lessgtr \xi_\mp \tag{1.35}$$

is a solution of the governing equations and can therefore be superposed on a uniform flow independently of the others. The circumferential wave number, $\kappa_{\eta,\mp\infty}$, of a disturbance is σG^{-1}. The axial wave number, $\kappa_{\xi,\mp\infty}$, of an entropic or vortical disturbance is determined by the relation $\omega = -\boldsymbol{\kappa}_{-\infty} \cdot \mathbf{V}_{-\infty}$. The attenuation constant, $\beta_{\mp\infty}$, and axial wave number, $\kappa_{\xi,\mp\infty}$, of an acoustic excitation also depend upon the inlet/exit free-stream velocity, the temporal frequency and the circumferential wave number of the excitation, but via more complicated relationships.

Equations (1.33)–(1.35) indicate the far-field information that must be supplied to determine a linearized unsteady flow solution. In particular, the interblade phase angle, σ, the temporal frequency, ω, the complex amplitudes (at $\mathbf{x} = \mathbf{0}$) of the entropic s and vortical ζ excitations at inlet, and the pressure excitations at inlet and exit must be prescribed. Usually only acoustic excitations that are of propagating type ($\beta_{\mp\infty} = 0$) are considered. Acoustic response disturbances that propagate away from the blade row, and entropic and vortical disturbances that are convected through the downstream boundary must be determined as part of the unsteady solution. It should be noted that if strong shocks are present in the flow, the far-downstream steady velocity will not be uniform, and an analytic solution for a pressure excitation from downstream [indicated by the subscript $+\infty$ in (1.35)] is unavailable in this case.

1.3.2 Isentropic and Irrotational Mean Flow

If any shocks that occur in the flow are at most of weak to moderate strength, then the steady background flow can be regarded as isentropic (with S = 0) and irrotational; i.e., $\mathbf{V} = \nabla\Phi$, where Φ is a steady velocity potential. In this case, the mass conservation equation for the underlying steady flow is

$$\nabla \cdot (\bar{\rho}\nabla\Phi) = 0 \tag{1.36}$$

and the momentum equation (1.25) can be integrated to yield the following (Bernoulli) relations for the mean-flow variables:

$$(M_{-\infty}A)^2 = \bar{\rho}^{\,\gamma-1} = (\gamma M_{-\infty}^2 P)^{(\gamma-1)/\gamma} = 1 - \frac{\gamma-1}{2} M_{-\infty}^2[(\nabla\Phi)^2 - 1]\,. \tag{1.37}$$

Here, M and $A = \gamma P/\bar{\rho}$ are the local steady Mach number and speed of sound propagation, respectively. The flow tangency condition, $\nabla\Phi \cdot \mathbf{n} = 0$, applies at the blade, (B_m), surfaces, and the steady pressure and normal velocity component must be continuous across blade wakes (W_m). Since entropy and vorticity changes across shocks are regarded as negligible, the mass, momentum and energy conservation laws cannot all be enforced at such surfaces. The usual practice is to require only that mass and tangential momentum be conserved, i.e., $[\![\bar{\rho}\mathbf{V}]\!] \cdot \mathbf{n} = 0$ and $[\![\tilde{\mathbf{V}}]\!] \cdot \boldsymbol{\tau} = 0$, respectively.

Numerical procedures for determining two-dimensional steady potential flows through cascades have been developed extensively, e.g., see Refs. [Cas83] and [WN85], particularly for flows with subsonic relative inlet and exit Mach numbers (i.e., $M_{\mp\infty} < 1$). In such calculations a Kutta condition is usually imposed at blade trailing edges in lieu of prescribing an exit freestream flow property. Also, the usual practice is to solve the conservative form of the mass-balance equation (1.36) throughout the entire fluid domain. Thus, shock-jump conditions are usually not imposed. Instead, as in nonlinear Euler calculations, shocks are captured through the use of special differencing techniques. In potential calculations the wake conditions are satisfied implicitly because the steady fluid properties are continuous and differentiable across wakes.

The Linearized Unsteady Flow

As indicated by Goldstein [Gol78] the system of field equations that govern the linearized unsteady perturbation of a potential mean flow can be cast into a very convenient form by decomposing the unsteady velocity into rotational ($\mathbf{v}_R$) and irrotational ($\nabla\phi$) parts. The rotational velocity, $\mathbf{v}_R$, is taken to be divergence-free far upstream of the blade row, i.e., $\nabla \cdot \mathbf{v}_R \equiv 0$ for $\xi < \xi_-$, and the unsteady pressure depends only upon the potential ϕ through the relation $p = -\bar{\rho}\bar{D}\phi/Dt$. However, for realistic blade profiles $\mathbf{v}_R \cdot \mathbf{n}$ and hence, $\nabla\phi \cdot \mathbf{n}$, will be singular along the mean blade and wake surfaces. Therefore, Atassi and Grzedzinski [AG89] introduced a modified form of the Goldstein decomposition, i.e.,

$$\mathbf{v} = \mathbf{v}_* + \nabla\phi = \mathbf{v}_R + \nabla\phi_* + \nabla\phi \ , \tag{1.38}$$

to facilitate the numerical resolution of the velocity potential, ϕ. Here, ϕ_* is a convected or pressure-less potential (i.e., $\bar{D}\phi_*/Dt = 0$) which satisfies the condition $\nabla\phi_* \cdot \mathbf{n} = -\mathbf{v}_R \cdot \mathbf{n}$ at blade and wake mean positions.

The system of field equations that governs the linearized unsteady flow variables, s, $\mathbf{v}_R$ and ϕ is determined by substituting the velocity decomposition (1.38) into the linearized Euler equations (1.27)–(1.29) to obtain

$$\frac{\bar{D}s}{Dt} = 0 \tag{1.39}$$

$$\frac{\bar{D}}{Dt}(\mathbf{v}_R - s\nabla\Phi/2) + [(\mathbf{v}_R - s\nabla\Phi/2) \cdot \nabla]\nabla\Phi = 0 \tag{1.40}$$

and

$$\frac{\bar{D}}{Dt}(A^{-2}\frac{\bar{D}\phi}{Dt}) - \bar{\rho}^{\ -1}\nabla \cdot (\bar{\rho}\nabla\phi) = \bar{\rho}^{-1}\nabla \cdot (\bar{\rho}\mathbf{v}_*) \ . \tag{1.41}$$

These equations are coupled only sequentially; hence, they can be solved in order to determine the complex amplitudes of the entropy (s), rotational velocity ($\mathbf{v}_R$) and velocity potential (ϕ), respectively. Moreover, closed form solutions [Gol78, HV91] can be determined for the entropy and rotational velocity fluctuations in terms of the known conditions at inlet. The velocity potential, ϕ, is governed by a second-order partial differential equation, along with conditions at the blade, wake, and shock surfaces and at the inlet and exit boundaries, and must be determined using numerical field methods.

We introduce the Lagrangian independent variable, $\mathbf{X} = \Delta\mathbf{e}_T + \Psi\mathbf{e}_N$, where

$$\Delta(\mathbf{x}) = \mathbf{x}_- \cdot \mathbf{e}_T + \int_{\mathbf{x}_- + [\Psi(\mathbf{x}) - \Psi(\mathbf{x}_-)]\mathbf{e}_N}^{\mathbf{x}} V^{-1} d\tau_\Psi$$

and

$$\Psi(\mathbf{x}) = \mathbf{x}_- \cdot \mathbf{e}_N + \int_{\mathbf{x}_-}^{\mathbf{x}} \bar{\rho}(\mathbf{e}_z \times \mathbf{V}) \cdot d\boldsymbol{\tau} \tag{1.42}$$

are the drift and stream functions, respectively, of the steady background flow. The unit vectors $\mathbf{e}_T$, $\mathbf{e}_N (= \mathbf{e}_z \times \mathbf{e}_T)$ and $\mathbf{e}_z$ point in the inlet freestream direction, normal to the inlet freestream direction, and out from the page, respectively; $\mathbf{x}_-$ is the position vector to the point of intersection (ξ_-, η_-) of the reference blade $(m = 0)$ stagnation streamline and the axial line $\xi = \xi_-$; $d\tau_\Psi$ is a differential element of arc length along a streamline; and $d\boldsymbol{\tau}$ is a differential vector tangent to the path of integration.

The solutions to the entropy (1.39) and rotational velocity (1.40) transport equations can be expressed in the form [Gol78]

$$s(\mathbf{x}) = s_{-\infty} \exp(i\boldsymbol{\kappa}_{-\infty} \cdot \mathbf{X}) , \tag{1.43}$$

and

$$\mathbf{v}_R(\mathbf{x}) = [\nabla(\mathbf{X} \cdot \boldsymbol{\mathcal{A}}_{-\infty}) + s_{-\infty}\nabla\Phi/2] \exp(i\boldsymbol{\kappa}_{-\infty} \cdot \mathbf{X}) , \tag{1.44}$$

where $\boldsymbol{\mathcal{A}}_{-\infty} = \mathbf{v}_{R,-\infty} - s_{-\infty}\mathbf{V}_{-\infty}/2$. Since the rotational velocity is divergence-free far upstream of the blade row, the vectors $\boldsymbol{\kappa}_{-\infty}$ and $\mathbf{v}_{R,-\infty}$ are orthogonal, and therefore,

$$\mathbf{v}_{R,-\infty} = i(\boldsymbol{\kappa}_{-\infty} \times \boldsymbol{\zeta}_{-\infty})/|\boldsymbol{\kappa}_{-\infty}|^2 , \tag{1.45}$$

where $\boldsymbol{\zeta}_{-\infty} = i\boldsymbol{\kappa}_{-\infty} \times \mathbf{v}_{R,-\infty}$ is a prescribed quantity, cf. (1.34).

For realistic configurations the steady background flow stagnates at blade leading edges. Therefore, the drift function has a logarithmic singularity, i.e., $\Delta \to a_0 \ln n$, at the mean blade and wake surfaces, and the rotational velocity is singular at such surfaces. A convected potential of the form

$$\phi_* = \left[-i\omega^{-1}\boldsymbol{\mathcal{A}}_{-\infty} \cdot \mathbf{V}_{-\infty} + F(\Psi)\right] \exp(i\boldsymbol{\kappa}_{-\infty} \cdot \mathbf{X}) , \tag{1.46}$$

where

$$F(\Psi) = \frac{\omega^{-1} G \cos\Omega_{-\infty}(\boldsymbol{\kappa}_{-\infty} \times \boldsymbol{\mathcal{A}}_{-\infty}) \cdot \mathbf{e}_z}{2\pi(1 - i a_0 \omega)} \sin\left[\frac{2\pi[\Psi(\mathbf{x}) - \Psi(\mathbf{x}_-)]}{G\cos\Omega_{-\infty}}\right] , \tag{1.47}$$

ensures that $\mathbf{v}_* \cdot \mathbf{n} = (\mathbf{v}_R + \nabla\phi_*) \cdot \mathbf{n} = 0$ at the mean blade and wake surfaces, and hence, that $\nabla\phi \cdot \mathbf{n}$ is finite at such surfaces [AG89].

The unsteady flow variables s, $\mathbf{v}_R$ and ϕ_* can be evaluated once the drift and stream functions of the steady background are determined. The unsteady potential (ϕ) can then be found numerically as a solution of the field equation (1.41) subject to the appropriate surface and far-field conditions. The flow tangency condition (1.30) with $\mathbf{V} = \nabla\Phi$ and $\mathbf{v} \cdot \mathbf{n} = \nabla\phi \cdot \mathbf{n}$ applies at the blade surfaces. Since the irrotational steady velocity and density are continuous downstream of the blade row, the linearized unsteady pressure and normal velocity component are continuous across the wakes, i.e., $[\![\bar{D}\phi/Dt]\!] = 0$ and $[\![\nabla\phi]\!] \cdot \mathbf{n} = 0$ for $\mathbf{x} \in W_m$. Finally, if we neglect changes in entropy $[\![s]\!]$ and vorticity $[\![\zeta]\!]$ across shocks and combine the conservation laws for mass and tangential momentum, we obtain the following condition for determining the jump in ϕ across a shock that terminates in the fluid

$$[\![\bar{\rho}(\nabla\phi + \mathbf{v}_* - A^{-2}\frac{\bar{D}\phi}{Dt}\nabla\Phi)]\!] \cdot \mathbf{n} = [\![\bar{\rho}]\!][i\omega + (\nabla\Phi \cdot \boldsymbol{\tau})\boldsymbol{\tau} \cdot \nabla](\mathbf{r} \cdot \mathbf{n})$$
$$+ (\mathbf{r} \cdot \mathbf{n})\boldsymbol{\tau} \cdot \nabla([\![\bar{\rho}]\!]\nabla\Phi \cdot \boldsymbol{\tau}) , \quad \mathbf{x} \in \mathrm{Sh}_{m,n} , \tag{1.48}$$

where $\mathbf{r} \cdot \mathbf{n} = -([\![\nabla\Phi]\!] \cdot \mathbf{n})^{-1}[\![\phi]\!]$ is the shock displacement normal to the mean shock locus.

The velocity potential fluctuations in the far upstream and far downstream regions depend upon the prescribed acoustic excitation as well as the acoustic and vortical response of the cascade. We can set $\phi(\mathbf{x}) = \phi_E(\mathbf{x}) + \phi_R(\mathbf{x})$ for $\xi \lessgtr \xi_{\mp}$, where ϕ_E accounts for acoustic excitations and is therefore prescribed, and ϕ_R is associated with the response of the blade row and therefore must be determined as part of the unsteady solution. Analytic solutions for ϕ_R can be determined [Ver89c], which satisfy the requirements that acoustic response disturbances either attenuate with increasing axial distance from the blade row or propagate carrying energy away from or parallel to the blade row and that vorticity must be convected downstream. These solutions can be matched to a near-field numerical solution to close the boundary-value problem for the unsteady potential.

1.3.3 Aeroelastic Response Parameters

Solutions to the foregoing nonlinear steady and linearized unsteady flow problems are required to determine the aerodynamic response information needed for blade-row aeroacoustic and aeroelastic predictions, i.e., the unsteady pressure response at inlet and exit and the unsteady pressures acting at the moving blade surfaces. We will proceed to describe briefly the local and global unsteady aerodynamic response parameters that are used in aeroelastic investigations, but refer the reader to Ref. [Ver89a] for more detailed information.

The pressure acting at the instantaneous position of the reference ($m = 0$) blade surface is given by

$$\tilde{P}(\tau_{\mathcal{B}},t) = P(\tau_{\mathrm{B}}) + \tilde{p}_{\mathcal{B}}(\tau_{\mathrm{B}},t) + \sum_n \tilde{P}_{Sh_n}(\tau_{\mathrm{B}},t) + \dots , \tag{1.49}$$

where $\tilde{p}_{\mathcal{B}} = \mathrm{Re}\{p_{\mathcal{B}} \exp(i\omega t)\}$, $p_{\mathcal{B}} = [p + (\mathbf{r}\cdot\nabla)P]_{\mathrm{B}}$, τ measures distance in the counterclockwise direction along the blade surface, and the subscripts $\mathcal{B}$ and B refer to the instantaneous and mean blade surfaces, respectively. The first two terms on the right-hand-side of (1.49) are the steady and the first-harmonic unsteady pressures acting at the moving blade surface. The third term is the anharmonic unsteady pressure associated with shock motion, and is determined by analytically continuing the steady and unsteady solutions from the mean to the instantaneous shock locations. Although the unsteady pressure response is not everywhere harmonic, its integral over a blade surface and, consequently, the first-order global unsteady aerodynamic loads are harmonic in time.

For aeroelastic calculations, it is convenient to introduce a set of generalized forces corresponding to a set of independent, but otherwise arbitrary, structural deformation modes. For this purpose, the reference blade displacement may be expressed in the form

$$\mathcal{R}_{\mathrm{B}}(\mathbf{x},t) = \sum_{i=1}^{I} \tilde{\delta}_i(t)\mathbf{R}_i(\mathbf{x}) = \sum_{i=1}^{I} \mathrm{Re}\{\delta_i e^{i\omega t}\}\mathbf{R}_i(\mathbf{x}) = \mathrm{Re}\{\mathbf{r}_{\mathrm{B}}(\mathbf{x})e^{i\omega t}\} , \tag{1.50}$$

where the generalized coordinate $\tilde{\delta}_i(t)$ is the displacement of the motion in the ith mode, $\mathbf{R}_i(\mathbf{x})$ is a real vector function which describes the shape and direction of this displacement, and we have used the subscript B to indicate blade displacement. The ith generalized force is then defined as

$$\tilde{Q}_i(t) = -\oint_{\mathcal{B}} \tilde{P}\mathbf{n}\cdot\mathbf{R}_i d\tau = Q_i + \tilde{q}_i + \dots , \tag{1.51}$$

where the unit normal vector $\mathbf{n}$ points out from the blade surface, and Q_i and $\tilde{q}_i = \mathrm{Re}\{q_i \exp(i\omega t)\}$ are the steady and the first-harmonic unsteady components, respectively, of the time-dependent generalized force $\tilde{Q}_i(t)$. After replacing the surface integration in (1.51) by an integration over the mean blade surface and performing some algebra [Ver89a], we find that

$$q_i = -\oint_{\mathrm{B}} [P\frac{\partial \mathbf{r}_{\mathrm{B}}}{\partial\tau}\times\mathbf{e}_z + p_{\mathcal{B}}\mathbf{n} - \sum_n r_{\mathrm{Sh}_n,\mathrm{B}}[\![P]\!]_{\mathrm{Sh}_n}\delta(\tau-\tau_{\mathrm{Sh}_n})\mathbf{n}]\cdot\mathbf{R}_i d\tau = \oint_{\mathrm{B}} q_{i,\tau}d\tau, \tag{1.52}$$

where $q_{i,\tau}$ is the distributed ith generalized force, $\mathbf{r}_{\mathrm{Sh}_n,\mathrm{B}} = (\mathbf{r}_{\mathrm{Sh}_n} - \mathbf{r}_{\mathrm{B}})\cdot\boldsymbol{\tau}_{\mathrm{B}}$ is the complex amplitude of the relative shock displacement along the moving blade surface, and the subscript Sh_n refers to the mean nth shock location.

Two time-independent, real parameters that are also useful in describing the aerodynamic response to prescribed blade vibrations are the work per cycle, W_C, and the pressure-displacement function, $w(\tau)$. The former is the work done by an airstream on a given blade over one cycle of its motion. Therefore, a prescribed blade motion is stable, neutrally stable or unstable according to whether the work per cycle is less than, equal to, or greater than zero, respectively. The pressure-displacement function describes the distribution of the work per cycle over a blade surface. If the blade motion is expressed in terms of generalized coordinates, as in (1.50), then

$$W_C = \oint_{\mathrm{B}} w(\tau)d\tau = \pi \oint_{\mathrm{B}} \mathrm{Im}\{\sum_{i=1}^{I} \delta_i^* q_{i,\tau}\}d\tau = \pi\mathrm{Im}\{\sum_{i=1}^{I} \delta_i^* q_i\} . \qquad (1.53)$$

where the superscript $*$ denotes the complex conjugate.

An important special case for turbomachinery aeroelastic calculations is one in which each incremental two-dimensional blade section vibrates in a rigid-body mode consisting of a translation in the y-direction and a counterclockwise rotation about an axis at $\mathbf{x} = \mathbf{x}_{\mathrm{P}}$. In this case $\mathbf{r}_{\mathrm{B}} = \delta_1 \mathbf{e}_y + \delta_2(\mathbf{e}_z \times \mathbf{R}_{\mathrm{P}})$, where the vector $\mathbf{R}_{\mathrm{P}}$ extends from the mean position of the reference blade axis of rotation, $\mathbf{x}_{\mathrm{P}}$, to a point on the mean reference blade surface. The relevant global response parameters are $q_1 = \mathbf{f} \cdot \mathbf{e}_y$, $q_2 = m_{\mathcal{P}} + \delta_2 M_{\mathrm{P}}$, and $W_C = \pi\mathrm{Im}\{\delta_1^*\mathbf{f} \cdot \mathbf{e}_y + \delta_2^* m_{\mathcal{P}}\}$, where, $\mathbf{f} \cdot \mathbf{e}_y$ and $m_{\mathcal{P}}$ are the complex amplitudes of the first-harmonic unsteady aerodynamic lift and moment about the moving pitching axis at $\mathbf{x} = \mathbf{x}_{\mathcal{P}}$, and M_{P} is the zeroth-order or steady moment about the mean pitching axis at $\mathbf{x} = \mathbf{x}_{\mathrm{P}}$.

1.3.4 Linearized Solution Methods

Numerical procedures for predicting unsteady perturbations of isentropic and irrotational steady background flows have received a great deal of attention over the past decade. Recent developments include the prediction of unsteady transonic flows excited by blade vibrations [UV90], and the prediction of the unsteady subsonic flows excited by vortical gusts [HV91, CD91]. Potential-based linearizations are very efficient, requiring only about ten CPU seconds or less on a CRAY YMP. At present, work is proceeding to apply such analyses in blade row aeroelastic [Smi90, SK91] and aeroacoustic [SS90] studies, and to couple linearized inviscid analyses to unsteady viscous-layer analyses to provide weak and strong inviscid/viscid interaction (IVI) solution capabilities for unsteady cascade flows [JEPC90, VBHA91].

Two potential-based unsteady aerodynamic linearizations, called FINSUP and LINFLO, have received wide application. The FINSUP analysis [Whi82, WN85, Whi90] applies at subsonic and supersonic inlet and exit Mach numbers, and to unsteady flows excited by prescribed blade motions and inlet and exit acoustic excitations. Here, the nonlinear steady and the first harmonic unsteady equations are approximated on a fixed, unstructured, triangular mesh using a finite element spatial discretization. Shock effects are captured in both the nonlinear steady and the linearized unsteady analyses. The steady equations are solved using a Newton iteration procedure in which each member of a sequence of approximations to the nonlinear steady solution are determined by direct matrix inversion via Gaussian elimination. The complex linear unsteady equations are solved directly using a similar inversion procedure.

The LINFLO analysis [VC84, UV90, HV91] applies at subsonic inlet and exit Mach numbers, and to unsteady cascade flows excited by prescribed blade motions, entropic and vortical excitations at inlet, and acoustic excitations at inlet and exit. LINFLO treats only the linearized unsteady flow problem; therefore, steady background flow information must be supplied as input for an unsteady flow calculation. For the most part, the cascade, full-potential analysis of Ref. [Cas83] has been used for this purpose. The linearized unsteady equations are discretized, using an implicit, least-squares, finite-difference approximation [CV81] on a fixed composite mesh [UV90], consisting of a blade-passage H-mesh and a local C-mesh which wraps around a blade. Although the H-mesh alone is usually sufficient for resolving subsonic flows, the local mesh allows an accurate modeling of leading-edge and shock phenomena. Shocks are captured in the steady analysis, but fitted in the linearized unsteady calculation via the implementation of the shock-jump condition (1.48).

A linearized Euler analysis was first proposed by Ni & Sisto [NS76] so that strong shock effects could be taken into account in unsteady flow calculations. However, because of the associated computing requirements, significant progress has been achieved only recently. In particular, Hall & Crawley [HC89] approximated the conservative forms of the nonlinear steady and the first-harmonic unsteady Euler equations on a fixed H-grid using a finite volume discretization. The steady equations are solved via Newton iteration. Successive approximations to the final steady solution as well as a direct linearized unsteady flow solution are obtained by applying Gaussion elimination to a large sparse system of linear algebraic equations. Results, determined by shock fitting, for a two-dimensional unsteady flow through a hyperbolic channel show an excellent definition of shock phenomena. In particular, the calculated moving shock remains normal to the channel walls at all times, and the amplitude of the shock displacement varies along the shock, decreasing with increasing shock strength.

Because of the difficulties anticipated in implementing a direct solution procedure for three-dimensional flows and in fitting shocks and wakes in complicated two- and three-dimensional unsteady cascade flows, Hall & Clark [HC91] have recently proposed a time-marching Euler procedure for steady and linearized unsteady flows. Here, the conservative forms of the nonlinear steady and first-harmonic unsteady Euler equations are solved by advancing solutions in pseudo-time to steady and periodic unsteady, states, respectively. Shocks and wakes are captured in both the steady and the unsteady flow calculations The linearized unsteady equations are solved on a harmonically deforming H-grid, thereby avoiding the need to approximate the mean-flow gradient terms [i.e., $\mathbf{r} \cdot \nabla \mathbf{V}$ in (1.30) and $\mathbf{r} \cdot \nabla P$ see (1.49)] that arise from transferring surface information from the instantaneous to the mean surface locations. The method requires approximately 1 CPU minute of CRAY YMP time for a representative, two-dimensional, subsonic calculation. Results for subsonic flat-plate and compressor cascades are in very good agreement with those of classical and full-potential based unsteady aerodynamic linearizations, respectively.

1.4 Example Response Predictions

Theoretical results will be presented to indicate the status of linearized unsteady aerodynamic prediction methods and to demonstrate several important features associated with the aerodynamic response of isolated two-dimensional blade rows to prescribed blade vibrations and vortical excitations. We consider two compressor-type cascades both operating at subsonic inlet conditions. The first, a high-speed compressor cascade, consists of cambered NACA 0006 airfoils and operates at high subsonic inlet Mach number; the second, a compressor exit guide vane (EGV), consists of highly cambered NACA 0012 blades and operates at an inlet Mach number of 0.3.

The steady background flows are assumed to be isentropic and irrotational and to satisfy a Kutta condition at blade trailing edges. Therefore, only information on the inlet free-stream conditions, e.g., inlet Mach number and flow angle, is required to calculate the steady flow fields. The unsteady flows through the NACA 0006 cascade are excited by prescribed single-degree-of-freedom (SDOF) blade motions at unit amplitude, ($|\delta_i| = 1$); those through the EGV, by prescribed vortical gusts with $\mathbf{v}_{R,-\infty} \cdot \mathbf{e}_N = (1, 0)$. These unsteady excitations are termed subresonant if all acoustic response waves attenuate with increasing axial distance from the blade row and superresonant (m, n) if m and n such waves persist far upstream and/or far downstream, respectively, and carry energy away from the blade row. An acoustic resonance occurs if at least one wave persists in the far field and carries energy along the blade row.

Theoretical results, including steady Mach number distributions and global and local unsteady aerodynamic response predictions, are presented below. The steady-flow predictions have been determined using the cascade full potential analysis of Ref. [Cas83]; the unsteady results, using the LINFLO analysis. The unsteady transonic solutions for the NACA 0006 cascade have been determined on a composite (H/C) mesh. Subsonic solutions for the EGV and a corresponding flat-plate cascade have been determined using the H-mesh alone.

More extensive results for the NACA 0006 and the EGV can be found in Refs. [Ver89a, UV90, HV91, VBHA91]. These configurations can serve as useful benchmarks for evaluating various unsteady aerodynamic prediction

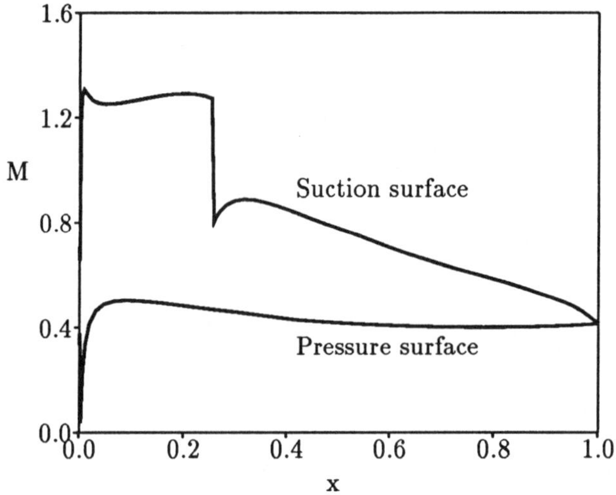

FIGURE 1.4. Surface Mach number distributions for the transonic NACA 0006 cascade: $M_{-\infty} = 0.8$, $\Omega_{-\infty} = 58$ deg.

methods. Indeed, the NACA 0006 cascade has been designated as one of the standard cascade configurations [FS83] for establishing a theoretical and experimental data base for unsteady cascade flows. Consequently, a number of investigators have applied linearized analyses, both potential-based and Euler, and time-marching Euler and Navier-Stokes simulations to this configuration. Thus far, the result comparisons have been very encouraging. In particular, Huff [Huf91] has applied a nonlinear Euler analysis to several of the transonic cases reported below and found very good agreement between the time-marching and the linearized response predictions.

1.4.1 Transonic NACA 0006 Cascade

This cascade has a stagger angle, Θ, of 45°, a gap/chord ratio, G, of unity, and operates at an inlet Mach number, $M_{-\infty}$, and flow angle, $\Omega_{-\infty}$, of 0.8 and 58 deg. respectively. The blades are constructed by superposing the thickness distribution of a NACA 0006 airfoil on a circular-arc camber line having a height at blade midchord of $y = 0.05$ [Ver89a]. The mean or steady flow through the cascade is transonic with a single normal shock occurring in each blade passage. The predicted blade-surface Mach number distributions are shown in Fig. 1.4. The transonic mean flow stagnates on the pressure surface at $x = 0.002$, and the normal shock impinges on the suction surface at $x = 0.258$. The Mach numbers at the base of the shock are 1.282 on the upstream side and 0.816 on the downstream side, and the exit Mach number and flow angle are 0.432 and 40.3 deg, respectively. The mean lift force acting on each blade is 0.42, and the jump in the steady pressure at the base of the shock, $[\![P_B]\!]$, is -0.477.

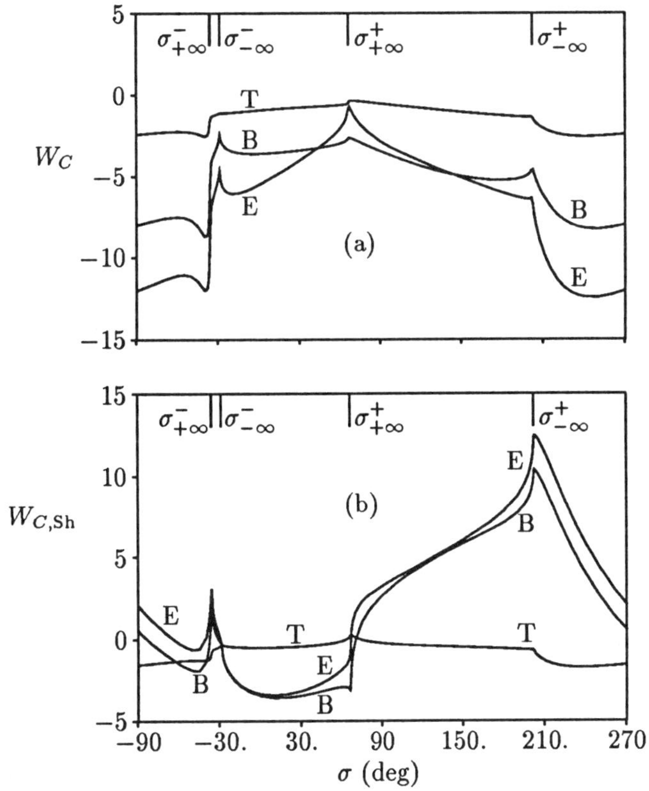

FIGURE 1.5. Work per cycle vs interblade phase for unit frequency, SDOF motions of the transonic NACA 0006 cascade: B, E and T denote bending, chordwise bending (elastic) and torsional blade motions, respectively: (a) Work per cycle; (b) Work per cycle due to shock motion.

Work per cycle predictions are given in Figure 1.5 for pure bending ($\mathbf{R} = \mathbf{e}_y$), pure torsion ($\mathbf{R} = \mathbf{e}_z \times \mathbf{R}_\mathrm{P}$) about midchord ($\mathbf{x}_\mathrm{P} = 0.5, 0$), and pure chordwise bending or elastic ($\mathbf{R} = \mathbf{e}_y \sin \pi x$) modes of blade vibration at unit amplitude ($|\delta| = 1$) and unit frequency ($\omega = 1$). The vertical lines above the curves denote the resonant interblade phase angles for a unit-frequency excitation. The curves in Figure 1.5a indicate that the SDOF bending, torsional and chordwise bending motions at unit frequency are stable ($W_C < 0$) and that the torsional vibrations have the lowest stability margin. They also indicate the rather complicated nature of the global unsteady aerodynamic response associated with realistic cascade flows, which results from the different acoustic responses (i.e., subresonant, superresonant (1,0), etc.) produced by unit frequency excitations at different interblade phase angles. Note that abrupt changes in global unsteady aerodynamic response behavior appear at the interblade phase angles at

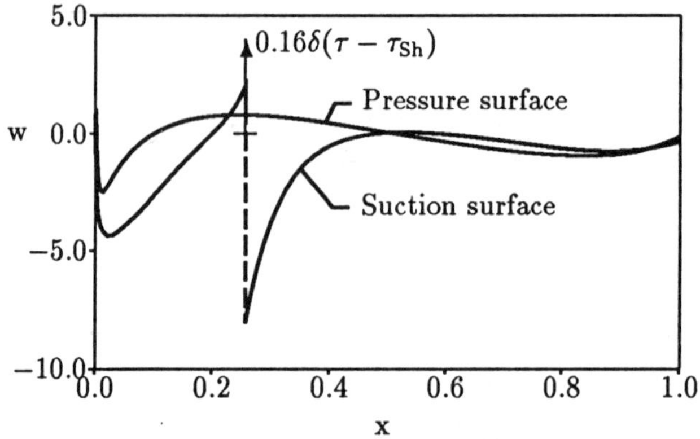

FIGURE 1.6. Pressure-displacement function distribution for the transonic NACA 0006 cascade due to an out-of-phase torsional vibration about midchord at $\omega = 1.5$.

which an acoustic resonance occurs.

As noted previously, there are two contributions to the unsteady airloads for a discontinuous transonic flow — one arising from the first harmonic unsteady pressure response; the other, from the anharmonic pressure response associated with shock motion. Shock-induced work per cycle responses, i.e., $W_{C,\mathrm{Sh}} = \mathrm{Im}\{r_{\mathrm{Sh,B}}([\![P_\mathrm{B}]\!]\mathbf{n}_\mathrm{B} \cdot \mathbf{r}_\mathrm{B}^*)_\mathrm{Sh}\}$, for the transonic NACA 0006 cascade are shown in Fig. 1.5b. The results for the bending, torsional and chordwise bending vibrations indicate that, except for the superresonant (1,1) blade motions [i.e., those occurring in the range $\sigma\epsilon(\sigma_{-\infty}^-, \sigma_{+\infty}^+)$], the shock loads tend to support the unit-frequency bending and chordwise bending motions. Also, these loads become quite severe for interblade phase angles near the upstream resonance condition, $\sigma = \sigma_{-\infty}^+$. The pure torsional motions at $\omega = 1$ give rise to relatively smaller shock loads, and these usually act to resist or stabilize the blade motions.

A benefit of using the work per cycle to describe global unsteady aerodynamic response behavior is that the responses to different modes of blade motion, including coupled modes, can be represented via this single, real, global, response parameter. The pressure displacement function serves the same purpose in describing the local response. A pressure displacement function distribution for the transonic NACA 0006 cascade is shown in Fig. 1.6 for a case in which the blades undergo a unit-amplitude, out-of-phase ($\sigma = 180$ deg), torsional motion about midchord at $\omega = 1.5$. This motion is stable with a work per cycle (W_C) response of -1.24. The work per cycle due to the motion of the shock ($W_{C,\mathrm{Sh}}$) is 0.16. The concentrated shock load, indicated by the arrow at $x = 0.256$ in Fig. 1.6, supports the torsional blade motion ($W_{C,\mathrm{Sh}} > 0$), but, over most of the blade, the harmonic unsteady loads, particularly those just downstream of the shock, tend to suppress this motion.

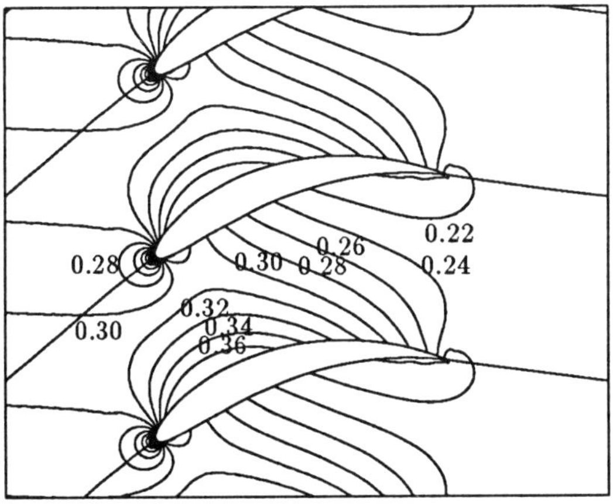

FIGURE 1.7. Mach number contours for steady flow at $M_{-\infty} = 0.3$ and $\Omega_{-\infty} = 40$ deg through the EGV cascade.

1.4.2 Compressor Exit Guide Vane

The compressor exit guide vane (EGV) has a stagger angle, Θ, of 15 deg, a blade spacing, G, of 0.6 and operates at a prescribed inlet Mach number, $M_{-\infty}$, and inlet flow angle, $\Omega_{-\infty}$, of 0.3 and 40 deg, respectively. The blades are constructed by superposing the thickness distribution of a NACA 0012 airfoil on a circular arc camber line having a height at midchord of $y = 0.13$. The calculated steady Mach number contours for this configuration are shown in Fig. 1.7. The exit Mach number and exit flow angle are 0.226 and -7.4 deg, respectively, and the mean lift force acting on each blade is 0.36.

We consider unsteady flows that are excited by inlet vortical disturbances. We will examine the behavior of the unsteady vorticity and pressure fields that result from the interaction between the EGV and a vortical gust defined by $\mathbf{v}_{R,-\infty} \cdot \mathbf{e}_N = (1,0)$, $\omega = 5$ and $\sigma = -2\pi$. In addition, we will examine the unsteady moment responses of the EGV and a corresponding flat plate cascade (i.e., $\Theta = \Omega = 40$ deg, $G = 0.6$ and $M = 0.3$) to vortical excitations at $\mathbf{v}_{R,-\infty} \cdot \mathbf{e}_N = (1,0)$, $\omega = 5$ and $-3\pi \leq \sigma \leq -\pi$. Note that the unsteady moments are taken about midchord ($\mathbf{x}_P = 0.5, 0$), and the mean positions of the flat-plate blades are aligned with the inlet free-stream flow direction.

Contours of the real parts of the complex amplitudes of the unsteady vorticity, $\mathrm{Re}\{\boldsymbol{\zeta}\} = \mathrm{Re}\{\zeta\}\mathbf{e}_z$, and the unsteady pressure, $\mathrm{Re}\{p\}$ for the EGV are shown in Fig. 1.8. As indicated, the vortical gust is distorted as it is convected by the nonuniform mean flow through the EGV blade row. The mean flow stagnates in the vicinity of the blade leading edges, and consequently, the vortex lines are severely stretched through thin regions adjacent to the mean blade and wake surfaces. In addition, since the vor-

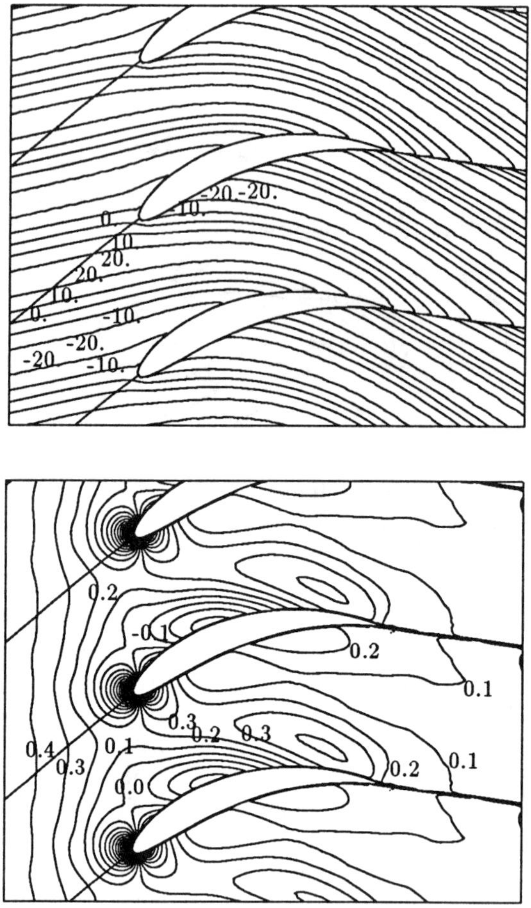

FIGURE 1.8. Contours of the in-phase components (real parts) of the unsteady vorticity and the unsteady pressure for the EGV cascade subjected to an incident vortical gust with $\mathbf{v}_g = (1, 0)$, $\omega = 5$ and $\sigma = -2\pi$.

ticity is convected at different mean velocities along the upper and lower surfaces of the EGV blades, an observer moving across the wake of an EGV blade would encounter strong and abrupt changes in vorticity. The unsteady pressure field that arises due to the interaction between this vortical gust and the EGV is depicted at the bottom of Fig. 1.8. This pressure response is superresonant (1,1). A relatively strong acoustic response wave persists far upstream and carries energy away from the blade row; a much weaker one persists far downstream.

The unsteady moments ($m_{\mathcal{P}}$) that act on the reference blades of the EGV and the corresponding flat-plate cascade are plotted versus interblade

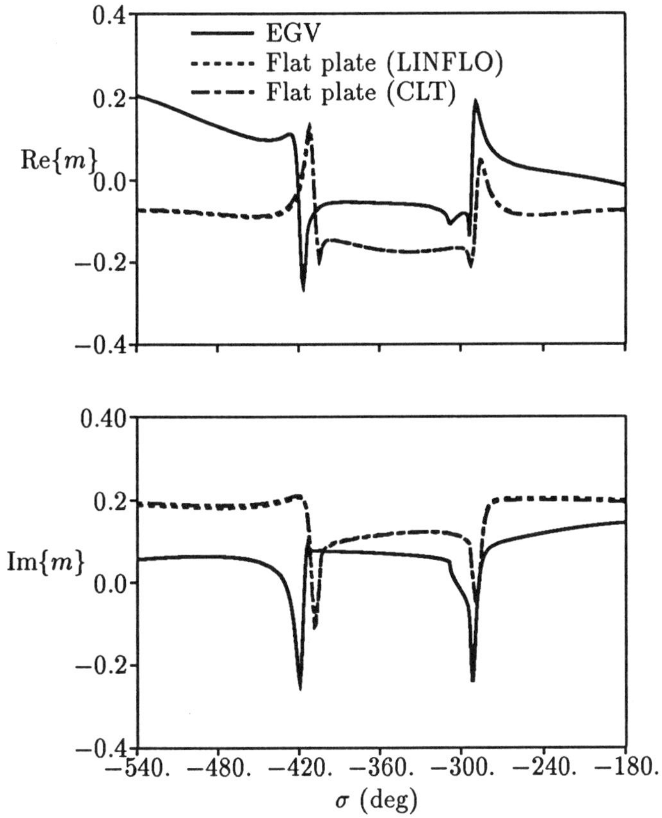

FIGURE 1.9. Unsteady moment about midchord vs interblade phase angle for the EGV and corresponding flat-plate cascades subjected to vortical gusts with $\mathbf{v}_{R,-\infty} \cdot \mathbf{e}_N = (1,0)$ and $\omega = 5$.

phase angle in Figure 1.9 for vortical excitations at $\mathbf{v}_{R,-\infty} \cdot \mathbf{e}_N = (1,0)$, $\omega = 5$ and $-3\pi(-540\,\text{deg}) \leq \sigma \leq -\pi(-180\,\text{deg})$ Here, the excitations at $\sigma = -404.2$ deg and -293.9 deg produce resonant acoustic response disturbances far upstream and far downstream of the flat-plate cascade and far upstream of the EGV; those at $\sigma = -414.3$ deg and -308.8 deg produce such responses far downstream of the EGV. The vortical gusts are convected without distortion by the uniform mean flow through the flat-plate cascade, i.e., the rectilinear vorticity contour patterns that exist far upstream of the EGV persist throughout the flat-plate flow field. The results in Fig. 1.9 indicate the effects of gust distortion, due to mean-flow nonuniformities, on blade response. In particular, mean-flow nonuniformities have a substantial effect on the unsteady moment response to high-wave-number vortical excitations, i.e., those occurring over the phase-angle range $-540\,\text{deg} \leq \sigma < -420\,\text{deg}$. Note that the LINFLO and classical linear theory (CLT) moment predictions for the flat-plate cascade are in excellent agreement.

1.5 Concluding Remarks

Two distinct approaches are being pursued actively to provide more accurate unsteady aerodynamic response information for turbomachinery aeroelastic and aeroacoustic design applications. In one, the time-dependent, nonlinear equations that describe viscid (i.e., the Navier-Stokes equations) or inviscid (the Euler equations) fluid motion are solved by advancing numerical calculations in time until a converged periodic unsteady state is determined. In the other or more traditional approach, inviscid ($Re \to \infty$) and small unsteady disturbance assumptions are invoked to obtain time-independent, nonlinear steady and linearized unsteady equations that are solved sequentially to determine the steady and first-harmonic, unsteady inviscid flow properties. A very encouraging sign, at present, is that good agreement is being achieved between the results of the nonlinear, time-marching and the linear frequency domain analyses for inviscid unsteady cascade flows driven by small-amplitude excitations.

The computational requirements associated with time-accurate numerical simulations of nonlinear unsteady flows will continue to prohibit the use of such simulations in the aeroelastic or aeroacoustic design process. Thus, the traditional approach must still be exploited so that useful aerodynamic response information can be provided. Inviscid unsteady aerodynamic linearizations offer great promise for meeting needs for efficient solutions that contain much of the essential physics of practical unsteady cascade flows. However, methods must be developed to include viscous displacement effects within the framework of a linearized inviscid analysis. A linearization of the viscous equations of motion may not be suitable for this purpose. If not, it will be necessary to couple linearized inviscid and nonlinear viscous-layer analyses, using an appropriate interaction law, to provide a useful inviscid/viscid interaction analysis for unsteady cascade flows. The traditional approach must also be extended for application to three-dimensional unsteady flows. Since strong-shock phenomena do not usually play a critical role, a potential-based linearization is probably adequate for two-dimensional applications. However, since vortical mean-flow effects are expected to be important, the linearized Euler approach is the natural one to consider for future extension to three-dimensional flows.

The major disadvantages of linearized inviscid and high Reynolds number inviscid/viscid interaction analyses are that they are difficult to develop, and they will have a limited range of application. Thus, research to develop accurate numerical simulations of fully nonlinear unsteady flows must also continue. In the near term, this work should be directed towards providing a reliable "numerical test facility" for unsteady cascade flows, which could be used to determine the importance of the nonlinear effects associated with shock behavior, viscous separation, finite-amplitude excitation, etc. This facility could also be used to indicate the regions of validity of approximate analyses, and to provide unsteady aerodynamic response information in parametric regions in which the approximate methods fail.

Acknowledgments: The preparation of this article was supported under the United Technologies Corporate Research Program. The author would like to thank Ms. R. Rudewicz for her help in this effort.

REFERENCES

[AA80] H. M. Atassi and T. J. Akai. Aerodynamic and aeroelastic characteristics of oscillating loaded cascades at low Mach number. *Trans. ASME A: J. Engng for Power*, 102(2):344, 1980.

[AG78] J. J. Adamczyk and M. E. Goldstein. Unsteady flow in a supersonic cascade with subsonic leading-edge locus. *AIAA J.*, 16(12):1248, 1978.

[AG89] H. M. Atassi and J. Grzedzinski. Unsteady disturbances of streaming motions around bodies. *J. Fluid Mech.*, 209:385, 1989.

[BF81] O. O. Bendiksen and P. Friedmann. Coupled bending-torsion flutter in supersonic cascade. *AIAA J.*, 19:774, 1981.

[BL78] B. S. Baldwin and H. Lomax. Thin-layer approximation and algebraic model for separated turbulent flow. Paper 78–257, AIAA 16th Aerospace Sciences Meeting, 1978.

[Cas83] J. R. Caspar. Unconditionally stable calculation of transonic potential flow through cascades using an adaptive mesh for shock capture. *Trans. ASME A: J. Engng for Power*, 105(3):504, 1983.

[CD91] J. E. Caruthers and W. N. Dalton. Unsteady aerodynamic response of a cascade to nonuniform inflow. Paper 91–GT–174, ASME Intl Gas Turbine and Aeroengine Congress and Expo., 1991.

[Chi80] M. R. Chi. Unsteady aerodynamics in stalled cascade and stall flutter prediction. Paper 80–C2/Aero–1, ASME, Century 2 Aerospace Conf., 1980.

[CS74] T. Cebeci and A. M. O. Smith. *Analysis of Turbulent Boundary Layers*, pages 211–239. Academic Press, 1974.

[CV81] J. R. Caspar and J. M. Verdon. Numerical treatment of unsteady subsonic flow past an oscillating cascade. *AIAA J.*, 19(12):1531, 1981.

[DDEM90] D. J. Dorney, R. L. Davis, D. E. Edwards, and N. K. Madavan. Unsteady analysis of hot streak migration in a turbine stage. Paper 90–2354, AIAA 26th Joint Prop. Conf., 1990.

[EAM77] J. I. Erdos, E. Alzner, and W. McNally. Numerical solution of periodic transonic flow through a fan stage. *AIAA J.*, 15(11):1559, 1977.

[FJ87] S. Fleeter and R. L. Jay. Unsteady aerodynamic measurements in flutter research. In M. F. Platzer and F. O. Carta, editors, *Unsteady Turbomachinery Aerodynamics, AGARD–AG–298*, 1987.

[Fle79] S. Fleeter. Aeroelasticity research for turbomachine applications. *J. of Aircraft*, 16(5):320, 1979.

[Fou86] A. Fourmaux. Unsteady flow calculation in cascades. Paper 86–GT–178, ASME Intl Gas Turbine Conf. and Exhibit, Dusseldorf, West Germany, 1986.

[FS83] T. H. Fransson and P. Suter. Two-dimensional and quasi three-dimensional experimental standard configurations for aeroelastic investigations in turbomachine-cascades. Report LTA–TM–83–2, Ecole Polytechnique Federale de Lausanne, Lausanne, Switzerland, 1983.

[GBA77] M. E. Goldstein, W. Braun, and J. J. Adamczyk. Unsteady flow in a supersonic cascade with strong in-passage shocks. *J. of Fluid Mech*, 83(Part 3):569, 1977.

[GBQ90] G. A. Gerolymos, E. Blin, and H. Quiniou. Comparison of inviscid computations with theory and experiment in vibrating transonic compressor cascades. Paper 90–GT–373, ASME 35th Intl Gas Turbine and Aeroengine Congress and Expo., Brussels, Belgium, 1990.

[GH91] M. B. Giles and R. Haimes. Validation of a numerical method for unsteady flow calculations. Paper 91–GT–271, ASME Intl Gas Turbine and Aeroengine Congress and Expo., 1991.

[Gil88] M. B. Giles. Calculation of unsteady wake rotor interaction. *J. Prop. and Power*, 4(4):356, 1988.

[Gil90a] M. B. Giles. Nonreflecting boundary conditions for Euler equation calculations. *AIAA J.*, 28(5):2050, 1990.

[Gil90b] M. B. Giles. Stator/rotor interaction in a transonic turbine. *AIAA J. Prop. and Power*, 6(5):621, 1990.

[Gol78] M. E. Goldstein. Unsteady vortical and entropic distortions of potential flows round arbitrary obstacles. *J. Fluid Mech.*, 89(Part 3):433, 1978.

[GR87] J. F. Groeneweg and E. J. Rice. Aircraft turbofan noise. *Trans. ASME, J. of Turbomachinery*, 109(1):130, 1987.

[HC89] K. C. Hall and E. F. Crawley. Calculation of unsteady flows in turbomachinery using the linearized Euler equations. *AIAA J.*, 27(6):777, 1989.

[HC91] K. C. Hall and W. S. Clark. Predictions of unsteady aerodynamic loads in cascades using the linearized Euler equations on deforming grids. Paper 91–3378, AIAA/SAE/ ASME/ASEE 27th Joint Prop. Conf., 1991.

[HD91] L. He and J. D. Denton. Inviscid-viscous coupled solution for unsteady flows through vibrating blades, part 1. description of the method; part 2. computational results. Papers 91–GT–125 and 91–GT–126, ASME Intl Gas Turbine and Aeroengine Congress and Expo., 1991.

[He90] L. He. An Euler solution for unsteady flows around oscillating blades. *Trans. ASME, J. of Turbomachinery*, 112(4):714, 1990.

[HR89] D. L. Huff and T. S. R. Reddy. Numerical analysis of supersonic flow through oscillating cascade sections by using a deforming grid. Paper 89–2805, AIAA 25th Joint Prop. Conf., 1989.

[HSR91] D. L. Huff, T. W. Swafford, and T. S. R. Reddy. Euler flow predictions for an oscillating cascade using a high resolution wave-split scheme. Paper 91–GT–198, ASME Intl Gas Turbine and Aeroengine Congress and Expo.,, 1991.

[Huf91] D. L. Huff, September 1991. Private communication.

[HV91] K. C. Hall and J. M. Verdon. Gust response analysis for cascades operating in nonuniform mean flows. *AIAA J.*, 29(9):1463, 1991.

[JEPC90] H. M. Jang, J. A. Ekaterinaris, M. F. Platzer, and T. Cebeci. Essential ingredients for the computation of steady and unsteady blade boundary layers. Paper 90–GT–160, ASME 35th Intl Gas Turbine and Aeroengine Congress and Expo., 1990.

[KG90] B. Krouthen and M. Giles. Numerical investigation of hot streaks in turbines. *J. Propulsion*, 6(6):769, 1990.

[KK85] M. Koya and S. Kotake. Numerical analysis of fully three-dimensional periodic flows through a turbine stage. *J. Engng for Gas Turbines and Power*, 107:945, 1985.

[KO70] S. Kaji and T. Okazaki. Propagation of sound waves through a blade row, ii. analysis based on the acceleration potential method. *J. Sound and Vibration*, 11(3):355, 1970.

[Lan57] F. Lane. Supersonic flow past an oscillating cascade with supersonic leading-edge locus. *J. Aeronautical Sciences*, 24:65, 1957.

[LF82] R. C. Lock and M. C. P. Firmin. Survey of techniques for estimating viscous effects in external aerodynamics. In P. L. Roe, editor, *Proc. IMA Conf. on Numerical Methods in Aeronautical Fluid Dynamics*, pages 337–430. Academic Press, 1982.

[MASS75] A. A. Mikolajczak, R. A. Arnoldi, L. E. Snyder, and H. Stargardter. Advances in fan and compressor blade flutter analysis and predictions. *J. Aircraft*, 12(4):325, 1975.

[Mel80] R. E. Melnik. Turbulent interactions on airfoils at transonic speeds — recent developments. In *AGARD Symp. on Computation of Viscous-Inviscid Flows*, 1980.

[Nam87] M. Namba. Three dimensional flows. In M. F. Platzer and F. O. Carta, editors, *Unsteady Turbomachinery Aerodynamics, AGARD-AG-298*, 1987.

[Ni79] R. H. Ni. A rational analysis of periodic flow perturbation in supersonic two-dimensional cascade. *Trans. ASME A: J. Engng for Power*, 101(3):431, 1979.

[NS76] R. H. Ni and F. Sisto. Numerical computation of nonstationary aerodynamics of flat plate cascades in compressible flow. *Trans. ASME: J. of Engng for Power*, 98:165, 1976.

[NW78] T. Nagashima and D. S. Whitehead. Linearized supersonic unsteady flow in cascades. R&M 3811, British Aeronautical Research Council, London, UK, 1978.

[Rai87] M. M. Rai. Navier-Stokes simulations of rotor-stator interaction using patched and overlaid grids. *AIAA J. Prop. and Power*, 3(5):387, 1987.

[Rai89] M. M. Rai. Three-dimensional Navier-Stokes simulations of turbine rotor- stator interaction; part 1 – methology, part 2 – results. *AIAA J. Prop. and Power*, 5(3):305, 1989.

[RD90] M. M. Rai and R. P. Dring. Navier-Stokes analyses of the redistribution of inlet temperature distortions in a turbine. *AIAA J. Prop. and Power*, 6(3):276, 1990.

[SA86] S. P. Surampudi and J. J. Adamczyk. Unsteady transonic flow over cascade blades. *AIAA J.*, 24(2):293, 1986.

[Sid91] L. D. G. Siden. Numerical simulation of unsteady viscous compressible flows applied to blade flutter analysis. Paper 91–GT–203, ASME Intl Gas Turbine and Aeroengine Congress and Expo., 1991.

[SK91] T. E. Smith and J. R. Kadambi. The effect of steady aerodynamic loading on the flutter stability of turbomachinery blading. Paper 91–GT–130, ASME Intl Gas Turbine and Aeroengine Congress and Expo., 1991.

[Smi71] S. N. Smith. Discrete frequency sound generation in axial flow turbomachines. R&M 3709, British Aeronautical Research Council, London, UK, 1971.

[Smi90] T. E. Smith. Aerodynamic stability of a high-energy turbine blade. Paper 90–2351, AIAA/SAE/ASME/ASEE 26th Joint Prop. Conf., 1990.

[SS90] A. Suddhoo and P. Stow. Simulation of inviscid blade-row interaction using a linearized potential code. Paper 90–1916, AIAA/SAE/ASME/ASEE 26th Joint Prop. Conf., 1990.

[UV90] W. J. Usab, Jr. and J. M. Verdon. Advances in the numerical analysis of linearized unsteady cascade flows. Paper 90–GT–11, ASME 35th Intl Gas Turbine and Aeroengine Congress and Expo., 1990.

[VBHA91] J. M. Verdon, M. Barnett, K. C. Hall, and T. C. Ayer. Development of unsteady aerodynamic analyses for turbomachinery aeroelastic and aeroacoustic applications. CR 4405, NASA Lewis Research Center, 1991.

[VC84] J. M. Verdon and J. R. Caspar. A linearized unsteady aerodynamic analysis for transonic cascades. *J. Fluid Mech.*, 149:403, 1984.

[Ver77] J. M. Verdon. Further developments in the aerodynamic analysis of unsteady supersonic cascades, i. the unsteady pressure field, ii. aerodynamic response predictions. *Trans. ASME A: J. Engng for Power*, 99(4):509, 1977.

[Ver89a] J. M. Verdon. The unsteady aerodynamic response to arbitrary modes of blade motion. *J. Fluids and Structures*, 3(3):255, 1989.

[Ver89b] J. M. Verdon. Unsteady aerodynamics for turbomachinery aeroelastic applications. In D. Nixon, editor, *Unsteady Transonic Aerodynamics, Progress in Astronautics and Aeronautics*, pages 287–347, 1989.

[Ver89c] J. M. Verdon. The unsteady flow in the far field of an isolated blade row. *J. Fluids and Structures*, 3(2):123, 1989.

[Ver90] J. M. Verdon. Linearized unsteady aerodynamics for turbomachinery aeroelastic applications. Paper 90–2355, AIAA/SAE/ ASME/ASEE 26th Joint Prop. Conf., 1990.

[Whi70] D. S. Whitehead. Vibration and sound generation in a cascade of flat plates in subsonic flow. Report CUED/A–Turbo/TR 15, Cambridge Univ. Engng Dept., Cambridge, UK, 1970.

[Whi82] D. S. Whitehead. The calculation of steady and unsteady transonic flow in cascades. Report CUED/A–Turbo/TR 118, Cambridge Univ. Engrng Dept, Cambridge, UK, 1982.

[Whi87] D. S. Whitehead. Classical two-dimensional methods. In M. F. Platzer and F. O. Carta, editors, *Unsteady Turbomachinery Aerodynamics, AGARD–AG–298*, 1987.

[Whi90] D. S. Whitehead. A finite element solution of unsteady two-dimensional flow in cascades. *Intl J. for Numerical Methods in Fluids*, 10:13, 1990.

[WN85] D. S. Whitehead and S. G. Newton. A finite element method for the solution of two-dimensional transonic flows in cascades. *Intl J. for Numerical Methods in Fluids*, 5(2):115, 1985.

An Innovative Algorithm for Periodic Flow Calculations Using a Parallel Architecture – Some Applications to Unsteady Aerodynamics

G. Carte[1]
P. Fraunié[1]
P. Dussouillez[1]

ABSTRACT An iterative scheme for the computation of periodic unsteady flows is presented, using a global space and time discretization, instead of a time marching procedure. A decomposition of domain is used to implement the algorithm on distributed architecture computers.
The preliminary results presented here concern two types of periodic flows: the first one is a pulsed boundary layer developing on a flat plate and the second one is a periodic wake behind a rectangular afterbody, when the flow generated Strouhal frequency is a new unknown.

1.1 Introduction

Different numerical methods for solving the Navier-Stokes equations are being used to aid in the study of unsteady flows. Using the fact that these equations are parabolic in the time direction, numerous explicit or implicit time marching schemes have been performed (see for example Peyret and Taylor, 1983 [1]). Studies concerning the stability, transition to turbulence and time evolution of the solution of the Navier-Stokes unsteady laminar equations are under progress concerning compressible and incompressible flows at low reynolds numbers [2, Karniadakis et al, 1990]. However, because of the non-linear advection term, rigorous second order temporal precision remains unachievable, although different implicit schemes increased this order from the first one. We can note furthermore that a time marching procedure involves an error cumulation but no convergence test.

Another case concerns the modeling of complex turbulent flows at high Reynolds numbers, when submitted to unsteady boundary conditions or

[1]Institut de Mécanique Statistique de la Turbulence, 12 avenue général Leclerc 13003 Marseille, France

when dominated by large structures. Large eddy simulations provide an accurate tool for investigation, although only a limited class of flows can be modelled using these computations. [3, Reynolds, 1989]. On the other hand, turbulence modelling needs further developments for complex situations such as separated or strongly unsteady flows [4, Launder and Tselepidakis, 1991], [5, Ha Minh, et al, 1989]. Concerning the turbulent closure, a numerical problem appears when unsteady (i.e. evoluting) Navier-Stokes equations have to be coupled together with unsteady closure equations.

A large simplification is avalaible when periodically time varying flows are considered. Then a statistical analysis can be used according to Hussain and Reynolds [6] who defined an ensemble averaging:

$$< \xi(\vec{X},t) >= \lim_{N\to\infty} \frac{1}{N} \sum_{i=1}^{N} \xi(\vec{X},t+iT) \tag{1}$$

Then, the mean Navier-Stokes equations becomes:

$$\frac{\partial < u_i >}{\partial t} + < u_j > \frac{\partial < u_i >}{\partial x_j} = -\frac{\partial < P >}{\partial x_i} + \frac{\partial}{\partial x_j}\left(\frac{1}{\mathcal{R}_e}\frac{\partial < u_i >}{\partial xj} - < u_i' u_j' >\right) \tag{2}$$

and the associated mean continuity equation is:

$$\frac{\partial < u_i >}{\partial x_i} = 0$$

The period T can be imposed by the unsteady boundary conditions as for example in the case of a profile in periodic motion, as well as flow generated, as the Strouhal frequency in the wake of a body. The assumption made here is that this period is stable and deterministic of the flow solution.

Furthermore, for real flows, the Navier-Stokes equations (2) can be coupled to a system of other non linear partial differential equations which are averaged in the same way. Examples of such equations are turbulent closure, thermal or chemical fields, and wall deformation equations.

We present here an iterative method that we expect [7, 8] to be able to solve such complex periodic problems encountered in spatio-temporal flows as in turbomachines or heat exchangers. Preliminary results are discussed in terms of validation as well as computing facilities. A simple unsteady self generated bidimensional flow behind an afterbody is computed to prove the convergence of the scheme, and a pulsed boundary layer is computed to study its implementation on a MIMD parallel architecture.

1.2 The "Periodic Convergence" Method

1.2.1 THE NUMERICAL PROCEDURE

An original numerical scheme is proposed here in order to obtain a converging solution of equation (2) in a time-space domain $\mathcal{D} = [0, T] \times \mathcal{R}^n$, where T is the temporal period and n the space dimension.

Considering initial conditions for all the unknowns in the whole $\mathcal{D}$ domain, the "periodically convergent scheme" provides a new approximation of the solution for each periodic iteration. So, for each iteration, all the quantities are approximated for each grid point and each time step.

Furthermore, the time boundary condition is cyclic and writes:

$$< \xi(\vec{X}, T) > = < \xi(\vec{X}, 0) > \tag{3}$$

where the quantity

$$\frac{\partial < \xi(\vec{X}, T) >}{\partial t}$$

is a new unknown of the problem.

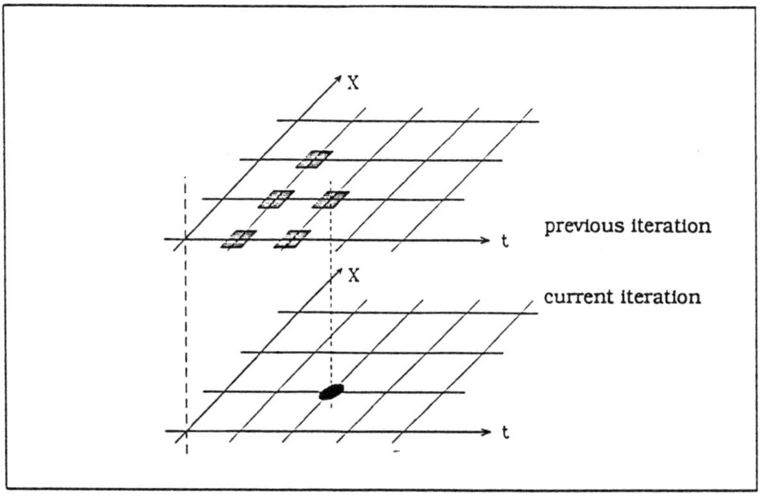

Figure 1: Periodic Convergence Concept

1.2.2 THE TIME DISCRETIZATION

A significant improvement available from this scheme is related to the time discretization which can be centered and written with higher order terms, to be consistant with the spatial discretization. Generally, introducing a second order time derivative $\partial^2 < u_i > / \partial t^2$ and a second order finite difference scheme, the time derivative becomes:

$$\left(\frac{\partial \xi}{\partial t}\right) = \frac{\alpha - \beta}{\Delta t}\xi_t + \frac{\beta \xi_{t+1} - \alpha \xi_{t-1}}{\Delta t} \tag{4}$$

with $\alpha + \beta = 1$.

First, when β equals zero, classical upwind time marching schemes can be tested by a converged numerical solution. Then the scheme can be explicit as well as implicit. In the presented examples, implicit schemes in finite and volume differences are presented, based upon the Keller box for a parabolic problem or an ADI scheme for an elliptic problem.

Secondly, improved time discretization is obtained when β takes non zero values. Furthermore, the numerical stability range concerning the time to space steps ratio can be increased for a given numerical scheme.

However, the preliminary tests performed show that the stability of the algorithm looks very sensitive to initial conditions.

1.3 Multitask Computational Procedure

1.3.1 METHODOLOGY

In the case of periodic time variations, when a convergent solution of a coupled system of equations is researched, the iterative algorithm described above uses a decomposition of the bounded space-time domain, for which implementation on distributed memory architectures is straight forward.

When assuming the space and time derivatives to be very similarily treated in the computational domain $[0, T] \times \mathcal{R}^n$, i.e. in a second order form over three consecutive points, calculations can be performed independently for each grid point $(\vec{X}, t)$, if an explicit iterative scheme is acting in the decomposition direction. This decomposition allows the implementation of the numerical method on a MIMD computer.

The choice of the decomposition direction depends evidently on the problem. Here two different (time and streamwise) directions have been chosen for two different problems.

Then, for each periodic iteration, the same solver is acting on the different processors associated to the subdomains, and a global convergence test is performed that synchronizes the multitask procedure, before data is exchanged between the nodes to allow a new run.

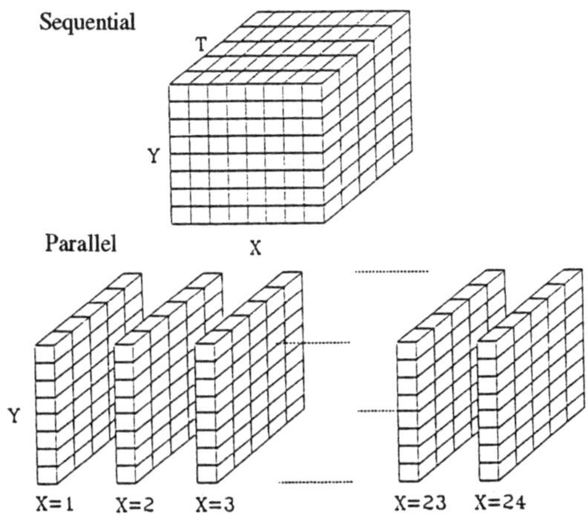

Figure 2: Domain Decomposition

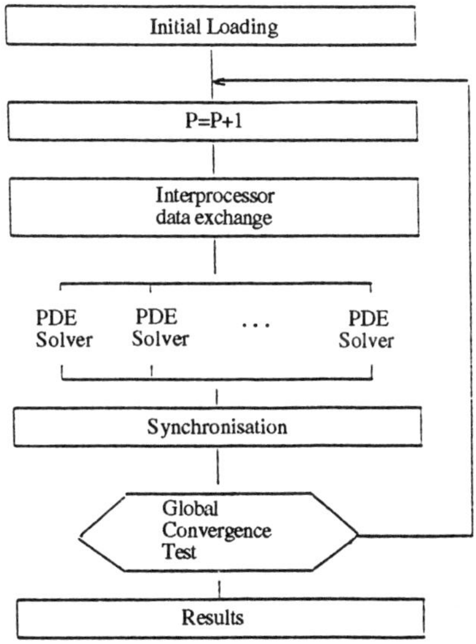

Figure 3: Parallel Algorithm

1.3.2 TECHNICAL CONSIDERATIONS

Choosing, for example, the time direction as the decomposition direction, each node will compute a part of the time steps in the period. The period is then divided by the number of processors.

Hopefully a good load balance between the processors enables a cpu acceleration that will depend on the adjusted domain decomposition.

Concerning the storage constraints, a lot of data needs to be duplicated. For instance, each node needs a copy of the entire spatial constant data.

Moreover, because of the temporal discretization, a storage overlapping of all the unknown values for the first and last time steps on each node is necessary. However the data flow exchange can be optimized when the network topology is consistent with the data distribution. Hence, a ring structure appears to be a good solution, reducing the communications to adjacent nodes. Note here that the cyclic boundary conditions (3) permit se the ring structure to close.

1.4 Applications

1.4.1 THE UNSTEADY BOUNDARY LAYER: A FORCED UNSTEADY FLOW

A simple unsteady problem concerns the well documented boundary layer developing on a flat plate placed in a sinusoidally pulsating flow, in the form $U_e(t) = \overline{U_e} + \Delta U_e \cos(\omega t)$.

In the laminar case, this flow is known as the unsteady Blasius flow, and refers to the Stokes analytical solution (when $\overline{U_e} = 0$). An unsteady effect occurs in the way of a shift advance on the first harmonic in a Fourier series decomposition, and this solution is well known to be periodically stable. The boundary layer thickness is then in order of magnitude of $6(\nu x/\overline{U_e})^{1/2}$.

The formulation of this problem is simple, using the Falkner-Skan parabolic equation:

$$(aF'')' + bFF' + c(F')^2 + d = x\left[F'\frac{\partial F'}{\partial x} - F''\frac{\partial F}{\partial x} + \frac{1}{U_e}\frac{\partial F'}{\partial t}\right] \qquad (5)$$

where the stream function Ψ identifies $(\overline{U_e}x)^{1/2}F(\xi,\eta,t)$, ξ equals x/L, η equals $y(\overline{U_e}/\nu x)^{1/2}$, and with

$$a = 1 + \frac{\nu}{\nu_t}, \quad b = \frac{1}{2}(1+c), \quad c = \frac{x}{u_e}\frac{du_e}{dx}, \quad d = \frac{x}{u_e^2}\left(\frac{\partial u_e}{\partial t} + u_e\frac{\partial u_e}{\partial x}\right)$$

Keller and Cebeci (1971) [9] proposed an accurate implicit scheme to solve this single equation in finite differences with second order space and first order time discretizations. A mixing-length closure model is introduced for the turbulent flow prediction.

Parallelism Procedure

An adapted algorithm has been implemented on a 24 transputers network (INMOS T800 associated to 1Mbyte) by considering a decomposition direction of the (x, y, t) domain following the "x" space direction [10, 11]. The Volvox transputers system was configured as an open ring, and 24 space steps have been kept for the boundary layer grid along the "x" streamwise direction, when each transputer is associated to one of the 24 "x" steps.

A (24x60x60) (x, y, t) mesh was necessary to obtain an optimal precision / CPU time ratio. For this configuration, 84 Kbytes where used by each transputer for a local treatement, to be compared to the 1.5 Mbytes necessary for the sequential algorithm in the same conditions.

Then for each period, synchronized data exchanges were necessary between adjacent memories in the optimized form as presented in figure 4.

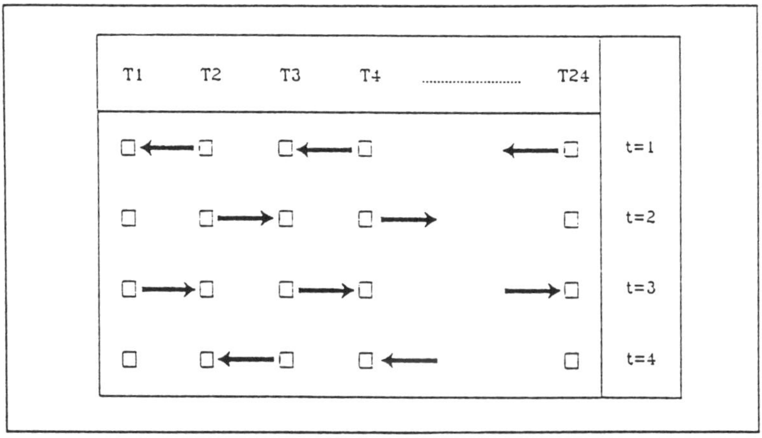

Figure 4: Local Data Exchanges

An overstorage of 40% was then necessary, in order to compute the derivative terms following the "x" direction.

The algorithm convergence has been successfully tested for forced periodic unsteady conditions in this case. However, the sensitivity of the stability of the numerical scheme to initial conditions has been pointed out. It could be connected to the low precision (first order) time discretization scheme used here.

Because of an advantageous computation versus data exchanges ratio, a good efficiency has been observed although no performance measurement tools were available. However, inhomogeneous task length (i.e. processors loading) has been observed between different "x" locations, especially depending on the turbulence occurence (computation of ν_t). In this way, a time direction decomposition would have provided better results.

Finally, despite the uncomfortable use of the "under progress" transputers systems and compilers (OCCAM then FORTRAN in our case), this simple application has shown that distributed memory systems can help in the near future for solving complex flow calculations with important storage and significant CPU time reductions.

Numerical Results

From an appropriate initial solution for each "x" and "t" discretization point, the "periodic convergence" scheme based on the fixed pulsating period T, equal to $2\pi/\omega$, has been tested in terms of the harmonic behaviour.

Equation (5) is parabolic in time and streamwise directions and an initial profile is generated at the first space step for all the time steps, when the temporal boundary condition (3) applies to all time steps. The first order upwind time discretization did not change but the unknown $\xi^p_{x,t}$ at the "x" location and the "t" time of the period at the pth iteration were related to the $\xi^{p-1}_{x\pm1,t\pm1}$ terms computed at the previous "p-1" iteration.

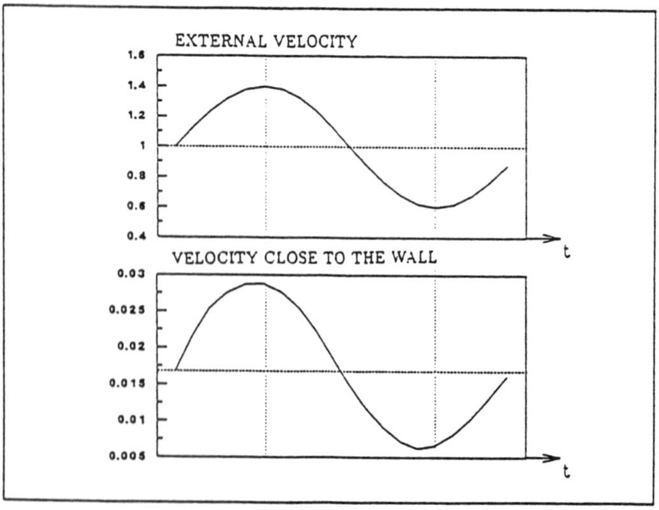

Figure 5: Unsteady Boundary Layer

This explicit iterative procedure permitted to find again the classical unsteady solution, as shown on Figure 5. Then, the convergence of the "periodic" algorithm and its ability to be implemented on a parallel computer were tested.

1.4.2 THE PERIODIC FLOW BEHIND AN AFTERBODY

The unsteady flow past a rectangular afterbody confined in a channel is a very classical test case concerning the flow induced periodicity. Numerous studies have been performed previously on this academic but very simple flow, using cartesian mesh generation techniques.

From this point of view, numerical preliminary tests have been performed, introducing the periodic convergence scheme with both explicit and implicit schemes.

A classical time marching implicit algorithm has been developed using a Douglas ADI scheme with a finite volume spatial discretization and a finite difference temporal discretization. The pressure and velocity formulation introduces an elliptic pressure term in the Navier-Stokes equations and a difficult convergent procedure is introduced to satisfy the continuity equation by resolution of the Poisson equation:

$$\nabla^2 \phi = div(\overrightarrow{U^*}) \tag{6}$$

where ϕ is a correction potential from the predicted velocity $\overrightarrow{U^*}$, as proposed by Braza et al, 1986 [12].

Downstream boundary conditions are fixed as:

$$\frac{\partial^2 u}{\partial x^2} = 0, \frac{\partial v}{\partial x} = 0, \frac{\partial^3 \phi}{\partial x^3} = 0, \tag{7}$$

for a convenient numerical behaviour. The computation iterative procedure leads to solving the system of the laminar Navier-Stokes equations when coupled to a Poisson equation for the pressure correction. For each time step, a convergence test is verified concerning the residue of equation (6).

The computation of such a flow provides a classical stable periodic solution at a Strouhal frequency for Reynolds numbers greater than a critical value.

An Adapted Scheme from ADI

Resolution of equations (2) in a $\mathcal{R}^2 \times [0, T]$ domain can be formulated by considering the three-dimensional problem where the spatial and temporal derivatives are treated in the same way. Then, the previous equation can be written:

$$\frac{\partial u_i}{\partial s} + \frac{\partial u_i}{\partial t} + u_j \frac{\partial u_i}{\partial x_j} = -\frac{\partial P}{\partial x_i} + \frac{1}{\mathcal{R}_e} \frac{\partial^2 u_i}{\partial x_j^2} \tag{8}$$

where "s" is a silent iterative variable. Note that in the periodic laminar case, $< \xi(t) >$ identifies $\xi(t)$. By analogy to quasi-steady numerical schemes, introducing an unsteady system of three dimensional partial difference equations in order to converge to a steady solution, we are interested here, in developing numerical schemes to get $\partial u_i / \partial s$ converging towards 0.

We note that the time derivative $\partial u_i / \partial t$ can be discretized in the same order than the spatial derivatives, and moreover that centered finite differences can be performed ($\beta \neq 0$ in equation (4)). However, we note a different structure of the transport equations for the spatial and time derivatives that lead us to consider the linear time derivative in an explicit way:

$$\left(\frac{\partial u_i}{\partial s}\right)^p + \left(u_j\frac{\partial u_i}{\partial x_j}\right)^p_{p-1} + \left(\frac{1}{\mathcal{R}_e}\frac{\partial^2 u_i}{\partial x_j^2}\right)^p_{p-1} = \left(\frac{\partial u_i}{\partial t}\right)_{p-1)} - \left(\frac{\partial P}{\partial x_i}\right)_{p-1} \quad (9)$$

where the $(\xi)^p$ quantities are explicitly discretized between the "p-1" and "p" iterations, the $(\xi)^{p-1}$ quantities are explicitly discretized at the "p-1" iteration, and the $(\xi)^p_{p-1}$ quantities can be implicitly discretized between the "p-1" and "p" iterations or explicitly discretized at the "p-1" iteration.

So, the bi-dimensional ADI numerical scheme can be adapted to the "periodic convergence" concept, by considering the time derivative term as a new unknown and a fictive "p" variable. So the ADI principle can be conserved in the form:

$$\frac{\xi^{p+1} - \xi^p}{\Delta p} + F(\xi^{p+1}, \xi^p) = S(\xi^p) \quad (10)$$

where F represents the first member implicit term including part of the spatial derivatives following the altered spatial direction, and where S represents the second member explicit term, including the spatial derivatives in the other altered direction and the time derivatives. In this way, the time derivatives (that can be centered) are always explicitly expressed, when space derivatives are sometimes implicitly expressed. Although it is not common, such a procedure can adapt very easily a classical ADI scheme for the "periodic convergence" procedure.

Numerical Procedure

Both explicit and implicit altered direction schemes have been developed to solve equation (9) when coupled to a Poisson pressure equation.

The numerical iterative procedure remains similar to the previous one, except for the additional unsteady term in the right member of equation (9).

For each space and time steps, a convergence test is applied to the residue of the velocity divergence (equation (6)), whose critical value has to be adjusted to increase the efficiency.

Every time a new scheme was tested, a range of values was to be optimized, concerning the spatial (Δx, Δy), temporal (Δt) and "iterative" (Δs) steps. The stability analysis of the different schemes has been extrapolated from time marching schemes experience for the spatial and temporal directions, when stability experience concerning the ADI schemes was translated from the temporal direction to the "s" iterative one. Note however that the time step Δt is free of numerical stability criteria, because it appears only for the temporal previously computed term, but that waves propagations in the solution can strongly depend on its value.

Different time discretization schemess, up to fourth order centered finite differences, have been tested in order to reduce the number of necessary iterations.

The first difficulty concerns the Strouhal number determination, whose approximation is iterated together with the periodic field. In this way, we chose to keep a constant number of time steps in the period and to iterate the time step value $\Delta t = k\Delta_o t$ from an initial value $\Delta_o t$ by introducing a relaxation parameter k. Two criteria have been selected to adjust this parameter from comparison between the first "t_1" and last "t_n" time steps in the period:

- the first test which is "qualitative", deals with the sign of the "v" spanwise component of the velocity at a point of the wake axis as "v" is oscillating in a sinusoidal way.
- the second test which is "quantitative", deals with the norm of the difference of a scalar field ξ between the two times "t_1" and "t_n". For example, we computed the quantity:

$$\Delta_\Sigma(\xi)^{t_1,t_n} = \sum_{x,y} \mid \xi_{x,y}(t_1) - \xi_{x,y}(t_n) \mid \tag{11}$$

Then, the parameter k can be adjusted for each "p" iteration from interpolation or extrapolation of $\Delta_\Sigma(\xi)^{t_1,t_{n-1}}$ and $\Delta_\Sigma(\xi)^{t_1,t_n}$, depending on the "qualitative" test. Of course, an unequal accuracy of this procedure has been observed and human intervention is often necessary.

At the same time, a convergence test will be computed for each iteration in terms of $\Delta_\Sigma(\xi(t_i))^{p-1,p} < \epsilon$.

Results

The numerical codes have been performed using a 4D320 Silicon Graphics station, with a mesh dimension of 240 x 30 space steps and 35 time steps in a period. Starting from a "classically" computed solution on the same mesh, and an estimated Strouhal frequency, the "iterative" explicit and implicit algorithms have been performed. First results show a possible convergence, although numerical oscillations still occur. The Reynolds number was 400, and the numerical parameters were: $\Delta t = 510^{-2}, \Delta s = 10^{-2}, \Delta x = \Delta y = 1/30$. The Strouhal frequency had to be adjusted up to the fourth order precision. The global convergence has been reached satisfactorily (figure 6), although numerical oscillations subsist due to an insufficient discretization precision or wave propagation or both. The pressure field (figure 7) shows a satisfying physical solution for this problem.

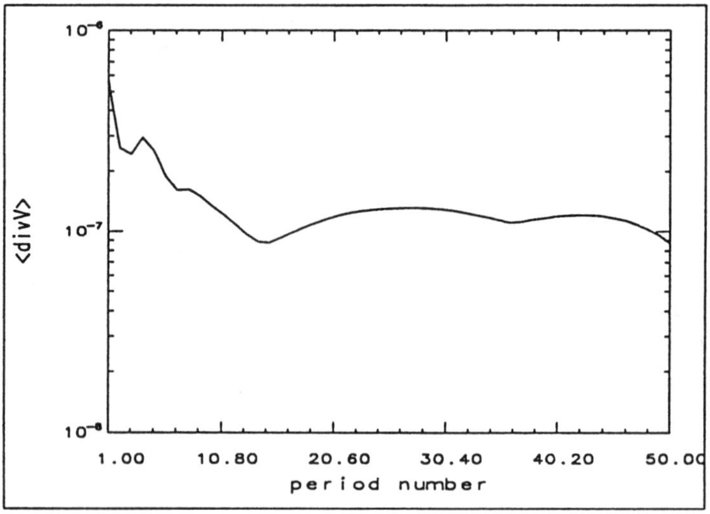

Figure 6: Periodic Evolution of $< div(\overrightarrow{U}) > at \quad t = 10$

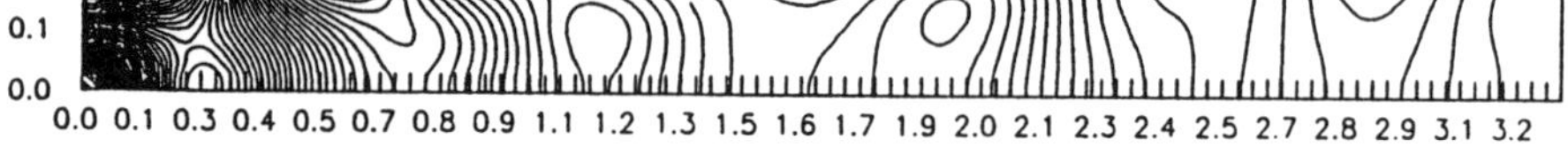

Figure 7: Pressure Field in the Wake of an Afterbody

At the end, the implementation of the code on a iPSC-860 hypercube is under progress, although delayed by necessary sequential algorithms developments.

1.5 Conclusion

The "periodic convergence" concept [7, Fraunié 1987] has been adapted to basic laminar unsteady flows, in forced and self generated periodic conditions, to demonstrate its ability to solve coupled partial differential equations.

In the case of finite difference time discretization, an adaptation of the ADI schemes is proposed, in order to solve bidimensional unsteady problems. At this point, the "periodic convergence" algorithm is shown to provide results as well as classical schemes using parabolic time marching. Only upwind temporal discretization schemes have been performed for comparison, and stability problems can occur. These problems will only be understood after much more experience is gained and a complete numerical analysis done. Especially, wall and downstream conditions are expected to generate low frequency instabilities impeding the periodic convergence. However, a good adjustement of the time and space grids would allow periodic problems to be solved in a satisfactory way.

Explicit iterative schemes have been developed, which permit solving coupled partial differential equation systems in periodic unsteady conditions. Such an approach could be of interest in future for the computation of fully turbulent periodic flows, using turbulent closure transport equations.

In another part, numerical performances have been partially investigated using parallel architecture computers, for a single equation test case, and it demonstrated encouraging performance. An overstorage (of 40 percent in our case) is still necessary, but is compensated for by an increased local data access performance. And a significant acceleration ratio has been observed.

Acknowledgements: This work was supported by the "Direction des Recherches Etudes et Techniques:, grant n 87 272.

G. Carte's fellowship was supported by the group "Transferts Thermiques, Aérodynamique" , Electricité de France, and the "Centre National de la Recherche Scientifique".

The authors thank the "Institut Mediterranéen de la Technologie" and "Région PACA" for disposal of a 24 transputers Volvox system.

The authors thank Mariana Braza, "Institut de Mécanique des fluides de Toulouse" for the benefit of her experience concerning Navier-Stokes solvers.

REFERENCES

[1] R. Peyret and T.D. Taylor. *Computational Methods for Fluid Flow.* 1983.

[2] G.E. Karniadakis and G.S. Triantafyllou. Three-dimensional dynamics and transition to turbulence in the wake of bluff objects. *submitted to J. Fluid Mech.*, 1990.

[3] W.C. Reynolds. The potential and limitations of direct and large eddy simulations. In *Whiter Turbulence workshop, Cornell Univ., March 22-24*, 1989.

[4] B.E. Launder and D.P. Tselepidakis. Progress and paradoxes in modelling near-wall turbulence. In *paper 29-1, 8th Symp. Turbulent Shear Flows, Munich*, 1991.

[5] M.W. Rubesin D.D. Vandromme H. Ha Minh, J.R. Viegas and P. Spalart. Physical analysis and second order modelling of an unsteady turbulent flow: the oscillating boundary layer on a flat plate. In *Proc. Turbulent Shear Flow 7, Stanford, U.S.A.*, 1989.

[6] A.K.M.F. Hussain and W.C. Reynolds. The mechanics of an organized wave in turbulent shear flow. *J. Fluid Mech., vol 41*, 1970.

[7] P. Fraunié. *Analyse des effets instationnaires sur un profil d'aile animé d'un mouvement de trajectoire circulaire.* PhD thesis, University Aix-Marseille II, 1987.

[8] P. Fraunié. Calcul d'écoulements en conditions aux limites périodiques sur architecture parallèle. application en aérodynamique instationnaire. In *Response of shear flows to imposed unsteadiness.* EUROMECH 272, Aussois, 14-18 jan, 1991.

[9] H.B. Keller and T. Cebeci. An accurate numerical method for boundary layer flows. *AIAA J. vol 10 n°9*, 1971.

[10] J.J. Casalot. Calcul d'écoulements périodiques sur architecture parallèle - application à la couche limite sur plaqe plane. Master's thesis, ESIM/IMT, 1989.

[11] E. Richalley. Parallélisation de la résolution des équations de navier-stokes sur un réseau de transputers, 1990.

[12] P. Chassaing M. Braza and H. Ha Minh. Numerical study and physical analysis of the pressure and velocity fields in the near wake of a circular cylinder. *J. Fluid Mech., vol 165*, 1986.

A Framework for Multi-Stage Unsteady Flow Calculations

Michael Giles[1]

ABSTRACT This paper describes a mathematical approach to the calculation of unsteady flow in multi-stage turbomachinery. An asymptotic formulation is used, with the small asymptotic parameter being the level of unsteadiness in each blade row. The baseline flow is the nonlinear steady flow that is computed by many existing multi-stage calculation methods. The first order linear perturbation is the unsteady flow field arising from stator/rotor interactions between neighboring blade rows. The second order correction contains the information about the time-averaged effect of unsteadiness on the mean flow. The advantage of this asymptotic approach is that it leads to a set of equations which can be solved numerically very much more cheaply than the full nonlinear unsteady flow equations, while still retaining the key features of the flow.

1.1 Introduction

1.1.1 A BRIEF REVIEW

Over the last twenty years considerable effort has been devoted to the calculation of unsteady flow in turbomachinery. All of the early work was based upon the approximation that the level of unsteadiness was sufficiently small that it could be considered to be a linear perturbation to some base steady-state flow field. One of the earliest noteworthy achievements was the development by Smith of a method of calculating the linear unsteady flow around a two-dimensional flat plate cascade [Smi71]. In this work the steady flow is taken to be a uniform subsonic flow, and the blade and wake are modelled with an unsteady vortex sheet. The theory of Smith was later extended by Ni [Ni78] to include supersonic steady-state flows. In addition to being computationally inexpensive, an advantage of these methods is the ability to analyze the unsteadiness due to wake/blade interaction, incoming potential waves, or structural bending or torsion. The weakness is the inappropriateness of the base steady flow solution.

[1]Associate Professor, Department of Aeronautics and Astronautics, M.I.T., Cambridge, MA 02139

The next major section of work was on linear potential methods, in which the steady flow is taken to be a solution of the nonlinear two-dimensional full potential equation. Whitehead [Whi82] and Verdon [VC84] developed quite similar numerical methods for solving the linear equations that result from assuming that the unsteadiness is sinusoidal in time. These linear potential methods have been the main technique used in the turbomachinery industry for the last ten years for the prediction of flutter associated with either bending or torsion. They have also been used recently to predict the unsteady potential interaction component of stator/rotor interaction [SS90]. The limitation of linear potential methods is the extreme difficulty involved in extensions to include three-dimensional rotational mean flows, flows with strong shocks, or flows with unsteady wakes.

Continuing with linear methods, in 1974, Ni did preliminary research on a numerical method for solving a modified form of the linear unsteady Euler equations [NS75]. Later, in 1987, Hall developed a finite element method for solving the linear Euler equations [HC87]. This work was for subsonic cascade flows, and transonic ducts with shock fitting. It also demonstrated the validity of the linear approximation up to quite substantial levels of unsteadiness. In 1988, Giles proposed that a modified form of Ni's method would be the most efficient way to solve the two-dimensional and three-dimensional linear Euler equations [Gil89]. At least four separate research groups are now working on such an approach. A key contribution by Lindquist is the proof that if correctly formulated and implemented, shock capturing can produce the same correct answers as shock-fitting [LG91]. This is important because shock capturing is very much simpler to implement, particularly in three dimensions.

Over the last ten years or so, there has also been a great deal of work on methods for calculating nonlinear unsteady flows by time-marching methods [AGEG91, Che88, EGSS89, EAM77, Fou86, Gil88, Gil90, Hod84, Rai87, Rai89]. The capabilities of the codes which have been developed vary widely. Some are two-dimensional, some have quasi-three-dimensional extensions, and some are fully three-dimensional. Some assume inviscid flow and use the Euler equations, while others adopt the Navier-Stokes equations with turbulence modelling. Some are able to treat arbitrary pitch ratios using techniques due to Erdos [EAM77] or Giles [Gil88] but most treat only very simple integer ratios (such as 2:3). Most can analyze stator/rotor interaction in which the unsteadiness is due to the relative blade motion. Some can also treat unsteadiness due to prescribed blade vibration or incoming wakes and potential disturbances. Depending on the capabilities of different methods, the required computer time can also vary greatly. Inviscid, two-dimensional computations can be performed in minutes on a CRAY, while viscous three-dimensional calculations can take days. It is clear however, that these nonlinear time-marching methods have given us very useful tools to investigate unsteady flows in isolated blade rows and single stages, and many papers are being written on the use of such codes to answer important engineering questions.

Having briefly discussed the history of both linear and nonlinear unsteady flow methods, the question that now arises is which is the most appropriate for the calculation of three-dimensional multi-stage unsteady flows. There are some who would argue that a multi-stage, nonlinear, unsteady, Navier-Stokes calculation is the only valid approach, and one should simply wait for computer hardware to improve sufficiently to make such calculations possible. I disagree strongly with this philosophy. Such calculations will take orders of magnitude more time than current viscous single-stage calculations. This is not only because there are more blade rows. It is also because the techniques used to treat arbitrary pitch ratios will not work in multi-stage situations and so the only correct treatment in many situations would be to include every single blade in the entire engine. This will remain an unreasonable option for many years.

In the early 1980's Adamczyk pointed out the impracticality of multi-stage unsteady Navier-Stokes calculations. One of the principal concerns about unsteadiness in turbomachinery flow is the effect of the unsteadiness in changing the mean flow properties such as mass flow and efficiency. Therefore, Adamczyk formulated a system of 'passage-averaged' equations in which the effect of deterministic periodic unsteadiness on the mean flow is included through terms which are similar in nature to the Reynolds stresses in the Reynolds-averaging of turbulent flow equations[Ada85]. These ideas of Adamczyk played a major role in stimulating my ideas for the quadratic terms in the approach in this paper.

The other major concerns about flow unsteadiness are the possibility of bending or torsional flutter, and the magnitude of unsteady loads and forced response due to various kinds of unsteady interactions. Flutter is by definition a linear problem, since it is concerned with the possible growth of infinitesimal unsteady perturbations. A growing body of evidence suggests that the forced response problem is also largely linear. Comparisons between linear and nonlinear calculations, for levels of unsteadiness typical of engine conditions, show that the linear calculations predict the forced response to within 5%. This level of accuracy is perfectly adequate for all practical engineering design purposes, and is probably within the level of other modelling errors. The great savings in computer time from linear methods therefore justify their use for this purpose.

1.1.2 Outline of the asymptotic approach

The approach which is proposed uses the first three terms in an asymptotic expansion in which the level of unsteadiness is the small asymptotic parameter. It is very closely related to the mathematical techniques used to analyze a phenomenon known as 'acoustic streaming', in which unsteady sinusoidal perturbations of a stationary fluid can generate a small second order mean flow [Bat67, Sch79].

The approach is best illustrated by considering the quasi-one-dimensional Euler equations for flow in a variable area duct

$$h\frac{\partial U}{\partial t} + \frac{\partial (hF)}{\partial x} - \frac{\partial h}{\partial x} P = 0, \tag{1.1}$$

where the vectors U, F, P are

$$U = \begin{pmatrix} \rho \\ \rho u \\ \rho E \end{pmatrix}, \quad F = \begin{pmatrix} \rho u \\ \rho u^2 + p \\ u\,(\rho E + p) \end{pmatrix}, \quad P = \begin{pmatrix} 0 \\ p \\ 0 \end{pmatrix}, \tag{1.2}$$

and h, ρ, u, E are the duct area, density, velocity, and total internal energy, respectively, and the pressure p is obtained through the assumption of an ideal, perfect gas from the equation

$$p = (\gamma - 1)\,\rho\left(E - \tfrac{1}{2}u^2\right). \tag{1.3}$$

The first key step is that the unsteady solution $U(x,t)$ is assumed to have an asymptotic form

$$U(x,t) = U^{(0)}(x) + \epsilon\, U^{(1)}(x,t) + \epsilon^2 U^{(2)}(x,t) + \cdots \tag{1.4}$$

ϵ is some measure of the level of unsteadiness. The first term is the steady term, representing a nonlinear steady-state base flow solution. The second term is the linear one, representing the leading order nature of the unsteady solution, and having zero time average. The third term is the quadratic term whose time-average will reveal the change to the mean flow due to the unsteadiness.

The second step is to carefully express $F(U)$ and $P(U)$ as corresponding asymptotic functions of ϵ by using Taylor series expansions.

$$F_i(U^{(0)} + \delta U) \approx F_i(U^{(0)}) + \frac{\partial F_i^{(0)}}{\partial U_j}\,\delta U_j + \frac{1}{2}\frac{\partial^2 F_i^{(0)}}{\partial U_j \partial U_k}\,\delta U_j\,\delta U_k + \cdots \tag{1.5}$$

$$\Longrightarrow \quad F_i(U^{(0)} + \epsilon\, U^{(1)} + \epsilon^2 U^{(2)} + \cdots) \approx \tag{1.6}$$
$$F_i(U^{(0)} + \epsilon \frac{\partial F_i^{(0)}}{\partial U_j} U_j^{(1)} + \epsilon^2 \left(\frac{\partial F_i^{(0)}}{\partial U_j} U_j^{(2)} + \frac{1}{2}\frac{\partial^2 F_i^{(0)}}{\partial U_j \partial U_k} U_j^{(1)}\, U_k^{(1)} \right) + \cdots$$

Similarly,

$$P_i(U^{(0)} + \epsilon\, U^{(1)} + \epsilon^2 U^{(2)} + \cdots) \approx \tag{1.7}$$
$$P_i(U^{(0)} + \epsilon \frac{\partial P_i^{(0)}}{\partial U_j} U_j^{(1)} + \epsilon^2 \left(\frac{\partial P_i^{(0)}}{\partial U_j} U_j^{(2)} + \frac{1}{2}\frac{\partial^2 P_i^{(0)}}{\partial U_j \partial U_k} U_j^{(1)}\, U_k^{(1)} \right) + \cdots$$

The third step is to substitute the expressions back into the equation of motion, and collect together all of the terms of a matching power of ϵ

to form a set of asymptotic equations. The terms which are $O(1)$ give the standard, nonlinear, steady state equation.

$$\frac{\partial}{\partial x}\left(hF(U^{(0)})\right) - \frac{\partial h}{\partial x}\,P(U^{(0)}) = 0. \tag{1.8}$$

The terms which are $O(\epsilon)$ produce an equation which is linear and homogeneous in $U^{(1)}$.

$$h\frac{\partial U_i^{(1)}}{\partial t} + \frac{\partial}{\partial x}\left(h\frac{\partial F_i^{(0)}}{\partial U_j}\,U_j^{(1)}\right) - \frac{\partial h}{\partial x}\,\frac{\partial P_i^{(0)}}{\partial U_j}\,U_j^{(1)} = 0 \tag{1.9}$$

Because this is a linear equation with coefficients that are not time-varying, $U^{(1)}$ can be represented as a sum of components of the form $\hat{U}(x)\exp(-i\omega t)$ and each of the different frequency components is mutually independent and satisfies the equation

$$-\,i\omega h\hat{U}_i + \frac{\partial}{\partial x}\left(h\,\frac{\partial F_i^{(0)}}{\partial U_j}\,\hat{U}_j\right) - \frac{\partial h}{\partial x}\,\frac{\partial P_i^{(0)}}{\partial U_j}\,\hat{U}_j = 0. \tag{1.10}$$

Gathering together terms which are $O(\epsilon^2)$ gives the following equation which is linear in $U^{(2)}$, but has an inhomogeneous term which is a quadratic function of $U^{(1)}$.

$$\begin{aligned} h\frac{\partial U_i^{(2)}}{\partial t} + \frac{\partial}{\partial x}\left(h\,\frac{\partial F_i^{(0)}}{\partial U_j}\,U_j^{(2)}\right) - \frac{\partial h}{\partial x}\,\frac{\partial P_i^{(0)}}{\partial U_j}\,U_j^{(2)} = \\ -\tfrac{1}{2}\frac{\partial}{\partial x}\left(h\,\frac{\partial^2 F_i^{(0)}}{\partial U_j \partial U_k}\,U_j^{(1)}U_k^{(1)}\right) + \tfrac{1}{2}\frac{\partial h}{\partial x}\,\frac{\partial^2 P_i^{(0)}}{\partial U_j \partial U_k}\,U_j^{(1)}U_k^{(1)} \end{aligned} \tag{1.11}$$

Taking the time-average of this equation yields

$$\begin{aligned} \frac{\partial}{\partial x}\left(h\frac{\partial F_i^{(0)}}{\partial U_j}\overline{U_j^{(2)}}\right) - \frac{\partial h}{\partial x}\,\frac{\partial P_i^{(0)}}{\partial U_j}\overline{U_j^{(2)}} = \\ -\tfrac{1}{2}\frac{\partial}{\partial x}\left(h\frac{\partial^2 F_i^{(0)}}{\partial U_j \partial U_k}\overline{U_j^{(1)}U_k^{(1)}}\right) + \tfrac{1}{2}\frac{\partial h}{\partial x}\,\frac{\partial^2 P_i^{(0)}}{\partial U_j \partial U_k}\,\overline{U_j^{(1)}U_k^{(1)}} \end{aligned} \tag{1.12}$$

The motivation for taking the time-average is that these second order terms are much smaller than the first order unsteady terms, so the unsteady portion of them is not important for the prediction of forced response loading and flutter. On the other hand, the time-average of the first order linear terms is zero, so it is precisely the time-average of the second order terms that reveals the change in the mean flow due to the average effect of the unsteadiness.

An alternate way of expressing this final equation for a particular value of ϵ, is to define a variable $\overline{U}$ which is equal to the steady state solution plus the time-average of the second order term,

$$\overline{U} = U^{(0)} + \epsilon^2 \overline{U^{(2)}} \tag{1.13}$$

and then recombine the steady state equation plus the time-average of the second order equation to form

$$\frac{\partial (hF_i(\overline{U}))}{\partial x} - \frac{\partial h}{\partial x} P_i(\overline{U}) = \tag{1.14}$$
$$-\frac{\epsilon^2}{2}\frac{\partial}{\partial x}\left(h \frac{\partial^2 F_i^{(0)}}{\partial U_j \partial U_k} \overline{U_j^{(1)} U_k^{(1)}} \right) + \frac{\epsilon^2}{2}\frac{\partial h}{\partial x} \frac{\partial^2 P_i^{(0)}}{\partial U_j \partial U_k} \overline{U_j^{(1)} U_k^{(1)}}.$$

This equation is very similar in nature to Adamczyk's passage-averaged equations in which the effect of unsteadiness enters into the equations as terms analogous to Reynolds stress terms in the Reynolds-averaged Navier-Stokes equations.

1.2 Steady-State Equations

1.2.1 Unsteady Euler equations

The unsteady Euler equations, describing the motion of an inviscid, compressible gas in three dimensions in a frame of reference rotating with angular velocity Ω about the x-axis, are

$$\frac{\partial U}{\partial t} = -\left(\frac{\partial F_x}{\partial x} + \frac{\partial F_y}{\partial y} + \frac{\partial F_z}{\partial z} \right) + S, \tag{1.15}$$

where the state vector U is a five-component vector,

$$U = \begin{pmatrix} \rho \\ \rho u_x \\ \rho u_y \\ \rho u_z \\ \rho E \end{pmatrix}, \tag{1.16}$$

the three flux vectors, F_x, F_y and F_z, are given by

$$F_x = \begin{pmatrix} \rho u_x \\ \rho u_x^2 + p \\ \rho u_x u_y \\ \rho u_x u_z \\ u_x(\rho E + p) \end{pmatrix}, \quad F_y = \begin{pmatrix} \rho u_y \\ \rho u_x u_y \\ \rho u_y^2 + p \\ \rho u_y u_z \\ u_y(\rho E + p) \end{pmatrix}, \quad F_z = \begin{pmatrix} \rho u_z \\ \rho u_x u_z \\ \rho u_y u_z \\ \rho u_z^2 + p \\ u_z(\rho E + p) \end{pmatrix}. \tag{1.17}$$

and the source term S containing both centrifugal and coriolis effects is

$$S = \begin{pmatrix} 0 \\ 0 \\ \rho\,(\Omega^2 y + 2\Omega u_z) \\ \rho\,(\Omega^2 z - 2\Omega u_y) \\ 0 \end{pmatrix}. \tag{1.18}$$

The total energy E is related to the pressure, density and velocities through the equation of state for an ideal gas.

$$E = \frac{p}{(\gamma - 1)\rho} + \tfrac{1}{2}(u_x^2 + u_y^2 + u_z^2) - \tfrac{1}{2}\Omega^2 r^2 \tag{1.19}$$

Integrating over a static control volume gives the integral form of the equations of motion.

$$\frac{d}{dt}\iiint_V U\,dV = -\iint_{\partial V} F_x\,dA_x + F_y\,dA_y + F_z\,dA_z + \iiint_V S\,dV. \tag{1.20}$$

It is this equation which is used for the spatial discretizations in most numerical time-marching methods. The time dependent term is retained to give a physically viable method of integrating in time towards the steady state solution. True time accuracy is often lost due to the use of acceleration techniques such as local time-stepping and multigrid, but the discrete steady-state equations are always a discretization of the steady terms on the r.h.s. of Eq. (1.20).

1.2.2 Boundary conditions

The boundary conditions on blade surfaces and endwalls is simply that there is no mass flux across the boundary, or equivalently that the normal component of the velocity is zero.

If the calculation is performed for a single blade-blade passage in a blade row with multiple blades, then the periodic boundary condition is that

$$U_p(x, \theta, r) = U_p(x, \theta + \Delta\theta), \tag{1.21}$$

where $\Delta\theta$ is the angular pitch and the 'primitive' variable vector U_p in cylindrical coordinates is defined as

$$U_p = \begin{pmatrix} \rho \\ u_x \\ u_\theta \\ u_r \\ p \end{pmatrix}. \tag{1.22}$$

At inflow and outflow boundaries, steady-state non-reflecting boundary conditions can be constructed [SG91]. In these, one specifies at the inflow circumferentially averaged values of total temperature, total pressure and flow angles as functions of radius. At the outflow circumferentially averaged values of pressure are specified, consistent with radial equilibrium. Circumferential harmonics of U or U_p are not set to zero but are allowed to evolve to values which are consistent with the existence of an infinite annular duct upstream or downstream of the computational domain.

At stator/rotor interfaces a very similar treatment is used, but the circumferentially averaged values of U or U_p are determined instead by the constraint the the total flux of mass, momentum and energy across the interface must match.

1.3 Linear Equations

1.3.1 Linear Euler equations

The construction of the linear equations begins with the integral form of the nonlinear, unsteady Euler equations, Eq. (1.20).

Next, following the SLiQ approach laid out in the Introduction, it is assumed that both the grid coordinates and the flow variables can be expressed as a sum of a steady-state part plus a small amplitude unsteady perturbation.

$$U(x,t) = U^{(0)}(x) + \epsilon U^{(1)}(x,t). \tag{1.23}$$

For convenience, the superscript $^{(0)}$ will be omitted unless it is absolutely necessary for clarity, and the superscript $^{(1)}$ will be replaced by $\widetilde{\ }$.

Following from the perturbations to U there are corresponding perturbations to all other terms. For an arbitrary scalar function $A(U)$ the linear perturbation is

$$\widetilde{A} = \frac{\partial A}{\partial U}\widetilde{U}. \tag{1.24}$$

After some algebra one finds that the linear perturbations to the velocities and pressure are

$$\begin{aligned} \widetilde{u}_x &= \rho^{-1}(\widetilde{\rho u}_x - u_x\widetilde{\rho}) \\ \widetilde{u}_y &= \rho^{-1}(\widetilde{\rho u}_y - u_y\widetilde{\rho}) \\ \widetilde{u}_z &= \rho^{-1}(\widetilde{\rho u}_z - u_z\widetilde{\rho}), \end{aligned} \tag{1.25}$$

$$\widetilde{p} = (\gamma-1)\left(\widetilde{\rho E} - \widetilde{\rho}\left(E - \frac{p}{(\gamma-1)\rho}\right) - \rho(u_x\widetilde{u}_x + u_y\widetilde{u}_y + u_z\widetilde{u}_z)\right), \tag{1.26}$$

the perturbations to the flux vectors are

$$\widetilde{F}_x = \begin{pmatrix} u_x\widetilde{\rho} + \widetilde{u}_x\rho \\ u_x\widetilde{\rho u}_x + \widetilde{u}_x\rho u_x + \widetilde{p} \\ u_x\widetilde{\rho u}_y + \widetilde{u}_x\rho u_y \\ u_x\widetilde{\rho u}_z + \widetilde{u}_x\rho u_z \\ u_x(\widetilde{\rho E} + \widetilde{p}) + \widetilde{u}_x(\rho E + p) \end{pmatrix} \tag{1.27}$$

$$\widetilde{F}_y = \begin{pmatrix} u_y \widetilde{\rho} + \widetilde{u}_y \rho \\ u_y \widetilde{\rho u_x} + \widetilde{u}_y \rho u_x \\ u_y \widetilde{\rho u_y} + \widetilde{u}_y \rho u_y + \widetilde{p} \\ u_y \widetilde{\rho u_z} + \widetilde{u}_y \rho u_z \\ u_y (\widetilde{\rho E} + \widetilde{p}) + \widetilde{u}_y (\rho E + p) \end{pmatrix} \tag{1.28}$$

$$\widetilde{F}_z = \begin{pmatrix} u_z \widetilde{\rho} + \widetilde{u}_z \rho \\ u_z \widetilde{\rho u_x} + \widetilde{u}_z \rho u_x \\ u_z \widetilde{\rho u_y} + \widetilde{u}_z \rho u_y \\ u_z \widetilde{\rho u_z} + \widetilde{u}_z \rho u_z + \widetilde{p} \\ u_z (\widetilde{\rho E} + \widetilde{p}) + \widetilde{u}_z (\rho E + p) \end{pmatrix}, \tag{1.29}$$

and the perturbation to the source term is

$$\widetilde{S} = \begin{pmatrix} 0 \\ 0 \\ \Omega^2 \widetilde{\rho}\, y + 2\Omega \widetilde{\rho u_z} \\ \Omega^2 \widetilde{\rho}\, z - 2\Omega \widetilde{\rho u_y} \\ 0 \end{pmatrix}. \tag{1.30}$$

The linear perturbation of the unsteady integral equation of motion is

$$\frac{d}{dt} \iiint_V \widetilde{U}\, dV = - \iint_{\partial V} \widetilde{F}_x dA_x + \widetilde{F}_y dA_y + \widetilde{F}_y dA_y + \iiint_V \widetilde{S}\, dV. \tag{1.31}$$

The fact that this equation is linear and all of the coefficients depend on U and geometric terms, which do not vary in time, means that there are harmonic solutions of the form $\widehat{U}(x) \exp(-i\omega t)$. The most general periodic solution can be expressed as the real part of a linear superposition of such solutions for a number of different real positive frequencies ω, all satisfying the necessary periodicity constraints.

$$\widetilde{U} = \mathcal{R}\left(\sum_\omega \widehat{U}_\omega \exp(-i\omega t) \right). \tag{1.32}$$

A key contribution of Ni [NS75] is to formulate the harmonic equation by retaining a time dependence in $\widehat{U}$. Substituting

$$\widetilde{U} = \widehat{U}(x, t) \exp(-i\omega t) \tag{1.33}$$

gives the unsteady, harmonic equation

$$\frac{d}{dt} \iiint_V \widehat{U}\, dV = - \iint_{\partial V} \widehat{F}_x dA_x + \widehat{F}_y dA_y + \widehat{F}_z dA_z + \iiint_V (\widehat{S} + i\omega \widehat{U})\, dV. \tag{1.34}$$

The advantage of retaining the time-dependent term is that numerical integration of this equation until a 'steady-state' solution is achieved will result in the desired harmonic solution. This allows one to employ any of a set of standard numerical techniques and tricks to accelerate the convergence.

1.3.2 Boundary conditions

The boundary conditions on blade surfaces and endwalls are again that there is no mass flux across the boundary. When this condition is linearized it produces the requirement that the normal component of the velocity perturbation is zero.

The periodic boundary condition is slightly more complicated than for the steady state flow due to the inclusion of the interblade phase angle ϕ.

$$\widehat{U}_p(x,\theta,r) = \widehat{U}_p(x,\theta+\Delta\theta)\exp(i\phi). \tag{1.35}$$

This is still a very much simpler boundary condition to implement numerically than the lagged periodic boundary condition in single stage nonlinear stator/rotor calculations [EAM77, Gil88]. Also, linearity allows one to superimpose two solutions with different interblade phase angles due to differing number of blades in upstream and downstream blade rows. This is not possible in nonlinear calculations and forces one to perform calculations for the entire annulus which is very much more expensive.

The most interesting boundary conditions are the inflow and outflow boundary conditions used to model unsteady stator/rotor interaction in a linear sense. Consider two neighboring blade rows, labelled 1 and 2, rotating with speeds Ω_1 and Ω_2. The key idea is that a stationary circumferential variation in one blade row becomes a rotating unsteady perturbation in the other blade row due to the relative rotation of the two rows. The steady-state primitive flow variables in cylindrical coordinates in blade row 1 can be expressed as the sum of an axisymmetric part plus circumferential harmonic terms, exactly as was done in constructing the steady boundary conditions.

$$U_p(x,\theta_1,r) = \bar{U}_p(x,r) + \mathcal{R}\left\{\sum_{m>0} U_{p\,m}\exp\left(\frac{2\pi m\, i\,\theta_1}{\Delta\theta_1}\right)\right\} \tag{1.36}$$

The angle θ_2 in neighboring blade row 2 is related through

$$\theta_1 + \Omega_1 t = \theta_2 + \Omega_2 t. \tag{1.37}$$

Hence,

$$\frac{2\pi m\,\theta_1}{\Delta\theta_1} = \frac{2\pi m\,(\theta_2 - (\Omega_1 - \Omega_2)\,t)}{\Delta\theta_1}. \tag{1.38}$$

and so the stationary mth mode in blade row 1 is equivalent to an unsteady mode in blade row 2 with the frequency

$$\omega_2 = \frac{2\pi m\,(\Omega_1 - \Omega_2)}{\Delta\theta_1}. \tag{1.39}$$

The harmonic boundary terms that go with this mode are

$$\widehat{U}_p(x,\theta_2,r) = U_{p\,m}(x,r)\,\exp\left(-\frac{i\,\theta_2\,\phi_2}{\Delta\theta_2}\right) \tag{1.40}$$

where the inter-blade phase angle is

$$\phi_2 = -\frac{2\pi m\,\Delta\theta_2}{\Delta\theta_1} \tag{1.41}$$

A slight complication is the desire to keep all frequencies ω positive. The above explanation is correct for the situation in which $\Omega_1 > \Omega_2$. If instead $\Omega_1 < \Omega_2$, then one must change the representation of the steady flow field to

$$U_p(x,\theta,r) = \bar{U}_p(x,r) + \mathcal{R}\left(\sum_{m<0} U_{p\,m} \exp\left(\frac{2\pi m\, i\,\theta}{\Delta\theta}\right)\right), \tag{1.42}$$

summing over negative integer values of m.

To obtain the forcing terms for blade row 1 due to the steady flow field in blade row 2, one simply swaps the roles of 1 and 2 in the above discussion.

1.4 Quadratic Source Terms

As with the derivation of the linear harmonic equations, the derivation of the quadratic terms starts with the integral form of the nonlinear, unsteady Euler equations.

$$\frac{d}{dt}\iiint_V U\,dV = -\iint_{\partial V} F_x dA_x + F_y dA_y + F_z dA_z + \iiint_V S\,dV. \tag{1.43}$$

The flow variables are represented by an asymptotic expansion in ϵ.

$$U(x,t) = U^{(0)}(x) + \epsilon U^{(1)}(x,t) + \epsilon^2 U^{(2)}(x,t) + O(\epsilon^3). \tag{1.44}$$

The next step is to construct the asymptotic form of all flow variables using Taylor series expansions. For an arbitrary scalar function $A(U)$,

$$\begin{aligned} &A(U^{(0)}+\epsilon U^{(1)}+\epsilon^2 U^{(2)}) \qquad (1.45)\\ &= A(U^{(0)}) + \frac{\partial A}{\partial U_j}(\epsilon U_j^{(1)} + \epsilon^2 U_j^{(2)}) + \tfrac{1}{2}\frac{\partial^2 A}{\partial U_j \partial U_k}(\epsilon U_j^{(1)})(\epsilon U_k^{(1)}) + O(\epsilon^3)\\ &= A(U^{(0)}) + \epsilon\frac{\partial A}{\partial U_j}U_j^{(1)} + \epsilon^2\left(\frac{\partial A}{\partial U_j}U_j^{(2)} + \tfrac{1}{2}\frac{\partial^2 A}{\partial U_j \partial U_k}U_j^{(1)}U_k^{(1)}\right) + O(\epsilon^3). \end{aligned}$$

Note that the second order term has two pieces, one due to the linear sensitivity to second order changes in U (which will be referred to as the second order linear perturbation) and the other due to the quadratic sensitivity to the first order changes in U (which will be referred to as the quadratic perturbation).

To avoid the use of superscripts, the following notation is introduced in addition to the notation used in the last section.

$$
\begin{aligned}
\tilde{\tilde{A}} &\equiv \frac{\partial A}{\partial U_j}\,\overline{U_j^{(2)}} \\
\overline{\overline{A}} &\equiv \tfrac{1}{2}\frac{\partial^2 A}{\partial U_j \partial U_k}\,\overline{U_j^{(1)} U_k^{(1)}}
\end{aligned}
\tag{1.46}
$$

The overlining on the r.h.s of these definitions denotes a time-average. The relationship between the second order linear perturbation $\tilde{\tilde{A}}$ and $\tilde{\tilde{U}}$ is exactly the same as the relationship between the first order perturbation $\tilde{A}$ and $\tilde{U}$, so the equations from the last section can be used to construct the linear second order perturbations to the flux vectors and the source term.

After considerable algebra the quadratic velocity perturbations are found to be

$$
\begin{aligned}
\overline{\overline{u}}_x &= -\rho^{-1}\,\overline{\tilde{\rho}\,\tilde{u}_x} \\
\overline{\overline{u}}_y &= -\rho^{-1}\,\overline{\tilde{\rho}\,\tilde{u}_y} \\
\overline{\overline{u}}_z &= -\rho^{-1}\,\overline{\tilde{\rho}\,\tilde{u}_z} \quad ,
\end{aligned}
\tag{1.47}
$$

and the quadratic pressure perturbation is

$$
\overline{\overline{p}} = -\tfrac{1}{2}(\gamma-1)\rho\left(\overline{\tilde{u}_x^2} + \overline{\tilde{u}_y^2} + \overline{\tilde{u}_z^2}\right). \tag{1.48}
$$

The quadratic perturbations to the flux vectors are

$$
\overline{\overline{F}}_x = \begin{pmatrix} 0 \\ \overline{\overline{u}}_x \rho u_x + \overline{\tilde{u}_x \widetilde{\rho u}_x} + \overline{\overline{p}} \\ \overline{\overline{u}}_x \rho u_y + \overline{\tilde{u}_x \widetilde{\rho u}_y} \\ \overline{\overline{u}}_x \rho u_z + \overline{\tilde{u}_x \widetilde{\rho u}_z} \\ \overline{\overline{u}}_x(\rho E + p) + u_x \overline{\overline{p}} + \overline{\tilde{u}_x \widetilde{\rho E}} + \overline{\tilde{u}_x \tilde{p}} \end{pmatrix} \tag{1.49}
$$

$$
\overline{\overline{F}}_y = \begin{pmatrix} 0 \\ \overline{\overline{u}}_y \rho u_x + \overline{\tilde{u}_y \widetilde{\rho u}_x} \\ \overline{\overline{u}}_y \rho u_y + \overline{\tilde{u}_y \widetilde{\rho u}_y} + \overline{\overline{p}} \\ \overline{\overline{u}}_y \rho u_z + \overline{\tilde{u}_y \widetilde{\rho u}_z} \\ \overline{\overline{u}}_y(\rho E + p) + u_y \overline{\overline{p}} + \overline{\tilde{u}_y \widetilde{\rho E}} + \overline{\tilde{u}_y \tilde{p}} \end{pmatrix} \tag{1.50}
$$

$$
\overline{\overline{F}}_z = \begin{pmatrix} 0 \\ \overline{\overline{u}}_z \rho u_x + \overline{\tilde{u}_z \widetilde{\rho u}_x} \\ \overline{\overline{u}}_z \rho u_y + \overline{\tilde{u}_z \widetilde{\rho u}_y} \\ \overline{\overline{u}}_z \rho u_z + \overline{\tilde{u}_z \widetilde{\rho u}_z} + \overline{\overline{p}} \\ \overline{\overline{u}}_z(\rho E + p) + u_z \overline{\overline{p}} + \overline{\tilde{u}_z \widetilde{\rho E}} + \overline{\tilde{u}_z \tilde{p}} \end{pmatrix} \tag{1.51}
$$

and the quadratic perturbation to S is zero.

Substituting the asymptotic expansions into the unsteady integral equation of motion, time-averaging, and removing the $O(1)$ terms which simply represent the steady-state equation, leaves

$$\iint_{\partial V} \overline{\overline{F}}_x dA_x + \overline{\overline{F}}_y dA_y + \overline{\overline{F}}_z dA_z - \iiint_V \overline{\overline{S}}\, dV = \text{QST} \tag{1.52}$$

where the quadratic source terms (QST) are

$$\text{QST} = -\iint_{\partial V} \overline{\overline{F}}_x dA_x + \overline{\overline{F}}_y dA_y + \overline{\overline{F}}_z dA_z . \tag{1.53}$$

Eq. (1.52) is a linear equation in $\overline{\overline{U}}$; the quadratic source terms are the inhomogeneity that drives this second order flow perturbation.

Following Ni's idea of adding an pseudo-unsteady term to allow the use of numerical time integration schemes, one obtains

$$\frac{d}{dt}\iiint_V \overline{\overline{U}}\, dV + \iint_{\partial V} \overline{\overline{F}}_x dA_x + \overline{\overline{F}}_y dA_y + \overline{\overline{F}}_z dA_z - \iiint_V \overline{\overline{S}}\, dV = \text{QST} \tag{1.54}$$

In this form, it is equivalent to the linear harmonic equation with zero frequency and the addition of the quadratic source terms. Thus it can be solved numerically with only trivial changes to the numerical procedure.

An alternative approach, outlined in the Introduction, is to define a time-averaged state

$$\overline{U} = U + \epsilon^2 \overline{\overline{U}} \equiv \overline{U^{(0)} + \epsilon U^{(1)} + \epsilon^2 U^{(2)}}, \tag{1.55}$$

and recombine the steady-state equation with second order equation to form

$$\frac{d}{dt}\iiint_V \overline{U}\, dV + \iint_{\partial V} \overline{F}_x dA_x + \overline{F}_y dA_y + \overline{F}_z dA_z - \iiint_V \overline{S}\, dV = \epsilon^2 \text{QST}. \tag{1.56}$$

This effectively introduces into the nonlinear steady-state equation the time-averaged effect of the linear unsteadiness through the simple inclusion of the quadratic source terms.

A final issue is the evaluation of the quadratic source terms for linear unsteady solution $\widetilde{U}$ which is represented as the real part of a sum of harmonic solutions of different frequencies.

$$\widetilde{U} = \mathcal{R}\left(\sum_\omega \widehat{U}_\omega \exp(-i\omega t)\right) \tag{1.57}$$

It turns out that this is very simple. For an arbitrary scalar function $A(U)$,

$$\overline{\overline{A}} \;=\; \frac{1}{2}\frac{\partial^2 A}{\partial U_j \partial U_k}\, \overline{U_j^{(1)} U_k^{(1)}}$$

$$= \frac{1}{2}\frac{\partial^2 A}{\partial U_j \partial U_k} \overline{\mathcal{R}\left(\sum_\omega \widehat{U}_{j\,\omega} \exp(-i\omega t)\right) \mathcal{R}\left(\sum_\omega \widehat{U}_{k\,\omega} \exp(-i\omega t)\right)}$$

$$= \frac{1}{4}\frac{\partial^2 A}{\partial U_j \partial U_k} \sum_\omega \mathcal{R}(\widehat{U}_{j\,\omega} \widehat{U}^*_{k\,\omega}). \tag{1.58}$$

Thus, the quadratic source terms can be expressed as a sum of the quadratic source terms for each distinct harmonic frequency

$$\text{QST} = \sum_\omega \text{QST}_\omega, \tag{1.59}$$

and each of the harmonic components is easily evaluated knowing $\widehat{U}$.

1.5 Concluding Remarks

This paper has presented an asymptotic approach toward the calculation of unsteady flows in multistage turbomachinery. The analysis proceeds in three phases. First one performs a steady-state multistage analysis in which the flow is assumed to be steady in each blade row in a frame of reference rotating with the blade row. The stator/rotor interface matching conditions ensure overall conservation of mass momentum and energy across the interface.

In the second phase the unsteadiness in each blade row is treated as a perturbation to the steady flow field. Assuming the perturbation is small it is represented by the linearized unsteady flow equations. The linear stator/rotor interaction is calculated by recognizing that a steady circumferential variation in a stator corresponds to a rotating unsteady disturbance in a rotor, and *vice versa*. Each circumferential harmonic produces a response of a different frequency. Due to linearity these can be computed independently and superimposed to form the composite unsteady flow.

The third phase calculates the change to the mean flow due to second order consequences of the unsteady flow field. This is achieved through the evaluation of quadratic source terms which are similar in origin to the Reynolds stress terms in the Reynolds averaged Navier-Stokes equations.

Through this process one can calculate the key features of the unsteady flow in turbomachinery, the unsteady forces and moments on blades due to interactions with neighboring rows, as well as the average effects on mass flow and efficiency. The efficiency changes come about as the result of wasted kinetic energy associated with the eventual dissipation of acoustic energy generated by unsteady stator/rotor interaction. A more complete treatment of efficiency prediction would require the application of the entire asymptotic approach to the Navier-Stokes equations. This is an algebraically complicated but otherwise straightforward extension which will be tackled in the future.

The motivation behind this work is that the whole asymptotic analysis can be achieved at a very small fraction of the cost of nonlinear multi-stage flow calculations because of the use of very efficient numerical methods designed for steady flow calculations. The proof of this statement will have to wait for future papers once the entire procedure has been implemented and validated. However, given the existing literature on multigrid methods for flows in turbomachinery and external aerodynamics it seems clear that these methods can be applied with equal success in this application.

REFERENCES

[Ada85] J. J. Adamczyk. Model equations for simulating flows in multistage turbomachinery. ASME Paper 95-GT-226, June 1985.

[AGEG91] R. S. Abhari, G. R. Guenette, A. H. Epstein, and M. B. Giles. Comparison of time-resolved turbine rotor blade heat transfer measurements and numerical calculations. ASME Paper 91-GT-268, June 1991.

[Bat67] G. K. Batchelor. *An Introduction to Fluid Dynamics.* Cambridge University Press, 1967.

[Che88] Y. S. Chen. 3-D stator-rotor interaction of the SSME. AIAA Paper 88-3095, 1988.

[EAM77] J. I. Erdos, E. Alzner, and W. McNally. Numerical Solution of Periodic Transonic Flow Through a Fan Stage. *AIAA Journal*, 15(11):1559–1568, Nov 1977.

[EGSS89] A. H. Epstein, M. B. Giles, T. Shang, and A. K. Sehra. Blade row interaction effects on compressor measurements. AGARD 74th Specialists Meeting on Unsteady Aerodynamic Phenomena in Turbomachines, Aug 1989.

[Fou86] A. Fourmaux. Unsteady flow calculation in cascades. ASME Paper 86-GT-178, 1986.

[Gil88] M. B. Giles. Calculation of unsteady wake/rotor interaction. *AIAA Journal of Propulsion and Power*, 4(4):356–362, July/August 1988.

[Gil89] M. B. Giles. Numerical methods for unsteady turbomachinery flow. VKI Lecture Series 1989-06, 1989.

[Gil90] M. B. Giles. Stator/rotor interaction in a transonic turbine. *AIAA Journal of Propulsion and Power*, 6(5), September-October 1990.

[HC87] K. C. Hall and E. F. Crawley. Calculation of unsteady flows in turbomachinery using the linearized Euler equations. Proceedings of Fourth Symposium on Unsteady Aerodynamics and Aeroelasticity of Turbomachines and Propellers, Sept 1987.

[Hod84] H. P. Hodson. An inviscid blade-to-blade prediction of a wake-generated unsteady flow. ASME Paper 84-GT-43, June 1984.

[LG91] D. R. Lindquist and M. B. Giles. On the validity of linearized unsteady Euler equations with shock capturing. AIAA Paper 91-1958, 1991.

[Ni78] R.-H. Ni. A rational analysis of periodic flow perturbations in supersonic two-dimensional cascade. ASME Paper 78-GT-176, 1978.

[NS75] R.-H. Ni and F. Sisto. Numerical computation of nonstationary aerodynamics of flat plate cascades in compressible flow. ASME Paper 75-GT-5, 1975.

[Rai87] M. M. Rai. Navier-Stokes simulations of rotor-stator interaction using patched and overlaid grids. *AIAA Journal of Propulsion and Power*, 3(5):387–396, 1987.

[Rai89] M. M. Rai. Three-dimensional Navier-Stokes simulations of turbine rotor-stator interaction. *AIAA Journal of Propulsion and Power*, 5(3):305–319, May-June 1989.

[Sch79] H. Schlichting. *Boundary Layer Theory.* McGraw-Hill, 1979.

[SG91] A. Saxer and M. Giles. Quasi-3D nonreflecting boundary conditions for Euler equation calculations. AIAA Paper 91-1603, June 1991.

[Smi71] S. N. Smith. Discrete frequency sound generation in axial flow turbomachines. University of Cambridge, Department of Engineering Report CUED/A-Turbo/TR 29, 1971.

[SS90] A. Suddhoo and P. Stow. Simulation of inviscid blade row interaction using a linearised potential code. AIAA Paper 90-1916, 1990.

[VC84] J.M. Verdon and J.R. Caspar. A linearized unsteady aerodynamic analysis for transonic cascades. *Journal of Fluids Mechanics*, 149:403–429, 1984.

[Whi82] D. S. Whitehead. The calculation of steady and unsteady transonic flow in cascades. University of Cambridge, Department of Engineering Report CUED/A-Turbo/TR 118, 1982.

Long Time Behavior of Unsteady Flow Computations

S. I. Hariharan[1]

ABSTRACT This paper addresses a specific issue of time accuracy in the calculations of external aerodynamic problems. The class of problems that is discussed here consists of inviscid compressible subsonic flows. These problems are inherently governed by a convective equation. This is readily seen by linearizing the Euler equation which results in a convective wave equation for the pressure. A key mathematical issue that is not well understood in literature for these problems is the long time behavior of the solution. This is an important aspect if one desires transient calculations of problems governed by the Euler equations or its derivatives such as the small disturbance equations or the potential formulations of the gust problem. In particular, difficulties arise for two-dimensional problems. In two dimensions the time decay rate of solutions of the wave equation is known to be rather slow. This applies to the convective wave equation as well. The consequences are rather severe if one focuses on the time accuracy of solutions of problems governed by the Euler equations.

In concert with the above mentioned problem, exterior flows require proper modeling of boundary conditions. Computations of these flow problems require truncation of infinite regions into finite regions with the aid of artificial boundaries. On these boundaries one must impose boundary conditions that are consistent with the physics as well as guarantee consistency with the original problem posed in the unbounded region. Moreover, these boundary conditions must have accuracy in time as well as space. Some of the well-known procedures do address the issues of spatial accuracy and have remedy for these conditions. Unfortunately, these procedures do not address the time accuracies, which are crucial for the transient problems. Our treatment is discussed in detail and examples are presented to verify the results.

---

[1]Department of Mathematical Sciences, The University of Akron, Akron, OH 44325. Research for the author was supported by the National Science Foundation, DMS-8921189 and by ICOMP, NASA Lewis Research Center.

1.1 Introduction

This paper deals with unanswered aspects of long time behavior of unsteady flow computations in two dimensions. Many of the available codes such as the FLO series deal with steady flows and their accuracies are well tested. State of the art grid generations and even complicated geometrical structures are included in these codes. While these developments reflect the maturity of computational fluid dynamics, the goal of the field is to understand aerodynamics in its entirety which includes the unsteady phenomena. It is important to understand this phenomena if one is to achieve successful calculations on by-products such as far field acoustics. A common tendency is to use steady fluid dynamics codes for unsteady calculations. Such calculations yield either unsatisfactory resolution of time accuracies or break down in finite time. The classical problem of subsonic-transonic compressible flows past a cylinder is an unresolved issue for this very reason. For example one can see a documentation of such computations in [15]. The key reason why this is difficult is due to lack of understanding in the modeling of far field boundary conditions that naturally plays an important if not an essential role in these computations.

It is an open issue whether one can predict solutions of unsteady Euler or Navier-Stokes equations in a satisfactory manner. A recent paper by [2] deals with numerical computations of viscous flows past a circular cylinder. The modeling of required far field approximate pressure condition was obtained using a theory proposed by [1] which is accurate for high frequencies. This limits the solutions to be accurate for short times. They also proposed approximate boundary conditions using characteristic variables which works better than the above said pressure boundary condition. It is not clear if these conditions maintain time accuracy of the solution and it needs to be addressed. This concern forms the basis of this paper. It should be emphasized the workers realized similar difficulties ahead of time and concentrated on linearized Euler equations and in the frequency domain. Namely, the work of Goldstein [12] stands out in the literature. Corrections to this model were proposed by Atassi [13] and subsequently he and his coworkers (e.g. see [14]) succeeded in predicting unsteady aerodynamic calculations through numerical procedures. These formulations and calculations are done through potential methods and the resulting equations are nonconstant coefficient second order linear partial differential equations. Calculations are accurate within the linearized limits. In certain cases, such as flat plate air foils, the theory behind these equations is well understood. As a result, these potential formulations have firm foundations. Recently several authors are beginning to solve linearized Euler equations directly. It is not clear if there are any real advantages to these procedures. It appears that the mathematical foundations do need careful work. Moreover, associated far field boundary conditions consistent with the linearized equations require a careful study. This is the other aspect this paper addresses.

Two-dimensional problems governed by any wave equation, unlike their one- or three-dimensional counterparts, suffer severe slow decay rates under certain conditions. Whether the problem is posed in the frequency domain or in the time domain, the problem of slow decay rates occurs. For example, for a very low frequency problem it would appear from a conceptual point of view that the terms multiplied by the frequencies could be dropped to consider a problem in the absence of those corresponding terms. A typical situation is the problem governed by the Helmholtz equation, where reduction is the Laplace equation. Similar arguments hold for the time dependent wave equation. If in the long time there is a steady state, the analogous situation is to consider the steady state part. These discussions imply that there is interplay between low frequencies and long time behavior. The researchers who looked at these questions considered only isolated issues. In this paper an attempt has been made to connect the theoretical issues behind these problems to aerodynamic problems.

In section two we discuss the behavior of the solution of reduced wave equations in exterior regions. In this section we point out the potential problems in considering the zero frequency limit or the steady state problems. Moreover, we show the proper form of the boundary operators one should consider in the prescription of radiation boundary conditions. In section three we consider a true time dependent model in which the corresponding corrections to the radiation boundary operator are discussed. Here we include available well-known procedures that are commonly used in the computational fluid dynamic literature to show these corrections in perspective. Numerical solutions are shown to establish the validity of the theory.

Finally in section four we provide extension of the theory proposed in section 3 for the treatment in the far field. This treatment should hold for the Euler equations and their linearized versions. The corresponding numerical treatment will be discussed elsewhere.

1.2 Low Frequency Behavior of the Reduced Wave Equation

Here we show a potential problem that arises with the reduced wave equation. In particular we shall discuss the low frequency behavior that arises from problems with Dirichlet boundary conditions. The reason for discussing such behavior is to establish the slow decay rates of the two dimensional wave equation. These decay rates play an important role in the construction of approximate boundary operators that are vital to the constructions of solutions. At the outset we want to emphasize that these decay rates and associated problems only apply to the two-dimensional wave and wave-like equations.

We shall begin with the two-dimensional problem that will motivate the decay rates in question. We consider a classical soft obstacle problem in two dimensions. Specifically we consider the following boundary value problem (P_k):

$$\begin{aligned} \Delta u + k^2 u &= 0 && \text{in } \Omega_e \subset \mathcal{R}^2 \\ u &= g_k && \text{on } \Gamma \\ \lim_{r\to\infty} (u_r - iku) &= 0 && r = \sqrt{x^2+y^2} \end{aligned} \tag{1.1}$$

where Ω_e is the exterior to a bounded region Ω and Γ is its boundary. For the soft obstacle problem $g_k = -u_{inc}$ where u_{inc} is the incident field.

The key question we address is what happens to the problem as $k \rightarrow 0$? To discuss this issue we consider an appropriate zero frequency problem. This problem is constructed in such a way that it preserves the zero frequency behavior on the equation and the boundary condition on Γ. However, the radiation condition at infinity is replaced by a boundedness condition at infinity and is as follows:

$$\begin{aligned} \Delta v &= 0 && \text{in } \Omega_e \subset \mathcal{R}^2 \\ v &= g_0 && \text{on } \Gamma \\ |v| &\leq M && \text{or } \mathrm{v} \rightarrow \mathrm{A}\log|\vec{\mathrm{x}}| \text{ as } |\vec{\mathrm{x}}| = \mathrm{r} \rightarrow \infty. \end{aligned} \tag{1.2}$$

We call this problem to be (P_0). It has been shown in the literature [10], [8] that in the limit as $k \rightarrow 0$ the solution of (P_k) approaches the solution of (P_0). However, this limit process is a slow one and we want to make a case for it as it has everything to do with constructions of boundary conditions for numerical simulations. Before we introduce the results let us focus our attention on the structure of $v_{\text{inc}} = -g_0$. The choice of v_{inc} depends on u_{inc}. Examples are as follows:

a) $u_{\text{inc}} = e^{ikx}$ (plane wave)

Then the corresponding zero frequency limit is readily seen and is $v_{\text{inc}} = 1$

b) $u_{\text{inc}} = -\frac{i}{4} H_0^{(1)}(k|\vec{x} - \vec{x}_0|)$ (singular source)

Then the corresponding zero frequency limit is taken as the principal leading spatial singularity $v_{\text{inc}} = \frac{1}{2\pi}\log|\vec{x} - \vec{x}_0|$. A discussion of this choice is seen in [7] and is motivated by several issues including an obvious one: the asymptotic behavior of the Hankel function for small arguments.

Now we state some of the key results:

a) (Crude) : $u \rightarrow v$ as $k \rightarrow 0$

b) (Sharp) : $u = v + O(\frac{1}{\ln k})$

c) (Sharper): $u = v + \frac{cf_0}{\beta - c_0} + O(k^2(\ln k)^2)$

The results in a) are due to an old work of MacCamy [8]. The results in b) are later observed by many others, namely Muravei [10], Werner [16] and Kress [17] which indicate the slow decay rates. The result c) is due to the author and his coworkers and the results appear in [9], [6] and [7]. To motivate the implication of these results first let us demonstrate the effect of these decays numerically. For this purpose we have taken a plane wave incident with unit amplitude. It is easily seen from problem (P_0) that the solution is $v_s = -1$ and the total field $v = v_{\text{inc}} + v_s = 0$. This solution is compared with the total solution for the nonzero frequency problem (P_k) with $u = u_{\text{inc}} + u_s$. Figure 1 shows magnitude of u for the frequencies 10^{-2}, 10^{-3} and 10^{-4} respectively. Clearly one can see that even for these very low frequencies the solution does not approach the correct low frequency limit of zero rapidly. The frequency or the wave number approaching zero corresponds to the long time limit of the solution to the corresponding time dependent wave propagation problem. In general for moderate times or for lower frequencies one cannot expect to use any properties of the steady state problem and in this case the solution of the zero frequency problem (P_0). This message has serious impacts on unsteady flow computations.

In light of the above discussion, we would like to outline a correction to this zero frequency computation. These ideas will later be transferred to an appropriate time domain problem and at the end we shall propose ideas as to how to extend the theory to the Euler equations. We outline a theory that appears in [6], [7] to show the sharp result b) and the sharper result c). The construction of these results involve potential theory. We use potential theoretic representation for the solution. Specifically we use single layer potential representations. Let $u(\vec{x})$ be the total field in (P_k), i.e.,

$$u(\vec{x}) = u_{inc}(\vec{x}) + u_s(\vec{x}). \tag{1.3}$$

Then, for the problem (P_k) the representation is as follows:

$$u(\vec{x}) = \int_\Gamma \sigma(\vec{y})\, G_k(\vec{x}, \vec{y})\, ds_{\vec{y}} + u_{inc}(\vec{x}), \quad \vec{x} \in \Omega_e \tag{1.4}$$

where

$$G_k(\vec{x}, \vec{y}) = \frac{-i}{4} H_0^{(1)}(k|\vec{x} - \vec{x}_0|) \tag{1.5}$$

is the free space Green's function. The scattered field satisfies the Sommerfeld's radiation condition due to this Green's function. One obtains a boundary integral equation for the determination of $\sigma(\vec{y})$ that is defined on the boundary Γ using the boundary condition on Γ.

$$\int_\Gamma \sigma(\vec{y})\, G_k(\vec{x}, \vec{y})\, ds_{\vec{y}} + u_{inc}(\vec{x}) = 0 \;, \vec{x} \in \Gamma \tag{1.6}$$

A similar construction with the free space Green's function for the Laplace's equation is made for the problem P_0 noting that any representation of the solution is subject to an arbitrary constant.

$$v(\vec{x}) = \int_\Gamma \mu(\vec{y})\, G_0(\vec{x},\vec{y})\, ds_{\vec{y}} + v_{inc}(\vec{x}) + c, \quad \vec{x} \in \Omega_e \tag{1.7}$$

with

$$\int_\Gamma \mu(\vec{y})\, ds_{\vec{y}} = -1 \tag{1.8}$$

where

$$G_0(\vec{x},\vec{y}) = \frac{1}{2\pi}\,(\log|\vec{x}-\vec{y}|). \tag{1.9}$$

The condition described by equation 1.8 takes care of the arbitrariness of the constant c and the boundedness of the solution at infinity. Again demanding the boundary condition on Γ we arrive at the boundary integral equation

$$\int_\Gamma \mu(\vec{y})\, G_0(\vec{x},\vec{y})\, ds_{\vec{y}} + v_{inc}(\vec{x}) + c = 0, \quad \vec{x} \in \Gamma. \tag{1.10}$$

Thus the complete solution of (P_0) is determined through solving the equations of the pair of equations 1.8 and 1.10 simultaneously. The key idea of the low frequency correction relies on relating σ and μ with appropriate low frequency corrections. Several of the quantities that arise can be related to the asymptotic formula for small values of z

$$\frac{-i}{4}\, H_0^{(1)}(z) = \frac{1}{2\pi}\,(\log(\frac{z}{2}) + \gamma) - \frac{i}{4} + O(z^2 \ln z) \tag{1.11}$$

where

$$\gamma = .5772156649\cdots \tag{1.12}$$

is the Euler constant. Using this expansion in equation 1.6 we obtain a related integral equation in the following form

$$\int_\Gamma \tilde{\sigma}\,(\vec{y})\, G_0(\vec{x},\vec{y})\, ds_{\vec{y}} + \beta(\int_\Gamma \tilde{\sigma}\,(\vec{y})\, ds_{\vec{y}} + 1) + v_{inc}(\vec{x}) = 0 \tag{1.13}$$

where

$$\beta = \frac{1}{2\pi}\,(\log(\frac{k}{2}) + \gamma) - \frac{i}{4}. \tag{1.14}$$

Successive approximation by considering equation (4) and (13) yields ([6])

$$\sigma = \tilde{\sigma} + O(k^2 \ln k). \tag{1.15}$$

Further it has been shown in [6] that

$$\tilde{\sigma}\,(\vec{x}) = \mu(\vec{x}) + \frac{c\, f_0(\vec{x})}{\beta - c_0}, \quad \vec{x} \in \Gamma \tag{1.16}$$

where $f_0(\vec{x})$ and the real constant c_0 satisfy the following equations :

$$\int_\Gamma f_0(\vec{y})\, G_0(\vec{x},\vec{y})\, ds_{\vec{y}} \, + \, c_0 \, = 0 \tag{1.17}$$

$$\int_\Gamma f_0(\vec{y})\, ds_{\vec{y}} \, = \, 1. \tag{1.18}$$

Use of this result leads to the results in c). There are two types of low frequency corrections that can be constructed. One is for $k|\vec{x}|$ small (the results in [6]) and another one that is proposed in the recent paper [7] for k small globally. The emphasis of the latter corrections are to include the low frequency corrections both in the near field as well as the far field. Those corrections are as follows

$$\tilde{w}\,(\vec{x}) \, = \, \int_\Gamma \tilde{\sigma}\,(\vec{y})\, G_0(\vec{x},\vec{y})\, ds_{\vec{y}} \, + \, \beta\,(\int_\Gamma \tilde{\sigma}\,(\vec{y})\, ds_{\vec{y}} \, + 1) \, + v_{inc}(\vec{x}) \tag{1.19}$$

which is the one that appears in [6].

$$w(\vec{x}) \, = \, \int_\Gamma \tilde{\sigma}\,(\vec{y})\, G_k(\vec{x},\vec{y})\, ds_{\vec{y}} \, + \, u_{inc}(\vec{x}) \tag{1.20}$$

This is the improved result that is valid both in the near field and in the far field. To illustrate the effectiveness of these approximations we solved for u, the full solution of Helmholtz equation, and the approximations $\tilde{w}$ and w for the case of a circular obstacle and a plane wave. All the results were calculated numerically. Figures 2 and 3 show the effect of these corrections as predicted by the theory for reduced frequencies $k = 10^{-2}$ and $k = 10^{-4}$ respectively.

The important information that is related to the computations of exterior aerodynamic calculations is as follows: Suppose one wants to compute solutions within a truncated region with the aid of an artificial boundary. Then for the well-posedness of the problem within the truncated region one must have normal derivatives prescribed on the boundary in an appropriate form. For example one can realize this form by considering the Sommerfeld's radiation condition. The operator that we propose is

$$\frac{\partial u}{\partial n}(\vec{x}) = \frac{\partial v}{\partial n}(\vec{x}) + \frac{1}{\beta - c_0}\int_\Gamma f_0(\vec{y})\frac{\partial G_0}{\partial n}\, ds_{\vec{y}} + \mathcal{O}(k^2(\log k)^2) \tag{1.21}$$

A similar relation holds if one is interested in specifying the normal derivative of the scattered field only. In this case u will be replaced by u_s and v will be replaced by v_s. The importance of this condition is that in the low frequency limits it is important to correct the zero frequency operator with the logarithmic terms which play a significant role. This will be evident in the next section where we shall deal with a true time dependent situation and the structure of these logarithmic corrections in the frequency translates to logarithmic corrections in time in the artificial boundary operators.

These constructions were made from a computational point of view but are consistent with the abstract theoretical results posed in [10].

A point of comfort and warning is that these corrections are not necessary in the case of a hard obstacle. That is, the Dirichlet condition on Γ is replaced with a Neumann condition $\frac{\partial u_s}{\partial n} = -\frac{\partial u_{inc}}{\partial n}$. It can be shown that the decay rates are much faster for this condition and does not involve any logarithmic corrections. Potential formulations that were introduced by Goldstein [12] and Atassi [13] fall into this category. As a result the computational results presented in [14] do behave well in the low frequency limits. However, issues dealing with the time dependent problems, such as problems governed by the Euler equations, are in need of further theoretical backing. We show a preliminary study in the next section.

1.3 Implications to a True Time Dependent Model

In this section we motivate a time dependent model problem that is used to indicate potential troubles in two dimensions. Specifically, we shall focus our attention on construction of absorbing boundary conditions on artificial boundaries. The model problem that we propose is governed by the wave equation exterior to a circle of radius r_0 with a time dependent data on r_0 which asymptotes to a constant value as $t \Rightarrow \infty$. For computational purposes we truncate the infinite region by an artificial boundary at $r = R$. At this boundary there is an exact condition (in the mathematical sense) that takes care of the effects of extending the region to infinity. The model problem is as follows:

$$u_{tt} = u_{rr} + \frac{1}{r} u_r, \qquad r_0 < r < R \tag{1.22}$$

$$u(r_0, t) = 1 + g(t) \tag{1.23}$$

$$u_r = \mathcal{F}(u) \qquad \text{at } r = R. \tag{1.24}$$

$$u(r, 0) = u_t(r, 0) = 0 \qquad r_0 < r < R \tag{1.25}$$

$$g(t) = O(\frac{1}{t^2}) \qquad \text{as } t \Rightarrow \infty \tag{1.26}$$

This model is designed to have the long time behavior of a solution $u \Rightarrow 0$ as $t \Rightarrow \infty$. The operator $\mathcal{F}(u)$ is the exact operator known only theoretically. It has a well defined structure in the transformed domain which we discuss now. The initial values suggest that it is easier to work with the Laplace transforms. Doing so one obtains

$$s^2 \overline{u} = \overline{u}_{rr} + \frac{1}{r} \overline{u}_r, \qquad r_0 < r < R \tag{1.27}$$

$$\overline{u}(r; s) = \overline{g} \tag{1.28}$$

$$\frac{\partial \overline{u}}{\partial r}(R; s) = \overline{\mathcal{F}}(\overline{u}) \tag{1.29}$$

The operator $\overline{\mathcal{F}}$ in the transformed space is readily verified by considering the solution of equation 1.27 in the exterior region and is as follows:

$$\overline{\mathcal{F}}(\overline{u}) = \frac{sK_0'(sR)}{K_0(sR)}\overline{u} \tag{1.30}$$

Available well-known boundary condition procedures rely on obtaining approximations of the symbol or the ratios of the modified Bessel fucnctions. Here we provide some of these approximations and operators. Let $\overline{B}$ be the approximation of $\overline{\mathcal{F}}$. Then

1. Engquist and Halpern [11]

$$\overline{B} = -s \tag{1.31}$$

which is the crudest approximation of $\overline{\mathcal{F}}$. Then the approximate boundary operator in the transformed space is

$$\frac{\partial \overline{u}}{\partial r} \approx \overline{B}(\overline{u}) \tag{1.32}$$

Inversion of the transform results in a time domain operator

$$\frac{\partial u}{\partial r} + \frac{\partial u}{\partial t} = 0 \tag{1.33}$$

The important part to consider here is the error in the approximation ?? using asymptotic properties of the modified Bessel functions for small and large arguments. This in turn will suggest the order of the error in real time. Doing so we obtain

$$\overline{\mathcal{F}} - \overline{B} = \begin{cases} \frac{1}{2R} & \text{as s} \rightrightarrows \infty \\ O(\frac{1}{\ln s}) & \text{as s} \rightrightarrows 0 \end{cases} \tag{1.34}$$

2. Bayliss and Turkel [4] first order

Analogous list in this are boundary operators in the transform domain

$$\overline{B} = -s - \frac{1}{2R} \tag{1.35}$$

the boundary operator in the time domain

$$\frac{\partial u}{\partial r} + \frac{\partial u}{\partial t} + \frac{1}{2R}u = 0 \tag{1.36}$$

and the errors behave like

$$\overline{\mathcal{F}} - \overline{B} = \begin{cases} O(\frac{1}{s}) & \text{as s} \rightrightarrows \infty \\ \\ -\frac{1}{2R} & \text{as s} \rightrightarrows 0 \end{cases} \tag{1.37}$$

3. Bayliss and Turkel [4] second order

Here the boundary operator in the transform domain is

$$\overline{B} = -\frac{s^2 + \frac{3s}{2R} + \frac{3}{8R^2}}{s + \frac{1}{R}} \tag{1.38}$$

and its translation in the time domain is

$$\left(\frac{\partial}{\partial r} + \frac{\partial}{\partial t} + \frac{5}{2R}\right)\left(\frac{\partial}{\partial r} + \frac{\partial}{\partial t} + \frac{1}{2R}\right)u = 0 \tag{1.39}$$

The errors in this case are

$$\overline{\mathcal{F}} - \overline{B} = \begin{cases} O(\frac{1}{s^2}) & \text{as s} \Rightarrow \infty \\ -\frac{3}{8R} & \text{as s} \Rightarrow 0 \end{cases} \tag{1.40}$$

4. Hagstrom [5] (part of current study)

The boundary operator in the transform domain is

$$\overline{B} = -\frac{s(s + \frac{1}{2R} + \delta)}{s + \delta} \tag{1.41}$$

and its translation in the time domain is

$$\left(\frac{\partial}{\partial t} + \delta\right)\left(\frac{\partial u}{\partial r}\right) + \left(\frac{\partial}{\partial t} + \frac{1}{2R} + \delta\right)\frac{\partial u}{\partial t} = 0 \tag{1.42}$$

Finally the errors behave like

$$\overline{\mathcal{F}} - \overline{B} = \begin{cases} O(\frac{1}{s}) & \text{as s} \Rightarrow \infty \\ O(\frac{1}{\ln s}) & \text{as s} \Rightarrow 0 \end{cases} \tag{1.43}$$

We propose to correct the errors in the low frequency limit, i.e., the long time behavior in constructions of these operators. This is the punch line of this paper and again we emphasize that this idea is crucial if one were to compute directly time dependent problems. For this purpose let us observe that if one were to include the next order correction to the asymptotics for small s, we obtain the following properties:

$$\overline{u}_r \sim \begin{cases} -(s + \frac{1}{2R})\overline{u} & \text{as } s \Rightarrow \infty \\ -\frac{1}{R\ln(Re^{\gamma}\frac{s}{2})}\overline{u} & \text{as } s \Rightarrow 0 \end{cases} \tag{1.44}$$

where γ is the Euler's constant. Now consider the case where

$$u \Rightarrow 1 \qquad \text{as } t \Rightarrow \infty \tag{1.45}$$

which has the behavior in the transform domain

$$\overline{u} \Rightarrow \frac{1}{s} \qquad \text{as } s \Rightarrow 0 \tag{1.46}$$

In particular

$$\overline{u}_r \sim -\frac{1}{Rs\ln(Re^{\gamma}\frac{s}{2})} \qquad \text{as } s \Rightarrow 0. \tag{1.47}$$

To obtain the corrected behavior in the time domain we look for a function $G(t)$ so that

$$\mathcal{L}\{G(t)u\} = -\frac{1}{Rs\ln(Re^{\gamma}\frac{s}{2})} \qquad \text{as } s \Rightarrow 0. \tag{1.48}$$

Construction of such a function will be found in [3]. The key result here is **Proposition:**

$$G(t) = \frac{1}{R[\ln\left(\frac{2t}{R}\right) + D]} \tag{1.49}$$

where D is a constant such that $D > e^2$. The proposed boundary operator here has the form

$$\left(\frac{\partial}{\partial t} + \delta\right)\frac{\partial u}{\partial r} = -\left(\frac{\partial}{\partial t} + \frac{1}{2R} + \delta\right)\frac{\partial u}{\partial t} - \delta G(t)u. \tag{1.50}$$

We compare this operator with the operator in category 4. When δ is a constant and $G = 0$ one retrieves this operator. The function $\delta(t)$ is determined by matching the next asymptotic correction in space (i.e., $O(\frac{1}{R^3})$). This yields

$$\delta = \frac{1}{4R}\frac{1}{1-2RG}. \tag{1.51}$$

The details of these calculations will be found in [3].

We demonstrate the effect of this proposition through a numerical simulation. For this simulation a specific choice of $g(t)$ is considered. It is modeled in such a way that it has oscillatory decay. The form is

$$u(r_0, t) = 1 - \frac{\cos 2\pi t}{1+t^2} \qquad t > 0. \tag{1.52}$$

Moreover, this problem does not have an explicit analytic solution. Therefore, the numerical simulation and comparison is made by doing the calculation in a large domain $r_0 < r < R$ with $r_0 = 1$ and $R = 252$. The value of R is chosen in such a way that the waves do not hit this artificial boundary so that no reflection in the computational domain is introduced in the time period (500 nondimensional units). The specific scheme that is used for computation is an explicit second order scheme (both in time and space). The graphs and associated results are given in Figure 4. In addition the results that are plotted in the figure represent the magnitude of error in time at a fixed point within the computational domain. This point in this case is chosen to be $r = 2$. The results are consistent with the predicted theory. The theory predicts that the results of Engquist and Halpern [11] have the largest error for short times and is demonstrated in the graph. However, for large times this error tends to decay but only at a very short rate which is consistent with the key emphasis of this paper. The second and third graphs are due to the computations using the Bayliss and Turkel [4] boundary operators of orders one and two respectively. Clearly one sees smaller errors at short times and the errors have a tendency of growing larger (they do indeed though not shown here). Finally the last curve is the current proposal [3] and has errors less than a percent for all times and it behaves well for short, long and in between times. These constructions are a guide to an abbreviated proposal that is given for aerodynamic problems, specifically for the Euler equations.

1.4 Extensions to Aerodynamics

We extend the ideas presented in the earlier sections to aerodynamic problems governed by the Euler equations. A similar analysis can be done for the viscous flows governed by the Navier-Stokes equations which is not presented here. It should be noted that to extend the analysis of the model problem it is vital to have a linear problem. Clearly the Euler equations or the Navier-Stokes equations are not linear. However, the analysis which we seek is for the far field boundary operators and is reasonable to assume linearity of the equations there. The form of the Euler equations we use here is as follows:

$$\frac{\partial \rho}{\partial t} + \operatorname{div}(\rho \underline{u}) = 0 \tag{1.53}$$

$$\rho \frac{D\underline{u}}{Dt} = -\nabla p \tag{1.54}$$

$$\rho \frac{De}{Dt} + p \operatorname{div} \underline{u} = 0 \tag{1.55}$$

There are a variety of ways to linearize these equations about a state. In this case we consider the state at infinity. Since we are dealing with a system of first order equations, additional boundary conditions are required according to the hyperbolic theory. The treatments that are presented here are for subsonic inflow as well as outflow. Along these lines Gustafsson [18] proposed a linearization with the variables $q = \operatorname{div}\underline{u}$, the divergence of the velocity field, and the entropy $S = \ln\left(\frac{p}{\rho^\gamma}\right)$. Noting that $\underline{\omega} = \operatorname{curl}\underline{u}$ the linearized equations look like

$$\frac{\partial S}{\partial t} + u_\infty \frac{\partial S}{\partial x} = 0 \tag{1.56}$$

$$\frac{\partial \omega}{\partial t} + u_\infty \frac{\partial \omega}{\partial x} = 0 \tag{1.57}$$

$$\begin{cases} \frac{\partial q}{\partial t} + u_\infty \frac{\partial q}{\partial x} + \frac{1}{\rho_\infty} \triangle p = 0 \\ \frac{\partial p}{\partial t} + u_\infty \frac{\partial p}{\partial x} + \rho_\infty c_\infty^2 q = 0 \end{cases} \tag{1.58}$$

In the last two equations, elimination of q yields a convective wave equation for the pressure

$$\left(\frac{\partial}{\partial t} + u_\infty \frac{\partial}{\partial x} \right)^2 p = c_\infty^2 \triangle p. \tag{1.59}$$

This is the place where the connection to our earlier analysis holds. The treatment of boundary conditions for the pressure is subject to the structure of the solutions of this equation. Before we proceed to discuss these equations we point out that the linearized equations for the entropy and vorticity suggest that at points of inflow one must specify S and ω. These are dictated by the hyperbolic theory as well.

Now we set up an appropriate initial boundary value problem that will be used to extend the analysis in the previous section. The model to be considered here is:

$$\psi_{tt} + 2U_\infty \psi_{xt} + U_\infty^2 \psi_{xx} = C_\infty^2 \psi_{xx} + C_\infty^2 \psi_{yy} \tag{1.60}$$

$$\psi = \psi_t = 0 \quad \text{at } t = 0 \tag{1.61}$$

where $\psi = p - p_\infty$. Taking the Laplace transform with respect to t yields:

$$s^2 \overline{\psi} + 2U_\infty s \; \overline{\psi}_x + U_\infty^2 \overline{\psi}_{xx} = C_\infty^2 \overline{\psi}_{xx} + C_\infty^2 \overline{\psi}_{yy}. \tag{1.62}$$

Now we make a dependent variable change from $\overline{\psi}$ to ϕ in the form

$$\overline{\psi} = e^{\lambda x} \phi. \tag{1.63}$$

This yields the following calculations:

$$\overline{\psi}_x = \lambda e^{\lambda x} \phi + e^{\lambda x} \phi_x \tag{1.64}$$

$$\overline{\psi}_{xx} = \lambda^2 e^{\lambda x} \phi + 2\lambda e^{\lambda x} \phi_x + e^{\lambda x} \phi_{xx} \tag{1.65}$$

$$\overline{\psi}_{yy} = e^{\lambda x} \phi_{yy} \tag{1.66}$$

Substituting (64-66) in the transformed equation we obtain

$$s^2\phi + 2U_\infty s\lambda\phi + 2U_\infty s\phi_x = (C_\infty^2 - U_\infty^2)(\phi_{xx} + 2\lambda\phi_x + \lambda^2\phi) + C_\infty^2 \phi_{yy}. \tag{1.67}$$

We pick the free parameter λ (to eliminate the first derivative term) so that

$$\lambda = \frac{U_\infty s}{C_\infty^2 - U_\infty^2}. \tag{1.68}$$

Let

$$\tilde{s} = \frac{s}{\sqrt{C_\infty^2 - U_\infty^2}} \tag{1.69}$$

$$\tilde{x} = \frac{x}{\sqrt{1 - M_\infty^2}} \tag{1.70}$$

$$\tilde{y} = y. \tag{1.71}$$

Then the equation for ϕ becomes

$$\phi_{\tilde{x}\tilde{x}} + \phi_{\tilde{y}\tilde{y}} = \tilde{s}^2 \phi. \tag{1.72}$$

Radially symmetric solutions satisfy

$$\phi_{\tilde{r}} = \tilde{s}\,\frac{K_0'(\tilde{s}\tilde{r})}{K_0(\tilde{s}\tilde{r})}\phi. \tag{1.73}$$

Like in our wave equation calculations for $\tilde{s}$ large we have

$$\phi_{\tilde{r}} = -\frac{(\tilde{s} + \frac{1}{2\tilde{r}} + \delta)\tilde{s}}{\tilde{s} + \delta}\,\phi \tag{1.74}$$

where

$$\tilde{r} = \sqrt{\tilde{x}^2 + \tilde{y}^2} = \sqrt{\frac{x^2}{1 - M_\infty} + y^2} = r\sqrt{\frac{\cos^2\theta}{1 - M_\infty} + \sin^2\theta}$$

$$= \frac{r}{\sqrt{1 - M_\infty}}\sqrt{1 - M_\infty \sin^2\theta}. \tag{1.75}$$

Therefore the chain rule yields

$$\frac{\partial\phi}{\partial\tilde{r}} = \frac{\sqrt{1 - M_\infty^2}}{\sqrt{1 - M_\infty^2 \sin^2\theta}}\,\frac{\partial\phi}{\partial r}. \tag{1.76}$$

Note that

$$\phi = \overline{\psi} e^{-\lambda x} = \overline{\psi} e^{-\lambda r \cos\theta}. \tag{1.77}$$

Therefore

$$\frac{\partial\phi}{\partial r} = \left(\frac{\partial\overline{\psi}}{\partial r} - \lambda\cos\theta\overline{\psi}\right) e^{-\lambda r\cos\theta}. \tag{1.78}$$

Noting

$$\lambda = \frac{U_\infty s}{C_\infty^2 - U_\infty^2} = \frac{M_\infty}{\sqrt{1 - M_\infty^2}}\tilde{s} \tag{1.79}$$

and letting $H(\theta) = \frac{\sqrt{1 - M_\infty^2}}{\sqrt{1 - M_\infty^2 \sin^2(\theta)}}$ we see that

$$H(\theta)\left(\frac{\partial\overline{\psi}}{\partial r} - \frac{M_\infty}{\sqrt{1 - M_\infty^2}}\tilde{s}\cos\theta\,\overline{\psi}\right) = -\frac{(\tilde{s} + \frac{1}{2r}H(\theta) + \delta)}{\tilde{s} + \delta}\tilde{s}\overline{\psi}. \tag{1.80}$$

Therefore

$$\frac{\partial \overline{\psi}}{\partial r} = -\frac{1}{H(\theta)} \left(\frac{\tilde{s} + \frac{1}{2r} H(\theta) + \delta}{\tilde{s} + \delta} \right) \tilde{s}\overline{\psi} + \frac{M_\infty}{\sqrt{1 - M_\infty^2}} \tilde{s} \cos\theta \overline{\psi} \tag{1.81}$$

$$\frac{\partial \overline{\psi}}{\partial r} = - \left(\frac{\mathcal{S}(\theta) + \frac{1}{2r} + \delta \mathcal{S}(\theta)}{\tilde{s} + \delta} \right) \tilde{s}\overline{\psi} \tag{1.82}$$

where

$$\mathcal{S}(\theta) = \frac{\sqrt{1 - M_\infty^2 \sin^2\theta} - M_\infty \cos\theta}{\sqrt{1 - M_\infty^2}} \tag{1.83}$$

Inclusion of the low frequency correction and inversion of Laplace transform yield a boundary operator for the difference of pressure $\psi = p - p_\infty$ in the form

$$\left(\frac{1}{d}\frac{\partial}{\partial t} + \delta \right) \psi_r = - \left(\frac{\mathcal{S}(\theta)}{d}\frac{\partial}{\partial t} + \frac{1}{2R} + \delta \mathcal{S}(\theta) \right) \cdot \frac{1}{d}\psi_t - \delta G(\theta, t)\psi \tag{1.84}$$

where $d = \sqrt{C_\infty^2 - U_\infty^2}$. Again one can establish the form of G as

$$G(\theta, t) = \frac{H(\theta)}{R \ln \left(\frac{2t\, H(\theta) d}{R} + D \right)}. \tag{1.85}$$

Corrections to $O(\frac{1}{R^3})$ yield in this case

$$\delta = \frac{1}{4R} \frac{\sqrt{1 - M_\infty^2}}{\sqrt{1 - M_\infty^2 \sin^2\theta}} \frac{1}{1 - 2RG}. \tag{1.86}$$

In conclusion, this paper points out potential problems in computing solutions of wave-like equations in exterior regions. Remedy of the problems dictates logarithmic correction terms in posing radiation conditions. An extension is proposed to the Euler equations in the far field.

References

[1] B. Engquist and A. Majda. Absorbing Boundary Conditions for the Neumerical Simulation of Waves *Math. Comp.,* Vol. 31, (1977), 629-651.

[2] Saul S. Abarbanel et. al. Secondary Frequencies in the Wake of a Circular Cylinder with Vortex Shedding *J. Fluid Mech.*, Vol. 225, (1991), 557-574.

[3] T. M. Hagstrom, S. I. Hariharan and R. C. MacCamy. Asymptotic Behaviors of Absorbing Boundary Conditions, (in preparation)

[4] A. Bayliss and E. Turkel. Radiation Boundary Conditions for Wave-like Equations *Comm. Pure Appl. Math.*, 33, (1980), 707-725.

[5] T. Hagstrom. Consistency and Convergence for Numerical Radiation Conditions, ICOMP Report No. 90-21, NASA TM-103262, (1990).

[6] S.I. Hariharan and R. C. MacCamy. Low Frequency Acoustic and Electromagnetic Scattering, *Appl. Num. Math.* Vol. 2., (1986), 29-35.

[7] S.I. Hariharan, R.C. MacCamy and C. Muralikrishna. Low Frequency Corrections for the Reduced Wave Equation, *to appear.*

[8] R.C. MacCamy. Low Frequency Acoustic Oscillations, *Quant. Appl. Math.*, Vol. 23, (1965), 247-256.

[9] S.I. Hariharan. Inverse Scattering for an Exterior Dirichlet Problem, *Quant. Appl. Math.*, No. 2, (1982), 273-286.

[10] L.A. Muravei. On the Asymptotic Behavior for Large Values of Time, of Solutions of Exterior Boundary Value Problems for the Wave Equation with Two Space Variables, *Math. USSR Sbornik*, Vol. 35, No. 3, (1979), 377-423.

[11] B. Engquist and L. Halpern. Far Field Boundary Conditions for Computation over Long Time, *Appl. Numer. Math.*, Vol. 4, No. 1 (1988), 21-25.

[12] M.E. Goldstein. Unsteady Vortical and Entropic Distortions of Potential Flows Round Arbitrary Obstacles, *J. Fluid Mech.*, Vol. 89, (1978), 433-468.

[13] H.M. Atassi. Unsteady Vortical Disturbances Around Bodies, *Proceedings of the Tenth U.S. National Congress of Applied Mechanics*, J.P. Lamb, Ed., ASME, (1986), 475-484.

[14] J.R. Scott and H.M. Atassi. Numerical Solution of Periodic Vortical Flows about a Thin Airfoil, *AIAA Paper 89-1691.*

[15] Eds. A. Dervieux, Brian Van Leer, J. Periaux and A. Rizzi. Numerical Simulation of Compressible Euler Flows: a GAMM Workshop, Braunschweig: Vieweg (1989).

[16] P. Werner. Low Frequency Asymptotics for the Reduced Wave Equation in Two Dimensional Exterior Spaces, *Math. Meth. in Appl. Sci.*, Vol. 8, (1986) 134-156.

[17] R. Kress. On the Low Wave Asymptotics for the Two-dimensional Exterior Dirichlet Problem for the Reduced Wave Equation, *Math. Meth. in Appl. Sci.*, Vol. 9, (1987), 335-341.

[18] B. Gustafsson. Far Field Boundary Conditions for Time Dependent Hyperbolic Problems, *SIAM J. Sci. Stat. Comp.*, 9, (1988), 812-828.

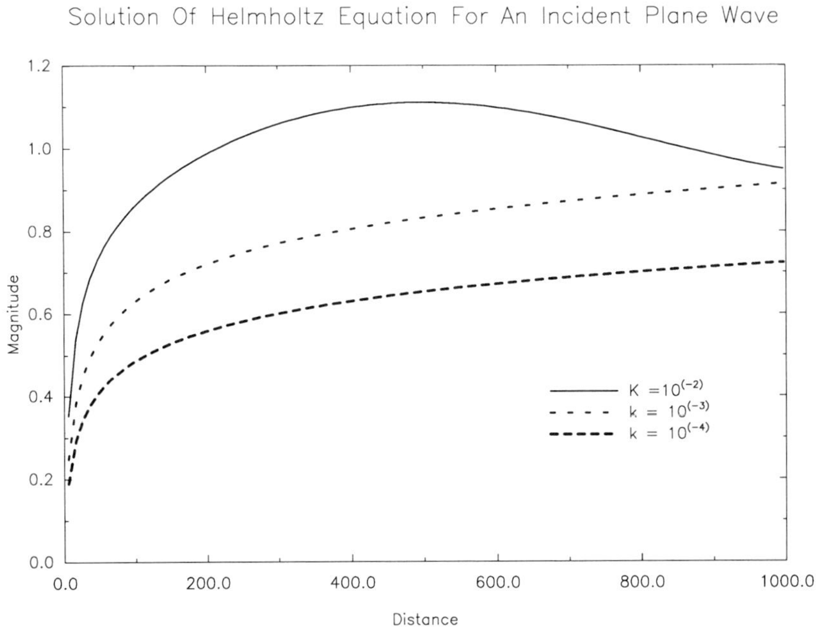

Figure 1.

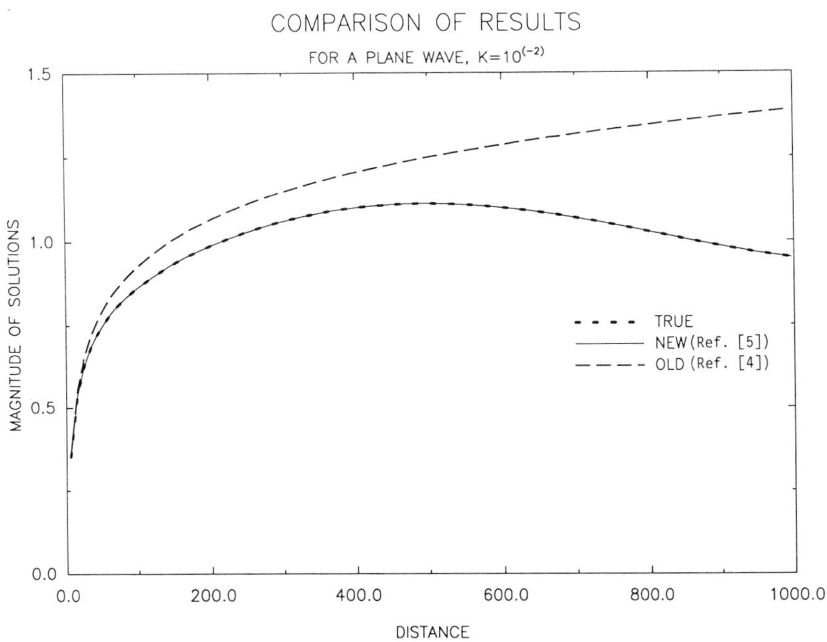

Figure 2.

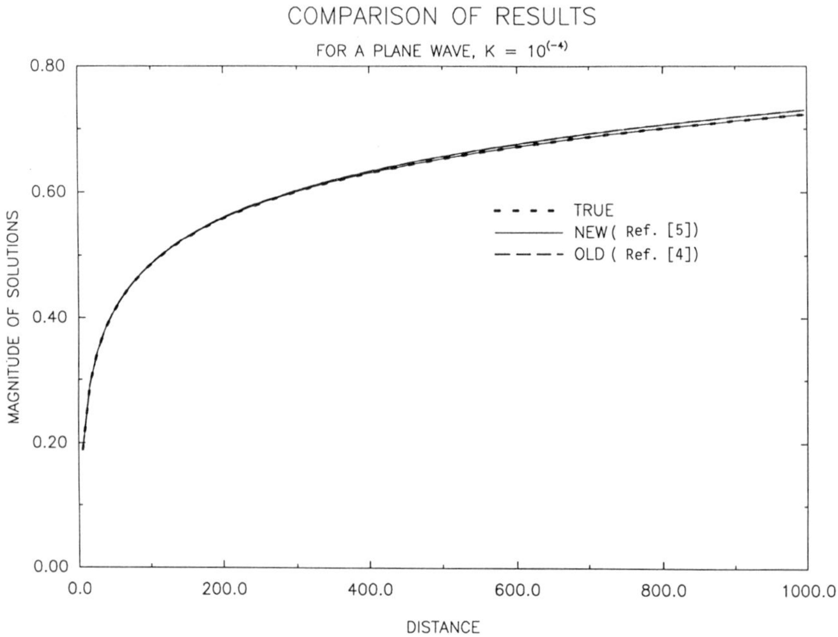

Figure 3.

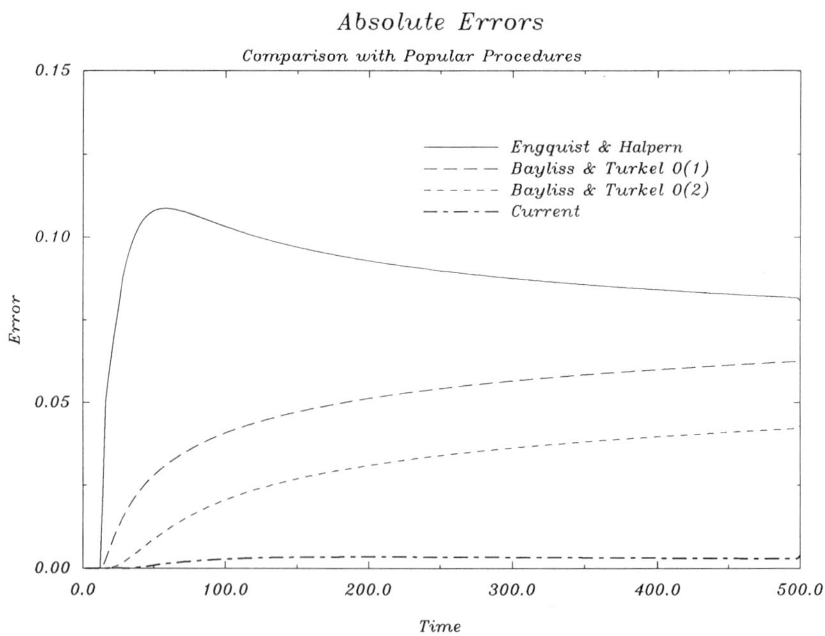

Figure 4.

Simulation of 3D-Unsteady Internal Flows

Najat Benbouta[1]
Pascal Ferrand[1]
Francis Leboeuf[1]

ABSTRACT This paper presents a numerical method to describe the unsteady flow field. This method solve the compressible, time-dependent, Euler and Navier-Stokes conservation equations for mass, momentum,and energy in curvilinear coordinates.The integration in time is performed by using an implicit single-step temporal scheme. The integration in space is peformed by using a finite-volume formulation coupling with an upwind scheme for the convective terms and a second order centered scheme for viscous terms.To validate and demonstrate the accuracy of this method, calculated results to obtain steady solutions in both inviscid or viscous, two or three dimensional flows are given.

1.1 Introduction

The internal flows in Turbomachinery are often 3D-unsteady, viscous and compressible. To compute such flows, the Navier-Stokes equations has to be solved. In order to reduce the computation time of complex flow calculations, most studies presently use implicit methods if a stationnary state is required(large time steps, thus less time iterations) or if the frequencies of the phenomena studied are sufficiently low(the time step is adapted to the cycle considered). Generally, these implicit methods use factorization in all space directions, based on the A.D.I technique; hence at each time we resolve a linear system with a tridiagonal[1,2] or bidiagonal[3] structure. The factorized schemes considerably reduce the CPU-time but they limit the time step and introduce numerical errors wich are difficult to evaluate: therefore, the non-factorizing schemes are more robust, though they are more expensive[4]. In the present study, the method used to compute the conservative Navier-stokes equations is implicit in time and makes no use of

[1]Groupe de Turbomachines, Laboratoire de Mecanique des Fluides et d'Acoustique,URA 263, Ecole Centrale de Lyon,36 Av Guy de Collongue, BP136, 69131 Ecully Cedex (France)

factorization in space. It is based on the "Delta" formulation of Pade. We implemented this method in terms of curvilinear coordinates, wich allows for a better descriptions of the flow behavior in complex geometries. In fact, since curvilnear coordinates are aligned with the geometrie, they permit a better treatment of the boundary conditions. A finite volume formulation has chosen in order to maintain the conservative character of the equations. For the spatial differences, an upwind scheme is used. This scheme is inherently dissipative and more expensive than a centred scheme, but it allows a correct treatment of strong shock waves. It is realized with the flux vector splitting technique of Van Leer[5]. In this technique, the convective fluxes are split into forward and backward contributions according to the sign of the eingenvalues. The spatial derivatives of splitted fluxes(backward and forward fluxes) were approximated as a flux balance in the control volume, using a MUSCL(Monotone Upstream-Centred Schemes of Conservation Laws) method[6]. The viscous terms are approximated by classical second order centred differences. To compute the linear system obtained after time and spatial discretizations, we use the iterative Gauss-Seidel's method whose convergence has been accelerated with the MMPE(Modified Minimal polynomial Extrapolation) method, developped by A.Sidi[7].To reduce the memory size, a particular storage of the matrix is adopted[8]. The purpose of the current work is to predict problems caused by flutter,through the modelisation of the 3D flows with a strong oscillations shock motion. The model described above will be applied to compute the unsteady transonic flow with emphasis on the numerical determination of the instantaneous shock position. First of all, it is necessary to validate the computed results for steady flow, because they are the starting point for the unsteady flow computations. The test case is the steady flow in a two dimnesional nozzle geometry.

1.2 Governing Equations

1.2.1 Basic Equations

The motion of compressible viscous flow is governed by the Navier-Stokes equations, wich express the conservation of mass, momentum, and energy. the conservative vector form of the dimensionless equations in cylindrical coordinate system (z,r,θ) is given by:

$$\partial_t q + \partial_z E_I + \partial_r F_I + \frac{1}{r}\partial_\theta G_I + H_I = Re^{-1}(\partial_z E_v + \partial_r F_v + \frac{1}{r}\partial_\theta G_v + H_v) + D \tag{1.1}$$

where, the unknown vector q is:

$$q = (\rho, \rho u, \rho v, \rho w, \rho e_0) \tag{1.2}$$

with ρ is the density, u, v, w are the components of the velocity, and e_0 is the total specific energy. E_I, F_I and G_I are the convective fluxes and E_v, F_v and G_v are the diffusive fluxes. H_I and H_v are the vectors associated with the curvature of the initial frame and D contains the rotation terms. The expression of theese fluxes may be found, for example, in [9]. The pressure P is related to the conseved variables through the perfect gas law:

$$P = (\gamma - 1)\rho(e_0 - \frac{1}{2}u^2 - \frac{1}{2}v^2 - \frac{1}{2}w^2) \tag{1.3}$$

where γ is the ratio of specific heats, taken as $\gamma = 1.4$ and the stokes hypothesis $3\lambda + 2\mu = 0$ is assumed.

1.2.2 Navier-Stokes Equations in Generalized Coordinates

In case of complex geometry problems, the use of the body-fitted curvilnear coordinate system (ξ,η,ζ) may be of interest. The equations system (1.1) has been generalized from cylindrical coordinate using a time-dependent transformation as:

$$\left\{ \begin{array}{l} \xi = \xi(t, z, r, \theta) \\ \eta = \eta(t, z, r, \theta) \\ \zeta = \zeta(t, z, r, \theta) \\ \tau = t \end{array} \right\} \tag{1.4}$$

Equation (1.1), subject to the general transformation (1.4), gives:

$$\partial_\tau \hat{q} + \partial_\xi \hat{E}_I + \partial_\eta \hat{F}_I + \partial_\zeta \hat{G}_I + \hat{H}_I = Re^{-1}(\partial_\xi \hat{E}_v + \partial_\eta \hat{F}_v + \partial_\zeta \hat{G}_v + \hat{H}_v) + \hat{D} \tag{1.5}$$

where:

$$\left\{ \begin{array}{ll} \hat{q} & = \frac{1}{J} q \\ \hat{D} & = \frac{1}{J} D \\ \hat{H}_I & = \frac{1}{J} H_I - r F_I \\ \hat{H}_v & = \frac{1}{J} H_v - r F_v \\ \hat{X}_I^\alpha & = \frac{1}{J}(\xi_t^\alpha q + \xi_z^\alpha E_I + \xi_r^\alpha F_I + \frac{1}{r}\xi_\theta^\alpha G_I) \\ \hat{X}_v^\alpha & = \frac{1}{J}(\xi_z^\alpha E_v + \xi_r^\alpha F_v + \frac{1}{r}\xi_\theta^\alpha G_v) \end{array} \right\} \tag{1.6}$$

with:

$$\left\{ \begin{array}{lll} \xi^1 = \xi & \hat{X}_I^1 = \hat{E}_I & \hat{X}_v^1 = \hat{E}_v \\ \xi^2 = \eta & \hat{X}_I^2 = \hat{F}_I & \hat{X}_v^2 = \hat{F}_v \\ \xi^3 = \zeta & \hat{X}_I^3 = \hat{G}_I & \hat{X}_v^3 = \hat{G}_v \end{array} \right\} \tag{1.7}$$

and J is the Jacobian determinant of the transformation (1.4).

1.3 Discretization

1.3.1 Finite Volume Formulation

Finite-volume-type discretization was chosen because of its general applicability and ability to preserve the conservation properties of the governed equations. The computational domain is subdivided into a set of control volumes and the equation (1.5) can be interpreted as describing the balance of mass, momentum and energy over an arbitrary control volume. Thus, the jacobian J represents the inverse of the cell volume and the quantities $\rho U/J$, $\rho V/J$ and $\rho W/J$ represent the mass flux crossing the cell interfaces in the ξ, η and ζ directions. The contravariant components of velocity U, V and W are defined as:

$$\left\{ \begin{array}{ll} U & = \xi_t + \xi_z u + \xi_r v + \frac{1}{r}\xi_\theta w \\ V & = \eta_t + \eta_z u + \eta_r v + \frac{1}{r}\eta_\theta w \\ W & = \zeta_t + \zeta_z u + \zeta_r v + \frac{1}{r}\zeta_\theta w \end{array} \right\} \tag{1.8}$$

1.3.2 Spatial Differencing

The governing equations are viewed as composed of inviscid (convective and pressure) and viscous(diffusive) terms. The spatial differencing reflects the physical nature of equations. Thus, upwind differencing is used for the convective and pressure terms and central differencing for the diffusive terms. The upwind differencing in the present work is effected through the technique of flux-vector splitting of Van Leer[5]. The three-dimensional splittings of Van Leer were originaly given for cartesian coordinates and the extension to the generalized coordinate is found in [6]. For the treatment of the moving grid, the splitting is based on the investigations of Parpia[10]. theese split fluxes have the advantage of first derivative continuity at eigenvalue sign changes, corresponding to the sonic and stagnation points, and can represent normal shocks with at most two, and generally one, transition zones.

Flux Vector Splitting:

Only the splitting for the flux in the ξ-direction is given, as the others can be obtained similarly. The flux vector $\hat{E}_I$ is split according to the contravariant Mach number in the ξ-direction, defined above as:

$$M_\xi = \frac{\bar{U}}{a} \tag{1.9}$$

with:

$$\bar{U} = \frac{U}{\|\vec{grad}\xi\|} \tag{1.10}$$

a: the sound velocity
U: the contravariant velocity in the ξ-direction

For supersonic flow, $|M_\xi| \geq 1$:

$$\begin{array}{llllll} \hat{E}_I^+ & = \hat{E}_I, & \hat{E}_I^- & = 0 & For & M_\xi \geq +1 \\ \hat{E}_I^+ & = 0, & \hat{E}_I^- & = \hat{E}_I & For & M_\xi \leq -1 \end{array} \quad (1.11)$$

For subsonic flow, $|M_\xi| < 1$:

$$\hat{E}^\pm = \frac{\|grad\vec{\xi}\|}{J} \left\{ \begin{array}{l} f^\pm_{mass} \\ f^\pm_{mass}\{\hat{\xi}_z(-\bar{U} \pm 2a)/\gamma + u)\} \\ f^\pm_{mass}\{\hat{\xi}_r(-\bar{U} \pm 2a)/\gamma + v)\} \\ f^\pm_{mass}\{\hat{\xi}_\theta(-\bar{U} \pm 2a)/\gamma + w)\} \\ f^\pm_{energy} \end{array} \right\} \quad (1.12)$$

where:

$$f^\pm_{mass} = \pm\frac{1}{4}\rho a(M_\xi \pm 1)^2 \quad (1.13)$$

$$\begin{aligned} f^\pm_{energy} = \; & f^\pm_{mass}.[\{-(\gamma-1)\bar{U}^2 \pm 2(\gamma-1)\bar{U}a + 2a^2\}/(\gamma^2-1) \\ & + \frac{1}{2}(u^2+v^2+w^2) - \hat{\xi}_t(-\bar{U} \pm 2a)/\gamma] \end{aligned} \quad (1.14)$$

and $\hat{\xi}_t$, $\hat{\xi}_z$, $\hat{\xi}_r$, $\hat{\xi}_\theta$ are given by:

$$\left\{ \begin{array}{lll} \hat{\xi}_t & = & \frac{\xi_t}{\|grad\xi\|} \\ \hat{\xi}_z & = & \frac{\xi_z}{\|grad\xi\|} \\ \hat{\xi}_r & = & \frac{\xi_r}{\|grad\xi\|} \\ \hat{\xi}_\theta & = & \frac{\xi_\theta}{\|grad\xi\|} \end{array} \right\} \quad (1.15)$$

$\hat{\xi}_z$, $\hat{\xi}_r$, $\hat{\xi}_\theta$ are the direction cosines of the directed interface in the ξ-direction and $(\bar{U} - \hat{\xi}_t)$is the velocity normal to a ξ=constant face. The fluxes in the other two directions are easily formed by interchanging ξ with either η or ζ.

Differencing of convective terms:

The generalized fluxes $\hat{E}_I$, $\hat{F}_I$ and $\hat{G}_I$ are split into forward and backward contributions according to the signs of the jacobian matrices $\partial\hat{E}_I/\partial\hat{q}$, $\partial\hat{F}_I/\partial\hat{q}$, $\partial\hat{G}_I/\partial\hat{q}$ and differencing accordingly. For example, the flux in the ξ-direction can be differenced as:

$$\partial_\xi \hat{E}_I = \partial_\xi^- \hat{E}_I^+ + \partial_\xi^+ \hat{E}_I^- \quad (1.16)$$

The split differences in equation(1.16) are implemented as a flux balance across a cell, corresponding to MUSCL type differencing (Monotone Upstream Centred Schemes for Conservation Laws)[6]. For example:

$$\partial_\xi^- \hat{E}_I^+ + \partial_\xi^+ \hat{E}_I^- = [\hat{E}_I^+(\hat{q}^-) + \hat{E}_I^-(\hat{q}^+)]_{i+\frac{1}{2}} - [\hat{E}_I^+(\hat{q}^-) + \hat{E}_I^-(\hat{q}^+)]_{i-\frac{1}{2}} \quad (1.17)$$

The notation $\hat{E}_I^+(\hat{q}^-)_{i+\frac{1}{2}}$ denotes $\hat{E}_I^+$ evaluated using the metric terms at the cell interface i+$\frac{1}{2}$ and the conserved state variable on the upwind side of the interface, obtained by the upwind state variable interpolations to cell interfaces:

$$\begin{aligned} \hat{q}^-_{i+\frac{1}{2}} &= \hat{q}_i + \tfrac{\varphi}{4}[(1+\kappa)\,\Delta\,\hat{q}_i + (1-\kappa)\,\nabla\,\hat{q}_i] \\ \hat{q}^+_{i+\frac{1}{2}} &= \hat{q}_{i+1} - \tfrac{\varphi}{4}[(1+\kappa)\,\nabla\,\hat{q}_{i+1} + (1-\kappa)\,\Delta\,\hat{q}_{i+1}] \end{aligned} \tag{1.18}$$

with:

$$\begin{aligned} \Delta\hat{q}_i &= \hat{q}_{i+1} - \hat{q}_i \\ \nabla\hat{q}_i &= \hat{q}_i - \hat{q}_{i-1} \end{aligned} \tag{1.19}$$

The upwind discretization above corresponds to an interpolation of $\hat{q}$ to cell interfaces followed by splitting rather differencing the split fluxes at cell centeres, corresponding to a more local splitting and leading to improvment results[6]. For first order accuracy, the control factor takes $\varphi = 0$. For higher than one order accuracy it takes $\varphi = 1$ and the parameter κ determines the accuracy of the spatial differencing: $\kappa = -1$ corresponds to a fully-upwind scheme, $\kappa = +1$ to a central difference scheme, and $\kappa = 1/3$ to the third-order upwind differencing and one for higher differencing. In general, some form of fluxes limiting is required for the scheme to avoid oscillations at discontinuities in the solution, such as shock waves. The limiting can be implemented by a locally modifying the difference quotients in the gradiants in order to ensure monotone interpolation. The first order scheme($\varphi = 0$) was used in the current computation.

Differencing of Viscous Terms:

The differencing for the viscous terms corresponds to second-order central differences, consistent with diffusive nature of the shear and heat flux terms. The second derivatives are treated consevatively as differences across cell interfaces of first derivative terms:

$$\hat{E}_v = \hat{E}_v(\hat{q}, \hat{q}_\xi, \hat{q}_\eta, \hat{q}_\zeta) = M.T(\hat{q}, \hat{q}_\xi, \hat{q}_\eta, \hat{q}_\zeta) \tag{1.20}$$

where:
M : contain the metric terms.
T : function of the variable and derivatives of variable.

$$\partial_\xi \hat{E}_v = [\hat{E}_v]_{i+\frac{1}{2}} - [\hat{E}_v]_{i-\frac{1}{2}} = M_{i+\frac{1}{2}}(T_{i+1} - T_i) - M_{i-\frac{1}{2}}(T_i - T_{i-1}) \tag{1.21}$$

1.4 Integration in Time

With the above mentioned "MUSCL" approach, a solution to equation(1.5) is obtained by using a single-step temporal scheme[11]:

$$\Delta\hat{q}^n = \frac{\theta\Delta\tau}{1+\beta}\partial_\tau\Delta\hat{q}^n + \frac{\Delta\tau}{1+\beta}\partial_\tau\hat{q}^n + \frac{\beta}{1+\beta}\Delta\hat{q}^{n-1} + \Theta[(\theta - \frac{1}{2} - \beta)\Delta\tau^2 + \Delta\tau^3] \quad (1.22)$$

where$\Delta\hat{q}^n = \hat{q}^{n+1} - \hat{q}^n$,$\hat{q}^n = \hat{q}(n\,\Delta\,\tau)$, and n denotes nth time step of computations. The implicit trapezoidal scheme takes $\theta = 1/2$ and $\beta = 0$ in(1.22) wich results in truncation error $\theta(\Delta\tau^3)$, and the implicit Euler's scheme takes $\theta = 1$ and $\beta = 0$ in(1.22) wich results in truncation error $\theta(\Delta\tau^2)$. The implicit trapezoidal scheme was used in the present computation. Applying the scheme (1.22) into equation (1.5), one obtain a solution to the unsteady Navier-Stokes equations:

$$\left\{\begin{array}{l} I + \frac{\theta\Delta\tau}{1+\beta}\{\partial_\xi^-[\hat{A}^+]^n + \partial_\xi^+[\hat{A}^-]^n + \partial_\eta^-[\hat{B}^+]^n + \partial_\eta^+[\hat{B}^-]^n \\ +\partial_\zeta^-[\hat{C}^+]^n + \partial_\zeta^+[\hat{C}^-]^n + \partial_\xi[\hat{P}]^n + \partial_\eta[\hat{R}]^n + \partial_\zeta[\hat{S}]^n \\ -\partial_{\xi\xi}^2[Q\hat{e}1]^n - \partial_{\eta\eta}^2[Q\hat{f}2]^n - \partial_{\zeta\zeta}^2[Q\hat{g}3]^n \\ +\partial_{\xi\eta}^2[\hat{L}]^n + \partial_{\xi\zeta}^2[\hat{M}]^n + \partial_{\eta\zeta}^2[\hat{N}]^n + [\hat{T}]^n] \end{array}\right\}\Delta\,\hat{q}^n = Res^n \quad (1.23)$$

where the residual at time level n is given by:

$$Res^n = \frac{\Delta\tau}{1+\beta}[-\partial_\xi^-\hat{E}^+ - \partial_\xi^+\hat{E}^- - \partial_\eta^-\hat{F}^+ - \partial_\eta^+\hat{F}^- - \partial_\zeta^-\hat{G}^+ - \partial_\zeta^+\hat{G}^- \quad (1.24)$$
$$+\partial_\xi\hat{E}_v + \partial_\eta\hat{F}_v + \partial_\zeta\hat{G}_v + \hat{H}]^n + \frac{\beta}{1+\beta}\Delta\,\hat{q}^{n-1}$$

The resolution scheme for the implicit part of (1.23) is based on the Gauss-Seidel's iterative method whose convergence have been accelerated with a plynomial extrapolation method and to reduce the size memory, a particular storage of matrix is adopted. This technique is detailed in[8].

1.5 Boundary Conditions

At the inlet boundary, the total temperature T_0, the total pressure P_0 and the transversal components of velocity v,w are imposed and the longitudinal velocity u is computed implicitly by solving the momentum equation in the ξ-direction. At the outlet boundary, the static pressure P is imposed and the unknown variables are determined implicitly from computations; the energy equation is replaced by an equation of this type: $dP = f(\tau)$. On the walls, the boundary conditions are applied explicitly. For inviscid computations, the normal velocity is set to zero and for viscous computations the velocity is set to zero. For both inviscid and viscous calculations, the pressure and temperature are determined by extrapolation from the interior.

1.6 Computational Results

The numerical method previously described was applied to the computation of a steady flow field. The method was applied to the computation of a two dimensional channel flow case. The geometry consists of a channel with a (non-dimensional) lenght of $L = 3.0$, a width of $l = 1.0$ and with a 10% thick sinusoidal "bump" on the upper wall; the lenght of the bump is identical to the channel width. This geometry is identical to the test case used in[12,13]. Figure 1 shows the computational mesh, composed of 23x17x7, used in both viscous and inviscid subsonic steady calculations, while figure 2 shows the computational mesh, composed of 37x11x7, used to compute the transonic inviscid flow. For all calculations, the flow in the channel is initially uniform. Figure 4,5 present respectively the convergence evolution of implicit and explicit parts in the subsonic viscous case; the method requires 400 steps with a $CFL = 10$ for the solution to reach viscous steady-state. In figures 5,6 are shown the isomach lines and longitudinal velocity in the viscous subsonic calculations. There is an acceleration of the flow at the throat of the channel. Also, at the upstream station a strongly marked boundary layer is present with a velocity profile that is advected up to the downstream station. The velocity profile is not consistent with the physical boundary conditions imposed at the inlet. Furthmore,in figure 7 we see that with respect to the uniform total pressure condition, the computation generates an errouneous static pressure at the inlet to compensate for the velocity deficit near the walls. The generation of the parabolic profile at the inlet is due to the deficiency of the flux-splitting technique wich is not adapted to a no-slip condition at the walls for viscous computation. This anomaly has been mentionned by Hanel et al[14]. Figures 8,9 permit the comparison between the 3D-results and 2D-results obtained by imposing a condition of symmetry in the third-direction. The parabolic form is more pronounced in the transversal direction wich is defined with only 7 regular points of mesh. Then, for viscous computation using the flux splitting, it is necessary to use a large number of grid points near the walls. For the 3D-computation the number of Mach at the inlet is $M_e = 0.08$, illustrating the capacity of the code to compute incompressible flows. Figures 10,11 show the isomach lines and static pressure contours for subsonic inviscid calculation at $M_{-\infty} = 0.6$. For this Eulerian calculation the static pressure has the correct behavior. For the transonic Euler calculation $M_{-\infty} = 0.675$, we use the mesh presented in figure 2. In this case a supersonic region appears in the solution (with a maximum value of mach is 1.3) that is determined by a shock as shown in figures 12,13,14. In this supersonic region, small oscillations occur due to the use of a centred scheme for calculating the Euler fluxes normal to the walls.

1.7 Conclusion

In this paper we have presented a time-accurate numerical method that can properly describe the steady or unsteady, viscous or inviscid, two or three dimensional flow fields. Numerical results in the subsonic and transonic steady cases are presented and indicate that the accuracy of the method is satisfactorily. However the results also show that upwind discretization cause additional numerical diffusion in viscous shear layers, where strong gradients are present. Thus the Euler fluxes must be well-treated in order to avoid numerical boundary layer effects at no-slip walls. Before unsteady calculations can be performed,it is necessary to make additional steady calculations to validate this method.

Acknowledgments: Support of this research by the METRAFLU Society,64 Chemin des Mouilles, 69134 Ecully Cedex FRANCE,is most gratefully acknowledged. The authors would like to present their thanks to Mr S. AUBERT for help in preparing the computer graphics.

REFERENCES

[1] R.M. Beam andR.F. Warming. An Implicit Factored Scheme For The Compressible Navier-Stokes Equations. In *AIAA Journal* Vol.16, pp. 393-402, 1978.

[2] W.R. Briely and H. Mac Donald. Solution Of The Three-Dimensional Compressible Navier-Stokes Equations By An Implicit Technique. In *Lectures Notes In Physics* Vol.35, pp. 105-110, 1975.

[3] R.W. Mac Cormack. A Numerical Method For Solving The Equations Of Compressible Viscous Flow. In *AIAA Journal* Vol.20, pp. 1275-1281, 1978.

[4] R.W. Mac Cormack. Current Status Of Numerical Solution Of The Navier-Stokes Equations. In *AIAA Paper* 85-0032, 1985.

[5] B. Van Leer. Flux Vector Splitting For The Euler Equations. In *Lectures Notes In Physics* Vol.170, pp. 501-502, 1982.

[6] W.K. Anderson, J.L. Thomas and B. Van Lerr. A Comparison Of Finite Flux-Vector Splitting For The Euler Equations. In *AIAA Paper* 85-0122, 1985.

[7] A. Sidi, JW.F. Ford and D.A. Smith. Acceleration Of Convergence Of Vector Sequences. In *SIAM J.NUM.ANAL.* VOL.23,N.1,Feb 1986.

[8] N. Benbouta, P. Ferrand and F. Leboeuf. Convergence Acceleration For Linear System Iterative Resolution And Application To Computational Fluid Mechanics. In *High Performance Computing J.L Delhaye and E. Gelenbe* (Edition Elsevier Sciences Publishers B.V North-Holland), 1989.

[9] R. Peyret, and H. Viviand. Computation Of Viscous Compressible Flows Based On Navier-Stokes Equations. In *AGARD* AG- 212, 1975.

[10] I.H. Parpia. Van Leer Flux Vector Splitting in Moving Coordinates. In *AIAA Journal* Vol.26, N.1, Jan.,1988.

[11] R.M. Beam and R.F. Warming. An Implicit Finite Differnce Algorithm For Hyperbolic Systems Conservation Law Form. In *J.Comput.Phys.* Vol.22, pp. 87-110, Sep.1976.

[12] G. Moretti. Experiments On Initial And Boundary Conditions. In *Symposium on Numerical and Physical Aspects of Aerodynamic Flow* Jan.,1981.

[13] R. Ni. A Multiple-Grid Scheme For Solving The Euler-Equations. In *AIAA Journal* Vol.20, N.11, Jan.,1982.

[14] D. Hanel, R. Schwane and Cr. Seider. On The Accuracy Of Upwind Schemes For The Solution of the Navier-Stokes Equations. In *AIAA Paper* 87-1105, 1988.

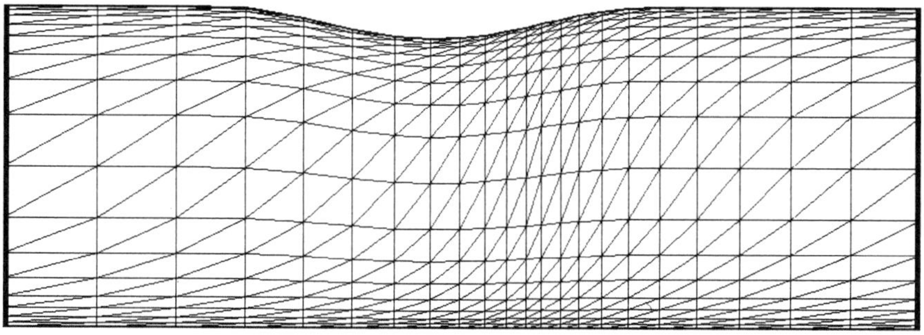

FIG. 1 Mesh : 23 x 17 x 7

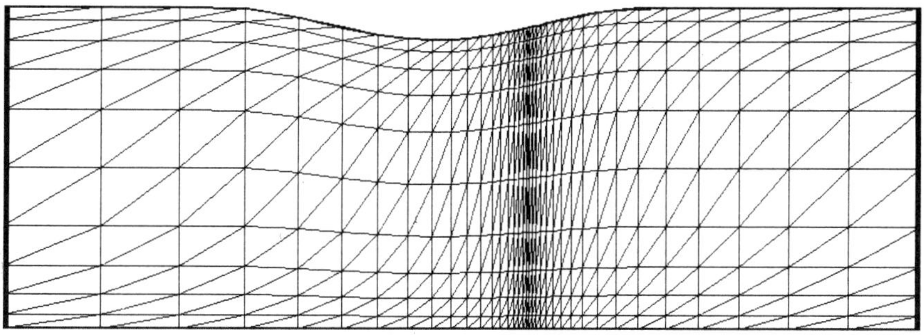

FIG. 2 Mesh : 37 x 11 x 7

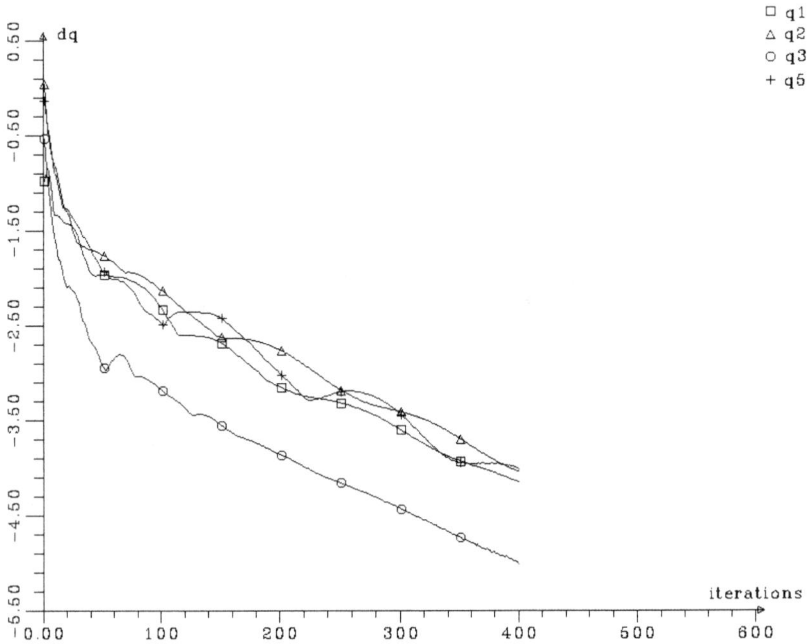

FIG. 3 Subsonic viscous flow convergence

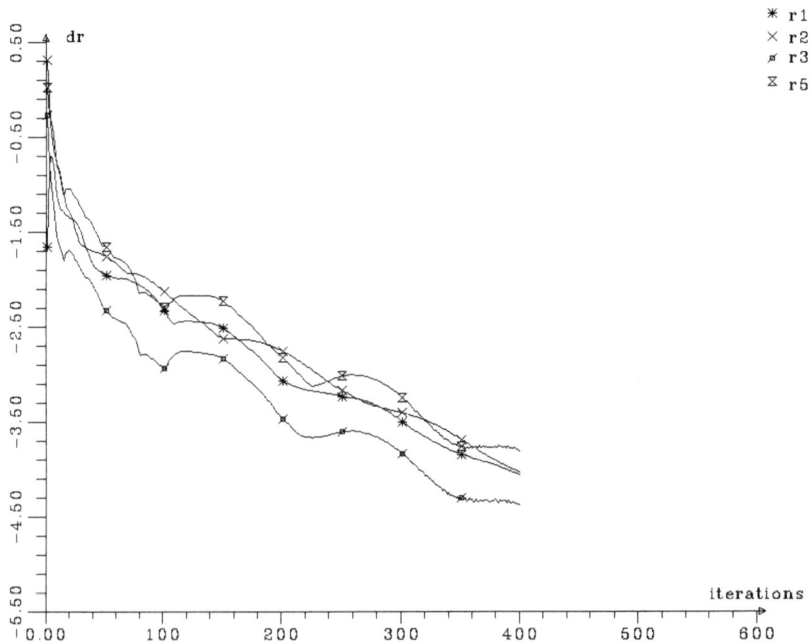

FIG. 4 Subsonic viscous flow convergence

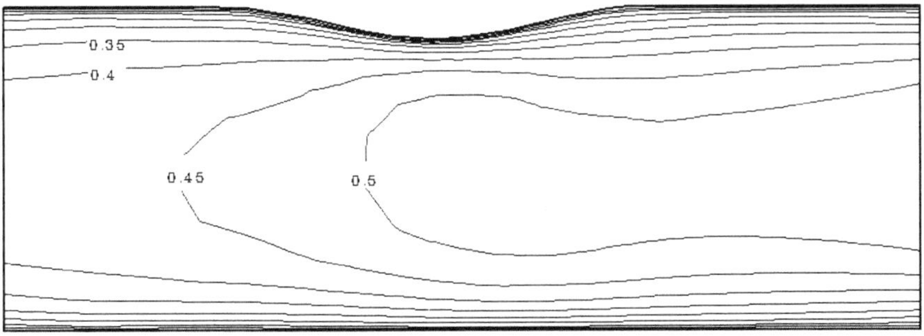

Mesh : 23 x 17 x 7

FIG. 5 Iso-Mach

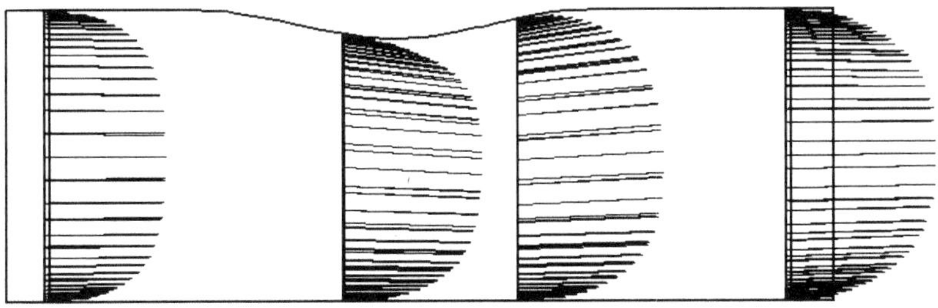

Mesh : 23 x 17 x 7

FIG. 6 Longitudinal velocity

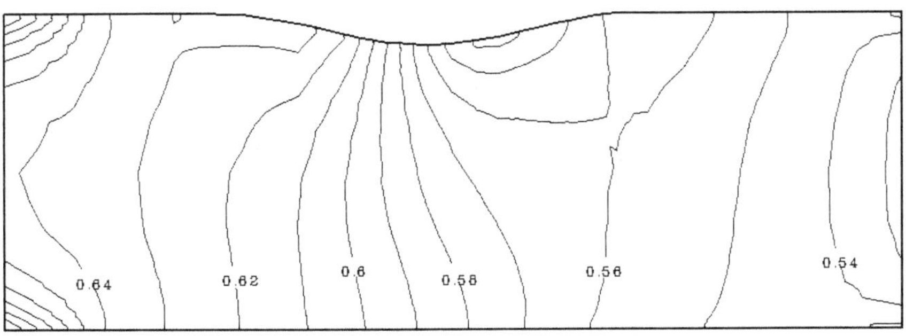

Mesh : 23 x 17 x 7

FIG. 7 Iso-pressure

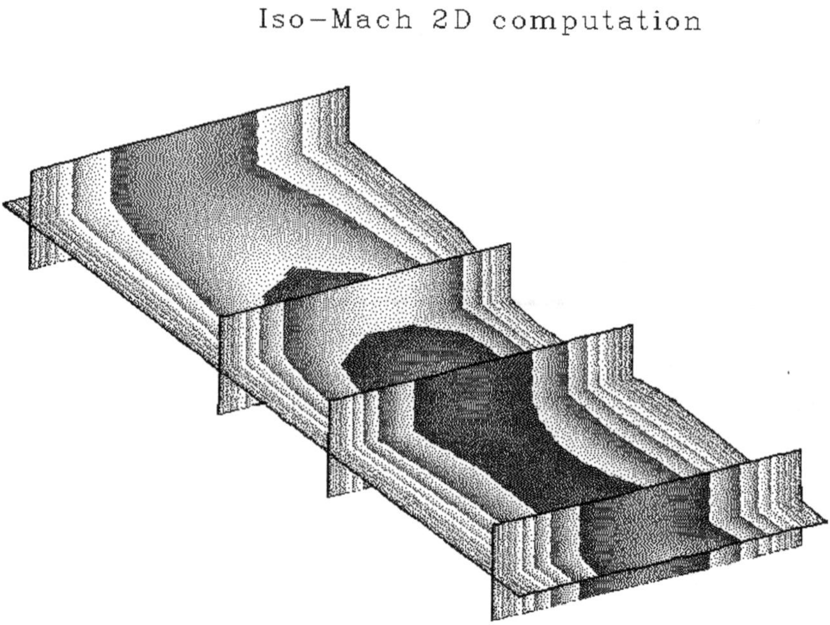

FIG. 8 Mesh : 23 x 17 x 7

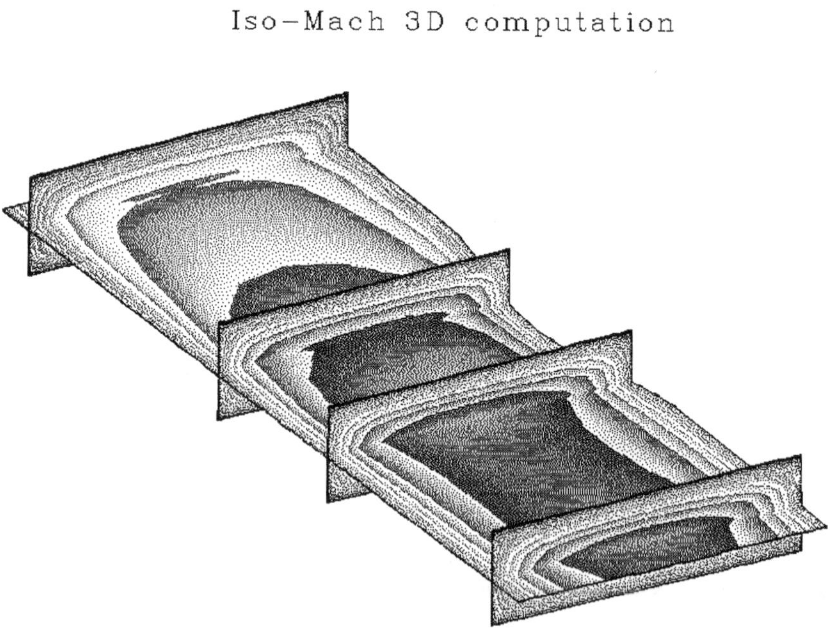

FIG. 9 Mesh : 23 x 17 x 7

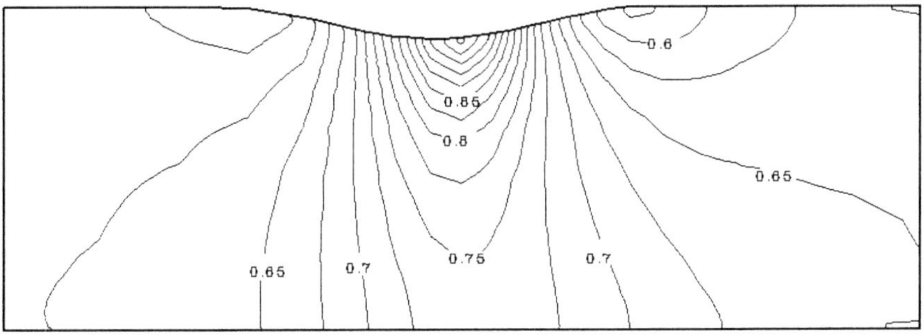

Mesh : 23 x 17 x 7

FIG. 10 Iso-Mach

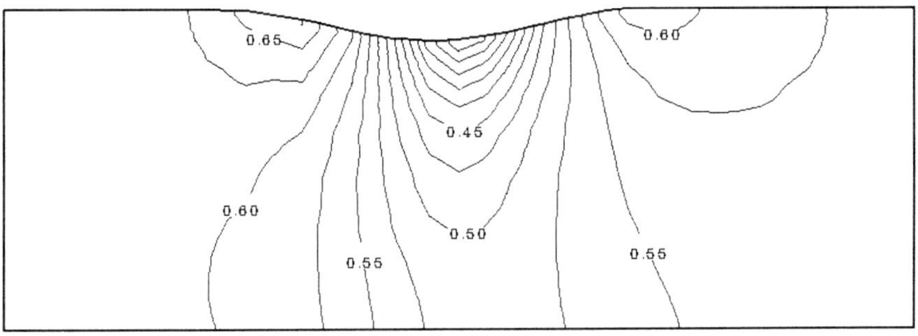

Mesh : 23 x 17 x 7

FIG. 11 Iso-pressure

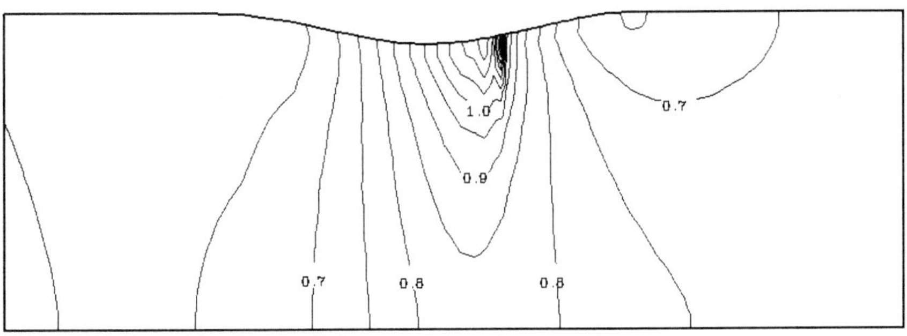

Mesh : 37 x 11 x 7

FIG. 12 Iso-Mach

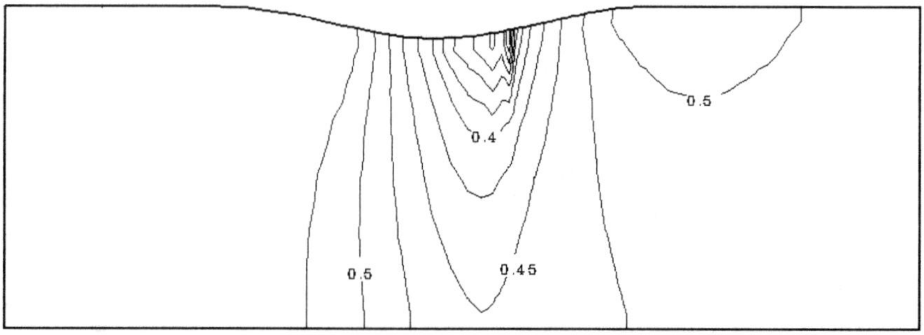

Mesh : 37 x 11 x 7

FIG. 13 Iso−pressure

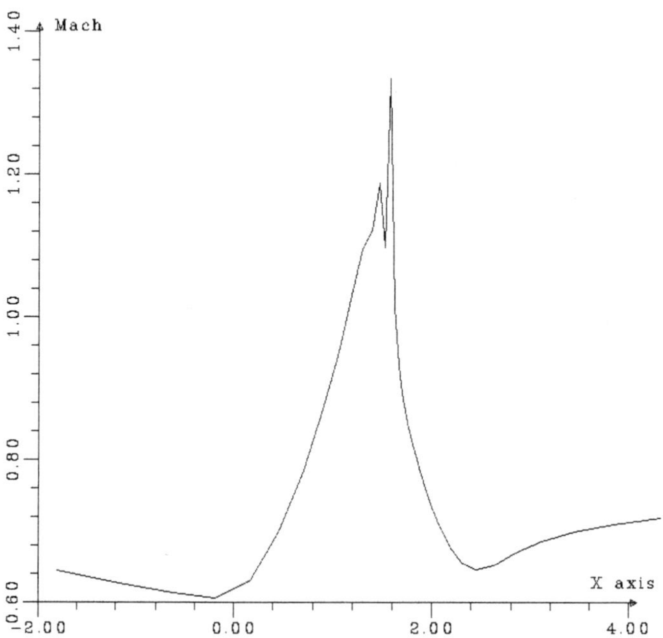

FIG. 14 Mach at the upper wall

II. Unsteady Cascade Flows

Time Linearized Euler Calculations for Unsteady Quasi-3D Cascade Flows

Gerhard Kahl[1]
Arno Klose[1]

ABSTRACT A method for calculating the quasi-3D unsteady flowfield in cascades is presented.
The unsteadiness may arise from oscillating blades (as in the flutter problem) or from inlet disturbances (as in the forced response problem).
In this method, the unsteady flow is assumed to be a small, harmonic perturbation of the steady flow. As long as the amplitude of the perturbations is moderate, the higher order terms in the governing equations derived under this assumption may be negelcted and the describing equations become linear. Thus every frequency component can be calculated separately and the results be obtained by superposition.
The steady-state flow, about which the unsteady part is linearized, is calculated by a finite-volume time-stepping Euler solver.
As the linearized unsteady equations are very similar to the basic equations for the steady solver, the discretization is almost identical for both solvers. Thus it is possible to use much of the steady code with little modification for the time-linearized unsteady code.
Because of the linearity of the unsteady equations, they can be solved very efficiently by an ADI-method.
The results obtained by this method compare favourably to flat plate theory as well as to nonlinear time stepping methods.
The approach presented is computationally very efficient, thus permitting application in the standard design procedure.

---

[1] MTU Motoren-und Turbinen-Union Munchen GmbH, Dachauer Strasse 665, 8000 Munchen 50, Germany.

Nomenclature

A, **B**, **C**	Jacobian matrices
B_n	streamsheet thickness
c	airfoil chord length
c_φ	absolute circumferential velocity
F, **G**	unsteady nonlinear flux in φ and m-direction, respectively
H	unsteady nonlinear source term
i, k	gridpoint indices
j	imaginary unit, $\sqrt{-1}$
m	meridional coordinate
r	radius
s	coordinate along airfoil surface
U	vector of unsteady nonlinear conservative variables
$\bar{\mathbf{u}}$	mean vector of conservative variables
u	perturbation vector of conservative variables
$\hat{\mathbf{u}}$	vector of time linearized complex amplitudes of conservative variable
w_m	relative meridional velocity
w_φ	relative circumferential velocity
δ	central difference operator
φ	circumferential coordinate
ω	angular frequency of oscillation
Ω	angular frequency of rotation

1.1 Introduction

The design of modern lightweight aeroengines leads to thin, highly loaded bladings in axial turbines and compressors. These bladings are increasingly susceptible to destructive vibrations, either self induced or induced by interactions between adjacent blade rows. This phenomenon necessitates the development of reliable and efficient computational methods for the prediction of the unsteady airloads acting on the airfoils. Current unsteady nonlinear time-stepping methods are computationally too expensive to be used routinely in the design process. It is thus necessary to introduce certain simplifications.

In this paper a time linearized Euler-method for the calculation of the unsteady airloads in inviscid quasi-3D flow is presented. The method fully accounts for the nonlinear effects of blade thickness and camber while restricting the unsteadiness in the flow to small perturbations of the steady flow. The theory is based on previous works by Hall [6] and Ni [8] while the discretization and large parts of the code are based on Happel's nonlinear steady Euler code [7].

To make best use of the linear properties of the problem, an aproximately factored implicit algorithm (ADI) is used for the interior of the computational domain. Exact nonreflecting boundary conditions based on the works of Acton [1] and Hall [6] are implemented at the far field boundaries. The theory, discretization and the method of solution are described and results for different test cases are shown.

1.2 Governing Equations

The unsteady flow is described by the two-dimensional inviscid and compressible Euler-equations on a streamsheet of revolution (blade-to-blade,S1) with varying radius and varying streamsheet thickness [7].

$$\frac{\partial}{\partial t}\mathbf{U} + \frac{1}{r}\left[\frac{\partial}{\partial \varphi}\mathbf{F}\right]_{S1,m} + \left[\frac{\partial}{\partial m}\mathbf{G}\right]_{S1,\varphi} = \mathbf{H} \tag{1.1}$$

with

$$\mathbf{U} = \begin{bmatrix} B_n r\rho \\ B_n r\rho\, w_m \\ B_n r\rho\, r c_\varphi \\ B_n r\rho\, u_{rot} \end{bmatrix}, \qquad \mathbf{F} = \begin{bmatrix} B_n r\rho\, w_\varphi \\ B_n r\rho\, w_m w_\varphi \\ B_n r\left(\rho\, r c_\varphi w_\varphi + rp\right) \\ B_n r\rho\, h_{rot} w_\varphi \end{bmatrix},$$

$$\mathbf{G} = \begin{bmatrix} B_n r\rho\, w_m \\ B_n r\left(\rho\, {w_m}^2 + p\right) \\ B_n r\rho\, r c_\varphi w_m \\ B_n r\rho\, h_{rot} w_m \end{bmatrix}, \qquad \mathbf{H} = \begin{bmatrix} 0 \\ B_n \rho {c_\varphi}^2 \frac{dr}{dm} + p\frac{d(B_n r)}{dm} \\ 0 \\ 0 \end{bmatrix},$$

and

$$\begin{aligned} c_\varphi &= w_\varphi + \Omega r \\ u_{rot} &= \frac{1}{\kappa - 1}\frac{p}{\rho} + \frac{1}{2}\left({w_m}^2 + {w_\varphi}^2 - (\Omega r)^2\right) \\ h_{rot} &= \frac{\kappa}{\kappa - 1}\frac{p}{\rho} + \frac{1}{2}\left({w_m}^2 + {w_\varphi}^2 - (\Omega r)^2\right) \end{aligned}$$

The fluid under consideration is assumed to be a perfect gas so that

$$RT = \frac{p}{\rho} \tag{1.2}$$

All variables are dimensionless, lengths are normalized by chord c, pressure and density by their respective values at the inlet, p_1 and ρ_1. Velocities are normalized by $\sqrt{p_1/\rho_1}$ and time by $c/\sqrt{p_1/\rho_1}$.

1.3 Time Linearization Approach

The basic idea of the linearization is to treat the unsteady flow as the sum of the underlying mean flow plus a small perturbation, i.e.:

$$\mathbf{U}(\varphi, m, t) = \bar{\mathbf{u}}(\varphi, m) + \mathbf{u}(\varphi, m, t) \tag{1.3}$$

Substitution of this expansion into eq. (1.1) and collection of zeroth and first order terms while neglecting second order terms (products of perturbation quantities) yields the steady state Euler equations (from the zeroth order terms), plus the time linearized Euler equations (from the first order terms). The steady state Euler equations equal eq. (1.1) with $\partial/\partial t = 0$. This shows that in case of small perturbations, the mean flow is equal to the unperturbed steady flow, hence the mean flow is not affected by the perturbations. If the steady state solution is known, only the first order time linearized Euler equations for the perturbation variables remain to be solved:

$$\tfrac{\partial}{\partial t}\mathbf{u} + \tfrac{1}{r}\tfrac{\partial}{\partial \varphi}\mathbf{A}\mathbf{u} + \tfrac{\partial}{\partial m}\mathbf{B}\mathbf{u} = \mathbf{C}\mathbf{u} \tag{1.4}$$

Here $\mathbf{A}$, $\mathbf{B}$ and $\mathbf{C}$ denote the Jacobian matrices:

$$\mathbf{A} = \left.\frac{\partial \mathbf{F}}{\partial \mathbf{U}}\right|_{\mathbf{U}=\bar{\mathbf{u}}}, \quad \mathbf{B} = \left.\frac{\partial \mathbf{G}}{\partial \mathbf{U}}\right|_{\mathbf{U}=\bar{\mathbf{u}}} \quad \mathbf{C} = \left.\frac{\partial \mathbf{H}}{\partial \mathbf{U}}\right|_{\mathbf{U}=\bar{\mathbf{u}}} \tag{1.5}$$

These equations could now be time marched to obtain the linearized unsteady flow. But for aeroelastic purposes, it is often sufficient to restrict the calculation to a single harmonic. It is then assumed that all perturbations are harmonic and only the complex amplitudes of the perturbations are calculated. Formally, this is equivalent to Fourier-transforming eq. (1.4) and determining the solution for one single harmonic. Due to the linearity of the system, interaction of two different harmonics is not possible. For a harmonic disturbance in the boundary conditions (i.e. harmonic oscillation of the airfoils with constant amplitude), the unsteady flow field will thus be harmonic with constant amplitudes. This means that the resulting complex amplitudes of the perturbations which we are interested in will be steady.

Nevertheless it is convenient for computational purposes to treat the amplitudes of the harmonic perturbations as time dependent. The reason for this seemingly over-complicated assumption is that the resulting equations are now very similar to the original nonlinear time-dependent Euler-equations that are used to calculate the steady flow via time-stepping. This renders it possible to use many parts of the steady solver for the unsteady time linearized solver with very little modification.

The expression for the complex, time dependent amplitudes of the perturbation variables then becomes:

$$\mathbf{u}(\varphi, m, t) = B_n r \hat{\mathbf{u}}(\varphi, m, t) e^{j\omega t} \tag{1.6}$$

Substitution of eq. (1.6) into eq. (1.4), evaluation of the time derivative of $\mathbf{u}$ and division by $B_n r e^{j\omega t}$ yields

$$\frac{\partial}{\partial t}\hat{\mathbf{u}} + \frac{1}{B_n r}\left[\frac{\partial}{\partial \varphi} B_n \mathbf{A}\hat{\mathbf{u}} + \frac{\partial}{\partial m} B_n r \mathbf{B}\hat{\mathbf{u}}\right] = (\mathbf{C} - j\omega \mathbf{E})\hat{\mathbf{u}} \qquad (1.7)$$

The only unknowns in these equations are the complex amplitudes $\hat{\mathbf{u}}$ of the perturbation variables $\mathbf{u}$, whereas the Jacobians $\mathbf{A}$, $\mathbf{B}$ and $\mathbf{C}$ depend only on the mean flow, which is assumed to be known in advance. In practice, the mean (=steady) flow is calculated using a time-stepping multigrid Euler-method. The problem is thus reduced to solving a set of linear first order PDEs with spatially varying coefficients.

1.4 Boundary Conditions

In the following section, the procedures for determining the perturbation amplitudes on the boundaries of the computational domain are described.

1.4.1 Nonreflecting Far-Field Boundary Conditions

The domain must be restricted in the axial direction while there is no physical boundary present at this location in the turbomachine. Thus the appropriate far-field boundary conditions must allow the perturbations emanating from the computational domain to pass out of the domain without causing reflections.

Sufficiently far up- or downstream of the cascade, it can be assumed that the steady flow is uniform in the axial and circumferential direction, if the streamsheet radius and thickness are both constant. In this case, exact nonreflecting boundary conditions can be derived (see for example [1, 6]).

The procedure will be only briefly outlined, a more detailed description can be found in the mentioned references.

If the steady flow is homogeneous, the Jacobians $\mathbf{A}$ and $\mathbf{B}$ are independent of φ and m and the source terms in $\mathbf{C}$ vanish. Eq. (1.7) then becomes a linear homogeneous PDE with constant coefficients, which can be cast into the form of an eigenvalue problem that can be solved analytically. The result is a set of four independent perturbation waves for each circumferential wave number. These waves can be identified as up- and downsteam travelling pressure waves (two downstream travelling pressure waves in axially supersonic flow), one entropy and one vorticity wave, both of which are convected downstream.

As these perturbation waves are independent of each other, it is possible to prescribe at the far field boundaries only the values of the waves entering the computational domain while leaving the ones passing out of the domain unaffected. Thus these waves will not be reflected at the boundaries. For the problem of vibrating airfoils, the amplitudes of the waves entering the flowfield will be set to zero whereas for the forced response problem they will be set to the incoming disturbance values.

The nonreflecting boundary conditions are implemented in primitive variables rather than using the conservative variables, as this reduces the necessary algebra considerably.

1.4.2 Flow Tangency Condition

On the airfoils, the inviscid flow remains tangent to the airfoil surfaces at all times. If we assume an arbitrary motion of the airfoils, the boundary condition on the airfoil surface is

$$\mathbf{V}\cdot\mathbf{N}-\frac{\partial\mathbf{R}}{\partial t}\cdot\mathbf{N}=0 \tag{1.8}$$

Here $\mathbf{V}$ is the velocity vector (w_φ, w_m), $\mathbf{R}$ is the position vector of the airfoil surface and $\mathbf{N}$ is the unit vector normal to the surface. Furthermore, $\mathcal{T}$ is the unit vector tangential to the surface and s is the coordinate along the surface.

We now assume that $\mathbf{R}$, $\mathbf{N}$ and $\mathcal{T}$ are, just like the flow variables, composed of a mean part plus a small, harmonic perturbation, i.e.:

$$\begin{aligned} \mathbf{R}(s,t) &= \bar{\mathbf{r}}(s)+\hat{\mathbf{r}}(s)e^{j\omega t}\\ \mathbf{N}(s,t) &= \bar{\mathbf{n}}(s)+\hat{\mathbf{n}}(s)e^{j\omega t}\\ \mathcal{T}(s,t) &= \bar{\tau}(s)+\hat{\tau}(s)e^{j\omega t}\end{aligned} \tag{1.9}$$

Defining

$$\begin{aligned} \hat{r}_n &= \hat{\mathbf{r}}\cdot\bar{\mathbf{n}}\\ \hat{v}_n &= \hat{\mathbf{v}}\cdot\bar{\mathbf{n}}\\ \bar{v}_n &= \bar{\mathbf{v}}\cdot\bar{\mathbf{n}}\\ \bar{v}_\tau &= \bar{\mathbf{v}}\cdot\bar{\tau}\end{aligned} \tag{1.10}$$

we find for the unit vector normal to the surface that

$$\hat{\mathbf{n}}=-\bar{\tau}\frac{\partial\hat{r}_n}{\partial s} \tag{1.11}$$

The linearized flow tangency condition then becomes

$$\hat{v}_n=j\omega\hat{r}_n+\bar{v}_\tau\frac{\partial\hat{r}_n}{\partial s} \tag{1.12}$$

Here the term $j\omega\hat{r}_n$ is due to the velocity of the airfoil surface whereas the term $\bar{v}_\tau\partial\hat{r}_n/\partial s$ is due to the rotation of the airfoil.

Eq. (1.12) is valid at the instantaneous position of the airfoil surface. This poses another problem, as the computational grid that is used is fixed in space. Therefore, a further extrapolation is needed to extrapolate from the instantaneous position of the airfoil surface to the mean position. The resulting condition for the normal velocity at the mean airfoil surface position is

$$\hat{v}_n = j\omega\hat{r}_n + \bar{v}_\tau \frac{\partial \hat{r}_n}{\partial s} - \hat{\mathbf{r}}\nabla\bar{v}_n \tag{1.13}$$

1.4.3 Periodicity Condition

For aeroelastic purposes, it can be assumed that the blades of the cascade are vibrating in a travelling wave mode. All airfoils then vibrate in the same modeshape with a constant time lag between the motions of adjacent blades(= constant interblade phase angle). In cases where this assumption is not immediately valid (i.e. clustered blades), the modeshapes can be decomposed into the sum of travelling wave modes due to the linearity of the system [3, 9]. The condition of all airfoils in the cascade vibrating in the same mode shape with a constant interblade phase angle introduces a similar condition on the complex amplitudes of the flow variables. All perturbation amplitudes at locations being one spacing apart in the circumferential direction will be equal in magnitude with a phase shift equal to the interblade phase angle. The application of this condition to the periodic boundaries allows the computational domain to be restricted to one blade passage.

1.5 Discretization

In the following section, the discretization of the time linearized Euler equations will be described as well as the method of solution and the implementation of the boundary conditions.

The Euler equations are solved on a structured H-type grid by a finite - volume - method, therefore they are first cast into an integral form [11]:

$$\int_V \frac{\partial}{\partial t}\hat{\mathbf{u}}\, dV + \oint_O \{B_n\, r\, \mathbf{B}\hat{\mathbf{u}}\, d\varphi - B_n\, \mathbf{A}\hat{\mathbf{u}}\, dm\} = \int_V (\mathbf{C} - i\omega\mathbf{E})\, \hat{\mathbf{u}}\, dV \tag{1.14}$$

These equations are discretized in time using a first order backward operator

$$\left.\frac{\partial \hat{\mathbf{u}}}{\partial t}\right|_{t+\Delta t} \approx \frac{\hat{\mathbf{u}}_{t+\Delta t} - \hat{\mathbf{u}}_t}{\Delta t} \tag{1.15}$$

This can be written as

$$\hat{\mathbf{u}}_{t+\Delta t} \approx \hat{\mathbf{u}}_t + \Delta\hat{\mathbf{u}} \tag{1.16}$$

The index t in $\hat{\mathbf{u}}_t$ will from now on be dropped for simplicity. Inserting eq. (1.16) into eq. (1.14) and putting all $\Delta\hat{\mathbf{u}}$ - terms on the left hand side while keeping the $\hat{\mathbf{u}}$ - terms on the right hand side leads to

$$\int_V [(\frac{1}{\Delta t} + i\omega) \cdot \mathbf{E} - \mathbf{C}]\Delta\hat{\mathbf{u}}\, dV - \oint_O B_n \mathbf{A}\Delta\hat{\mathbf{u}}\, dm + \oint_O B_n r\mathbf{B}\Delta\hat{\mathbf{u}}\, d\varphi = \int_V [-i\omega\mathbf{E} + \mathbf{C}]\hat{\mathbf{u}}\, dV + \oint_O B_n \mathbf{A}\hat{\mathbf{u}}\, dm - \oint_O B_n r\mathbf{B}\hat{\mathbf{u}}\, d\varphi \quad (1.17)$$

It can be seen that the right hand side depends only on the known values of the steady flow and the amplitudes $\hat{\mathbf{u}}$ and can thus be calculated explicitly, whereas the left hand side contains the unknown corrections $\Delta\hat{\mathbf{u}}$. Therefore, the left hand side will be called the "implicit part" and the right hand side will be referred to as the "explicit part" of the calculation.

Clearly, as $\hat{\mathbf{u}}$ approaches its steady state value, $\Delta\hat{\mathbf{u}} \to 0$ and the entire left hand side goes to zero as well.

This renders it possible to use a simple discretization for the left hand side (the implicit part) thus reducing the bandwidth of the matrices. At the same time a high precision of the obtained solution is maintained by using a more sophisticated discretization for the right hand side.

We use two types of elements for this discretization: the "basic element", consisting of four nodes and the "Euler element" consisting of the four basic elements that are adjacent to one node (fig. 1.1). The "Euler element" thus contains nine nodes.

The nodes of the H-type mesh are numbered by two indices: the index $i\,(i = 1, \ldots, i_{max})$ increases in the meridional direction whereas the index $k\,(k = 1, \ldots, k_{max}$ increases in the tangential direction.

1.5.1 Discretization of the Explicit Part

The discretization of the right hand side is taken from Happel's steady Euler solver [7]. It utilizes all nine nodes of each Euler element in determining the integrals.

Each integral is formed by summing up the contributions of all four basic elements belonging to the Euler element. Accordingly, all integrals are first evaluated for the basic elements.

To evaluate the volume integrals, the averages of the values of the nodes along the circumferential gridlines (i =const) are multiplied by the appropriate fraction of the control volume and the results are summed up.

The surface integrals are evaluated by averaging the fluxes of the adjacent nodes for each cell face. The average is then multiplied by the appropriate cell face length and the contributions of all four cell faces are summed up.

The flux balance is then set up for the basic element. Finally, the flux balances of all four basic element elements belonging to one Euler-element are summed up to yield the discrete value RHS for the right hand side of eq.(1.17).

1.5.2 Discretization of the Implicit Part

The discretization of the left hand side uses only five of the nine nodes belonging to an Euler element. For the element that centers at (i, k), these are the nodes $(i, k+1)$, $(i-1, k)$, $(i, k-1)$ and $(i+1, k)$ (fig. 1.1).

The volume integrals are evaluated by multiplying the value at the cell center by the cell volume $V_{i,k}$:

$$\int_{V_{i,k}} [\cdots]\, dV \approx [\cdots]_{i,k} V_{i,k} \tag{1.18}$$

To evaluate the surface integrals, the values at the center nodes of each cell face are taken to be valid for the whole cell face.

The first surface integral is approximated as

$$\begin{aligned} \oint_{O_{i,k}} B_n \mathbf{A} \Delta \hat{\mathbf{u}}\, dm \quad &\approx \quad (B_n \mathbf{A}\hat{\mathbf{u}})_{i,k+1}(m_{i-1,k+1} - m_{i+1,k+1}) \\ &+ \quad (B_n \mathbf{A}\hat{\mathbf{u}})_{i-1,k}(m_{i-1,k-1} - m_{i-1,k+1}) \\ &+ \quad (B_n \mathbf{A}\hat{\mathbf{u}})_{i,k-1}(m_{i+1,k-1} - m_{i-1,k-1}) \\ &+ \quad (B_n \mathbf{A}\hat{\mathbf{u}})_{i+1,k}(m_{i+1,k+1} - m_{i+1,k-1}) \end{aligned} \tag{1.19}$$

Defining

$$\begin{aligned} \Delta_i m_{i,k} &= m_{i+1,k} - m_{i-1,k} \\ \Delta_k m_{i,k} &= m_{i,k+1} - m_{i,k-1} \end{aligned} \tag{1.20}$$

we have

$$\begin{aligned} \oint_{O_{i,k}} B_n \mathbf{A} \Delta \hat{\mathbf{u}}\, dm \quad \approx \quad & -(B_n \mathbf{A}\hat{\mathbf{u}}\Delta_i m)_{i,k+1} \\ & -(B_n \mathbf{A}\hat{\mathbf{u}}\Delta_k m)_{i-1,k} \\ & +(B_n \mathbf{A}\hat{\mathbf{u}}\Delta_i m)_{i,k-1} \\ & +(B_n \mathbf{A}\hat{\mathbf{u}}\Delta_k m)_{i+1,k} \end{aligned} \tag{1.21}$$

Introducing the delta - formulation with

$$\begin{aligned} (\delta_i[\ldots])\hat{\mathbf{u}}_{i,k} &\equiv [\ldots\hat{\mathbf{u}}]_{i+1,k} - [\ldots\hat{\mathbf{u}}]_{i-1,k} \\ (\delta_k[\ldots])\hat{\mathbf{u}}_{i,k} &\equiv [\ldots\hat{\mathbf{u}}]_{i,k+1} - [\ldots\hat{\mathbf{u}}]_{i,k-1} \end{aligned} \tag{1.22}$$

this can be rewritten as

$$\oint_{O_{i,k}} B_n \mathbf{A} \Delta \hat{\mathbf{u}}\, dm \approx [-\delta_k(B_n \mathbf{A}\Delta_i m) + \delta_i(B_n \mathbf{A}\Delta_k m)]\hat{\mathbf{u}}_{i,k} \tag{1.23}$$

Defining $\Delta_i \varphi$ and $\Delta_k \varphi$ analogous to $\Delta_i m$ and $\Delta_k m$

$$\begin{aligned} \Delta_i \varphi_{i,k} &= \varphi_{i+1,k} - \varphi_{i-1,k} \\ \Delta_k \varphi_{i,k} &= \varphi_{i,k+1} - \varphi_{i,k-1} \end{aligned} \tag{1.24}$$

the second surface integral can be approximated as

$$\oint_{O_{i,k}} B_n r \mathbf{B} \Delta \hat{\mathbf{u}}\, d\varphi \approx [-\delta_k(B_n r \mathbf{B} \Delta_i \varphi) + \delta_i(B_n r \mathbf{B} \Delta_k \varphi)] \hat{\mathbf{u}}_{i,k} \tag{1.25}$$

The complete left hand side in its discrete form is now:

$$\begin{aligned} \int_{V_{i,k}} \mathbf{Q} \Delta \hat{\mathbf{u}}\, dV \quad & - \oint_{O_{i,k}} B_n \mathbf{A} \Delta \hat{\mathbf{u}}\, dm + \oint_{O_{i,k}} B_n r \mathbf{B} \Delta \hat{\mathbf{u}}\, d\varphi \approx \\ (\mathbf{Q} V \quad & + \delta_i[-B_n \mathbf{A} \Delta_k m + B_n r \mathbf{B} \Delta_k \varphi] \\ & + \delta_k[+B_n \mathbf{A} \Delta_i m - B_n r \mathbf{B} \Delta_i \varphi]) \Delta \hat{\mathbf{u}}_{i,k} \end{aligned} \tag{1.26}$$

where $\mathbf{Q}$ is defined as

$$\mathbf{Q} \equiv \mathbf{E} \cdot (\frac{1}{\Delta t} + i\omega) - \mathbf{C} \tag{1.27}$$

The discretization is clearly of central-difference type, thus it contains no dissipation. To eliminate spurious sawtooth-modes, second order artificial dissipation is added to the implicit part and second and fourth order artificial dissipation is added to the explicit part.

The equations for all interior points ($i = 2 \ldots i_{max} - 1,\ k = 2 \ldots kmax - 1$) of the computational domain are assembled in a matrix. The left hand side (the implicit part) can be approximately factored as

$$\begin{array}{llllll} (\mathbf{Q}V & + & \delta_i[-B_n \mathbf{A} \Delta_k m & +B_n r \mathbf{B} \Delta_k \varphi] & & \\ & + & \delta_k[B_n \mathbf{A} \Delta_i m & -B_n r \mathbf{B} \Delta_i \varphi]) & \Delta \hat{\mathbf{u}}_{i,k} \approx & \\ (\mathbf{E} & +(\mathbf{Q}V)^{-1} & \delta_i[-B_n \mathbf{A} \Delta_k m & +B_n r \mathbf{B} \Delta_k \varphi]) & & \\ \cdot(\mathbf{Q}V & + & \delta_k[B_n \mathbf{A} \Delta_i m & -B_n r \mathbf{B} \Delta_i \varphi]) & \cdot \Delta \hat{\mathbf{u}}_{i,k} & \\ & & \equiv Factor_1 \cdot Factor_2 \cdot \Delta \hat{\mathbf{u}}_{i,k} & & & \end{array} \tag{1.28}$$

This approach decouples the finite differences along the two types of gridlines. By introducing

$$Factor_2 \Delta \hat{\mathbf{u}}_{i,k} \equiv \Delta \hat{\mathbf{u}}'_{i,k} \tag{1.29}$$

this set of equations is solved in two sweeps by first solving

$$(\mathbf{E} + (\mathbf{Q}V)^{-1}) \delta_i[-B_n \mathbf{A} \Delta_k m + B_n r \mathbf{B} \Delta_k \varphi] \Delta \hat{\mathbf{u}}' = RHS. \tag{1.30}$$

for $\Delta \hat{\mathbf{u}}'$, and then solving eq. (1.29) for $\Delta \hat{\mathbf{u}}$.

Each of these sweeps only contains finite differences along one type of gridlines with each equation coupling the values at three gridpoints along one of the gridlines. Thus the resulting matrices are reduced to block-tridiagonal.

The discretization described above is valid for all regular Euler elements. To obtain expressions for the nodes along the periodic boundaries and the airfoil surfaces, values are defined at the centers of the basic elements along these boundaries. The flux balance for these basic elements is set up as described, the result is taken as $\Delta\hat{u}/\Delta t$ at the center of these elements (fully explicit treatment). After all the values at the centers of these elements have been calculated, the results are interpolated to give the values at the nodes along the boundaries.

The flow tangency condition is then applied to the cell face centers at the boundaries along the airfoils by setting the normal velocity to the prescribed value while leaving density, pressure and tangential velocity unaffected. Along the periodic boundaries, the phase-corrected complex amplitudes of the points on the upper and lower boundary are averaged.

Similarly, the values at the inlet and outlet far field boundaries are first calculated through the use of the basic elements along the boundaries. The nonreflecting boundary conditions are then applied to these values. Due to the finite number of gridpoints along the far field boundaries, only a finite number of circumferential wavenumbers (equal to the number of gridpoints in the circumferential direction k_{max} minus one) need to be considered.

The explicit treatment of the boundary conditions probably lowers the stability limit of the method, thus reducing the usable time step, but the resulting simplification of the code seems to justify this sacrifice.

1.6 Results

To assess the capabilties of the method and to identify potential problems and limitations, test cases of varying complexity have been analyzed.

1.6.1 Flat Plate Cascade

The first test case is a staggered flat plate cascade in subsonic flow. The stagger angle is $\gamma = 30^o$, the space /chord ratio is 1., the mach number is $M = .5$ and the reduced frequency is $k = 1$.

Figures 1.2–1.5 show the amplitude and phase of the resulting pressure difference coefficients Δc_p, for torsional (fig. 1.2,1.3) and plunging (fig. 1.4,1.5) oscillation at a phase angle of $\sigma = 90$ deg.

Together with the computed results, the results obtained by D.S. Whitehead's LINSUB-program [10] for flat plate cascades are shown (dashed lines). The results of the two methods are seen to agree almost perfectly.

1.6.2 Gostelow's Cascade

The second test case is the Gostelow-cascade [5]. Fig. 1.6 shows the computational grid and the steady flow isobars for this cascade.

The airfoils are assumed to oscillate in a torsional mode about their midchords with a phase angle of 180 deg. This case has also been analyzed by Atassi & Akai [2] by means of an integral method. Please note that the gradient terms in the boundary conditions have been neglected in this example as well as in the Atassi & Akai calculation.

In fig. 1.7 and 1.8, the solid lines show the real (fig. 1.7) and imaginary (fig. 1.8) part of the unsteady pressure preturbation for this test case as calculated by the time linearized Euler method. The dashed lines show the results of Atassi & Akai, the agreement between the two methods is seen to be quite good.

The third test case is a forced response calculation for the same cascade. Hall [6] has done calculations for this case using Giles' UNSFLO code [4]. In this example, the incoming flow is assumed to contain a harmonically varying axial velocity with a reduced frequency of $k = 1.257$. The phase angle is $\sigma = -90^o$. As the blades are not vibrating in this case, the phase angle of 90° here refers to a quarter sine wave of the inlet disturbance per cascade spacing.

Our results for this test case are shown in fig. 1.9 and 1.10 (solid lines) along with the UNSFLO results of Hall and Giles (dashed lines). The agreement is not as good as in the former test cases, but still the qualitative agreement is acceptable.

1.7 Conclusions

A method for calculating the unsteady airloads acting on airfoils in cascade due to airfoil oscillation or incoming disturbances has been presented.

The governing equations are time linearized and the perturbations are assumed to be harmonic in time. This restricts the validity of the method to small perturbations while considerably simplifying the problem and thus reducing the computer time needed for its solution.

The far field boundaries are modeled by nonreflecting boundary conditions which allow the perturbations emanating from the computational domain to pass out of the domain without causing reflections.

The first results calculated by the program show that the presented method is capable of predicting the unsteady forces acting on the airfoils of a cascade with reasonably good accuracy while being computationally far more efficient than unsteady nonlinear time stepping methods.

References

[1] Acton,E.; Cargill,A.M.: Non-Reflecting Boundary Conditions for Computations of Unsteady Flows in Turbomachines, *UAATP Proceedings of the Fourth Symposium*, pp. 211-228, Aachen, 1987.

[2] Atassi,H.;Akai,T.J.: Aerodynamic Force and Moment on Oscillating Airfoil in Cascade, *ASME-paper 78-GT-181*, 1978.

[3] Gerolymos,G.A.: Periodicity, Superposition and 3D Effects in Super-

sonic Compressor Flutter Aerodynamics, *Int. Jnl. of Turbo and Jet Engines*, No. 7, pp. 143-152, 1990

[4] Giles,M.: *UNSFLO: A Numerical Method For Unsteady Inviscid Flow in Turbomachinery*, Technical Report CFDL-TR-86-6, Massachusetts Institute of Technology, Cambridge, Massachusetts, December 1986.

[5] Gostelow,J.P.: *Cascade aerodynamics*, chapter 5, Pergamon Press, 1984.

[6] Hall,K.C.: *A Linearized Euler Analysis of Unsteady Flows in Turbomachinery*, NASA-CR.180987, Massachusetts Institute of Technology, June 1987.

[7] Happel,H.W.: Computation of Transonic 2D Cascade Flow and Comparison with Experiments, *AGARD-CP-401*, 1986.

[8] Ni,R.H.: *Non-Stationary Aerodynamics of Arbitrary Cascades in Compressible Flow*, Stevens Institute of Technology, June 1974.

[9] Nixon,D.: Computation of Unsteady Transonic Cascade Flows, *AIAA Journal*, vol. 21, no. 5, pp.790-792, 1982.

[10] Whitehead,D.S.: Classical Two-Dimensional Methods, *AGARD Manual on Aeroelasticity in Axial-Flow Turbomachines. Vol.1: Unsteady Turbomachine Aerodynamics.*, AGARD-AG-298, vol. 1, chapter 3,1987.

[11] Zirkelbach,J.: *Lösung der zeitlinearisierten Euler-Gleichungen für schwingende Schaufelgitter in Quasi-3D Strömung*, Diploma thesis, Techn. Univ. München, 1990, unpublished.

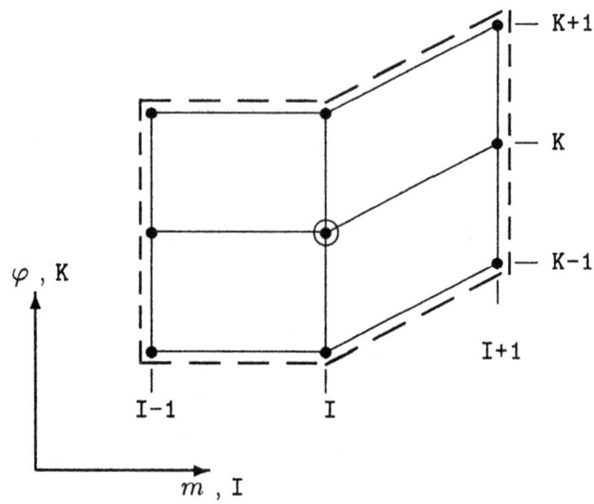

FIGURE 1.1. Euler-element (i,k) consisting of the basic elements $(i-1,k-1)$, $(i,k-1)$, $(i-1,k)$, (i,k).

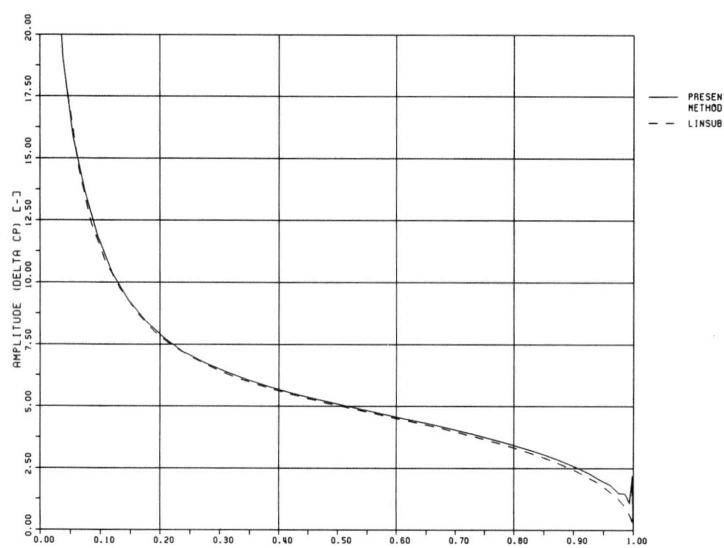

FIGURE 1.2. Flat plate cascade: Amplitude of Δc_p for torsional oscillation about midchord, reduced frequency $k = 1$, phase angle $\sigma = 90^\circ$

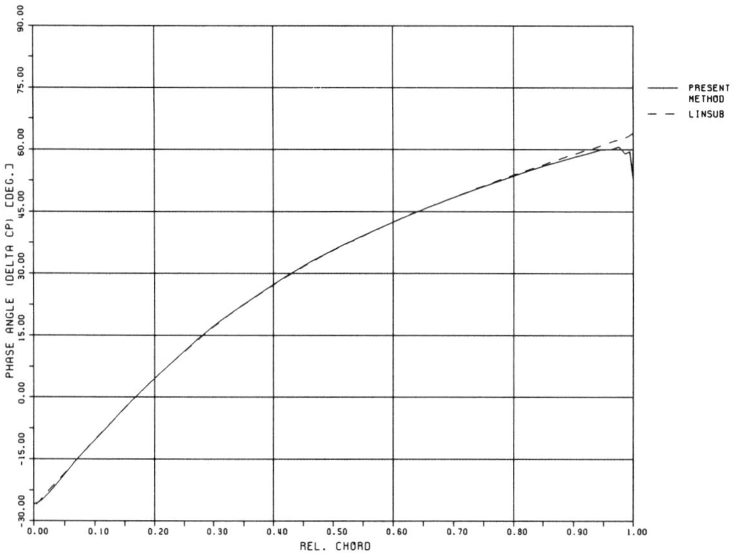

FIGURE 1.3. Flat plate cascade: Phase angle of Δc_p for torsional oscillation about midchord, reduced frequency $k = 1$, phase angle $\sigma = 90^o$

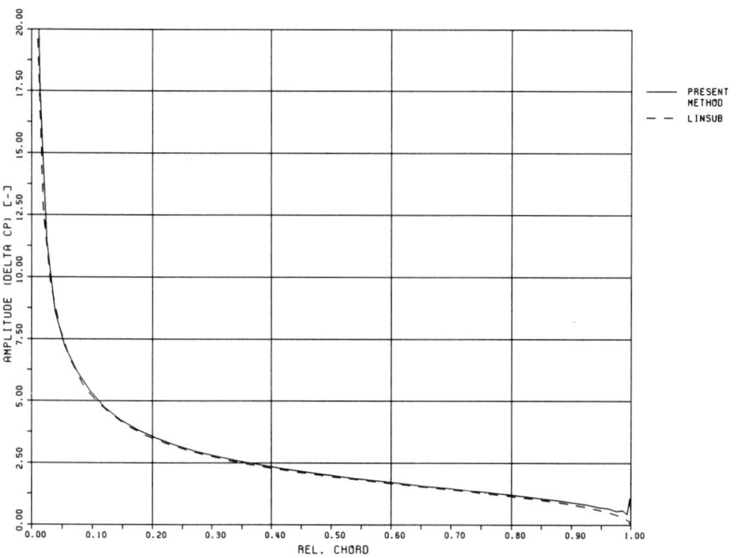

FIGURE 1.4. Flat plate cascade: Amplitude of Δc_p for plunging oscillation, reduced frequency $k = 1$, phase angle $\sigma = 90^o$

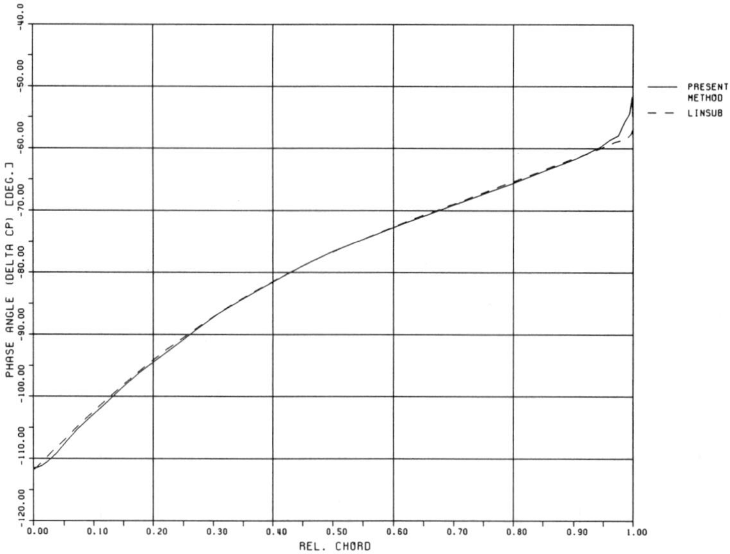

FIGURE 1.5. Flat plate cascade: Phase angle of Δc_p for plunging oscillation, reduced frequency $k = 1$, phase angle $\sigma = 90^o$

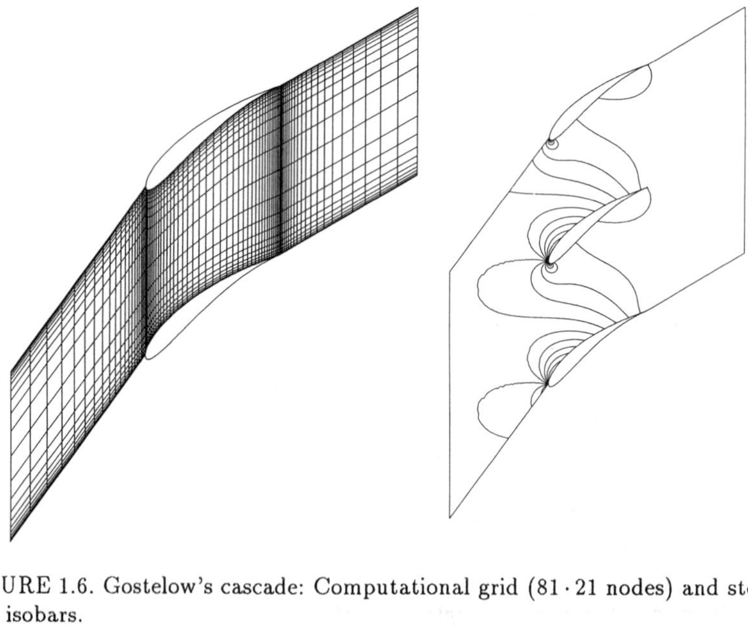

FIGURE 1.6. Gostelow's cascade: Computational grid ($81 \cdot 21$ nodes) and steady flow isobars.

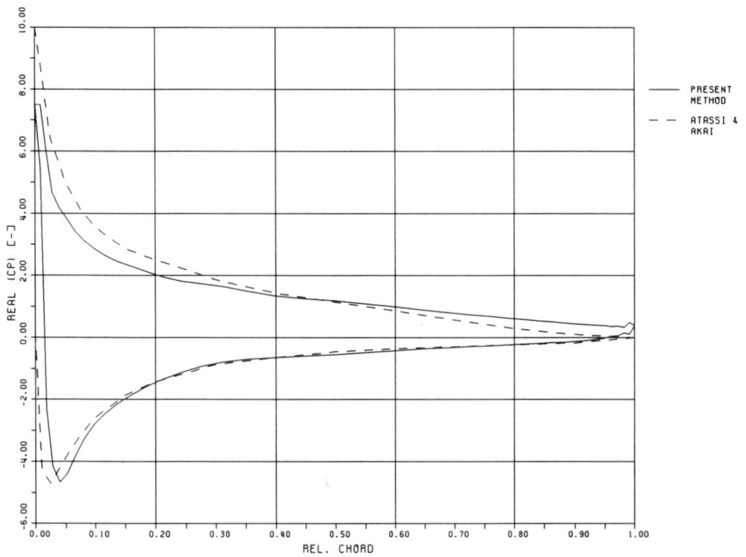

FIGURE 1.7. Gostelow's cascade: Real part of c_p for torsional oscillation about midchord, reduced frequency $k = .4$ phase angle $\sigma = 180^o$

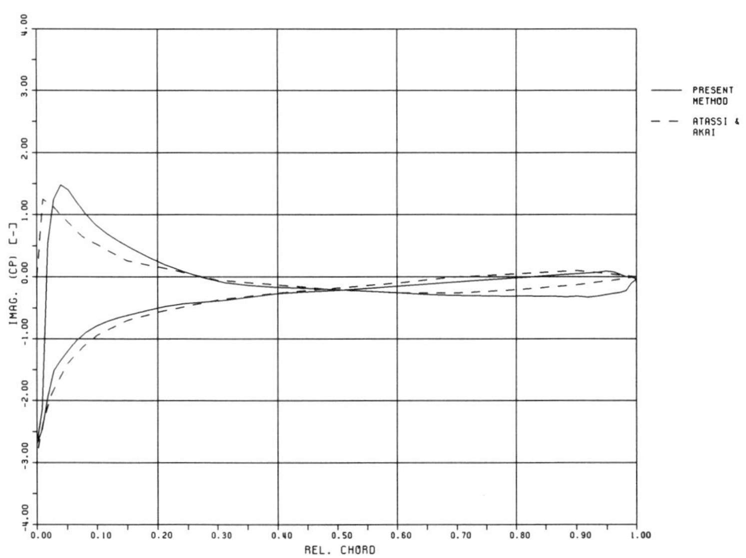

FIGURE 1.8. Gostelow's cascade: Imaginary part of c_p for torsional oscillation about midchord, reduced frequency $k = .4$ phase angle $\sigma = 180^o$

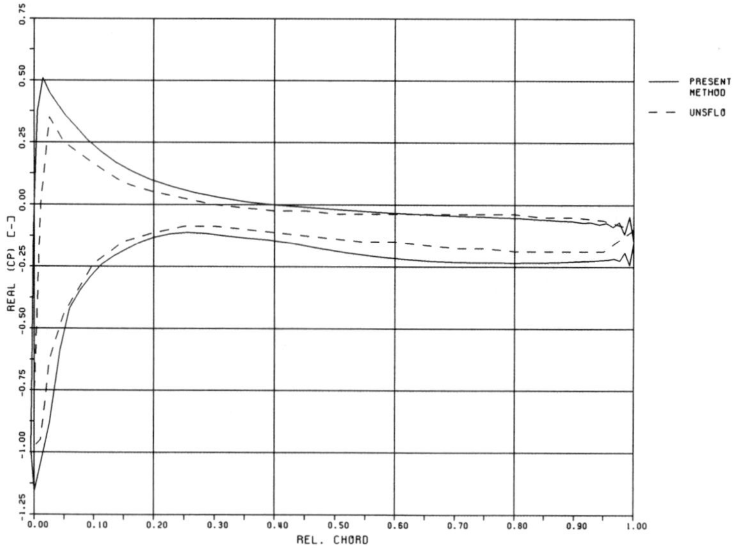

FIGURE 1.9. Gostelow's cascade: Real part of c_p for inlet disturbance, reduced frequency $k = 1.257$ phase angle $\sigma = -90^o$

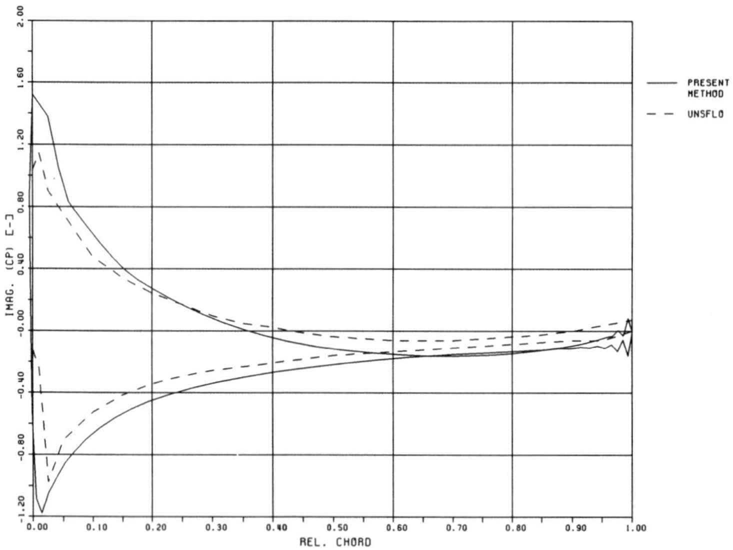

FIGURE 1.10. Gostelow's cascade: Imaginary part of c_p for inlet disturbance, reduced frequency $k = 1.257$ phase angle $\sigma = -90^o$

Unsteady Flow Field Predictions for Oscillating Cascades

Dennis L. Huff [1]

Abstract

The unsteady flow field around an oscillating cascade of flat plates with zero stagger is investigated using a time-marching Euler code. This case has an exact solution based on linear theory and serves as a model problem to study pressure wave propagation in the numerical solution. The importance of using proper unsteady boundary conditions, grid resolution and time step size is demonstrated for a moderate reduced frequency. Results show that an approximate non-reflecting boundary condition based on linear theory does a good job of minimizing reflections from the inflow and outflow boundaries and allows the placement of the boundaries to be closer than cases using reflective boundary conditions. Stretching the boundary to dampen the unsteady waves is another way to minimize reflections. Grid clustering near the plates does a better job of capturing the unsteady flow field than cases using uniform grids as long as the CFL number is less than one for a sufficient portion of the grid. Finally, a solution based on an optimization of grid, CFL number and boundary conditions shows good agreement with linear theory.

Nomenclature

c	chordlength, speed of sound or unsteady characteristic variables
ΔCP	pressure difference coefficient, $(p_- - p_+)/(\rho_1 V_1^2 \vert h_1 \vert k)$
e	energy
F,G	flux vectors

[1]NASA Lewis Research Center, MS 77-6, 21000 Brookpark Road, Cleveland, OH 44135.

h_1	amplitude of oscillation for plunging motions based on chordlength
Im{ }	imaginary part of { }
k	semi-chord reduced frequency, $\omega c/(2V_1^2)$
M	Mach number
p	static pressure
Q	dependent variable vector
Re{ }	real part of { }
t,τ	time
u,v	velocities in the x,y directions, respectively
V	total velocity
x,y	spatial coordinates
ξ, η	curvilinear coordinate directions
ρ	fluid density
σ	interblade phase angle
ω	plate oscillation frequency

Subscripts

1,2	conditions at inlet/exit
+,-	upper/lower surfaces on plate
L,R	left/right of an interface

Introduction

The accurate prediction of the unsteady flow field in turbomachinery remains a critical challenge for predicting noise and aeroelastic response. Advanced designs of ducted propellers for large commercial aircraft have relatively large shroud diameters with swept composite blades. The blades are expected to be more flexible than the blades used in current turbofans. Blade row interaction effects will be important for both aeroelasticity and aeroacoustics. Analytical and experimental predictions of the unsteady aerodynamics associated with these designs are needed to ensure safe operation and lower noise.

While three-dimensional methods are needed for a complete simulation of the unsteady flows associated with swept blades, two-dimensional analyses of linear cascades are used in practice. Aeroelastic and aeroacoustic analyses commonly model a single blade row responding to a gust or blade vibration.

There are a number of numerical and analytical methods available for addressing the two-dimensional, oscillating cascade problem. Time linearized analysis has been used the most since it does a reasonably good job near design conditions and is inexpensive to compute. Also, exact specification of the inflow and outflow boundary conditions is possible. An assumption made with all of these schemes is that the unsteady flow field is a small perturbation of the mean flow. Work in this category includes Smith (ref.1), Whitehead (ref.2), Verdon (ref.3), Hall/Crawley (ref.4) and Fang/Atassi (ref.5). Time-marching, non-linear schemes are also available that solve the full-potential (ref.6), Euler (refs.7-12), and Navier-Stokes (refs.8,11) equations. These methods are needed for flows where small-perturbation approaches are no longer valid. However, these methods require larger amounts of computer time compared to linearized methods. In addition, proper treatment of the inflow and outflow boundary conditions is difficult, but essential for accuracy.

The objective of this paper is to fundamentally investigate the unsteady flow field around an oscillating cascade using a time-marching Euler code. The code solves the non-linear Euler equations using a high resolution wave-split scheme. Only linear solutions are sought for a simple cascade of flat plates. Using an Euler code is overkill for this problem, but an exact solution is known from linear theory that can be used as a baseline for code validation. Furthermore, fundamental studies of pressure wave propagation using a time-marching CFD code are needed to better understand the limitations of numerical solutions. Choosing the proper grid resolution, time step size and boundary conditions are important for resolving the unsteady flow field. Modeling the linear flow for oscillating flat plates is a first step toward validating a numerical solution used for aeroelasticity and aeroacoustics and needs to be modeled correctly before solutions for more complex flow fields can be considered relevant. This is particularly true for moderate blade passing frequencies associated with forced response problems in aeroelasticity and blade row interaction problems in aeroacoustics.

In the present work, the numerical predictions of the unsteady aerodynamics of an unstaggered, oscillating cascade of flat plates is studied in detail. An unstaggered cascade is used so that the computational grid is perfectly orthogonal. An oscillation frequency is chosen that simulates a representative blade passing frequency in an advanced ducted propeller design. The importance of using the proper inflow and outflow boundary conditions is demonstrated. The effects of the boundary distance from the cascade, grid clustering and time step size on the solution are investigated.

Governing Equations and Numerical Algorithm

The unsteady two-dimensional Euler equations written in conservative differential form are used and are transformed from a Cartesian to a time-dependent curvilinear reference frame. This transformation process and complete details of the numerical algorithm are presented in references 13-17. A brief overview is given below.

The transformed equations can be written in vector form as

$$\frac{\partial Q}{\partial \tau} + \frac{\partial F}{\partial \xi} + \frac{\partial G}{\partial \eta} = 0 \tag{1}$$

where

$$Q = Jq \tag{2a}$$

$$F = J(\xi_t q + \xi_x f + \xi_y g) \tag{2b}$$

$$G = J(\eta_t q + \eta_x f + \eta_y g) \tag{2c}$$

and

$$q = [\rho, \rho u, \rho v, e]^T \tag{3a}$$

$$f = [\rho u, \rho u^2 + p, \rho u v, u(e + p)]^T \tag{3b}$$

$$g = [\rho v, \rho u v, \rho v^2 + p, v(e + p)]^T \tag{3c}$$

$\xi_t, \xi_x, \xi_y, \eta_t, \eta_x$ and η_y are the metrics and J is the Jacobian of transformation.

These equations are discretized and solved using a finite volume method. As discussed in reference 13, a method to allow the fluxes to be accurately represented at cell faces is based on the one-dimensional approximate Riemann solver of Roe (ref.18) at cell interfaces for each coordinate direction. This approach extracts flow field information from characteristically dictated directions and is commonly referred to as flux difference splitting (FDS) and is applicable to multidimensional space, so long as the assumption is made that all wave propagation occurs normal to a particular cell interface. To provide higher order spatial accuracy, a corrective flux is appended to the first order fluxes. In addition, in order to control dispersive errors commonly encountered with higher order schemes, so-called "limiters" are used to limit components of the interface flux resulting in total variation diminishing (TVD) schemes. All solutions presented herein were obtained using the basic algorithm developed in reference 13, which is third-order accurate spatially and second-order accurate in time.

Grid

For the simple flat plate cascades considered in this paper, the flow equations are solved on Cartesian H-grids. Uniformly spaced grids are generated algebraically. Grids with cells clustered near the plates or coarsened near the inflow/outflow boundaries were generated using a two-dimensional version of the IGB code developed by Beach (ref.19) for turbomachinery. This grid generator uses a hyperbolic tangent function to determine the spacing upstream and downstream of the cascade and ensures smooth variation in grid spacing. Once a grid is generated for a single passage, it is stacked to form a cascade with multiple passages.

All applications are for plunging motions of plates with an interblade phase angle, $\sigma = 90$ degrees. Based on the direct store method presented in reference 12, this requires four blade passages. The Euler code is blocked such that the passage-centered H-grids and the corresponding dependant variables are stored on a SSD (Solid-State Storage Device) on the CRAY computer and retrieves the information as needed. Within one block, the lower computational boundary contains the upper surface of one blade in the cascade, while the upper computational boundary contains the lower surface corresponding to the adjacent blade. The grid from each passage is deformed such that the grid follows the plunging motion near the plates and the grid near the center of the passage remains fixed. This is done using weighting functions and is described in more detail in reference 12. The code time-marches the grid and the flow solution for harmonic oscillations of the plates.

Boundary Conditions

Solid Surface and Periodic Boundary Conditions

The solid surface boundary conditions used in these studies implement zero pressure gradient conditions. Other forms of boundary conditions have been investigated by Janus (ref.20) and others at Mississippi State University and found that the zero pressure gradient boundary conditions are sufficient as long as the grid near the surface is adequately resolved. The boundaries that extend upstream and downstream of the plates are called "periodic boundaries" for plates with harmonic motion and a specified interblade phase angle. The Euler code stores phantom cells on either side of these boundaries and uses injection of the dependant variables to specify the boundary conditions for a given passage.

Steady Inflow and Outflow Boundary Conditions

The far-field, steady boundary conditions are based on characteristic variables and assume a locally one dimensional flow at the boundary and is derived in reference 20. For cascades, the direction of the flow and the characteristics at the boundaries are known. Let the subscript "a" denote approaching a boundary, "b" refers to on the boundary, and "l" means leaving the boundary. For a subsonic inflow, the characteristics approaching the boundary are set equal to the characteristics on the boundary, assuming that the metrics at the boundary are equal to the metrics at "a" and "l", and gives the following information for the flow variables at the inflow boundary:

$$p_b = 1/2\{p_a + p_l + \rho_1 c_1[\xi_x(u_a - u_l) + \xi_y(v_a - v_l)]\} \tag{4a}$$

$$\rho_b = \rho_a + \frac{p_b - p_a}{c_1^2} \tag{4b}$$

$$u_b = u_a + \xi_x \frac{p_a - p_b}{\rho_1 c_1} \tag{4c}$$

$$v_b = v_a + \xi_y \frac{p_a - p_b}{\rho_1 c_1} \tag{4d}$$

With the non-conservative flow variables known, the conservative dependant variables (Eq. 3a) can be easily computed at the boundary.

The flow variables at the outflow boundary are found in a similar manner. The characteristics on the boundary are set equal to the characteristics leaving the boundary. In this case, the incoming characteristic is the upstream running pressure wave and therefore the pressure is set equal to the pressure at "l". For internal steady flows, this is usually specified by the user. The non-conservative flow variables become:

$$p_b = p_l \tag{5a}$$

$$\rho_b = \rho_a + \frac{p_b - p_a}{c_2^2} \tag{5b}$$

$$u_b = u_a + \xi_x \frac{p_a - p_b}{\rho_2 c_2} \tag{5c}$$

$$v_b = v_a + \xi_y \frac{p_a - p_b}{\rho_2 c_2} \tag{5d}$$

These characteristic variable boundary conditions are only valid for steady flows when applied to cascades since the outflow is assumed to have uniform static pressure. There are a number of flow solvers today that mistakenly use similar boundary conditions for unsteady flows. For comparison purposes, these boundary conditions will be call "reflective" boundary conditions when they are later used for unsteady flows. A better approximation for the correct unsteady boundary conditions are examined next.

Unsteady Inflow and Outflow Boundary Conditions

For unsteady flows in turbomachinery, the exact specification of the unsteady boundary conditions for non-linear problems is not feasible. Approximate "non-reflecting" boundary conditions are developed by assuming that linear theory can be applied. A very useful publication by Giles (Ref.21) presents the formulation of boundary conditions for internal flows. Much of the original theoretical work for this topic was done by Kreiss (Ref.22). Giles has transformed the theory into a form that can be easily implemented in general Euler solvers for turbomachinery. The basic implementation of this formulation determines the steady flow using the boundary conditions derived above and then solves the time-linearized form of Euler equations at the boundary to determine the perturbation flow variables in terms of the characteristic variables. This allows time variations of static pressure at the outflow and reduces reflections from the boundaries. Due to space limitations, only the final form of the unsteady boundary conditions are presented based on reference 21.

A set of characteristic variables representing an entropy wave, a vorticity wave and upstream/downstream running pressure waves are written in terms of the perturbation variables as:

$$\begin{pmatrix} c_1 \\ c_2 \\ c_3 \\ c_4 \end{pmatrix} = \begin{pmatrix} -1 & 0 & 0 & 1 \\ 0 & 0 & 1 & 0 \\ 0 & 1 & 0 & 1 \\ 0 & -1 & 0 & 1 \end{pmatrix} \begin{pmatrix} \delta\rho \\ \delta u \\ \delta v \\ \delta p \end{pmatrix} \tag{6}$$

where $\delta\rho$, δu, δv and δp are the perturbations from the steady flow.

For a subsonic inflow, the amplitudes of the incoming unsteady characteristics (c_1, c_2, c_3) are set to zero and the outgoing characteristic (c_4) is computed using Eq. 6. For subsonic outflow, $c_4 = 0$ and the remaining characteristics are computed. Once the characteristics are known, the perturbation variables are found using an inverse transform:

$$\begin{pmatrix} \delta\rho \\ \delta u \\ \delta v \\ \delta p \end{pmatrix} = \begin{pmatrix} -1 & 0 & \frac{1}{2} & \frac{1}{2} \\ 0 & 0 & \frac{1}{2} & -\frac{1}{2} \\ 0 & 1 & 0 & 0 \\ 0 & 0 & \frac{1}{2} & \frac{1}{2} \end{pmatrix} \begin{pmatrix} c_1 \\ c_2 \\ c_3 \\ c_4 \end{pmatrix} \tag{7}$$

The primitive flow variables can be found using the following relationships:

$$\rho = \rho_0 + \delta\rho \tag{8a}$$

$$u = u_0 + \delta u \tag{8b}$$

$$v = v_0 + \delta v \tag{8c}$$

$$p = p_0 + \delta p \tag{8d}$$

where ρ_0, u_0, v_0 and p_0 are determined from the steady solution. Equation 3a is then used to determine the conservative flow variables at the boundaries.

Results and Discussion

A cascade of flat plates is used for studying the unsteady flow field due to small amplitude plunging motions. The purpose of the plunging motions is to propagate pressure waves upstream and downstream to study the accuracy of the CFD code to resolve the unsteady aerodynamics. Results are presented for one flow condition and various numerical conditions, such as inflow/outflow boundary conditions, grids and time steps (CFL numbers). While the blade vibration problem is usually associated with flutter analysis, the forcing frequencies considered will be high enough to simulate forced response problems. The small disturbances introduced by the plate motion allows for comparisons with small-perturbation theory.

Model Problem

The flat plate cascade geometry used in the present study is shown in Figure 1a. The stagger angle is zero and the gap-to-chord ratio is one. Uniform flow with $M_1 = 0.50$ is used for the steady flow. Small plunging oscillations with $h_1 = 0.001$ and $\sigma = 90$ degrees are used for all of the cases presented. A semi-chord reduced frequency (k) of four was chosen to simulate a realistic rotor/stator frequency expected in advanced ducted propeller designs. According to linear theory, this flow condition propagates single mode pressure waves both upstream and downstream in a direction not aligned with the Cartesian grids. This defines a model problem with a known solution that can be used to study wave dynamics using CFD.

The procedure used in the time-marching Euler code is to first determine a steady solution with no blade motion. For flat plates with zero incidence, this reduces to uniform flow and can be specified analytically. The unsteady solutions are started with each plate moving according to the specified interblade phase angle. Unless specified otherwise, the maximum CFL number used in each of the following solutions is unity. The solution is ran until the unsteady flow field reaches periodicity. For the present study, the unsteady pressures on the plate are Fourier Transformed into their real and imaginary parts after each cycle of oscillation. The solution is stopped after the coefficients from the first harmonic do not vary significantly. This usually takes anywhere from 3 to 20 cycles of plate oscillation. The solutions presented below are done on the CRAY-YMP computer and take anywhere from 20 minutes to 5 hours of CPU time, depending on the grid and time step.

Visualization of the Effects of Boundary Conditions

A set of runs are done to visually demonstrate pressure wave propagation and the importance of using correct inflow and outflow unsteady boundary conditions. Consider a uniformly spaced grid (dx=dy=.05 chordlengths) with the inflow and outflow boundaries located six chordlengths away from the leading and trailing edges, respectively. The grid resolution is too coarse for accurate computations of surface pressures, but should be sufficient for visual demonstration of pressure wave propagation to the far-field. Figure 1 (b-f) show snapshots of pressure contours just after the plates are impulsively moved from uniform flow. The amount of time between snapshots is approximately constant and is arbitrarily denoted ΔT, which gives five instantaneous plots of pressure over enough time for the downstream running wave to reach the boundary. The left block uses the "reflective" boundary conditions and the right block uses the "non-reflective" boundary conditions. Since the "reflective" boundary conditions enforce constant static pressure at the exit boundary, the downstream running wave is reflected (Figure 1 e and f). As the solution continues, the downstream domain is contaminated with an upstream-running wave that is non-physical for the desired solution. With proper grid resolution, this wave enters the cascade and affects the unsteady pressures on the plates.

Figure 2 shows the unsteady pressure distributions on the plate after the solution has ran twenty plate oscillation cycles. There are four solutions shown; two of which are solutions using the uniform grid with the "reflective" and "non-reflective" boundary conditions shown in Figure 1. The remaining two solutions are using the "reflective" boundary conditions on grids that coarsen as they extend to the inlet and exit boundaries. Coarsening the grid introduces numerical dissipation into the solution that dampens the waves as they propagate. (This has been commonly used for isolated airfoil analysis). The grid has been coarsened such that the resolution in the passage is the same (dx=dy=.05 chordlengths), but the aspect ratio of the cells gradually change to 3:1 (dx=.15,dy=.05) at the inflow and outflow boundaries in one case and 10:1 (dx=.50,dy=.05) in the other case. There is a significant difference between the "reflective" case and the "non-reflective" case using a uniform grid, as would be expected based on the contour plots in Figure 1. Also, the solutions using the grid coarsening technique and "reflective" boundary conditions tend toward the "non-reflective" boundary conditions case using a uniform grid. It is expected that grid coarsening will only work if the grid spacing variation is smooth, as abrupt changes in the grid may cause internal reflections.

Effect of Boundary Distance From the Cascade

An investigation of the effect of inflow and outflow boundary distance from the cascade is done using both the "reflective" and "non-reflective" boundary conditions. The upstream and downstream boundaries are placed 0.25, 0.5, 1.0, 1.5, and 2.0 chordlengths away from the leading and trailing edges of the plates, respectively. The grid density is refined to dx=dy=.02 chordlengths. Each case was run through 20 cycles of oscillation to give the unsteady pressure predictions shown in Figures 3 and 4. The results shown in Figure 3 use the "reflective" boundary conditions and never reach a solution that is independent of the boundary distance. In contrast, the cases using the "non-reflective" boundary conditions (Figure 4) are nearly identical when the boundaries are 1.5 and 2.0 chordlengths away from the cascade. This demonstrates the advantage of using "non-reflective" boundary conditions for reducing the computational domain.

Results from linear theory (ref.1) are also plotted in Figures 3 and 4. The results using the "non-reflective" boundary conditions that are independent of the boundary distance are in general, but not exact agreement with the theory (short dashed lines in Figure 4). Since the theory is exact for flat plates, the differences between the solutions are due to numerical error. It will be demonstrated later that the discrepancies are due to grid resolution. For now, the study on the effect of boundary distance using different inflow and outflow boundary conditions should still be valid.

Effect of Boundary Conditions With Grid Coarsening

The results shown in Figure 2 have already demonstrated the effects of grid coarsening using the "reflective" boundary conditions on a coarse grid. Based on these results, damping the pressure waves by coarsening the grid should render a solution insensitive to either the "reflective" or "non-reflective" boundary conditions. To verify this, another set of solutions are considered using a finer grid near the plates that smoothly coarsens to the inflow and outflow boundaries. The same boundary distances of six chordlengths upstream and downstream are used. The grid density is $dx=dy=.02$ chordlengths in the passage and coarsens to $dx=.50$ and $dy=.02$ chordlengths at the inflow and outflow boundaries. The unsteady pressure distributions are plotted in Figure 5 for solutions using both the "reflective" and "non-reflective" boundary conditions, along with the exact solution from linear theory. The solutions are identical with each other and differ from the theory by the same amount found in Figure 4 using the "non-reflective" boundary conditions. Therefore, a solution has been found that is not being influenced by reflections from the inflow and outflow boundaries.

Effect of Time Step (CFL Number)

All of the previous solutions were ran with a maximum CFL number of unity. For cases with varying grid spacing, this means that the smallest cell does not exceed a CFL number of one and the larger cells have a CFL number less than one. The grids with uniform spacing have a CFL number of one for every cell. Specifying a maximum CFL number of one preserves, at least formally, the second-order temporal accuracy of the numerical solution. However, many unsteady solutions using a time-marching code run higher CFL numbers to reduce the run times. When this is done, an assumption is made that the higher CFL numbers occur only at the smaller cells and do not significantly effect the global solution. The following solutions show how bad this assumption can be if care is not taken to ensure that the CFL number is small enough to resolve the unsteady flow field.

A study of the effects of CFL number is done on two types of grids, including a uniformly spaced grid and a grid with points clustered near the plates. The "non-reflecting" boundary conditions are used in all of the remaining results. The uniformly spaced grid (dx=dy=.02 chordlengths) is identical to the one used in the study of boundary distance effects with the boundary located 0.5 chordlengths away from the cascade (Figures 3 and 4). Unsteady pressure predictions using three different values of CFL number are shown in Figure 6. The results using a CFL number of one are the same as those shown in Figure 4. The good agreement with linear theory is fortuitous since the predictions with a boundary distance of 0.5 chordlengths was found to be insufficient. Increasing the CFL number to values of two and three introduces temporal error, as expected. Using a CFL number of three completely destroys the character of the unsteady wave forms. For this case, two solutions have been plotted after twenty and forty cycles of plate oscillation to verify that the solution has reached periodicity.

A similar study is shown in Figure 7 for a grid that is clustered in the y-direction. The spacing near the plates is dx=.05, dy=.02 chordlengths and smoothly varies to a spacing of dx=dy=.05 chordlengths at the center of the passage. Hence, the maximum CFL number occurs near the plates. The boundary distance is 0.5 chordlengths upstream and downstream of the cascade. From the predictions shown in Figure 7, clustering the grid near the surface has helped preserve the character of the pressure waves, but still quickly loses the accuracy for maximum CFL numbers of two and three. It should be noted that for a maximum CFL number of three, the minimum CFL number is 0.4 on this grid. This means that there is a portion of the grid that should convect disturbances correctly near the center of the passage. However, in the region of the grid near the plates, which is where the unsteady disturbances are being generated, the correct magnitude and phase are not being computed. This is why the nature of the disturbance is captured with a clustered grid, but the accuracy is poor for higher CFL numbers. As demonstrated in the next section, it is possible for highly clustered grids to do a reasonable job of predicting the unsteady flow field providing that an adequate portion of the grid has a CFL number less than unity.

An Optimal Solution

Finally, using the studies presented above, an optimal solution is presented in Figure 8. A highly clustered grid near the leading and trailing edges (dx=.003, dy=.001 chordlengths) is used. The 121 x 41 grid (x by y) is stretched six chordlengths upstream and downstream of the cascade using the grid coarsening technique. The maximum CFL number is ten, which occurs near the leading and trailing edges of the plates. The CFL number quickly diminishes to less than one away from the plates, leaving a large percentage of the grid with an adequate time step to resolve the unsteady flow. Comparisons of the unsteady pressure distributions with linear theory in Figure 8 shows good agreement for most of the chordlength.

Conclusions

A fundamental investigation of the unsteady flow field around an oscillating cascade of flat plates with zero stagger is done using a time-marching Euler code. The study demonstrates the importance of using proper unsteady boundary conditions, grid resolution and time step size to accurately model the unsteady flow field. Imposing constant static pressure across the outflow causes reflections from the downstream boundary and is shown by plotting pressure contours at various times. These reflections contaminate the unsteady pressure distributions on the plates. A study on the effect of boundary distance shows that the unsteady plate pressures never reach a solution that is independent of boundary distance. Using approximate non-reflecting boundary conditions based on linear theory improves the solutions and allows the boundaries to be placed closer to the cascade (approximately 1.5 to 2 chordlengths away from the cascade). Coarsening the grid by stretching the boundaries away from the cascade introduces numerical damping to the solutions and is also an effective way to obtain correct unsteady surface pressures on the plates, although it is only valid for unsteady flows due to plate motion. These solutions are insensitive to "reflecting" or "non-reflecting" boundary conditions. Several solutions are presented that vary the time step size on both a clustered grid and a uniformly spaced grid. The results show that the unsteady flow field becomes significantly inaccurate as the CFL number exceeds one on a uniformly spaced grid. The clustered grid results do a better job of capturing the flow, but are still inaccurate as fewer grid cells have a CFL number less than one. Good agreement with linear theory is found using "non-reflective" boundary conditions on a highly clustered grid that stretches six chordlengths upstream and downstream from the cascade. For this case the maximum CFL number is ten near the leading and trailing edges and quickly diminishes to less than one away from the plates.

Acknowledgments

The author would like to thank Dr. Timothy Swafford, Dr. David Whitfield and Dr. Mark Janus, of Mississippi State University, for the helpful discussions concerning the flow algorithm. Also, Dr. John Adamczyk, of NASA Lewis Research Center, for his encouragement during the course of this investigation.

References

1. Smith, S.N., "Discrete Frequency Sound Generation in Axial Flow Turbomachines," ARC-R/M-3709, 1971.
2. Whitehead, D.S., "Classical Two-Dimensional Methods," *AGARD Manual on Aeroelasticity in Axial-Flow Turbomachines Volume 1 Unsteady Turbomachinery Aerodynamics, AGARDograph No. 298*, M.F. Platzer and F.O. Carta, editors, London, 1987, pp. (3-1) - (3-30).
3. Verdon, J.M. and Casper, J.R., "A Linearized Unsteady Aerodynamic Analysis for Transonic Cascades," *Journal of Fluid Mechanics*, Vol. 149, December 1984, pp. 403-429.
4. Hall, K.C. and Crawley, E.F., "Calculation of Unsteady Flows in Turbomachinery Using the Linearized Euler Equations," Fourth Symposium on Unsteady Aerodynamics and Aeroelasticity of Turbomachines and Propellers, Aachen, West Germany, September 1987.
5. Fang, J. and Atassi, H.M., "Compressible Flows with Vortical Disturbances Around a Cascade of Loaded Airfoils," Sixth Symposium on Unsteady Aerodynamics and Aeroelasticity of Turbomachines and Propellers, University of Notre Dame, U.S.A., September 1991.
6. Kao, Y.F., "A Two-Dimensional Unsteady Analysis for Transonic and Supersonic Cascade Flows," Ph.D. Thesis, Purdue University, May 1989.
7. Fransson, T.H. and Pandolfi, M., "Numerical Investigation of Unsteady Subsonic Compressible Flows Through an Oscillating Cascade," ASME Paper 86-GT-304, June 1986.
8. Huff, D.L., "Numerical Analysis of Flow Through Oscillating Cascade Sections," AIAA Paper 89-0437, Jan. 1989.
9. He, L., "An Euler Solution for Unsteady Flows Around Oscillating Blades," ASME Paper 89-GT-279, June 1989.
10. Gerolymos, G.A., Blin, E., Quiniou, H., "Comparison of Inviscid Computations with Theory and Experiment in Vibrating Transonic Compressor Cascades," ASME Paper 90-GT-373, June 1990.
11. Giles, M. and Haimes, R., "Validation of a Numerical Method for Unsteady Flow Calculations" ASME Paper 91-GT-271, June 1991.
12. Huff, D.L., Swafford, T.W. And Reddy, T.S.R., "Euler Flow Predictions for an Oscillating Cascade Using a High Resolution Wave-Split Scheme," ASME Paper 91-GT-198, June 1991.
13. Whitfield, D.L., Janus, J.M., and Simpson, L.B., "Implicit Finite Volume High Resolution Wave-Split Scheme for Solving the Unsteady Three-Dimensional Euler and Navier-Stokes Equations on Stationary or Dynamic Grids," Mississippi State Engineering and Industrial Research Station Report No. MSSU-EIRS-ASE-88-2, February, 1988.

14. Prewitt, Nathan C., "Two-Dimensional Euler Code for the Prediction of Pressure Distributions About an Airfoil," MS Thesis, Mississippi State University, Mississippi State, MS, December 1988.
15. Janus, J.M., "Advanced 3-D CFD Algorithm for Turbomachinery," PhD Dissertation, Mississippi State University, Mississippi State, MS, May 1989.
16. Arabshahi, A., "A Dynamic Multiblock Approach to Solving the Unsteady Euler Equations About Complex Configurations," PhD Dissertation, Mississippi State University, Mississippi State, MS, May 1989.
17. Whitfield, D.L., Janus, J.M., and Arabshahi, A., "Unsteady Euler Solutions on Dynamic Blocked Grids for Complex Configurations," AGARD Conference Proceedings, No.464, 1990.
18. Roe, P.L., "Approximate Rieman Solvers, Parameter Vectors, and Difference Schemes," *Journal of Computational Physics*, Vol. 43, pp. 357-372, 1981.
19. Beach, T.A., "An Interactive Grid Generation Procedure for Axial and Radial Flow Turbomachinery," AIAA Paper 90-0344, 1990.
20. Janus, J.M., "The Development of A Three-Dimensional Split Flux Vector Solver with Dynamic Grid Applications," MS Thesis, Mississippi State University, Mississippi State, MS, August 1984.
21. Giles, M., "Non-Reflecting Boundary Conditions for Euler Equation Calculations," AIAA Paper 89-1942-CP, 1989.
22. Kreiss, H.O., "Initial Boundary Value Problems for Hyperbolic Systems," *Communications on Pure and Applied Mathematics*, 23:277-298, 1970.

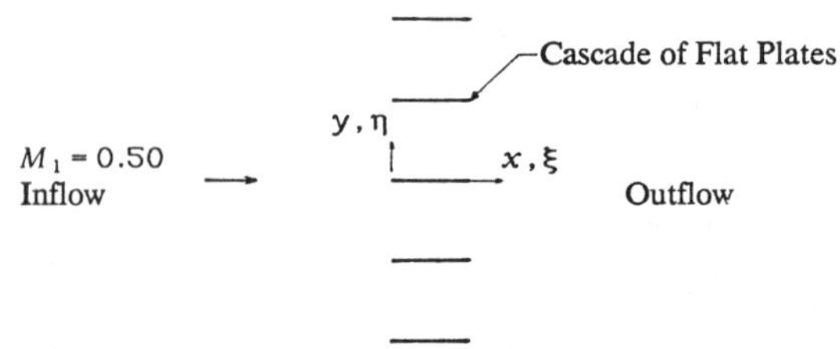

a) Cascade geometry for model problem

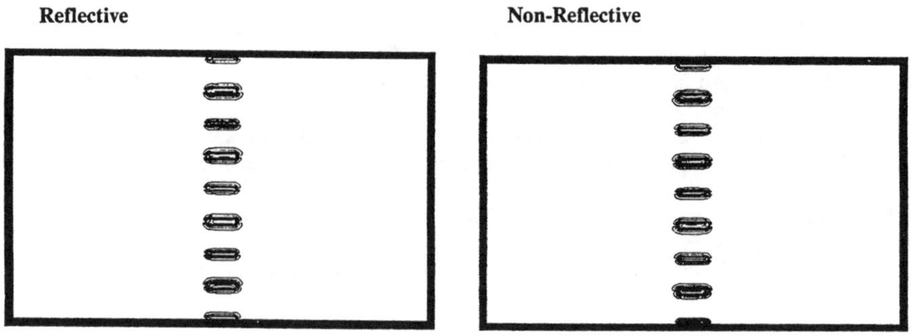

b) $t > 0$

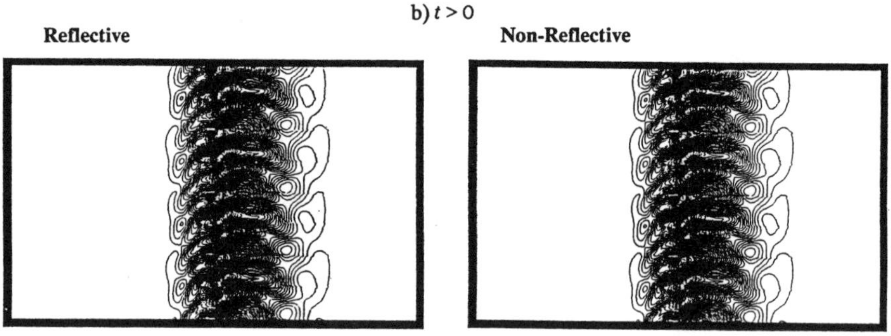

c) $t = \Delta T$

Figure 1. Cascade geometry and instantaneous static pressures for model problem.

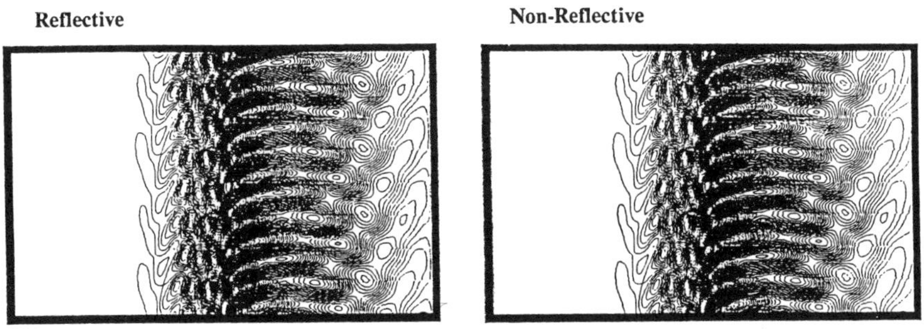

d) $t = 2\Delta T$

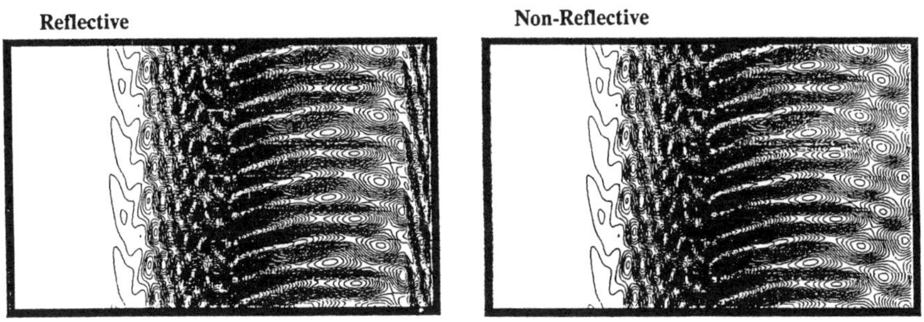

e) $t = 3\Delta T$

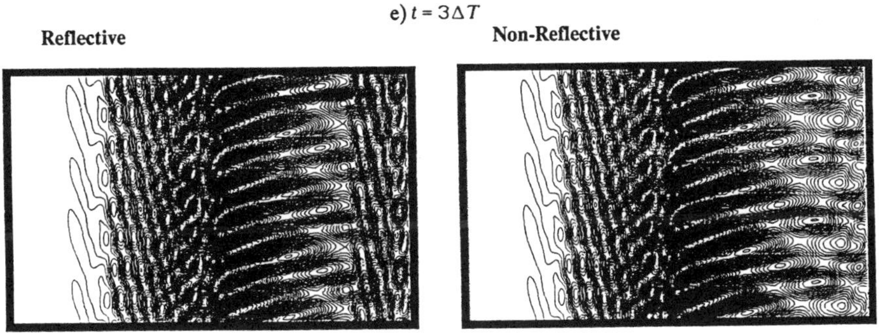

f) $t = 4\Delta T$

Figure 1. (cont.) Cascade geometry and instantaneous static pressures for model problem.

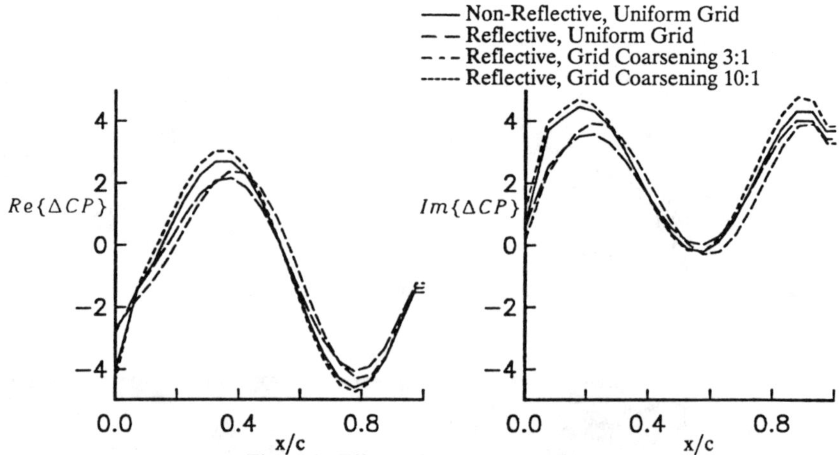

Figure 2. Effects of boundary conditions and grid on unsteady pressure distributions.

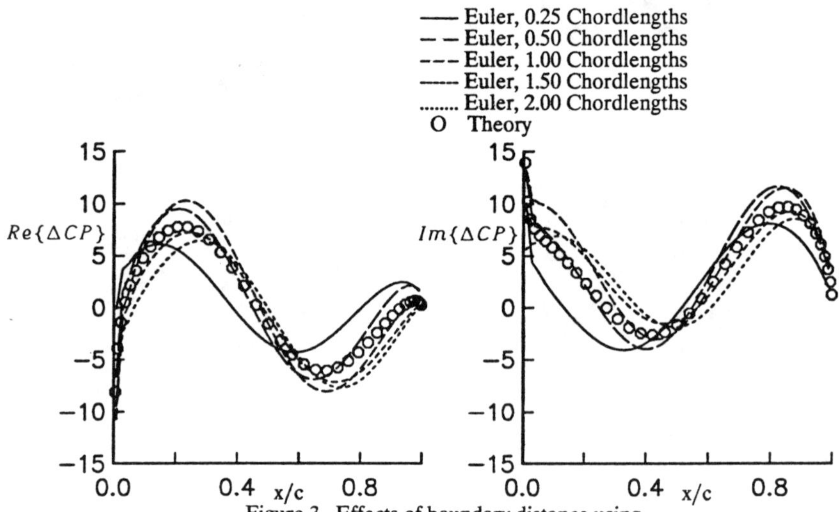

Figure 3. Effects of boundary distance using reflective boundary conditions.

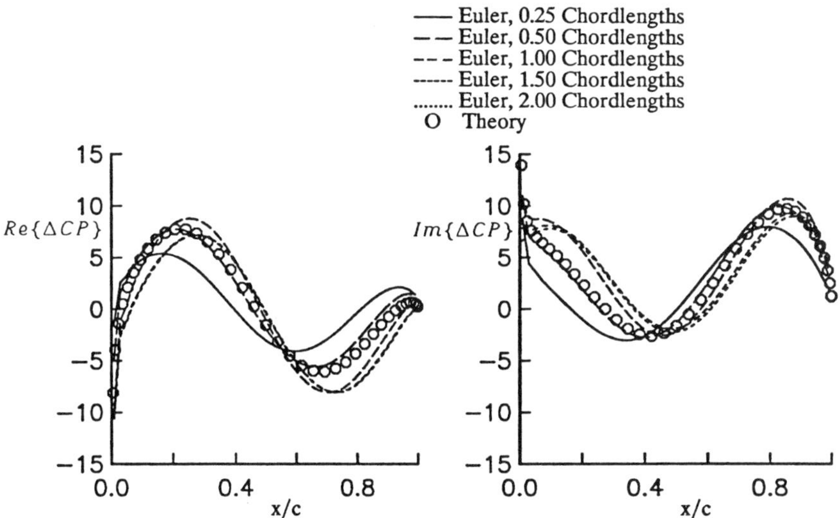

Figure 4. Effects of boundary distance using non-reflective boundary conditions.

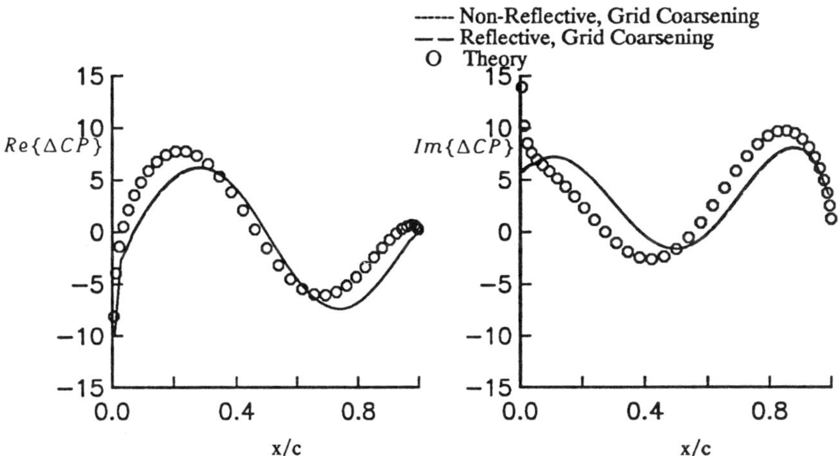

Figure 5. Effects of boundary conditions on the unsteady pressure distributions using grid coarsening.

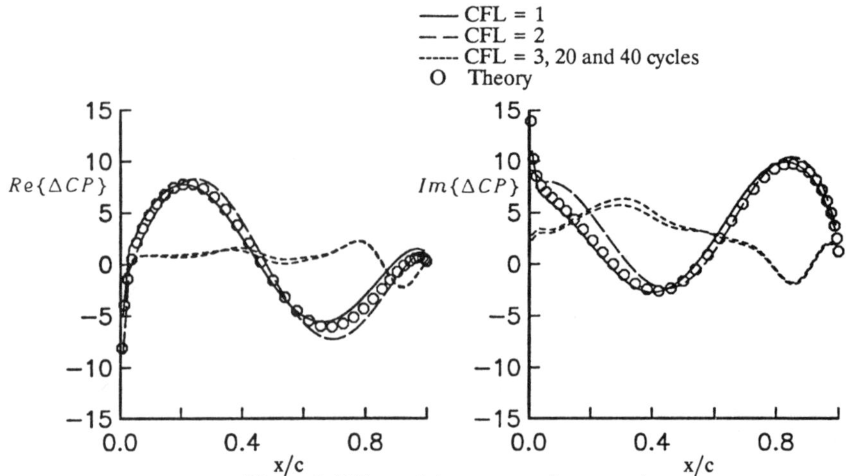

Figure 6. Effects of time step on the unsteady pressure distributions using a uniform grid.

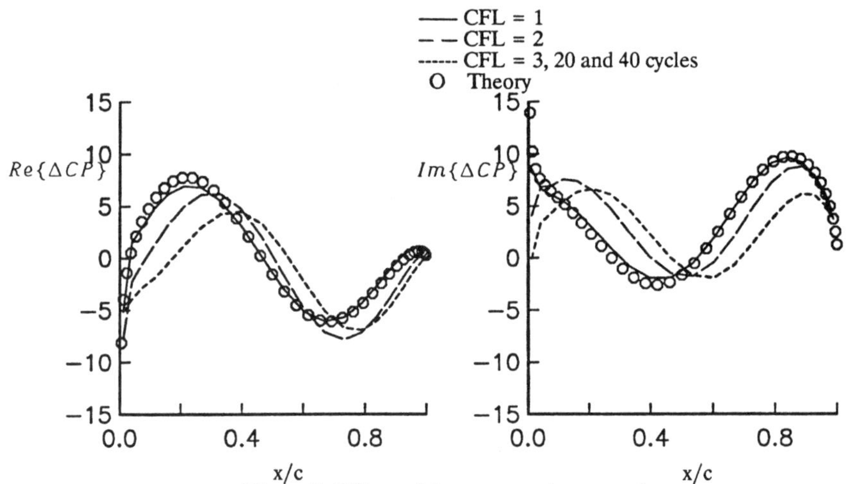

Figure 7. Effects of time step on the unsteady pressure distributions using a clustered grid.

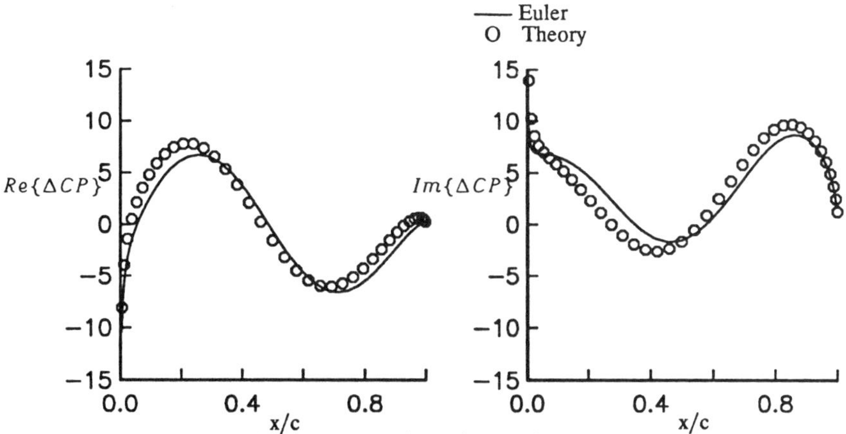

Figure 8. Comparison of unsteady pressure distributions with linear theory for an optimal Euler solution.

Compressible Flows with Vortical Disturbances Around a Cascade of Loaded Airfoils

Jisheng Fang[1]
Hafiz M. Atassi[2]

ABSTRACT A highly efficient numerical method is developed for three-dimensional, periodic, vortical flows around a cascade of loaded airfoils. The method uses the approximation of the rapid distortion theory and thus fully accounts for the effects of distortion of the vortical disturbances as they propagate through the spatially varying mean flow. The numerical scheme is based on the splitting of the unsteady velocity into a vortical part which is a known function of the upstream flow conditions and the Lagrangian coordinates of the mean flow, and a potential part satisfying a nonconstant-coefficient, inhomogeneous, convective wave equation. A new downstream out-flow boundary condition is derived to complete the unsteady boundary value problem. By writing the downstream condition in terms of the unsteady pressure, the difficulties associated with the singular behavior of the unsteady vortical velocity are avoided. This condition is relatively simple and suitable for numerical computations. By using a body-fitted coordinate system, the unsteady potential is computed in the frequency domain. Results are presented to demonstrate the effects of mean blade loading on the aerodynamic response of a cascade to vortical gust excitations.

1.1 Introduction

For most turbomachines and propellers, flow unsteadiness is the result of upstream nonuniformities which are caused by a variety of sources, such

[1]Research Associate, Aerospace and Mechanical Engineering Department, University of Notre Dame, Notre Dame, IN 46556

[2]Professor, Aerospace and Mechanical Engineering Department, University of Notre Dame, Notre Dame, IN 46556

as atmospheric turbulence, inlet distortion, installation effects, blade tip vortex, rotor-stator wake interaction etc. In the blade frame of reference, these nonuniformities appear as convected vortical waves, known as *gusts* [Ata86]. Moreover, they are usually periodic and dominated by the blade passing frequency and its harmonics.

Such gusts can cause a number of undesirable effects. For example, they induce unsteady forces on the blade surface which may cause forced vibrations and radiate noise. When the frequency of the aerodynamic excitation is close to the machine natural frequency, the forced vibration could be destructive. Since it is not possible to eliminate all kinds of nonuniformities, it is important to determine those irregular patterns yielding large amplitude of induced forces and eliminate them.

The earliest attempts to analyze unsteady flows around cascades go back to the 1950's when turbojet engines began equipping high speed aircrafts. The early work dealt with inviscid flows around two-dimensional cascades of oscillating flat plates in uniform upstream flows of zero mean incidence. Assuming small disturbances, the flow is linearized about the uniform mean flow. The problem is then formulated in terms of an integral equation which is solved using a collocation technique. From 1960's to 1980's, highly efficient semi-analytic solutions were developed for subsonic [Whi60, Whi70, Ven80] and supersonic [Kur74, Ver77, NW78] cascade flows. For a full review of these early works we refer the readers to Whitehead's review paper [Whi87].

As the technology of turbomachine blading evolved toward higher blade loading, the classical linearization about a uniform mean flow becomes inadequate. The coupling between the mean and unsteady flows becomes important and must be account for to obtain accurate predictions of the unsteady aerodynamics of the blades. Models accounting for the effects of mean flow incidence and blade and cascade geometry were then developed for a cascades of oscillating blades by Atassi and Akai [AA80] for low Mach number flows, Verdon and Caspar [VC80], and Whitehead and Grant [WG80] for subsonic flows, and Verdon and Caspar [VC84], and Whitehead and Newton [WN85] for transonic flows. The unsteady disturbances were regarded as small-amplitude harmonic fluctuations about a nonuniform mean flow. The mean flow was determined from a solution of the potential flow equations, and the unsteady flow was governed by a set of linear equations with non-constant coefficients that depended upon the underlying mean flow. These equations were homogeneous and the unsteady flow was potential. A review of these models was given by Verdon [Ver87].

For a cascade in a vortical upstream flow (the gust problem), the unsteady flow is not potential. However, for a uniform mean flow, a splitting theorem can be derived [Gol76, Ata86] and the vortical velocity can be identified with the upstream vortical disturbances. Thus, the problem can be reduced to that of finding the potential part of the unsteady velocity. In that regard, the problem becomes similar to that of a cascade of oscillating blades except for the form of the blade surface boundary condition.

For loaded cascades, because of the nonlinear distortion of the unsteady flow by the mean flow, it is not possible to entirely split the unsteady velocity field into vortical and potential parts. Goldstein and Atassi [GA76] were the first to recognize the importance of the distortion of the unsteady field by the mean flow. They introduced the concept of "rapid distortion" [Pra33, BP54] for high-speed large-structure disturbances in unsteady aerodynamics. This approach accounts for the distortion of the incoming large eddies by the mean flow and neglects the inter-eddy interaction. This leads to a linearization of the unsteady governing equations about a nonuniform mean flow. Using this approach, Goldstein and Atassi [GA76] and Atassi [Ata84] studied the case of single lifting airfoils in a periodic gust in incompressible flows and derived analytical expressions for the lift function to account for the effects of thickness and camber as well as mean flow incidence. Their results show a strong influence of the mean airfoil loading on the unsteady pressure, lift and moment coefficients.

This approach was later generalized by Goldstein [Gol78] for unsteady three-dimensional disturbances with potential mean flows. Goldstein split the unsteady velocity field into vortical, entropic and potential parts. The expressions of the vortical and entropic parts are known functions in terms of the upstream gust conditions and the mean flow Lagrangian coordinates. The potential part is governed by a non-constant coefficient, inhomogeneous, convective wave equation.

For real flows around bodies, there is always a stagnation point near the blade leading edge. At this point, the Lagrangian coordinates are infinite. As a result, Goldstein's vortical velocity becomes singular on the blade surface and its wake. To avoid this difficulty, Atassi and Grzedzinski [AG89] proposed a modification of Goldstein's decomposition which eliminates that singular behavior and results in a regular boundary value problem for the unsteady potential. Moreover, Atassi and Grzedzinski's splitting yields a vortical velocity whose normal and tangential components vanish on the blade surface and its wake. This feature makes their splitting particularly suitable for numerical computations.

Atassi and Grzedzinski's splitting was recently used by Scott [Sco90], and Scott and Atassi [SA90] for unsteady vortical flows around a single airfoil, Hall and Verdon [HV91] and Fang [Fan91] around a cascade of airfoils.

In the present paper, a numerical method based on Atassi and Grzedzinski's formulation is presented for subsonic flows around a cascade of loaded airfoils subject to a three-dimensional gust.

Because the computational domain is finite, the boundary-value problem for the cascade in a gust should be complemented with suitable out-flow conditions upstream and downstream. These conditions must allow outgoing acoustic waves to propagate out of the computational domain. Verdon [Ver89] derived an anlytic far-field solution which was used by Hall and Verdon [HV91]. In the present work, we expand the unsteady pressure in plane waves near the out-flow boundaries. This leads to a relatively simple and numerically suitable out-flow boundary condition which can be coupled with the computational solutions.

To examine the effect of mean blade loading on the unsteady response, two cascade configurations were studied at different Mach numbers and frequencies.

1.2 Mathematical Formulation

1.2.1 Linearized Euler Equations

We consider an unsteady, inviscid, compressible, subsonic flow past a stationary linear cascade of airfoils with velocity and entropy disturbances imposed on a uniform upstream velocity $U_\infty \vec{i}_1$ and entropy s_∞ (See Figure 1.1). All lengthes is normalized by half a chord $\frac{c}{2}$, velocities by upstream mean flow speed U_∞, and time by $\frac{c/2}{U_\infty}$. The governing equations are the continuity, momentum and entropy equations as follows:

$$\frac{D\rho}{Dt} + \rho \vec{\nabla} \cdot \vec{U} = 0 \quad (1.1)$$

$$\rho \frac{D\vec{U}}{Dt} = -\vec{\nabla} p \quad (1.2)$$

$$\frac{Ds}{Dt} = 0 \quad (1.3)$$

where

$$\frac{D}{Dt} = \frac{\partial}{\partial t} + \vec{U} \cdot \vec{\nabla}$$

t denotes time, and ρ, $\vec{U}$, p and s denote the fluid density, velocity, pressure and entropy respectively. We assume that the unsteady flow is linearized about a mean potential flow, so that

$$\vec{U}(\vec{x},t) = \vec{U}_0(\vec{x}) + \vec{u}(\vec{x},t) \quad (1.4)$$

$$p(\vec{x},t) = p_0(\vec{x}) + p'(\vec{x},t) \quad (1.5)$$

$$\rho(\vec{x},t) = \rho_0(\vec{x}) + \rho'(\vec{x},t) \quad (1.6)$$

$$s(\vec{x},t) = s_0(\vec{x}) + s'(\vec{x},t) \quad (1.7)$$

where the subscript 0 denotes the steady mean flow quantities, $\vec{u}$, p', ρ' and s' are the small unsteady perturbations of velocity, pressure, density and entropy, respectively. The coordinate system is shown in Figure 1.2.

Substituting (1.4) to (1.7) into equations (1.1) to (1.3) yields the zero order and the first order equations of conservation of continuity, momentum, and energy. We assume that the solutions to the zero order equations, which are $\vec{U}_0$, p_0, ρ_0 and s_0 in equations (1.4) to (1.7), are known and obtained by a steady flow solver.

Neglecting the higher order terms, we then obtain the first order equations for $\vec{u}$, p', ρ' and s':

$$\frac{D_0\rho'}{Dt} + \rho'\vec{\nabla}\cdot\vec{U}_0 + \vec{\nabla}\cdot(\rho_0\vec{u}) = 0 \tag{1.8}$$

$$\rho_0(\frac{D_0\vec{u}}{Dt} + \vec{u}\cdot\vec{\nabla}\vec{U}_0) + \rho'\vec{U}_0\cdot\vec{\nabla}\vec{U}_0 = -\vec{\nabla}p' \tag{1.9}$$

$$\frac{D_0 s'}{Dt} = 0 \tag{1.10}$$

where

$$\frac{D_0}{Dt} = \frac{\partial}{\partial t} + \vec{U}_0\cdot\vec{\nabla}$$

is the linearized material derivative associated with the mean flow.

Equations (1.8) to (1.10) form a set of partial differential equations for the unknown unsteady quantities $\vec{u}$, p', ρ' and s', which are functions of the spatial coordinates $\vec{x}$ as well as time t. In the remainder of this paper we assume a potential mean flow velocity field

$$\vec{U}_0 = \vec{\nabla}\Phi. \tag{1.11}$$

We further assume that upstream there are only vortical disturbances without entropy disturbances ($s' = 0$), and no far field acoustic sources.

For a cascade, out-going acoustic waves may not vanish far upstream, therefore the upstream total velocity must be of the form

$$\vec{U}(\vec{x},t) = U_\infty\vec{i}_1 + \vec{u}_g(\vec{x} - \vec{i}_1 U_\infty t) + \vec{u}_a(\vec{x},t) \text{ as } x_1 \longrightarrow -\infty \tag{1.12}$$

where $\vec{u}_a$ denotes the unsteady velocity associated with upstream acoustic, and $\vec{u}_g$ is the upstream imposed vortical disturbance satisfying

$$\vec{\nabla}\cdot\vec{u}_g = 0. \tag{1.13}$$

Because of the linear nature of the problem, we take, without loss of generality, one Fourier component as the upstream vortical disturbance, so that

$$\vec{u}_g = \vec{a}e^{i(\vec{k}\cdot\vec{x} - k_1 t)}, \tag{1.14}$$

where the gust amplitude vector $\vec{a} = (a_1, a_2, a_3)$ satisfies $|\vec{a}| << U_\infty$, and $\vec{k} = (k_1, k_2, k_3)$ is the wave number vector which gives the direction of propagation of the gust. Note that k_1 is the reduced frequency defined by

$$k_1 = \frac{\omega c}{2U_\infty}.$$

According to (1.13), $\vec{a}$ and $\vec{k}$ must satisfy

$$\vec{a} \cdot \vec{k} = 0. \tag{1.15}$$

The upstream condition for $\vec{u}$ is

$$\vec{u} \longrightarrow \vec{u}_g + \vec{u}_a \quad \text{as} \quad x_1 \longrightarrow -\infty. \tag{1.16}$$

1.2.2 Decomposition of the Unsteady Velocity

Golstein [Gol78] proposed a unified approach to streaming motion with small unsteady vortical and entropic disturbances imposed on the upstream flow. The unsteady velocity is decomposed into the sum of (i) a vortical part $\vec{u}^{(I)}$ which is known in terms of the upstream disturbances and the mean flow Lagrangian coordinates,which does not contribute to the unsteady pressure p', and (ii) a potential part $\vec{\nabla}\phi^*$ so that

$$\vec{u} = \vec{u}^{(I)} + \vec{\nabla}\phi^*. \tag{1.17}$$

ϕ^* satisfies a single non-constant-coefficient, convective wave equation

$$\frac{D_0}{Dt}(\frac{1}{c_0^2}\frac{D_0\phi^*}{Dt}) - \frac{1}{\rho_0}\vec{\nabla}.(\rho_0\vec{\nabla}\phi^*) = \frac{1}{\rho_0}\vec{\nabla}.(\rho_0\vec{u}^{(I)}). \tag{1.18}$$

The vortical velocity $\vec{u}^{(I)} = (u_1^{(I)}, u_2^{(I)}, u_3^{(I)})$ is given by

$$u_l^{(I)} = \vec{u}_g \cdot \frac{\partial \vec{X}}{\partial x_l}, \quad (l = 1, 2, 3) \tag{1.19}$$

where the components of $\vec{X} = (X_1, X_2, X_3)$ are defined as follows. $X_2 = X_2(x_1, x_2, x_3)$ and $X_3 = X_3(x_1, x_2, x_3)$ are independent integrals of the equations

$$\frac{dx_1}{U_{01}} = \frac{dx_2}{U_{02}} = \frac{dx_3}{U_{03}} \tag{1.20}$$

such that

$$X_2 \longrightarrow x_2 \quad \text{and} \quad X_3 \longrightarrow x_3 \quad \text{as} \quad x_1 \longrightarrow -\infty, \tag{1.21}$$

where $U_{0l}(l = 1, 2, 3)$ are the three components of the mean velocity $\vec{U}_0$. $X_1 = X_1(x_1, x_2, x_3)$ is defined as $X_1 = U_\infty \Delta$, where Δ is the Lighthill's 'drift' function

$$\Delta(x_1, x_2, x_3) = \frac{x_1}{U_\infty} + \int_{-\infty}^{x_1} [\frac{1}{U_{01}} - \frac{1}{U_\infty}]\, dx_1'. \tag{1.22}$$

The integral is carried out along a streamline. For a two-dimensional mean flow around a linear cascade, X_2 becomes the mean flow stream function Ψ_0, and X_3 becomes the spanwise coordinate x_3.

The boundary condition for ϕ^* on the blade surface Σ is

$$\vec{n}\cdot\vec{\nabla}\phi^* = -\vec{n}\cdot\vec{u}^{(I)} \qquad \text{for} \qquad \vec{x}\in\Sigma. \tag{1.23}$$

For a real fluid motion around a blade of cascades, there is a frontal stagnation point $\vec{x}_{st}$ on the blade surface. Atassi and Grzedzinski [AG89] showed that the vortical velocity $\vec{u}^{(I)}$ given by (1.19) becomes singular on the entire surface of the blade and its wake. As a result, equation (1.23) shows that ϕ^* has a singular gradient along the blade surface.

To remove this singularity, Atassi and Grzedzinski [AG89] showed that it is possible to construct a potential velocity field that cancels the dominant singular behavior of $\vec{u}^{(I)}$ along the blade surface Σ and the wake W. In their decomposition of the unsteady velocity, $\vec{u}$ is split into a known vortical part $\vec{u}^{(R)}$ and a potential part $\vec{\nabla}\phi$, so that

$$\vec{u} = \vec{u}^{(R)} + \vec{\nabla}\phi, \tag{1.24}$$

where ϕ satisfies

$$\frac{D_0}{Dt}\left(\frac{1}{c_0^2}\frac{D_0\phi}{Dt}\right) - \frac{1}{\rho_0}\vec{\nabla}.(\rho_0\vec{\nabla}\phi) = \frac{1}{\rho_0}\vec{\nabla}.(\rho_0\vec{u}^{(R)}). \tag{1.25}$$

$\vec{u}^{(R)}$ is given by

$$\vec{u}^{(R)} = \vec{u}^{(I)} + \vec{\nabla}\tilde{\phi}, \tag{1.26}$$

and $\tilde{\phi}$ is a known function in terms of the gust condition and the mean flow Lagrangian coordinates. The expression for $\tilde{\phi}$ for a linear cascade was given by [AG89] as

$$\tilde{\phi} = \frac{i}{k_1}\{a_1 + \frac{i\tilde{s}(a_2k_1 - a_1k_2)}{2\pi(1 + ia_0U_\infty k_1)}$$

$$\times sin[\frac{2\pi}{\tilde{s}U_\infty\rho_\infty}(\Psi_0 - \Psi_0^{(0)})]\}e^{i(\vec{k}\cdot\vec{x} - k_1t)} \tag{1.27}$$

where ρ_∞ is the upstream mean flow density, $\Psi_0^{(0)}$ is the value of the mean flow stream function taken at the blade surface, $\tilde{s}$ is the component of the cascade spacing vector $\vec{s}$ perpendicular to the upstream mean velocity, a_0 is defined by

$$a_0 = -\left(\frac{\partial U_0}{\partial n_0}\right)_{st}^{-1} \tag{1.28}$$

where $U_0 = |\vec{U}_0|$, n_0 denotes the derivative in the outward direction normal to the blade surface, and the subscript st denotes the value taken at the leading edge stagnation point.

The unsteady pressure is given by

$$p' = -\rho_0(\vec{x})\frac{D_0\phi}{Dt}. \tag{1.29}$$

Atassi and Grzedzinski [AG89] showed that $\vec{u}^{(R)}$ has zero normal and streamwise components on the blade surface and in the wake, which leads to a regular blade surface condition

$$\vec{\nabla}\phi \cdot \vec{n} = 0 \qquad \text{for} \qquad \vec{x} \in \Sigma. \tag{1.30}$$

This zero-gradient condition not only eliminate the singular behavior of condition (1.23) but also it is suitable for accurate numerical computations since it is not sensitive to the inaccuracies of numerically calculated mean flow gradients near the stagnation point.

The other boundary conditons include the following: (i) The cascade quasi-periodicity condition:

$$\phi(\vec{x} + \vec{s}, t) = \phi(\vec{x}, t)e^{i\sigma} \tag{1.31}$$

or

$$p'(\vec{x} + \vec{s}, t) = p'(\vec{x}, t)e^{i\sigma} \tag{1.32}$$

where

$$\sigma = \vec{k} \cdot \vec{s} \tag{1.33}$$

is the interblade phase angle. (ii) The unsteady potential ϕ is not continuous across the wake but the unsteady pressure and the normal velocity are continuous, which can be expressed as follows:

$$\frac{D_0}{Dt}(\Delta\phi) = 0 \;\; \text{and} \;\; \Delta(\vec{\nabla}\phi \cdot \vec{n}) = 0 \;\; \text{for} \quad \vec{x} \in W \tag{1.34}$$

where Δ denotes the jump in the value across the wake. In addition, we assume that the the unsteady pressure is continuous at the trailing edge of the blades (Kutta condition). (iii) At the out-flow boundaries, the unsteady pressure must satisfy non-reflective acoustic radiation conditions. This out-flow boundary condition is very important for accurate predictions of the unsteady response especially for moderate and high Mach numbers and reduced frequencies where the acoustic radiation becomes significant. The discussion concerning the out-flow boundary condition will be given in the next section.

1.3 Out-Flow Boundary Condition

1.3.1 Governing Equation in the Far Field

Upstream and downstream of the cascade, the mean flow is uniform, thus the coefficients of the governing equation are constants. By taking the material derivative of both sides of (1.25), we get

$$\frac{1}{c_0^2}\frac{D_0^2 p'}{Dt^2} - \nabla^2 p' = 0 \tag{1.35}$$

upstream and downstream of the cascade. Note that p' is regular and continuous upstream and downstream.

We now introduce (see figure 1.2) the upstream and downstream frames of reference $(\vec{j}_1, \vec{j}_2)$ and $(\vec{j}'_1, \vec{j}'_2)$ with origins o and o', respectively. The corresponding coordinate systems are (y_1, y_2) and (y'_1, y'_2), respectively. $\vec{j}_1$ is in the upstream mean flow direction, and $\vec{j}'_1$ downstream direction.

We can expand p' in plane wave series using the coordinate system of (y_1, y_2, x_3) for upstream and (y'_1, y'_2, x_3) for downstream

$$p' = \sum_{n=-\infty}^{\infty} c_{n+} p'_{n+} + \sum_{n=-\infty}^{\infty} c_{n-} p'_{n-}, \tag{1.36}$$

where the coefficients $c_{n\pm}$ are complex constants. $p'_{n\pm}$ has the general form

$$p'_{n\pm} = e^{i[(a_{n\pm}y_1 + b_{n\pm}y_2 + k_3 x_3) - k_1 t]} \quad \text{for upstream} \tag{1.37}$$

$$p'_{n\pm} = e^{i[(a_{n\pm}y'_1 + b_{n\pm}y'_2 + k_3 x_3) - k_1 t]} \quad \text{for downstream} \tag{1.38}$$

Substituting (1.37) and (1.38) into (1.35) and (1.32), the complex wave number for every single mode $\vec{k}_{n\pm} = (a_{n\pm}, b_{n\pm}, k_3)$ is determined. Note that $\vec{k}_{n\pm}$ depends on the mean flow conditions and thus hss different values upstream and downstream for a loaded cascade. Depending on the values of $a_{n\pm}$ and $b_{n\pm}$, (1.37) or (1.38) may represent an unsteady pressure field with the following far field charateristics: (i) an exponentially decaying field, (ii) an exponential increasing field, (iii) waves propogating away from the row of the blades with constant amplitude, or (iv) waves propogating toward the cascade with constant amplitude. Those non-wave-type terms of (1.37) or (1.38) which exponentially increase in the far field are physically not acceptable. For a gust problem without far-field acoustic sources, those wave-type terms of (1.37) or (1.38) which represent waves propagating toward the row of blades do not exist physically. Consequently, p' on the boundary can be constructed as a sum of all decaying terms (i) and propogating waves (iii). The coefficients $c_{n\pm}$ in (1.37) or (1.38) can be calculated numerically by matching the far field with the near field.

This out-flow boundary condition is valid for both the upstream and downstream, though, it is particularly suitable for the downstream. It avoids the singular behavior associated with the unsteady vortical velocity field which becomes infinity and indeterminate in the wake. This singularity makes it impossible to construct a downstream condition directly for ϕ without approximation. Since the total flow is regular upstream, it is then more convenient to use ϕ directly instead of p'. In this case, a derivation similar to what have been given in (1.35), (1.36) and (1.37) can be obtained for ϕ^*.

1.3.2 THE EXPRESSION FOR $p'_{n\pm}$

Substituting (1.37) and (1.38) into the governing equation (1.35) and the cascade quasi-periodicity condition (1.32) yields the equations for $a_{n\pm}$ and $b_{n\pm}$ in the upstream or the downstream

$$M^2(a_{n\pm} - k'_1)^2 + a_{n\pm}^2 + b_{n\pm}^2 + k_3^2 = 0 \tag{1.39}$$

$$b_{n\pm} - \frac{2n\pi + \sigma - a_{n\pm}s^*}{\tilde{s}} = 0, \tag{1.40}$$

where

$$k'_1 = \frac{k_1}{U_0}, \tag{1.41}$$

s^* and $\tilde{s}$ are the components of the spacing vector $\vec{s}$ parallel and perpendicular to the mean flow in the frame of $(\vec{j}_1, \vec{j}_2)$ for the upstream or $(\vec{j}'_1, \vec{j}'_2)$ for the downstream. Note that s^* and $\tilde{s}$ have different values upstream and downstream for a loaded cascade. M and U_0 refer to the upstream or the downstream mean flow Mach number and speed, respectively.

$a_{n\pm}$ can be readily obtained from (1.39) and (1.40)

$$a_{n\pm} = \frac{-k'_1 M^2 \tilde{s}^2 + (2n\pi + \sigma)s^* \pm d_n \tilde{s}}{d^{\dagger 2}} \tag{1.42}$$

where

$$d_n = \sqrt{\beta^2(\kappa^2 d^{\dagger 2} - \Gamma_n^2)} \tag{1.43}$$

$$\kappa = \sqrt{\frac{k'^2_1 M^2}{\beta^4} - \frac{k_3^2}{\beta^2}} \tag{1.44}$$

$$\Gamma_n = 2n\pi + \sigma + \frac{s^* k'_1 M^2}{\beta^2} \tag{1.45}$$

$$d^\dagger = \sqrt{s^{*2} + \beta^2 \tilde{s}^2} \tag{1.46}$$

$$\beta = \sqrt{1 - M^2}, \tag{1.47}$$

The branch cut for d_n is taken such that

$$d_n = |d_n| \quad \text{for} \quad d_n^2 > 0 \tag{1.48}$$
$$d_n = i\,|d_n| \quad \text{for} \quad d_n^2 < 0. \tag{1.49}$$

For $d_n^2 > 0$, (1.37) and (1.38) represent plane waves propagating in the direction of $\vec{k}_{n\pm} = (a_{n\pm}, b_{n\pm}, k_3)$. The acoustic cut-on condition is

$$d_n^2 = 0. \tag{1.50}$$

1.3.3 Behavior of Acoustic Modes Near Cut-On

When the system is close to the acoustic cut-on, the unsteady response of the cascade is very sensitive to flow conditions and exhibits a large variation. This indicates that the coupling between the near field and the far field is strong. It is thus interesting to investigate the behavior of acoustic modes near the acoustic cut-on.

We now introduce an upstream and a downstream frame of references $(\vec{e}_1, \vec{e}_2, \vec{e}_3)$ and $(\vec{e}_1{}', \vec{e}_2{}', \vec{e}_3{}')$ with origins at o and o', respectively, where $\vec{e}_1$ and $\vec{e}_1{}'$ are in the machine axis direction pointing to the downstream, and $\vec{e}_3$ and $\vec{e}_3{}'$ are in the spanwise direction (see figure 1.2).

We write $\vec{k}_{n\pm}$ with respect to these frames of reference upstream and downstream

$$\vec{k}_n = (\pm\frac{d_n S}{d^{\dagger 2}} + \delta_n, \frac{2n\pi + \sigma}{S}, k_3) \tag{1.51}$$

where $S = |\vec{s}|$ and

$$\delta_n = \frac{M^2 \tilde{s}(-k'_1 S^2 + (2n\pi + \sigma)s^*)}{S d^{\dagger 2}} \tag{1.52}$$

When the the system approaches the acoustic cut-on condition, i.e. $d_n^2 \rightarrow 0+$, the wave number vector for a single model $\vec{k}_n^c$ can be written in the frame $(\vec{e}_1, \vec{e}_2, \vec{e}_3)$ or $(\vec{e}_1{}', \vec{e}_2{}', \vec{e}_3{}')$ as

$$\vec{k}_n^c = (\delta_n, \frac{2n\pi + \sigma}{S}, k_3). \tag{1.53}$$

(1.53) shows that when a new mode cuts in, it starts to propagate in the direction $\vec{k}_n^c$, which may not be in the direction of the stagger line since δ_n may not be zero as $d_n^2 \rightarrow 0+$. In addition, when $|d_n|$ is small and $d_n^2 > 0$, δ_n may dominate $\pm d_n S/d^{\dagger 2}$ in (1.51). Consquently, depending upon the value of δ_n, both p'_{n+} and p'_{n-} in (1.36) may physically exist but propagate in different directions as out-going waves. Note that these two waves have the same circumferential component of wave number vector in the cascede stagger line direction because of the quasi-periodicity condition (1.32).

For a cascade of flat plates, since the upstream and the downstream mean flow conditions are the same, acoustic cut-on always occur at either upstream or downstream, but not both, depending upon the value of δ_n. However, this is not true for a cascade of loaded airfoils, where downstream and upstreammean flow conditions are different due to the mean blade loading.

1.4 Numerical Scheme

The present boundary-value problem depends on the time t and the spanwise coordinate x_3. However, because of the form of the upstream gust condition (1.14) and the two-dimensional geometry of the cascade, it is possible to factor out these two variables. We introduce a dependent variable ψ such that

$$\phi = \psi(x_1, x_2) e^{i(k_3 x_3 - k_1 t)}. \tag{1.54}$$

Substituting (1.54) into (1.25), (1.30), (1.31), (1.34) and the far-field boundary condition, we obtain governing equation and boundary conditions for $\psi(x_1, x_2)$.

The boundary value problem for ψ was solved using finite differencing approximations in a body-fitted coordinate system. One coordinate is the stream function Ψ_0 of the mean flow and the other is a family of straight lines parallel to the cascade stagger line. The computational domain is an extended single passage of the cascade.

The mean flow fields, which are entirely subsonic, were obtained by using McFarland's [McF84] subsonic steady cascade flow solver. The advantage of this solver is its efficiency. The mean flow properties were then calculated along stream lines using a spatial marching technique.

1.5 Results and Discussion

The numerical scheme was first validated by comparison with a flat plate cascade code based on an integral method [Ven80]. We then examined the effects of mean blade loading, blade and cascade geometry, and upstream gust condition on the unsteady lift coefficient C_L and the normalized unsteady pressure P', P' and C_L are defined by:

$$P' = \frac{\overline{p'}}{\rho_\infty U_\infty \, |\vec{a}| \, e^{ik_1 t} e^{ik_3 x_3}}$$

$$C_L = \frac{\overline{L}}{\pi \rho_\infty c U_\infty \, |\vec{a}| \, e^{ik_1 t} e^{ik_3 x_3}}$$

where L is the unsteady aerodynamic lift and the overbar indicates the complex conjugate.

1.5.1 Results for Cascades of Flat Plates

The results for three cascades of flat plates with Mach numbers of 0.3, 0.5 and 0.7 are presented below. The mean flow is uniform with zero incidence. The upstream vortical disturbances described by equation (1.14) are imposed far upstream with $\vec{a} = (0, 1, 0)$ and $\vec{k} = (k_1, 0, 0)$ where k_1 varies. The blade spacing is 1 for all of the cases.

For a cascade of flat plates with Mach number of 0.3, stagger of 45°, the unsteady lift coefficients predicted by present analysis are presented in Figure 1.3 and compared with those of an integral method analysis [Ven80]. The reduced frequency k_1 varies from 0 to 5. For these flow conditions, the first acoustic mode cut-on occurs at $k_1 = 2.86$. The results show a excellent agreement for a large range of the reduced frequency including the frequency near the acoustic cut-on, where the unsteady lift is very sensitive and exhibits large variation.

Figure 1.4 shows the unsteady lift coefficients for a cascade of flat plates with Mach number of 0.5 and stagger of 30°. The results are compared with those of integral method. For these flow conditions, the first acoustic mode cut-on is at $k_1 = 2.5$ and second at $k_1 = 5.0$. Again, in the whole range of reduced frequencies from 0 to 5 including two acoustic modes cut-on, the results show excellent agreement between the present analysis and the integral method.

At a higher Mach number of 0.7 a stagger of 15°, Figure 1.5 compares our results with those of the integral method. The first acoustic cut-on occurs at a very low frequency, and the second one at $k_1 = 2.07$. In the whole range, including the peaks of the unsteady lift caused by acoustic cut-on, the present analysis gives highly accurate results.

1.5.2 Effects of Cascade Geometry

The unsteady responses for two compressor cascades in a gust are studied to demonstrate the effects of the mean blade loading. The cascade No. 1 has a relative thin configuration ($\theta = 0.06$), small camber (m $= 0.05$), moderate relative spacing of 1 and stagger (χ) of 45°. The second cascade has a larger relative thickness and camber ($\theta = 0.12$, m $= 0.13$), smaller relative spacing of 0.6 and stagger of 15°. Both cascades consist of Joukowski airfoils. The upstream flow angle α_1 is 55° for cascade No. 1, and 40° for cascade No. 2. The reduced frequency k_1 varies from 0 to 5. The geometries of these two cascades are given in Figure 1.6.

Cascade of Loaded Airfoils No. 1

For this cascade, the inlet Mach number takes values of 0.3, 0.5 and 0.7. The corresponding outlet Mach numbers are 0.22, 0.35, 0.45, respectively. The upstream flow angles are constant of 55°, the downstream flow angles are 39.6^o, 39.8^o and 40.1^o, respectively. The vortical excitation is due to a gust with $\vec{k} = (k_1,0,0)$, $\vec{a} = (0,1,0)$, where k_1 ranges from 0 to 5. The interblade phase angle σ is determined by $\sigma = \vec{k} \cdot \vec{s}$. The real and imaginary parts of the unsteady lift coefficients versus the reduced frequency are shown in figures 1.7, 1.8 and 1.9. These results show that near the cut-on condition, the unsteady lift exhibits large variation. Figure 1.10 and 1.11 show the unsteady pressure field (Real part) for M=0.7 and k_1=1, and k_1=2, respectively. For k_1=1, there is no acoustic mode in the far field, but the system is close to cut-on condition. It can be seen that the unsteady pressure is building up in the near field but decays in the far field. For k_1=2, one acoustic mode cuts on in the upstream and one in the downstream. The unsteady pressure is reduced in the near field as the acoustic waves propagate in the far field.

Cascade of Loaded Airfoils No. 2

The inlet Mach numbers are 0.01, 0.3 and 0.5, the corresponding outlet Mach numbers are 0.007, 0.23 and 0.36, respectively. The upstream flow angles are constant at 40°, while the downstream flow angle is almost constant at -6.68^o. The gust is $\vec{k} = (k_1, 0, 0), \vec{a} = (0, 1, 0)$, where k_1 ranges from 0 to 5. The interblade phase angle σ is determined by $\sigma = \vec{k} \cdot \vec{s}$. For this cascade, figures 1.12, 1.13 and 1.14 show the real and imaginary part of the unsteady lift coefficients. These results show that the mean blade loading strongly affect the gust response. For small reduced frequency ($k_1 \leq 1$), the mean blade loading reduces the unsteady lift, which is consistent with the results observed for a single airfoil by of Goldstein and Atassi [GA76], and Atassi [Ata84], and Scott and Atassi [SA90]. For moderate and high reduced frequencies ($1 \leq k_1 \leq 5$), the mean blade loading effect strongly depends on the combination of the upstream flow conditions and the cascade parameters. The unsteady pressure field is shown in Figure 1.15 for $M = 0.5$ and $k_1 = 3.5$ with no acoustic mode in the far field; and Figure 1.16 for $M = 0.5$ and $k_1 = 4$ with one acoustic modes in the upstream. The figures show a strong unsteady pressure build-up in the near field before the acoustic cut-on, and its dissipation to the far field after cut-on.

Effects of Two- and Three-Dimensional Gust

To demonstrate the effects of two- and three-dimensional gusts, we take $M = 0.5$ and consider the four cases of cascade No. 2: 1) one-dimensional gust, 2) two-dimensional gust, 3) three-dimensional gust without sweep and 4) three-dimensional gust with sweep. Figure 1.17 show the velocity triangles for these four cases. The gust parameters for 1) are: $\vec{k} = (k_1,0,0)$, $\vec{a} = (0,1,0)$; For 2): $\vec{k} = (k_1,k_2,0)$, $\vec{a} = (-0.707,0.707,0)$, where $k_2 = k_1$; For 3): $\vec{k} = (k_1, k_2, 1)$, $\vec{a} = (-0.707, 0.707, 0)$, where $k_2 = k_1$; For 4): $\vec{k} = (k_1, k_2, k_3)$, $\vec{a} = (-0.47, 0.82, 0.33)$, where $k_2 = k_1$ and k_3 is determined by the condition $\vec{a} \cdot \vec{k} = 0$. In all cases k_1 varies from 0 to 5. Figure 1.18 shows the amplitudes of unsteady lift coefficient for these cases. Note the strong effects of the gust wave numbers and sweep on the unsteady response.

1.6 Conclusions

A linearized analysis and a numerical scheme have been developed for subsonic compressible flows around a cascade of loaded airfoils subject to a three-dimensional gust.

The analysis is based on Goldstein's [Gol78] and Atassi and Grzedzinski's [AG89] formulations for unsteady vortical flows. The numerical scheme developed yields the unsteady cascade response for a wide variety of unsteady vortical flow conditions and for a broad range of reduced frequencies and Mach numbers.

Another contribution of the present work is the derivation of a new far field acoustic analysis. It avoids the difficulties associated with the singular behavior of the unsteady vortical velocity downstream. This condition is relatively simple and does not require the development of far field analytical solutions. It is also found that an accurate, regular far field condition is very important for accurate solutions especially for higher Mach numbers and/or higher reduced frequencies.

The results obtained by the present analysis were compared with those of an integral method. For a large range of the reduced frequency (0 to 5), and a large range of Mach number (0 to 0.7), the comparison shows excellent agreement.

The present paper also examined the effect of mean blade loading on the unsteady responses due to a gust. Detailed results were given for two compressor cascades. The unsteady pressure fields were also given in contour form to demonstrate the effect on the near field unsteady pressure as well as the far field acoustic due to the mean blade loading. These contours show that the present far field conditions satisfy the non-reflective acoustic radiation condition.

The results show that for low reduced frequencies ($k_1 < 1$), the mean blade loading caused by blade camber and thickness reduces the unsteady lift, this is consistent with the results for single airfoils in a gust. For moderate and high reduced frequencies ($1 \le k_1 \le 5$), the mean flow loading effect strongly depends on the combination of upstream flow conditions and cascade parameters.

The numerical scheme development in present work is computationally

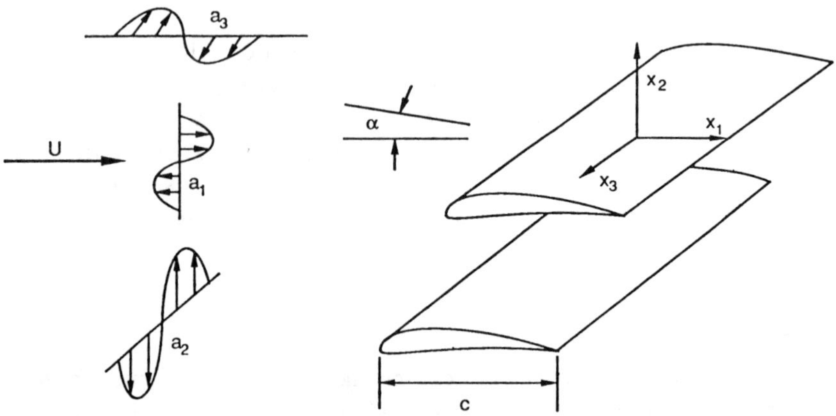

FIGURE 1.1. Cascade of loaded airfoils in a three-dimensional gust

efficient. Therefore this scheme should be useful for turbomachinery aerodynamic and aeroacoustic design investigations. Currently, typical CPU time per frequency on a scientific workstation DEC-3100 is about 2 minutes, depending upon the frequency and the Mach number.

Acknowledgements

This work is supported by NASA Lewis Research Center under Grant NAG 732. The authors are grateful to Mr. Dennis Huff for his assistance and support.

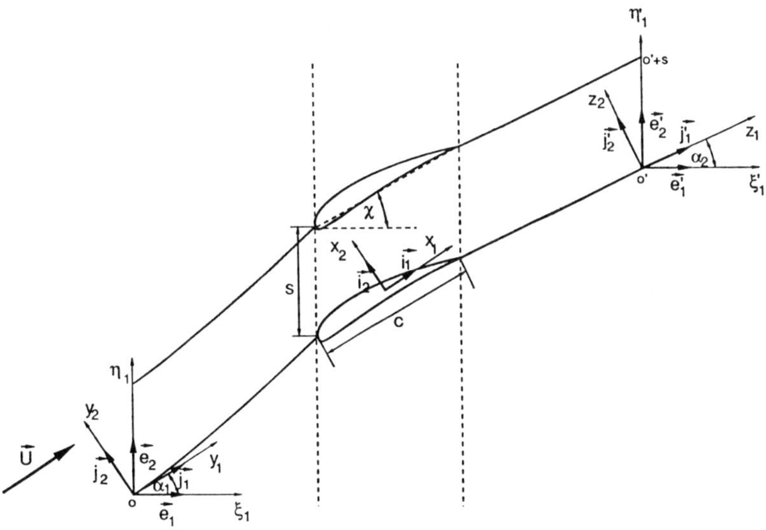

FIGURE 1.2. Two-dimensional cascade of airfoils and the computational coordinate system

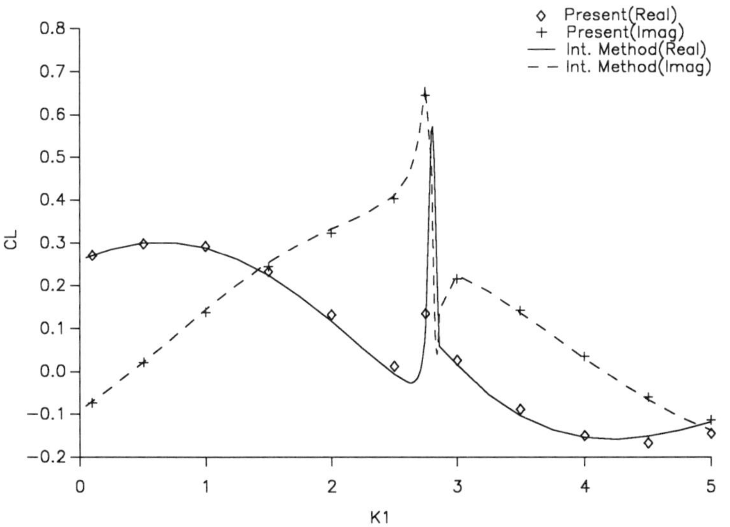

FIGURE 1.3. Unsteady lift coefficient vs. reduced frequency for a cascade of flat plates. Present results are compared with those of an integral method. M=0.3, $\chi = 45^\circ$, S/C=1.

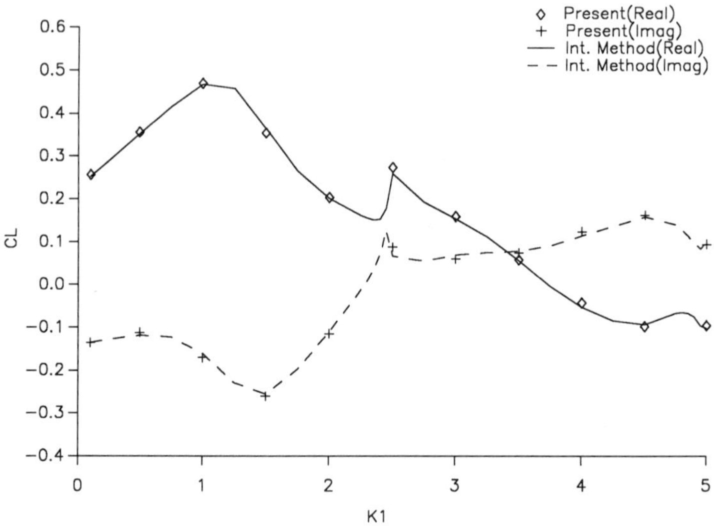

FIGURE 1.4. Unsteady lift coefficient vs. reduced frequency for a cascade of flat plates. Present results are compared with those of an integral method. M=0.5, $\chi = 30^o$, S/C=1.

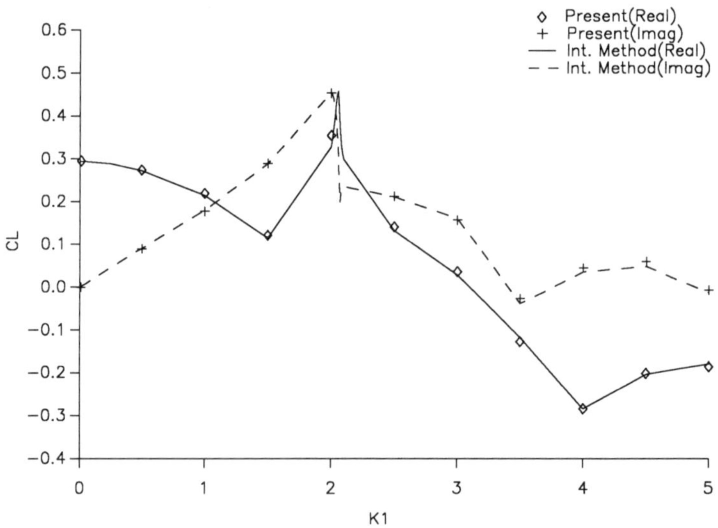

FIGURE 1.5. Unsteady lift coefficient vs. reduced frequency for a cascade of flat plates. Present results are compared with those of an integral method. M=0.7, $\chi = 15^o$, S/C=1.

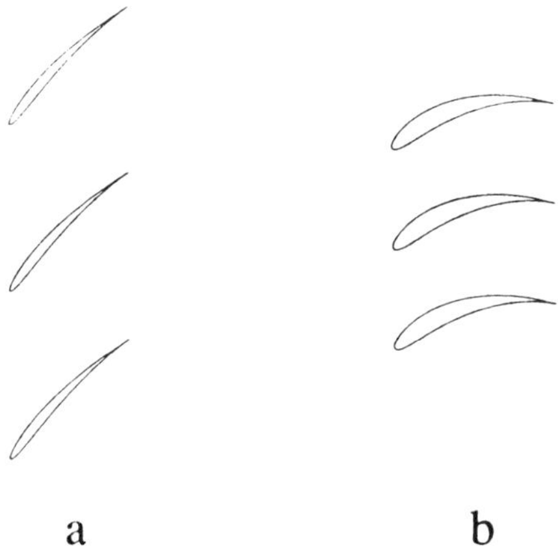

FIGURE 1.6. Geometries of Cascades. a) Cascade No. 1, b) Cascade No. 2

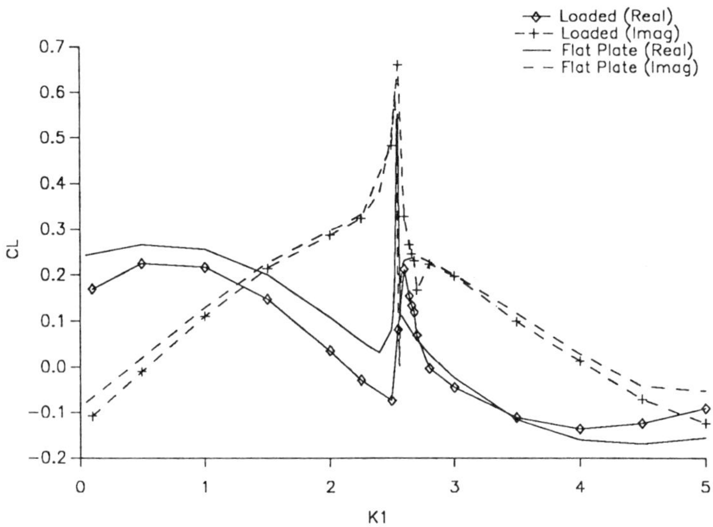

FIGURE 1.7. Unsteady lift coefficient vs. reduced frequency for the cascade No.1. Present results are compared with those for a corresponding flat plate cascade. $M = 0.3,\ k_2 = k_3 = a_1 = a_3 = 0,\ a_2 = 1.$

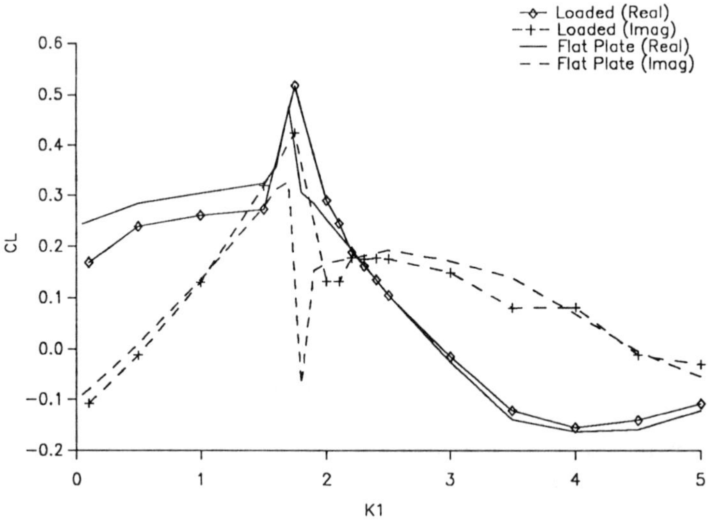

FIGURE 1.8. Unsteady lift coefficient vs. reduced frequency for the cascade No. 1. Present results are compared with those for a corresponding flat plate cascade. $M = 0.5$, $k_2 = k_3 = a_1 = a_3 = 0$, $a_2 = 1$.

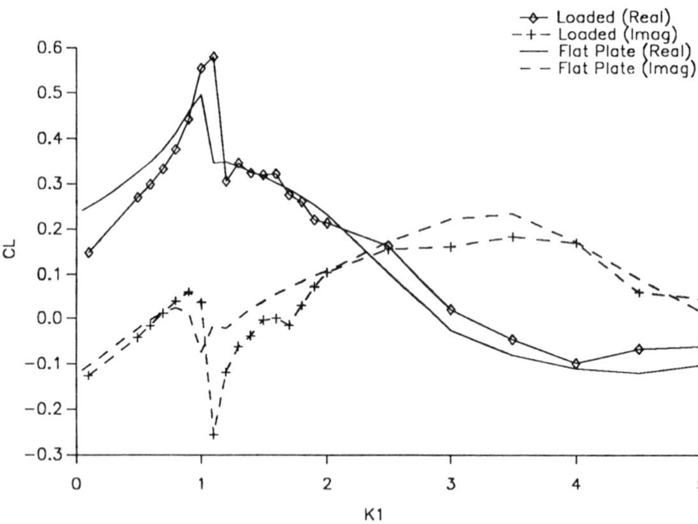

FIGURE 1.9. Unsteady lift coefficient vs. reduced frequency for the cascade No. 1. Present results are compared with those for a corresponding flat plate cascade. $M = 0.7$, $k_2 = k_3 = a_1 = a_3 = 0$, $a_2 = 1$.

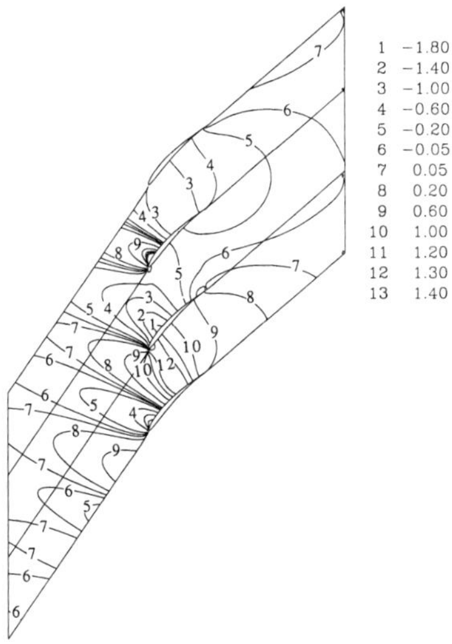

FIGURE 1.10. Contour of the unsteady pressure (real part) for the cascade No. 1. M=0.7, $\vec{a}$ = (0,1,0), $\vec{k}$ = (1,0,0).

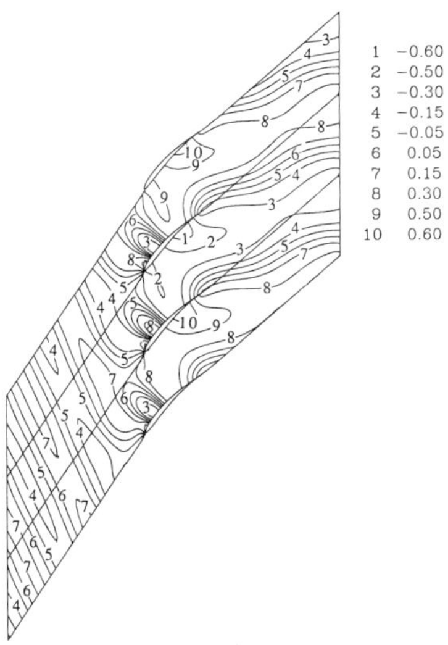

FIGURE 1.11. Contour of the unsteady pressure (real part) for the cascade No. 1. M=0.7, $\vec{a}$ = (0,1,0), $\vec{k}$ = (2,0,0).

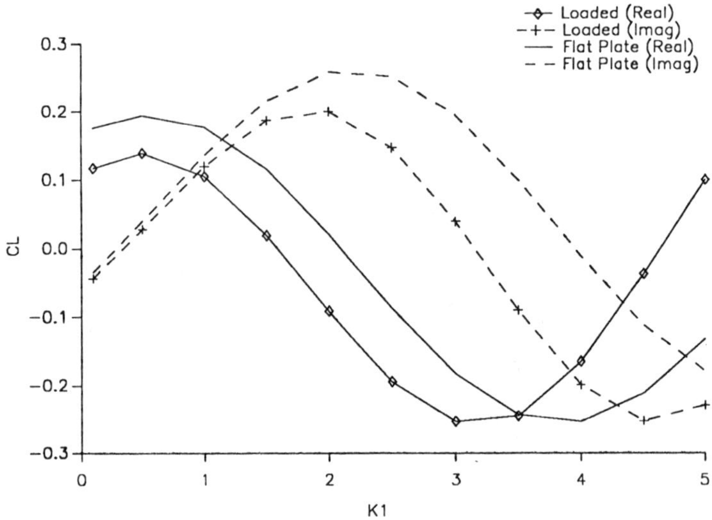

FIGURE 1.12. Unsteady lift coefficient vs. reduced frequency for the cascade No. 2. Present results are compared with those for a corresponding flat plate cascade. $M = 0.01,\ k_2 = k_3 = a_1 = a_3 = 0,\ a_2 = 1.$

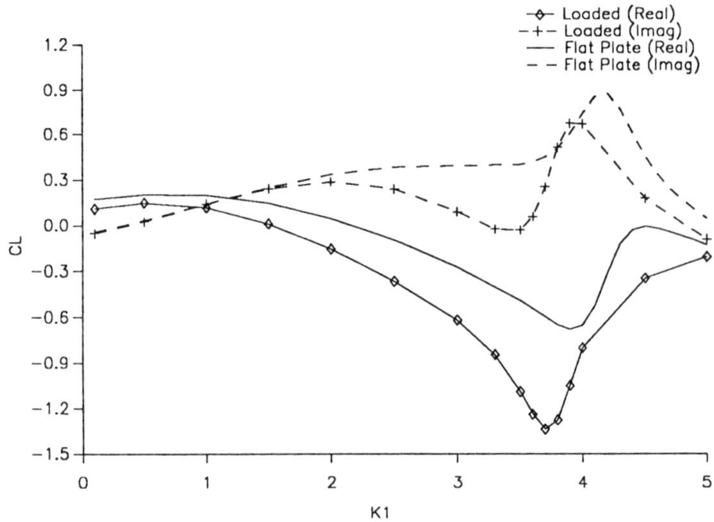

FIGURE 1.13. Unsteady lift coefficient vs. reduced frequency for the cascade No. 2. Present results are compared with those for a corresponding flat plate cascade. $M = 0.3,\ k_2 = k_3 = a_1 = a_3 = 0,\ a_2 = 1.$

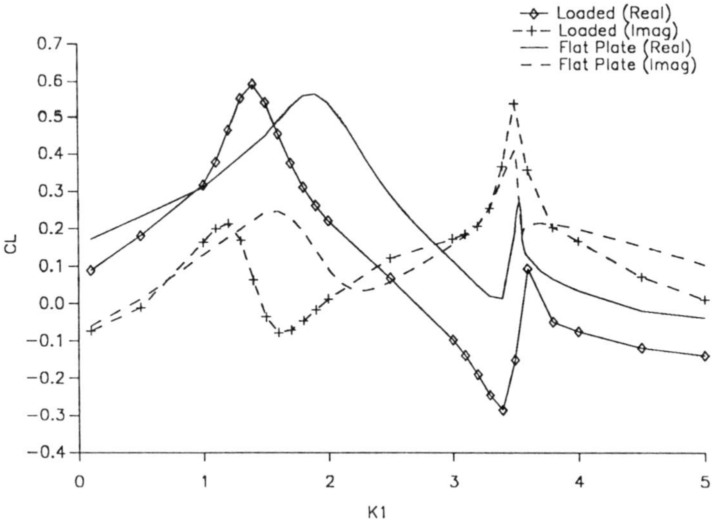

FIGURE 1.14. Unsteady lift coefficient vs. reduced frequency for the cascade No. 2. Present results are compared with those for a corresponding flat plate cascade. $M = 0.5$, $k_2 = k_3 = a_1 = a_3 = 0$, $a_2 = 1$.

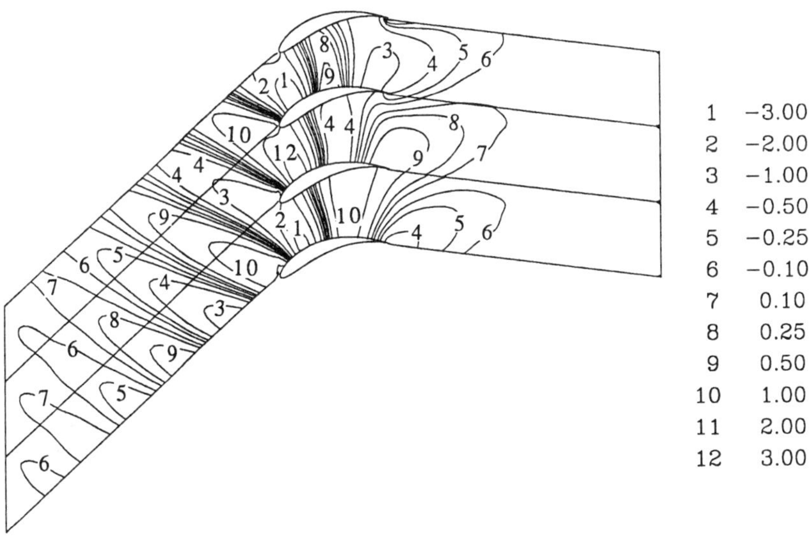

FIGURE 1.15. Contour of the unsteady pressure (real part) for the cascade No. 2. M=0.5, $\vec{a} = (0,1,0)$, $\vec{k} = (3.5,0,0)$.

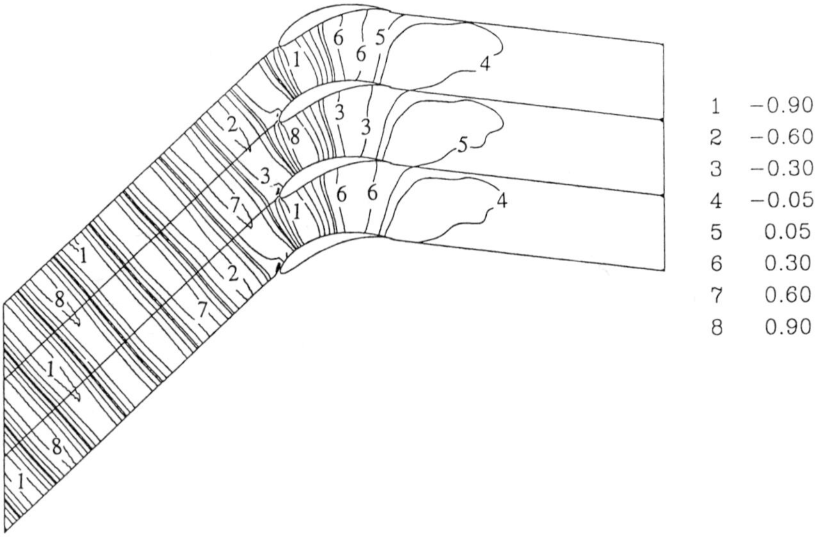

FIGURE 1.16. Contour of the unsteady pressure (real part) for the cascade No. 2. M=0.5, $\vec{a} = (0,1,0)$, $\vec{k} = (4,0,0)$.

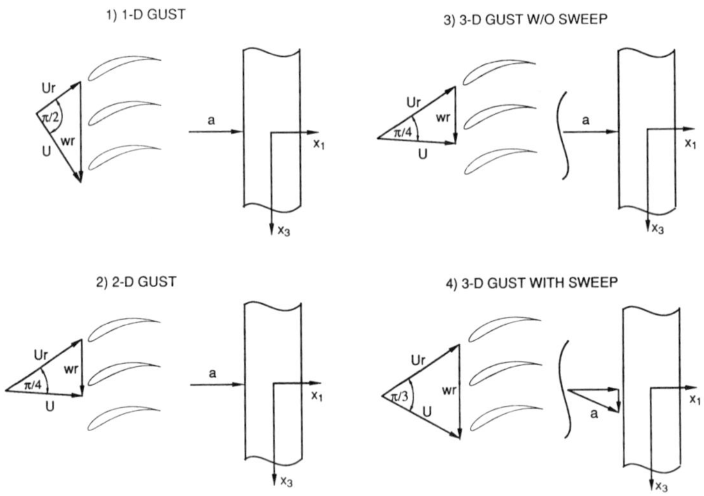

FIGURE 1.17. Velocity triangles for 1) one-dimensional gust, 2) two-dimensional gust, 3) three dimensional gust without sweep and 4) three-dimensional gust with sweep.

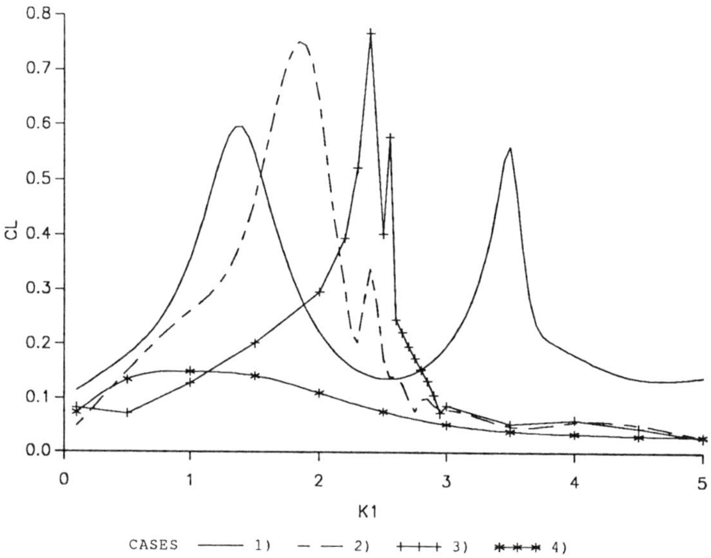

FIGURE 1.18. Amplitudes of unsteady lift coefficient vs. reduced frequency for the cascade No. 2 with vortical excitation due to 1) one-dimensional, 2) two-dimensional, 3) three-dimensional gust without sweep and 4) three-dimensional gust with sweep. M = 0.5.

References

[AA80] T. J. Akai and H. M. Atassi. Aerodynamic and aeroelastic characteristics of oscillating loaded cascades at low mach number, part 2: Stability and flutter boundaries. *Trans. ASME A: Journal of Engineering for Power*, 102:524–356, 1980.

[AG89] H. M. Atassi and J. Grzedzinski. Unsteady disturbances of streaming motions around bodies. *J. Fluid Mech.*, 209:385–403, 1989.

[Ata84] H. M. Atassi. The sears problem for a lifting airfoil revisited-new results. *J. Fluid Mech.*, 141:109–122, 1984.

[Ata86] H.M. Atassi. Unsteady vortical disturbances about bodies. In *Proceedings of the Tenth U. S. National Congress of Applied Mechanics.* J.P. Lamb, ed., ASME, 1986.

[BP54] G. K. Batchelor and I. Proudman. The effect of rapid distortion of a fluid in turbulent motion. *Quart. J. Mech. Appl. Math.*, 1:83–103, 1954.

[Fan91] J. Fang. *Compressible Flows with Vortical Disturbances Around A Cascade of Airfoils.* PhD thesis, University of Notre Dame, April 1991.

[GA76] M. E. Goldstein and H. M. Atassi. A complete second-order theory for the unsteady flow about an airfoils due to a periodic gust. *J. Fluid Mech.*, 74:741–765, 1976.

[Gol76] Marvin E. Goldstein. *Aeroacoustics.* McGraw-Hill, 1976.

[Gol78] M. E. Goldstein. Unsteady vortical and entropic distortions of potential flows round arbitrary obstacles. *J. Fluid Mech.*, 89:433–468, 1978.

[HV91] K.C. Hall and J. M. Verdon. Gust response analysis for cascade operating in nonuniform mean flows. *AIAA Journal*, 29(9):1463–1471, September 1991.

[Kur74] M. Kurosaka. On the unsteady supersonic cascade with a subsonic leading edge - an exact first order theory: Part 1 and 2. *Trans. ASME A: Journal of Engineering for Power*, 96:13–31, 1974.

[McF84] E. R. McFarland. A rapid blade-to-blade solution for use in turbomachinery design. *Journal of Engineering for Gas Turbines and Power*, 106:376–382, 1984.

[NW78] T. Nagashima and D. S. Whitehead. Linearized supersonic unsteady flows in cascade. Technical Report 3811, A.R.C., 1978.

[Pra33] L. Prandtl. Attaining a steady air stream in wind tunnels. Technical Report Tech. Memo., No. 726, N.A.S.A., 1933.

[SA90] J. S. Scott and H. M. Atassi. Numerical solutions of the linearized euler equations for unsteady vortical flows around lifting airfoils. In *28th Aerospace Sciences Meeting, Reno.* AIAA Paper, 1990.

[Sco90] J. S. Scott. *Compressible Flows with Periodic Vortical Disturbances around Lift Airfoils.* PhD thesis, University of Notre Dame, April 1990.

[VC80] J. M. Verdon and J. R. Caspar. Subsonic flow past an oscillating cascade with finite mean flow deflection. *AIAA J.*, 18:524–356, 1980.

[VC84] J. M. Verdon and J. R. Caspar. A linearized unsteady aerodynamic analysis for transonic cascade. *J. Fluid Mech.*, 149:403–429, 1984.

[Ven80] C.S. Ventres. A computer program to calculate cascade 2-d kernel. In *NASA Technic Memorandum.* NASA, 1980.

[Ver77] J. M. Verdon. Further development in the aerodynamic analysis of unsteady supersonic cascade, part 1: The unsteady pressure field, and part 2: Aerodynamic response predictions. *Trans. ASME A: Journal of Engineering for Power*, 99:509–525, 1977.

[Ver87] J. M. Verdon. Linearized unsteady aerodynamic theory. In *AGARD Manual on Aeroelasticity in Axial-Flow Turbomachinery*, volume 1. M. F. Platzer and F. O. Carta, Editors, 1987.

[Ver89] J. M. Verdon. The unsteady flow in the far field of an isolated blade row. *Journal of Fluids and Structures*, 3:123–149, 1989.

[WG80] D. S. Whitehead and R. J. Grant. Force and moment coefficients for high deflection cascade. In *2nt Intl. Symp. on Aeroelasticity in Turbomachines.* P. Suter, Editors, 1980.

[Whi60] D.S. Whitehead. Force and moment coefficient for vibrating aerofoils in cascade. In *A.R.C. 22133.* Cambridge Univ. Engineering Department, 1960.

[Whi70] D.S. Whitehead. Vibration and sound generation in a cascade of flat plates in subsonic flow. In *A.R.C. 32017.* Cambridge Univ. Engineering Department, 1970.

[Whi87] D. S. Whitehead. Classical two-dimensional method. In *AGARD Manual on Aeroelasticity in Axial-Flow Turbomachinery*, volume 1. M. F. Platzer and F. O. Carta, Editors, 1987.

[WN85] D.S. Whitehead and S.G. Newton. A finite element method for the solution of two-dimensional transonic flows in cascades. *International Journal for Numerical Method in Fluids*, 5:115–132, 1985.

Development of a Steady Potential Solver for Use with Linearized Unsteady Aerodynamic Analyses

Daniel Hoyniak[1]
Joseph M. Verdon[2]

Abstract

The need for accurate and numerically efficient response predictions for vibrating turbomachinery blading has resulted in the development of linearized unsteady aerodynamic analyses for cascade flows. One such model is the linearized inviscid flow (LINFLO) analysis developed by Verdon and Caspar, (1982,1984). This model assumes that the unsteady flow can be represented as a perturbation of the underlying nonuniform mean flow. Thus, in order to predict the unsteady aerodynamic response of a cascade using this technique, the steady flow field must first be calculated. A full potential solver developed explicitly for use with the LINFLO analysis is described herein. The solver uses the nonconservative form of the nonlinear potential flow equations together with an implicit, least-squares, finite-difference approximation to solve for the steady flow field. The difference equations are developed on a composite mesh which consists of a C-grid embedded in a rectilinear (H-grid) cascade mesh. The composite mesh is capable of resolving blade-to-blade and farfield phenomena on the H-grid, while accurately resolving local phenomena on the C-grid. The resulting system of algebraic equations are arranged in matrix form using a sparse matrix package and solved by Newton's method.

Steady and unsteady results are presented for two cascade configurations, a high speed compressor, and a turbine with high exit Mach number. The compressor configuration is composed of airfoils having a NACA 6%

[1]NASA Lewis Research Center, MS 77-6, 21000 Brookpark Road, Cleveland, OH 44135.

[2]United Technologies Research Center, MS 129-20, 411 Silver Lane, East Hartford, CT 06108.

compressor configuration is composed of airfoils having a NACA 6% thickness distribution superimposed on a circular arc camber line. The turbine cascade is the fourth standard configuration reported on by Bölcs and Fransson (1986).

Introduction

The need for accurate and numerically efficient techniques for predicting the aeroelastic and aeroacoustic response behavior of turbomachinery blading has resulted in the development inviscid unsteady aerodynamic linearization procedures (see Atassi & Akai,1980; Whitehead, 1982; Verdon & Caspar, 1982,1984; Hall & Clark, 1991; and Caruthers and Dalton, 1991). In general these techniques address the unsteady response of an isolated, two dimensional cascade. They include the effects of realistic design features such as blade geometry, mean blade loading, and operation at transonic Mach numbers. The unsteady fluctuations are regarded as a small-amplitude harmonic (in time) disturbances superimposed on a fully nonuniform mean or steady flow. The steady flow is determined as a solution of the nonlinear inviscid equations, while the unsteady flow is governed by linear equations with variable coefficients that depend on the underlying steady flow. Thus, in order to predict the unsteady response it is first necessary to determine the steady flow field.

Recently, a number of investigators have used Newton's method to obtain solutions to the equations of fluid dynamics. Giles (1985), and Hall & Crawley (1989), used the Euler equations to solve for the transonic flow through a two dimensional cascade of airfoils. Bender & Khosla (1988), and Bailey & Beam (1989), both used this technique to obtained solutions of the two dimensional inviscid and viscous equations associated with flow over an isolated airfoil.

This paper describes the application of Newton's method to the solution of the nonlinear steady potential equation describing the flow through a two dimensional cascade of airfoils. This method was selected for two reasons. The first is that it essentially transforms the solution of the nonlinear velocity potential equation into a series of linear equations that are similar to those obtained for the linearized unsteady problem, thus, making it possible to use the numerical algorithm developed for the solution of the linearized unsteady method has the potential for obtaining a converged solution very quickly.

The first section of the paper describes the partial differential equation and boundary conditions for the nonlinear steady velocity potential. Next, Newton's method is applied and the resulting linear differential equation and boundary conditions related to those obtained for the linearized unsteady flow. The last section describes the application of this model to the solution of the steady and unsteady flows associated with a high speed compressor cascade, and the turbine cascade designated as the fourth standard configuration by Bölcs and Fransson, (1986).

Governing Equations

The cascade geometry used in this mathematical model is shown in Figure 1. It consists of an isolated, infinite array of identical airfoils. The reference airfoil for the cascade defines the origin of a Cartesian coordinate system with the ξ axis oriented along the cascade axial direction, and the η axis oriented along the tangential direction. The gap distance between blades is defined by the magnitude of the gap vector G which is directed along the η axis. The inlet and exit flow angles are defined by, $\Omega_{\mp\infty}$, respectively. The steady state positions of the airfoil chord lines coincide with the line segments $0 \le \xi \le \cos\Theta, \eta = \xi \tan\Theta + mG$, where m is a blade index number, m = 0 denotes the reference blade, and Θ is the cascade stagger angle. The blade and cascade physical parameters have been normalized by the blade chord.

Field Equations

The equations that govern the steady, two dimensional, adiabatic flow, with negligible body forces, of an inviscid non-heat-conducting perfect gas are obtained from the conservation of mass, momentum, and energy. It is also assumed that at some distance upstream ($\xi < \xi_{-\infty}$) and downstream ($\xi > \xi_{+\infty}$) the flow is at most a small isentropic and irrotational perturbation to the steady uniform flow. Because of these assumptions the flow is isentropic and irrotational throughout the domain, and the velocity field can be defined in terms of a velocity potential function. The steady flow field can now be

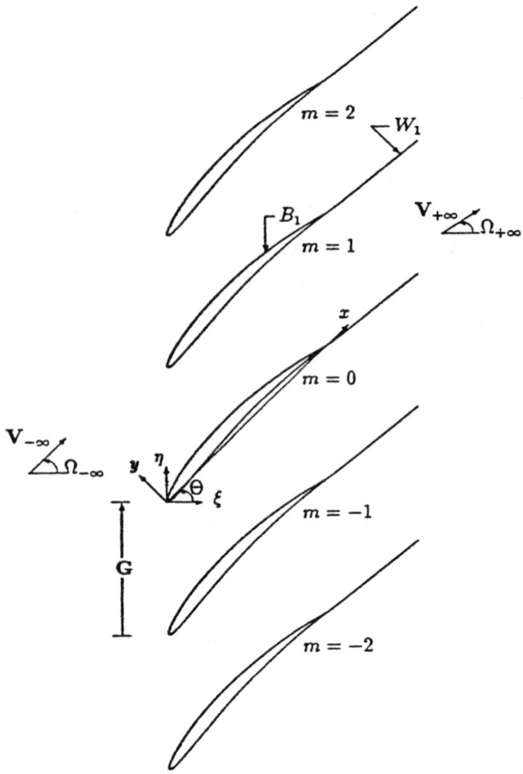

Figure 1 Two-dimensional cascade geometry.

in terms of a velocity potential function. The steady flow field can now be defined by the conservation of mass equation,

$$\nabla\cdot\left(\tilde{\rho}\,\nabla\tilde{\Phi}\right) = 0 \tag{1}$$

where $\tilde{\rho}$ is the density and $\tilde{\Phi}$ is the velocity potential. In addition to this mass conservation equation, the following relationships between the mean flow variables can be obtained from the momentum equation

$$\tilde{\rho}^{\gamma-1} = \left\{M_{-\infty}\tilde{A}\right\}^2 = 1 - \left(\frac{\gamma-1}{2}\right)M_{-\infty}^2\left\{(\nabla\tilde{\Phi})^2 - 1\right\} \tag{2}$$

where, $\tilde{A}$, is the local speed of sound propagation, and $M_{-\infty}$ is the inlet Mach number. Equations 1 and 2 can be combined to determine a single nonconservative equation for the potential, i.e.,

$$\tilde{A}\,\nabla^2\tilde{\Phi} - \nabla\tilde{\Phi}\cdot\frac{\nabla(\nabla\tilde{\Phi})^2}{2} = 0 \tag{3}$$

Boundary Conditions

In addition to the field equation (Eq. 3), the flow is subject to the following boundary conditions in order to obtain a solution for the steady flow field. If we assume that the flow remains attached to the blade surfaces, the velocity potential must satisfy the flow tangency condition,

$$\nabla\tilde{\Phi}\cdot\vec{n} = 0\,,\ \vec{X}\in B_m \tag{4}$$

where $\vec{n}$ is a unit normal vector to the airfoil surface, $\vec{x}$ is a position vector, and B_m refers to the mth blade surface.

For the steady flows being considered here, the inlet and exit velocities are subsonic. The inlet and exit boundary conditions require that three of the four uniform freestream components (e.g $V_{\pm\infty}$ or $\Omega_{\pm\infty}$) be specified. The fourth is determined so that the global form of the mass conservation is satisfied. In addition, conditions can be imposed at the blade edges [e.g., a zero load condition at a sharp trailing edge (Kutta condition)]. In this case, the flow at the trailing edge of the airfoil must satisfy the following relationship,

$$\left|\frac{\partial\tilde{\Phi}^{+}_{T.E.}}{\partial S^{+}}\right| = \left|\frac{\partial\tilde{\Phi}^{-}_{T.E.}}{\partial S^{-}}\right| \tag{5}$$

where $S^{\pm}$ is the arc distance along the upper (+), and lower (-) surface of the airfoil.

Since the airfoils in this cascade are assumed to be identical, and the inlet flow is uniform, the following blade-to-blade periodicity condition can be applied along the upstream periodic boundary,

$$\tilde{\Phi}(\xi,\eta+G) - \tilde{\Phi}(\xi,\eta) = V_{-\infty}\sin\Omega_{-\infty}G \tag{6}$$

and,along the downstream periodic boundary

$$\tilde{\Phi}(\xi,\eta+G) - \tilde{\Phi}(\xi,\eta) = V_{+\infty}\sin\Omega_{+\infty}G \tag{7}$$

The location of these periodic boundaries are arbitrary, however in this model the upstream periodic boundary is chosen so that it is parallel to the inlet velocity vector, and intersects the airfoil at the leading edge point. The downstream periodic boundary is arbitrarily chosen. Upon completion of the steady flow calculation, the downstream stagnation streamline is determined since this will be required for the unsteady analysis.

These boundary conditions allow the steady flow to be calculated in a region represented by a single extended flow passage. This region is defined by the upstream and downstream periodic boundaries, the upstream and downstream far-field boundaries, the upper surface of reference airfoil (m = 0), and the lower surface of the airfoil defined by m = 1.

Newton's Method

Rewriting the velocity potential as,

$$\tilde{\Phi}(\vec{X}) = \Phi(\vec{X}) + \phi(\vec{X}) \tag{8}$$

where Φ is an approximation to the solution $\tilde{\Phi}$, $\phi = \tilde{\Phi} - \Phi$ is a correction term, and $\vec{x}$ is a position vector. After substituting Eq. 8 into Eq. 3 and performing the necessary algebra we find that,

$$\begin{aligned} A^2\nabla^2\phi - (\gamma-1)\nabla^2\Phi\frac{\bar{D}\phi}{Dt} - \frac{\bar{D}^2\phi}{Dt^2} - \nabla\phi\cdot\frac{\nabla(\nabla\Phi)^2}{2} \\ = -A^2\nabla^2\Phi + \nabla\Phi\cdot\frac{\nabla(\nabla\Phi)^2}{2} \end{aligned} \tag{9}$$

where the material derivative operator is defined as,

$$\frac{\bar{D}}{Dt} = \nabla\Phi\cdot\nabla \tag{10}$$

and,

$$A^2 = 1 - \left(\frac{\gamma-1}{2}\right)M_{-\infty}^2[(\nabla\Phi)^2 - 1] \tag{11}$$

Eq. 9 can be written in the form,

$$L(\Phi)\phi = N(\Phi) \tag{12}$$

where L is a linear differential operator which depends on the steady velocity potential, Φ, and N is a nonlinear differential operator that also depends on Φ. The initial distribution for Φ is some estimate of the converged solution, e.g. $\Phi^0 = V_{-\infty}\xi$. The solution to Eq. 9 than proceeds as follows. At the nth Newton iteration Eq. 9 is solved for the correction ϕ^n using the estimate of the converged solution Φ^n. The $(n+1)^{th}$ i.e. Φ^{n+1}, approximation to $\tilde{\Phi}$ becomes $\Phi^{n+1} = \Phi^n + \phi^n$. This process is continued until $|\phi^n| < \epsilon_{max}$, where ϵ_{max} is some preselected convergence tolerance.

In addition to rewriting the field equation into a form that will facilitate the application of Newton's method, it is also necessary to convert the boundary conditions to this form. This is accomplished by substituting Eq.8 into the boundary conditions Eqs.5 through 8 and performing the necessary algebra. The resulting expressions are:

Flow tangency:

$$\nabla\phi\cdot\vec{n} = -\nabla\Phi\cdot\vec{n} \tag{13}$$

Periodicity:

$$\phi(\xi,\eta+G) - \phi(\xi,\eta) = V_{\pm\infty}\sin\Omega_{\pm\infty}G - [\Phi(\xi,\eta+G) - \Phi(\xi,\eta)] \tag{14}$$

Kutta condition:

$$\left|\frac{\partial\phi^+_{T.E.}}{\partial s^+}\right| - \left|\frac{\partial\phi^-_{T.E.}}{\partial s^-}\right| = -\left(\left|\frac{\partial\Phi^+_{T.E.}}{\partial s^+}\right| - \left|\frac{\partial\Phi^-_{T.E.}}{\partial s^-}\right|\right) \tag{15}$$

The differential equation for the linearized unsteady potential is,

$$A^2\nabla^2\phi - (\gamma-1)\nabla^2\Phi\frac{\bar{D}\phi}{Dt} - \frac{\bar{D}^2\phi}{Dt^2} - \nabla\phi\cdot\frac{\nabla(\nabla\Phi)^2}{2} = 0 \tag{16}$$

where,

$$\frac{\bar{D}}{Dt} = i\omega + \nabla\Phi \cdot \nabla \tag{17}$$

A detailed derivation of the this equation is given by Verdon and Caspar (1982,1984). The only difference between the left hand sides of Eqs. 9 and 16 is that the unsteady potential, ϕ, is a complex value. If the blade oscillation frequency, ω, is set to zero in Eq.17, the left hand sides of the two expressions are identical. The advantages of using the forms for the field equation and boundary conditions given by Eq. 9 and Eqs. 13-15 respectively, is that the differencing schemes and matrix inversion procedure used in the numerical algorithm for the linearized unsteady potential flow can, with little modification, be used to determine the steady potential flow field. However, since the steady problem is nonlinear an iterative solution, such as that outlined above, is needed. Once the steady flow has been determined, the linearized unsteady potential equation can be solved for the unsteady velocity potential, and ultimately for the unsteady pressures and global unsteady airloads.

The difference approximations to the linear operator in Eq. 9 are obtained using an implicit, least-squares, interpolation procedure. This leads to a nine point "centered" difference stencil at field points within the solution domain, and a nine point one sided difference stencil at blade surface points. A detailed description of this differencing approximation can be found in Caspar & Verdon (1981).

Composite Calculation Mesh

Following Usab and Verdon (1989), the difference approximations are applied on a computational mesh consisting of a sheared H type mesh which will capture aerodynamic phenomena over the extended blade-passage solution domain, and a local surface-fitted C-mesh which will resolve high gradient phenomena in the region of blunt leading edges. Details of this composite mesh construction can be found in Usab & Verdon (1989), while details on the construction of the global and local meshes can be found in Caspar &

Verdon (1981). Only a brief outline of the composite mesh procedure will be given here.

The composite mesh used in this model is constructed by overlaying the local surface fitted mesh over the global cascade mesh as shown in Figure 2. Points on the composite mesh can be identified as cascade-mesh solution points, local-mesh solution points, cascade-mesh coupling points, and local-mesh coupling points depending on where they lie with respect to a user defined overlap zone. The overlap region is defined in terms of the local mesh by stepping in n cells from the outer boundary of the local mesh. Composite mesh points are identified as cascade-mesh solution points if they lie outside the interior boundary of the overlap zone, and cascade-mesh coupling points if they are needed to complete the differencing stencil for cascade-mesh solution points lying within the overlap zone. Local-mesh solution points are those which lie inside the outer boundary of the local mesh. While local-mesh coupling points are those that lie on the outer boundary of the local mesh.

The inclusion of these coupling points into the system of discrete equations destroys the block tri-diagonal structure that exists when the H-grid and C-grid are considered separately. Although each coupling equation involves points that are spatially close to the coupling point under consideration, the former are not necessarily neighbors in the composite equation system. Therefore, the final system of equations contains a sparse coefficient matrix of large band width. Consequently, special storage and inversion techniques are required to achieve an efficient solution. The sparse matrix solver used is that developed by Eisenstat, et al. (1977).

Model Verification

The mathematical model for the steady flow calculation was verified using two cascade configurations. Composite-mesh steady flow solutions were determined for both cascades. In both cases converged solutions were obtained in six or less iterations. Once the steady calculations were obtained, the linearized unsteady aerodynamic model, LINFLO, developed by Verdon and Caspar (1982,1984) was used to calculate the unsteady response of the cascade to a prescribed blade motion.

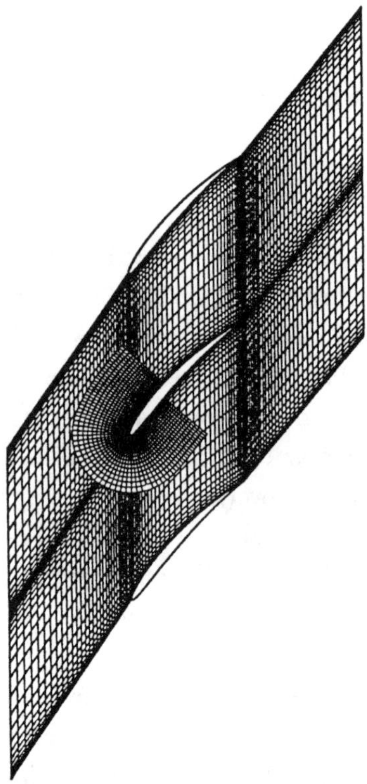

Figure 2 Composite mesh for NACA 0006 compressor cascade.

NACA 0006

The first geometry chosen was one studied by Usab & Verdon (1990). The blades are constructed by superposing the thickness distribution of a modified NACA four digit series airfoil on a circular-arc camber line. The cascade geometry has a solidity of unity (i.e. G = 1.0), and a stagger angle, Θ, of 45 degrees. The inlet Mach number, $M_{-\infty}$, was set at 0.7, and the inlet flow angle, $\Omega_{-\infty}$, was 55 degrees. The steady flow is assumed to satisfy a Kutta condition

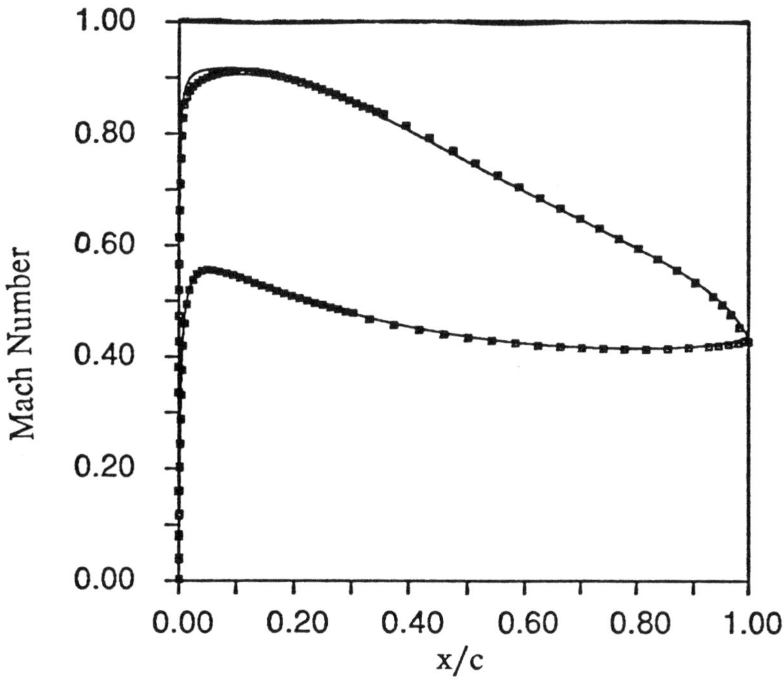

Figure 3 Steady surface Mach number distribution for NACA 0006 compressor cascade, $M_{-\infty} = 0.7$, $\Omega_{-\infty} = 45$ Deg; ——— SFLOW; □ CASPOF.

at the trailing edge, therefore only the inlet Mach number and flow angle were specified.

The composite mesh used to calculate the steady flow through this cascade geometry is shown in Figure 2. It is made up of a global H-mesh consisting of 96 axial lines and 21 tangential mesh lines. The local C-mesh consists of 71 radial lines and 21 circumferential lines. The predicted blade surface Mach number distribution for this mesh is shown in Figure 3. Also shown is a prediction obtained from the steady velocity potential model, CASPOF, developed by Caspar, et al. (1980). It should be note that CASPOF also uses both a global H-mesh and a local C-mesh; however, the two meshes do not overlap and the governing equations are not solved simultaneously on both meshes as it is in the present analysis. Instead, (Caspar et al. 1980) the H-mesh was used to obtain a global solution which then provided the outer boundary condition for the C-mesh calculation. The exit Mach number, $M_{+\infty}$, and flow angle, $\Omega_{+\infty}$, calculated by the present method were found to be 0.447, and 40.3 degrees respectively. Those obtained from CASPOF were $M_{+\infty} = 0.446$, and $\Omega_{+\infty} = 40.2$ deg. Referring to Figure 3 it can be seen that there is excellent agreement between the two predictions. There are however some small differences in the leading and trailing edge regions. These differences are attributed to the differences in the meshes, and the solution techniques used in both models.

Having obtained a steady solution, an unsteady calculation was made to demonstrate the coupling to the unsteady flow solver. The LINFLO code was used to calculate the unsteady blade surface pressures resulting from a harmonic torsion mode oscillation of the blades. The reduced frequency of the blade motion based on chord is unity and the elastic axis is placed at midchord. Motions at two interblade phase angles are considered such that one corresponds to a subresonant motion, and the other to a superresonant motion. Figure 4 shows the real and imaginary components of the predicted unsteady pressure as a function of blade chord for the subresonant, $\sigma = 180$ deg, and superresonant, $\sigma = 30$ deg motions. Also shown is the prediction from LINFLO using the steady results obtained from CASPOF. As can be seen the two predictions agree quite well, the small differences in the leading and trailing edge regions are attributed to the noted differences in the two steady predictions.

Turbine Cascade

The second test case selected for the verification of the steady flow solver is a turbine cascade, the fourth standard configuration reported by Bölcs and Fransson (1986). In the present study, the airfoil geometry was modified so that the profile closed in a wedge shaped trailing edge. This modification was done in such a manner that the chord of the blade was not changed.The cascade geometry consists of a set of airfoils which have a stagger angle, Θ, of 56.6 deg, and a blade spacing of, $G = 0.76$. Once the steady flow solution is obtained the unsteady aerodynamic response at the blade surface resulting

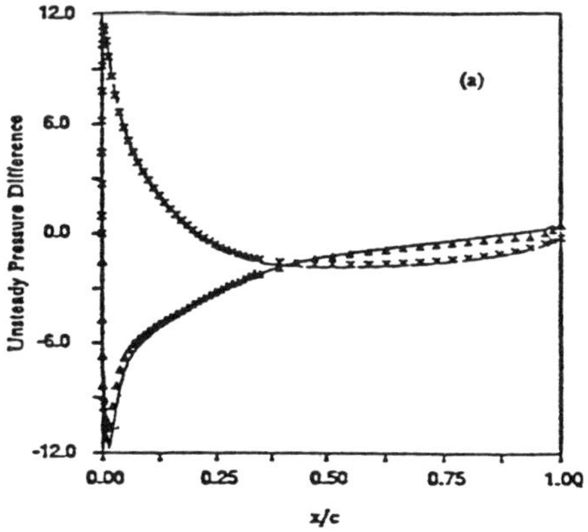

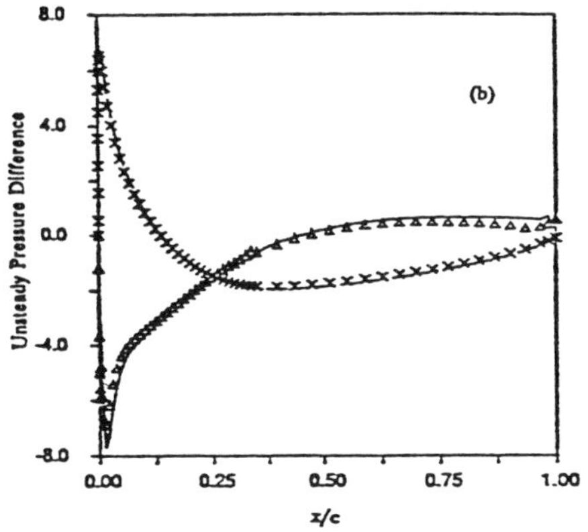

Figure 4 Unsteady aerodynamic response to torsion mode oscillation of the NACA 0006 cascade, α = (1,0), ω = 1; ——— in phase (real) component of ΔP, ----- out of phase (imaginary) component of ΔP, SFLOW & LINFLO calculation; Δ in phase (real) component of ΔP, x out of phase (imaginary) component of ΔP, CASPOF & LINFLO calculation; (a) σ = 180 deg: (b) σ = 30 deg.

from a prescribed blade motion can be determined using the LINFLO code.

Bōlcs and Fransson (1986), and He (1989), both report that in order to match the experimentally determined exit conditions, the effect of streamtube height variation needs to be included in the steady flow calculations. He (1989) used a linear variation with an exit height to inlet height ratio of 1.1. Since the steady flow model used is this report does not presently have this capability, the steady flow calculations were performed by matching the experimentally determined exit Mach number, M_{+-} = 0.90. A comparison of

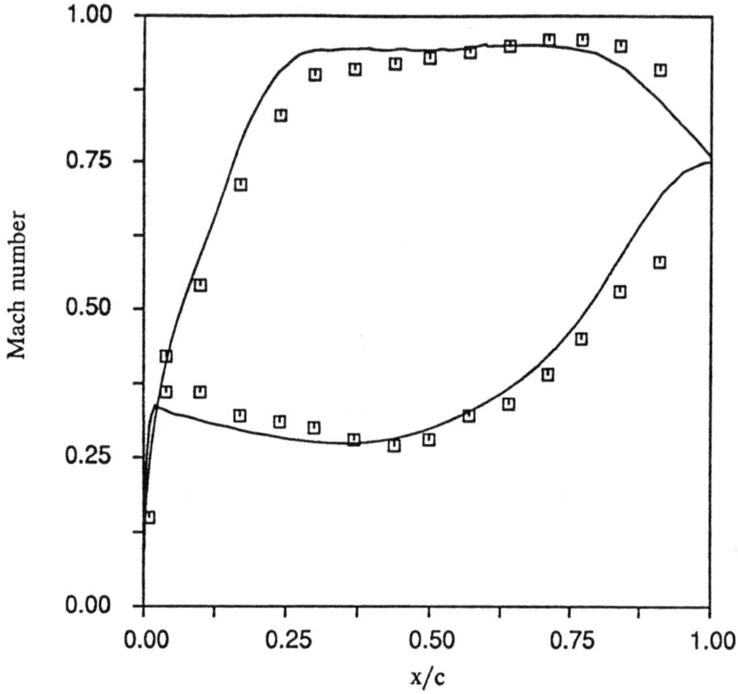

Figure 5 Steady surface Mach number distribution for turbine cascade, $M_{-\infty}$ = 0.26, $\Omega_{-\infty}$ = 45 deg; ——— SFLOW; □ measurement.

the predicted blade surface Mach number distribution with the experimental data for this exit condition is shown in Figure 5. The inlet Mach number was determined to be 0.26, the inlet flow angle 45 deg, and the exit flow angle 71.8 deg. The experimentally determined exit flow angle was 71.0 deg. Referring to Figure 5 it can be seen that the agreement between the prediction and data is quite good from the leading edge to about 85% chord. The discrepancies over the last 15% of the blade can probably be related to the blade modification, and the constant streamtube height used in the calculation.

Unsteady response predictions for a translational motion with displacement amplitudes of h_x = 0.0016 and h_y = 0.0029 in the x and y directions, a reduced frequency based on chord of 0.24, and oscillating at an interblade phase angle, σ, of -90 deg. is shown in Figure 6. This figure shows the magnitude and

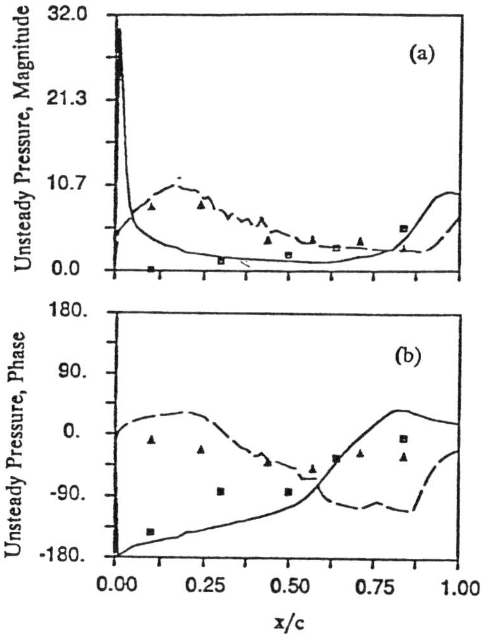

Figure 6 Unsteady aerodynamic response to bending displacement of turbine cascade: h = (0.0016,0.0029), ω = 0.24, σ = -90 Deg; ----- suction surface SFLOW & LINFLO, ▲ measurement; ——— pressure surface SFLOW & LINFLO, □ measurement: (a) magnitude, (b) phase.

phase of the unsteady pressure on the suction and pressure surfaces of the reference airfoil as a function of chordwise distance. It can be seen that there is reasonable agreement between the prediction and experiment for the magnitude of the unsteady response on the suction surface. The pressure surface comparison shows good agreement except in the leading edge region. The comparison of the phase information shows that the experimental trend on the pressure surface is well predicted while that on the suction surface is not predicted in the trailing edge region. The oscillations observed in the predictions can be related to the blade surface definition.

Figure 8 shows a plot of the aerodynamic damping, Ξ, versus interblade phase angle. The calculated values of aerodynamic damping are in reasonable agreement with the measured values for σ = 0 deg and σ = 180 deg, but wide discrepancies exist for σ = ±90 deg. These discrepancies are probably

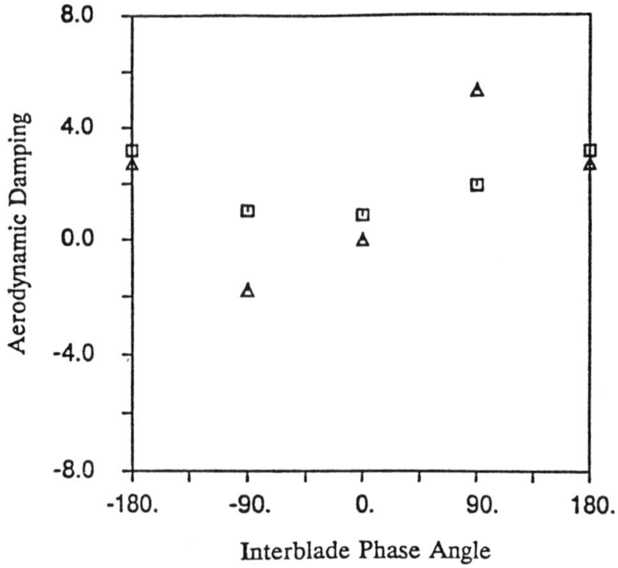

Figure 7 Aerodynamic damping vs. interblade phase angle; □ prediction, ▵ measurement.

related to the noted differences between the predicted and measured steady behavior in the trailing egde region of the blade.

Concluding Remarks

The steady potential analysis described in this paper has resulted in two improvements in the ability to calculate the steady potential field of a two dimensional cascade. The first is the implementation of a composite mesh solution procedure which allows detailed flow information to be obtained in the vicinity of blunt leading edges. The second is the use of Newton's method for the solution of a nonlinear equation. This results in a numerical procedure that converges rapidly. The coupling of this solver to an existing linearized unsteady solver results is a computationally efficient model which can be used to study the aeroelastic and aeroacoustic behavior of advanced turbomachinery configurations.

The steady portion of the model has been verified by comparison with other predictions for a compressor cascade, and with experimental data for a turbine cascade. The results have been used as input data for the linearized unsteady analysis. In the case of the compressor geometry the steady calculation agrees quite well with predictions made using another steady flow solver. The steady predictions for the turbine geometry showed good agreement with the experimental results of Bölcs & Fransson (1986) despite the fact that no streamtube height variation was assumed in the steady solver. In general the unsteady predictions for the turbine are in good agreement with the data for both the magnitude and phase of the unsteady pressure response on the pressure surface of the airfoil. However, although the suction surface predictions showed reasonable agreement for the magnitude of the unsteady pressure, the predictions and measurements for the phase are in poor agreement. The aerodynamic damping prediction were in agreement with the measurements for interblade phase angles of ± 180 degs, but wide discrepancies exists for $\sigma = \pm 90$ deg.

References

Atassi, H., and Akai, T.J., 1980, "Aerodynamic and aeroelastic Characteristics of Oscillating Loaded Cascades at Low Mach Number. I Pressure Distribution, Forces and Moments," *Transactions of American Society of Mechanical Engineering, Journal of Engineering for Power*, Vol 102,No. 2, pp. 344-351.

Bailey, H.E. and Beam, R.M., 1991, "Newton's Method Applied to Finite-Difference Approximations for the Steady-State Compressible Navier-Stokes Equations", *Journal of Computational Physics*, Vol 93., pp 108-127.

Bender, E.E., and Khosla, P.K., 1988, "Application of Sparse Matrix Solvers and Newton's Method to Fluid Flow Problems", AIAA paper number 88-3700-CP.

Bölcs, A., and Fransson, T.H., 1986, "Aeroelasticity in Turbomachines - Comparison of Theoretical and Experimential Cascade Results", Communication du Laboratorie de Thermique Appliquèe et de Turbomachines, No. 13, Lausanne, EPFL.

Caruthers, J.E., and Dalton, W.N., 1991, "Unsteady Aerodynamic Response of a Cascade to Nonuniform Inflow," ASME Paper 91-GT-174, Presented at the International Gas Turbine and Aeroengine Congress and Exposition, Orlando, Florida, June 3-6, 1991.

Caspar, J.R., Hobbs, D.E., and Davis, R.L., 1980, "Calculation of Two-Dimensional Potential Cascade Flow Using Finite Area Methods," *AIAA Journal*, Vol. 18, pp. 103-109.

Caspar, J.R., and Verdon, J.M., 1981, "Numerical Treatment of Unsteady Subsonic Flow Past an Oscillating Cascade," *AIAA Journal*, Vol. 19, pp 1531-1539.

Eisenstat, S.C., Grusky, M.C., Schultz, M.H., and Sherman, A.H., 1977, "The Yale Sparse Matrix Package. II. The Non-symmetric Codes," Research Report No. 114, Yale University, Department of Computer Science, New Haven, CT.

Giles, M., 1985, "Newton Solution of Steady Two-Dimensional Transonic flows", PhD thesis, Massachusetts Institute of Technology, Cambridge, Massachusetts.

Hall, K.C., and Clark W.S., 1991, "Prediction of Unsteady Aerodynamic Loads in Cascades Using Time Linearized Euler Equations on Deforming Grids," AIAA Paper 91-3378, Presented at the 27th AIAA Joint Propulsion Conference, 1991, Sacramento Calf.

Hall, K.C., and Crawley, E.F., 1989, "Calculations of Unsteady Flows in Turbomachinery Using The Linearized Euler Equations," *AIAA Journal*, Vol. 27, No. 6, pp 777-787.

Verdon, J.M., and Caspar, J.R., 1982, "Development of a Linear Unsteady Aerodynamic Analysis for Finite-Deflection Subsonic Cascades," *AIAA Journal*, Vol. 20, pp 1259-1267.

Verdon, J.M., and Caspar, J.R., 1984, "A Linearized Unsteady Aerodynamic Analysis for Transonic Cascades," *Journal of Fluid Mechanics*, Vol. 149, pp 403-429.

Whitehead, D.S., 1982, "The Calculation of Steady and Unsteady Transonic Flows in Cascades," Cambridge University Engineering Dept. Rept. CUED/A-Turbo/TR 118, Cambridge, UK.

Usab, W.J., and, Verdon, J.M., 1989, "Advances in the Numerical Analysis of Linearized Unsteady Cascades Flows," ASME Paper 90_GT-11, Presented at the International Gas Turbine and Aeroengine Congress and Exposition, Brussels, Belgium.

Calculation of Unsteady Linearized Euler Flows in Cascades Using Harmonically Deforming Grids

Kenneth C. Hall[1]
William S. Clark[2]

ABSTRACT A method for calculating unsteady, inviscid, compressible flows in cascades is presented. Using the linearized Euler technique, the flow is decomposed into a steady or mean flow plus a harmonically varying small disturbance flow. The equations that describe the small disturbance flow are linear variable coefficient equations, and are solved using a pseudo-time time marching Lax-Wendroff technique. Unlike previous linearized methods, however, the solution is computed on a harmonically deforming computational grid that conforms to the motion of the vibrating airfoils. The mean flow and perturbation flow solutions are defined in the deforming coordinate system rather than in a coordinate system fixed in space. Hence, no extrapolation terms are required to implement the upwash boundary conditions at the airfoil surfaces significantly improving the accuracy of the method. For transonic flow calculations, unsteady shock motions are modelled using shock capturing. The unsteady loads due to the shock motion are then seen as pressure impulses. Representative computational results are presented for transonic channel flows and subsonic and transonic cascade flows.

1.1 Introduction

To understand and predict flutter and forced response in turbomachinery blade rows, aeroelasticians require accurate and efficient models of the unsteady flow fields that arise from blade motion and incident gusts. Over the past two decades, many computational methods have been developed

---

[1]Assistant Professor, Department of Mechanical Engineering and Materials Science, Duke University, Durham, NC 27706.
[2]Research Assistant, Department of Mechanical Engineering and Materials Science, Duke University, Durham, NC 27706.

to analyze unsteady flows in turbomachinery. For example, the linearized potential methods have been developed extensively [WG81, VC82, VC84, VU88, HV91]. Linearized potential analyses are computationally very efficient, and for aeroelastic applications, the small disturbance assumption is not a severe restriction. However, because of the inherent assumption of irrotational and isentropic flow, the linearized potential analyses are only valid for subsonic and low transonic speeds and flows without inlet swirl. To model the more complex flow fields found in transonic compressors and turbines, and as computer speeds and storage have improved, several time-accurate time-marching Euler [Hod84, FLM87, Gil88, HSR91] and Navier-Stokes [Rai89a, Rai89b, GH91] codes have been developed. These codes, unlike the linearized potential codes, account for both steady and unsteady vorticity and entropy. However, despite the recent advances in computer hardware, the time-marching approach is still computationally too expensive for routine design use.

Recently, several investigators have begun to develop linearized Euler analyses [NS76, HC89, HC91, LG91]. As in the linearized potential technique, the unsteady flow is assumed to be composed of a nonlinear mean flow plus a harmonically varying small perturbation flow. The unsteady perturbation flow is governed by the linearized Euler equations. Using this model of the flow accounts for entropy and vorticity generation and shocks, and permits analysis of unsteady flows about rotational nonisentropic mean flows. Furthermore, the solution of the linearized Euler equations, while not as efficient as solution of the linearized potential equations, is one to two orders of magnitude more efficient than computing the full nonlinear flow field while still retaining the essential physics of unsteady flows associated with forced response and the onset of flutter.

In this paper, a recently developed Euler analysis is presented. The linearized Euler equations are discretized and solved using a conservative pseudo-time time marching Lax-Wendroff scheme [NS76, Ni82, Dan87, HC91]. Two important features of the present analysis are the use of a deforming computational grid for flutter calculations, and shock capturing for transonic flow calculations. For flutter calculations, the use of a deforming computational grid eliminates large error producing gradient terms that would otherwise appear in the upwash boundary conditions and in the evaluation of the unsteady pressure on the airfoil surfaces. For transonic flow calculations, shock capturing is used to model the pressure impulse resulting from small harmonic motion of the shock. Shock capturing is considerably easier to implement than shock fitting. However, to predict correctly the unsteady loads due to the shock impulse, one must use a discretization scheme for the unsteady linearized flow solver that is a faithful linearization of the nonlinear flow solver [LG91]. Representative computational results are presented that demonstrate the capability of the method to predict unsteady subsonic and transonic flows.

1.2 Theory

1.2.1 The Linearized Euler Equations

For compressible flows in which viscous forces may be neglected, the unsteady flow is governed by the Euler equations representing the conservation of mass, momentum, and energy, i.e,

$$\frac{\partial \hat{\mathbf{U}}}{\partial t} + \frac{\partial \hat{\mathbf{F}}}{\partial x} + \frac{\partial \hat{\mathbf{G}}}{\partial y} = 0 \tag{1.1}$$

Here, $\hat{\mathbf{U}}$ is the vector of conservation variables and $\hat{\mathbf{F}}$ and $\hat{\mathbf{G}}$ are the so-called flux vectors. These are given by

$$\hat{\mathbf{U}} = \begin{bmatrix} \hat{\rho} \\ \hat{\rho}\hat{u} \\ \hat{\rho}\hat{v} \\ \hat{e} \end{bmatrix}, \quad \hat{\mathbf{F}} = \begin{bmatrix} \hat{\rho}\hat{u} \\ \hat{\rho}\hat{u}^2 + \hat{p} \\ \hat{\rho}\hat{u}\hat{v} \\ \hat{\rho}\hat{u}\hat{h}_0 \end{bmatrix}, \quad \hat{\mathbf{G}} = \begin{bmatrix} \hat{\rho}\hat{v} \\ \hat{\rho}\hat{u}\hat{v} \\ \hat{\rho}\hat{v}^2 + \hat{p} \\ \hat{\rho}\hat{v}\hat{h}_0 \end{bmatrix}$$

Where $\hat{\rho}$ is the static density, $\hat{u}$ and $\hat{v}$ are the x- and y-components of the velocity, $\hat{p}$ is the pressure, $\hat{e}$ is the internal energy, and $\hat{h}_0$ is the total enthalpy.

Most previous linearized analyses have used computational grids fixed in space. To apply the upwash boundary condition at the airfoil surface, the boundary conditions must be extrapolated from the instantaneous position of the airfoil to the mean airfoil position, that is, to the boundary of the fixed grid. The additional extrapolation terms appearing in the upwash boundary condition contain velocity gradient terms that can be large and difficult to evaluate accurately, especially near the leading and trailing edges of the airfoils. Whitehead and Grant [WG81] overcame this difficulty for rigid-body airfoil motion by using a transformed velocity potential that can be viewed as equivalent to using a computational grid that undergoes rigid-body motion. Hall [Hal92] has recently developed a linearized potential analysis based on a variational principle that includes the effects of a grid undergoing arbitrary deformations thus allowing both rigid-body and flexible blade motion.

In the present linearized Euler analysis, we also make use of a deforming grid. The physical coordinates x, y, t are related to the computational coordinates ξ, η, τ by the transformation

$$x(\xi,\eta,\tau) = \xi + f(\xi,\eta,\tau) \tag{1.2}$$
$$y(\xi,\eta,\tau) = \eta + g(\xi,\eta,\tau) \tag{1.3}$$
$$t(\xi,\eta,\tau) = \tau \tag{1.4}$$

The grid motion functions f and g are chosen so that the motion of the grid conforms to the motion of the airfoils. A typical example of unsteady

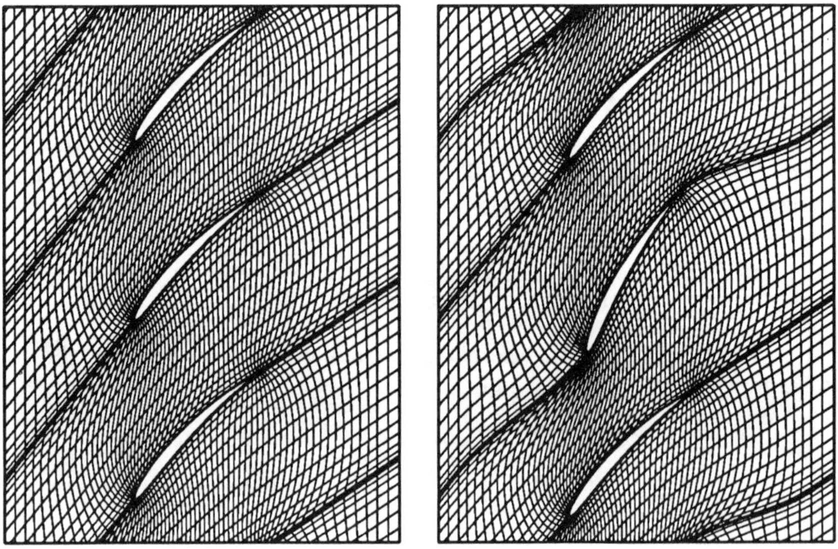

FIGURE 1.1. Left: Grid in computational coordinate system, ξ, η. Right: Grid in physical coordinate system, x,y, for the case of a cascade of airfoils plunging with an interblade phase angle, σ, of 90 deg. The cascade shown is Standard Configuration No. 10.

grid motion is shown in Fig. 1.1. Shown is a cascade of airfoils designated Standard Configuration No. 10 [Fra91] vibrating in pitch with an interblade phase angle, σ, of 90 deg. In the computational coordinate system, the airfoils and the grid appear to be stationary; in the physical domain, the airfoils and grid deform unsteadily.

Integrating the differential form of the Euler equations [Eq. (1.1)] over a control volume D that is fixed in the computational coordinate system – but deforming in the physical coordinate system – and applying the divergence theorem gives the Euler equations in integral form, i.e.,

$$\iint_D \frac{\partial}{\partial t}\hat{\mathbf{U}}\, dx\, dy + \oint_{\partial D} \left(\hat{\mathbf{F}}\, dy - \hat{\mathbf{G}}\, dx\right) = 0 \tag{1.5}$$

With some manipulation, Eq. (1.5) can be expressed as

$$\frac{\partial}{\partial \tau}\iint_D \hat{\mathbf{U}}\, dx\, dy + \oint_{\partial D}\left[\left(\hat{\mathbf{F}} - \hat{\mathbf{U}}\frac{\partial f}{\partial \tau}\right)\frac{\partial y}{\partial \eta} - \left(\hat{\mathbf{G}} - \hat{\mathbf{U}}\frac{\partial g}{\partial \tau}\right)\frac{\partial x}{\partial \eta}\right] d\eta$$
$$+ \oint_{\partial D}\left[\left(\hat{\mathbf{F}} - \hat{\mathbf{U}}\frac{\partial f}{\partial \tau}\right)\frac{\partial y}{\partial \xi} - \left(\hat{\mathbf{G}} - \hat{\mathbf{U}}\frac{\partial g}{\partial \tau}\right)\frac{\partial x}{\partial \xi}\right] d\xi = 0 \tag{1.6}$$

Equation (1.6) is valid for both small and large disturbance unsteady flows. Large amplitude flows, which exhibit nonlinear behavior, could be computed by time marching Eq. (1.6). However, such computations are very expensive. Fortunately, for many flows of interest in aeroelastic applications, the unsteadiness in the flow is temporally harmonic and small compared with the mean flow. Therefore, to simplify the problem and also reduce the required computational time, we assume that the flow is decomposed into a zeroth-order mean or steady flow plus a forst-order unsteady perturbation flow. Hence,

$$\hat{\mathbf{U}}(\xi,\eta,\tau) = \mathbf{U}(\xi,\eta) + \mathbf{u}(\xi,\eta)e^{j\omega\tau} \tag{1.7}$$

Here, $\mathbf{U}$ is the vector of conservation variables representing the mean flow field, and $\mathbf{u}$ is the vector of small perturbation harmonic amplitudes of the conservation variables. The motion of the airfoils, and hence the grid, is assumed to be small and harmonic so that Eqs. (1.2)-(1.4) are replaced by

$$x(\xi,\eta,\tau) = \xi + f(\xi,\eta)e^{j\omega\tau} \tag{1.8}$$

$$y(\xi,\eta,\tau) = \eta + g(\xi,\eta)e^{j\omega\tau} \tag{1.9}$$

$$t(\xi,\eta,\tau) = \tau \tag{1.10}$$

where now f and g represent the small perturbation harmonic amplitudes of the grid motion.

Substitution of the perturbation assumptions into the nonlinear Euler equations, Eq. (1.6), and collection of terms of zeroth and first-order results in the mean flow Euler equations and the linearized Euler equations respectively. The mean flow Euler equations in integral form are given by

$$\oint_{\partial D} (\mathbf{F}\,d\eta - \mathbf{G}\,d\xi) = 0 \tag{1.11}$$

The linearized Euler equations are given by

$$\begin{aligned} j\omega \iint_D \mathbf{u}\,d\xi\,d\eta + \oint_{\partial D} \left(\frac{\partial \mathbf{F}}{\partial \mathbf{U}}\mathbf{u}\,d\eta - \frac{\partial \mathbf{G}}{\partial \mathbf{U}}\mathbf{u}\,d\xi \right) = \\ - j\omega \iint_D \mathbf{U}\,(df\,d\eta + d\xi\,dg) \;+\; j\omega \oint_{\partial D} (f\mathbf{U}\,d\eta - g\mathbf{U}\,d\xi) \\ - \oint_{\partial D} (\mathbf{F}\,dg - \mathbf{G}\,df) \end{aligned} \tag{1.12}$$

The Jacobian matrices $\partial\mathbf{F}/\partial\mathbf{U}$ and $\partial\mathbf{G}/\partial\mathbf{U}$ appearing in Eq. (1.12) are evaluated using the mean flow solution. For example, for an ideal gas with constant specific heats, one finds that the Jacobian $\partial\mathbf{F}/\partial\mathbf{U}$ is given by

$$\frac{\partial \mathbf{F}}{\partial \mathbf{U}} = \begin{bmatrix} 0 & 1 & 0 & 0 \\ \frac{\gamma-3}{2}U^2 + \frac{\gamma-1}{2}V^2 & -(\gamma-3)U & -(\gamma-1)V & \gamma-1 \\ -UV & V & U & 0 \\ U\left(\frac{\gamma-1}{2}{V_T}^2 - h_0\right) & h_0 - (\gamma-1)U^2 & -(\gamma-1)UV & \gamma U \end{bmatrix} \tag{1.13}$$

where $V_T^2 = U^2 + V^2$.

The mean flow [described by (1.11)] is independent of the unsteady perturbation flow. The linearized Euler equations, Eq. (1.12), are linear, inhomogeneous, variable coefficient equations. The variable coefficients, i.e., the Jacobians, are functions of the mean flow field. The inhomogeneous portion of Eq. (1.12), that is, the right-hand side, is a function of the mean flow and the prescribed grid motion.

1.2.2 NEAR-FIELD BOUNDARY CONDITIONS

Periodicity

Because the unsteady flow is governed by linear equations, the response of the cascade to blade motion or a gust may be found by decomposing the disturbance into a sum of travelling wave modes each with a different frequency, ω, and interblade phase angle, σ. The total response of the cascade to the sum of modes is equal to the sum of the responses to each of the individual modes. Without loss of generality, we consider here a single travelling wave mode. The complex periodicity condition upstream and down stream of the cascade is then

$$\mathbf{u}(\xi, \eta + G) = \mathbf{u}(\xi, \eta)e^{j\sigma} \tag{1.14}$$

where G is the blade-to-blade gap. This periodicity condition allows the computational domain to be reduced to a single blade passage.

Flow Tangency

The upwash boundary condition on the airfoil surface is

$$\mathbf{v} \cdot \mathbf{n} = \left(j\omega\mathbf{r} + \mathbf{V} \cdot \tau \frac{\partial \mathbf{r}}{\partial s} \right) \cdot \mathbf{n} \tag{1.15}$$

where $\mathbf{V}$ and $\mathbf{v}$ are the mean perturbation flow velocities, $\mathbf{n}$ and τ are unit vectors normal to and tangent to the mean airfoil surface, s is the arc length along the airfoil surface, and $\mathbf{r}$ is the perturbation displacement of the airfoil surface. The two terms on the right-hand side represent the upwash due to the local translation and rotation of the airfoil surface. Note that because we use a deforming computational grid, no extrapolation terms are required.

1.2.3 NUMERICAL INTEGRATION

To integrate numerically the linearized Euler equations, we use the pseudo-time time marching approach suggested by Ni and Sisto [NS76]. The unsteady perturbation amplitude $\mathbf{u}$ is assumed to vary with time so that

Eq. (1.7) is replaced by

$$\hat{\mathbf{U}}(\xi,\eta,\tau) = \mathbf{U}(\xi,\eta) + \mathbf{u}(\xi,\eta,\tau)e^{j\omega\tau} \tag{1.16}$$

Substitution of Eq. (1.16) into Eq.(1.6) and collection of terms of equal order results in the pseudo time-dependent linearized Euler equations,

$$\frac{\partial}{\partial\tau}\iint_D \mathbf{u}\,d\xi\,d\eta + j\omega\iint_D \mathbf{u}\,d\xi\,d\eta + \oint_{\partial D}\left(\frac{\partial\mathbf{F}}{\partial\mathbf{U}}\mathbf{u}\,d\eta - \frac{\partial\mathbf{G}}{\partial\mathbf{U}}\mathbf{u}\,d\xi\right) = \mathbf{b} \tag{1.17}$$

where $\mathbf{b}$ is the same as the right-hand side of the original (time-invariant) linearized Euler equation, Eq. (1.12). Note that the first term on the left-hand side is the only term that does not appear in the original linearized Euler equation. This pseudo-time dependent term is driven to zero by time marching the equation to steady state.

Note the similarity of the linearized Euler equations, Eq. (1.17), to the nonlinear time-dependent Euler equations, Eq. (1.5). One advantage of the pseudo-time time-marching approach is that several nonlinear Euler solvers have been developed that can be applied with little modification to the linearized Euler equations. For example, the conservation variables $\hat{\mathbf{U}}$ in the nonlinear Euler solver are replaced by the perturbation conservation variables $\mathbf{u}$ in the linearized Euler solver. The nonlinear flux vectors $\hat{\mathbf{F}}$ and $\hat{\mathbf{G}}$ are replaced by the perturbation flux vectors $(\partial\mathbf{F}/\partial\mathbf{U})\mathbf{u}$ and $(\partial\mathbf{G}/\partial\mathbf{U})\mathbf{u}$. Furthermore, since only the steady state solution of the perturbation solution $\mathbf{u}$ is desired, acceleration techniques such as multi-grid and local time stepping may be used to reduce dramatically the required computational time.

For subsonic flow calculations, any consistent discretization of the linearized Euler equations is sufficient to produce consistent unsteady flow solutions. For transonic flows with shock capturing, however, the discretized perturbation equations should be a faithful linearization of the discretized unsteady nonlinear equations used to compute the steady flow [LG91]. In the present study, we use Ni's conservative finite volume implementation of the Lax-Wendroff method [Ni82, Dan87] to compute both the steady and unsteady flow fields. Local time stepping and Ni's multiple-grid acceleration technique are used to speed convergence in both the steady nonlinear Euler and unsteady linearized Euler codes. For subsonic flows, a fourth-difference smoothing operator is used to eliminate sawtooth modes. For transonic flows, a combination of fourth-difference and second difference smoothing is used to capture shocks.

1.2.4 Far-Field Boundary Conditions

Because the computational domain extends a finite distance in the axial direction (see Fig. 1.1), far-field boundary conditions are required at the far-field boundaries to prevent spurious reflections of outgoing waves. Previous investigators have found analytical expressions for the behavior of the linearized equations in the far field, and have matched these far-field solutions to the near-field numerical solutions at the upstream and downstream far-field boundaries. These analytical solutions were found for the special case of uniform mean flow in the far field. For transonic Euler flows, however, the mean flow in the far field is in general nonuniform. For example, the presence of a shock or multiple shocks in the blade passage will produce nonuniform total pressure in the downstream region. No general analytical expression has been developed for this situation.

To overcome this difficulty, the authors have developed an alternative treatment the far-field boundary conditions. Instead of finding the analytical eigenmodes of the linearized Euler equations in the far-field, we find the eigenmodes of the *discretized* linearized Euler equations. The numerical eigenmodes are then matched to the numerical near-field solution at the far-field boundaries. This procedure, although computationally less efficient, produces numerically exact nonreflecting boundary conditions. Reference [HC91] describes the implementation of these boundary conditions in more detail.

1.3 Results

1.3.1 Transonic Channel Flow

The first test case to be presented is of transonic flow through a diverging channel. This case is presented to demonstrate the ability of the linearized Euler method to model accurately shock motion using shock capturing.

The channel considered here has a height, A, given by

$$A(x) = A_{\text{inlet}} \left\{ 1.10313 + .10313 \tanh\left[5\left(x - \frac{1}{2}\right)\right]\right\}, \quad 0 \leq x \leq 1 \quad (1.18)$$

So that the results can be compared to a one-dimensional shock-fitting theory, A_{inlet} is taken to be small compared with the channel length ($A_{\text{inlet}} = 0.01$). The inflow total pressure, P_T, total density, ρ_T, and flow velocity, U, are 1.0, 1.364, and 1.0, respectively. The back pressure, P_{exit} is 0.7422. Shown in Fig. 1.2 is the Mach number and pressure distribution as computed by the present nonlinear steady Euler solver using a 129×5 node computational grid. Also shown for comparison is the solution found using a steady quasi-one-dimensional, shock-fitting, Euler solver using 1001 grid nodes in the x direction. Note the excellent agreement between the two approaches. The only noticeable differences occur at the shock where the present nonlinear Euler solver smears the shock over about five grid nodes.

Next, we consider the case of a quasi-steady perturbation in the back pressure. The perturbation solution was computed using three different methods. We first computed the solution using the present linearized solver; then using a quasi one-dimensional, shock-fitting, linearized Euler solver;

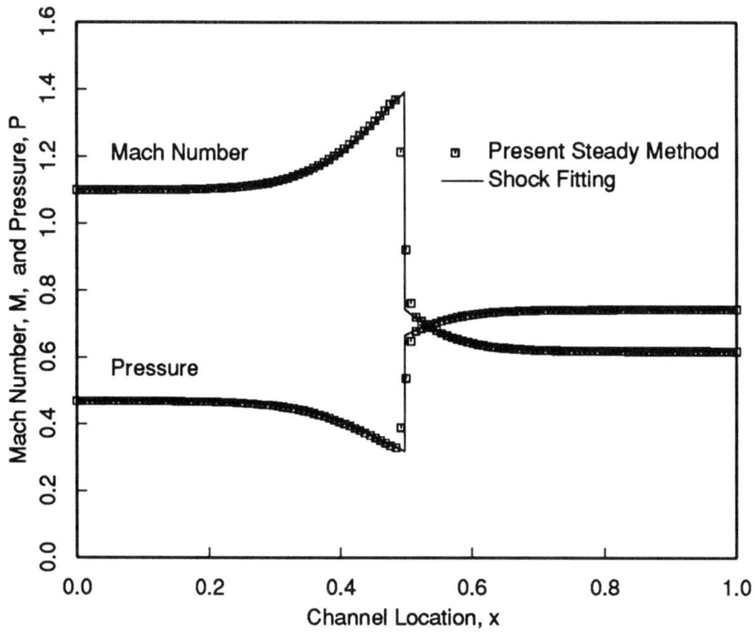

FIGURE 1.2. Steady transonic flow in a diverging channel.

and finally, by differencing the solutions found using the present nonlinear steady Euler solver at two slightly different back pressures. The results are shown in Fig. 1.3. The three solutions are seen to be in very good agreement away from the shock. At the shock, the two shock-capturing techniques produce an impulse of pressure. This impulse represents the load exerted on the channel walls due to the displacement of the shock. Note that the present linearized Euler solution is nearly identical to the differenced nonlinear solutions.

We now consider the case of an unsteady variation in the back pressure. Shown in Fig. 1.4 are the computed real and imaginary parts of the unsteady pressure distribution due to a unit perturbation in the back pressure with an excitation frequency, ω, of 1.0. The results of the present linearized Euler code agree well with the quasi one-dimensional, shock-fitting, linearized Euler solver.

To determine whether the present linearized Euler solver correctly predicts the unsteady loads induced by the motion of the shock, we integrated the pressure along the lower wall to obtain the "wall force." These results are tabulated in Table 1.1 for several different excitation frequencies. Also given in Table 1.1 is the wall force computed using the shock-fitting code. For our purposes, the grid is sufficiently fine that the shock-fit solution

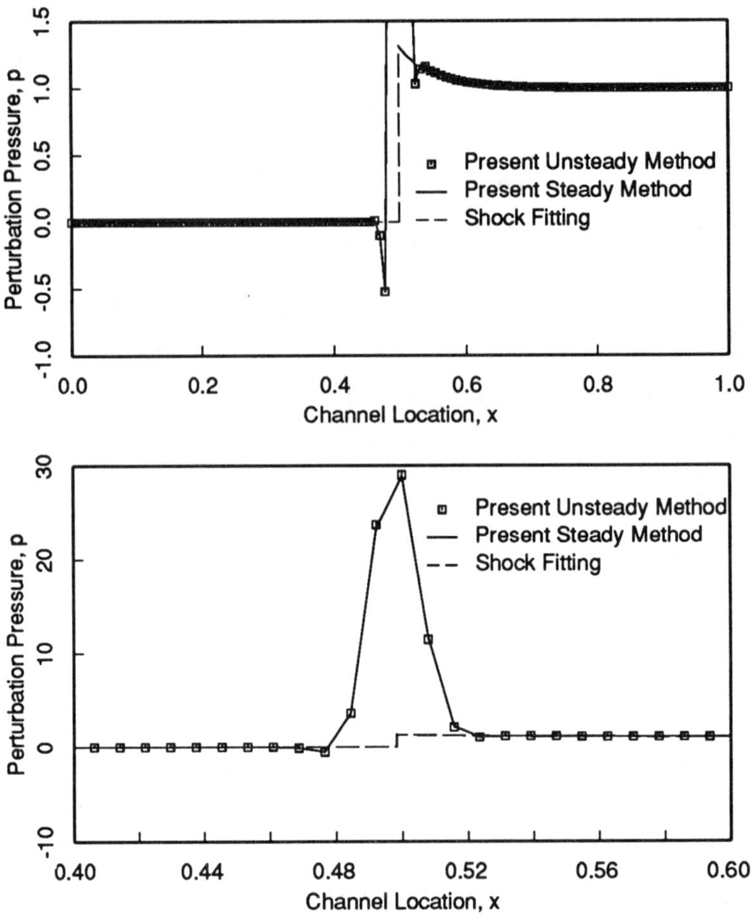

FIGURE 1.3. Perturbation pressure in a diverging channel due to a steady perturbation in back pressure. Bottom figure shows detail of shock impulse.

may be considered exact. For low frequencies, the two codes are in almost perfect agreement suggesting that the shock impulse found using shock capturing is properly modelled. For the higher frequency cases, the agreement is still good, but there is a slight error in the phase of the wall force. We believe that these differences are due to dispersion errors in regions away from the shock rather than a limitation in shock capturing at high frequencies. Finally, For the $\omega = 1$ case, we deliberately introduced two types of inconsistency into the linearized Euler solver to determine their influences on the solutions. In the first case, we used a different level of smoothing in the linearized unsteady Euler code than was used in the nonlinear steady

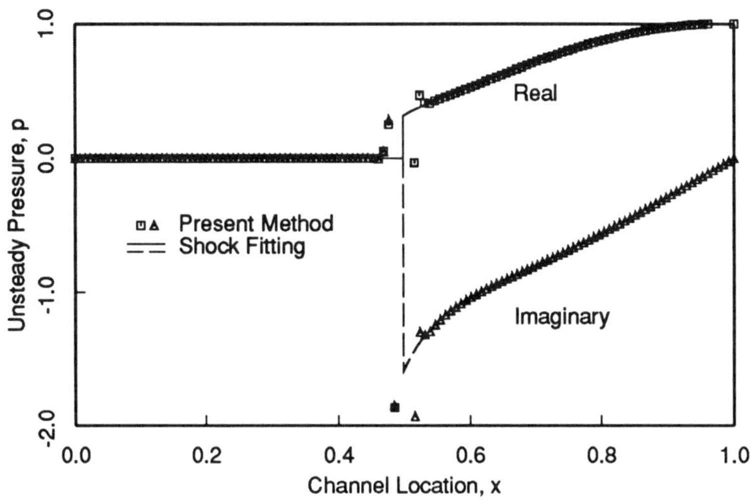

FIGURE 1.4. Unsteady pressure in a diverging channel due to an unsteady perturbation in back pressure. The reduced frequency, $\overline{\omega}$, based on channel length and inflow velocity is 1.0.

Euler code. In the second case, global time stepping was used in the steady Euler code and local time stepping was used in the linearized unsteady code. In both cases, the inconsistency between steady and unsteady codes is seen to increase the error in the unsteady wall force. The errors are much more pronounced, however, for the latter case. These results are consistent with the claim by Lindquist and Giles that the linearized code should be a faithful linearization of the nonlinear code [LG91].

1.3.2 Subsonic Flow in a Compressor

Having demonstrated the accuracy of the method for transonic channel flow, we next consider unsteady flows in a compressor geometry to demonstrate the ability of the method to accurately compute the unsteady flow about a cascade of loaded airfoils. The cascade considered is the newly designated Standard Configuration No. 10 [Fra91]. The airfoils have a circular arc camber distribution with a maximum height of 5 percent of the chord with a slightly modified NACA 0006 thickness distribution. The stagger angle, Θ, is 45 deg and the gap-to-chord ratio, G, is 1.0. For the subsonic case considered here, the steady inflow angle, $\Omega_{-\infty}$, is 55 deg and the inflow Mach number, $M_{-\infty}$, is 0.7. Under these conditions, the flow is entirely subsonic with a maximum Mach number on the suction surface of about 0.92.

Figure 1.5 shows the computed unsteady pressure distribution for the case where the airfoils pitch about points near their midchords with an

TABLE 1.1. Predicted pressure loads in a transonic diverging channel due to an unsteady perturbation in back pressure.

Frequency, ω	Scheme	Wall Force	Shock Force
0.0	1D Shock Fitting	$1.0305 + 0.0000j$	$0.5133 + 0.0000j$
	Nonlinear Euler[a]	$1.0326 + 0.0000j$	—
	Present Method	$1.0320 + 0.0000j$	—
0.5	1D Shock Fitting	$0.6866 + 0.6115j$	$0.1796 - 0.4426j$
	Present Method	$0.6863 + 0.6100j$	—
1.0	1D Shock Fitting	$0.1249 - 0.6267j$	$-0.2543 - 0.2797j$
	Present Method	$0.1293 - 0.6222j$	—
	Present Method[b]	$0.1315 - 0.6184j$	—
	Present Method[c]	$0.1778 - 0.6133j$	—
2.0	1D Shock Fitting	$-0.0807 - 0.1801j$	$-0.1182 + 0.1667j$
	Present Method	$-0.0750 - 0.1831j$	—

[a] Results from the steady analysis were found for two slightly different back pressures. The two solutions were then differenced and normalized by ΔP_{exit}.

[b] Smoothing coefficient in steady analysis is 0.15; smoothing coefficient in unsteady analysis is 0.30.

[c] Global time stepping used in steady analysis; local time stepping used in unsteady analysis.

interblade phase angle, σ, of 90 deg and a reduced frequency, $\overline{\omega}$, of 0.5 (Standard Configuration No. 10, Case 2). Here, the pressure is nondimensionalized by the upstream steady flow quantity ρV_T^2. To assess the accuracy of the analysis, solutions were computed on both a 65×17 and a 129×33 node grid. The good agreement between the coarse and fine grid solutions indicates that the fine grid solution is nearly grid converged. Also shown for comparison is the solution found using a deforming grid linearized potential method [Hal92]. Note that the solutions are well behaved in the neighborhood of the leading and trailing edges. The use of a deforming grid greatly improves the accuracy of the solution in these regions.

The linearized Euler solver is very efficient. The unsteady solution computed on the 65×17 node grid required 565 seconds of computer time on an IBM RISC System/6000 model 320 computer. For comparison, the steady solution required 110 seconds.

Having computed the unsteady pressure distribution on the airfoil surface, one can integrate to obtain the pitching moment. Shown in Figure 1.6 is the imaginary part of the pitching moment computed for a range of interblade phase angles, $-180 \leq \sigma \leq 180$, at a reduced frequency, $\overline{\omega}$, of 1.0. Also shown is the pitching moment computed using Hall's deforming grid linearized potential method. The peaks in the solution are acoustic resonances. We were unable to obtain converged solutions near one of the upstream acoustic resonance points ($\sigma = 117.1$ deg). Note that there is generally good agreement between the linearized Euler and linearized potential

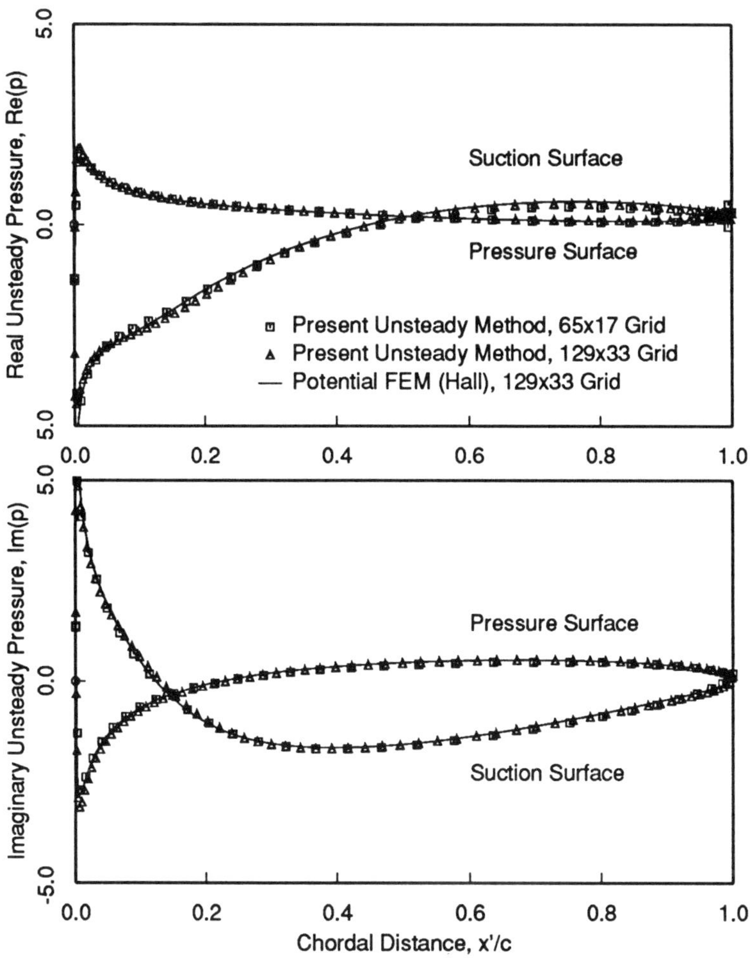

FIGURE 1.5. Unsteady pressure on surface of subsonic compressor blade due to pitching motion (Standard Configuration No. 10, Case 2). The reduced frequency, $\overline{\omega}$, is 0.5, and the interblade phase angle, σ, is 90 deg.

solution. One noticeable difference is seen at $\sigma = -20$ deg. This point is very near one of the upstream acoustic resonance points ($\sigma = -26.93$). Nevertheless, the good agreement elsewhere in the superresonant regions demonstrates the effectiveness of the numerical far-field boundary conditions.

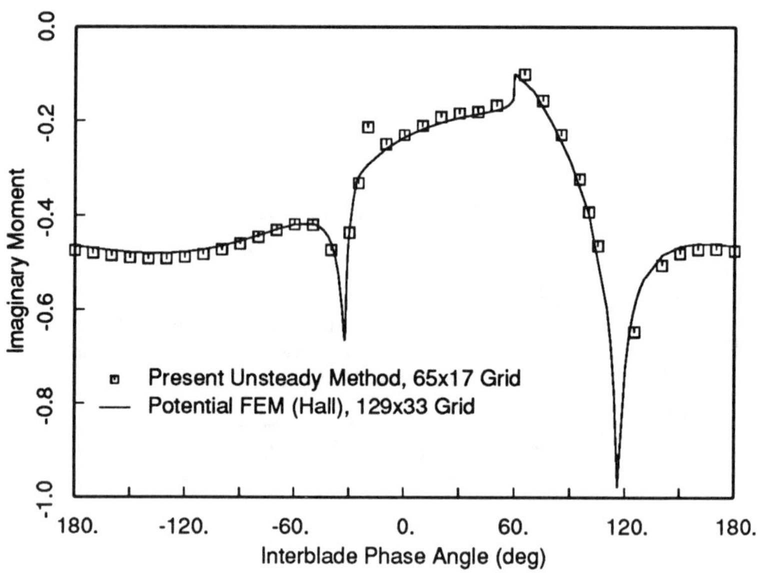

FIGURE 1.6. Imaginary pitching moment acting on subsonic compressor blades due to pitching motion (Standard Configuration No. 10). The reduced frequency, $\overline{\omega}$, is 0.5.

1.3.3 Transonic Flow in a Compressor

The final case to be considered here is flow through a transonic compressor. The cascade is again Standard Configuration No. 10. However, the flow conditions are such that there is a supersonic patch on the suction surface of the airfoil. The mean inflow angle, $\Omega_{-\infty}$, is 58 deg and the inflow Mach number, $M_{-\infty}$, is 0.8. Shown in Fig. 1.7 is the computed isentropic Mach number distribution on the surface of the airfoil. Also shown is the Mach number distribution computed using a full potential technique [VC84, VU88]. Note that for the solution computed with the present nonlinear Euler solver, the shock is smeared over about five grid points.

Having computed the steady flow solution, we consider the case of the airfoils vibrating in pitch with an interblade phase angle, σ, of 90 deg with a reduced frequency, $\overline{\omega}$, of 1.0 (Standard Configuration No. 10, Case 20). Figure 1.8 shows the computed unsteady pressure distribution on the airfoil surface. The shock impulse representing the unsteady load due to unsteady shock motion is clearly visible on the suction surface. Also shown is the pressure distribution computed using Verdon's linearized potential code, LINSUB code [VU88]. The linearized Euler and linearized potential solutions agree quite well on the pressure surface. However, the agreement is not as good on the suction surface, particularly ahead of the shock. This

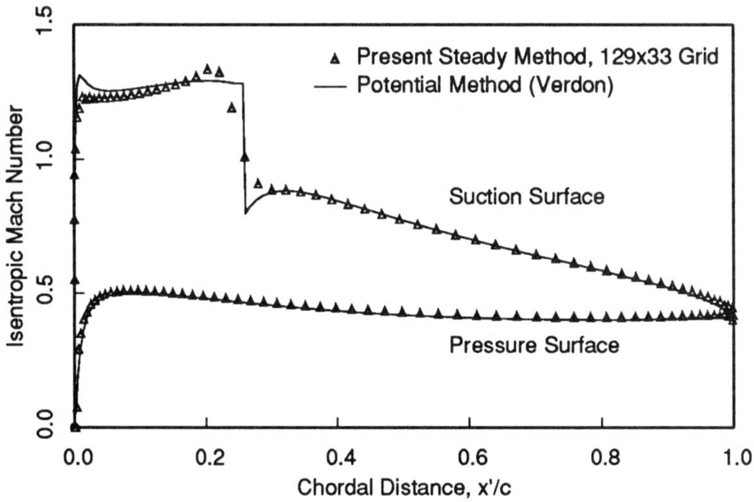

FIGURE 1.7. Steady pressure distribution on transonic compressor blade (Standard Configuration No. 10). The inflow Mach number, M, is 0.8, the gap-to-chord ratio, G, is 1.0, the stagger angle, Θ, is 45 deg., and the inflow angle, $\Omega_{-\infty}$, is 58 deg.

difference could arise from neglecting entropy and vorticity generation in the steady and unsteady potential analyses.

1.4 Summary

A two-dimensional linearized Euler solver has been developed that is capable of computing unsteady small disturbance flows in cascades. Two important features of the analysis are the use of a deforming grid to improve the accuracy of flutter calculations, and the use of shock capturing to model shock motion. Using the pseudo-time time-marching technique, one can solve the linearized equations with any of a number of well known schemes originally developed to solve the nonlinear Euler equations. Furthermore, since only the steady state harmonic perturbation solution is desired, convergence acceleration techniques may be used to speed convergence. Results presented in the paper clearly demonstrate the ability of the analysis to predict unsteady subsonic and transonic flows accurately and efficiently.

Acknowledgments: This work was supported by a grant from NASA Lewis Research Center, NASA Grant NAG3-1192, with Dr. Daniel Hoyniak serving as technical monitor. The authors wish to acknowledge helpful discus-

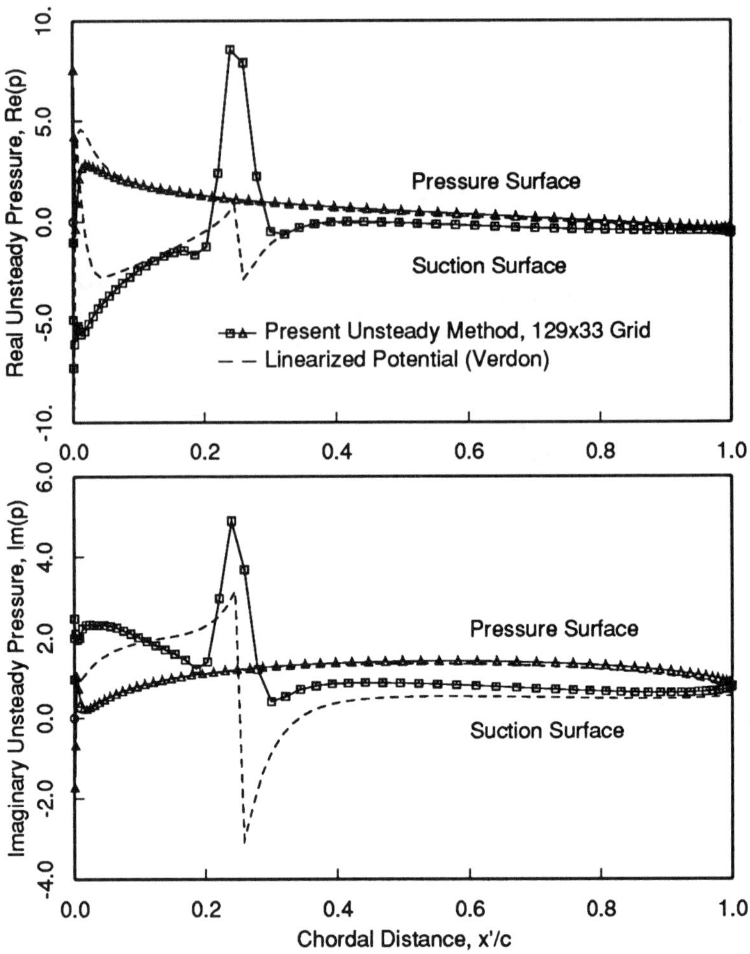

FIGURE 1.8. Unsteady pressure on surface of transonic compressor blade due to pitching motion (Standard Configuration No. 10, Case 20). The reduced frequency, $\overline{\omega}$, is 1.0, and the interblade phase angle, σ, is 90 deg.

sions with Dr. Graham Holmes, Dr. Michael Giles, and Ms. Dana Lindquist. In addition the authors are indebted to Dr. Joseph M. Verdon for providing unsteady pressure distribution predictions for Standard Configuration No. 10.

REFERENCES

[Dan87] J. D. Dannenhoffer, III. *Grid Adaptation for Complex Two-Dimensional Transonic Flows.* ScD thesis, Massachusetts Institute of Technology, Cambridge, Massachusetts, August 1987.

[FLM87] A. Fourmaux and A. Le Meur. Computation of unsteady phenomena in transonic turbines and compressors. In H.E. Gallus and S. Servaty, editors, *Unsteady Aerodynamics and Aeroelasticity of Turbomachines and Propellers, Proceedings of the Fourth International Symposium*, pages 87–108, Aachen, West Germany, September 6-10, 1987.

[Fra91] T. H. Fransson, June 1991. Private communication.

[GH91] M. B. Giles and R. Haimes. Validation of a numerical method for unsteady flow calculations. ASME paper 91-GT-271, presented at the International Gas Turbine and Aeroengine Congress and Exposition, Orlando, Florida, June 3-6, 1991.

[Gil88] M. B Giles. Calculation of unsteady wake/rotor interaction. *AIAA Journal of Propulsion*, 4(4):356–362, July-August 1988.

[Hal92] K. C. Hall. A deforming grid variational principle and finite element method for computing unsteady small disturbance flows in cascades. AIAA Paper 92-0655, to be presented at the AIAA 30th Aerospace Sciences Meeting, Reno, Nevada, January 6-9, 1992.

[HC89] K. C. Hall and E. F. Crawley. Calculation of unsteady flows in turbomachinery using the linearized Euler equations. *AIAA Journal*, 27(6):777–787, June 1989.

[HC91] K. C. Hall and W. S. Clark. Prediction of unsteady aerodynamic loads in cascades using the linearized Euler equations on deforming grids. AIAA Paper 91-3378, presented at the AIAA/SAE/ASME/ASEE 27th Joint Propulsion Conference, Sacramento, California, June 24-26, 1991.

[Hod84] H. P. Hodson. An inviscid blade-to-blade prediction of a wake generated unsteady flow. ASME paper 84-GT-43, 1984.

[HSR91] D. L. Huff, T. W. Swafford, and T. S. R. Reddy. Euler flow predictions for an oscillating cascade using a high resolution wave-split scheme. ASME paper 91-GT-198, presented at the International Gas Turbine and Aeroengine Congress and Exposition, Orlando, Florida, June 3-6, 1991.

[HV91] K. C. Hall and J. M. Verdon. Gust response of a cascade operating in a nonuniform mean flow. *AIAA Journal*, 29(9):1463–1471, September 1991.

[LG91] D. R. Lindquist and M. B. Giles. On the validity of linearized unsteady Euler equations with shock capturing. AIAA Paper 81-1598-CP, presented at the AIAA 10th Computational Fluid Dynamics Conference, Honolulu, Hawaii, June 24-27, 1991.

[Ni82] R. H. Ni. A multiple-grid scheme for solving the Euler equations. *AIAA Journal*, 20(11):1565–1571, November 1982.

[NS76] R. H. Ni and F. Sisto. Numerical computation of nonstationary aerodynamics of flat plate cascades in compressible flow. *Transactions of the ASME: Journal of Engineering for Power*, 98:165–170, April 1976.

[Rai89a] M. M. Rai. Three-dimensional Navier-Stokes simulations of turbine rotor-stator interaction; part I – methodology. *AIAA Journal of Propulsion and Power*, 5(3):307–311, May-June 1989.

[Rai89b] M. M. Rai. Three-dimensional Navier-Stokes simulations of turbine rotor-stator interaction; part II – results. *AIAA Journal of Propulsion and Power*, 5(3):312–319, May-June 1989.

[VC82] J. M. Verdon and J. R. Caspar. Development of a linear unsteady aerodynamic analysis for finite-deflection subsonic cascades. *AIAA Journal*, 20(9):1259–1267, September 1982.

[VC84] J. M. Verdon and J. R. Caspar. A linearized unsteady aerodynamic analysis for transonic cascades. *Journal of Fluid Mechanics*, 149:403–429, December 1984.

[VU88] J. M. Verdon and W. J. Usab, Jr. Advances in the numerical analysis of linearized unsteady cascade flows. Technical Report UTRC Report R88-957685-1, United Technologies Research Center, 1988.

[WG81] D. S. Whitehead and R. J. Grant. Force and moment coefficients of high deflection cascades. In P. Suter, editor, *Proceedings of the 2nd International Symposium on Aeroelasticity in Turbomachines*, pages 85–127, Zurich, 1981. Juris-Verlag.

2D Linearized Harmonic Euler Flow Analysis for Flutter and Forced Response

D. G. Holmes[1]
H. A. Chuang[2]

ABSTRACT
A two dimensional, unsteady, linearized Euler solver has been developed, and applied to both flutter and forced response problems. Solutions are obtained at a single frequency, with the time derivatives $\partial Q/\partial t$ replaced by $-i\omega Q$. The unsteady solver is derived from an existing steady flow Euler solver that uses adaptive triangular grids. The solver is not restricted to blade row geometries. The solution technique is the false time marching of Ni, used in conjunction with a Runge-Kutta scheme. As with most steady flow Euler solvers, shocks are captured, rather than fitted. The inlet and outlet boundary conditions are the two dimensional single frequency non-reflective boundary conditions due to Giles. For flutter problems, the grid moves with the blade: grid points on the blade surface remain fixed to the blade, and the motion of interior grid points is computed using a Laplacian smoother. Results are shown for three flutter calculations and one forced response calculation.
Potential methods are clearly more efficient than linearized Euler methods, but are restricted to shock free flows, or to flows with weak shocks and shock geometries simple enough to permit shock fitting. This and other investigations have shown that Euler methods can produce results comparable to potential methods. For transonic flows, the unsteady force applied to a blade by an oscillating shock – the "shock foot force" – can depend significantly on the way the shock is modeled. Improvements in the way unsteady Euler solvers treat shocks are required, possibly even extending to making the grid oscillate with the shock. Once the issue of the correct treatment of shocks can be resolved, Euler methods will be able to produce useful solutions in regimes inaccessible to potential methods.

---

[1] Corporate Research and Development Center, General Electric Company, ES207, P.O. Box 8, Schenectady, NY 12301.
[2] General Electric Aircraft Engines, One Neumann Way, MD A334, Cincinnati, OH 45215.

1.1 Introduction

The are two major reasons for an interest in unsteady flow in turbomachines: blade flutter and forced response. Effective engineering analysis of flutter and forced response problems requires both structural and aerodynamic calculations. For the aerodynamic part of the analysis, the state of the art (in the sense of the most advanced methods in regular use) is probably linearized potential flow analysis, as exemplified by [Ver77]. Following in the footsteps of steady flow analysis methods – where the progression has been from panel methods to potential flow analysis, to inviscid (Euler) flow analysis, to viscous (Navier-Stokes) flow analysis – the next natural step for unsteady flows is unsteady Euler flow analysis. Many efforts to perform unsteady Euler analysis of turbomachinery flows are underway – see, for example, [He89]. The usual approach is to observe that most steady flow Euler solvers reach a steady state solution by time marching, and that an unsteady flow Euler solver can thus be derived very easily from a steady flow solver. All such Euler solvers solve the *non-linear* unsteady Euler equations.

Giles [Gil89] has strongly advocated an intermediate step between linearized potential and non-linear Euler: *linearized* Euler. The benefit of linearized Euler is that we can take advantage of the linearity, and compute relatively inexpensive *frequency domain* solutions. Ni [Ni74][NS75] reported the earliest such effort. Hall [Hal87][Hal89] has done more recent work. Ni solves an isentropic form of the Euler equations; Hall solves the full Euler equations. Ni uses a false time marching solution technique to compute frequency domain solutions; Hall uses a least squares solution technique. Hall treats shocks using a shock fitting technique.

This paper describes the development of a 2D linearized Euler code for turbomachinery blade rows, and other geometries. The code is based on an unstructured grid, adaptive Euler and Navier-Stokes solver for steady flows, applicable to almost any geometry [HLC88][HC89]. A finite volume interpolation scheme on an unstructured triangular mesh is used. Ni's false time marching solution technique is used, with a multi-step Runge-Kutta scheme. Shocks are treated by capturing, with a partial linearization of the non-linear adaptive second and fourth order smoother. Either 1D or 2D non-reflecting boundary conditions are used, following Giles [Gil88].

We report here on the application of this code to four test cases. Two are flutter test cases reported by Bolcs and Fransson [BF86] – one a subsonic compressor blade row, the other a transonic turbine blade row. The third is a supersonic flat plate flutter test case reported by Verdon [Ver77]. The fourth test case is a forced response problem reported by Hall [Hal87]. We conclude with a discussion of some outstanding problems, primarily the issue of shock capture.

1.2 The Linearized Euler Equations on a Moving Grid

The derivation of the linearized Euler equations on a moving grid is given here at some length, since the authors are unaware of any convenient reference in which the equations can be found.

1.2.1 The Euler equations on a moving grid

The 2D Euler equations are written in terms of the density ρ, the absolute fluid velocity (v_x, v_y), the pressure p, the total internal energy per unit mass E, and the total enthalpy H. Application of the boundary condition at the surface of a moving blade is simplified if we write the equations on a moving grid: at the blade surface the grid velocity can be chosen so that grid points remain fixed to the moving blade. On a grid moving with velocity (u_x, u_y) the equations, in conservation form, are

$$\frac{\partial Q}{\partial t} - u_x \frac{\partial Q}{\partial x} - u_y \frac{\partial Q}{\partial y} + \frac{\partial F}{\partial x} + \frac{\partial G}{\partial y} = 0 \tag{1.1}$$

where

$$Q = \begin{bmatrix} \rho \\ \rho v_x \\ \rho v_y \\ \rho E \end{bmatrix} \qquad F = \begin{bmatrix} \rho v_x \\ \rho v_x^2 + p \\ \rho v_x v_y \\ \rho v_x H \end{bmatrix} \qquad G = \begin{bmatrix} \rho v_y \\ \rho v_x v_y \\ \rho v_y^2 + p \\ \rho v_y H \end{bmatrix} \tag{1.2}$$

and where

$$E = \frac{p}{(\gamma - 1)\rho} + \frac{1}{2}(v_x^2 + v_y^2), \qquad H = \frac{\gamma p}{(\gamma - 1)\rho} + \frac{1}{2}(v_x^2 + v_y^2) \tag{1.3}$$

and γ is the specific heat ratio.

We put the Euler equations (1.1) in control volume form by integrating over a fixed control volume that is instantaneously coincident with a moving control volume attached to the grid. Applying Gauss' theorem we obtain

$$\int \left[\frac{\partial Q}{\partial t} - u_x \frac{\partial Q}{\partial x} - u_y \frac{\partial Q}{\partial y} \right] dV + \oint F\, dy - \oint G\, dx = 0 \tag{1.4}$$

This control volume form can also be obtained by applying the conservation laws directly to a moving control volume – see, for example, reference [Bat90].

1.2.2 Linearization

We define mean and perturbation coordinates as follows

$$x = \bar{x} + \epsilon x'(\bar{x}, \bar{y}, t), \qquad y = \bar{y} + \epsilon y'(\bar{x}, \bar{y}, t) \tag{1.5}$$

where $(\bar{x}, \bar{y})$ are the *mean* coordinates of the moving grid, and (x', y') are the *time varying* coordinates of the grid motion, multiplied by a small parameter ϵ. The grid velocity components become

$$u_x = \epsilon u'_x, \qquad u_y = \epsilon u'_y \tag{1.6}$$

since the grid motion is of order ϵ: there is no order one motion of the grid. The flow variables can be broken up into a mean part and a small, time varying part, e.g.,

$$Q = \bar{Q}(\bar{x}, \bar{y}) + \epsilon Q'(\bar{x}, \bar{y}, t) \tag{1.7}$$

The control volume can also be broken up into a mean part and a small perturbation.

The volume and surface integrals in Equation (1.4) can be linearized to give

$$\begin{aligned} \int A\,dV &= \int \bar{A}\,d\bar{V} + \epsilon \int A'\,d\bar{V} + \epsilon \int \bar{A}\,dV' + \mathcal{O}(\epsilon^2) \\ &= \int \bar{A}\,d\bar{V} + \epsilon \int A'\,d\bar{V} + \epsilon \oint \bar{A}(x'\,d\bar{y} - y'\,d\bar{x}) + \mathcal{O}(\epsilon^2) \end{aligned} \tag{1.8}$$

$$\oint A\,dx = \oint \bar{A}\,d\bar{x} + \epsilon \oint A'\,d\bar{x} + \epsilon \oint \bar{A} \left[\frac{\partial x'}{\partial \bar{x}}\,d\bar{x} + \frac{\partial x'}{\partial \bar{y}}\,d\bar{y} \right] + \mathcal{O}(\epsilon^2) \tag{1.9}$$

$$\oint A\,dy = \oint \bar{A}\,d\bar{y} + \epsilon \oint A'\,d\bar{y} + \epsilon \oint \bar{A} \left[\frac{\partial y'}{\partial \bar{x}}\,d\bar{x} + \frac{\partial y'}{\partial \bar{y}}\,d\bar{y} \right] + \mathcal{O}(\epsilon^2) \tag{1.10}$$

We now have the machinery we need to linearize Equation (1.4). For the mean flow

$$\oint \bar{F}\,d\bar{y} - \oint \bar{G}\,d\bar{x} = 0 \tag{1.11}$$

and for the perturbation

$$\begin{aligned} &\int \left[\frac{\partial Q'}{\partial t} - u'_x \frac{\partial \bar{Q}}{\partial \bar{x}} - u'_y \frac{\partial \bar{Q}}{\partial \bar{y}} \right] d\bar{V} + \oint F'\,d\bar{y} - \oint G'\,d\bar{x} + \\ &\qquad \oint \bar{F} \left[\frac{\partial y'}{\partial \bar{x}}\,d\bar{x} + \frac{\partial y'}{\partial \bar{y}}\,d\bar{y} \right] - \oint \bar{G} \left[\frac{\partial x'}{\partial \bar{x}}\,d\bar{x} + \frac{\partial x'}{\partial \bar{y}}\,d\bar{y} \right] = 0 \end{aligned} \tag{1.12}$$

Write

$$A = \left[\frac{\partial F}{\partial Q} \right]_0, \qquad B = \left[\frac{\partial G}{\partial Q} \right]_0 \tag{1.13}$$

Thus,

$$F' = AQ', \qquad G' = BQ' \tag{1.14}$$

We can now arrange equation (1.12) into a left hand side involving Q', and a right hand side involving all other terms

$$\int \frac{\partial Q'}{\partial t}\, d\bar{V} + \oint A Q'\, d\bar{y} - \oint B Q'\, d\bar{x} =$$
$$\int \left[u'_x \frac{\partial \bar{Q}}{\partial \bar{x}} + u'_y \frac{\partial \bar{Q}}{\partial \bar{y}} \right] d\bar{V} -$$
$$\oint \bar{F} \left[\frac{\partial y'}{\partial \bar{x}}\, d\bar{x} + \frac{\partial y'}{\partial \bar{y}}\, d\bar{y} \right] + \oint \bar{G} \left[\frac{\partial x'}{\partial \bar{x}}\, d\bar{x} + \frac{\partial x'}{\partial \bar{y}}\, d\bar{y} \right] \tag{1.15}$$

We now assume that the time dependence of the grid perturbations is harmonic, i.e.,

$$u'_x(\bar{x}, \bar{y}, t) = \hat{u}_x(\bar{x}, \bar{y}) e^{-i\omega t}, \qquad u'_y(\bar{x}, \bar{y}, t) = \hat{u}_y(\bar{x}, \bar{y}) e^{-i\omega t} \tag{1.16}$$

We also assume that the time variation of the flow variables is harmonic, but with an amplitude that depends on time, e.g.,

$$Q'(\bar{x}, \bar{y}, t) = \hat{Q}(\bar{x}, \bar{y}, \tau) e^{-i\omega t} \tag{1.17}$$

where we have used the variable τ to represent the time dependence of the amplitude. All quantities such as $\hat{u}_x$ and $\hat{Q}$ are complex.

Equation (1.15) becomes

$$\int \frac{\partial \hat{Q}}{\partial \tau}\, d\bar{V} - i\omega \int \hat{Q}\, d\bar{V} + \oint A \hat{Q}\, d\bar{y} - \oint B \hat{Q}\, d\bar{x} =$$
$$\int \left[\hat{u}_x \frac{\partial \bar{Q}}{\partial \bar{x}} + \hat{u}_y \frac{\partial \bar{Q}}{\partial \bar{y}} \right] d\bar{V} -$$
$$\oint \bar{F} \left[\frac{\partial \hat{y}}{\partial \bar{x}}\, d\bar{x} + \frac{\partial \hat{y}}{\partial \bar{y}}\, d\bar{y} \right] + \oint \bar{G} \left[\frac{\partial \hat{x}}{\partial \bar{x}}\, d\bar{x} + \frac{\partial \hat{x}}{\partial \bar{y}}\, d\bar{y} \right] \tag{1.18}$$

We propose time marching equation (1.18) to a steady state, i.e. until the amplitudes of the flow perturbations have stabilized. This "false time marching" approach to the solution of the linearized Euler equations was first suggested by Ni [NS75]. The advantage of this approach is that all of the methods invented for time marching the non-linear Euler equations to a true steady state can easily be modified for this application.

1.3 The Solution Algorithm

1.3.1 THE INTERPOLATION SCHEME

We use a finite volume interpolation scheme on triangles. The interpolation scheme is identical to that used in the steady flow Euler and Navier Stokes solver described in [HLC88]. The solver in [HLC88] is used to create the mean flow solutions needed by this unsteady flow solver. The mean flow solutions are made on mixed grids composed of both triangular and quadrilateral cells. For the unsteady flow solver each quadrilateral cell is split into two triangular cells across its shortest diagonal.

The finite volume interpolation scheme can be interpreted as a Galerkin finite element scheme with first order elements, but with a lumped mass matrix. One advantage of the linearized false time marching is that the finite element approach can be used to treat the physical time terms consistently, but the false time terms can be left lumped to make an explicit solution scheme possible. Numerical experiments indicate that the consistent treatment of the physical time terms gives better solutions at high wave numbers. This benefit of using a consistent mass matrix has also been noted in [Sid91]. Note that an explicit *non-linear* Euler solution scheme cannot treat the unsteady terms in a consistent way: the mass matrix must always be lumped.

1.3.2 The time marching scheme

The time marching scheme is the multi-step Runge-Kutta scheme popularized by Jameson, *et al.* [JST81]. A three step scheme proves to be better in this linearized unsteady application than either a four or five step scheme.

1.3.3 The partially linearized smoothing algorithm

The smoothing algorithm is a partial linearization of an adaptive second / fourth order scheme, of the type proposed by Jameson, *et al.* [JST81]. The non-linear scheme on which the linearization is based uses a two stage fourth order smoother composed of two successive second order operators, with the first stage designed to be as close to a Laplacian as possible [HLC88]. The linearization is *partial* in the following sense. The smoothing operator at node j can be written as

$$L_j(Q_k) = L_2\left(Q_k, \nu_2(Q_k)\right) + L_4\left(Q_k, \nu_4(Q_k)\right) \tag{1.19}$$

where the nodes k are the nodes that form the difference stencil for the smoothing operator at node j. Note that both ν_2 and ν_4 are, because of the adaptive nature of the smoothing algorithm, strongly non-linear functions of Q_k. After linearization

$$L'(Q') = \left(\frac{\partial L_2}{\partial Q} + \frac{\partial L_4}{\partial Q} + \frac{\partial L_2}{\partial \nu_2}\frac{\partial \nu_2}{\partial Q} + \frac{\partial L_4}{\partial \nu_4}\frac{\partial \nu_4}{\partial Q}\right) Q' \tag{1.20}$$

Computation of $\partial L_2/\partial Q$ and $\partial L_4/\partial Q$ is easy. Computation of $\partial \nu_2/\partial Q$ and $\partial \nu_4/\partial Q$ is hard. For some values of the mean flow $\bar{Q}$ (e.g., a uniform mean flow), ν_2 and ν_4 may be non-differentiable functions of Q. Even if

$\partial\nu_2/\partial Q$ and $\partial\nu_4/\partial Q$ could be computed without difficulty, these terms would produce a linearized smoother with a difference stencil so large that the cost of computing the smoothing operator would far exceed the cost of computing the Euler operator, which seems somewhat unsatisfactory. There are several ways in which one could proceed:

1. Drop the terms that are difficult and expensive to evaluate, i.e., write
$$L'(Q') = \left(\frac{\partial L_2}{\partial Q} + \frac{\partial L_4}{\partial Q}\right) Q' \tag{1.21}$$
2. Devise an alternative adaptive smoothing algorithm that has coefficients ν_2 and ν_4 that are always differentiable. This has been tried in one dimension, but with only limited success: the alternative smoothing algorithms always had poorer shock capturing properties than the standard algorithm, or
3. Do the differentiation numerically, i.e., compute the small changes in ν_2 and ν_4 in response to small changes in Q. This has been tried in one dimension, but not in two dimensions. Even if this approach were successful in two dimensions, the issue of computational cost would remain.

We have elected to use the first approach, and done only a partial linearization, dropping the difficult terms. The terms that are dropped are of importance only near shocks. We discuss below whether this approach produces results that are valid for flows with shocks.

1.3.4 Boundary conditions

We have attempted to implement the exact, 2D, single frequency non-reflecting boundary conditions of Giles [Gil88]. We assume that the M nodes across an inlet or outlet boundary are uniformly spaced, and at each partial time step perform a digital Fourier transform of the unsteady solution to obtain $M-1$ distinct wave number components. (Because of periodicity, only $M-1$ boundary nodes are independent.) The original nodal values can be reconstructed, by a reverse Fourier transform, from any set of consecutive wave numbers consistent with the specified interblade phase lag. Since we can only apply the boundary conditions to wave number components that are physically meaningful and free of aliasing, we pick a set of tangential wave numbers (k_y) centered on the wave number, consistent with the phase lag, that has the lowest magnitude, i.e., $M-1$ wave numbers of the form
$$k_{y_n} = \frac{\sigma \pm 2n\pi}{P} \qquad n = 0, 1, 2, \ldots \tag{1.22}$$
where σ is the interblade phase lag and P is the blade pitch. Giles' boundary condition is applied to each of these wave number components, and the wave

number components are then transformed back to the nodal values. Giles suggests that in the context of false time marching this boundary condition has to be lagged, but we have not found this to be necessary.

We also have available the options of applying the boundary condition as if the boundary disturbance were at a single, specified wave number, and of applying a simple, one dimensional non-reflecting boundary condition. Some simple test problems have demonstrated the necessity of using the full, two-dimensional non-reflecting boundary conditions.

In [Gil88], the 2D single-frequency boundary condition is applied by solving a system of differential equations in terms of the 1D characteristic variables. There were some misprints in [Gil88] that resulted in non-orthogonality of the right (column) eigenvectors corresponding to the third and fourth roots, but these eigenvectors were not used in his 2D formulation. We decided to orthonormalize the eigenvectors, so that we can construct the transformation matrices from primitive to 2D characteristic variables and vice versa. Because of the normalization, the product of these two matrices is the identity matrix; and this also allows us to apply the boundary condition directly in terms of the 2D characteristic variables. Below is the transformation matrix from characteristic to primitive variables (in dimensional form) after orthonormalization of the corrected right eigenvectors:

$$\left(\begin{array}{cccc} -\dfrac{1}{c^2} & 0 & \dfrac{1}{2c^2(1-v_y\lambda)S} & \dfrac{1}{2c^2(1-v_y\lambda)S} \\ 0 & -\dfrac{v_x\lambda}{\rho c[(v_x\lambda)^2+(1-v_y\lambda)^2]} & \dfrac{(cS-v_x)}{2\rho cS(1-v_y\lambda)(c-v_xS)} & -\dfrac{(cS+v_x)}{2\rho cS(1-v_y\lambda)(c+v_xS)} \\ 0 & \dfrac{(1-v_y\lambda)}{\rho c[(v_x\lambda)^2+(1-v_y\lambda)^2]} & \dfrac{\lambda(c^2-v_x^2)}{2\rho cS(1-v_y\lambda)^2(c-v_xS)} & \dfrac{\lambda(c^2-v_x^2)}{2\rho cS(1-v_y\lambda)^2(c+v_xS)} \\ 0 & 0 & \dfrac{1}{2(1-v_y\lambda)S} & \dfrac{1}{2(1-v_y\lambda)S} \end{array} \right) \tag{1.23}$$

where

$$S = \sqrt{1 - \frac{\lambda^2(c^2-v_x^2)}{(1-v_y\lambda)^2}}, \qquad \text{and} \qquad \lambda = \frac{k_y}{\omega} \tag{1.24}$$

1.3.5 SETTING THE GRID MOTION

Given the motion of the grid boundary, it is necessary to compute values for the grid velocity (u_x, u_y) at all of the interior nodes of the grid. The exact velocities assigned to the interior nodes of the grid are not important, although the grid velocity should probably vary smoothly. We have chosen to use a simple Laplacian smoother to assign velocities to the interior grid points, i.e.,

$$\nabla^2 \hat{u}_x = 0, \qquad \nabla^2 \hat{u}_y = 0 \tag{1.25}$$

These equations are solved on the triangular grid by a simple point Jacobi iterative scheme. A high degree of convergence is not necessary.

1.3.6 CONVERGENCE ACCELERATION

Since the goal of the time marching is to achieve a steady state, local time stepping can be used to accelerate convergence. Use of local time steps is permissible since the time is "false" time. Other conventional acceleration techniques such as implicit residual averaging have not been implemented.

Only one other convergence acceleration technique is used: Aitken extrapolation [DB74]. Since use of Aitken extrapolation for time marching codes has not been reported before, it is described here.

We have observed that convergence of the time marching scheme is sometimes very slow, and that, when it is slow, it is usually very *linear*. The successive estimates of the solution $Q^{(n)}$ form a series converging linearly on the converged solution $\tilde{Q}$, i.e.,

$$\frac{Q^{(n)} - \tilde{Q}}{Q^{(n-1)} - \tilde{Q}} = \frac{Q^{(n-1)} - \tilde{Q}}{Q^{(n-2)} - \tilde{Q}} = \alpha \tag{1.26}$$

where α is the convergence rate. In this application α is complex, and will, for slowly converging cases, have a magnitude very close to unity. Solving for $\tilde{Q}$

$$\tilde{Q} = Q^{(n)} - \frac{\left(Q^{(n)} - Q^{(n-1)}\right)^2}{Q^{(n-2)} - 2Q^{(n-1)} + Q^{(n)}} \tag{1.27}$$

We can apply Equation (1.27) to each of the four Euler equations, at each node, using three successive estimates of Q spaced 50 or 100 iterations apart. If convergence is linear, $\tilde{Q}$ is a dramatically improved estimate of the solution. $\tilde{Q}$ can be used as the starting point of continued iterations, and eventually the extrapolation process in Equation (1.27) can be repeated, and so on.

This process is very simple in principle: the hard part is making sure that the convergence is linear before doing it. If the convergence is not close to linear, Aitken extrapolation only makes things worse. One approach we have found to be successful is to compute the convergence rate for each equation at each node, using Equation (1.26). Aitken extrapolation is only performed if the standard deviation of these estimates of α, normalized on the mean, is less than, say, 10%.

It is not known why Aitken extrapolation is often effective in this application. Clearly linear convergence implies that a single mode of the solution is persisting long after all other modes have been attenuated. It is not known what it is about the current algorithm that causes a single mode to be so persistent. It is not known, finally, whether Aitken extrapolation might be useful for any other time marching code.

1.3.7 COMPUTER PLATFORMS

The code runs on a variety of platforms – on Unix workstations such as the Apollo, HP, and Sun, on the Cray Y/MP, and on the Intel iPSC/860 hypercube.

On the Cray vectorization is facilitated by sorting the edge list so that batches of 64 edges do not reference any common node [JBW86].

The Intel hypercube is the most interesting platform. For a typical flutter calculation, solutions will be sought over the full 0° to 360° range of interblade phase lags. These calculations are essentially independent of each other, and can readily be dealt out to the processors of a hypercube by a simple host program. There is no communication between processors, and communication between the processors and the host is minimal. High parallel efficiencies are readily achieved, and we have run cases on a 16 node iPSC/860 at twice the speed on one processor of a Cray Y/MP.

1.4 Illustrative Test Cases

1.4.1 LOW SPEED COMPRESSOR BLADE ROW

The first test case is a low speed compressor cascade, which is the first standard configuration tested by Bolcs and Fransson [BF86]. This cascade has 10° of camber and a stagger angle of 55°. The cascade was tested at an inlet Mach number of 0.18. The blades pitch about mid-chord at a reduced frequency of 0.122. Figure 1.1 shows the aerodynamic work, as defined by Bolcs and Fransson, plotted as a function of interblade phase angle. The figure shows the experimental data, calculations made using the linearized potential code FINSUP (FINSUP seems to be the most accurate and reliable of the computational schemes reported in [BF86]), and calculations made using the linearized Euler solver reported here. The agreement between the current method and FINSUP is good. This is as it should be: at these subsonic Mach numbers there should be no difference between the linearized Euler and the linearized potential solutions. Both calculation methods do a reasonable job of predicting the experimental data.

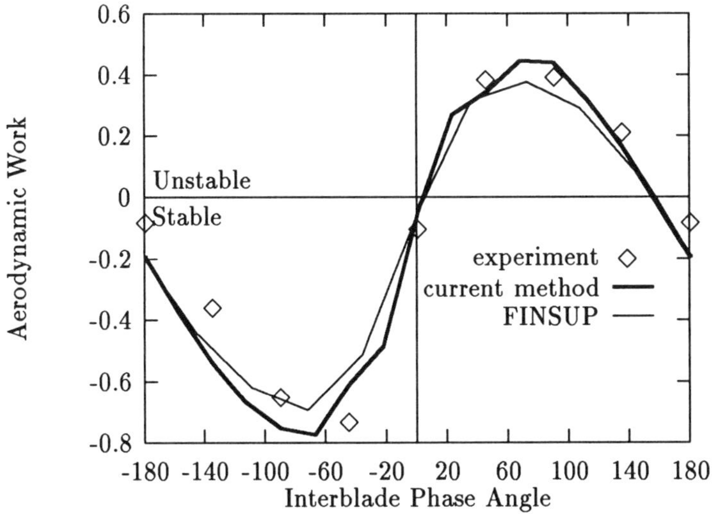

FIGURE 1.1. Unsteady aerodynamic work due to torsional blade vibrations for a low speed compressor

1.4.2 High speed turbine blade row

The second test case is a high speed turbine cascade, which is the fourth standard configuration tested by Bolcs and Fransson [BF86]. This cascade has 45° of camber and a stagger angle of 55°. The cascade was tested at inlet Mach numbers ranging from 0.19 to 0.29, and isentropic exit Mach numbers from 0.58 to 1.19. The blades translate in a direction that is close to pitchwise at a reduced frequency that varies with Mach number from 0.168 to 0.082. Although the inviscid algorithm that we use for computing the mean flow can handle the true rounded trailing edge shape without difficulty, we elected to put a short wedge on the trailing edge, in order to give a more realistic pressure distribution at the trailing edge. With a rounded trailing edge the flow accelerates rapidly around the trailing edge, and decelerates again via small, strong shocks. Figure 1.2 shows the aerodynamic work plotted as a function of interblade phase angle, at an exit Mach number of 0.9. The figure shows the experimental data, calculations made using the linearized potential code FINSUP, and calculations made using the linearized Euler solver reported here. The agreement between the current method and FINSUP is fairly good, and both methods show the same trend as the data. We also show the aerodynamic work, at a −90° phase lag, versus Mach number in Figure 1.3. Both calculation methods show a tendency towards stability with increasing Mach number, as shown

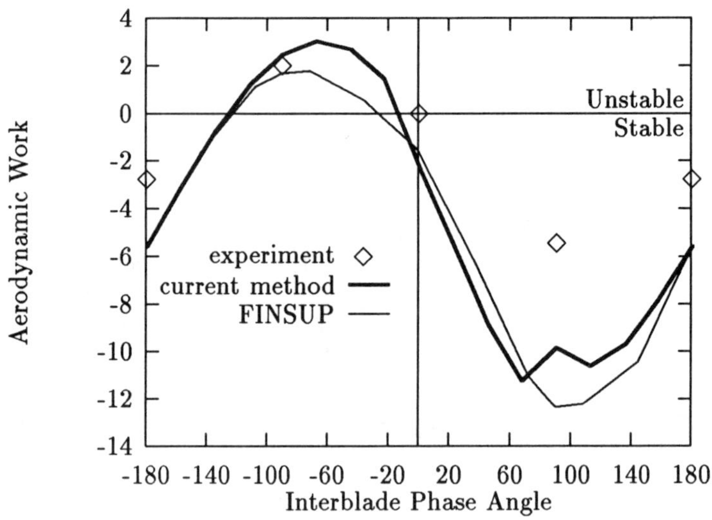

FIGURE 1.2. Effects of phase angle on unsteady aerodynamic work due to bending blade vibrations of a high speed turbine

by the data. It might be argued that the current linearized Euler method is doing a better job in this case than the linearized potential method.

1.4.3 Supersonic flat plate cascade

The third test case is a supersonic flat plate cascade reported by Verdon [Ver77]. The comparison in this case is purely computational: i.e., between our linearized Euler calculation and Verdon's linearized potential algorithm. The cascade has a pitch-to-axial-chord ratio of 1.665, and a stagger angle of 59.53°. The incident flow is at a zero angle-of-attack, and at a Mach number of 1.345. The blades pitch about mid-chord with reduced frequencies (k) of 0.5 and 2.0. The grid for the linearized Euler solution contains 890 nodes. Figure 1.4 shows a plot of the dimensionless real part of the work input to the blade *vs.* interblade phase lag, for the current linearized Euler code and Verdon's linearized potential code. Resonant interblade phase angles are $-9.7°$ and $-42.8°$ for $k = 0.5$, and $-38.8°$ and $-171.0°$ for $k = 2.0$. The linearized Euler and linearized potential results are very close. The linearized Euler code has some trouble converging near resonances: this needs to be investigated further.

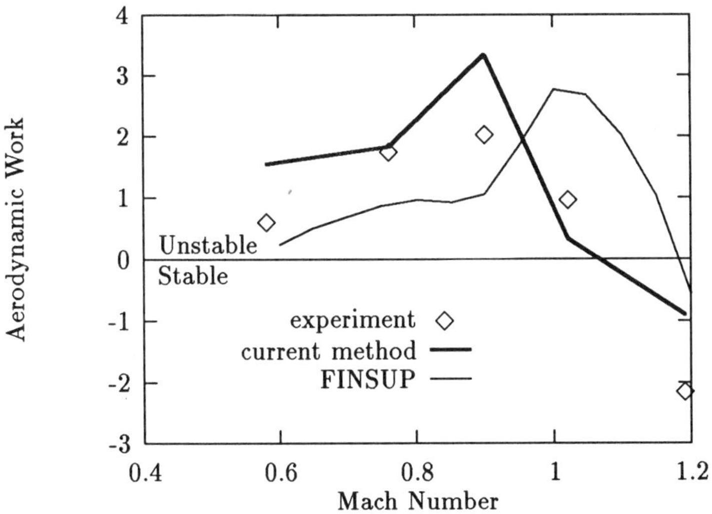

FIGURE 1.3. Effects of exit Mach number on unsteady aerodynamic work due to bending blade vibrations of a high speed turbine

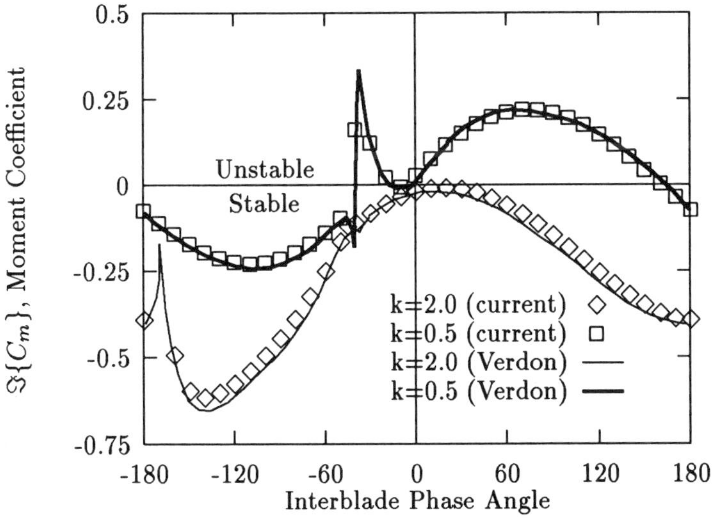

FIGURE 1.4. Unsteady aerodynamic moment coefficient due to torsional blade vibrations for a supersonic flat-plate cascade

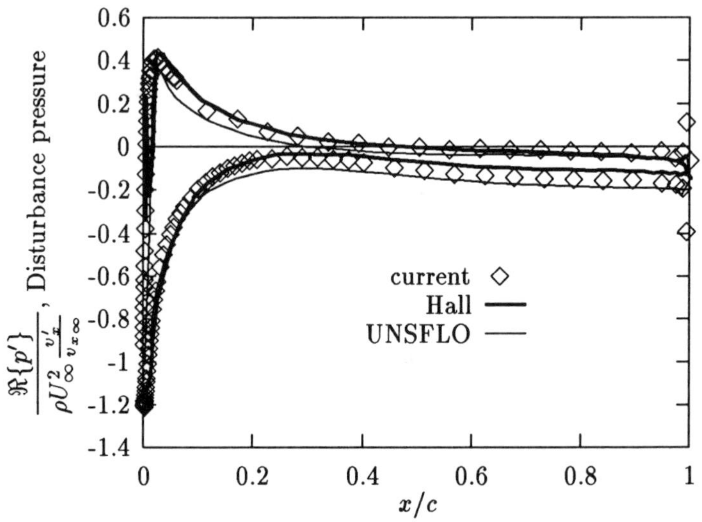

FIGURE 1.5. Real part of the nondimensional disturbance pressure on a Gostelow airfoil due to a vortical gust

1.4.4 Vortical gust incident on the Gostelow cascade

The fourth test case is the subsonic Gostelow cascade reported by Hall [Hal87]. As with the third test case, the comparison made here is purely computational. The Gostelow cascade of Joukowski airfoils has a pitch-to-chord ratio of 0.99, and a stagger angle of 37.5°. The incident flow is at 16° with respect to the blade chord, and the Mach number is 0.25 upstream and 0.17 downstream. The unsteady response of a cascade due to an upstream sinusoidal velocity defect (vortical gust) is analyzed. The velocity defect approximates obstructions upstream of a rotor. The sinusoidal variation has a wavelength equal to four times the pitch, equivalent to an interblade phase angle of $-90°$. The reduced frequency is 1.257. Figure 1.5 shows the real part of the nondimensional pressure on the airfoil surface due to the incoming vortical gust. The numerical solutions of Hall and ours show good agreement, as expected. Our results also compared favorably with that of UNSFLO [Gil86], which is a non-linear Euler solver.

1.5 Problems Outstanding

1.5.1 The treatment of shocks

We have chosen to use shock *capturing* via adaptive smoothing, rather than shock *fitting*, even though successful shock fitting for linearized unsteady Euler solutions has been reported by Hall [Hal89]. This is because shock geometries in real turbomachines can be complex and shocks can appear in unexpected locations, requiring much more intelligence of a shock fitting algorithm than is at present possible.

For steady flows, shock capturing is far more widely used than shock fitting, but this does not mean that one can use shock capturing for unsteady flows without some thought. For steady flows it is easy to create shock capturing schemes that are conservative, so that the jump conditions across the shock are satisfied. Any shock capturing scheme that satisfies the jump conditions at a shock is giving us most of what we need: any other feature, such as capturing a shock within a couple of cells, or producing a shock profile that is strictly monotone, is a bonus. For unsteady flows we cannot afford to be so casual: the shock profile can be important. Unsteady variations in the profile of a shock attached to a blade influence the unsteady "shock foot force" applied to the blade by the oscillating shock, and this shock foot force can be a large part of the total unsteady force on the blade. For a large amplitude, strongly non-linear Euler analysis, in which the shock oscillates back and forth, sweeping over several grid cells, variations in the shock profile are probably not important. For a small amplitude non-linear Euler analysis, in which the shock oscillates with an amplitude less than a cell width, and for a *linearized Euler analysis*, variations in the shock profile are all important. This is a problem that is beginning to be addressed by those working in linearized unsteady Euler flows [LG91], but which seems to have been totally ignored by the non-linear unsteady Euler community.

We have done some numerical experiments which indicate that the shock capturing problem is real. One set of experiments involved using our current 2D linearized Euler code, as well as a 1D analog, to compute harmonic perturbations of a shock in a parallel duct, caused by acoustic waves incident on the shock from upstream: the computed force on the wall of the duct compared poorly with the "exact" solutions that can be computed with the help of Landau [Lan59]. Another set of experiments involved a non-linear one-dimensional analysis of perturbations of a shock in a diverging duct: the computed values of the unsteady force on the walls of the duct, when normalized on the perturbation amplitude, became highly irregular as the amplitude became small. Note that a small amplitude non-linear analysis should be functionally equivalent to a linearized analysis, suggesting that a full linearization of the shock capturing algorithm (Section 1.3.3 above) will *not* be enough to ensure valid solutions.

It is far from clear how to proceed. One could admit defeat and only do large amplitude non-linear Euler calculations, where "large" means calculations in which the shock motion is large compared to the width of the local grid cells. The hope would be that "large" would be large enough to eliminate the shock problem, but small enough to allow meaningful flutter boundaries to be calculated. It would, however, be preferable to make linearized Euler solvers produce accurate solutions! Two possible approaches are:

1. Modify the smoother (in both the steady flow solver and in the linearized unsteady solver), so that the shock is smeared over, say, ten grid cells. When the shock is highly smeared, even small displacements of the shock can be represented accurately. This approach has been explored by Lindquist and Giles [LG91].

2. The grid could be made to move with the shock. This should solve the shock problem, if it can be done. The grid would move with the velocity of the boundary on blade surfaces, with the velocity of the shock near the shock, and a Laplacian smoother would set the grid velocity elsewhere. It is not clear how this should be done. Some method of sensing the shock and computing its velocity is needed, and this begins to look like shock fitting! One approach would be to note that when the grid does *not* move with the shock, the perturbation solution shows a strong spike at the shock: if a grid velocity can be chosen such that the spike disappears, the grid is moving with the shock.

Resolving this issue of the treatment of shocks is vital if linearized Euler methods are to be as useful.

1.5.2 Acceleration

We have already shown (Section 1.3.7 above) that the code vectorizes and parallelizes readily. The only common acceleration technique we have not exploited is multigrid. Multigrid has been shown by Mavripilis, *et al.*, to work effectively on unstructured grids [MJM89]. We are currently incorporating multigrid into the code.

1.5.3 Three dimensions

In a real turbomachinery blade row the flow is rarely 2D. Much of the analysis of mean flow in a blade row is based on the assumption that the flow is *quasi-3D*, i.e., that the flow is confined to an axisymmetric streamsheet whose thickness varies in the (r, z) plane, but not in the θ direction. There is something to be gained in modifying the code to solve the linearized quasi-3D equations – and we are doing this – but the main benefit is to make it easier to make use of existing codes that create quasi-3D, rather than 2D, mean flow solutions. There is absolutely no reason to believe that unsteady disturbances obligingly confine themselves to the quasi-3D mean flow streamsheets! Cases in which 2D or quasi-3D unsteady flow codes give accurate results will be relatively rare. Unsteady flow codes in 3D must be developed.

Acknowledgements: The authors would like to thank the General Electric Company, and in particular R. E. Kielb, and R. Mani, for supporting this work and for permission to publish this paper. We have had many helpful discussions with Professor K. C. Hall of Duke University, to whom we also extend our thanks.

References

[Bat90] J. T. Batina. Implicit flux-split Euler schemes for unsteady aerodynamic analysis involving unstructured dynamic meshes. AIAA Paper 90-0936, 1990.

[BF86] A. Bolcs and T. H. Fransson. Aeroelasticity in turbomachines: Comparison of theoretical and experimental cascade results. Communication du Laboratoire de Thermique Appliquée et de Turbomachines Nr. 13, L'École Polytechnique Fédérale de Lausanne, 1986.

[DB74] G. Dahlquist and A. Bjorck. *Numerical Methods.* Prentice-Hall, New Jersey, 1974.

[Gil86] M. B. Giles. UNSFLO: A numerical method for unsteady inviscid flow in turbomachinery. Computational Fluid Dynamics Laboratory Report CFDL-TR-86-6, MIT, 1986.

[Gil88] M. B. Giles. Non-reflecting boundary conditions for the Euler equations. Computational Fluid Dynamics Laboratory Report CFDL-TR-88-1, MIT, 1988.

[Gil89] M. B. Giles. Numerical methods for unsteady turbomachinery flow. In *Numerical Methods for Flows in Turbomachinery, VKI Lecture Series 1989-06*, 1989.

[Hal87] K. C. Hall. *A Linearized Euler Analysis of Unsteady Flows in Turbomachinery.* PhD thesis, M.I.T., 1987.

[Hal89] K. C. Hall. Calculation of unsteady flows in turbomachines using the linearized Euler equations. *AIAA Journal*, 27(6):777–787, 1989.

[HC89] D. G. Holmes and S. D. Connell. Solution of the 2D Navier-Stokes equations on unstructured, adaptive grids. AIAA 9th Computational Fluid Dynamics Conference, AIAA Paper 89-1932-CP, 1989.

[He89] L. He. An Euler solver for unsteady flows around oscillating blades. ASME Paper 89-GT-279, 1989.

[HLC88] D. G. Holmes, S. H. Lamson, and S. D. Connell. Quasi-3D solutions for transonic, inviscid flows by adaptive triangulation. ASME Paper 88-GT-83, 1988.

[JBW86] A. Jameson, T. J. Baker, and N. P. Weatherill. Calculation of inviscid transonic flow over a complete aircraft. AIAA paper 86-0103, 1986.

[JST81] A. Jameson, W. Schmidt, and E. Turkel. Numerical solutions of the Euler equations by finite-volume methods using Runge-Kutta time-stepping scheme. AIAA paper 81-1259, 1981.

[Lan59] L. M. Landau. *Fluid Mechanics.* Pergamon Press, London, 1959.

[LG91] D. R. Lindquist and M. B. Giles. On the validity of linearized unsteady Euler equations with shock capturing. AIAA 10th Computational Fluid Dynamics Conference, AIAA Paper 91-1598-CP, 1991.

[MJM89] D. J. Mavripilis, A. Jameson, and L. Martinelli. Multigrid solution of the Navier-Stokes equations on triangular meshes. AIAA Paper AIAA-89-0120, 1989.

[Ni74] R.-H. Ni. *Non-Stationary Aerodynamics of Flat Plate Cascades in Compressible Flow.* PhD thesis, Stevens Institute of Technology, 1974.

[NS75] R.-H. Ni and J. Sisto. Non-stationary aerodynamics of flat plate cascades in compressible flow. ASME Paper 75-GT-5, 1975.

[Sid91] L. D. G. Siden. *Numerical Simulation of Viscous Compressible Flows Applied to Turbomachinery Blade Flutter.* PhD thesis, Chalmers University of Technology, Goteborg, Sweden, 1991.

[Ver77] J. M. Verdon. Further developments in the aerodynamic analysis of unsteady supersonic cascades, parts 1 & 2. ASME Paper 77-GT-44 & 45, 1977.

Prediction of Unsteady 2D-Flow in Turbomachinery Bladings

D. Peitsch[1]
H.E. Gallus[1]
H.-P. Kau[2]

Abstract

A numerical algorithm for calculating the unsteady flow in turbomachinery bladings is presented. The unsteadiness can be caused by oscillating cascades and incoming wakes of preceding blade rows simultaneously. The analysis of the time-dependent flow field yields the resulting blade forces and moments on the described profile.

The flow is assumed to be two-dimensional, time-dependent, inviscid and subsonic at the boundaries. Since the describing Euler-equations are not linearized, it is possible to handle with large oscillations and to compute non-harmonic phenomena, such as wakes, in only one computational run. The solution of the hyperbolic system of equations is carried out on a moving computational grid. At interior points, Mac-Cormack's explicit predictor-corrector-scheme is used, whereas the boundary conditions are formulated by means of a post-correction method based on one-dimensional characteristics. Emphasis was given to the formulation of a periodicity condition, which reduces the computational domain to only one representative channel even in the case of simultaneous unsteady phenomena of different frequencies. Results are shown for oscillating compressor and turbine cascades in comparison with experimental data. Furthermore, special test cases of the Fourth Standard Configuration have been investigated. At last, a result is shown for a flow with non-uniform inflow conditions.

---

[1]Institut fur Strahlantriebe und Turboarbeitsmaschinen, RWTH , D-5100 Aachen, Germany.
[2]BMW-Rolls Royce, Munich, Germany.

Introduction

The increased concentration of power per unit in the recent years lead to a high loading of turbomachinery bladings. The reliability of the whole machine can therefore be endangered by the impact of unsteady effects on the material fatigue. Since these phenomena have great influence on the efficiency of the machine, the prediction of the blade loads due to the unsteadiness of the flow is under intensive investigation, which is proven by the great number of publications dealing with this subject /1-6/.

Periodic disturbances of the flow in turbomachines are unavoidable due to the power transformation in rotating systems. They are caused by the potential interaction of blade rows, forced and self-excited blade vibrations, turbulence and stall. In addition, unsteadiness is generated by wakes of upstream blade rows. The investigation of the flutter instability problem is of interest in an early state of turbomachinery design to receive the aerodynamic forces and moments just in critical regions of the operating map. This can not only be done by experiments. Due to the high costs, modern computational methods have to be developed. Since the computing power until now is not yet sufficient to take into account all effects occurring in complex turbomachinery flows, simplifications have to be made. In the presented method, the flow is two-dimensional, compressible and inviscid.

Mathematical Model

The two-dimensional, compressible, unsteady and inviscid flow is described by a set of partial differential equations, consisting of the continuity equation, the equations of momentum and the energy equation. Today this set is commonly called the Euler-equations. Written in conservative formulation, the Euler-equations take the form:

$$\frac{\partial}{\partial t}\boldsymbol{U} + \frac{\partial}{\partial x}\boldsymbol{E}(\boldsymbol{U}) + \frac{\partial}{\partial y}\boldsymbol{F}(\boldsymbol{U}) = \boldsymbol{0} \qquad (1)$$

with:

$$U = \begin{pmatrix} \rho \\ \rho u \\ \rho v \\ e \end{pmatrix} \qquad E(U) = \begin{pmatrix} \rho u \\ \rho u^2 + p \\ \rho u v \\ (p+e)u \end{pmatrix} \qquad F(U) = \begin{pmatrix} \rho v \\ \rho u v \\ \rho v^2 + p \\ (p+e)v \end{pmatrix}$$

This set is closed by the assumption of a perfect gas:

$$e = \frac{p}{\gamma - 1} + \frac{\rho}{2}(u^2 + v^2) \tag{2}$$

The system is solved by finite differences on a curvilinear contour-conform grid as shown in Fig. 1. The discretization takes into account the local stretching of cells according to the transformation introduced by Thompson, Thames and Mastin /7/. Since the grid is time-dependent in case of oscillating blades, elliptical smoothing in the vicinity of the profile takes place at every time step to minimize discretization errors. The movement of the computational grid and the curvature terms are integrated in the Euler-equations. This leads to the transformed set of equations:

$$\frac{\partial}{\partial \tau}\hat{U} + \frac{\partial}{\partial \xi}\hat{E} + \frac{\partial}{\partial \eta}\hat{F} = 0 \tag{3}$$

with:

$$\hat{U} = U\,(x_\xi y_\eta - x_\eta y_\xi)$$

$$\hat{E} = U \cdot U_k + p \begin{pmatrix} 0 \\ y_\eta \\ -x_\eta \\ u y_\eta - v x_\eta \end{pmatrix} \qquad \hat{F} = U \cdot V_k + p \begin{pmatrix} 0 \\ -y_\xi \\ x_\xi \\ v x_\xi - u y_\xi \end{pmatrix}$$

where:

$$U_k = x_\eta(y_\tau - v) + y_\eta(u - x_\tau) \qquad V_k = x_\xi(v - y_\tau) + y_\xi(x_\tau - u)$$

This system is structured by analytical methods with the result, that even though formulated on a curved grid, the computational effort for the solution is not much higher than for the solution in the non-transformed region. This quasi-linear, hyperbolic partial differential equation system forms a properly posed initial/boundary-value problem together with an appropriate set of boundary conditions. Since no simplifying assumptions are made (like small perturbations), a time-step method has to be used.

Solution Algorithm

Equation (2) is solved by means of a combined method. Since interior points possess an entire system of 8 neighbouring points, the well known explicit two-step-scheme of MacCormack /8/ can be used here. The conditions at the boundaries shown in Fig. 1 are formulated by a post-correction-method based on one-dimensional characteristics /9/.

Inner Points

At interior points, equation (3) is directly formulated with finite-differences. The MacCormack scheme consists of the predictor-step, which gives a first approximation of the flow values at the new time step t+dt

$$\overline{U^{n+1}_{i,j}} = J^{n+1}_{i,j} \left\{ \frac{U^n_{i,j}}{J^n_{i,j}} - \frac{\Delta t}{\Delta \xi}\left[\hat{E}^n_{i+1-\delta,j} - \hat{E}^n_{i-\delta,j} \right] - \frac{\Delta t}{\Delta \eta}\left[\hat{F}^n_{i,j+1-\epsilon} - \hat{F}^n_{i,j-\epsilon} \right] \right\} \tag{4}$$

and the corrector-step, which uses this approximation for updating the fluxes E and F. Averaging the results gives the new values, which are of second order accurate, both in time and space

$$U_{i,j}^{n+1} = \frac{1}{2} J_{i,j}^{n+1} \left\{ \frac{\overline{U_{i,j}^{n+1}}}{J_{i,j}^{n+1}} + \frac{U_{i,j}^{n}}{J_{i,j}^{n}} \right. \tag{5}$$

$$\left. - \frac{\Delta t}{\Delta \xi} \left[\hat{E}_{i+\delta,j}^{n+1} - \hat{E}_{i-1+\delta,j}^{n+1} \right] - \frac{\Delta t}{\Delta \eta} \left[\hat{F}_{i,j+\epsilon}^{n+1} - \hat{F}_{i,j-1+\epsilon}^{n+1} \right] \right\}$$

The direction of the one-sided differences in the first step is cycled through the time steps. The second step uses the opposite direction for the computation, which is controlled by the variables δ and ϵ in equations (4) and (5). The rotation of the algorithm produces numerical fluctuations of high frequency. Since no interaction between numerical and physical effects in the solution of the non-linear set of equations may occur, the difference between both frequencies has to be very large. In the present method, this is ensured anyway by using the CFL-criterion for the maximum time step. Even in the case of high-frequent excitation of the profiles, a physical period consists of some hundred time steps, whereas the numerical period consists of four time steps.

Boundary Points

Since the influence of different unsteady phenomena is investigated simultaneously, special emphasis was given to the formulation of the boundary conditions. They are formulated by means of a post-correction method based on one-dimensional characteristics, which has the advantage to need only a short computation time. As a first step, a lightly modified MacCormack step is used to estimate the values at the boundaries for a new time step. These values are corrected in a second step using one-dimensional characteristics to satisfy the boundary conditions. Important for the application is, whether the eigenvectors of the characteristic equations point out of the computational domain or not. In the first case, they have to be replaced by boundary conditions. A further description of the governing equations is given in /1/. The added boundary conditions in the presented method are (Fig. 1):

At the subsonic upstream boundary total temperature T_t , total pressure p_t and the flow angle α are prescribed. In case of incoming wakes, p_t and α are given as a function of time.

At the downstream boundary either the static pressure p_2 or a non-reflecting formulation can be given.

On the profile the flow velocity normal to the surface is equal to the velocity of the profile. This gives the well known slip-condition in the non-vibrating case.

A non-permeable, fully flexible slipline is attached to the trailing edge point of the profile to account for outgoing unsteady vortices in the rather coarse mesh used for the computation. Each point of the slipline keeps two values for tangential velocity and density, one for the upper and one for the lower side of the slipline. Static pressure and normal velocity are equal on both sides.

The periodic lines in circumferential direction are handled by a special periodicity condition, which enables the reduction of the computational domain to only one channel even in the unsteady case.

Oscillation of profiles

The profile, which is assumed to be a rigid body, can move with both a sinusoidal bending and torsion vibration. The oscillation is determined by frequency, amplitude and the phase lag of two adjacent profiles. For the bending motion, a direction relative to the chord is prescribed, whereas for the torsional motion the axis of rotation is given. In addition the phase lag between bending and torsion is at choice. Before computing the unsteady flow problem, a stationary solution has to be evaluated starting from any given flow values. At the time t=0 the oscillation starts. This leads to the problem of a non-differentiable velocity at this time. The profile's moment of inertia prevents in reality the infinite acceleration. Hence, in the present method, the transition from the steady to the unsteady boundary condition is made differentiable (n=1):

$$\underline{t<0:} \qquad s_p = 0 \qquad v_p = 0 \qquad \lim_{t\to 0} v_p = 0$$

$$\underline{0 \le t \le nT:} \qquad s_p = A\,\frac{t}{nT}\,\sin(\omega t) \qquad v_p = \frac{A\omega t}{nT}\cos(\omega t) \qquad \lim_{t\to 0} v_p = 0$$

$$\underline{t>nT:} \qquad s_p = A\sin(\omega t) \qquad v_p = A\,\omega\cos(\omega t)$$

Fig. 2 shows a comparison of the pressure course over some periods of computation between a non-differentiable start of motion and the formulation above. In the first case, extremely high gradients in the flow values appear. This has a negative influence on the convergence of the undamped algorithm. In contrast, the overshoot of the pressure is completely avoided with the new formulation and the periodic course is obtained very fast.

Wake Model

The wakes of rotating cascades induce very high pressure fluctuations on the downstream profile's surfaces. Experiments have shown, that wakes are essentially characterized by a defect of the relative velocity in the center of the wake accordingly to the defect of total pressure p_t, caused by the viscous effects on the surface of preceding blades. The velocity distribution at the upstream boundary can be described by an exponential function as shown in Fig. 3. The depth of the wake h_D is given by per cent of the velocity in the relative frame of reference. Superposition of the velocity in circumferential direction yields the fluctuations of velocity and flow angle in the absolute frame of reference. The distribution of the total pressure is computed each time step from the actual velocity distribution with the assumptions of constant total temperature and equalized averaged static pressure. The loss of the total pressure is a criterion for the profile losses of the upstream wake generator.

Periodicity Condition

The usage of a periodicity condition reduces the computational domain to only one channel. In the steady flow, each channel is similar, hence the periodicity condition is very easy to formulate. In contrast, the computation of the unsteady flow requires a very expensive handling of storaging the values at the periodic lines, especially in the case of simultaneous oscillation of blades and incoming wakes. The time of the whole system used for one period has to be obtained by the periodical similarity both of the channel's geometry and the upstream boundary conditions. This procedure gets very complex in the case of unequal spacing of wake generator and the considered channel.

Convergence Criterion

The development of the presented time step method was carried out with special emphasis on the efficiency in the industrial usage. To avoid a further computation after obtaining the periodic state, a criterion has been developed, which analyzes the course of convergence at the end of each period. In the case of little changes of the fluctuations in comparison with the preceding period the computation stops properly. The extensive analysis of the received flow values can follow immediately.

Results

Steady Cascade Flow

Before the presentation of results with oscillating blades, a computed steady-state solution is presented to show the prediction capability of the algorithm. Fig. 8 shows the time averaged pressure coefficients for the "Fourth Standard Configuration" /10/ using a computational grid of 51x17 points. The turbine cascade described in this configuration operates under high subsonic flow conditions. The agreement between measured and calculated results is fairly good, only in the near sonic region there exist some minor deviation.

Oscillating Compressor Cascade

Since the available reliable experimental data about aerodynamic pressure distributions caused by unsteady flow in turbomachines are poor, expensive investigations took place in the annular test rig of the Institut für Strahlantriebe und Turboarbeitsmaschinen at Aachen, Germany. The results contained in /11/ were obtained from measurements on a cascade, which was excited to oscillations in his whole entirety (Travelling-Wave-Mode, TWM). The great moving masses resulted in little amplitudes at low frequencies. In the succeeding experiments only one blade was excited, while the remaining blades are idle (Influence-Wave-Mode, IWM). The costs for this procedure are much lower than for the above mentioned experiments. Contemporarily, the reduced moving mass and larger volume for the installation of more powerful electrodynamic exciters lead to the possibility of oscillations with higher amplitudes and frequencies, both in bending and torsion mode. The IWM does not exist in the real machine, but it is easy to generate in the experiment and it is possible to overcome it by calculation with a minor modification of the program in spite of not existing periodicity of adjacent channels. The purpose of the experimental tests was the establishing of a data base for the verification of the computational method. The Mach number realized and the span of the profile were not comparable to endangered cases in installed engines. The resulting blade forces and moments were uncritical. But both experimental and numerical investigations show, that just growing Mach numbers and amplitudes have a great influence on the aerodynamic loading.

As a first result, Fig. 4 contains a comparison of the dynamic pressure between calculation and experiment for the introduced compressor cascade. A bending oscillation in TWM with the frequency $\upsilon = 132$ Hz and the interblade phase angle $\sigma = 180°$ was given. The unsteady pressure was measured in five positions D1 to D5 on the pressure side and six positions S1 to S6 on the suction side. The agreement in amplitude and phase is very good. Fig. 5 shows the pressure fluctuations due to a coupled bending-torsion motion in IWM. The bending motion was carried out perpendicular to the chord with an amplitude of $A = 0.5$ mm, whereas the torsion motion had an amplitude of $\alpha = 0.65°$ and a phase lag of 10° to the bending. In this case some discrepancies occur, especially on the pressure side. At the location D2 the computation shows only half of the amplitude of the measurement, whereas the predicted value at D3 exceeds the measurement by 20 per cent. There seems to be some deviation in the measured results, because there is no physical explanation for the high amplitude at D3, which is located near the torsional axis. On the whole, the accuracy is fairly well and will increase with a better discretization and a better analysis of the data from the moving pressure transducers.

Oscillating Turbine Cascade

For the investigated subsonic turbine cascade, Fig. 6 and Fig. 7 contain results for two different Mach numbers and frequencies of the performed bending oscillation. In Fig. 6 the agreement of computation and experiment is good at most kulite locations. The relative coarse discretization at the leading edge yields strong fluctuations on the suction side. Averaging the amplitudes (dashed line) shows only a small deviation of calculation and experiment. The reduction of the fluctuation from leading to trailing edge is not predicted as measured in the experiment, but the deviation of ca. 30 Pa is nearly the accuracy of the pressure transducers. A higher frequency and Mach number results in much higher pressure fluctuations in spite of a smaller amplitude (Fig. 7). The agreement of the results is very well, also on the pressure side.

Fourth Standard Configuration

The presented method has been applied to the "Fourth Standard Configuration", described in /10/. The experimental tests have been performed in the annular cascade facility at the Lausanne Institute of Technology /12/. The presented method is validated with the same test cases used for the publication on the UAATP-Symposium 1987, held in Aachen /13/. The constant outlet Mach number ($Ma_{2is}=0.9$) and different interblade phase angles were prescribed for the four considered aeroelastic test cases. The time averaged pressure distribution (Fig. 8) was already mentioned above.

The predicted unsteady blade surface pressure coefficients are summarized in magnitude and phase lead in Figs. 9 to 12 for four different phase angles. Results from the presented method are compared with the experimental data and a linearized numerical method, indicated as "Method 7" /14,15/. The agreement of both amplitude and phase angle of the pressure coefficient along the blade chord is good. The predicted amplitudes agree in all cases. In comparison with the results presented 1987, the amplitudes are lightly increased, especially on the pressure side. This effect is caused by the new undamped boundary treatment in contrary to the damped bicharacteristics-method used in /13/. The required computation time for the test cases using the post-correction method is clearly decreased compared with the earlier used method. This opens the possibility for a wide range of applications in the industrial design process.

Fig. 13 shows the predicted damping coefficient. The agreement is very good for all interblade phase angles, except for $\sigma = 90°$. In this case, the phase lead has a bigger discrepancy especially at the locations of the last two transducers on the pressure side and the fifth transducer on the suction side (Fig. 11). This leads to a wrong contribution to the aerodynamic work, whereas for example in the case of $\sigma = -90°$ (Fig. 9) the averaged discrepancies of the phase lead are much smaller, so the aerodynamic damping coefficient is computed correctly.

This result enables the prediction of aerodynamic work resp. damping coefficient in endangered, i.e. the unstable region of oscillation.

Flows with Non-Uniform Inflow Conditions

The computation of unsteady flows with non-uniform inlet conditions was performed with very different frequencies and pitches of the wake. Simultaneous excitation by oscillating profiles and wakes can be considered as well. As an example, Fig. 14 shows a case with a wake of high frequency. In experiment and calculation the depth of the wake amounts about 20 per cent, the relative width about 50 per cent. The agreement is good. In reality the wakes are mixed out by viscous effects, which are not included in the computational method. This results in too high predicted amplitudes in relation to the measured data towards the trailing edge.

Conclusion

The presented method solves the non-linear, two-dimensional, unsteady Euler-Equations for subsonic turbomachinery flows. Simultaneous excitation by oscillating profiles and unsteady non-uniform inlet conditions can be considered. The required computational time is small in comparison with previous methods due to the changed boundary treatment.

The obtained results show good agreement with experimental data for compressor and turbine cascades. Especially the application on the Fourth Standard Configuration verifies the prediction capability of the calculation method.

References

1 H.-P. Kau, H.E. Gallus:
"Numerical Investigation of Unsteady Flow in Oscillating Turbine and Compressor Cascades", AGARD-CP-468/469, 1989

2 H.J. Joubert:
"Flow Field Calculation in Compressor Operating with Distorted Inlet Flow", ASME-Paper 90-GT-212, 1990

3 T.H. Fransson:
"Analysis of Experimental Time-Dependent Blade Surface Pressures from an Oscillating Turbine Cascade with the Influence-Coefficient Technique", ASME-Paper 90-GT-225, 1990

4 G.A. Gerolymos:
"Comparison of Inviscid Computations with Theory and Experiment in Vibrating Transonic Compressor Cascades", ASME-Paper 90-GT-373, 1990

5 J.M. Verdon:
"Linearized Unsteady Aerodynamics for Turbomachinery Aeroelastic Applications", AIAA-Paper 90-2355, 1990

6 K. Rao, R. Delaney et. al.:
"Investigation of Unsteady Flow Through Transonic Turbine Stages", Part I + II, AIAA-Paper 90-2408 / 2409

7 J.F. Thompson, F.C. Thames, C.W. Mastin:
"Boundary-Fitted Curvilinear Coordinate Systems for Solutions of Partial Differential Equations on Field Containing Any Number of Arbitrary Two-Dimensional Bodies", NASA CR-2729, July 1976

8 R.W. MacCormack:
"Basic Explicit and Implicit Finite-Difference and Finite-Volume Schemes for Hyperbolic and Parabolic Systems", VKI-Lecture-Series 1985-04, 1985

9 A. Fourmaux:
"Unsteady Flow Calculation in Cascades", ASME-Paper 86-GT-178, 1986

10 A. Bölcs, T.H. Fransson (eds.):
"Aeroelasticity in Turbomachines - Comparison of Theoretical and Experimental Cascade Results.", Communication de Laboratoire du Thermique Appliquée et de Turbomachines, No. 13, Lausanne, EPFL, 1986

11 W.B. Knauf:
"Experimentelle Untersuchungen zur Ermittlung der aerodynamischen Schaufeldämpfung an einem schwingenden Axialverdichtergitter.", Diss., RWTH Aachen, 1985

12 A. Bölcs:
"A Test Facility for the Investigation of Steady and Unsteady Transonic Flows in Annular Cascades", ASME-Paper 83-GT-34, 1983

13 S. Servaty, H.E. Gallus, H.-P. Kau:
"Computation of Aerodynamic Blade Loads Due to Wake Influence and Aerodynamic Damping of Turbine and Compressor Cascades", Proceedings of the Fourth Symposium On Unsteady Aerodynamic and Aeroelasticity of Turbomachines and Propellers, Aachen, 1987

14 S. Newton, D.S. Whitehead:
"A Finite Element Method for the Solution of Two-Dimensional Transonic and Supersonic Flows", International Journal for Numerical Methods in Fluids, Vol. 5, pp. 115-132, 1985

15 R.D. Cedar, P. Stow:
"The Addition of the Quasi-Three-Dimensional Terms into a Finite Element Method for Transonic Turbomachinery Blade-to-Blade Flows", International Journal for Numerical Methods in Fluids, Vol. 5, pp. 101-114, 1985

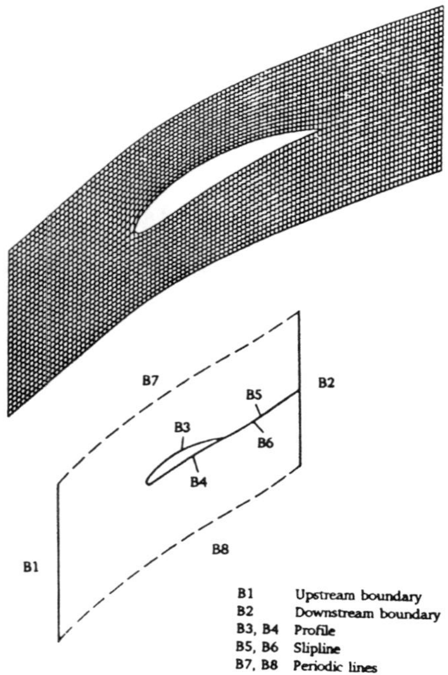

Figure 1:
Computational grid and boundaries

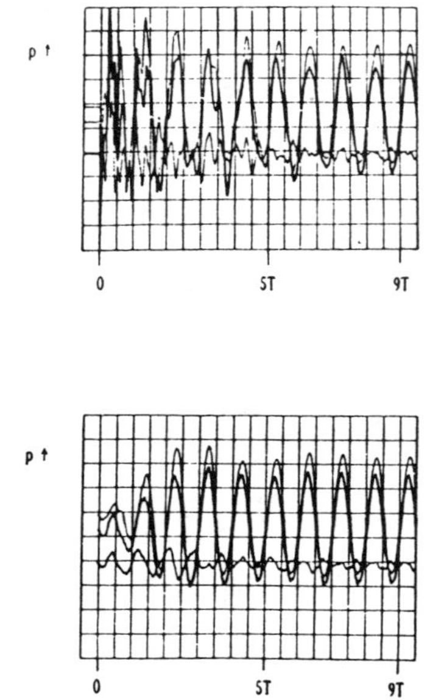

Figure 2:
Comparison of dynamic surface pressure for non differentiable and differentiable start of motion

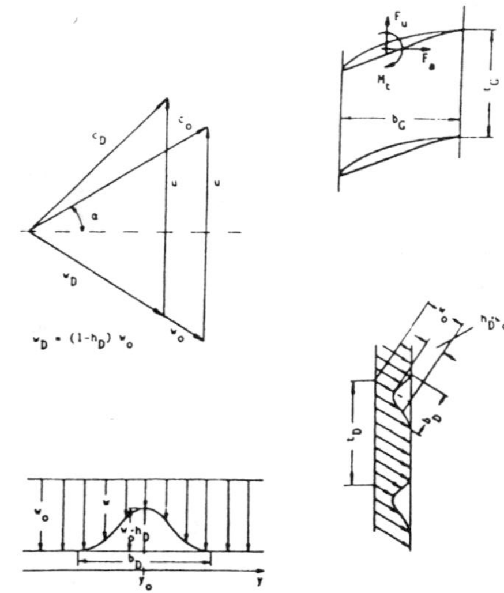

$$\frac{w(y)}{w_o} = 1 - h_D\, e^{-16\frac{(y-y_o)^2}{b_D^2}}$$

Figure 3:
Wake model in the relative and the absolute frame of reference

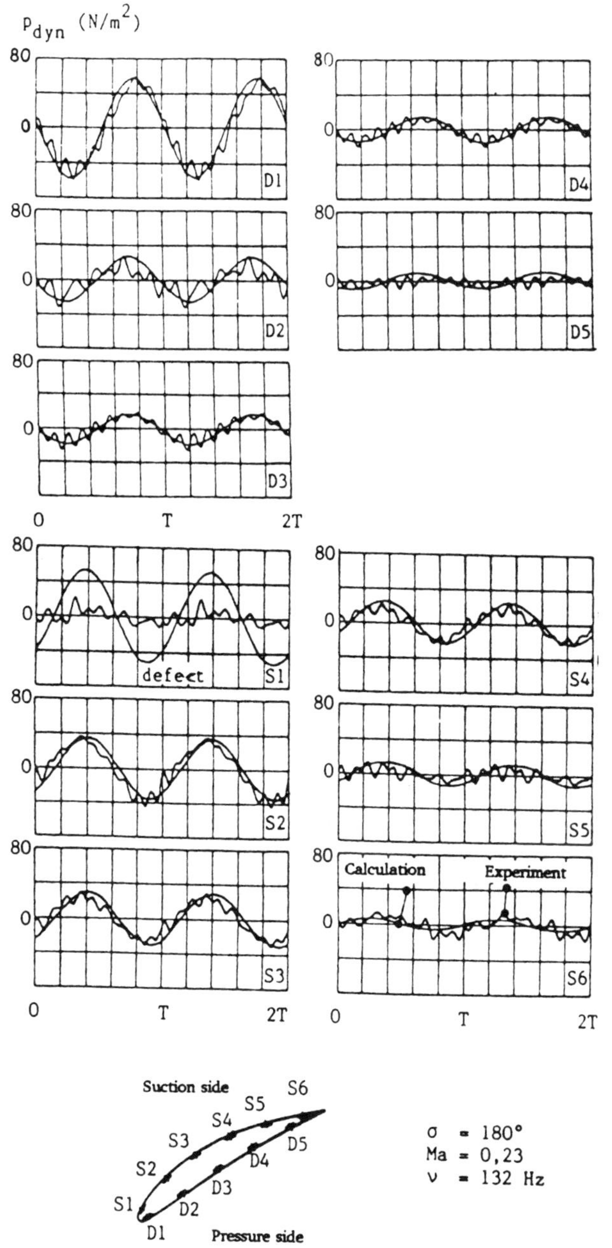

Figure 4:
Comparison of calculated and measured pressure fluctuations on a compressor profile in bending oscillation (TWM)

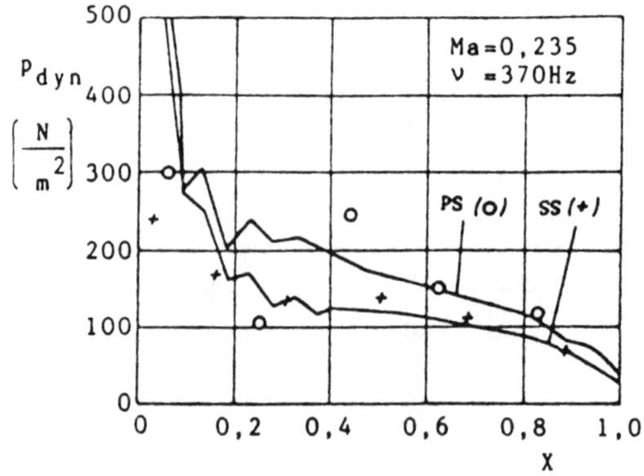

Figure 5:
Compressor cascade in coupled bending-torsion oscillation (IWM)

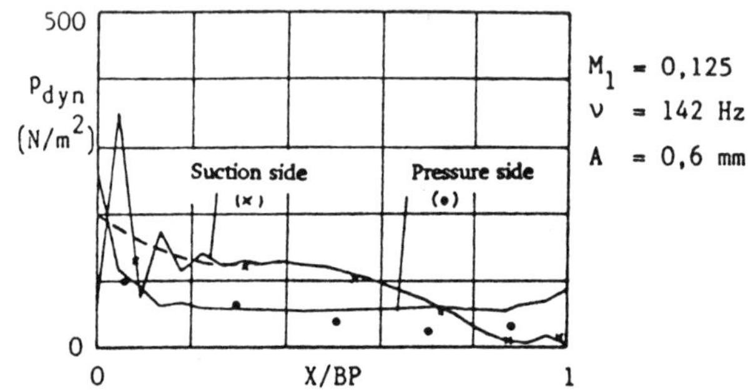

Figure 6:
Turbine cascade in bending oscillation (IWM, 142 Hz)

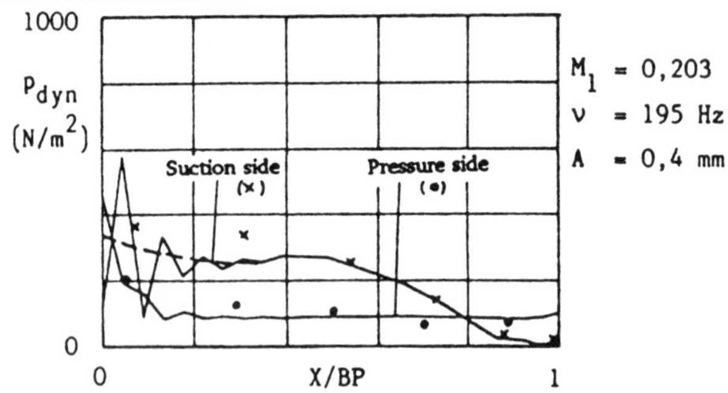

X, • Experiment —— Calculation

Figure 7:
Turbine cascade in bending oscillation (IWM, 195 Hz)

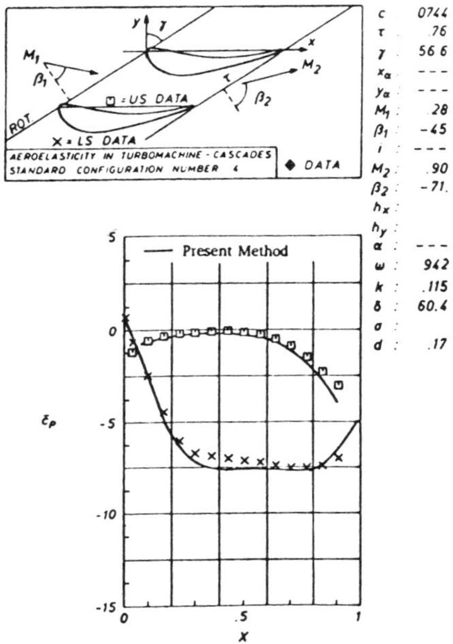

Figure 8:
Fourth Standard Configuration
Time averaged blade surface pressure distribution, Case 3, 6-8

Case 3

$\sigma = -90°$

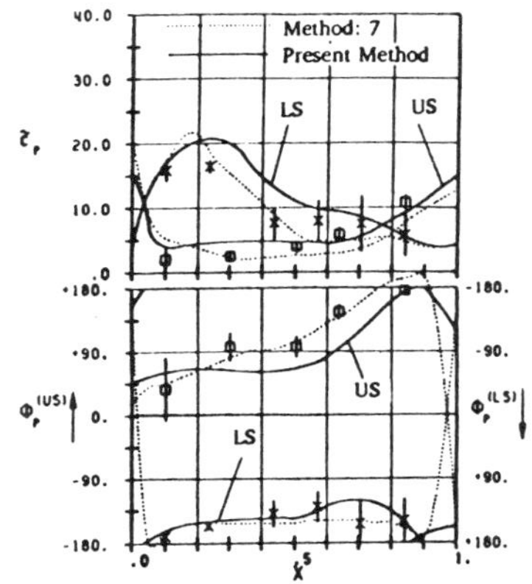

Figure 9:
Fourth Standard Configuration
Magnitude and phase lead of unsteady blade surface pressure coefficient
Case 3

Case 6

$\sigma = 180°$

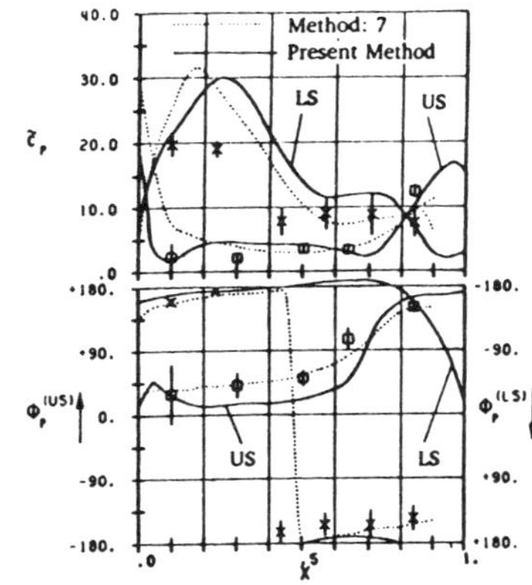

Figure 10:
Fourth Standard Configuration
Magnitude and phase lead of unsteady blade surface pressure coefficient
Case 6

Case 7

$\sigma = 90°$

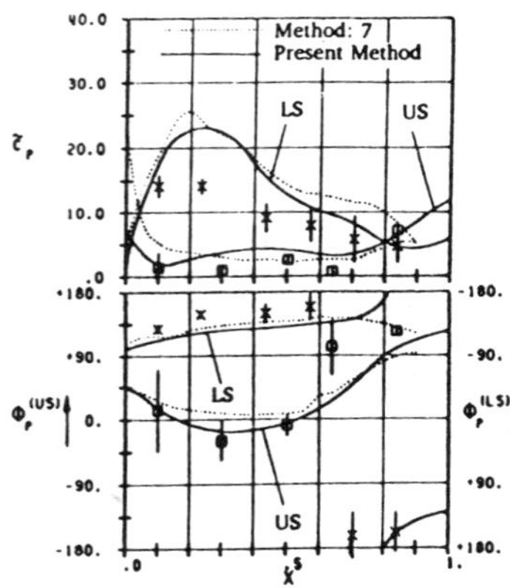

Figure 11:
Fourth Standard Configuration
Magnitude and phase lead of unsteady blade surface pressure coefficient
Case 7

Case 8

$\sigma = 0°$

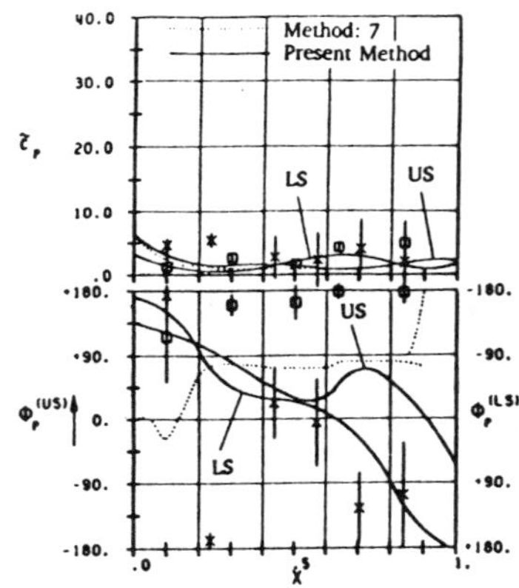

Figure 12:
Fourth Standard Configuration
Magnitude and phase lead of unsteady blade surface pressure coefficient
Case 8

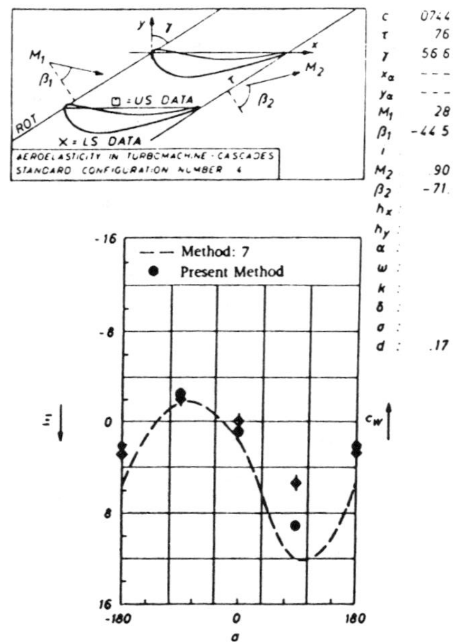

Figure 13:
Fourth Standard Configuration
Aerodynamic work and damping coefficient
Case 3, 6-8

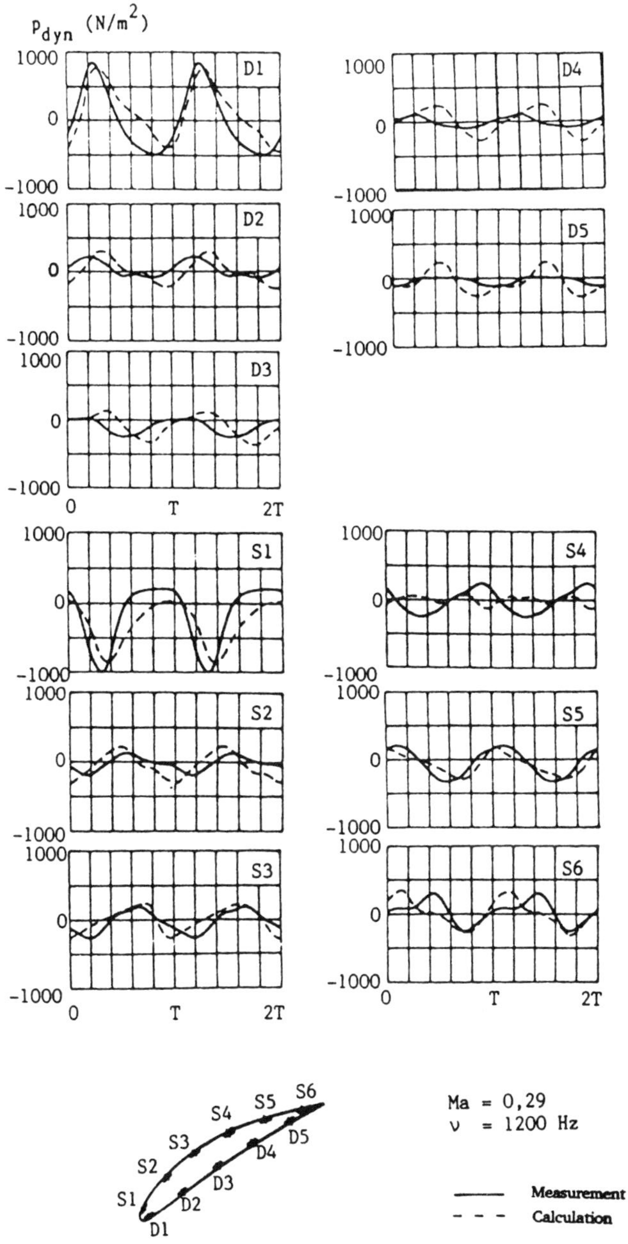

Figure 14:
Compressor cascade with non-uniform, unsteady inlet conditions

Axial-Gap Effects on the Propagation of Unsteady Flow in Cascades

Theodosios Korakianitis[1]

ABSTRACT This paper investigates the effects of stator-to-rotor axial gap on the two-dimensional propagation of pressure disturbances due to potential-flow interaction and viscous-wake interaction from upstream blade rows in axial-turbine-blade rotor cascades. Results are obtained by modeling the effects of the upstream stator viscous wake and potential-flow field on the downstream rotor flow field and computing the unsteady flow fields in the rotor frame. The amplitudes of the two types of interaction for typical turbines are based on a review of available experimental and computational data. The potential-flow field is modeled as a sinusoidal pressure disturbance of amplitude 4% of the local pressure across the stator trailing edges that decays downstream. The velocity wake is modeled as a Gaussian velocity defect of varying amplitude and width, depending on the axial gap between the blade rows. The wake amplitudes and widths are based on conservation of loss of incoming momentum to the rotor due to the wake. The axial gap between rotor and stator is varied to show how the two interactions propagating in different directions reinforce or counteract each other at different axial gaps. The corresponding forces on rotor blades are computed for typical values of reduced frequency.

1.1 Nomenclature

a	acoustic velocity (eq. 1.4)
B	amplitude of potential variation (eq. 1.6)
b	axial chord of rotor or stator cascade
C_L	tangential loading (lift) coefficient
c	local absolute-flow velocity (eq. 1.3)
c_0	rotor-inlet total sonic velocity (eq. 1.9)
c_n	stator-outlet absolute-flow averaged velocity
D	wake amplitude, fraction of c_n (eq. 1.3)
d_{ax}	stator-rotor axial gap (figure 1.1)

[1]Department of Mechanical Engineering, Washington University, One Brookings Drive, Campus Box 1185, St.Louis, MO 63130

F	force (eq. 1.9)
j	$\equiv \sqrt{-1}$ (eq. 1.6)
M_z	moment (in the z direction) (eq. 1.9)
N	number of blades (eq. 1.1)
R	($\equiv N_{rb}/N_{sb}$) stator-to-rotor-pitch ratio
S	pitch of a cascade
t	time (nondimensionalized by y/S_{sb})
u	velocity component in the x direction
V	component of velocity
v	velocity component in the y direction
W	characteristic wake width, fraction of S_{sb} (eq. 1.3)
(x, y, z)	cartesian coordinates (in subscripts also)
α	flow angle
α_n	angle of wake propagation at inlet, along c_n
Δ	perturbation operator (on p, u and v)
ς	parameter for the y period of Φ (eq. 1.6)
ξ	parameter for the x decay of Φ (eq. 1.6)
ρ	relative total density at rotor inlet
Φ	velocity potential (defined by eq. 1.5)
ω	rotor passing frequency (eq. 1.2)
$\tilde{\omega}$	reduced frequency parameter (eq. 1.2)
Subscripts	
i, o	cascade inlet, outlet respectively
ip	flow property for potential-flow model
iw	flow property in the wake
q	suction (s) or pressure (p) side (eq. 1.3)
rb, sb	rotor and stator blade row respectively
ϕ	perturbation value due to potential (eq. 1.5)

1.2 Introduction

Two important parameters in unsteady stator-rotor interactions are the reduced frequency and the stator-to-rotor-pitch ratio R, defined by:

$$R \equiv \frac{N_{rb}}{N_{sb}} = \frac{S_{sb}}{S_{rb}} \tag{1.1}$$

The effect of R on geometry is illustrated in figure 1.1. The stators are shown dashed in the figure because the approach models the stator-disturbance/rotor interaction and not the full stator-rotor interaction. The velocity diagrams, the rotors, and the nondimensional geometry of the upstream stator are identical, but R is varied by changing the value of S_{sb}. R is inversely proportional to the reduced frequency parameter, defined by:

$$\tilde{\omega} \equiv \frac{\omega \cdot b_{rb}}{c_x} = \frac{2\pi\, V_{rb}\, b_{rb}}{S_{rb}\, c_x} \cdot \frac{1}{R} \tag{1.2}$$

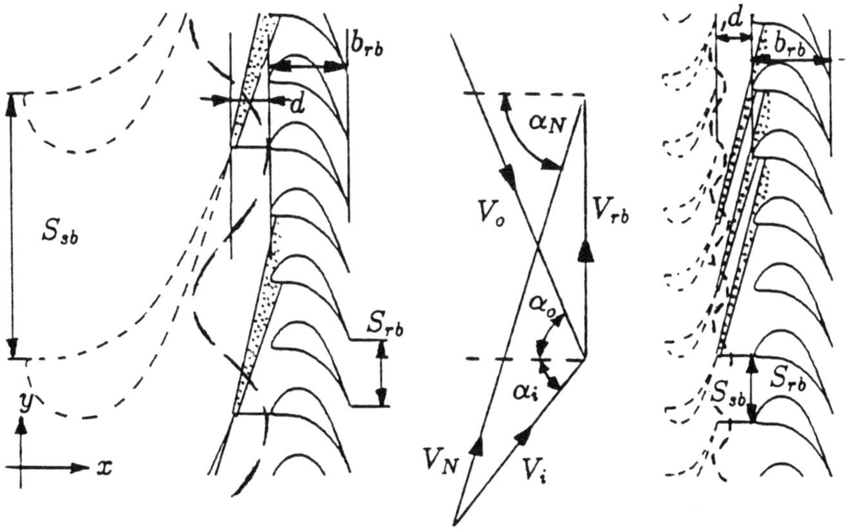

(a) $R = S_{sb}/S_{rb} = 4.0$ (b) $R = S_{sb}/S_{rb} = 1.0$

FIGURE 1.1. The effect of R on the stage geometry. The velocity diagrams and the downstream rotors (flow from left to right) are identical.

The sources of two-dimensional flow unsteadiness between the rotor and the stator are: viscous velocity wakes shed from the trailing edge of the stator; inviscid potential-flow interaction because of the relative motion of the lifting surfaces; two-dimensional vortices shed at the stator trailing edge; flutter of both cascades; and the effect of flow changes due to cooling flows in high-pressure high-temperature turbine stages. In figure 1.1 the velocity wake is illustrated in the dotted region starting at the stator trailing edges; the potential-flow interaction is illustrated as a sinusoidal variation in potential (and pressure) with a maximum at the stator trailing edge. As R increases the potential-flow interaction relative to the rotor increases, and its frequency and decay rate decrease. As R increases the amplitude of the velocity defect in the wake remains about constant, and one may observe a modest increase in the width of the wake because of the increased Reynolds numbers and larger lengths over which the boundary layers develop on the larger stators.

1.3 Background

The amplitude of the harmonics of the spatially analyzed unsteady forces on rotor blades as a function of R for a series of typical turbine cascades were computed in past investigations [Kor87,Kor88a,Kor88b,Kor90a,Kor90b]. The effect of the viscous wake interaction and of the potential-flow interaction from upstream stators were modeled as incoming flow disturbances into the computational flow field of the downstream rotors. The resulting unsteady flows in the rotor fields were computed, using the wake and potential-flow models as unsteady inlet-boundary conditions. The effects of two-dimensional-vortex shedding (which result in much higher harmonics of much lower amplitude in the forcing functions) were neglected, and infinitely rigid blades without cooling were assumed. The mechanisms of propagation of these interactions into the rotor flow fields were also investigated [Kor91c,Kor91a] in various rotor cascades of low, intermediate and high aerodynamic loading (tangential-lift coefficients of 0.8, 1.0 and 1.2). In these studies (past and the present one) we used Giles' [Gil88] computer program UNSFLO to compute the flow field and the forces. Initially UNSFLO was an Euler solver for the two-dimensional, unsteady, compressible, inviscid flow around rotor blades. Although later versions of the program can compute viscous stator-rotor flows (see Giles' contribution to this volume), we have concentrated on identifying the propagation of the disturbances in the core of the flow using the inviscid version of the program. This program was chosen because it can handle arbitrary values of R with reasonable CPU and storage requirements due to a novel "tilting" of the time domain. The accuracy of the computations has been checked by comparing results of calculations using UNSFLO with the results of numerous steady and unsteady-flow cases of known theoretical and experimental output [Kor87,Kor88a,Kor90a,Gil88,GH91].

The unsteady-pressure disturbances from the wake and potential-flow interactions from upstream propagate via different mechanisms and in different directions in the downstream cascade. In this paper we examine the flow fields at four axial gaps and illustrate how one may influence (and in some cases may minimize) the unsteady flows and forcing functions. Although in the following we have assumed an upstream stator influencing a downstream rotor, the conclusions can also be used for an upsteam rotor influencing a downstream stator.

1.4 The rotor-inlet boundary

The two disturbances from the stator (viscous wake and potential-flow interaction) have been modeled as inputs to the computational rotor-inlet boundary. This simplification provides accurate computational results only if one is extremely careful to specify the correct boundary conditions to the problem. Details of the following derivations have been published elsewhere [Kor87,Kor90a,Kor91c,Gil88]. A few important equations have been included here for clarity and completeness because they are essential to understanding the model of the rotor-inlet boundary, and the following results and discussion.

The velocity disturbance due to the wake is characterized by the max-

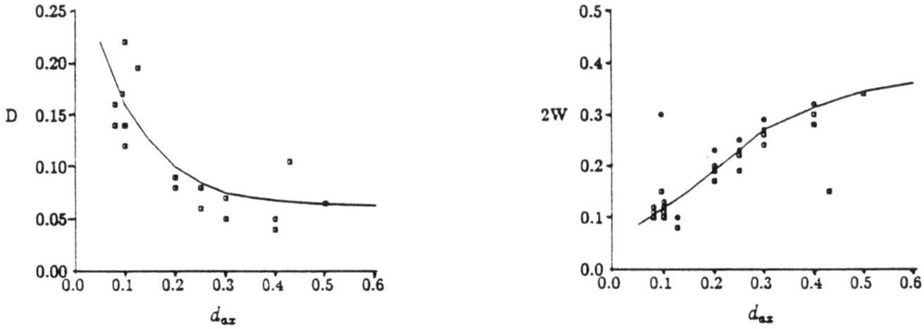

FIGURE 1.2. Experimental turbine-wake data as a function of axial gap. Left, amplidudes D. Right, full widths 2W

imum amplitude of the velocity defect D, expressed as a fraction of the undisturbed velocity, and by the "width" W of the velocity defect. Most velocity wakes observed in experimental data have velocity distributions which resemble Gaussian distributions. For the wake model we assumed that: the velocity defect is a Gaussian distribution with chatacteristic width expressed as a fraction of the pitch of the blade cascade that generates the wake; in the stator frame the flow vectors in the wake are parallel to the undisturbed flow (a velocity deficit with no angle variation); the static pressure is constant across the wake; and the total enthalpy is constant across the wake. The above are modeled by:

$$\begin{aligned} u_{iw,q} &= \left[1 - D \cdot exp\frac{1}{2}\left(\frac{tan(\alpha_n)x - y}{S_{sb}W_q}\right)^2\right] c_n cos(\alpha_n) \\ v_{iw,q} &= \left[1 - D \cdot exp\frac{1}{2}\left(\frac{tan(\alpha_n)x - y}{S_{sb}W_q}\right)^2\right] c_n sin(\alpha_n) \end{aligned} \tag{1.3}$$

where q is either s (for the suction surface) or p (for the pressure surface).

The axial gap between blade rows in modern engines is between 0.2 and 0.5 of the axial chords. Narrower gaps result in shorter engines but increase the unsteadiness between the blade rows. Experimental wake data for isolated airfoils, for fan or compressor cascades, and for turbines have been published in [Kor90a]. The data points in figure 1.2 show turbine wake data extracted from [DJHW82], [Hod83,Hod85a,Hod85b], [Che85], [MA85], [Hub86], [BR82], [BFKR84], [Bin85], [BFMR87], and [WDSH87]. Typically the suction side of the wakes has larger width than the pressure side. The solid lines represent the values of D and $2W$ that were used in this investigation. They were chosen so that they represent the same loss of momentum (4.74%) due to wake at the rotor-inlet boundary, and they are shown in table 1.1. A common width has been used for both sides of the wake. (The different widths of the suction and pressure sides have a sec-

TABLE 1.1. Wake amplitudes D and widths $2W$ as a function of axial gap d_{ax}

d_{ax}	D	$2W$
0.05	0.2200	0.0862
0.10	0.1600	0.1185
0.15	0.1250	0.1518
0.20	0.1000	0.1912
0.25	0.0860	0.2266
0.30	0.0750	0.2700
0.35	0.0700	0.2991
0.40	0.0680	0.3134
0.45	0.0660	0.3300
0.50	0.0645	0.3442
0.60	0.0630	0.3604

ondary effect on the unsteadiness-propagation mechanisms downstream).

The model for the potential-flow disturbance was developed by observing experimentally measured and computed static pressure fields of various turbine-blade cascades. Across the line of the trailing edges there is a variation of static pressure with maxima at (or very near) the trailing edges and minima at (or very near) the middle of the passage. The exact location and shape of the pressure variation depends on the geometric shape of the passage. It is nearly sinusoidal for cascades of various geometries and tangential lift-coefficients, and the amplitude of the pressure disturbance decays very fast with distance downstream. Numerous examples of this static pressure variation have been reviewed in the experimental data listed in [Kor87,Kor90a]. For example, measured cascade data such as those shown in figure 6 of [Son85], figure 4 of [YY85], and figure 7 of [BS84] and figure 6 of [SVHB83], show the near-sinusoidal shape and the rapid decay of the potential-flow interaction. The sinusoidal model is a good representation for many cascade geometries, and for high and low lift coefficients (0.8 to 1.2) [Kor91a].

The potential-flow model (in the stator frame) is derived as a two-dimensional, linear, isentropic, irrotational perturbation to uniform flow (see page 198 of [LR57]):

$$(u^2 - a^2)\frac{\partial u}{\partial x} + (v^2 - a^2)\frac{\partial v}{\partial y} + u\, v\left(\frac{\partial u}{\partial y} + \frac{\partial v}{\partial x}\right) = 0 \tag{1.4}$$

The velocity-potential relationship is defined by:

$$\frac{\partial \Phi}{\partial x} \equiv u \qquad\qquad \frac{\partial \Phi}{\partial y} \equiv v \tag{1.5}$$

For subsonic flows one expects that the potential-flow interaction is periodic in the y (circumferential or tangential) direction and that it decays exponentially in the x (axial) direction. Thus the general solution of equation 1.4 is of the form:

$$\Phi(x,y) = B \cdot exp(j\varsigma y + \xi x) \tag{1.6}$$

Substituting these in equation 1.4 we derive solutions for the corresponding velocity perturbations Δu and Δv. The latter are coupled with the conditions of no variation in entropy and total enthalpy to give a cosinusoidal pressure perturbation Δp (later unsteady flow figures show the resulting shape of Δp in the rotor frame) of the form:

$$\begin{aligned} \Delta p &= \Delta p_\phi \cos\left\{2\pi(\epsilon + \Delta\epsilon) - tan^{-1}\left[\frac{tan(\alpha_n)}{\sqrt{1-M^2}}\right]\right\} \\ \Delta p_\phi &= \frac{\rho c M \Delta v_\phi}{\sqrt{1-M_x^2}} \end{aligned} \tag{1.7}$$

where M is given by $M = \sqrt{M_x^2 + M_y^2}$, M_x and M_y are the axial and circumferential Mach numbers in the stator frame, and ϵ and $\Delta\epsilon$ are phase-shifting constants that locate the maximum pressure perturbation and the wake centerline at the stator trailing edge. Δv_ϕ is the maximum perturbation in circumferential velocity v at the inlet boundary. The perturbations are functions of the axial distance from the computational rotor-inlet boundary. The direction of propagation of the potential field is given by:

$$tan(\alpha_{ip}) = -\frac{M_x M_y}{1 - M_x^2} \ . \tag{1.8}$$

Experimental and computational data [Kor87] indicate that across the line of the trailing edges Δp_ϕ is typically between 4% and 4.5% of the average pressure. The corresponding values of Δv_ϕ for values of V_{rb} typical of modern engines are between 4.52% and 5.31% of c_n. In the computations presented in this paper we used $\Delta p_\phi = 0.04$.

The combined disturbance of the potential-flow interaction and the wake interaction from the upstream stator at its oulet boundary is found by adding the values of the two disturbances. These are input at the computational rotor-inlet boundary via a coordinate transformation for V_{rb}.

1.5 Disturbance propagation

The potential-flow and wake interaction models propagate in the rotor passages via different mechanisms, presented in [Kor91c]. These mechanisms are the same for a variety of blade-loading and flow conditions [Kor91a].

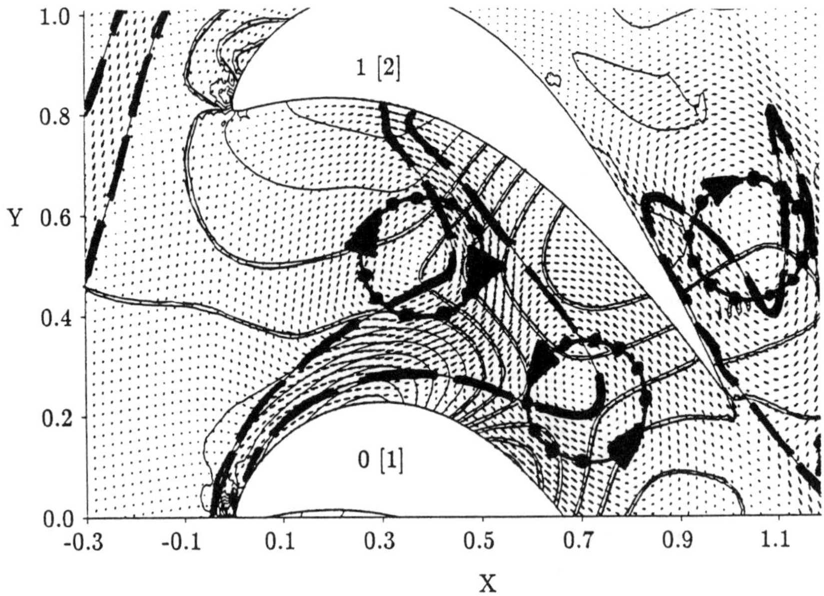

FIGURE 1.3. Unsteady flow vectors from wakes, illustrating regions of increased and decreased pressure upstream and downstream of the wake centerline (from [Kor91a])

1.5.1 POTENTIAL-FLOW INTERACTION

As the potential-flow interaction from the stator extends into the rotor cascade, large portions of it enter the passages when the direction of propagation of the potential is aimed near the center of the rotor passage. The potential flow field of the stator is cut by the advancing potential flow field of the rotor. After it is cut it moves downstream according to equation 1.7 as a potential-flow disturbance in the rotor-relative frame. The frequency (with respect to the rotor) and the decay rate of the potential-flow interaction in the rotor cascade are affected by the rotor geometry and performance, and the value of R.

1.5.2 WAKE INTERACTION

This has been explained in detail in [Kor91c]. The stator-wake centerlines are first bent by the potential flow field of the advancing rotor. As the leading edge in the rotor stagnation region interacts with the lower-momentum fluid in the wake recirculating-flow patterns are established in the stagnation region of the leading edge of the rotor. These recirculating flow patterns are generated as the wake is being cut; once generated they result in a counter-clockwise rotating unsteady-flow pattern downstream of the wake centerline, and a clockwise rotating unsteady-flow pattern upstream of the wake centerline. The wakes are cut by the passing rotor into individual segments that are acting in each passage. After the stator wake is cut to produce a segment of a wake in the rotor passage, the two ends of the wake segment travel at the local speeds: the portion attached to the pressure side moves downstream into the passage much slower than the portion attached to the suction side, because the local flow velocities are higher on the suction side of the cascade. At the same time lower momentum fluid moves from the wake end near the pressure side to the wake end near the suction side. The last two phenomena cause a thinner wake on the pressure side, a thicker wake on the suction side, and a counter-clockwise rotation of the centerline of the wake as it moves through the passage.

Figure 1.3 (from [Kor91a]) shows the cut wake centerline as dashed entropy contours in a rotor passage, superimposed on: unsteady flow vectors marking the counter rotating recirculating-flow patterns; and superimposed on unsteady pressure contours (with respect to the average flow). The negative unsteady pressure regions are marked with double-line contours and the positive unsteady pressure regions are marked with single-line contours. The recirculating flows result in regions of increased and decreased unsteady pressure upstream and downstream of the wake centerline respectively. As the recirculating-flow patterns move downstream into the rotor cascade the wake is sheared, distorted and enlarged, while the amplitude of the unsteady pressure maximum or minimum is decreased and its region of influence increased.

1.5.3 Combined Potential and Wake Interaction

The corresponding unsteady pressure maxima and minima due to the combined interaction are the combined (additive) effects of the two interactions described above. The above mechanisms are confirmed by results of this and past studies.

The value of R has a significant effect on which type of interaction dominates the unsteady flow field. For low values of R (less than 1.5) the stator potential-flow interaction decays very fast downstream so that it hardly affects the flow in the rotor cascade. The unsteady pressure contours inside the cascade passage are due to the wake interaction only. For high values of R (greater than 2.5) the stator potential-flow interaction decays very slowly downstream and it dominates the flow in the rotor cascade.

1.6 Results

Using the above explanations of the mechanisms of generation of unsteady pressure regions due to the wake and due to the potential we can identify (and with insight predict) the regions of influence of the wake and the potential-flow interactions, and the resulting shape of the unsteady forcing functions.

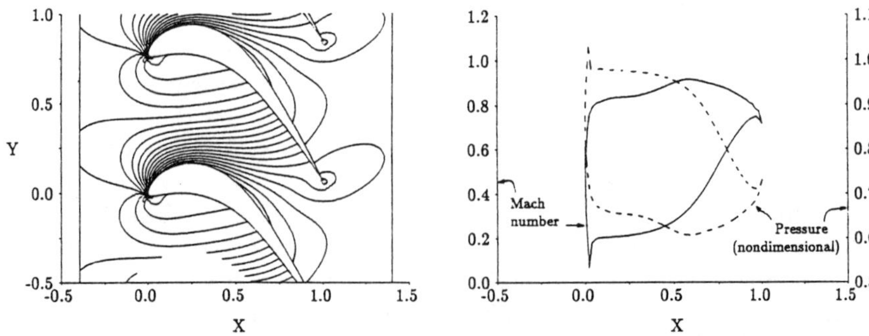

FIGURE 1.4. Steady flow performance of the sample cascade. Left, nondimensinal pressure contours of increment 0.02. Right, surface pressure and Mach number distributions

TABLE 1.2. Some information on the sample cascades. The corresponding reduced frequencies can be evaluated using equation 2

inlet flow angle α_i	40.00^o
outlet flow angle α_o (< 0)	60.00^o
stagger angle (< 0)	35.00^o
loading coefficient C_L	1.00
cascade solidity S_{rb}/b_{rb}	0.7779
inlet Mach number	0.405
outlet Mach number	0.800
steady-flow x-force $F'_{x,av}$	0.1179
steady-flow y-force $F'_{y,av}$	0.1945
steady-flow z-moment $T'_{z,av}$	0.1050
nozzle angle α_n	70.94^o
V_{rb} for α_n	0.6300

Results have been obtained for an intermediately-loaded cascade ($C_L = 1.0$) designed with a prescribed-curvature distribution method [Kor91b]. Its design characteristics are shown in table 1.2 and its steady-flow performance is shown in figure 1.4.

The wake and potential-flow interactions have been studied on the sample rotor cascade moving past a stator that has double the pitch of the rotor ($R = 2$). In production engines designers would normally try to choose unequal integers for all blade rows, and R would not be an integer. This unlikely geometry is useful for the purposes of this paper because it provides a symmetry that facilitates the discussion, because it is in a range where both types of interaction are of comparable magnitude so that neither one dominates the generation of the unsteady part of the force, and because it is close to the value of R of many turbine stages in production engines. The origin (zero) of the nondimensional period of the unsteadi-

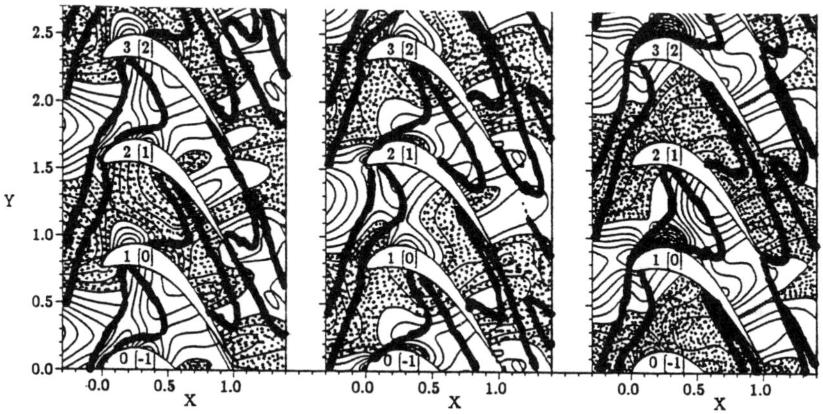

FIGURE 1.5. Unsteady flow fields for $d_{ax} = 0.20$

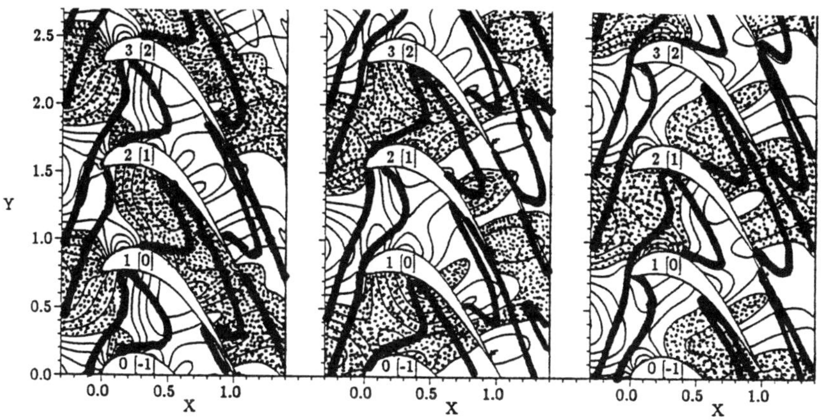

FIGURE 1.6. Unsteady flow fields for $d_{ax} = 0.30$

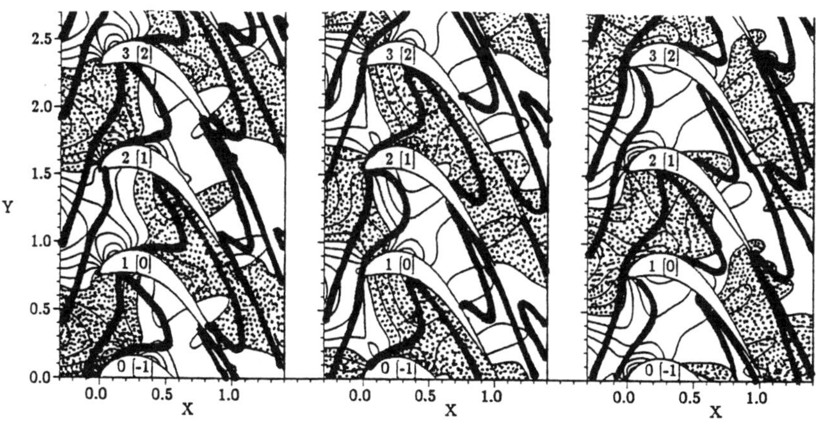

FIGURE 1.7. Unsteady flow fields for $d_{ax} = 0.40$

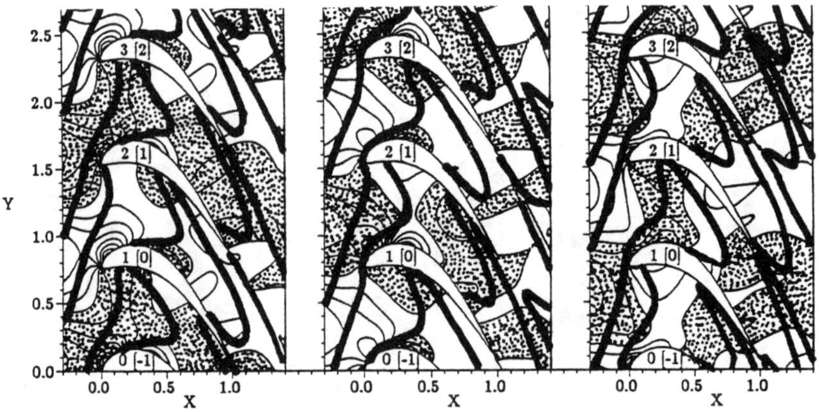

FIGURE 1.8. Unsteady flow fields for $d_{ax} = 0.50$

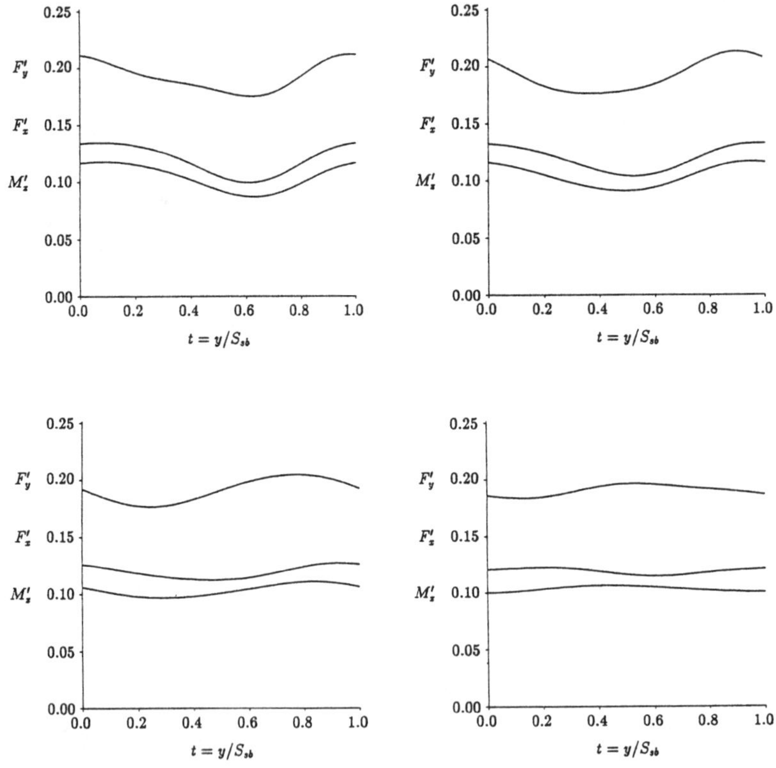

FIGURE 1.9. Forcing functions on the rotor. Top left for $d_{ax} = 0.20$. Top right for $d_{ax} = 0.30$. Bottom left for $d_{ax} = 0.40$. Bottom right for $d_{ax} = 0.50$.

ness in each case corresponds to the non-dimensional time $t = y/S_{sb}$ and y location at which the centerline of a stator velocity wake touches the leading edge of a rotor blade (numbered blade 0 in the unsteady flow figures), and the next stator wake touching the leading edge of blade 2. The end of the period corresponds to the velocity wake from the next stator blade touching the leading edge of the same rotor blade. The figures of unsteady flow fields show the rotor-relative frame, with the inlet boundary (modeling the stator cascade) moving towards the negative y direction as time increases. Figures 1.5, 1.6, 1.7 and 1.8 show the resulting unsteady pressure contours (instantaneous minus average pressure) superimposed on entropy contours for stator-to-rotor axial gaps $d_{ax} = 0.20$, 0.30, 0.40 and 0.50 of the rotor axial chord respectively. Because $R = 2$ the same figures show in the immediately above or below passage the flow half a period later, so that the flow at six instants during the period can be seen in each figure. The flow for $t = y/S_{sb} = 0.00$ and for $[t = y/S_{sb} = 0.50]$ is shown on the left. The flow for $t = y/S_{sb} = 0.33$ and for $[t = y/S_{sb} = 0.83]$ is shown in the middle. The flow for $t = y/S_{sb} = 0.67$ and for $[t = y/S_{sb} = 0.17]$ is shown on the right. The blade numbers in square brackets correspond to flow conditions at times in square brackets. The regions of positive values of unsteady pressure contours (pressure increased above the average) are dotted; the regions of negative unsteady pressure contain contours only.

The dimensional forces per unit length of blade span (such as F_x in the x direction) are related to the nondimensional forces (F'_x) by:

$$F'_x \equiv \frac{F_x/z}{\rho\, b_{rb}\, c_0^2} \tag{1.9}$$

(The moment M'_z is taken about the leading edge, and it is divided by b_{rb}^2). The period of the unsteadiness is one full stator-cascade pitch. The forcing functions for one stator pitch for the four axial gaps are shown in figure 1.9.

1.7 Discussion

The wakes illustrated by the width of the incoming entropy contours in figures 1.5, 1.6, 1.7 and 1.5 become thicker with increasing axial gap. The figures show the cutting and shearing of the wake by the rotor. Cascades with higher tangential-lift coefficient C_L have higher suction-surface and lower pressure-surface velocities. The centerlines of the cut wakes attached to the suction and pressure sides of the passages respectively travel at the local flow velocities downstream. The wakes are sheared more: in cascades with higher values of C_L; when they enter from lower α_n; and when they are wider (larger d_{ax}).

The effect of the potential-flow field of the stator on the potential-flow field of the rotor can be seen by the distortion (existence) of the unsteady pressure lines at the rotor inlet boundaries ($x = -0.30$) of all unsteady-flow figures and at all times. The figures show the positive unsteady pressure at the (modeled) trailing edges of the stators, the negative unsteady pressure between the trailing edges of the stators, and the rapid decay of the potential-flow interaction downstream of the stator trailing edge. The weaker potential-flow field of the rotor blades propagates upstream, distorts, and eventually cuts the stronger potential-flow field of the stator. The portion cut in the passage propagates downstream as a potential-flow disturbance (in this study equations 1.7, transformed in the rotor-relative frame and superimposed on the rotor flow field). The effect of the potential-flow field of the rotor upstream can also be seen by the distortion (bending) of the wake centerlines near the leading edge of the rotors.

At corresponding times in figures 1.5, 1.6, 1.7 and 1.8 the wakes are at the same locations (except they are thicker at larger axial gaps). The potential-flow interaction unsteady-pressure maxima and minima are at different locations, reflecting that they started from different trailing-edge locations (originated at different values of x). The regions of positive and negative unsteady pressures from the wake and the potential are superimposed. For example in figure 1.5 at time $t = 0.0$ and near $(x, y) = (0.4, 0.5)$ the negative unsteady pressure from the potential is superimposed on the negative unsteady pressure from the wake, resulting in large negative values of pressure. One passage upstream, near $(x, y) = (0.5, 1.2)$, the positive unsteady pressure from the potential is superimposed on the positive unsteady pressure from the wake, resulting in large positive values of pressure. The combination of "in phase" unsteady pressures around blade 0 results in substantial variations in F'_y. As d_{ax} increases the location of maxima and minima due to the potential change and eventually become "out of phase" with those due to the wake. This occurs (at corresponding times and places to the above) in figure 1.8. At time $t = 0.0$ and near $(x, y) = (0.4, 0.5)$ the positive unsteady pressure from the potential is superimposed on the negative unsteady pressure from the wake, resulting in small (local) positive values of pressure. One passage upstream, near $(x, y) = (0.5, 1.2)$, the negative unsteady pressure from the potential is superimposed on the positive unsteady pressure from the wake, resulting in small (local) positive values of pressure. The combination of unsteady pressures around blade 0 results in very small variations in F'_y. The effects of the potential and the wake become "in phase" again at $d_{ax} \approx 0.70$, resulting in large variations in F'_y. For these larger values of axial gap the potential-flow interaction onto the rotor is reduced due to its decay downstream. The influence of the wake is reduced (due to spreading of the wake downstream) but to a much lesser extent.

The above prove that by varying R and d_{ax} one may find geometries in which for most of the period the influence of wake and potential counteract each other to minimize the unsteady forces. This occurs to large parts of the period in figure 1.8.

1.8 Conclusions

The unsteady forces on two-dimensional gas-turbine rotors due to potential-flow interaction and viscous-wake interaction from upstream blade rows are computed using a compressible, two-dimensional, inviscid rotor / stator-disturbance interaction program. The viscous wake and the potential-flow interaction from the upstream stator are modeled at the rotor-inlet boundary. The mechanisms of generation of unsteady forces on the rotor blades are reviewed.

We conclude the following:

1. The results used in this paper to facilitate the discussion have been compared with the results from numerous other cascades of a variety of loading distributions and geometries. The reasoning presented in this paper is confirmed by all cases we have studied to date.
2. The potential flow field of the rotor cuts into the potential flow field of the stator and vice versa. After the potential-flow disturbance from the stator is cut into a rotor cascade, it propagates into the relative flow field of the rotor passage as a potential-flow disturbance superimposed on the rotor flow field.
3. The potential flow field of the rotor near the leading edge and the leading edge itself cut into the wake. The cutting action generates two counter-rotating vortical patterns flanking the wake-segment centerline in the passage. The vortical pattern upstream of the wake centerline generates an increase in local pressure (and forces acting on the sides of the passage). The vortical pattern downstream of the wake centerline generates a decrease in local pressure (and forces acting on the sides of the passage). The wake-segment ends attached to the sides of the passage travel at the local flow velocities: the wake-segment end near the suction surface moves downstream faster than the wake-segment end near the pressure surface; and the centerline of the wake segment rotates counter-clockwise as it propagates downstream. The wake is sheared more at higher values of tangential-loading coefficient and at lower values of nozzle angles.
4. The resulting unsteady forces on the blades are generated by the combined (additive) interaction of the two disturbances.
5. Since the mechanisms and locations of generation of unsteady pressure distributions due to the potential-flow and wake interactions are different, designers could use the above arguments to influence the shape of the unsteady forcing functions and minimize the magnitude of the unsteadiness by varying the stator-to-rotor pitch ratio and the axial gap between blade rows.

Acknowledgements: The author thanks Professor Michael B. Giles and Rolls Royce plc. for their permission to use UNSFLO in this investigation.

REFERENCES

[BFKR84] A. Binder, W. Forster, H. Kruse, and H. Rogge. *An experimental investigation into the effects of wakes on the unsteady turbine rotor flow.* ASME paper 84-GT-178, 1984.

[BFMR87] A. Binder, W. Forster, K. Mach, and H. Rogge. *Unsteady flow interaction caused by stator secondary vortices in a turbine rotor.* Transactions of the ASME, Journal of Turbomachinery, Volume 109, No. 2, pp. 251-257, April 1987. ASME paper 86-GT-302, 1986.

[Bin85] A. Binder. *Turbulence production due to secondary vortex cutting in a turbine rotor.* ASME paper 85-GT-193, 1985.

[BR82] A. Binder and R. Romey. *Secondary flow effects and mixing of the wake behind a turbine stator.* ASME paper 82-GT-46, 1982.

[BS84] E. Boletis and C. H. Sieverding. *Experimental study of the flow field behind an annular turbine nozzle guide vane with and without downstream rotor.* ASME Paper 84-GT-15, 1984.

[Che85] S. Chen. *Application of the rotating water table to nozzle wake excitation in low pressure turbines.* ASME paper 85-IGT-45, 1985.

[DJHW82] R. P. Dring, H. D. Josyln, L. W. Hardin, and J. H. Wagner. *Turbine rotor-stator interaction.* ASME paper 82-GT-3, 1982.

[GH91] M. B. Giles and R. Haimes. *Validation of a numerical method for unsteady flow calculations.* ASME paper 91-GT-271. Accepted for the Transactions of the ASME, 1991.

[Gil88] M. B. Giles. Calculation of unsteady wake/rotor interactions. *AIAA Journal of Propulsion and Power. AIAA paper 87-0006,* 4(4), July/August 1988.

[Hod83] H. P. Hodson. *Unsteady boundary-layers on axial-flow turbine rotor blades.* PhD thesis, University of Cambridge, United Kingdom, February 1983.

[Hod85a] H. P. Hodson. *An inviscid blade-to-blade prediction of a wake-generated unsteady flow.* Transactions of the ASME, Journal of Engineering for Gas Turbines and Power, Volume 107, pp. 337-344, 1985. ASME Paper 84-GT-43.

[Hod85b] H. P. Hodson. *Measurements of wake-generated unsteadiness in the rotor passages of axial flow turbines.* Transactions of the ASME, Journal of Engineering for Gas Turbines and Power, Volume 107, pp. 467-476, April 1985. ASME paper 84-GT-189, 1984.

[Hub86] M. Hubensteiner. *Einflub der Schaufelprofilform auf die instationaren Schaufelkrafte eines mittleren Turbinengitters (Influence of the blade-figure on the unsteady blade-forces of a middle turbine stage).* VDI Berichte pp. 209-232, 1986.

[Kor87] T. Korakianitis. *A design method for the prediction of unsteady forces on subsonic, axial gas-turbine blades.* Doctoral thesis (Sc.D.) in Mechanical Engineering, Massachusetts Institute of Technology, Cambridge, MA, USA, 1987.

[Kor88a] T. Korakianitis. *On the prediction of unsteady forces on gas-turbine blades. Part 1: Typical results and potential-flow-interaction effects.* ASME paper 88-GT-89, 1988.

[Kor88b] T. Korakianitis. *On the prediction of unsteady forces on gas-turbine blades. Part 2: Viscous-wake-interaction and axial-gap effects.* ASME paper 88-GT-90, 1988.

[Kor90a] T. Korakianitis. On the prediction of unsteady forces on gas turbine blades: part 1: description of the approach. *Accepted for the Transactions of the ASME, Journal of Turbomachinery. Washington University report WU/CCM 90-5*, 1990.

[Kor90b] T. Korakianitis. On the prediction of unsteady forces on gas turbine blades: part 2: analysis of the results. *Accepted for the Transactions of the ASME, Journal of Turbomachinery. Washington University report WU/CCM 90-6*, 1990.

[Kor91a] T. Korakianitis. *Blade-loading effects on the propagation of unsteady flows and on forcing functions in axial turbine cascades.* Washington University report WU-ICE 91-4, 1991.

[Kor91b] T. Korakianitis. *Hierarchical development of three direct-design methods for two-dimensional axial-turbomachinery cascades.* ASME paper presented in the 1991 International Joint Power Generation Conference and Exposition, October 6-10, San Diego, CA. Washington University report WU-ICE 91-3, 1991.

[Kor91c] T. Korakianitis. On the propagation of viscous wakes and potential-flow in axial-turbine cascades. *Accepted for the*

Transactions of the ASME, Journal of Turbomachinery. ASME paper 91-GT-373, 1991.

[LR57] H. W. Liepmann and A. Roshko. *Elements of gasdynamics.* John Wiley & Sons, 1957.

[MA85] J. Moore and R. Y. Adhye. *Secondary flows and losses downstream of a turbine cascade.* Transactions of the ASME, Journal of Engineering for Gas Turbines and Power, Volume 107, No. 4, pp. 961-968, October 1985. ASME paper 85-GT-64, 1985.

[Son85] T. Sonoda. *Experimental investigation on a spatial development of streamwise vortices in a turbine inlet guide vane cascade.* ASME paper 85-GT-20, 1985.

[SVHB83] C. H. Sieverding, W. Van Hove, and Boletis. *Experimental study of the three-dimensional flow field in an annular turbine nozzle guidevane.* ASME paper 83-GT-120, 1983.

[WDSH87] S. Wittig, K. Dullenkopf, A. Schulz, and R. Hestermann. *Laser-doppler studies of the wake-effected flow field in a turbine cascade.* Transactions of the ASME, Journal of Turbomachinery, Volume 109, No. 2, pp. 170-176, April 1987. ASME paper 86-GT-160, 1986.

[YY85] A. Yamamoto and R. Yanagi. *Production and development of secondary flows and losses within a three dimensional turbine.* ASME paper 85-GT-217, 1985.

III. Unsteady Transonic and Supersonic Flows

Unsteady Lifting Surface Theory for Supersonic Through-Flow Fan

Masanobu Namba[1]
Toshiya Hanada[2]

ABSTRACT The purpose of the present study is to predict analytically the unsteady loading on vibrating blades of a supersonic through-flow fan and investigate the aerodynamic instability. It is assumed that each blade operates with zero steady loading, and vibrates with small displacement amplitude, and a linearized unsteady lifting surface theory for a supersonic through-flow fan on the basis of the finite radial eigenfunction series approximation is developed. Numerical results of the unsteady loadings and unsteady aerodynamic works are presented. The bending vibration with zero mean loading is always stable, and the torsional instability heavily depends upon relative locations of the leading edge Mach wave reflection point and the torsion axis.

1.1 Introduction

Recently a propulsion system equipped with a single-stage supersonic through-flow fan, which operates at supersonic axial Mach numbers, is considered as one of the promising candidates that would be suitable for long-range supersonic cruise aircrafts (Refs. [1, 2]). But, previous studies related to cascades operating at supersonic axial velocity are incomplete because they are based on simplified linear cascade models.

The first author has developed the linearized steady and unsteady lifting surface theories for the rotating annular cascades operating at subsonic axial Mach numbers (Refs. [3, 4, 5]). This theoretical method is based on the technique of the finite radial eigenfunction series approximation which possesses an advantage described below.

The disturbance pressure field generated by a rotating cascade in an

[1]Professor, Deptartment of Aeronautical Engineering, Kyushu University, 6–10–1 Hakozaki Higashiku, Fukuoka, 812 Japan
[2]Graduate Student (Doctoral Course), ibid.

infinitely long annular duct with rigid walls can be expressed as a double infinite series of acoustic modes in the annular duct, which are referred to by the circumferential wave number n and the radial node number ℓ. In general such a disturbance pressure field is expected to have a rather small radial variation. This fact implies that the contribution from acoustic modes of higher radial orders will be very small. Therefore the series expansion with respect to the radial order ℓ can be approximated by a finite series. As a result of this mathematical treatment, one can obtain not only a clear insight into the three-dimensional structure of the disturbance flow field, but also mathematical expressions appropriate for numerical work.

In this paper an unsteady lifting surface theory for a supersonic through-flow fan is developed on the basis of the finite radial eigenfunction series technique. The theory is applied to vibrating blades, and numerical calculation is conducted to study the aerodynamic instability. It is found that the bending vibration under zero mean loading is always stable, and that the torsional instability heavily depends upon relative locations of the leading edge Mach wave reflection point and the torsion axis and is in general enhanced as the torsion axis moves off the unsteady pressure center.

1.2 Theoretical Model and Basic Assumptions

The model considered herein is an isolated rotor composed of N blades rotating at a constant angular velocity Ω^* in an infinitely long annular duct as shown in Fig. 1.1. Each blade vibrates harmonically with an angular frequency ω^* and an interblade phase angle $2\pi\sigma/N(\sigma = 0, 1, \cdots, N-1)$ under zero mean steady loading. For this flow model the following basic assumptions are made.

1. The fluid is an inviscid perfect gas.
2. The undisturbed flow is of a uniform supersonic axial velocity W_a^* and a uniform density ρ_0^*.
3. The perturbation to the flow is small and isentropic.

In the following discussion, dimensionless quantities are denoted by symbols without asterisk. Lengths, time, density, velocity and pressure are scaled with respect to r_T^* (rotor tip radius), r_T^*/W_a^*, ρ_0^*, W_a^* and $\rho_0^* \, {W_a^*}^2$ respectively. Furthermore, the dimensionless rotational speed of the rotor and the reduced frequency are difined as $\Omega = \Omega^* r_T^*/W_a^*$ which denotes the ratio of the circumferential rotor tip speed to the axial flow velocity and $\omega = \omega^* r_T^*/W_a^*$ respectively.

Let t a time coordinate and (r, θ, z) be a cylindrical coordinate system which is fixed to the rotor as shown in Fig. 1.1. Then the position of blade

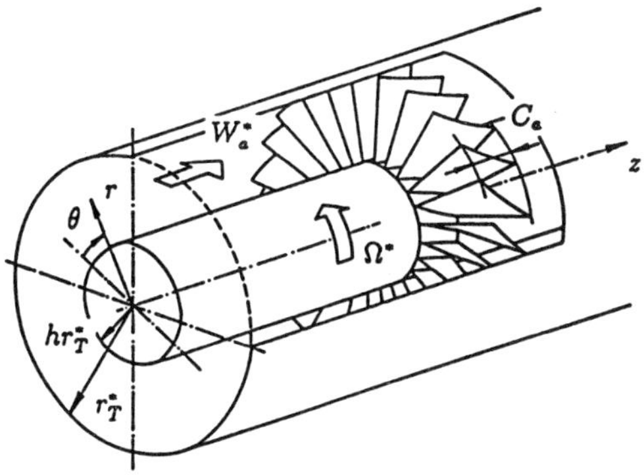

FIGURE 1.1. Annular cascade and duct geometry.

surfaces can be expressed by

$$\left.\begin{array}{l} \eta = \dfrac{2m\pi}{N} + \dfrac{f(r,z)}{r} e^{i\omega t + i2m\pi\sigma/N} \quad (m = 0, 1, \ldots, N-1) \\ \quad -C_a/2 \leq z \leq C_a/2 \quad \text{and} \quad h \leq r \leq 1 \end{array}\right\} \tag{1.1}$$

Here $\eta = \theta - \Omega z =$ constant denotes an undisturbed helical streamsurface, $f(r,z)e^{i\omega t + i2m\pi\sigma/N}$ denotes a circumferential displacement of the m-th blade surface from the helical streamsurface $\eta = 2m\pi/N$. Further C_a denotes the dimensionless axial chord length, and is assumed to be constant along the span to simplify the mathematical treatment. In the cases of the j-th order pure bending vibration where the circumferential displacement of the zeroth blade is given by $HC_a a_j^{(b)}(r)e^{i\omega t}$, and the k-th order pure torsional vibration where the twist angle of the zeroth blade is given by $\Theta a_k^{(t)}(r)e^{i\omega t}$ (See Fig. 1.2), $f(r,z)$ can be expressed by

$$f(r,z) = HC_a a_j^{(b)}(r) \quad \text{and} \quad f(r,z) = \Theta a_k^{(t)}(r) q_0^2(r)(z - z_e) \tag{1.2}$$

respectively, where $q_0(r) = \sqrt{1 + \Omega^2 r^2}$ is the undisturbed relative flow velocity with respect to the rotor fixed coordinate system (r, θ, z), and z_e denotes the location of the torsion axis.

1.3 Disturbance Flow Field

As is generally known, the disturbance pressure field is governed by the linearized acoustic wave equation as follows:

$$\left(\boldsymbol{\nabla}^2 - M_a^2 \frac{D^2}{Dt^2}\right) p = \boldsymbol{\nabla} \cdot \boldsymbol{F} - \frac{DQ}{Dt}. \tag{1.3}$$

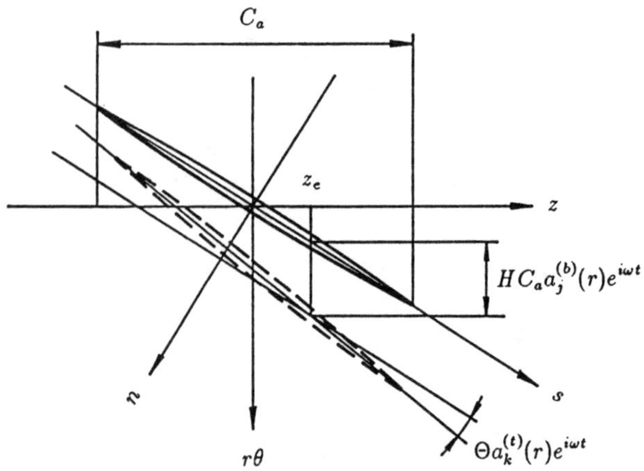

FIGURE 1.2. Cylindrical coordinate system fixed to the rotor

Here $\boldsymbol{F}$ denotes the body force acting on the fluid equivalent to a counteraction of aerodynamic forces on blades, Q denotes the mass source which accounts for the blade thickness effect, and M_a is the undisturbed axial Mach number. Further D/Dt represents the substantial derivative defined by

$$\frac{D}{Dt} = \frac{\partial}{\partial t} + \frac{\partial}{\partial z} + \Omega\frac{\partial}{\partial \theta} \tag{1.4}$$

and $\boldsymbol{\nabla}$ is the differential operator defined by

$$\boldsymbol{\nabla} = \boldsymbol{e}_r\frac{\partial}{\partial r} + \boldsymbol{e}_\theta\frac{1}{r}\frac{\partial}{\partial \theta} + \boldsymbol{e}_z\frac{\partial}{\partial z} \tag{1.5}$$

where $\boldsymbol{e}_{(\ )}$ denotes a unit vector in the direction indicated by its subscript.

Let the pressure difference across the m-th blade be denoted by $\Delta p(r,z) \times e^{i\omega t+2m\pi\sigma/N}$. Then the body force $\boldsymbol{F}$ can be expressed as follows:

$$\boldsymbol{F} = \sum_{m=0}^{N-1}\int_h^1 d\rho \int_{-C_a/2}^{C_a/2}\Big[-\boldsymbol{e}_n\Delta p(\rho,\zeta)e^{i\omega t+i2m\pi\sigma/N}$$
$$\times\frac{1}{r}\delta(r-\rho)\delta(\theta-\phi-2m\pi/N)\delta(z-\zeta)\Big]_{\phi=\Omega\zeta} q_0(\rho)d\zeta. \tag{1.6}$$

Here $\delta(\ )$ is the Dirac delta function, and $\boldsymbol{e}_n$ denotes a unit vector in the n-wise direction normal to the undisturbed helical streamsurface $\eta =$ constant. In the present model the mass source effect is absent. Then substituting Eq. (1.6) into Eq. (1.3), one can obtain the wave equation governing the unsteady disturbance pressure field $p(r,\eta,z)e^{i\omega t}$ as follows:

$$\left[\nabla^2 - M_a^2\left(i\omega + q_0(r)\frac{\partial}{\partial s}\right)^2\right] p(r,\eta,z)$$
$$= \sum_{m=0}^{N-1}\int_h^1 d\rho \int_{-C_a/2}^{C_a/2} e^{i2m\pi\sigma/N}\left[-\Delta p(\rho,\zeta)\frac{\partial}{\partial n}\right.$$
$$\left.\times\frac{1}{r}\delta(r-\rho)\delta(\theta-\phi-2m\pi/N)\delta(z-\zeta)\right]_{\phi=\Omega\zeta} q_0(\rho)d\zeta. \tag{1.7}$$

Here

$$\frac{\partial}{\partial n} = \frac{1}{q_0(r)}\left(\frac{1}{r}\frac{\partial}{\partial\theta} - \Omega r\frac{\partial}{\partial z}\right), \tag{1.8}$$

$$\frac{\partial}{\partial s} = \frac{1}{q_0(r)}\left(\frac{\partial}{\partial z} + \Omega\frac{\partial}{\partial\theta}\right), \tag{1.9}$$

where s denotes a coordinate parallel to undisturbed helical streamlines. The boundary condition at the inner and outer duct walls is

$$\frac{\partial p}{\partial r} \quad \text{at } r = h \text{ and } 1. \tag{1.10}$$

As the formal solution to Eq. (1.7), one can obtain the unsteady disturbance pressure field in the following form:

$$p(r,\eta,z) = \int_h^1 d\rho\int_{-C_a/2}^{C_a/2}\Delta p(\rho,\zeta)K_p(r,\eta,z-\zeta|\rho)d\zeta. \tag{1.11}$$

Here K_p is the pressure kernel function and given by

$$K_p(r,\eta,z-\zeta|\rho) = -q_0(\rho)\left[\frac{\partial}{\partial n}G(r,\theta,z|\rho,\phi,\zeta)\right]_{\phi=\Omega\zeta} \tag{1.12}$$

where G denotes a Green function which implies the pressure field induced by a row of monopoles the strength of which is unity. As shown in Ref. [4], G for supersonic axial flows is expressed in the form

$$G(r,\theta,z|\rho,\phi,\zeta) = -\mathrm{H}(z-\zeta)\frac{N}{4\pi\alpha_a^2}\sum_{\nu=-\infty}^{\infty}\sum_{\ell=0}^{\infty}R_\ell^{(n)}(r)R_\ell^{(n)}(\rho)$$
$$\times\frac{1}{\gamma_\ell^{(n)}}\left[E_{\ell+}^{(n)}(\theta-\phi,z-\zeta) - E_{\ell-}^{(n)}(\theta-\phi,z-\zeta)\right] \tag{1.13}$$

where $\alpha_a^2 = M_a^2 - 1$, $n = \nu N + \sigma$ and $\mathrm{H}(z-\zeta)$ denotes the Heaviside step function. Furthermore

$$E_{\ell\pm}^{(n)}(\theta-\phi,z-\zeta)$$
$$= \frac{1}{i}\exp\left[in(\theta-\phi) - i(n\Omega+\omega)(z-\zeta)\frac{M_a^2}{\alpha_a^2} \pm i\gamma_\ell^{(n)}(z-\zeta)\right] \tag{1.14}$$

$$\gamma_\ell^{(n)} = \sqrt{\left\{ k_\ell^{(n)^2} + (n\Omega + \omega)^2 \frac{M_a^2}{\alpha_a^2} \right\} \frac{1}{\alpha_a^2}} \tag{1.15}$$

Further $R_\ell^{(n)}(r)$ and $k_\ell^{(n)}$ are orthogonal radial eigenfunctions and eigenvalues of the following Sturm-Liouville boundary value problem:

$$\left.\begin{aligned} &\left\{ \frac{d^2}{dr^2} + \frac{1}{r}\frac{d}{dr} + \left(k_m^{(n)^2} - \frac{n^2}{r^2} \right) \right\} R_m^{(n)}(r) = 0, \\ &\frac{d}{dr} R_m^{(n)}(r) = 0, \quad \text{at } r = h \text{ and } 1, \\ &\int_h^1 r R_m^{(n)}(r) R_\ell^{(n)}(r) dr = \delta_{m\ell}. \end{aligned}\right\} \tag{1.16}$$

The expresstion of the unsteady disturbance velocity field $\boldsymbol{q}(r, \eta, z)e^{i\omega t}$ is derived by integrating the linearized Euler equation of motion

$$\left[i\omega + q_0(r)\frac{\partial}{\partial s} \right] \boldsymbol{q}(r, \eta, z) = -\boldsymbol{\nabla} p(r, \eta, z) \tag{1.17}$$

as follows:

$$\boldsymbol{q}(r, \eta, z) = \int_h^1 d\rho \int_{-C_a/2}^{C_a/2} \Delta p(\rho, \zeta) \boldsymbol{K_q}(r, \eta, z - \zeta | \rho) d\zeta \tag{1.18}$$

where $\boldsymbol{K_q}$ is the velocity kernel function given by

$$\boldsymbol{K_q} = -e^{-i\omega z} \int_{-\infty}^{z} e^{i\omega z} \left[\boldsymbol{\nabla} \cdot [K_p]_{\eta=\theta-\Omega z} \right]_{\theta=\eta+\Omega z} dz. \tag{1.19}$$

It should be noted that the series expansions expressing the kernel functions are in general not uniformly convergent. Therefore, before conducting numerical computations, the regular part and singular part involved in kernel functions should be separated. By use of the relations on the basis of the finite radial mode expansion, the velocity kernel function can be resolved into four parts and can be written in the following form:

$$\boldsymbol{K_q} = \boldsymbol{K_q^{(R)}} + \boldsymbol{K_q^{(SRR)}} + \boldsymbol{K_q^{(SRS)}} + \boldsymbol{K_q^{(JUMP)}}. \tag{1.20}$$

The first and second terms are regular parts in nonclosed and closed forms respectively, which are bounded everywhere in the field. The third term denotes the singular part which involves the Dirac delta function. The last term denotes the part of jump discontinuity accros $\eta = 0$ plane. Note that not all of four parts exist in some component of the kernel functions. In particular K_{q_n} necessary for determination of unsteady blade loadings is composed of three parts as follows:

$$K_{q_n} = K_{q_n}^{(R)} + K_{q_n}^{(SRR)} + K_{q_n}^{(SRS)}. \tag{1.21}$$

As $K_{q_n}^{(SRS)}$ involves the delta function terms, one can partially conduct integration with respect to ζ in the integral representation (1.18) of the n-wise disturbance velocity, obtaining

$$q_n(r,\eta,z)$$
$$= \int_h^1 d\rho \int_{-C_a/2}^{C_a/2} \Delta p(\rho,\zeta)\left\{K_{q_n}^{(R)}(r,\eta,z-\zeta) + K_{q_n}^{(SRR)}(r,\eta,z-\zeta)\right\} d\zeta$$
$$+ \int_h^1 \frac{\rho r}{q_0(r)} \sum_{\ell=0}^{L-1} R_\ell^{(\infty)}(r) R_\ell^{(\infty)}(\rho) \sum_{\mu=\mu^-}^{\mu^+} T_\ell^{(\mu)}(\eta)\Delta\bar{p}\left(\rho, Z_\ell^{(\mu)}(z,\eta)\right) d\rho \tag{1.22}$$

where

$$T_\ell^{(\mu)}(\eta) = -\frac{\alpha_a^2}{2}\hat{\gamma}_\ell^{(\infty)} \exp\left[i2\mu\pi\sigma/N - i\frac{\omega M_a^2}{\hat{\gamma}_\ell^{(\infty)}\alpha_a^2}|\eta - 2\mu\pi/N|\right], \tag{1.23}$$

$$Z_\ell^{(\mu)}(z,\eta) = z + \frac{\eta - 2\mu\pi/N}{A_\ell}, \tag{1.24}$$

$$A_\ell = \begin{cases} A_\ell^+ & : \eta - 2\mu\pi/N < 0, \\ A_\ell^- & : \eta - 2\mu\pi/N > 0, \end{cases} \tag{1.25}$$

$$A_\ell^\pm = -\frac{\Omega}{\alpha_a^2} \pm \hat{\gamma}_\ell^{(\infty)}, \tag{1.26}$$

$$\hat{\gamma}_\ell^{(\infty)} = \sqrt{\left\{\kappa_\ell^{(\infty)\,2} + \frac{\Omega^2 M_a^2}{\alpha_a^2}\right\}\frac{1}{\alpha_a^2}} \tag{1.27}$$

As shown in Fig. 1.3, Eq. (1.24) implies that the Mach line of the ℓ-th radial order passing through a field point $z = z'$, $\eta = \eta'$ in (z,θ) plane starts from a point of z-wise coordinate equal to $Z_\ell^{(\mu)}(z,\eta)$ on the μ-th blade surface. Furthermore μ_+ and μ_- denote, respectively, the upper and lower limits of the integer μ that satisfies

$$-C_a/2 \le Z_\ell^{(\mu)}(z,\eta) \le C_a/2. \tag{1.28}$$

1.4 Determination of Blade Loadings

The flow tangency condition on a blade surface is expressed as

$$q_n(r,0,z) = \frac{1}{q_0(r)}\left[i\omega + \frac{\partial}{\partial z}\right] f(r,z). \tag{1.29}$$

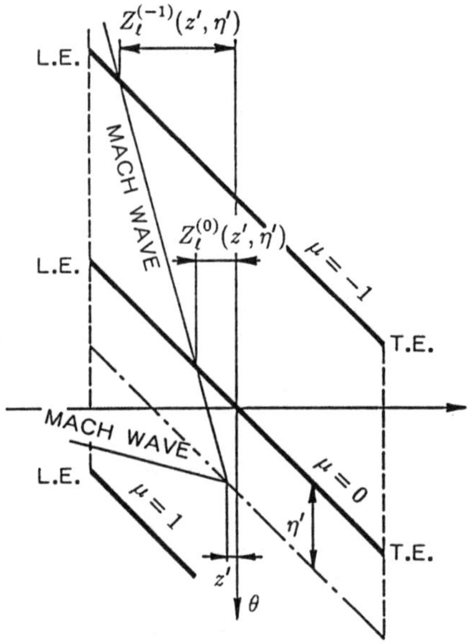

FIGURE 1.3. Mach wave and definition of $Z_\ell^{(\mu)}(z,\eta)$

Substituting Eq. (1.22) into Eq. (1.29), one obtains an integral equation for the pressure difference amplitude $\Delta p(r,z)$ across the blade as follows:

$$\int_h^1 d\rho \int_{-C_a/2}^{C_a/2} \Delta p(\rho,\zeta)\left\{K_{q_n}^{(R)}(r,0,z-\zeta)+K_{q_n}^{(SRR)}(r,0,z-\zeta)\right\}d\zeta$$

$$+\int_h^1 \frac{\rho r}{q_0(r)}\sum_{\ell=0}^{L-1} R_\ell^{(\infty)}(r)R_\ell^{(\infty)}(\rho)\sum_{\mu=\mu^-}^{\mu^+} T_\ell^{(\mu)}(0)\Delta\bar{p}\left(\rho, Z_\ell^{(\mu)}(z,0)\right)d\rho$$

$$= \frac{1}{q_0(r)}\left[i\omega+\frac{\partial}{\partial z}\right]f(r,z). \tag{1.30}$$

Let Express $\Delta p(r,z)$ in terms of radial and chordwise mode function series as follows:

$$\Delta p(r,z) = \sum_{k=0}^{L-1} R_k^{(\infty)}(r)G_k(z). \tag{1.31}$$

As shown in Ref. [4], $1/\kappa_k^{(\infty)}(k=0,1,\ldots,L-1)$ constitute a finite sequence of numbers which are equally spaced between 1 and h, and each $R_k^{(\infty)}(r)$ shows a highest extremum at $r \simeq 1/\kappa_k^{(\infty)}$. Then Eq. (1.31) implies that the pressure difference distribution $\Delta p(r,z)$ near the span station $r \simeq 1/\kappa_k^{(\infty)}$

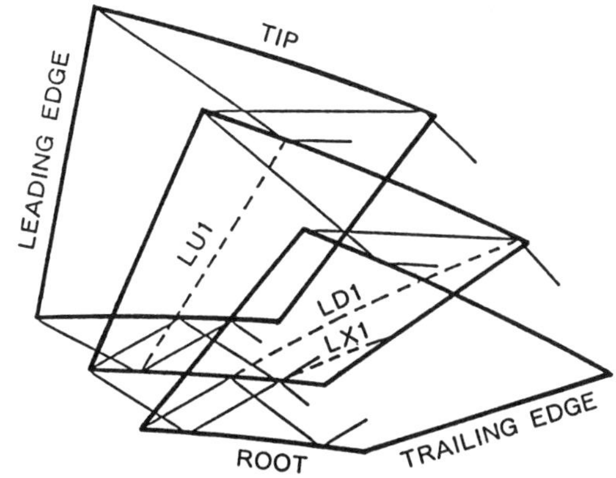

FIGURE 1.4. Reflection pattern of the leading edge Mach wave

is essentially dominated by the k-th term $R_k^{(\infty)}(r)G_k(z)$, and as the result $G_k(z)$ is discontinuous at reflection points of the k-th mode Mach waves emanating from the leading edges of the adjacent blades. The z coordinates of the reflection points are given by

$$\left.\begin{aligned} z_{LUm}^{(k)} &= -C_a/2 + (m-1)s_z^{(k)+} - ms_z^{(k)-}, \\ z_{LDm}^{(k)} &= -C_a/2 + ms_z^{(k)+} - (m-1)s_z^{(k)-}, \\ z_{LXm}^{(k)} &= -C_a/2 + ms_z^{(k)+} - ms_z^{(k)-}, \end{aligned}\right\} \tag{1.32}$$

where $m = 1, 2, \ldots$, and

$$s_z^{(k)\pm} = \frac{2\pi}{N} \frac{\alpha_a^2}{{\kappa_k^{(\infty)}}^2 + \Omega^2} \left(\frac{\Omega}{\alpha_a^2} \pm \hat{\gamma}_k^{(\infty)} \right). \tag{1.33}$$

In order to deal with the integral equation for the discontinuous functions $G_k(\zeta)$, Nagashima and Whitehead's technique [6] is applied.

1.5 Aerodynamic Coefficients

The present paper deals with pure bending and pure torsional vibrations. The natural mode shapes of a uniform cantilever beam are assumed for spanwise distributions of vibration amplitudes, $a_j^{(b)}(r)$ (the j–th order bending mode) and $a_k^{(t)}(r)$ (the k–th order torsional mode).

Local unsteady lift coefficient for the j-th order bending vibration is

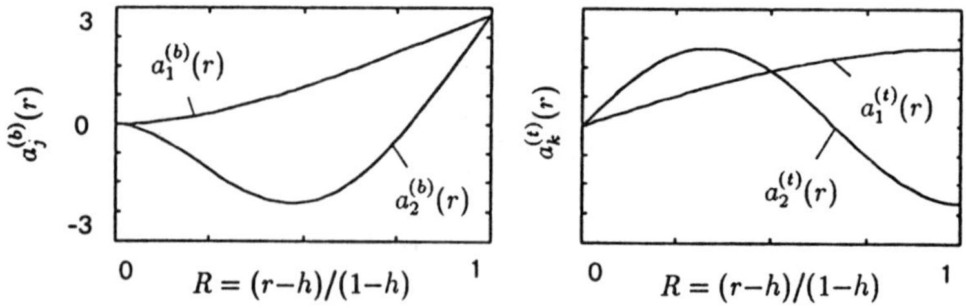

FIGURE 1.5. Spanwise distributions of vibration amplitudes

defined by

$$C_{L_j}^{(b)}(r) = \frac{\hat{L}_j^{(b)}(r)/C_a}{i\pi\omega C_a a_j^{(b)}(r)}, \tag{1.34}$$

were $q_0(r)\hat{L}^{(b)}$ is the lift force per unit span and $\hat{L}^{(b)}$ is given by

$$\hat{L}^{(b)} = \int_{-C_a/2}^{C_a/2} \Delta p^{(b)}(r,z)dz. \tag{1.35}$$

On the other hand, local moment coefficient for the k-th order torsional vibration about $z = z_e$ is defined by

$$C_{M_{ek}}^{(t)}(r) = \frac{\hat{M}_{e_k}^{(t)}(r)/C_a^2}{\pi q_0^2(r) a_k^{(t)}(r)}, \tag{1.36}$$

where $q_0^2(r)\hat{M}_e^{(t)}$ is the moment of the force per unit span about $z = z_e$ and $\hat{M}_e^{(t)}$ is given by

$$\hat{M}_e^{(t)} = \int_{-C_a/2}^{C_a/2} \Delta p^{(t)}(r,z)(z - z_e)dz. \tag{1.37}$$

The local aerodynamic work per cycle per unit span is given by

$$W(r) = \frac{\pi}{2}\int_{-C_a/2}^{C_a/2} \left[-i\bar{f}(r,z)\Delta p(r,z) + if(r,z)\Delta\bar{p}(r,z)\right]dz \tag{1.38}$$

where $\Delta\bar{p}(r,z)$ and $\bar{f}(r,z)$ denote the complex conjugates of $\Delta p(r,z)$ and $f(r,z)$ respectively. Then one can write

$$W(r) = \pi C_a^2 H^2 C_{W_j}^{(b)}(r) \quad \text{and} \quad W(r) = \pi C_a^2 \Theta^2 C_{W_{ek}}^{(t)}(r) \tag{1.39}$$

where

$$C_{W_j}^{(b)}(r) = \pi\omega C_a \left\{a_j^{(b)}(r)\right\}^2 Re\left[C_{L_j}^{(b)}(r)\right], \tag{1.40}$$

$$C_{W_{ek}}^{(t)}(r) = \pi q_0^4(r) \left\{ a_k^{(t)}(r) \right\}^2 Im \left[C_{M_{ek}}^{(t)}(r) \right]. \tag{1.41}$$

Finally, the total aerodynamic work per cycle is given by

$$\begin{aligned} \bar{W} &= \int_h^1 W(r)dr \\ &= \pi C_a^2 H^2 \bar{C}_{W_j}^{(b)} \quad \text{or} \quad \pi C_a^2 \Theta^2 \bar{C}_{W_{ek}}^{(t)} \end{aligned} \tag{1.42}$$

where

$$\bar{C}_{W_j}^{(b)} = \int_h^1 C_{W_j}^{(b)}(r)dr \quad \text{and} \quad \bar{C}_{W_{ek}}^{(t)} = \int_h^1 C_{W_{ek}}^{(t)}(r)dr. \tag{1.43}$$

1.6 Numerical Examples and Discussions

In the following Figures, the real and imaginary parts of unsteady aerodynamic coefficients are denoted by $Re[\ ]$ and $Im[\ ]$ respectively.

Figure 1.6 demonstrates the unsteady loading distribution across the blade for the first order bending vibration (the upper figure) and the first order torsional vibration with torsion axis at the half chord line (the lower one). One can find that the unsteady loading for both vibrations are discontinuous at the line LU1 where the leading edge Mach wave of the pressure-side blade reflects, and from Fig. 1.5 and Fig. 1.6, it appears that the spanwise distribution of both vibration amplitude and that of unsteady loading have the same profile. It should be noted that the imaginary component of the unsteady loading, *i.e.*, the local aerodynamic work for the torsional vibration is exclusively concentrated upon the region between the reflection line LU1 and the trailing edge. Therefore it can be expected that the relative locations of the line LU1 and the torsion axis will be crucial for determination of the flutter boundary.

In Figs. 1.7, 1.8 and 1.9, the influence of design parameters on unsteady local aerodynamic force and work for the first order bending vibrations is demonstrated.

Figure 1.7 shows that the unsteady aerodynamic force and work heavily depends upon the interblade phase angle σ/N. The absolute value of the aerodynamic work is highest at $\sigma/N = 0.5$. From Figs. 1.8 and 1.9, it can be observed that the absolute value of the unsteady lift coefficient decreases as the reduced frequency ωC_a and axial Mach number M_a increases. For various combinations of prescribed design parameters, one can find that the unsteady aerodynamic force always acts against the blade motion, and consequently the unsteady local aerodynamic work coefficient becomes always negative. In other words, the bending vibration is aerodynamically stable.

Figure 1.10 demonstrates the influence of interblade phase angle σ/N on unsteady local aerodynamic moment and work for the first order torsional vibrations with the torsion axis at the half chord line. This figure indicates that the unsteady local aerodynamic moment also heavily depends upon the interblade phase angle. In this case, however, the absolute value of the unsteady aerodynamic work becomes highest at $\sigma/N = \pm 0.25$.

Figures 1.11, 1.12, 1.13 and 1.14 demonstrate the influence of the design parameters on the total aerodynamic work for the first and second order torsional vibrations.

From Fig. 1.11, one can find that the variation of the total aerodynamic work with the interblade phase angle σ/N for the torsional vibration with the torsion axis at the three quarters line ($z_e/C_a = 0.25$) takes the opposite direction against that for the torsional vibration with the torsion axis at the quarter or the half chord line ($z_e/Ca = -0.25$ or 0.0). The reason for such a tendency is clear from the unsteady loading distribution for the torsional vibration shown in Fig. 1.6, which indicates that the twisting moment is minimum if the torsion axis is located near the unsteady pressure center, *i.e.*, the mid-point between the reflection line LU1 and the trailing edge. Therefore, as shown in Fig. 1.12, which shows the dependence of the total aerodynamic work for the interblade phase angle $\sigma/N = 0.25$ or -0.25 on the torsion axis location z_e/C_a, the absolute value of the total aerodynamic work increases as the torsion axis moves away from the unsteady pressure center, but the total work for the torsion axis located ahead of the unsteady pressure center takes the opposite sign against that for the torsion axis located behind. Consequently, when the interblade phase angle is positive, the torsional vibration with the torsion axis located ahead of the unsteady pressure center is stable, and the torsional instability is enhanced as the torsion axis moves towards the trailing edge. On the other hand, when the interblade phase angle is negative, the instability is enhanced as the torsion axis moves off the unsteady pressure center towards the leading edge.

Figure 1.13 shows the effect of axial Mach number M_a on the total aerodynamic work for the torsional vibration with the torsion axis at the quarter, the half and the three quarters chord lines. The leading edge Mach wave reflects only at line LU1 up to $M_a = 2.3$, but for $M_a > 2.3$ no reflection of the leading edge Mach wave occurs on the blade surfaces. Since the reflection line LU1 moves towards the trailing edge as axial Mach number M_a, the region between the reflection line LU1 and the trailing edge decreases, and hence the total aerodynamic work also decreases with increasing M_a. But at last the total work becomes negative and almost invariant with M_a because of no reflection. It is worth to note that the stability of the torsional vibration with the torsion axis at the three quarters chord line is not necessarily increased with increasing Mach number as long as the Mach wave reflection occurs.

Figure 1.14 indicates that the critical reduced frequency can be reduced by increasing the axial Mach number, since the blade surface region, which the local unsteady work concentrates upon, is reduced by the fact that the reflection line LU1 moves towards the trailing edge as the axial Mach number increases.

Finally no substantial difference in the total aerodynamic work is observed between the first order and second order torsional vibrations. In other words the three-dimensional effect is small.

1.7 Conclusions

To predict the unsteady aerodynamic instability of a supersonic through-flow fan, an unsteady lifting surface theory is developed. From some numerical examples the following conclusions are drawn out.

1) The bending vibration with zero mean loading is always stable.

2) The torsional instability heavily depends upon the relative locations of the leading edge Mach wave reflection points and the torsion axis. When the interblade phase angle is positive, the torsional vibration with the torsion axis located ahead of the unsteady pressure center, *i.e.*, the mid-point between the reflection line of the leading edge Mach wave of the pressure-side blade and the trailing edge, is stable, and the torsional instability is enhanced as the torsion axis moves towards the trailing edge. On the other hand, when the interblade phase angle is negative, the instability is enhanced as the torsion axis moves off the unsteady pressure center towards the leading edge.

3) The torsional vibration becomes stable as the reduced frequency and the axial Mach number increase. The critical reduced frequency is reduced with increasing the axial Mach number.

4) In general the three-dimensional effect on the unsteady aerodynamic work is small.

References

[1] Trucco, H. *Study of Variable Cycle Engines Equipped With Supersonic Fans.* Final Report, ATL TR 201 Advanced Technology Laboratries Inc., NAS3–17559, NASA CR–134777, 1975.

[2] Franciscus, L. C. *The Supersonic-Through Flow Turbofan for High Mach Propulsion.* AIAA Paper, 87-2050, 1987.

[3] Namba, M. *Lifting Surface Theory for a Rotating Subsonic or Transonic Blade Row.* A.R.C., R&M 3740, 1972.

[4] Namba, M. *Three–Dimensional Flows*, AGARD Manual on Aeroelasticity in Axial–Flow Turbomachines. AGARD–AG–298, Vol.1.

[5] Namba, M. and Ishikawa, A. *Three-Dimensional Aerodynamic Characteristics of Oscillating Supersonic and Transonic Annular Cascades.* Transaction the ASME, Journal of Engneering for Power, Vol.105, 1983.

[6] Nagashima, T. & Whitehead, D. S. *Linearlized Supersonic Unsteady Flow in Cascade.* A.R.C., R&M 3811, 1977.

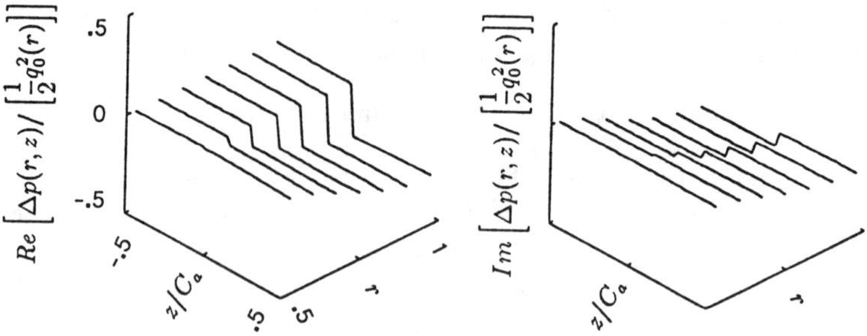

The first order bending vibration

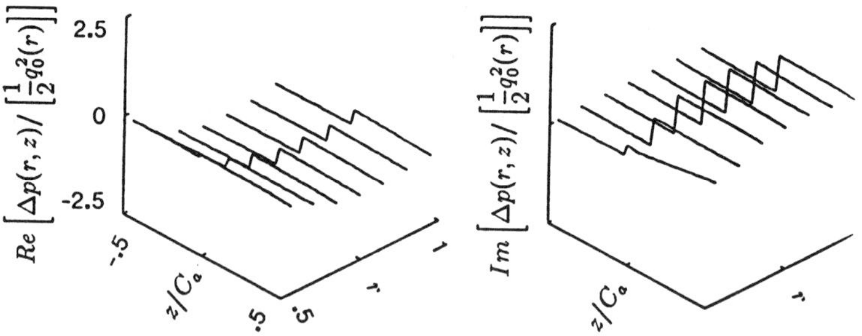

The first order torsional vibration ($z_e/C_a = 0.0$)
($N = 40$, $NCa = 4.0$, $\omega C_a = 0.2$, $\Omega = 1.5$, $h = 0.5$, $M_a = 1.5$, $\sigma/N = -0.25$)

FIGURE 1.6. Unsteady blade loading distributions

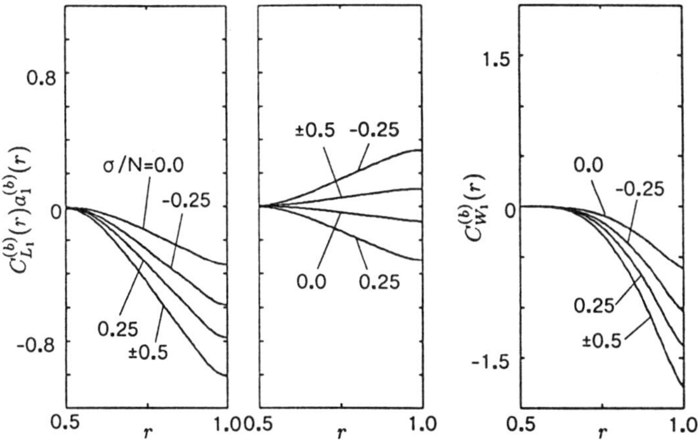

($N = 40$, $NCa = 4.0$, $\omega C_a = 0.2$, $\Omega = 1.5$, $h = 0.5$, $M_a = 1.5$)

FIGURE 1.7. Influence of the interblade phase angle on the unsteady aerodynamic force coefficients for the first order bending vibration

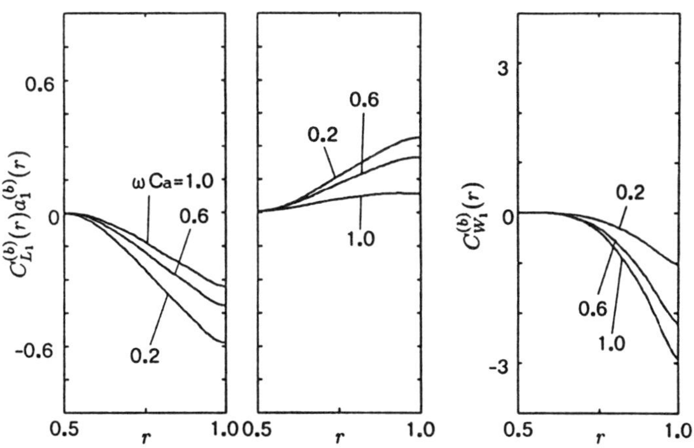

($N = 40$, $NCa = 4.0$, $\Omega = 1.5$, $h = 0.5$, $M_a = 1.5$, $\sigma/N = -0.25$)

FIGURE 1.8. Influence of the reduced frequency on the unsteady aerodynamic force coefficients for the first order bending vibration

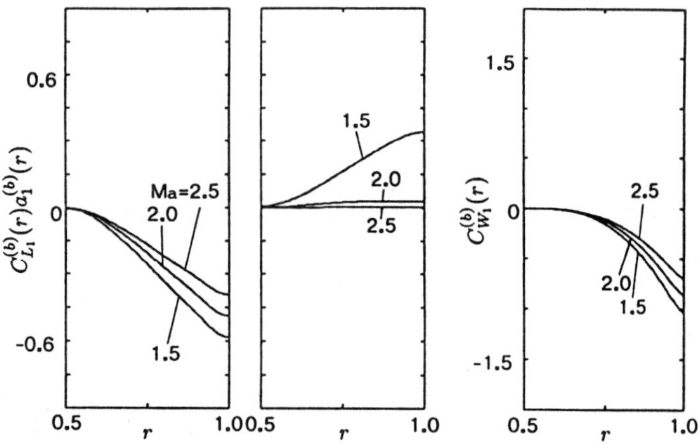

($N = 40$, $NCa = 4.0$, $\omega C_a = 0.2$, $\Omega = 1.5$, $h = 0.5$, $\sigma/N = -0.25$)

FIGURE 1.9. Influence of the axial Mach number on the unsteady aerodynamic force coefficients for the first and order bending vibration

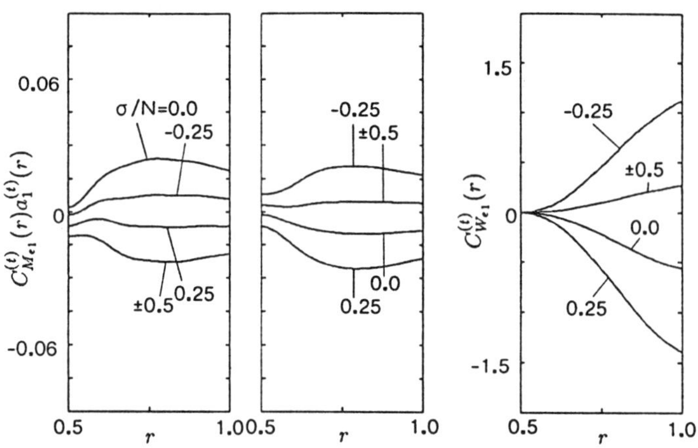

($N = 40$, $NCa = 4.0$, $\omega C_a = 0.2$, $\Omega = 1.5$, $h = 0.5$, $M_a = 1.5$, $z_e/C_a = 0.0$)

FIGURE 1.10. Influence of the interblade phase angle on the unsteady aerodynamic moment coefficients for the first order torsional vibration

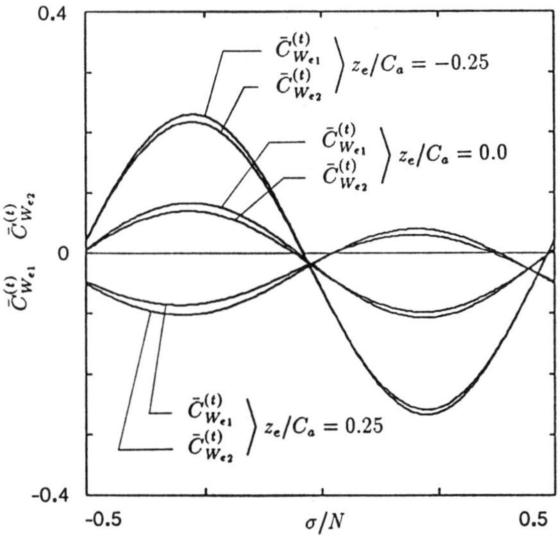

($N = 60,\ NCa = 6.0,\ \omega C_a = 0.2,\ \Omega = 1.5,\ h = 0.5,\ M_a = 1.5$)

FIGURE 1.11. Influence of the interblade phase angle on the total aerodynamic work coefficients for the first and second order torsional vibrations with the torsion axis at the quarter, the half and the three quarters chord lines

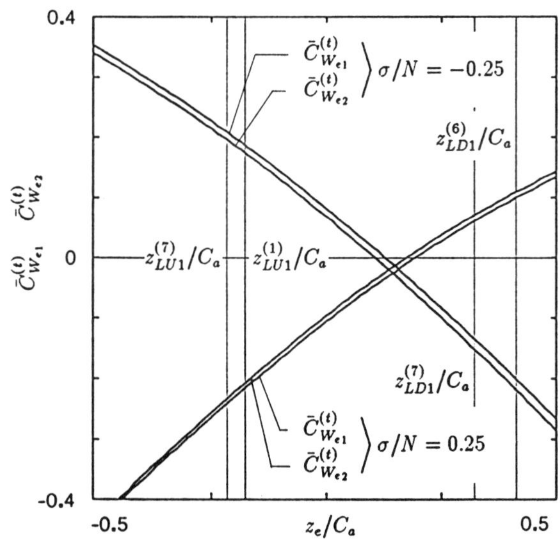

($N = 60,\ NCa = 6.0,\ \omega C_a = 0.2,\ \Omega = 1.5,\ h = 0.5$)

FIGURE 1.12. Influence of the location of the torsion axis on the total aerodynamic work coefficients for the first and second order torsional vibrations. Vertical lines indicate the location of the leading edge Mach wave reflection.

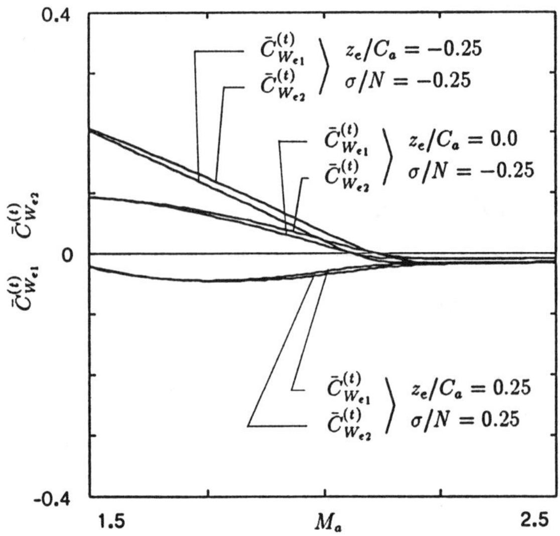

($N = 40$, $NCa = 4.0$, $\omega C_a = 0.2$, $\Omega = 1.5$, $h = 0.5$, $M_a = 1.5$)

FIGURE 1.13. Influence of the axial Mach number on the total aerodynamic work coefficients for the first and second order torsional vibrations with the torsion axis at the quarter, the half and the three quarters chord lines

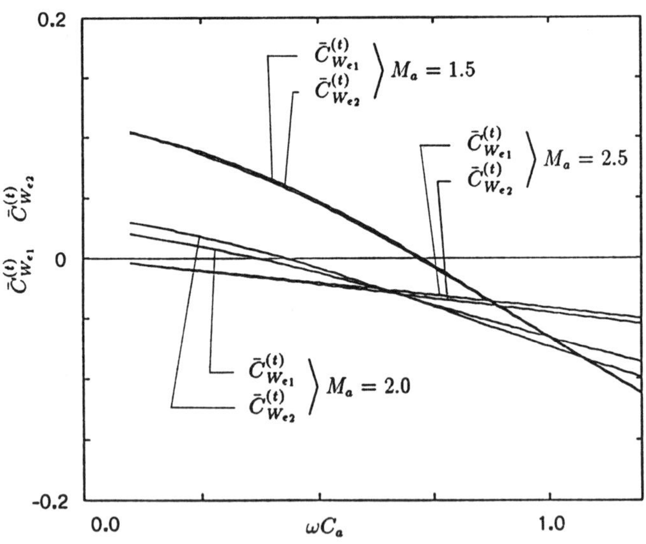

($N = 40$, $NCa = 4.0$, $\Omega = 1.5$, $h = 0.5$, $\sigma/N = -0.25$)

FIGURE 1.14. Influence of the reduced frequency on the total aerodynamic work coefficients for the first and second order torsional vibrations

Time-Dependent Predictions and Analysis of Turbine Cascade Data in the Transonic Flow Region

A. Bölcs[1]
A. Cargill[2]
T.H. Fransson[1,3]
A. Suddhoo[2]
K. Vogeler[4]

ABSTRACT

Experimental and numerical model cascade time-dependent results have been obtained, under mainly attached flow conditions, in the travelling wave vibration mode on two model steam turbine cascades. Both cascades oscillate in the first bending vibration mode at realistic reduced frequencies.

Time-dependent blade surface pressure coefficients are studied, both in the travelling wave mode and with the influence coefficient technique. It is concluded that the prediction model accurately captures the measured steady-state and time-dependent flow phenomena under subsonic flow conditions for both cascades. The travelling wave mode results indicate that the time-dependent blade pressure surface pressure coefficient is almost independent of the outlet flow velocity and similar in phase between the two cascades.

For supersonic outlet flow conditions, both the steady-state and the time-dependent blade pressure surface pressure coefficient are similar to those at high subsonic flow conditions. In addition, experiments show aerodynamic shock waves on the blade surface as localized unsteady pressure peaks with an associated phase angle shift.

It is concluded that trustworthy predictions can be obtained from the numerical model employed for subsonic flow conditions on cascades of the type presented. At supersonic outlet flow conditions major discrepancies between the data and the predictions are found in the vicinity of the shock waves. From the experiments it is also concluded that an aerodynamic shock wave can have a significant, but usually localized, influence on the unsteady pressures on the blade surface. The amplitudes can become fairly large, and a phase shift can appear in the blade pressure response. This is not always clearly demonstrated for results in the travelling wave mode as these depend on the interblade phase angle of the cascade vibrating mode.

[1]Ecole Polytechnique Federale de Lausanne, Switzerland.
[2]Rolls-Royce plc, Derby, United Kingdom.
[3]Presently at Royal Institute of Technology, Stockholm, Sweden.
[4]Asea Brown Boveri, Baden, Switzerland.

NOMENCLATURE

Symbol	Explanation	Dimension
c	Blade chord length	m
$\tilde{c}_p(x,t)$	Unsteady perturbation pressure coefficient in travelling wave mode $\tilde{c}_p(x,t) = \frac{1}{h}\frac{\tilde{p}(x,t)}{p_{w1} - p_1} = \tilde{c}_p(x)\, e^{i(\omega t + \Phi_p)}$	-
e	Distance between blades and measuring station	m
f	Blade vibration frequency	Hz
h	Dimensionless (with chord) bending vibration amplitude	-
k	Reduced frequency, $k=(\omega c)/(2 u_2)$	-
M	Mach Number	-
p	Pressure	mbar
$\tilde{p}$	Unsteady perturbation pressure	mbar
Re	Reynolds number, $Re = (u_1 c)/\nu$	-
t	Time	sec
u	Velocity	m/sec
t	Blade pitch	m
x	Coordinate along chord	m
β	Flow angle	deg
β_g	Stagger angle	deg
δ	Blade vibration direction	deg
σ	Interblade phase angle. Positive when blade "n+1" leads blade "n"	deg
Φ_s	Phase angle. Positive when disturbance leads blade "n"	deg
ν	Kinematic viscosity	m^2/sec

Sub- and Superscripts

ic	Influence coefficient	p	Pressure
s	Isentropic	w	Stagnation values
1	Upstream flow conditions	2	Downstream flow conditions
1B	First bending mode		
m, n	Influence of blade "m" on blade "n"		

INTRODUCTION

It is today generally known that low velocity subsonic attached unsteady flow conditions can be well predicted for blades with small thickness and small camber. The predictions do however not agree so well for transonic flows and on curved profiles.

The objective of the present work is to compare two sets of experimental data, obtained in an annular test facility, with predictions from a small perturbation potential flow solver, and to analyse the results based on an interpretation of quasi-steady flow and influence coefficients.

EXPERIMENTAL SET-UP

The tests on the present cascade were performed in the non-rotating annular cascade tunnel (Fig.1) at the Swiss Federal Institute of Technology, Lausanne [Bölcs, 1983]. The velocity and flow angle profiles in the test section are obtained by the regulation of two separate inlet valves ("1" and "2") and two independent inlet guide vanes ("5" and "6"), together with two separate suction slots immediately downstream of the preswirl vanes. Four independent suction slots up- and downstream of the test cascade can act on the velocity profile as well as on the boundary layer on the outer and inner tunnel walls.
The flow conditions are measured with two aerodynamic probes ("8") which can traverse over the channel height as well as in the circumferential direction. The test facility is also equipped with Schlieren optic and laser holography for flow visualization ("9"), as well as a Laser-Two-Focus system for determination of velocities in the blade passages.
Each blade is elastically suspended on a spring-mass system (Fig. 2), designed to reproduce the first three eigenfrequencies, as well as the vibration direction of the first bending mode, as determined from the real blading in the turbine. The inner wall of each blade ("2") is polished for shadowgraph visualization, with an overlapped gap between the blades in order to allow blade vibrations.

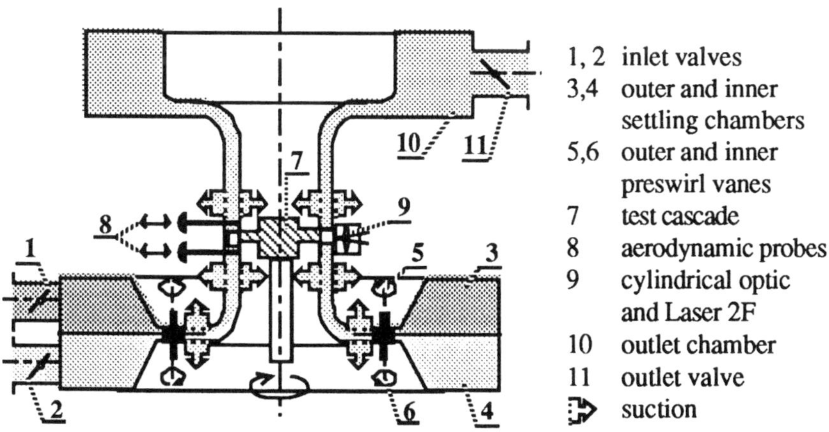

Fig. 1: *Schematic view of annular test facility.*

The blades are forced into vibration, in the travelling wave mode, by means of individual electromagnetic exciters, each with its own feed-back loop and by transformer plates ("8") fixed on the foot of the blade. The blade vibration is measured by inductive displacement transducers ("9"), fixed on an impact ring ("10") which is also used to clamp the blades for steady-state tests. The interblade phase angle and blade vibration amplitude can thus be regulated individually and almost any cascade vibration pattern can be realized.

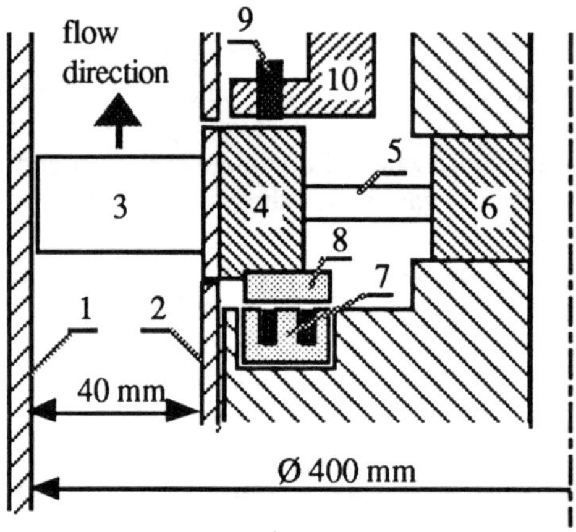

1 Outer wall
2 Inner wall
3 Blade
4 Mass element
5 Spring element
6 Central body
7 Electromagnet
8 Transformer plate
9 Vibration transducer
10 Impact ring

Fig. 2: *Blade vibration system*

PREDICTION MODEL

The numerical model employed in the study (FINSUP) was originally developed by Dr. Whitehead at Cambridge University [Whitehead, 1990; Whitehead and Newton, 1985] and extended in collaboration with Rolls Royce.
The model is based on the linearized unsteady quasi three-dimensional potential flow in turbine and compressor cascades and uses a numerical mesh of triangular finite elements. The program calculates both sub- and supersonic steady-state and time-dependent flow conditions, where the unsteadiness is introduced through blade vibrations.

CASCADE GEOMETRIES

The first cascade "RAX" (Fig.3) is a section from the second-to-last, free-standing, stage of a large steam turbine. It is used as the "Fourth International Standard Configuration for Aeroelasticity in Axial-Flow Turbomachine Cascades" [Bölcs and Fransson, 1986] and has as such been calculated by different prediction models. In general the agreement between data and predictions is reasonable for subsonic flow conditions. The experimental data can thus be qualified as good. The second cascade "AFT" is taken from a tip section of a large steam turbine last stage.

The investigated type of airfoil has, for cascade one, a non-negligable blade thickness and camber and operates under high subsonic flow conditions. Cascade two has low camber and thickness.

The cascade geometries, nominal flow conditions are given in Table 1.

Table 1: Geometry and nominal flow conditions for cascade "RAX" and "AFT".

Cascade Model	**"RAX"**	**"AFT"**
Blade span	40 mm	40 mm
Number of blades	20	20
Chord (c)	74.4 mm	78.1 mm
Camber	45 °	6.5 °
Thickness/chord	0.17	0.051
Hub diameter	320 mm	320 mm
Tip diameter	400 mm	400 mm
Pitch (t) (midspan)	56.5 mm	56.5 mm
t/c (midspan)	0.76	0.724
e_1/c	0.073	0.333
e_2/c	0.892	0.834
Minimal section (midspan)	17.4 mm	20 mm
Stagger angle (β_g)	56.6 °	65 °
Eigenfrequency (f_{1B}) (bend.)	150 Hz	114±2 Hz
Bending mode direction (d_{1B})	60.4 °	53 °
Nominal flow conditions:		
M_1	0.31	0.46
M_2	0.90	1.38
β_1	-44.1 °	-62 °
β_2	-72.4 °	-66 °
Reynolds number (Re)	$6.8 \cdot 10^5$	$1.2 \cdot 10^6$

COMPARISON OF EXPERIMENTAL DATA AND PREDICTIONS

Cascade "RAX"

In general it can be concluded that the time-averaged agreement in the blade surface Mach number distribution between the two-dimensional "FINSUP" predictions and the data is good for the subsonic flow conditions, both for the large and the small inlet flow angles (Fig. 4a).Also for the supersonic outlet flow

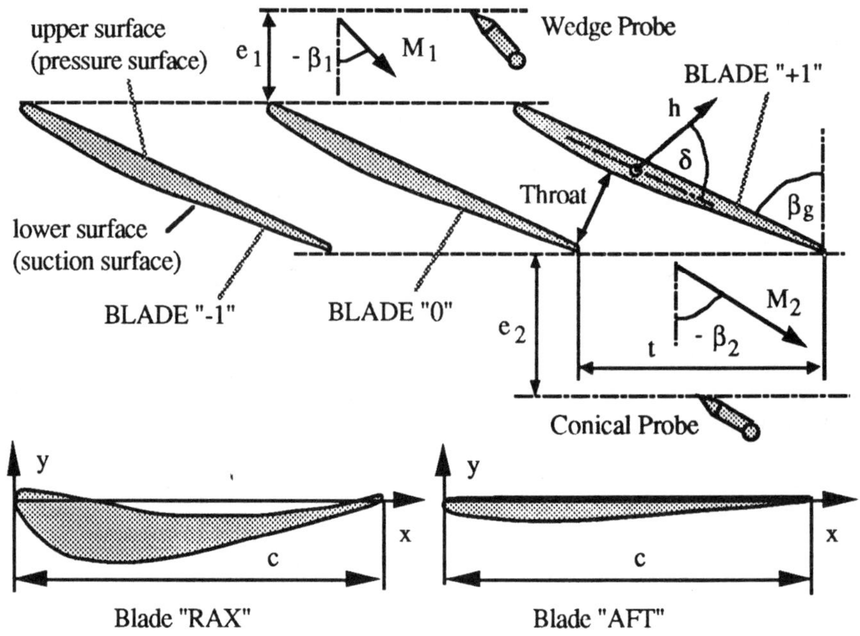

Fig. 3: Geometry for cascades "RAX" and "AFT" at midspan.

cases the correlation corresponds fairly well with the experimental data, apart from in the vicinity of the shocks (Fig. 4b).

For subsonic outlet flow at low inlet flow angles (Figs. 5) both the data and the predictions indicate an increase in unsteady blade suction surface pressure amplitude with increasing interblade phase angle, with similar trend and magnitude. Moreover, both the data and the predictions indicate an increase in the amplitude of the unsteady blade surface pressure coefficient towards the pressure surface trailing edge. This increase is larger at the higher than the lower outlet flow velocities, and is thought to come about as the cascade is overlapped in the blade pressure surface trailing edge region.For M_{s2}=0.71 (Figs. 5) the experimental suction surface pressure coefficient amplitude data are lower than the prediction for all interblade phase angles. The agreement in pressure amplitude and phase angle on the lower blade surface (pressure side) is excellent, especially at large interblade phase angles, along the whole chord, whereas the suction surface phase angle is different for the aft 60% chord. It is noted that all suction surface transducers, downstream of x/c~0.4, lie outside the overlapped part of the blade channel.

For closer to sonic outlet flow velocity (M_{s2}=0.95, Fig. 6) the suction surface dimensionless pressure coefficient amplitude has decreased somewhat in theupstream part of the blade whereas it has increased downstream of the 40%

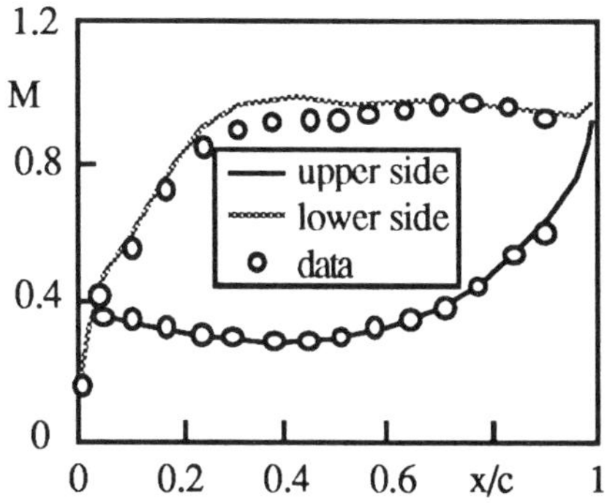

a: M_1=0.19, β_1=-18°, M_{s2}=0.95

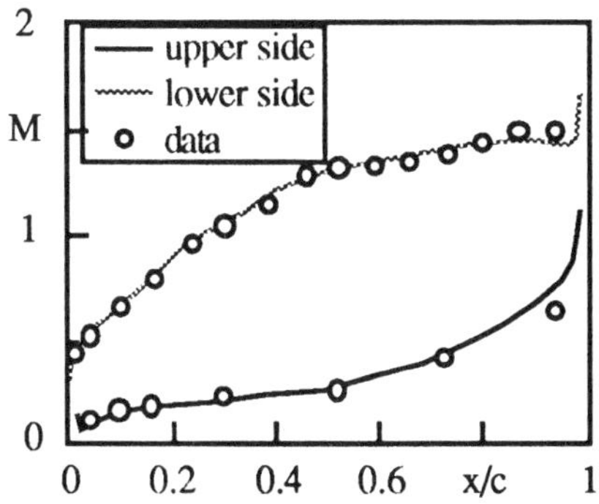

b: M_1=0.19, β_1=-14°, M_{s2}=1.26

Fig.4: *Cascade "RAX": Time-averaged blade surface Mach number distribution. Two-dimensional predictions.*

chord, both for the data and the prediction. The decrease is explained by the variation in inlet flow angle, and thus the increase in the inlet dynamic pressure, whereas the increase towards the trailing edge comes about as the flow velocity in the blade passage is closer to sonic. A small area change because of the blade vibration has thus a larger influence on the unsteady pressure amplitude.

At a supersonic outlet flow velocity of M_{s2}=1.13 (Fig.7) a good agreement is found in the steady-state blade surface Mach number distribution, apart from close to the shock. Schlieren flow visualizations from a linear test facility indicate an aerodynamic shock wave around x/c≈0.45 and a second around x/c≈0.75.

An excellent agreement between data and prediction of the time-dependent pressure coefficient is again found on the pressure surface, both in amplitude and in phase. The agreement between the data and the predictions is also excellent in the suction surface leading edge region, i. e. in the part of the flow field which is subsonic. However, the magnitude of the unsteady effects of the aerodynamic shock wave does not agree between data and prediction. The shock wave at around 45% chord is only seen by the pressure transducer at x/c=0.44 at some interblade phase angles, but not by the prediction. The experiment indicate a high unsteady pressure around 71% chord, whereas the prediction shows a much smaller increase at 90% chord. The 71% chord value agrees fairly well with the shock position from the linear cascade tests, whereas the predicted increase at 90% chord can not be explained.

Moreover, the experiments indicate, for all interblade phase angles, a considerable change in the pressure coefficient phase angle Φ_p for the pressure transducers in the vicinity of the shock waves. Such a tendency is not at all showed by the predictions. The implication of these differences in Φ_p is that the 71% chordwise position may show a local aerodynamic damping of opposite sign between the prediction and the experiment, which in turn can significantly influence the

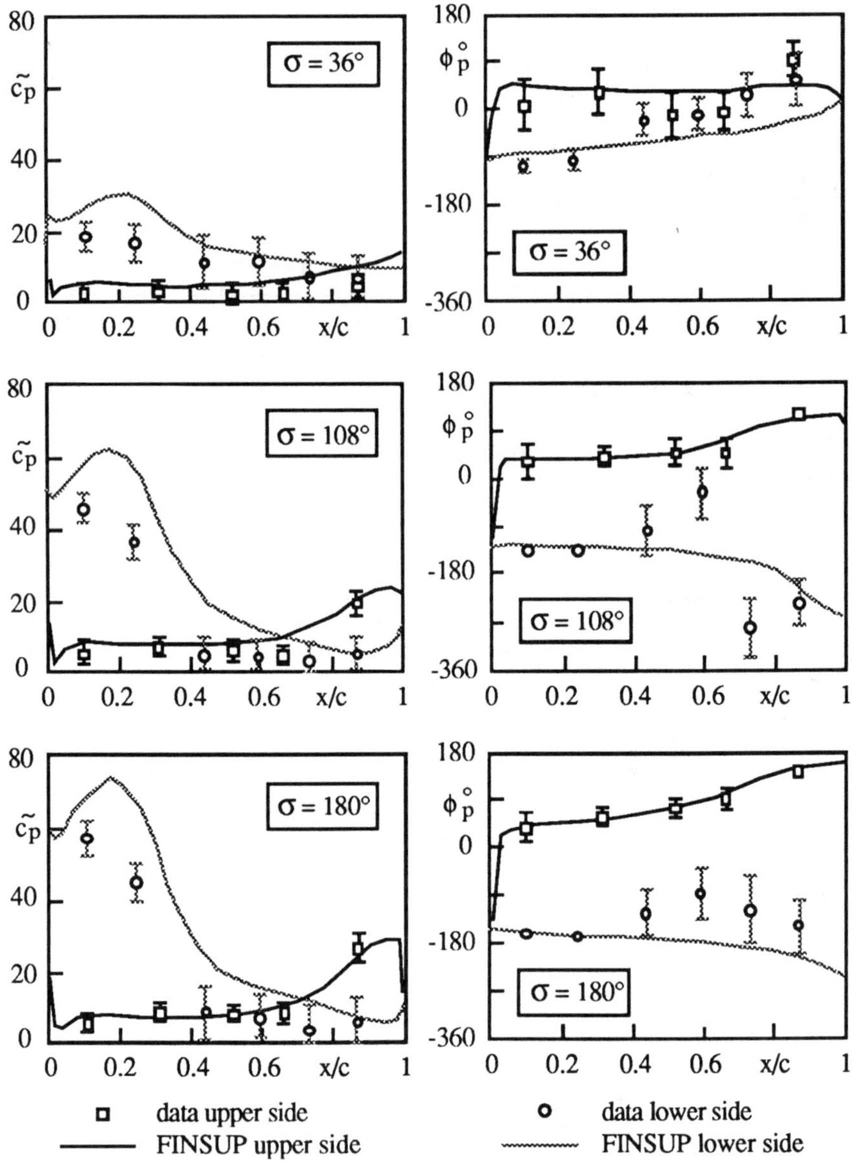

Fig. 5: *Cascade "RAX": Time-dependent blade surface pressure distribution for different interblade phase angles at low inlet flow angle and subsonic outflow (β_1=-13, M_{s2}=0.71).*

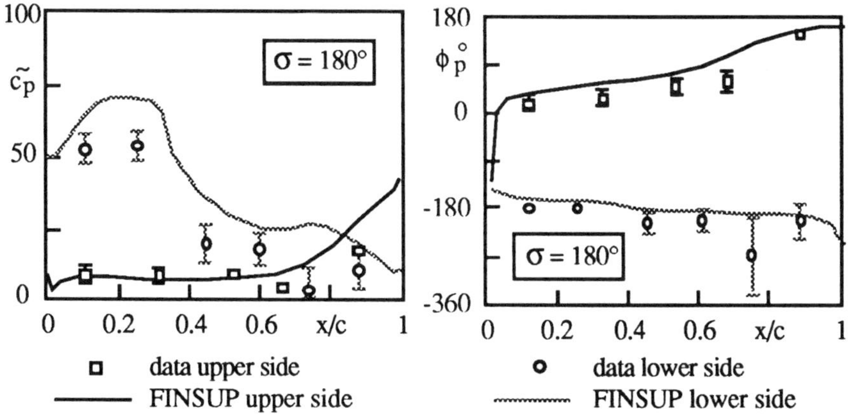

***Fig. 6:** Cascade "RAX": Time-dependent blade surface pressure distribution at low inlet flow angle and subsonic outflow. (β_1=-18, M_{s2}=0.95)*

global aerodynamic damping integrated along the blade surface.

The comparison of interblade phase angle σ=180° with the corresponding results at the lower outlet flow velocities shows that the predicted unsteady supersonic results do not significantly differ from the subsonic ones, neither for the pressure nor the suction surface.

Also at a higher supersonic outlet flow velocity of M_{s2}=1.26 the measured and predicted steady-state blade surface Mach number distributions show a good agreement. As for the previously discussed sample cases the amplitude and phase angle of the time-dependent blade surface pressure coefficient shows a good agreement between the two-dimensional "FINSUP" predictions and the experimental data on the pressure surface, both as regards to the phase angle and the amplitude. In the subsonic region it is good, whereas the influence of the shock wave does not correlate, especially in the phase.

It is concluded that for subsonic flow conditions the correlation between the unsteady data and the prediction is good to excellent for all interblade phase angles, in the travelling wave vibration mode. Major differences appear in the supersonic part of the flow field. These differences at supersonic outlet flow conditions can have significant influence on the, along the blade surface, integrated aerodynamic damping for some operating conditions.

Cascade "AFT"

The experimental blade surface Mach number distributions, as well as Schlieren flow visualizations, show, because of the circular leading edge a large overexpansion at the leading edge suction surface for the nominal inlet flow angle and all investigated outlet flow velocities (Fig. 9). For outlet flow Mach numbers

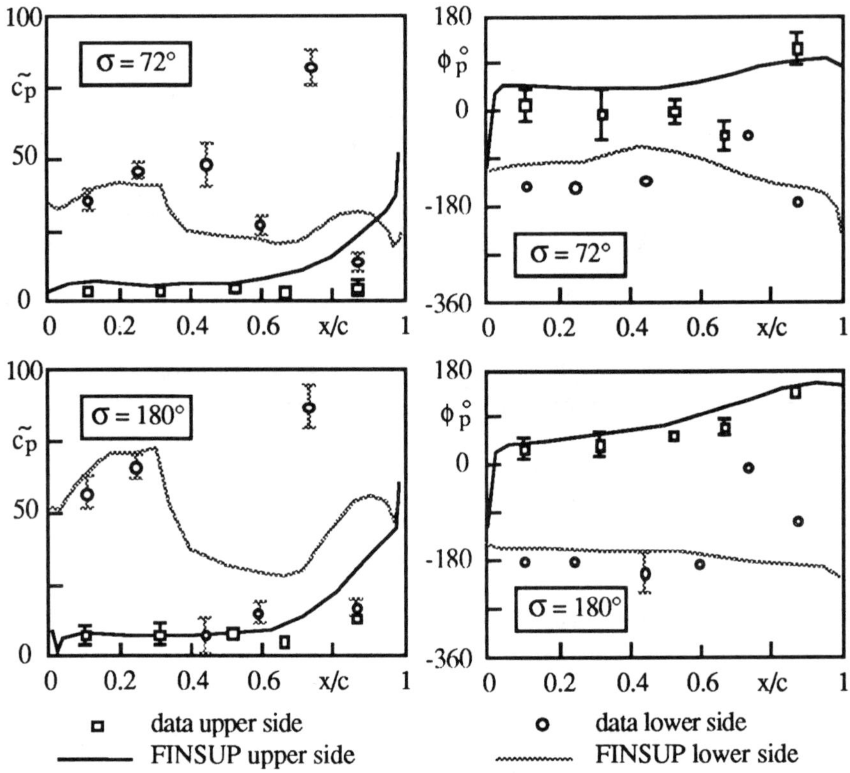

Fig. 7: Cascade "RAX": Time-dependent blade surface pressure distribution at low inlet flow angle and supersonic outflow (β_1 =-18°, M_{s2} =1.13).

larger than M_{s2}≈0.7 this overexpansion zone terminates with a normal shock wave. This shock wave is highly unstable, and in this region some differences in the Mach number distribution is found between the pressure taps on different blades.

The predicted shape of the Mach number distribution on the pressure surface agrees well with the experimental data, but the two-dimensional prediction model indicate, for all outlet flow Mach numbers, a higher Mach number towards the trailing edge than the measured one. The stream tube thickness was therefore varied in the calculations to compensate for factors such as side wall boundary layer growth and tip leakage, and it was found that a ratio of 0.9 gave the best agreement with the steady-state data. This ratio was thus used for the subsequent calculations.

For the subsonic Mach numbers the predictions indicate a slightly higher expansion, around x/c≈0.30, before the recompression zone starts. For the supersonic sample case "FINSUP" predicts the position of the shock wave at

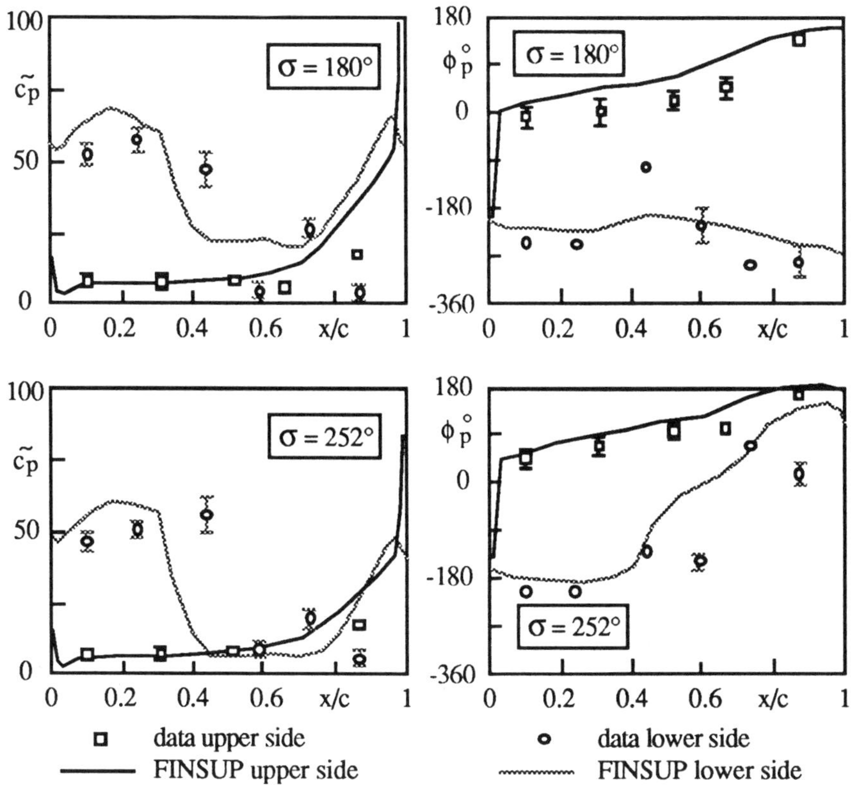

Fig. 8: Cascade "RAX": Time-dependent blade surface pressure distribution at low inlet flow angle and supersonic outflow (β_1 =-14°, M_{s2} =1.26).

x/c≈0.6 fairly well, but neither the magnitude of this shock nor the post-expansion or the second shock wave impinging on the blade suction surface around x/c≈0.8 correlates well with the data.
At nominal inlet flow angle and M_{s2}=0.60 good agreement is found between the data and the prediction along the pressure surface for most interblade phase angles (Figs. 10). The measured increase in pressure coefficient amplitude towards the blade pressure surface trailing edge is predicted, and the phase angles along the whole pressure surface correlate well.
The same type of results are found at a higher subsonic outlet Mach number at nominal incidence as well as for β_1≈-57°.
The correlation between the suction surface time-dependent amplitude data and predictions, for different steady-state loadings, is also acceptable although some differences exist in the magnitude for some interblade phase angles.
As for the "RAX"-cascade, it can be concluded that a very good agreement

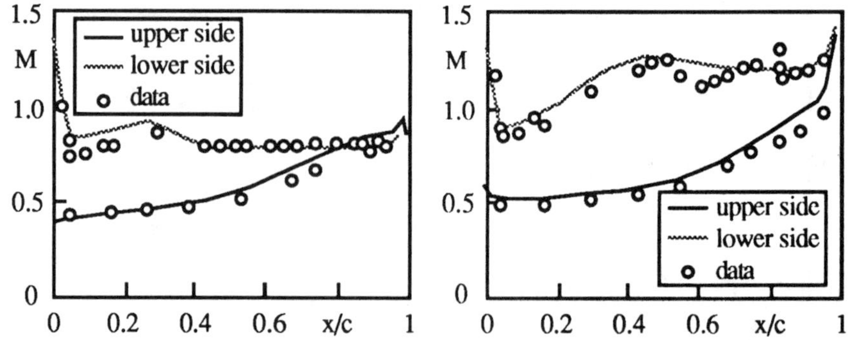

a: M_1=0.44, β_1=-62°, M_{s2}=0.79   b: M_1=0.49, β_1=-62°, M_{s2}=1.20

Fig. 9: *Steady-state blade surface Mach number distribution for the cascade "AFT". (Predictions with a quasi 3D ratio of 0.9)*

between the data and the "FINSUP"-predictions are found on the blade pressure surface for most interblade phase angles. This holds for both the amplitude and the phase angle.
The suction surface data do not correlate as well with the theory. Especially the phase angle shows some larger discrepancies over a large range of interblade phase angles. This will briefly be discussed in the section on influence coefficients.

DISCUSSION BASED ON THE INFLUENCE COEFFICIENT TECHNIQUE

In order to improve both the prediction model and the experimental procedure it is necessary to analyze the results in detail and to find out from where the differences appear as well as their physical origin.

Cascade "RAX"

From the above it has been seen that all experimental data and the predictions show that the unsteady complex blade surface pressures are, for the pressure surface on cascade "RAX", largely independent of outlet Mach number at inlet flow angles close to $\beta_1 \approx -15°$. A considerable variation exists in the absolute value of the amplitude for the different outlet flow velocities, as the blade surface pressure coefficient is made dimensionless with the upstream dynamic pressure. The pressure surface phase angles are almost constant, at a given interblade phase angle, for the different outlet flow velocities.
In the cases where the aerodynamic shock waves are situated close to a pressure transducer, a considerable unsteady effect is measured on the blade suction surface, both in the amplitude and the phase. The influence depends on the interblade

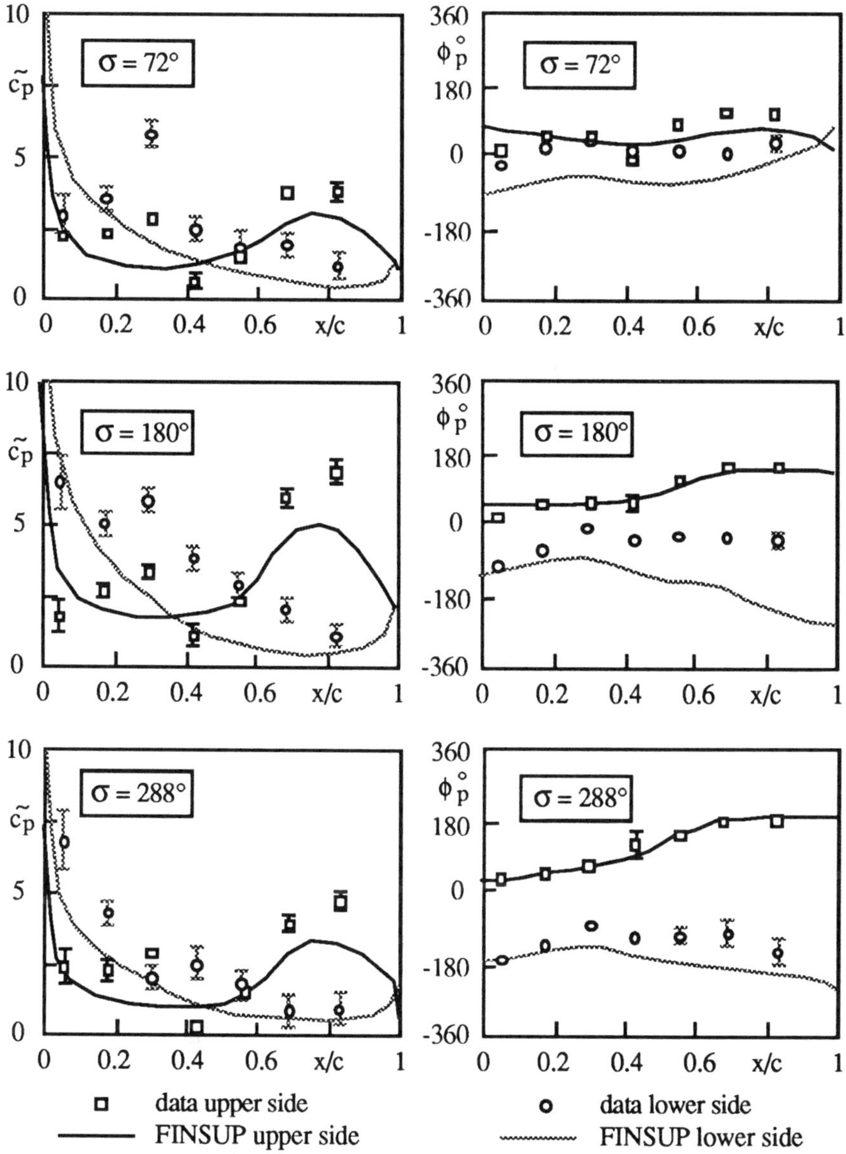

Fig. 10: Cascade "AFT": Correlation between experimental data and "FINSUP" analysis of the time-dependent blade surface pressure coefficient at $M_{s2}=0.61$, $k=0.133$.

phase angle, which indicates that the shock influence comes both from the reference blade itself and one or several of its neighbors.

These influences are studied in the form of amplitude and phase angle influences from blades "-1", "0" and "+1" on blade "0" in Figs. 11. Details about how these influences are obtained have been given by Schläfli [1989] and are thus not repeated here. The notation is such that:

- $\tilde{c}^{n,0}_{p,ic}$ indicates the pressure coefficient amplitude influence of the vibrating blade "n" on the non-vibrating reference blade "0". The influence of blade "0" on itself is called the eigen-influence, and is the influence when only blade "0" vibrates.
- $\tilde{\Phi}^{n,0}_{p,ic}$ indicates the pressure coefficient phase angle influence of the vibrating blade "n" on the reference blade "0".

From the pressure surface amplitude influences (Fig. 11) the experiments show that, towards the trailing edge, the eigen-influence of blade "0" is somewhat larger than the influence of the suction surface neighbor (blade "+1"), whereas the prediction indicates influences from both blades "0" and "+1" of similar magnitude. Apart from this the agreement between prediction and experiment is excellent for all studied outlet flow velocities. Both the experiments and the predictions indicate that the influence from blade "-1" is negligible.

The results also indicate that the amplitude, in absolute value, of the unsteady forces increases considerably as the outlet flow velocity approaches sonic.

The pressure surface phase angle influence-plots (Fig. 11) show an excellent agreement between the predicted and measured eigen-influence. Both the predicted and measured influences of blades "0" and "+1" are independent of outlet flow velocity. This is to be expected as the steady-state flow is very similar on the pressure surface for all studied cases. However, a systematic difference between the prediction and the experiment is found in the influence of blade "+1" on blade "0". For all flow conditions the measured influence is about 20°-40° lower than the predicted one.

It is also noted that the phase angle influence of blades "0" and "+1" are about 180° apart, which indicates that the cascade effects on the blade pressure surface can be considered to be largely quasi-steady at the reduced frequencies under study (k=0.082-0.134, depending on the outlet flow velocity). This is seen from both the experiments and the predictions. We refer to this as a "quasi-steady cascade effects".

It is thus concluded that, for the "RAX"-cascade, the difference between the predicted and measured unsteady pressures on the blade pressure surface found in the travelling wave vibration mode, appears because of a systematic disagreement as regards to the trailing edge influence of blade "+1" (=pressure surface neighbor blade) on blade "0".

It is also concluded that the blade pressure surface is always "eigen-damped" for all outlet flow velocities studied, both from the predictions and from the experiments.

The trend and magnitude of the experimental and predicted suction surface influences is in good agreement for subsonic flow conditions, with the exceptions that the leading edge region influence of blade "-1" is over predicted and that the

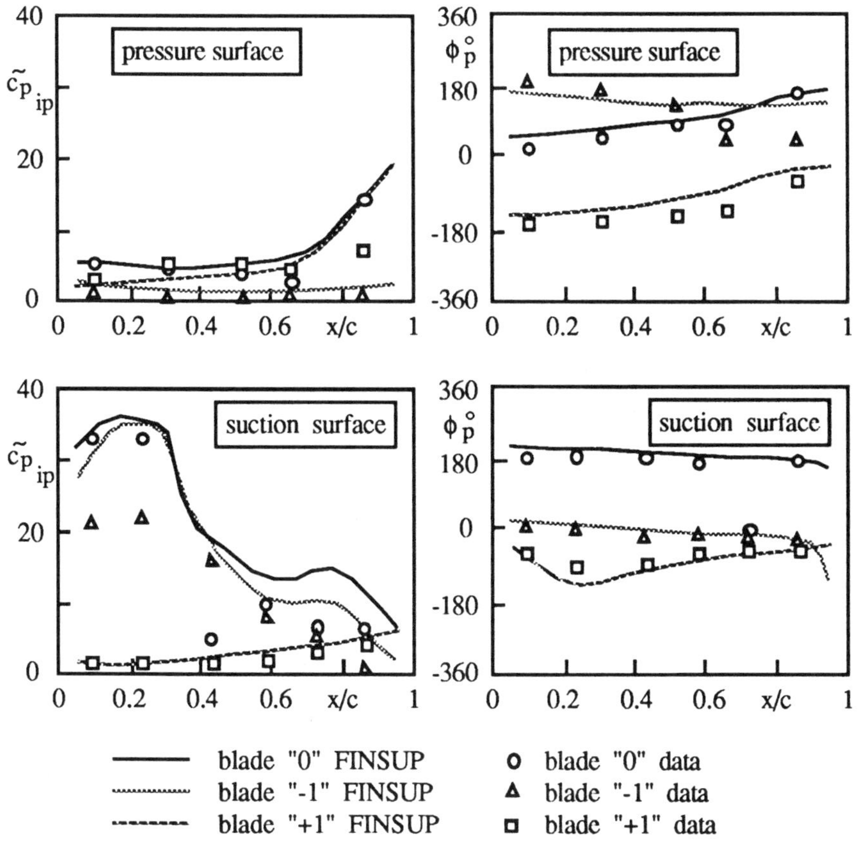

Fig. 11: Cascade "RAX": Time-dependent pressure amplitude and phase angle influence from blades "-1", "0" and "+1" on the suction surface and the pressure surface of blade "0". (M_1 =0.19, β_1 =-18°, M_{s2} =0.95).

44% pressure transducer sees a much smaller eigen-influence than the computation predicts.

This overprediction, especially of the influence of blade "-1", corresponds to the similar information in the pressure surface trailing edge region and is to be expected as the blade passage is overlapped for approximately x/c= 0.0 to 0.37 on the suction surface and x/c= 0.63 to 1.0 on the pressure surface.

As on the pressure surface the predicted phase difference between blades "0" and "-1" is almost constant at 180°. The trend of the experiment is similar, with an excellent agreement.

At the low supersonic flow velocity, M_{s2}=1.13, some larger differences are found between the prediction and the data (Fig. 12). An eigen-influence phase shift is

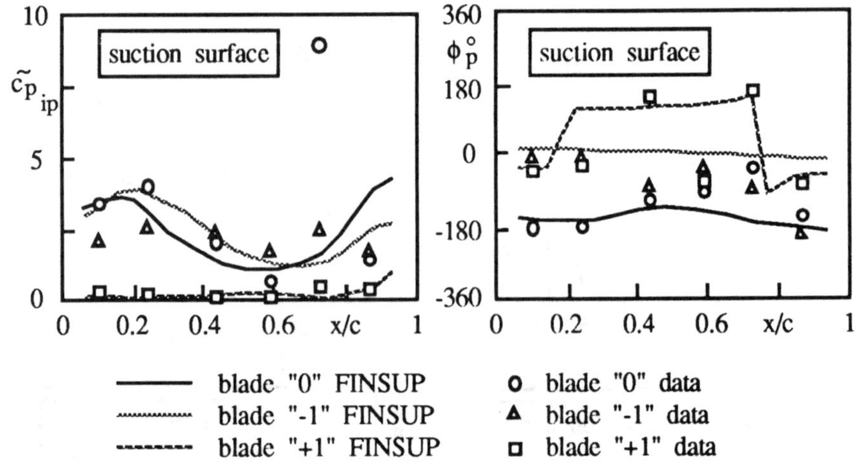

Fig. 12: Cascade "RAX": Time-dependent pressure amplitude and phase angle influence from blades "-1", "0" and "+1" on the suction surface of blade "0". (M_1=0.19, β_1=-18, M_2=1.13).

measured but not predicted between x/c≈0.4 to the shock position at x/c≈0.7. On the suction surface, for supersonic flow conditions, it is concluded that the underestimation in the prediction of the shock wave can be dangerous for the overall stability of the blade. Furthermore, the predictions show a large increase in amplitude eigen-influence towards the suction surface trailing edge, which is not seen so clearly in the data. As the phase angles are not well predicted in this region, and as the amplitude influence from blade "-1" on blade "0" is small, it is not certain that this over-prediction will always give a conservative value of global aerodynamic damping.

Cascade "AFT"

An example of the influences of the time-dependent blade surface pressure coefficient amplitude on the blade pressure surface of cascade "AFT" is presented in Figs. 13. The tendency is very similar to the corresponding figures for the cascade "RAX" . First of all it is noted that the scale of the amplitude of the dimensionless pressure coefficient is much smaller for cascade "AFT" than for "RAX". This is however not so in absolute pressure as the dynamic pressures are different. This is in line with the previously expressed discussion that a higher blade surface Mach number gives a higher absolute unsteady pressure level.
Similar to the "RAX" cascade the steady-state "AFT"-Mach number on the blade pressure surface increases considerably towards the trailing edge. The time-dependent experimental data indicate, as for "RAX", that the eigen-influence in the trailing edge region is larger than the influence from blade "+1", whereas the

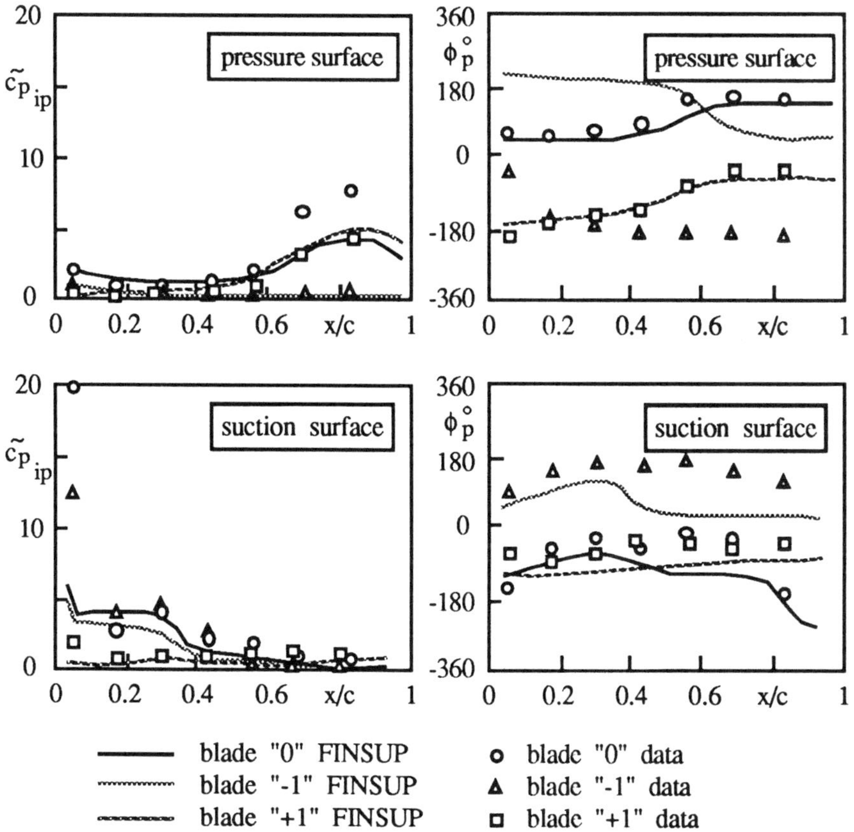

Fig. 13: Cascade "AFT": Time-dependent pressure amplitude and phase angle influence from blades "-1", "0" and "+1" on the suction surface and the pressure surface of blade "0". (M_1=0.44, β_1=-62°, M_{s2}=0.79)

predictions indicate that the influences from blades "0" and "+1" are of similar magnitude. In most cases the predicted influence of blade "+1" is larger than from blade "0".

Contrary to the pressure surface results for "RAX", the leading edge shows a non negligible unsteady eigen-influence. This can probably be explained by the circular leading edge shape of the "AFT" blade and the corresponding expansion around the nose at the flow angles studied.

The influence of the phase angle on the pressure surface is almost identical for the two cascades "RAX" (Figs. 11,12) and "AFT" (Figs. 13), although the geometry is quite different. The blade suction surface (Figs. 13) shows a somewhat different time-dependent behavior of the influences from neighbor blades than the cascade "RAX". The region of significantly higher unsteady pressure amplitude close to

the leading edge is much smaller for "AFT" than for "RAX", both from the predictions and from the experiments, although the peak becomes lower with increasing negative β_1, as expected from the steady-state Mach number distribution.
For the subsonic sample cases the correlation between the data and the prediction is good. The amplitude of the dimensionless pressure coefficient decreases steadily towards the trailing edge, and stays in the same order of magnitude as for the cascade "RAX" as far as the absolute unsteady pressure level is concerned.
The unsteady pressure phase angles on the suction surface are represented in Figs. 13. The tendency of the predictions agree well with the data, and the difference in phase angle between the influences from blades "0" and "-1" lies around 180° in the overlapped part of the blade passage, both for the prediction and the data.

CONCLUSIONS

Experimentally determined steady-state and time-dependent blade surface pressures, obtained in an annular test facility, have been compared to predictions by a finite element potential flow solver, "FINSUP", for two model turbine cascade sections.
The studied cascades are geometrically different (camber, thickness), but have a well defined interblade passage and are overlapped for about 35% of the chord.
Most time-dependent predictions were obtained with "FINSUP" in the two-dimensional mode, and no special attempt was made to refine the computational mesh in the vicinity of the expected shock positions.
From the results the following conclusions, believed to be of general nature for this kind of turbine blades, are drawn:

- The predicted and measured steady-state blade surface Mach number distributions correlate well for subsonic flow.
- The pressure surface steady-state Mach number distributions correlate also well for supersonic outlet flow conditions, whereas the suction surface show some deviation in the neighborhood of the shock waves.
- The prediction model gives a good agreement, for subsonic flow conditions, with the time-dependent travelling wave data, on both blade surfaces.
- The pressure surface of the blade is "eigen-damped" for all studied sample cases, for both cascades. This has been seen also on two other turbine cascades [Schläfli, 1989] and is believed to be a general trend for turbine cascades with attached flow.
- "FINSUP" predicts a higher magnitude influence from the immediate neighbor blade than actually measured. This indicates that the time-dependent perturbations attenuates faster in the experiment than for the inviscid linearized theory. Some differences, between prediction and data, found in results represented in the travelling wave mode can be explained by this fact.
- For both cascades the influence from blades "0" and "+1" (on the pressure surface), and blades "-1" and "0" (on the suction surface), are about 180° out of phase, both as regards to the prediction and the experiment, in the overlapped part

of the blade passage. This indicates that the unsteady cascade effects are quasi-steady for the reduced frequencies studied on the two steam turbine cascades. The magnitude of damping or excitation will thus depend largely on the unsteady effects on the reference blade itself, as well as on the amplitude of the influence from the neighbor blades.

- The measured change in the unsteady suction surface complex pressure coefficient in the vicinity of a shock wave is in general not captured by "FINSUP" (and not always seen in the experiment neither unless a transducer is situated close to the foot of the shock). It is thus believed to be important to well capture, both numerically and experimentally, the local effects of the aerodynamic shock waves.
- The results show that very trustworthy predictions can be obtained with "FINSUP" for steam- and industrial gas turbine sections with subsonic flow. For supersonic flow some major differences between the data and the predictions are found.

From the experimental results the conclusions are drawn that an aerodynamic shock wave has a significant influence on the unsteady pressures on the blade surface. The amplitude can become very large, and a phase shift can appear in the pressure. Results obtained in the travelling wave mode do not always, as they depend on the interblade phase angle, show this clearly. It is thus concluded that the physical understanding of shock influences may be fund more easily if the influence coefficients are studied in detail, both as regards to the amplitude and the phase angle.

ACKNOWLEDGEMENT

The experiments presented in the study have been performed under different contracts from ASEA BROWN BOVERI and " Commission pour l'encouragement de la recherche scientifique ", Switzerland (Cascade "RAX") and ANSALDO COMPONENTI S.P.A. FRANCO TOSI, University of Florence and ENEA, Italy (Cascade "AFT"). The program "FINSUP" has been developed at Cambridge University and at Rolls Royce plc.

The authors express their thanks to all the above institutions for the permission to publish the findings. The views expressed are the author's alone and do not necessarily correspond with the views of the above mentioned companies.

Most of the experimental results have been obtained by Dr. D. Schläfli, Mr. M. Degen, P. Ott, A. Ezzat, F. Jolles and S. Abbassi. Without their support and many fruitful discussions the study could not have been made.

REFERENCES

Bölcs, A.; 1983 "A Test Facility for the Investigation of Steady and Unsteady Transonic Flows in Annular Cascades", *ASME Paper 83-GT-34, 1983.*

Bölcs, A.; Fransson, T. H.; 1986 "Aeroelasticity in Turbomachines: Comparison of theoretical and experimental cascade results". *Communication du Laboratoire de thermique appliquée et de turbomachine, EPF-Lausanne, Switzerland, No. 13, 1986.*

Schläfli, D.; 1989 "Experimentelle Untersuchung der instationären Strömung in oszillierenden Ringgittern", *Thèse No. 800, EPF-Lausanne, Switzerland.*

Whitehead, D. S.; 1990 "A Finite Element Solution of Unsteady Two-Dimensional Flow in Cascades." *International Journal for Numerical Methods in Fluids, Vol. 10, pp. 10-34, 1990.*

Whitehead, D. S.; Newton, S. G.; 1985 "A Finite Element Method for the Solution of Two-Dimensional Transonic Flows in Cascades."*International Journal for Numerical Methods in Fluids, Vo. 5, pp. 115-132, 1985.*

Supersonic Stator-Rotor Interaction in a Turbine Stage

Gilles Billonnet[1]

ABSTRACT

This paper describes the unsteady phenomenon which occurs between stator and rotor in an impulse supersonic turbine. Shock and expansion waves appear upstream of the rotor characterized by an inlet supersonic flow. Generally, the flow is started and the flow rate is determined by the unique incidence phenomena i.e. the axial velocity remains subsonic and the flow direction depends only on the blade entrance shape for a given Mach number. Shock waves are propagating upstream of the rotor. The wave system moves with the rotor and propagates upstream into the stator. As the leading edge thickness is large and the relative inlet Mach number is high, the shock wave intensity is significant. A complex interaction phenomenon between the waves and the walls is then generated, and it is very difficult to asses their intensity experimentally. The present study suggests a theoretical analysis of this phenomenon.

A two-dimensional inviscid flow calculation is performed. The numerical method used here is able to compute every unsteady features of the flow and captures strong shock waves.

In order to describe this unsteady interaction phenomenon, a simplified geometrical model consistent with the particular properties of supersonic flows has been chosen.

The stator outflow is fully supersonic and there are any upstream effects of the outflow conditions. The internal stator shape is generally designed by using a supersonic nozzle. By the same way, the inlet relative flow is started, and thus the rotor upstream conditions do not depend on the relative flow field after the leading edge region. Therefore, only the downstream geometry of the stator and the upstream geometry of the rotor have an effect on the phenomena, the absolute upstream Mach number and the rotational speed being given. Consequently, the computation is restricted to the case of two simplified grids moving with respect to each other. The first one simulates the stator exit shape and consists of straight walls without thickness. The second one which simulates the rotor inlet shape is realistic compared to the real turbine leading edges.

The computation is carried out on a domain which includes one blade-to-blade passage of the stator and four blade-to-blade passages of the rotor. As the azimutal extensions of the two blade-to-blade passage groups are the same, we imposed a flow periodicity condition on the extremal boundaries of each group.

The results emphasize that the flow field is highly unsteady and periodic as expected. The periods differ depending on whether we are located in the fixed

[1]Office National d Etudes et de Recherches Aerospatiales BP72,
92322 Chatillon Cedex, France.

frame or in the rotating frame. The unsteady effects are very significant on the upper wall of the nozzle where impinge the shock waves coming from the rotor . The amplitude of static pressure fluctuations is maximum near the trailing edge. On the contrary, the unsteady effects are weak within the rotor in terms of static pressure fluctuations. The shock losses are compared with the case of an isolated rotor and the effect of the axial stator-rotor gap is described.

Even though the geometry is simplified, the inviscid computation shows clearly that the unsteady effects cannot be neglected in supersonic turbines.

In purpose of designing such turbines, the choice of the axial gap value turns out to be crucial to control as much as possible the interaction phenomenon.

1. INTRODUCTION

In general case, flow in turbomachines is tridimensional, unsteady and viscous.

At the present time, it is difficult to emphasize all aspects of the flow in turbomachines and in particular unsteady effects.

Since a few number of years, available computers allow us to calculate unsteady flow aspect in a turbine or compressor stage. The internal flow of a complete turbomachine is governed by the complete Navier-Stokes equations but the resolution of these equations would need too much time even on the biggest computers.

Nevertheless, a two-dimensional flow simulation can give a good estimation of unsteady flows. Assuming that there are no external unsteady effects such as time depending upstream or downstream condition, or blade vibration, the main phenomenon is led by the rotation effect.

ONERA has been working on this interaction phenomenon for several years. At the same time, supersonic axial turbine has been dealt with in aerodynamic studies.

The specific stator-rotor interaction occurring in supersonic turbine stage, is described in this paper. An inviscid flow calculation is performed in the case of two simplified grids moving with respect to each other.

Before describing the aerodynamic phenomena, it is necessary to identify the particular flow features which characterize such turbines. It allows us to simplify the blade geometry and to restrict the study to the specific interaction between stator and rotor.

2. FLOW PROPERTIES IN SUPERSONIC TURBINE STAGE

Typical impulse stages are usually encountered in advanced turbines of rocket engine turbopumps. The stages includes a supersonic stator followed by a full supersonic rotor. The stator is generally made up by an annular row including several supersonic nozzles. These two-dimensional nozzles are inclined to the axial direction. The flow downstream of the stator is largely supersonic with a subsonic axial velocity. Upstream of the rotor, the relative flow is supersonic and usually remains supersonic across the moving blade with a strong deflection.

A shock and expansion waves system appears at the rotor leading edges. That system propagates upstream from the rotor and moves with the blades. The waves are then reflected on the stator walls and a complex interaction phenomenon is generated.

Let us first consider the flow upstream of the isolated rotor. The blade shape is characterized by large rounded leading edge. Detached shocks appear near the leading edge associated with a subsonic flow area. The shock intensity is all the more important as the leading edge thickness reaches 4 % to 7 % of the blade pitch. Generally, as the inlet flow is started and periodic, it is led by the unique incidence phenomenon. Thus, the mass flow is only determined by the blade entrance shape for a given upstream Mach number. The blade entrance shape which influences on the phenomenon is limited by the last characteristic line coming from the upper blade leading edge as shown on figure 1a. The influence of leading edge thickness and suction surface angle on the "unique incidence" value is well-known. Some works give "unique incidence" relations which connect flow angle to upstream Mach number for different values of leading edge thickness and suction surface angle (1). The "unique incidence" condition is valid for started supersonic blades with subsonic axial flow.

This specific feature of flow allows us to simplify the rotor blade geometry. This one is limited to the entrance region. It consists of a flat plate cascade with rounded leading edges. The plates tilt is realistic compared to the actual rotor blade one. However, the blading suction surface curvature is not taken into account but as it is small, its effect can be neglected. The leading edge thickness value corresponds to 7 % of the pitch. This value is encountered on actual rotor blades.

Let us consider the flow downstream of the isolated stator. Since the flow is started and fully supersonic in the nozzle, there is no influence of the exit flow upon the upstream flow. More particularly, the mass flow is fixed by the sonic throat for given total pressure and total temperature. The region influenced by the exit flow can be determined by the characteristic line coming from the lower wall trailing edge as shown on figure 1b. In the same way, this property allows us to simplify the stator shape. Then, the stator shape consists of inclined straigth walls. The tilt of the walls is the same as the supersonic nozzle. Assuming that the upstream Mach number delivered by the stator is constant, the uniform flow direction corresponds to the stator tilt. Finally, the walls are assumed to be without thickness because the phenomena in the trailing edge region (separated flow) can not be accurately simulated by means of inviscid flow calculation.

The simplified geometry of the stage is shown on figure 2.

3. COMPUTATION METHOD

Although viscous effects can modify locally the main phenomenon, we assumed that their influences remain negligeable. A two-dimensional inviscid flow computation is then used. The computation is performed on a domain which can include several blade-to-blade passages for the stator and rotor.

The rotor mesh grid is shown on figure 3. This O mesh type allows us to accurately restitute the leading edge geometry. The domain includes the rotor-stator interface which is very close to the leading edge. Special techniques are used to elaborate the mesh grid such as an optimization technique developed at ONERA (2).

As we describe earlier, the geometry is limited to the entrance blade cascade as previously described. The mesh for one channel of the rotor blade domain includes 2208 points.

The stator mesh grid is a simple H mesh type. For building out this kind of grid, the optimization technique can be avoid. The stator mesh includes 16335 points.

The Euler computation method is characterized by an explicit centered difference scheme. The equations are discretized using a two-step Lax-Wendroff type scheme. The method is derived from the method developed by Cambier, Couaillier and Veuillot for the two-dimensional Navier-Stokes equations (3). The time step used in the computation is constant and is determined by the CFL criteria. Artificial viscosity terms are added in order to stabilize the numerical scheme. The boundary conditions are treated using the characteristic method. This technique is based on the mathematical properties of hyperbolic system which gives characteristic relations. On each boundary, the flow is then determined by the applied conditions and the characteristic relations. The boundary conditions must be adequally chosen in order to get physical meaningful results.

Before perfecting the unsteady computation, the method is first applied and tested on the isolated rotor grid. A simple H mesh domain is then added upstream from the rotor blade domain (see figure 3). This calculation simulates the upstream flow of the rotor and gives results about the "unique incidence" phenomenon and the periodic shock and expansion waves system. The isobaric lines of the calculated flow is given on figure 4. Such computations allow us to check "unique incidence" values obtained from another source. The flow pattern emphasizes that the shock waves intensity decreases from the leading edge towards the upstream boundary. The waves intensity becomes smaller and smaller until the upstream boundary where the flow remains almost uniform.

Then, we performed the full calculation of the stator followed by the rotor, taking into account the rotational motion of the rotor. Nevertheless, it is impossible to calculate all the blades of both rotor and stator. The calculation is carried out on a domain which includes a reduced number of blades (4). More precisely, since the flow computation of N_1 channels of the first blade-row and N_2 of the second is excluded, a smaller computational domain is determined. This domain contents K_1 and K_2 channels such that the ratio K_1/K_2 is as close as possible to the ratio N_1/N_2, while the blade number N_1 and N_2 are usually prime numbers with respect to each other. In a practical case of supersonic turbine, the stator has N_1 blades and the rotor N_2 blades ; the ratio N_1/N_2 is 0.213.

To get as close as possible the real case, two numbers K_1 and K_2 must be found which satisfy the following dual condition : they must be as small as possible but with a ratio close to 0.213.

The sets of numbers giving a ratio closest to this value are :

$$\frac{K_1}{K_2} = \frac{1}{1} = 1 \;; \qquad \frac{K_1}{K_2} = \frac{1}{2} = 0.5 \;; \qquad \frac{K_1}{K_2} = \frac{1}{4} = 0.25$$

Let us note that if the ratio $N_1/N_2$was 0.25, the last solution $K_1 = 1$, $K_2 = 4$ would have been precise. Hence, these simplifying assumptions that are valid in an exact case constitute a satisfactory approximation. Obviously, the case $K_1 = K_2 = 1$ where the unsteady computations are carried out on only one blade channel for each blade row is much less realistic than the case $K_1 = 1$, $K_2 = 4$.

The periodicity has to be applied to the restricted number of channels rather than the whole rotor.

The application for current supersonic turbine leads to choose one stator blade and four rotor blades. Although the technique suggests not to change the blade pitch, in this specific case, the blade numbers are assumed to have a highest common factor equal to 4. Therefore, the azimutal extension of both stator and rotor domains is the same, and the flow becomes periodic every one blade in the stator and every four blades in the rotor.

The stator domain and the rotor domain are separated by an artificial interface on which we applied a flow continuity condition at each time step. This condition takes into account the relative motion between the stator and the rotor. On the extremal boundaries of the rotor and the stator, the flow periodicity condition is applied at each time step.

The conditions applied on the others boundaries are the following:

- a slide condition on the solid walls

- on the upstream boundary, constant total pressure, total temperature and tangential velocity are imposed in order to generate uniform flow ;

- on each rotor downstream boundary, no condition is applied because on this boundary, the perpendicular component of the velocity remains supersonic ;

- a flow continuity on the boundaries shared by two adjacent channels.

4. UNSTEADY FLOW SIMULATION

A first calculation is performed on the domain described before. The rotational speed of the rotor is chosen according to the velocity triangle of supersonic turbine. The absolute flow conditions lead to a relative Mach number value and a relative flow angle according to the unique incidence phenomenon. The axial gap between the stator trailing edge and the rotor leading edge is equal in value to 30 % of the rotor pitch. This value is also realistic compared to the one encountered in realistic cases. It corresponds to about 20 % of the axial chord of actual rotors.

The results prove that the flow is periodic and indicate that the period of the phenomena differs depending on whether we are located in the fixed frame or in the rotating frame. The rotor period corresponds to the time required by the rotor to go over one stator pitch. According to the blade number ratio, the rotor period is then equal to the quarter of the stator period.

Figure 6 shows isobaric lines of the instantaneous flow field. One can distinguish the region which is affected by the interaction of shocks and expansion waves with the stator upper wall. Upstream from this region, the flow remains steady and uniform because it is not influenced by the rotor as expected. In particular, the static pressure as well as the Mach number are constant on the upper wall. The instantaneous static pressure distribution on the upper wall is plotted as a function of the axial abscissa on figure 7, for a given position of the rotor. It indicates that the static pressure amplitude increases with the axial position. The amplitude reaches its maximum near the trailing edge.

In order to find out the main unsteady effects, we study the time-evolution of the flow at different locations of the computation domain. These numerical probes are embedded in the fixed or rotating frame. A first set of probes is located on the stator-rotor interface (middle of axial gap) in the fixed frame. Their positions are given on figure 8.

The flow evolution of the probe n° 1 (figure 9) shows large fluctuations of the static pressure. This probe is located in the middle of the stator blade-to-blade channel and it goes through the moving expansion and shock waves. Figure 10 shows the corresponding absolute Mach number and flow angle evolutions. The shock waves and the expansion waves generate large deflections of the absolute flow. These time-evolutions are periodic and the period corresponds to the rotor period Tr. In the same way, the time-evolutions of probes attached to the rotor, are periodic and the period corresponds to the stator period Ts. After one period Tr, the rotor is in the same position again with respect to the stator.

For all time-evolutions, the time scale is dimensionless and equal in value of the ratio T/T_S. The probes n° 2 and n° 3 are also located on the stator-rotor interface near the stator upper wall. Without the stator, these probes would have the same evolution as the previous one, except of the difference in phase. The static pressure evolution as well as the absolute Mach number and flow angle direction ones (figures 11 and 12) demonstrate that the flow is quite different. These probes are set in the disturbed flow area where a reflecting waves system occurs. In particular, the probe n° 2 is characterized by a sudden variation of static pressure compared to the probe n° 1. Consequently, the absolute flow is highly unsteady on the interface.

A second group of probes is located on the stator walls. Most of them are located on the upper wall in the disturbed flow area. The corresponding static pressure evolutions are shown on figure 13 to 16. The probe n°7 points out a small variation of the pressure. The intensity of oblic shocks reflecting on the wall is reduced at that point. The more the probe is located near the trailing edge, the more the static pressure amplitudes are important. At the trailing edge, the intensity of the shock waves is maximum and the corresponding amplitude of the static pressure variation reaches 33 % of the mean pressure (probe n° 4). On the lower wall, some probe evolutions confirms that the flow is uniform and steady up to the trailing edge.

The next series of probes (n° 8 to 10) is attached to the rotor. In the moving frame, the flow period corresponds to the rotor period Tr. The pressure evolutions of the probes n° 8, 9 and 10 are given on figure 17. These probes are located on the suction side of the lower blade and in the middle of the corresponding blade-to-blade passage near the downstream boundary. The evolutions clearly show that the static presure fluctuations are reduced within the rotor. The amplitudes become maximum when the blade turns through the disturbed flow area, after about two rotor periods. The moving blade cascade has then covered two pitches of the rotor. Subsequently, unsteady effects are weak within the rotor. It means that the effect of the stator on the rotor is very moderated.

The probes n° 11 to 13 are similar to the previous ones but they are located on the upper blade-to-blade passage. Their flow evolutions are the same as the previous one with phase difference. The shift in phase between the two groups of probes corresponds to two rotor periods (equal to an half of the stator period). All the blade passages have the same flow evolution and each one is out of phase with respect to the others. Therefore, the analysis of one rotor blade is then sufficient to know the whole rotor flow.

The analysis concerns also the losses which are generated through the shock waves. The thickness of the rotor leading edge has a great influence on losses. It is interesting to investigate the interaction effect on losses in that case. Figure 18 shows the time-evolution of the mean losses generated from upstream to downstream. The losses value is expressed as the ratio of the absolute total pressure to the isentropic total pressure*.

The mean value is computed on the downstream boundary where the flow is fully supersonic. Two curves are plotted on figure 18. The curve A is related to one blade-to-blade passage and the curve B is related to the whole rotor. The total pressure evolutions can be compared to the losses corresponding to the isolated rotor. One can notice the stator influence on losses. The largest fluctuations occur when the unsteady effects are the most intense within the rotor. The losses could probably change whether including viscous effect. Strong interactions between the shocks and the boundary layers could appear. Consequently, the results must be interpreted cautiously by recalling that Euler calculation method is only able to restitute shock waves losses.

Effect of axial gap

The same computation is performed for a different gap value. The stator-rotor gap is equal to 50 % of the rotor pitch compared to the

---

* the isentropic total is calculated in assuming an isentropic flow through the stage

previous one (30 % of the rotor pitch). The other computation conditions remain the same. According to the results analysis, it is obvious that an increase of the gap has mainly two consequences :

- significant decrease in unsteady effect on the stator wall ;
- slight increase in total pressure recovery.

Many factors may change the aerodynamic phenomena and a complete assesment of their influences would need a parametric analysis. In order to prevent significant pressure fluctuations, it is important to design carefully the axial gap. The gap choice varys according to the aerodynamic conditions and the geometrical configuration. Generally, the aerodynamic conditions are fixed by the turbine design and the axial gap also depends on technological criteria. The designer has to optimize as well as possible the gap between each row.

5. CONCLUSION

This paper describes the supersonic stator-rotor interaction in a simplified turbine stage. Although flow is assumed to be inviscid and two-dimensional, the shock waves system coming from the rotor leading edges is accurately predicted. The flow is highly unsteady upstream of the rotor and a large variation of the static pressure occurs on the stator wall. We focus on the effect of the axial gap between the two blade cascades in supersonic turbine stage. A good estimation of such flow behaviour is then possible by means of numerical simulation. This kind of analysis would certainly help the designer to understand aerodynamic phenomena. Nevertheless, viscous and 3D effects should add other complexity to this interaction phenomenon.

In the future, it will probably become possible to compute viscous flow in a complete stage. These more complex simulation will need experimental validation to be validated.

ACKNOWLEDGEMENT

The author would like to express his thanks to the society S.E.P. (Société Européenne de Propulsion) and CNES (Centre National d'Etudes Spatiales) for their financial support.

REFERENCES

(1) Starken H., Schreiber H., Yong Xing Z., 1984, "Mass Flow Limitation of Supersonic Blade Row Due to Leading Edge Blockage", ASME Paper N° 84-GT-233.

(2) Jacquotte O.P. and Cabello J., "Three Dimensional Grid Generation Method Based on a Variation Principle", La Recherche Aérospatiale, N° 1990-4, pp. 7-11 (English edition).

(3) Cambier L., Couaillier V. and Veuillot J.P., "Numerical Solution of the Navier-Stokes Equations by a Multigrid Method", La Recherche Aérospatiale, N° 1988-2, pp. 23-42 (English edition).

(4) Fourmaux A. and Le Meur A. "Computation of Unsteady Phenomena in Transonic Turbines and Compressor", 4th Symposium on Unsteady Aerodynamics and Aeroelasticity of Turbomachines and Propellers, Aachen, September 87.

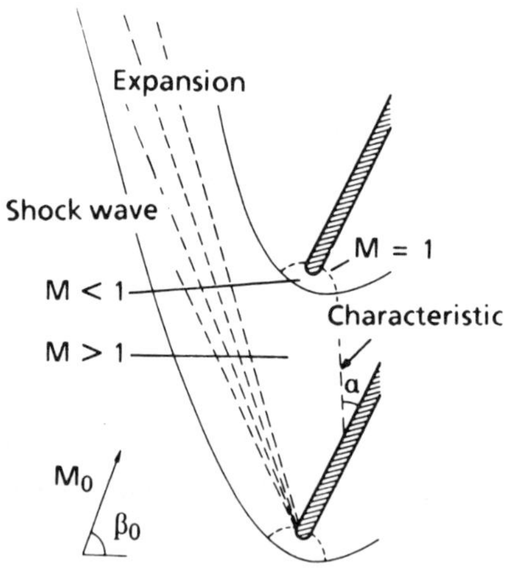

Fig. 1a -Entrance shape of rotor

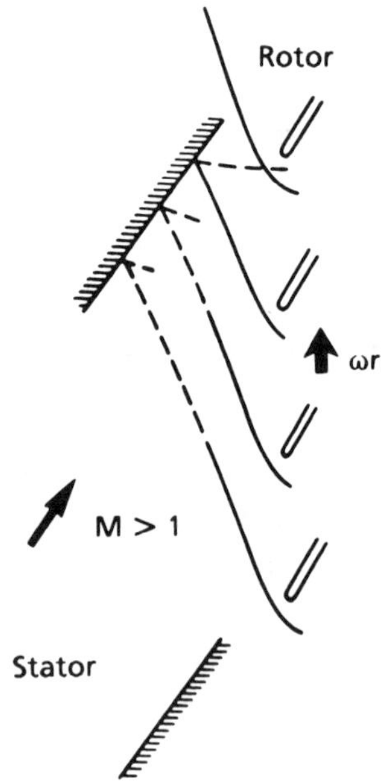

Fig. 2 - Simplified geometry of the turbine stage.

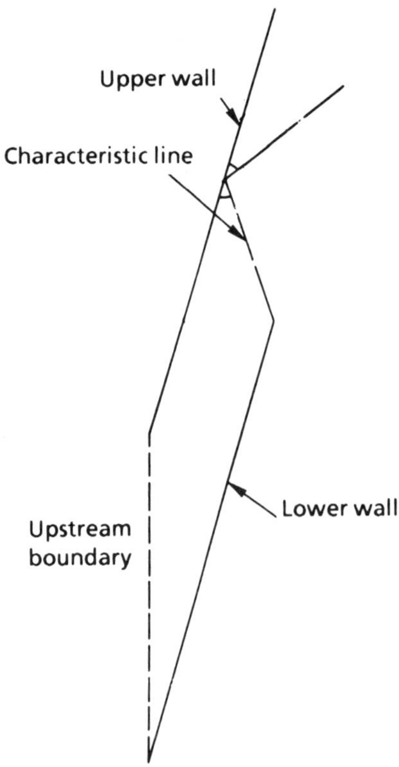

Fig. 1b - Exit shape of stator.

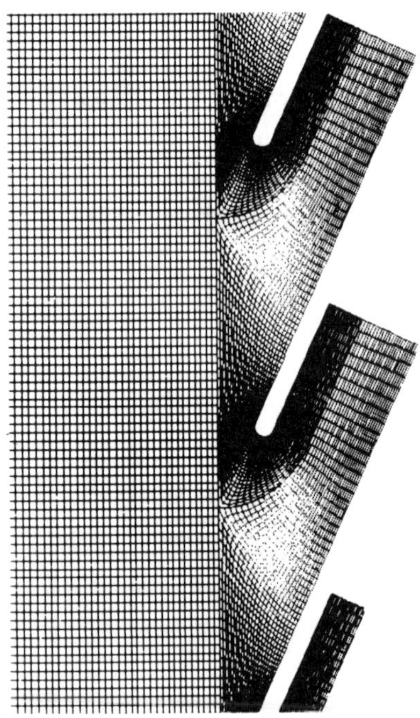

Fig. 3 - Rotor mesh grid.

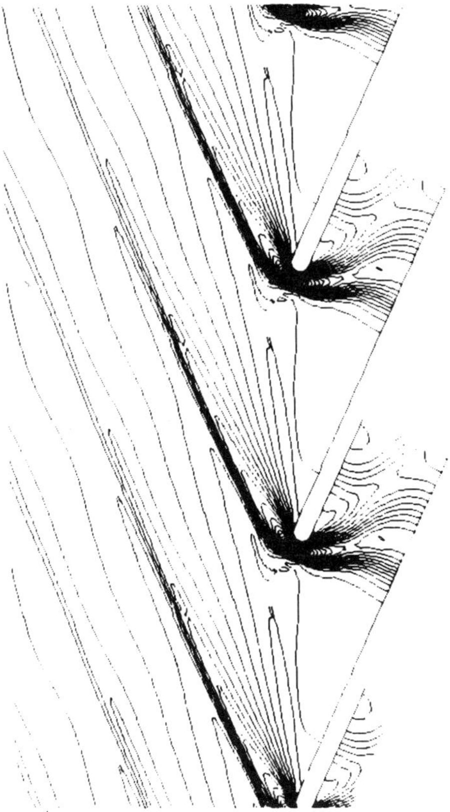

Fig. 4 - Isobaric lines of the isolated rotor.

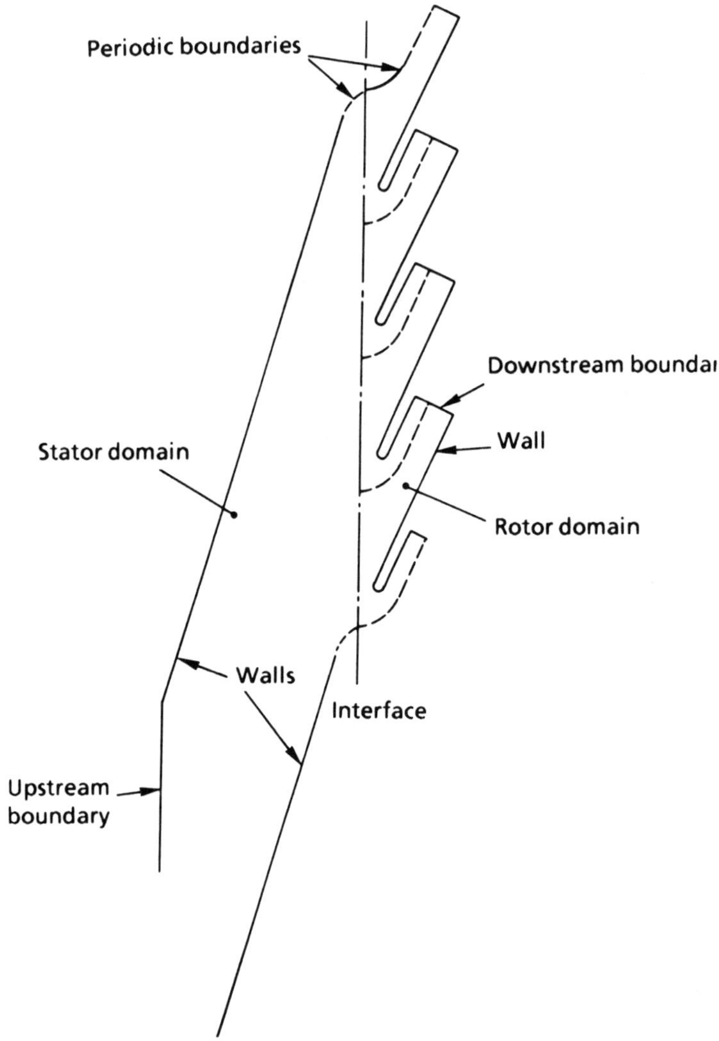

Fig. 5 - Computation domain.

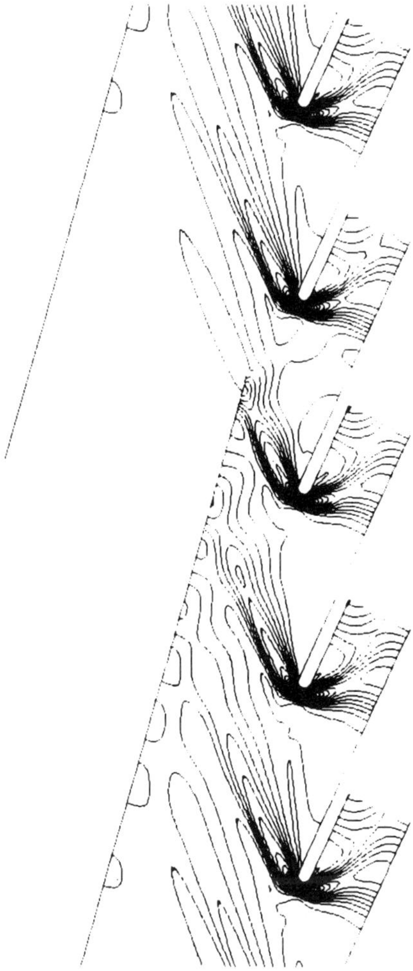

Fig. 6 - Isobaric lines of instantaneous flow in the stage.

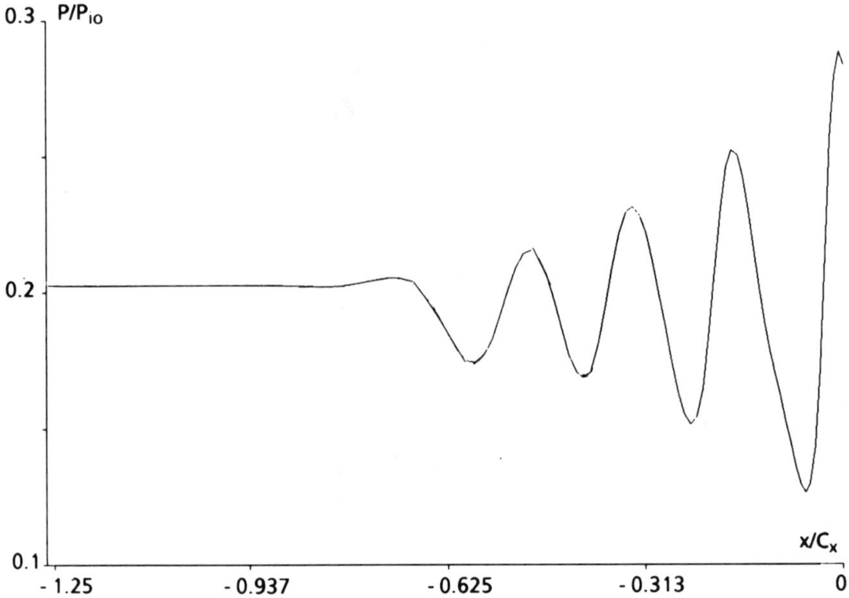

Fig. 7 - Instantaneous static pressure evolution on the upper wall of the stator.

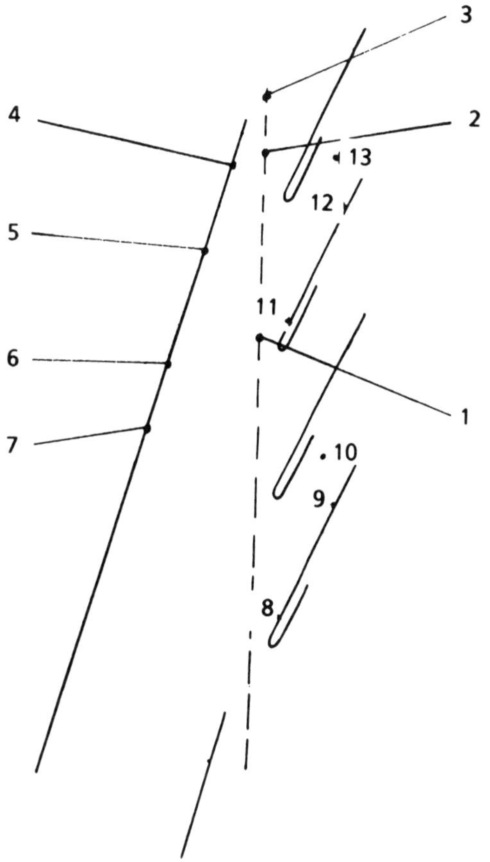

Fig. 8 - Position of the probes.

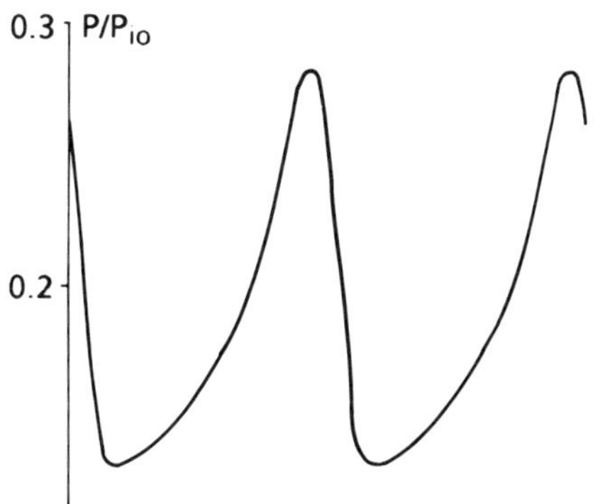

Fig. 9 - Static pressure evolution of the probe n° 1.

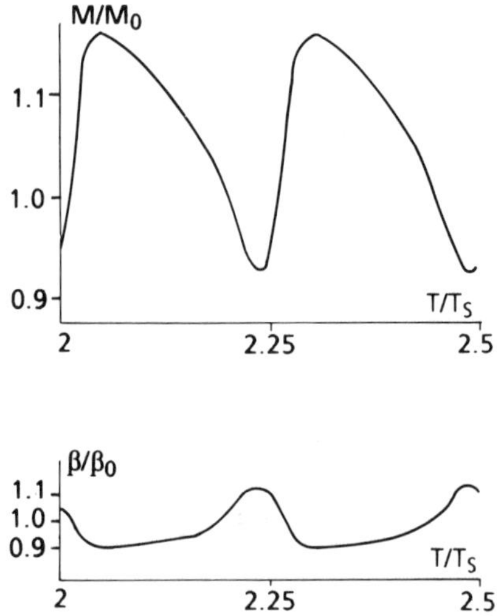

Fig. 10 - Mach number and flow direction of the probe n° 1.

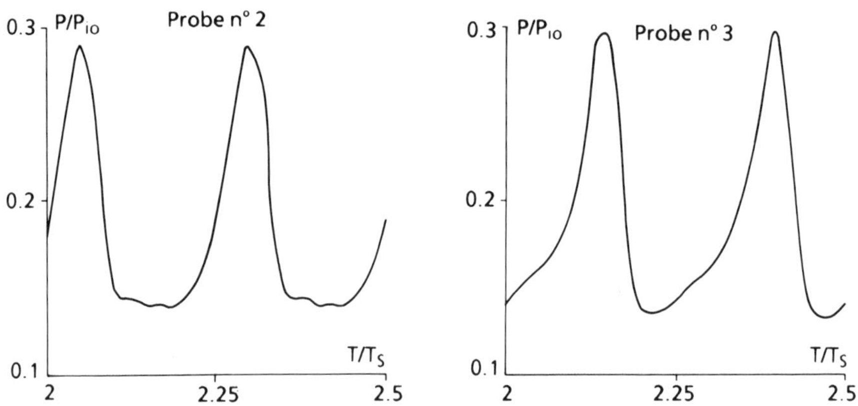

Fig. 11 - Static pressure evolution of the probes n° 2 and 3.

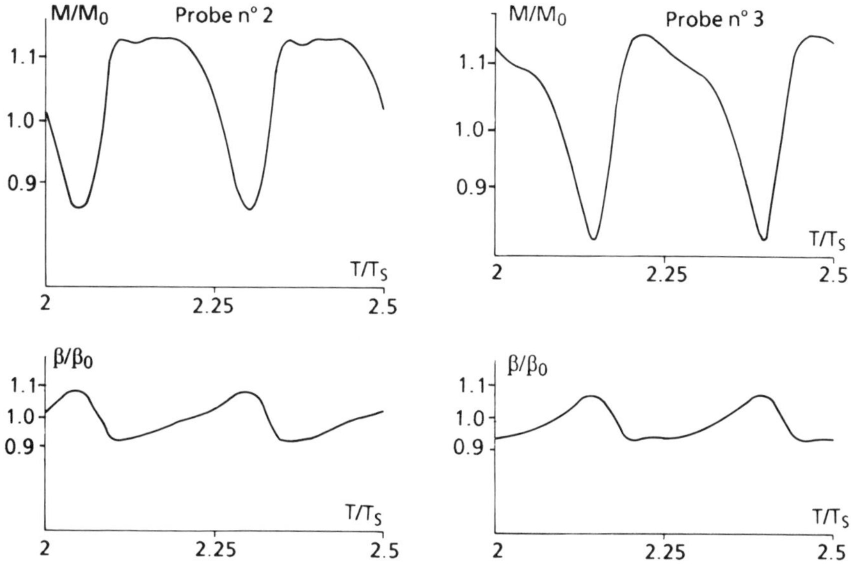

Fig. 12 - Mach number and flow direction of the probes n° 2 and 3.

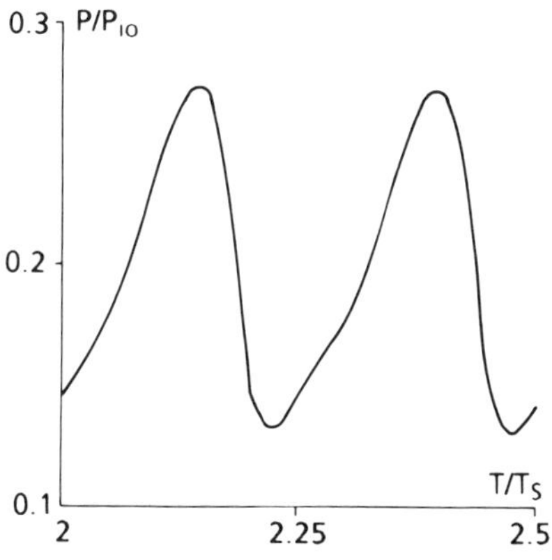

Fig. 13 - Static pressure evolution of the probe n° 4.

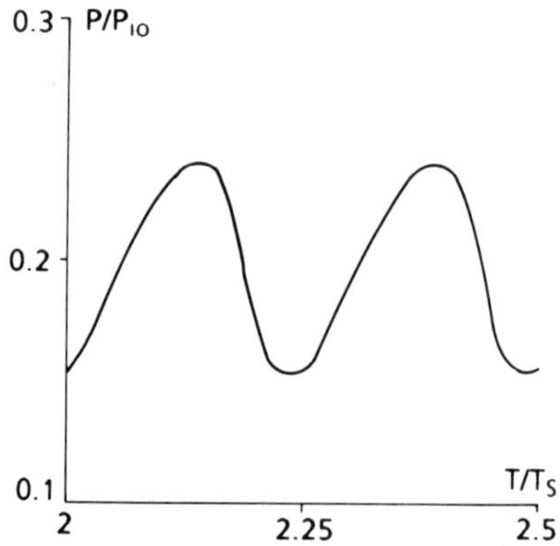

Fig. 14 - Static pressure evolution of the probe n° 5.

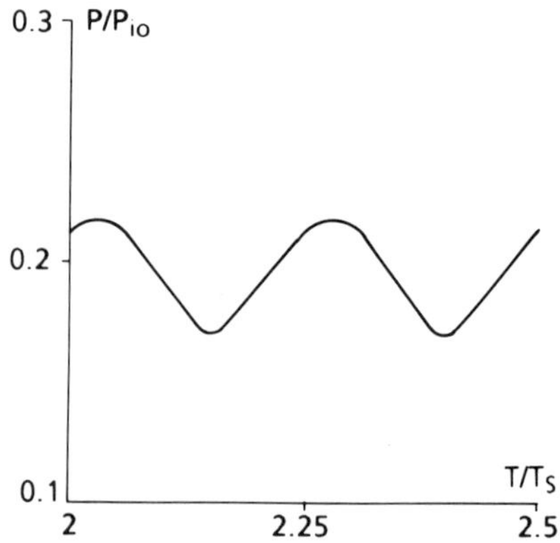

Fig. 15 - Static pressure evolution of the probe n° 6.

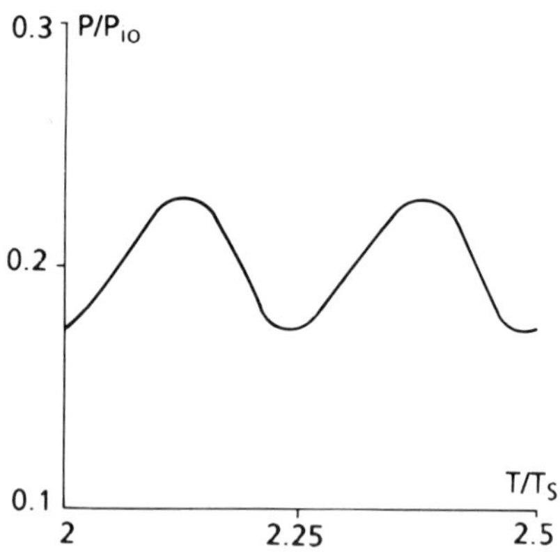

Fig. 16 - Static pressure evolution of the probe n° 7.

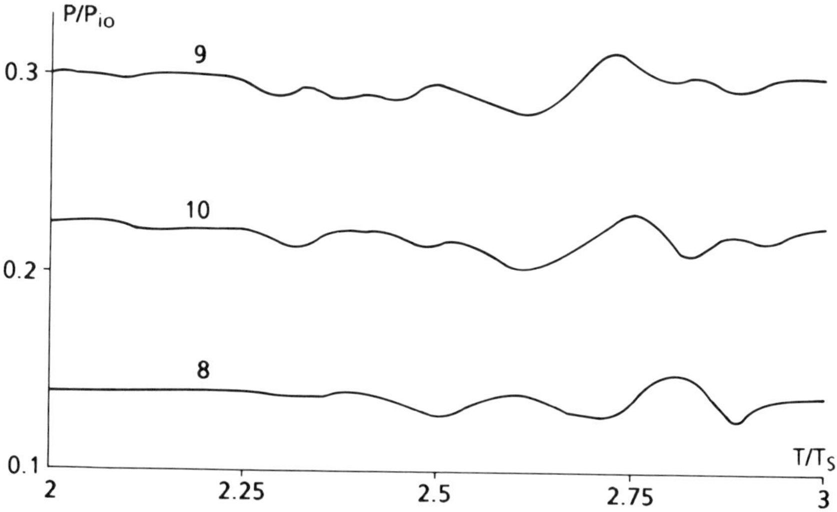

Fig. 17 - Static pressure evolution of the probes n° 8 to 10.

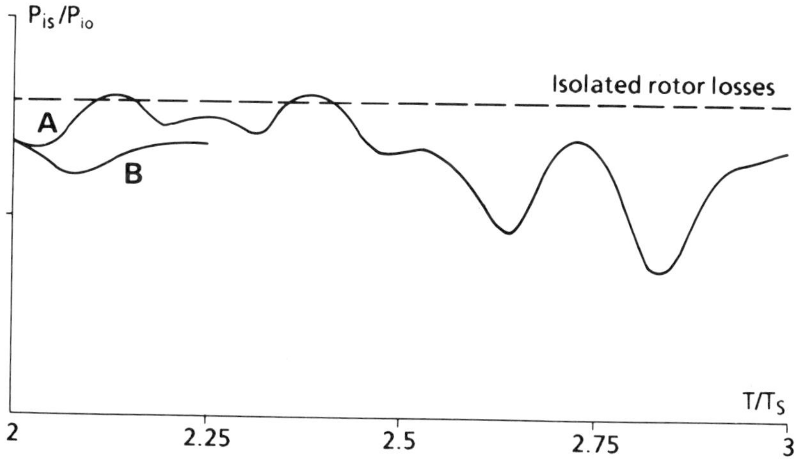

Fig. 18 - Time-evolution of the mean losses.

Oscillation of an Airfoil Behind a Detached Normal Shock

Shojiro KAJI [1]
Masanori OGAWA [2]
Hirofumi DOI [1]

ABSTRACT The effect of detached bow shock waves on airfoils in bending oscillation was investigated. For this purpose the response of a normal shock wave exposed to disturbances was studied. By the incident pressure disturbance the shock wave is motivated, and, at the same time, the pressure wave, vorticity wave, and entropy wave are generated. It is shown that when the pressure wave hits the normal shock wave at an angle, the reflected pressure wave increases in combination with flow Mach number. The oscillating cascaded airfoils accompanied with the detached bow shock waves are analysed based on the semi-actuator disk method. Results show that the existence of detached bow shock waves destabilize significantly the bending oscillation, and that the flutter boundary extends dramatically to high frequency regions.

1.1 Introduction

In the design of high speed fans and compressors, unstarted supersonic flutter is one of the most serious problems of blade vibration [1][2]. This type of flutter occurs in bending mode, near the stall line on the compressor characteristics map [3][4][5]; that is to say on high back pressure conditions of supersonic cascades with a detached bow shock wave on each blade. Although a couple of calculations have been performed based on the unsteady Euler equations [6][7], the essential feature of unstarted supersonic flutter is not fully clarified. The unsteady motion of detached bow shock waves and accompanying flow separation on blade surfaces may be responsible for this flutter [3].

This paper treats, as the first step to understand unstarted supersonic flutter, the motion of a detached bow shock wave induced by the oscilla-

[1] Department of Aeronautics, University of Tokyo, Hongo, Bunkyo-ku, Tokyo 113, Japan.
[2] Engine Experiment Department, Nissan Motor Co., Ltd., 6-1 Daikoku-cho, Tsurumi-ku, Yokohama, Japan.

tion of a single airfoil, based on the linearized theory. The unsteady aerodynamic characteristics of the airfoil are also investigated. Furthermore, cascaded airfoils in bending oscillation with detached bow shock waves are investigated based on the semi-actuator disk theory.

1.2 Formulation of the Problem

Figure 1 shows the situation of the problem. A single flat plate airfoil is placed on the x-axis with zero angle of attack in the uniform subsonic flow of velocity U_1. This subsonic flow is resulted from the uniform supersonic flow of velocity U_0 by the normal shock wave which is located a finite distance L ahead of the leading edge of the airfoil. The normal shock wave therefore sits parallel to the y-axis, i.e., normal to the airfoil chord. The distance L may depend on the inlet flow Mach number, the leading edge radius of the airfoil, and also the back pressure condition in the case of compressor cascades, however, it is just a variable parameter in this analysis.

When the airfoil makes a harmonic oscillation, the pressure wave disturbances are emitted by the airfoil which propagate upstream and hit upon the normal shock. Then, the normal shock wave is motivated relative to the flow and the pressure disturbances are reflected back to downstream. At the same time the vorticity and the entropy disturbances are produced due to deformation of the normal shock and due to the variation of shock strength, and these are convected downstream. The flow field around the oscillating airfoil is in turn influenced by these disturbances.

Before proceeding to the formulation, therefore, we had better make clear what is the response of the normal shock wave against the incident pressure disturbance.

1.2.1 Response of Normal Shock Wave against Disturbances

Figure 2 shows the normal shock wave which is motivated by incident disturbances of angular frequency ω. Its velocity amplitude is denoted by $q(y)$, therefore, its displacement around the equilibrium position can be expressed by $\frac{q(y)}{i\omega}e^{i\omega t}$. The relation of the mean flow between the upstream and the downstream of the shock wave can be derived by the equations for the conservation of mass flow, momentum, and energy as follows:

$$\rho_0 U_0 = \rho_1 U_1 \tag{1.1}$$

$$p_0 + \rho_0 U_0^2 = p_1 + \rho_1 U_1^2 \tag{1.2}$$

$$\frac{\kappa}{\kappa - 1}\frac{p_0}{\rho_0} + \frac{1}{2}U_0^2 = \frac{\kappa}{\kappa - 1}\frac{p_1}{\rho_1} + \frac{1}{2}U_1^2 \tag{1.3}$$

where the suffixes 0 and 1 denote the upstream and the downstream of the shock wave, respectively.

When disturbances are present, the velocity, the pressure, and the density can be expressed as $\left(U_0+\hat{u}_0 e^{i\omega t}, \hat{v}_0 e^{i\omega t}\right), p_0+\hat{p}_0 e^{i\omega t}, \rho_0+\hat{\rho}_0 e^{i\omega t}$ for the upstream of the normal shock wave, and $\left(U_1+\hat{u}_1 e^{i\omega t}, \hat{v}_1 e^{i\omega t}\right), p_1+\hat{p}_1 e^{i\omega t}, \rho_1+\hat{\rho}_1 e^{i\omega t}$ for the downstream. The quantities with $\hat{}$ denote small perturbations induced by disturbances. These perturbation quantities should satisfy the linearized equations of mass, momentum and energy.

$$\frac{D\hat{\rho}}{Dt}+\rho\left(\frac{\partial \hat{u}}{\partial x}+\frac{\partial \hat{v}}{\partial y}\right)=0 \tag{1.4}$$

$$\frac{D\hat{u}}{Dt}=-\frac{1}{\rho}\frac{\partial \hat{p}}{\partial x} \tag{1.5}$$

$$\frac{D\hat{v}}{Dt}=-\frac{1}{\rho}\frac{\partial \hat{p}}{\partial y} \tag{1.6}$$

$$\frac{D}{Dt}\left(\frac{\hat{p}}{p}-\kappa\frac{\hat{\rho}}{\rho}\right)=0 \tag{1.7}$$

where

$$\frac{D}{Dt}=\frac{\partial}{\partial t}+U\frac{\partial}{\partial x}$$

and κ denotes the ratio of the specific heats. From these four equations of mass flow, momentum, and energy, we can find out four waves of disturbance; two pressure waves, one propagating upstream and another propagating downstream, one vorticity wave and one entropy wave.

If we assume the disturbances have the phase velocity V_S in the direction along the shock, i.e., in the y-direction, then, two pressure waves can be expressed in the form

$$e^{i\omega\left(t-\frac{y}{V_S}\right)+i\lambda x}$$

and the x-wise wave number λ can be obtained as

$$\lambda=\frac{-U\pm a\sqrt{1+\frac{U^2-a^2}{V_S^2}}}{U^2-a^2}\omega \tag{1.8}$$

where a denotes the velocity of sound. Since we know that the plane pressure wave in the uniform flow can be expressed by

$$e^{i\omega\left(t-\frac{x\cos\beta+y\sin\beta}{a+U\cos\beta}\right)}$$

where $\frac{\pi}{2}<\beta_1<\pi, 0<\beta_2<\frac{\pi}{2}$, we can write

$$V_S=\frac{a+U\cos\beta}{\sin\beta} \tag{1.9}$$

From the equation of motion we get

$$\hat{u} = \frac{\cos\beta}{\rho a}\hat{p} \tag{1.10}$$

$$\hat{v} = \frac{\sin\beta}{\rho a}\hat{p} \tag{1.11}$$

The vorticity wave is convected downstream by the main flow velocity U without accompanying the pressure fluctuation nor density fluctuation, thus we get

$$\hat{u} = \hat{u}_V e^{i\omega\left(t - \frac{x}{U} - \frac{y}{V_S}\right)} \tag{1.12}$$

$$\hat{v} = \hat{v}_V e^{i\omega\left(t - \frac{x}{U} - \frac{y}{V_S}\right)} \tag{1.13}$$

From the equation of continuity we find

$$\hat{v}_V = -\frac{V_S}{U}\hat{u}_V \tag{1.14}$$

The entropy wave is also convected by the main flow at velocity U, without accompanying the pressure fluctuation nor velocity fluctuation, therefore, the entropy fluctuation takes the form of the density fluctuation.

$$\hat{\rho} = \hat{\rho}_e e^{i\omega\left(t - \frac{x}{U} - \frac{y}{V_S}\right)} \tag{1.15}$$

Figure 3 shows the control surface including the normal shock wave. In the figure, ϵ is of the same order of magnitude as the perturbation quantities but still larger than the displacement of the shock wave. The angle of inclination of the shock wave, α, can be expressed by

$$\alpha = \frac{\partial}{\partial y}\frac{q(y)}{i\omega}e^{i\omega t} \tag{1.16}$$

Applying the conservation law of mass with respect to the control volume, we obtain

$$\hat{\rho}_0 U_0 + \rho_0 \hat{u}_0 - (\hat{\rho}_1 U_1 + \rho \hat{u}_1) + (\rho_1 - \rho_0)\, q = 0 \tag{1.17}$$

The conservation law of momentum can be expressed in the x-direction by

$$\hat{p}_0 + 2\rho_0 U_0 \hat{u}_0 + \hat{\rho}_0 U_0^2 - \left(\hat{p}_1 + 2\rho_1 U_1 \hat{u}_1 + \hat{\rho}_1 U_1^2\right) = 0 \tag{1.18}$$

and in the y-direction by

$$\rho_0 U_0 \hat{v}_0 - \rho_1 U_1 \hat{v}_1 + (p_0 - p_0)\, \alpha = 0 \tag{1.19}$$

The conservation law of energy is expressed by

$$\rho_0 U_0 \{\frac{\kappa}{\kappa - 1}\frac{p_0}{\rho_0}\left(\frac{\hat{p}_0}{p_0} - \frac{\hat{\rho}_0}{\rho_0}\right) + U_0 \hat{u}_0\}$$
$$-\rho_1 U_1 \{\frac{\kappa}{\kappa - 1}\frac{p_1}{\rho_1}\left(\frac{\hat{p}_1}{p_1} - \frac{\hat{\rho}_1}{\rho_1}\right) + U_1 \hat{u}_1\} - (p_1 - p_0)\, q = 0 \tag{1.20}$$

From these four equations we can determine four unknown quantities relative to the incident disturbance. For example, if the pressure wave is incidental from downstream of the shock wave, the reflected pressure wave, the vorticity wave, the entropy wave, and the velocity of shock movement are determined. If the pressure wave is present as the incident disturbance in the upstream field of the shock wave, then, the pressure wave propagating downstream, the vorticity wave, the entropy wave, and the velocity of shock movement are to be determined. In the same way, the convective disturbances such as vorticity wave and entropy wave existing in the upstream field of the shock wave can be the incident disturbance, respectively.

Figures 4(a) to (d) show the results for the case when the pressure wave is incidental from downstream. When the plane pressure wave strikes normal to the shock wave, very limited amount of pressure wave is reflected back to downstream. However, when it strikes the normal shock wave at an angle, the reflected pressure wave increases considerably (see Fig. 4(a)). Figure 4(b) and (c) show that the entropy wave and the vorticity wave increase monotonically as the Mach number ahead of the normal shock wave, M_0, increases, although much dependence on the angle of incidence is found in the vorticity wave amplitude. Figure 4(d) shows the velocity of shock movement. This quantity increases monotonically with M_0, too, and shows significant dependence on the angle of incidence of the pressure wave.

Figures 5(a) to (d) show the results when the pressure disturbance is incidental from the upstream of the shock wave. We see that the amplitudes of the pressure wave, the entropy wave, and the velocity of shock movement increases monotonically against the upstream flow Mach number, M_0. We must recognize that the pressure and the velocity of shock movement are both one order of magnitude greater than the incident pressure wave when the Mach number is high. This signifies that the small pressure fluctuation in the high Mach number flow contains high energy, and thus, it is transformed into high pressure amplitude in the downstream of the shock wave, and at the same time, it induces large movement of the shock wave. We notice the entropy wave is relatively not so significant and the vorticity wave is very weak.

Figures 6(a) to (d) show the results for the case of the incident vorticity wave. We see that considerable pressure waves and shock movements are induced downstream of the shock wave, and that their amplitudes increase monotonically with M_0.

1.2.2 Analysis of a Single Airfoil Oscillating behind the Normal Shock Wave

As is usual in unsteady aerodynamics, oscillation of the airfoil can be represented by the distribution of pressure dipoles whose axis is perpendicular to the airfoil chord. On the other hand, oscillation of the normal shock wave can be represented by the superposition of two distributions of pressure dipoles as the mean shock position [8]: one is the distribution of dipoles whose axis is perpendicular to the normal shock wave, and another is that of dipoles with the axis parallel to the shock wave. The former represents the variation of shock strength along the shock wave, and the latter indicates the effect of shock inclination which is responsible for the shed vorticity due to shock motion. These distributions of pressure dipoles induce velocity fluctuations at the airfoil position. We should note here that the solution to the flow field obtained in this way is valid only in the field downstream from the normal shock wave.

From the above discussion, now we have four unknowns; the distribution of pressure dipoles on the airfoil, two kinds of distributions of pressure dipoles in place of the normal shock wave, and the distribution of the velocity of shock motion. For these four unknowns boundary conditions are as follows:

(i) The upwash velocity induced by all of pressure dipoles should satisfy the condition of airfoil oscillation.

(ii) The conservation condition of mass flow at the normal shock wave, Eq.(1.17).

(iii) The conservation condition of momentum in the x-direction at the normal shock wave, Eq.(1.18).

(iv) The conservation condition of momentum in the y-direction at the normal shock wave, Eq.(1.19).

(v) The conservation condition of energy at the normal shock wave, Eq.(1.20).

In the description of conservation conditions across the normal shock wave, the upstream perturbations $\hat{p}_0, \hat{\rho}_0, \hat{u}_0, \hat{v}_0$ are not present in this problem. Although we have four conditions at the normal shock wave, they are reduced to three conditions with respect to the perturbation pressure and two components of velocity by eliminating the perturbation density $\hat{\rho}_1$. In this way we obtain four boundary conditions in total, and can determine four kinds of unknowns.

Figure 7 shows the shock wave motion every 45 degrees of the phase angle of oscillation. The movement is anti-phase for the upper part and the lower part of the oscillating airfoil, and has twin peaks nearest to the leading edge of the airfoil. The distribution of dipoles with axes normal to the shock front is shown in Fig. 8. The dipole strength decreases gradually as the distance from the airfoil plane increases, and also the phase delay increases almost linearly along the distance from the x-axis.

The shock wave motion changes considerably depending on the stand-off distance of the normal shock wave, L, the frequency of oscillation, k, and the upstream Mach number, M_0. Figure 9(a) is the one similar to Fig. 7, but the stand-off distance, L, is ten times larger than that for Fig. 7. We see that when L increases the peaky character of shock motion found in Fig. 7 will be lost gradually. Figure 9(b) shows the case for a higher frequency of four times larger than that for Fig. 7. We recognize the almost wavy variation of shock motion along its distance.

As is shown in Fig. 10, when the upstream Mach number, M_0, is reduced, the shock movement increases significantly. This fact suggests the effect of detached bow shock wave becomes more and more important as the transonic flow conditions are realized.

1.2.3 Experimental Observation

In order to investigate the shock motion and unsteady airfoil characteristics, an experiment for a single airfoil in bending oscillation has been performed. As the first step an optical measurement was made. Bending oscillation was given by use of a high speed motor cycle engine(YAMAHA 10X01) in which the crank arm radius was changed from 45mm to 4mm. The maximum frequency of oscillation was about 150 Hz. The test section is 180mm high and 50mm wide. One side wall is composed of an optical mirror with a slot through which the airfoil was inserted and oscillated. The schlieren photograph was taken through another side glass window by high speed camera at the speed about 500 frames per second.

Figure 11 shows a series of photographs at different phases of oscillation. The experimental conditions are the inlet Mach number, 1.15, the frequency of oscillation, 40 Hz, the number of frames of the camera, 520/s. The bow shock wave stands about 15% of chord ahead the leading edge of the airfoil, and oscillates about 5% of chord in amplitude. Figure 12 shows the variation of stand-off distance of the bow wave during one cycle of oscillation. Two of different cycles were analysed. We see that near 180 degrees of the phase of oscillation, which corresponds to the lowest quasi-steady loading position of the airfoil, the bow wave approaches nearest to the airfoil. However, the phase of oscillation where the bow wave goes farthest from the airfoil is not clear.

Figure 13 shows the case of a higher Mach number of 1.4. The bow shock wave looks to be attached to the leading edge of the airfoil, and to perform quasi-steady oscillation. But the detailed observation by use of the frame analyzer indicates, as shown in Fig. 14, that the bow wave has a small deviation from its quasi-steady position accompanied with a phase variation along its distance. The maximum deviation is about 1% of chord, which corresponds to 12.5% of oscillation amplitude.

1.3 Flutter Analysis for an Unstarted Supersonic Cascade Based on the Semi-Actuator Disk Model

1.3.1 Basic Assumption in the Formulation

In order to extend the concept of a single airfoil oscillating behind the normal shock wave to cascaded airfoils, a simplified assumption was adopted in combination with the semi-acutuator disk theory.

As shown in Fig. 15, in actual cascaded airfoils, when the flow condition is unstarted supersonic, one branch of the bow shock wave of one airfoil may reach the suction surface of the adjacent airfoil. This means the subsonic flow behind the detached bow shock wave is reaccelerated to be supersonic on the suction side of the airfoil and again decelerated by this suction surface shock wave. This process is highly dependent on the airfoil profile. In the present analysis, we assume that the bow shock wave does not reach the suction surface of the adjacent airfoil, so that the airfoil behind the bow wave is situated in a completely subsonic flow field. This situation is schematically shown in Fig. 16 for the semi-actuator disk model. Here, we have a series of segmented bow shock waves a finite distance ahead of each airfoil.

In the present analysis, the treatment of the flow fields behind the leading edge plane is the same as that in the usual actuator disk theory [9]. On the other hand, the upstream field is quite different. The field upstream from the segmented shock wave plane has no disturbances because the flow is supersonic. On the other hand, the field between the segmented shock wave plane and the leading edge plane has various disturbances: The pressure wave propagating upstream which is induced by airfoil oscillation, the pressure wave reflected back to downstream by the segmented shock waves, the vorticity wave and the entropy wave produced by shock motion and convected downstream. Although the quasi-steady shock segment is normal to the inlet flow, the unsteady motion includes the slight inclination of the shock segment, and this inclination differs segment by segment because of the interblade phase angle. Therefore, we should permit the existence of the vorticity wave in this region.

1.3.2 Results and Discussion

Numerical calculations were performed for the cascade of stagger angle $\theta = 60°$. The steady flow conditions are determined by the upstream Mach number, M_0, and the inlet flow angle, β. When the upstream Mach number is $M_0 = 1.2$, the cascade inlet Mach number is reduced to M_1=0.842 by the normal shock wave. If the inlet flow angle $\beta = 70°$, then the flow is turned at the leading edge plane by 10 degrees, and the cascade outlet Mach number becomes $M_2 = 0.446$. The distance between the leading edge plane and the plane of segmented shock waves, L, is selected to be $L = 0.1$, specified otherwise. The angle of oscillation, ψ, is also a variable parameter. But in almost calculations it is fixed to $\psi = 45°$, specified otherwise.

Figure 17 shows the typical variation of the aerodynamic force which is in phase with the oscillation velocity against the reduced frequency, $k = \omega c/W_1$, where W_1 is the inlet velocity at the leading edge plane of the cascade. If this component of the aerodynamic force becomes positive, the bending oscillation will be excited and flutter may result. From the figure we see that when $k < 0.1$ flutter is inevitable. When k is increased further the oscillation will be damped, but we have the region of excitation again in $0.4 < k < 1.0$. The range of excitation is fairly wide and locates in a high frequency region. Therefore, flutter may be encountered in practical designs. This positive humps of the curve will subside below the coordinate axis as the absolute value of the interblade phase angle parameter, σ, increases ($\sigma < -0.8$), σ is defined as $\sigma = \omega c/V_S$, and V_S is the phase velocity of flutter modes along the cascade. Flutter occurs for negative values of σ, which indicates the unstable mode should be the forward traveling wave.

To make clear the effect of the normal shock wave oscillation, the unsteady aerodynamic force was calculated for the subsonic inlet flow which is just correspondent to that behind the normal shock wave. Figure 18 shows the typical results. We confirm the flutter boundary is given by a very small k, around 0.02. Furthermore, we have no positive hump of the aerodynamic force in subsonic inlet flow cases.

Figures 19(a) to (d) show the effect of inlet flow Mach number on the unsteady aerodynamic force. When we increase the inlet flow Mach number, M_0, to 1.3 or 1.4, the second hump of the curve discussed in Fig. 17 would not enter the positive side for almost all values of the inter-blade phase angle parameter, σ. The flutter boundary in terms of k for these supersonic cases is around $k = 0.12$, which is the value extremely higher than that for the correspondent subsonic cases.

On the other hand, when we reduce the inlet flow Mach number, M_0, from $M_0 = 1.2$ to 1.1, the second hump of the curve for the aerodynamic force versus k enters the positive side, and the positive region spreads significantly. Therefore, in such cases flutter becomes inevitable.

It is important to note the effect of the stand-off distance of the normal shock wave , L, on the unsteady aerodynamic force. Figures 20(a) to (d) show the effect of stand-off distance on the unsteady aerodynamic force. Generally speaking, when L increases the oscillation tends to be stabilized. For example, in the case of $L = 0.3$, the second hump of the aerodynamic force curve becomes positive in the small range of k for the limited range of σ near -0.1. On the other hand, when L decreases to 0.05, the range of excited oscillation spreads significantly in terms of k, as well as σ.

The angle of incidence has less significant effects on airfoil oscillation. As shown in Figs.21(a) to (d), at zero angle of incidence (Fig.21(a)), the flutter boundary in terms of k is aroud 1.6, near $\sigma = -0.2$. As the angle of incidence increases, this flutter boundary increases up to $k = 2.2$, near $\sigma = -0.2$ (see Fig.21(b)). Further increase in angle of incidence would not push the flutter boundary to higher k, but rather shows the tendency of saturation (Fig.21(c)). Finally, as shown in Fig.21(d), the flutter boundary goes down to about 0.7. At present the reason why the oscillation should be stabilized at excessively high angles of incidence, but the fact that the Mach number downstream from the cascade is significantly reduced as the angle of incidence increases may explain the tendency of the results.

One more point we should note is that a considerably large region of k for destabilized oscillation exists even at zero angle of incidence (Fig.21(a)). Since we do not observe such destabilizing effect in case of purely subsonic or supersonic zero incidence inlet flow, this destabilization should be attributed to the existence of the detached normal shock wave ahead of each airfoil.

1.4 Conclusion

In order to elucidate the mechanism of unstarted supersonic flutter, the effect of detached bow shock waves on airfoils in bending oscillation was investigated. First, the response of a normal shock wave exposed to disturbances was studied. Then, the problem of a single airfoil oscillating behind the normal shock wave was treated. Finally, cascaded airfoils in bending oscillation accompanied by detached bow shock waves were analysed based on the semi-acutuator disk method. Following conclusions were obtained.

(1) When pressure disturbances hit upon the normal shock wave at an angle, considerable amount of pressure waves are reflected back to downstream.

(2) When disturbances hit upon the normal shock wave from upstream, disturbances of significant amplitude are induced downstream from the normal shock wave.

(3) The normal shock wave shows wavy motion along its distance induced by airfoil oscillation.

(4) The unsteady aerodynamic force is much influenced by the existence of the detached bow shock wave. The degree of influence changes depending on the stand-off distance of the bow shock wave.

(5) Existence of detached bow shock waves destabilize significantly the bending oscillation of cascaded airfoils. The flutter boundary extends considerably to high frequencies. The flutter mode in this case is a forward traveling wave.

(6) The unstable frequency range shrinks as the Mach number ahead of the shock increases, and it widens as the Mach number decreases. This flutter, therefore, is characteristic to transonic flows.

(7) The stand-off distance of the bow shock wave has significant influence on the occurrence of flutter. On the other hand, the angle of incidence has less significant effect.

REFERENCES

[1] H.Joubert,"Supersonic Flutter in Axial Flow Compressors ," *Unsteady Aerodynamics of Turbomachines and Propellers* ,Cambridge,Sept.1984.

[2] E.Széchényi et al,"A Straight Cascade Wind-Tunnel Study of Fan Blade Flutter in Started Supersonic Flow ," *Unsteady Aeronautics of Turbomachines and Propellers* ,Cambridge,Sept.1984.

[3] F.Sisto,"Introduction and Overview," *AGARD Manual on Aeroelasticity in Axial-Flow Turbomachines* ,vol.1,1987.

[4] F.O.Carta,"Aeroelasticity and Unsteady Aerodynamics," Chapter 7, *Aircraft Propulsion Systems Technology and Design. AIAA* ,1989.

[5] J.M.Verdon,"Unsteady Aerodynamics for Turbomachinery Aeroelastic Applications," Chapter 7, *Unsteady Transonic Aerodynamics, Progress in Astronautics and Aeronautics* ,vol.120,1989.

[6] H.Joubert et al,"Aircraft Compressor Flutter Analysis," *Unsteady Aerodynamics and Aeroelasticity of Turbomachines and Propellers* ,Aachen,Sept.1987.

[7] G.A.Gerolymos,"Numerical Integration of the Blade-to-Blade Surface Euler Equations in Vibrating Cascades," *AIAA J.*,vol.26, no.12, Dec.1988.

[8] M.Ogawa,"Study on Unstarted Flutter in Transonic Compressor Cascades," *Master Thesis, Department of Aeronautics, Univ. of Tokyo,* 1991.

[9] S.Kaji,"Stall Flutter of Linear Cascade in Compressible Flow,"C 282/20 *IMechE Vibrations in Rotating Machinery*,Cambridge,Sept.1980.

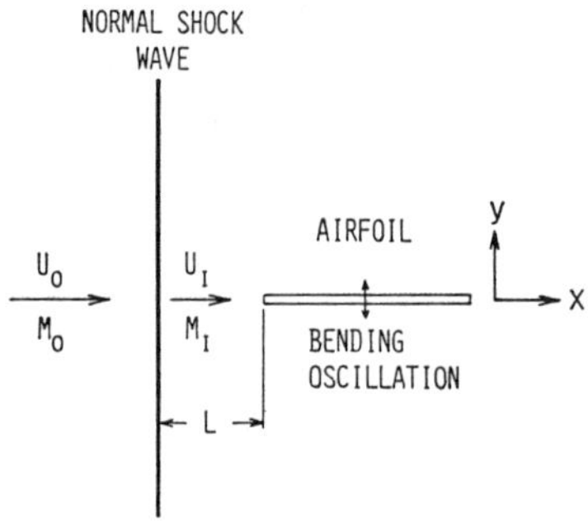

Fig.1 Oscillating airfoil behind normal shock wave.

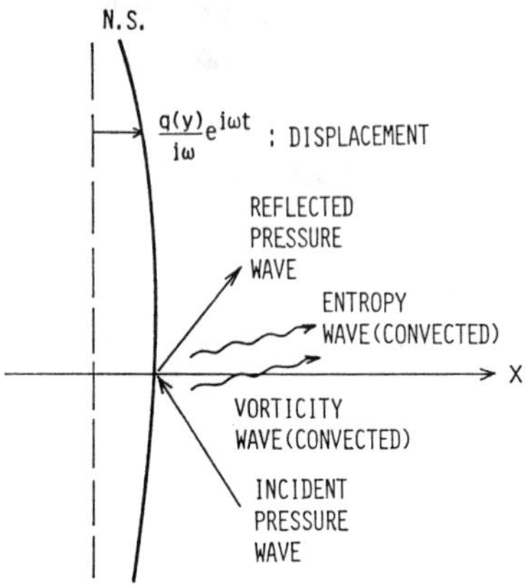

Fig.2 Shock motion due to disturbances.

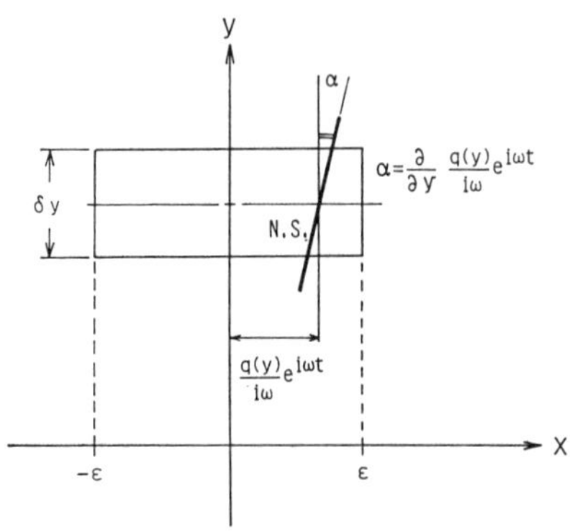

Fig.3 Control surface across normal shock wave.

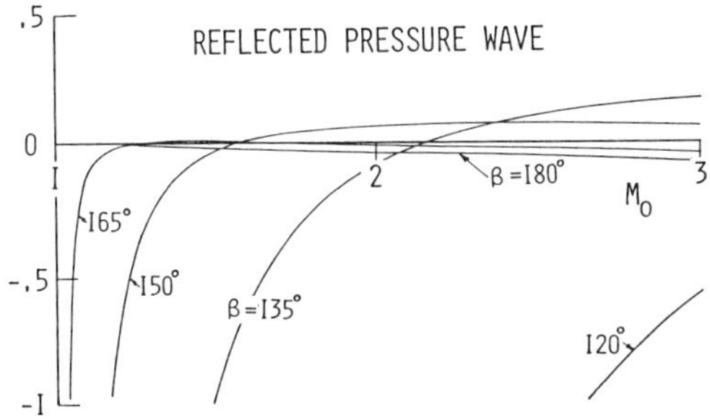

Fig.4

Case for incident pressure wave $\hat{p}_i$ from downstream.

(a) Pressure wave

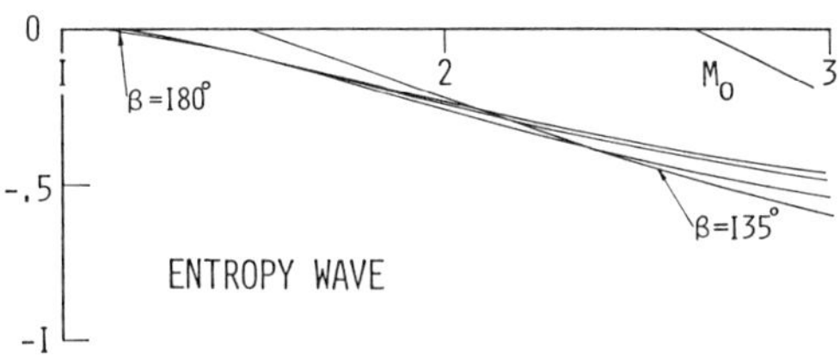

(b) Entropy wave

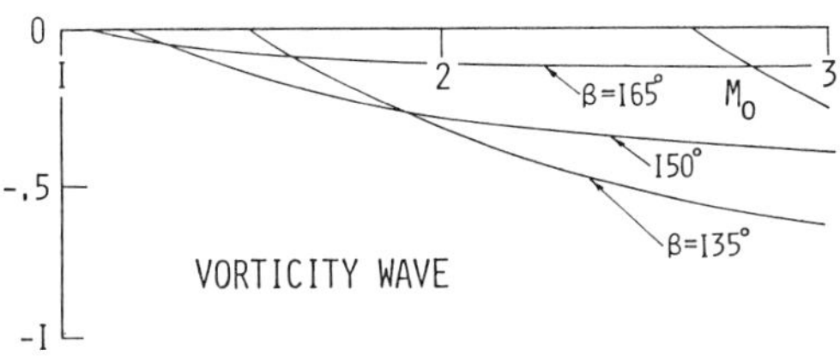

(c) Vorticity wave

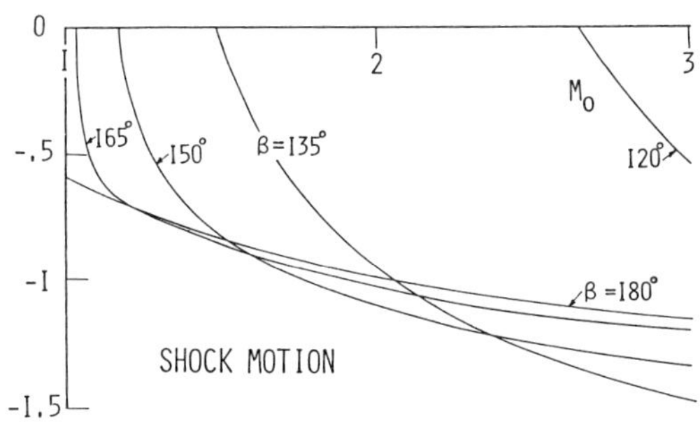

(d) Shock motion

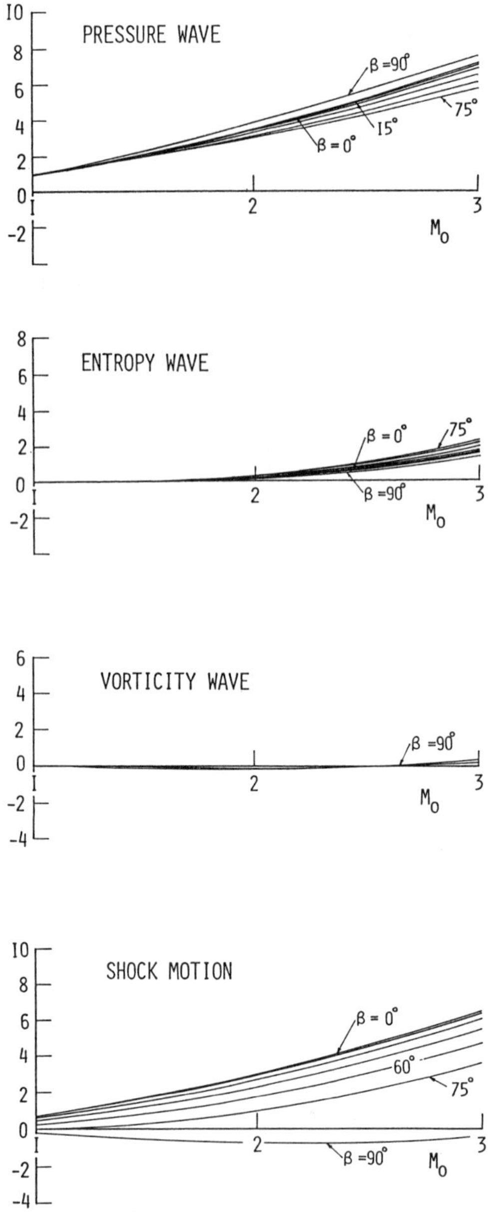

Fig.5

Case for incident pressure wave $\hat{p}_i$ from upstream.

(a) Pressure wave

(b) Entropy wave

(c) Vorticity wave

(d) Shock motion

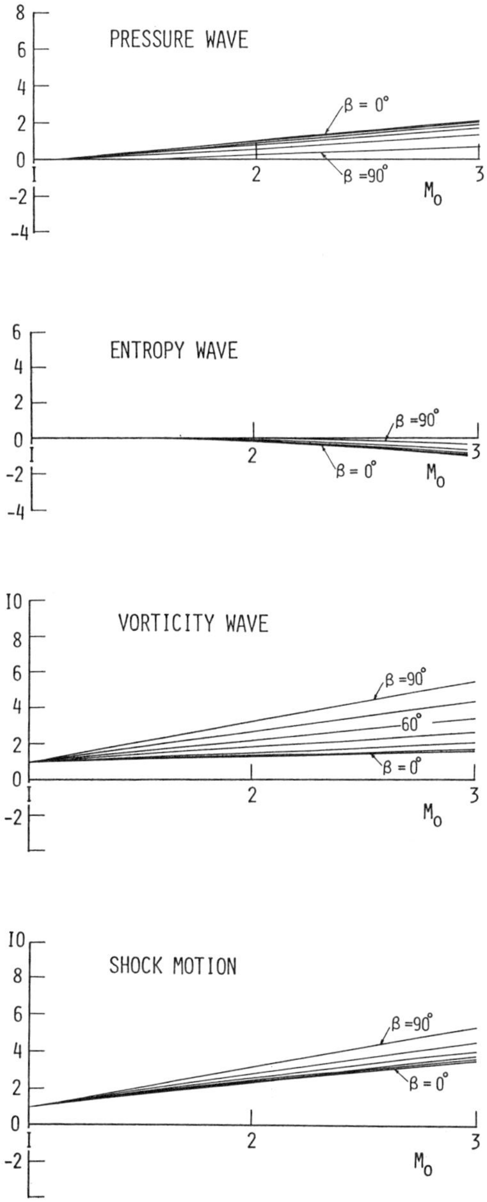

Fig.6

Case for incident vorticity wave $\rho_0 a_0 \hat{u}_v$ from upstream.

(a) Pressure wave

(b) Entropy wave

(c) Vorticity wave

(d) Shock motion

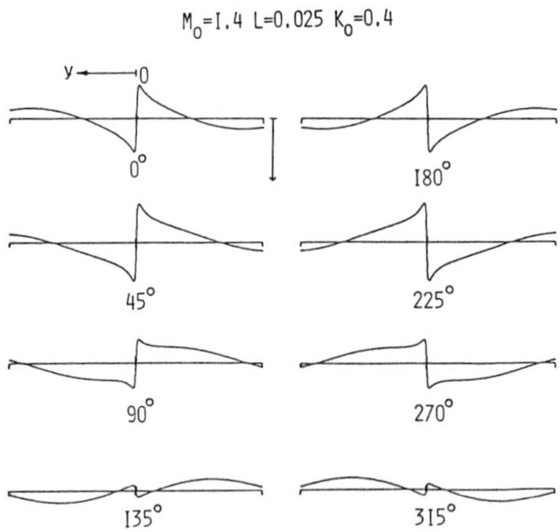

Fig.7 Shock motion at every 45° of phase of oscillation. Vertical arrow: Amplitude of airfoil oscillation.

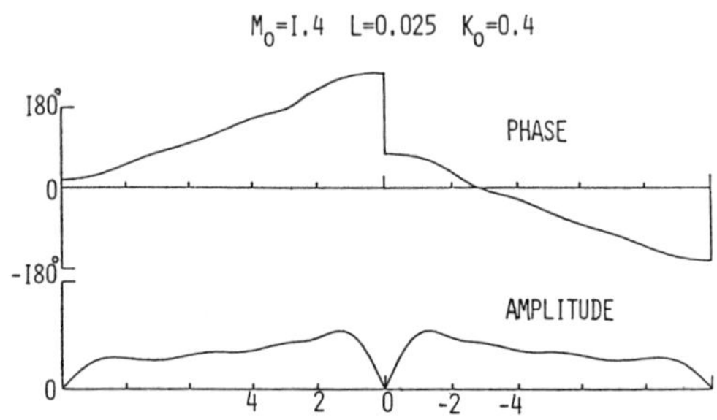

Fig.8 Distribution of axial dipoles along normal shock wave.

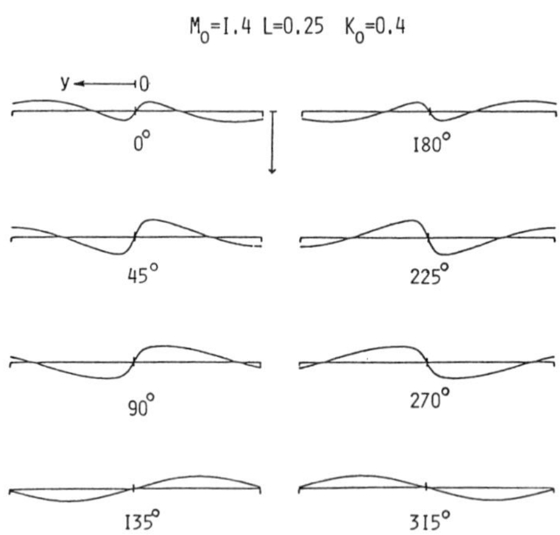

Fig.9(a) Shock motion for large stand-off distance case, L=0.25.

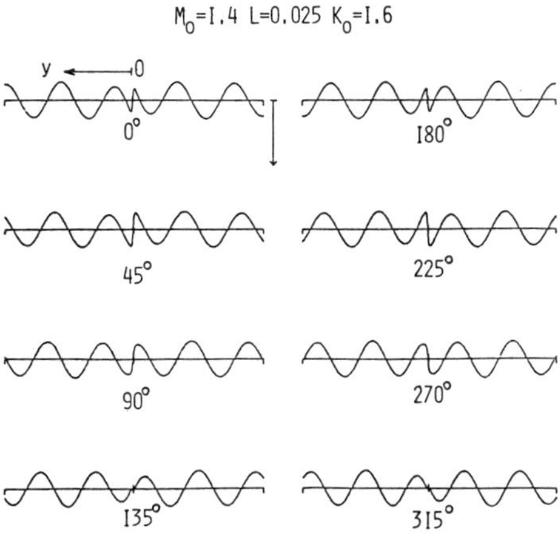

Fig.9(b) Shock motion for high frequency case, K_0=1.6.

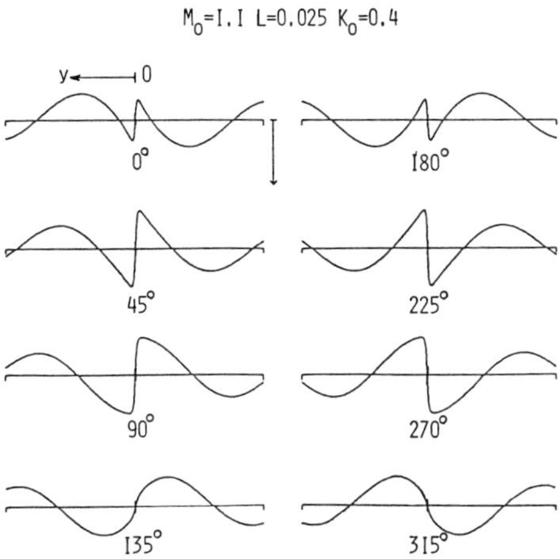

Fig.10 Shock motion for transonic Mach number case, M_0=1.1.

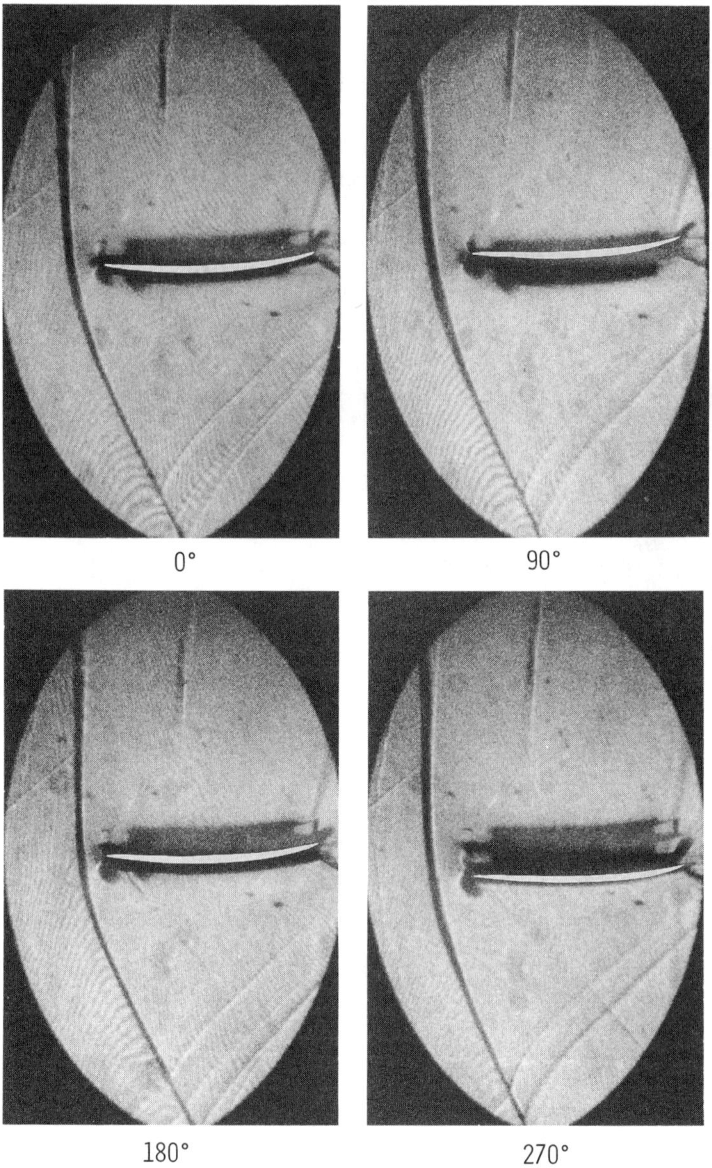

Fig.11 Schlieren photograph at every 90° of phase of oscillation, M_0=1.15, f=40Hz.

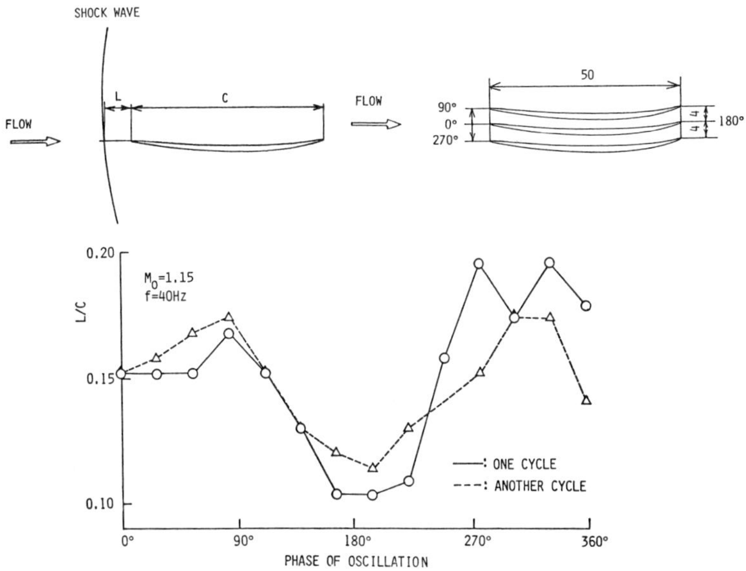

Fig.12 Variation of stand-off distance of shock wave during one cycle of oscillation.

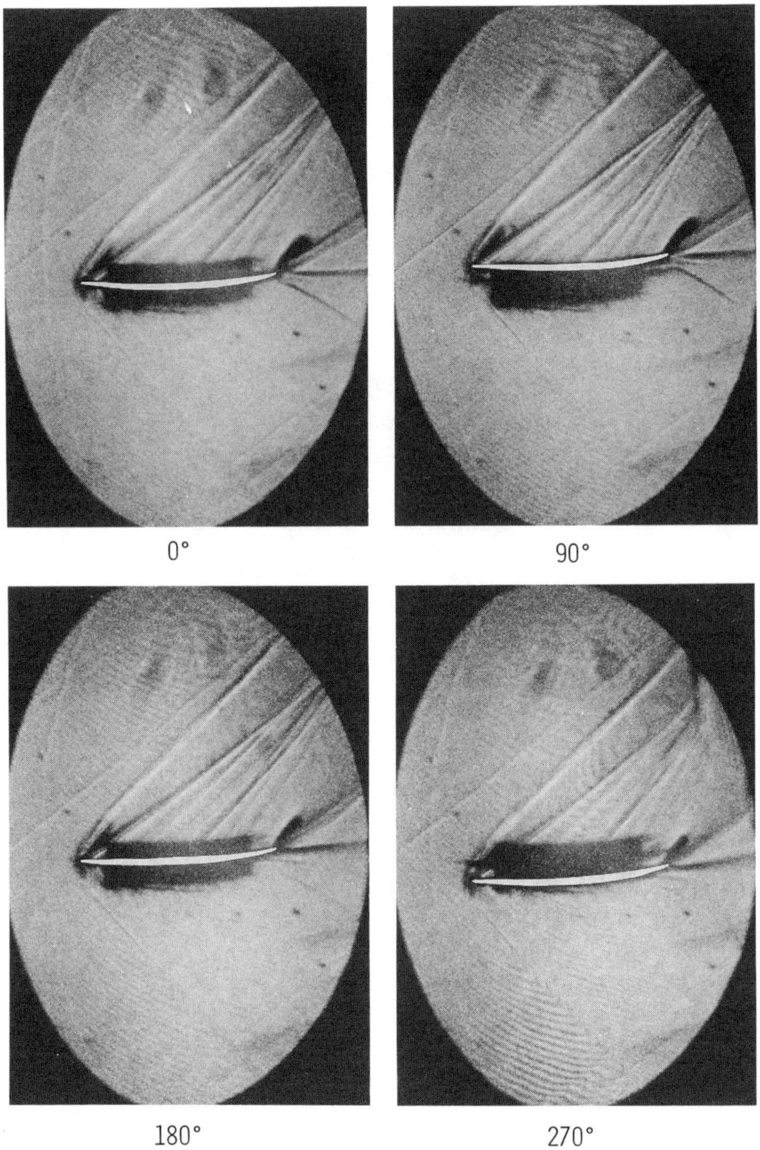

Fig.13 Schlieren photograph at every 90° of phase of oscillation, M_0=1.4, f=40Hz.

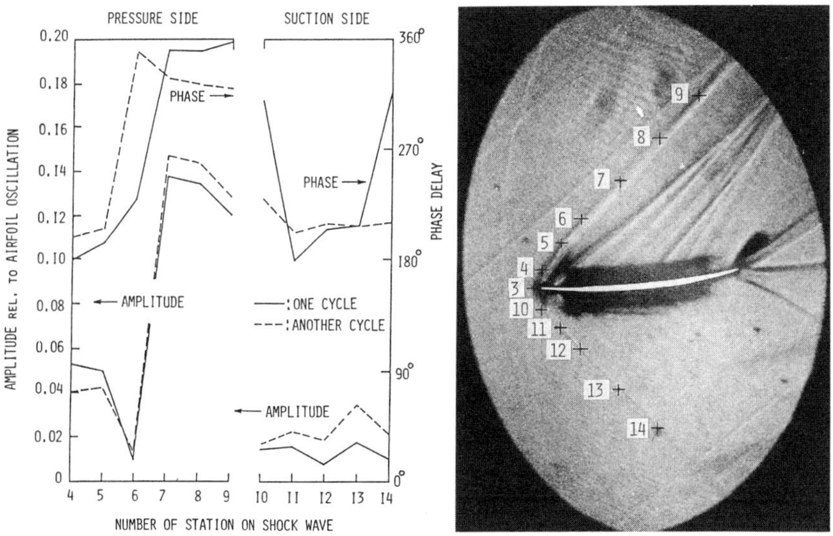

Fig.14 Deviation of bow shock wave from quasi-steady position.

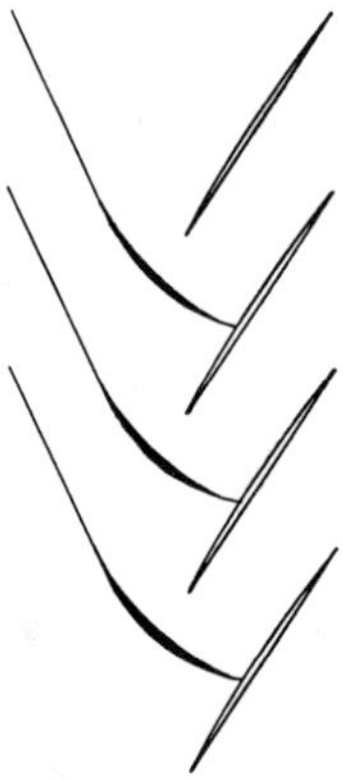

Fig.15
Unstarted supersonic cascade.

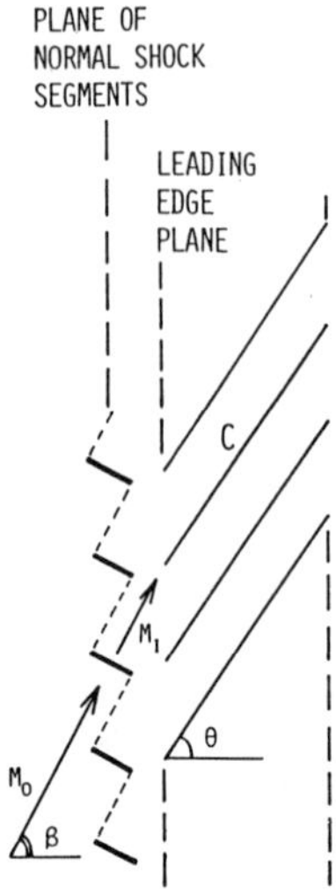

Fig.16
Semi-actuator disk model combined with segmented, detached bow shock waves.

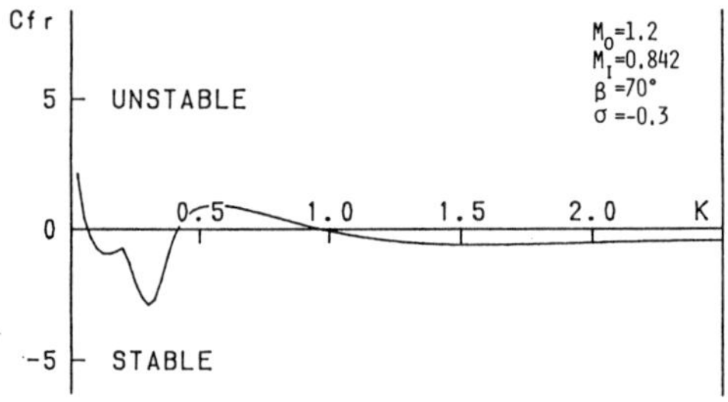

Fig.17 Aerodynamic force versus reduced frequency.
C_{fr}: Component in phase with oscillation velocity.
Stagger angle, θ=60°
Angle of oscillation, ψ=45°
Stand-off distance, L=0.1

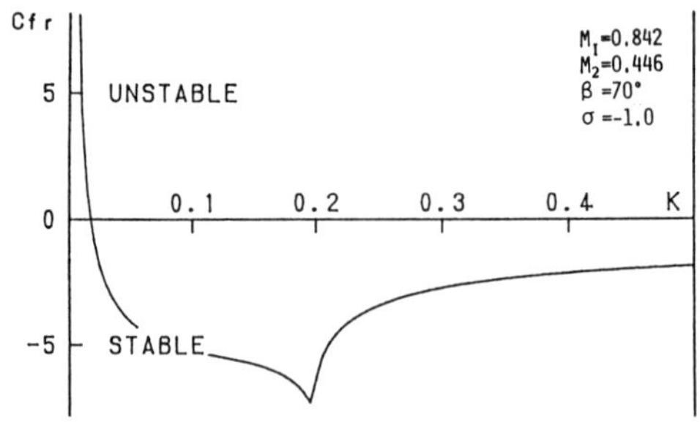

Fig.18 Aerodynamic force versus reduced frequency; subsonic case.

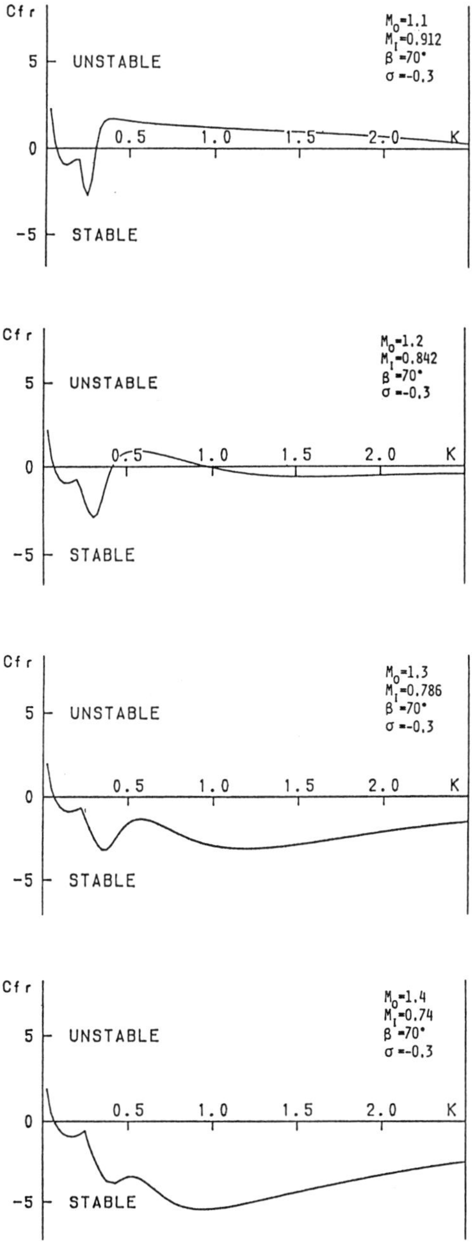

Fig.19 Effect of inlet flow Mach number on aerodynamic force.

(a) M_O=1.1

(b) M_O=1.2

(c) M_O=1.3

(d) M_O=1.4

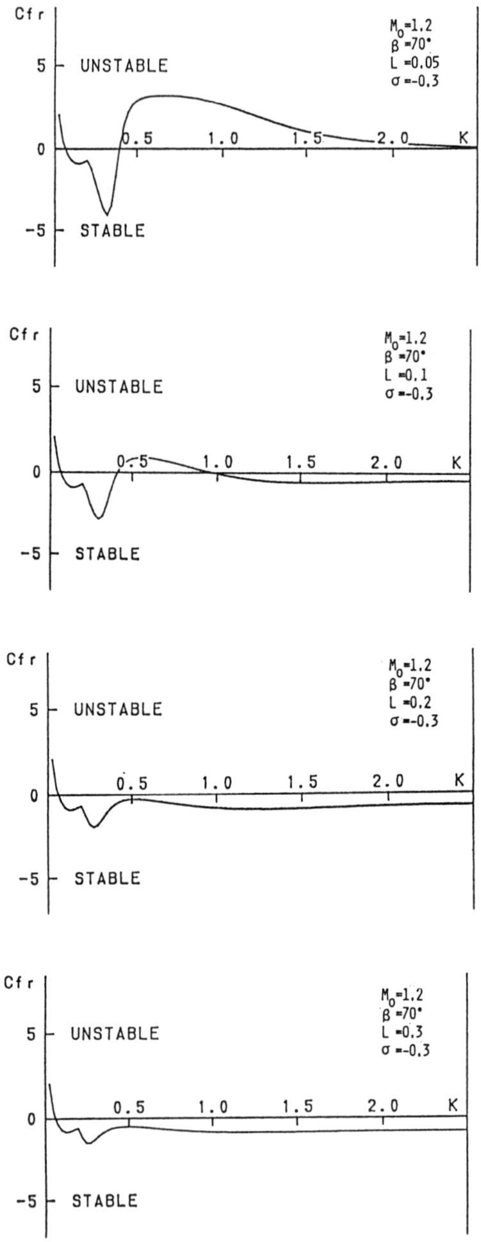

Fig.20 Effect of stand-off distance on aerodynamic force.

(a) L=0.05

(b) L=0.1

(c) L=0.2

(d) L=0.3

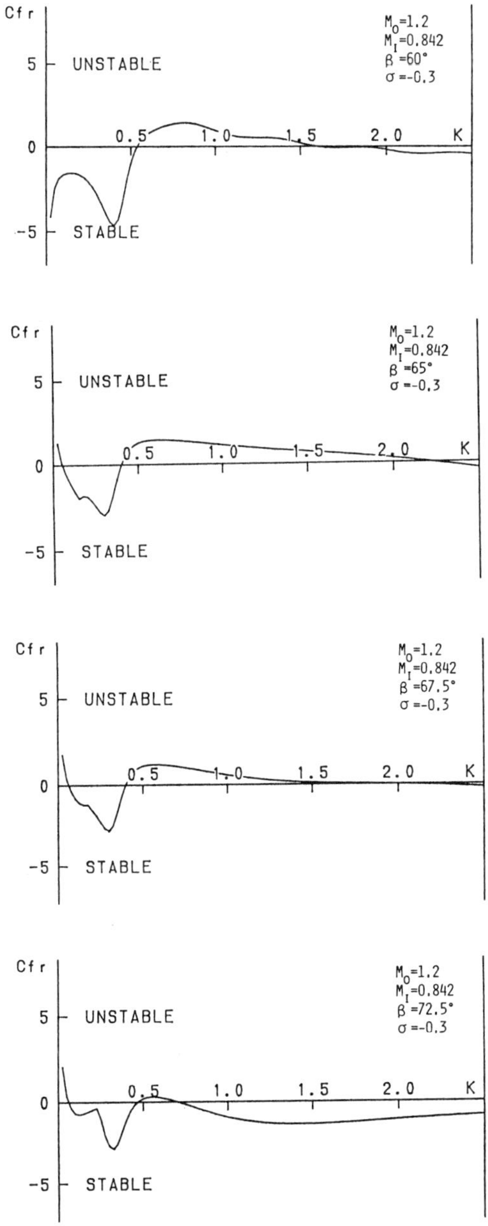

Fig.21 Effect of inlet flow angle on aerodynamic force.

(a) β=60°

(b) β=65°

(c) β=67.5°

(d) β=72.5°

IV. Unsteady Viscous Effects

Experimental Determination of Nonlinear Transfer Functions Characterizing Shock Wave/Turbulent Boundary Layer Interactions

F.O. Thomas[1]
H.C. Chu[1]
C.M. Putnam[1]

I. Introduction

Shock wave/turbulent boundary layer interactions have been the topic of numerous investigations and yet many aspects of the physical mechanisms associated with the interaction remain poorly understood. The extension of low speed turbulence closure models to the interaction region between a shock wave and turbulent boundary layer can lead to predictions of wall pressure distribution, skin friction, mean velocity profile, and turbulence quantities which are far from satisfactory[1]. The failure of equilibrium based turbulence models is generally attributed to the extreme rapid distortion of the flow field. Nonequilibrium turbulence closure models based upon rapid distortion theory have met with some success. Anyiwo and Bushnell[2] noted, however, that the physics of the turbulent stress amplification which has been experimentally demonstrated to occur across the shock needs to be incorporated into the current generation of rapid distortion theories. A simple linearized model by Anyiwo and Bushnell[2] theoretically demonstrates three primary modes of turbulence amplification across the shock: (1) a "direct" inviscid vorticity amplification, (2) conversion of incident acoustic and entropy modes into vortical modes and (3) turbulence production due to shock wave oscillation.

In the work to be reported here, the unsteady wall pressure fluctuations associated with the interaction of a fully turbulent boundary layer with a compression ramp-generated shock is experimentally investigated. The primary objective of the research is to apply a novel digital signal processing technique to digitized time-series wall pressure fluctuation data in order to investigate the local nonlinear spectral energy transfer associated with the modification of the turbulent field across the shock interaction region.

---

[1] Aerospace and Mechanical Engineering, Hessert Center Aero Lab, University of Notre Dame, Notre Dame, IN 46556.

II. Experimental Apparatus and Measurement Techniques

All experiments were conducted in the continuous supersonic in-draft wind tunnel facility located at the Hessert Center for Aerospace Research at the University of Notre Dame. This facility is described in some detail in reference[3] and therefore only essential details will be reported here. The tunnel is powered by three 125 HP Allis-Chalmers rotary vacuum pumps, which draw ambient room air through a contraction inlet which provides a smooth and parallel entrance flow to a converging-diverging nozzle. The C-D nozzle discharges into a rectangular duct which provides for boundary layer growth and additional flow acceleration to Mach 1.5 before reaching the test section. The test section is located 117 cm downstream of the nozzle exit and is 12.7 cm in width and 10.9 cm in height. At this streamwise position provision was made to flush mount 10.3 cm x 12.7 cm aluminum compression ramp models directly into the tunnel floor. In order to avoid significant condensation effects, experiments were run only when the absolute humidity ratio was less than 0.16% and for which the tunnel was observed to be clear of condensation fog.

Unswept compression ramp models were used in the experiments in order to acquire both static and fluctuating wall pressure data. The ramp angle was limited to 6° in each case and all of the models spanned the tunnel floor with no aerodynamic fences used. The effect of the side wall boundary layers was thought to be negligible based on the fact that the dynamic pressure measurements were confined to the centerplane of the 12.7 cm tunnel width. Spanwise static pressure measurements and surface flow visualization using titanium dioxide verified the substantial two-dimensionality of the shock interaction over the middle 10 cm of the test section. For obtaining wall pressure fluctuation measurements two aluminum 6° wedge models had a 2.54 cm hole bored either upstream of the wedge apex or along the ramp face in order to accommodate a flush mounted aluminum plug which contained five dynamic pressure transducers. Kulite dynamic differential pressure transducers (model XCQ-062-50D) were used. These transducers were each d_t=1.6 mm in outer diameter and had an active sensing element d_s= 0.71 mm in diameter. In terms of inner wall variables these dimensions correspond to a d_t^+=485 and d_s^+=215 , respectively. The transducers have protective screens above the diaphragm that limit the frequency response to approximately 50 kHz. The operating differential pressure range of transducers was 0-345 kPa. The transducers were spaced Δx=2.5 mm apart (center-to-center) due to spatial limitations imposed by the transducer mount. The transducers were adjustable in the direction normal to the wall via a micrometer adjustment so they could be slightly recessed as recommended Dolling and Dussauge[4]. In the case of the of the model with the transducer plug placed in front of the compression corner, the ramp could be moved over a streamwise distance of Δx=2.3 cm. The change in the boundary layer thickness over the wedge's range of motion was not significant. In the model with the plug fitting flush onto the wedge face, the ramp was fixed. The output of each of the transducers was amplified prior to digitization via amplifiers built in house. The analog signals from the amplifiers were AC coupled via a high-pass filter (with a cutoff set at 0.1 Hz) and were each passed through anti-alias filters that had a roll-off of 60 dB/decade. The conditioned signals were subsequently digitized with the 12-bit multi-channel A/D converter of a Masscomp MC-5550 data acquisition system with a sample-and-hold for true simultaneous sampling. All subsequent data processing was performed on the MC-5550 computer.

III. Basic Nature of the Interaction

A. Boundary Layer Characteristics

Table 1: Undisturbed Boundary Layer (M=1.5)

Parameter	SI units	English units
M_∞	1.50±0.03	
U_∞	424±6 m/s	1390±20 ft/s
Re_∞ (x10^{-6})	12.5 /m	3.81 /ft
P_T	98.6 kPa	14.3 psi
T_T	290±3 K	522±5 R
P	26.2±0.7 kPa	3.8 ± 0.1 psi
T	200±1 K	360±2 R
δ (.99u/U_∞)	14.3±0.5 mm	0.56±0.02 inches
δ*	2.5 ±0.1mm	0.098±0.004 in.
θ	1.4±0.1 mm	0.055±0.004 in
H	1.8	
Re_θ	17500	
β	-0.47	
c_f (x10^3)	2.0 ±0.1	

A detailed series of measurements were performed in order to characterize the boundary layers upstream of the shock interaction. The nominal properties of the incoming freestream flow and boundary layer are presented in Table 1. It should be noted that the boundary layer transition occurred naturally and that the resulting turbulent boundary layer develops under the influence of a mild favorable pressure gradient upstream of the shock interaction.

B. Mean Characteristics of the Interaction

Figure 1a presents the mean static pressure distribution which characterizes the shock interaction. The data shown was obtained along the centerline of the tunnel and the streamwise coordinate x is centered at the wedge corner and has

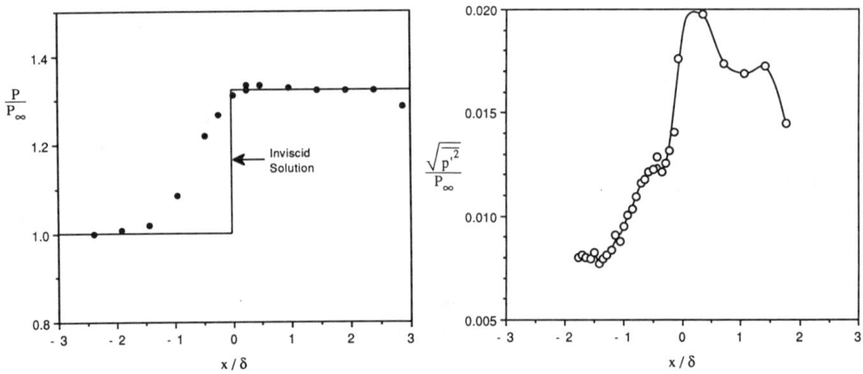

Figure 1a, b. Mean Static Pressure Distribution

been non-dimensionalized by the incoming boundary layer thickness, δ. The static pressure jump extends from approximately $-2 < x/\delta < 0.5$ where it reaches the inviscid theoretical value. Figure 1b presents the streamwise variation in the wall pressure fluctuation intensity, $\sqrt{\overline{p'^2}} / P_\infty$ through the interaction. This figure clearly indicates the increase in pressure fluctuation intensity that is associated with the turbulence stress amplification across the shock. The incoming boundary layer intensity is nominally 0.79% in general agreement previous published values. The amplification of fluctuations is observed to commence near $x/\delta=-1.4$ with maximum intensity occurring on the wedge face near $x/\delta=0.4$. Downstream of that location the pressure intensity decays which is indicative of turbulent stress relaxation downstream of the shock interaction.

C. Spectral Development of Fluctuations Through the Interaction

The spectral development of wall pressure fluctuations through the shock interaction was investigated by application of standard Fast Fourier Transform techniques. The power spectra were obtained by ensemble averaging 200 blocks of 1024 time series points which provide a frequency resolution of $\Delta f=127$ Hz. The power spectra are broadband and their interpretation is facilitated by recognizing that

$$\sigma_p^2(f) = \int_0^\infty G(f)df = \int_0^\infty f^*G(f)d(\ln f) \qquad (1)$$

In this manner the area under each spectral curve is directly proportional to the fluctuation energy. A streamwise sequence of power spectra spanning the shock interaction region is presented in Figure 2. This figure shows that initially (e.g. $x/\delta= -1.8$) the greatest energy and turbulence amplification occurs within a

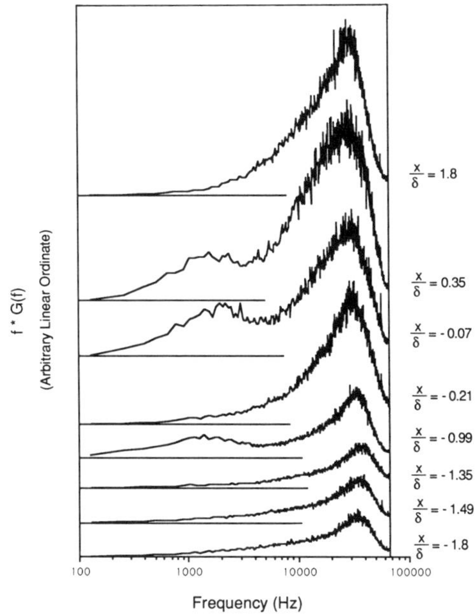

Figure 2. Pressure Fluctuation Spectral Development

frequency band centered near 35 kHz which is quite close to the characteristic frequency of the turbulence for the incoming boundary layer flow, f $\approx U_\infty/\delta$. Figure 2 shows that this characteristic frequency associated with peak fluctuation energy decreases throughout the interaction. At the last measurement station shown, $x/\delta=1.8$, the peak is centered near 30 kHz which again is of the order U_∞/δ for the characteristic conditions behind the shock interaction. Also of interest in Figure 2 is the development by $x/\delta = -0.99$ of a secondary spectral peak centered near 2 kHz, which is a much lower characteristic frequency on the order of f $\approx 0.1U_\infty/\delta$. This peak decays but reappears even more prominently downstream near the wedge apex at $x/\delta \approx -0.07$. It is well known that such frequencies are characteristic of both shock oscillation and separation bubble motion as reported in several unswept compression ramp experiments.

The contribution of fluctuations within the frequency bands 0-5 kHz and 5-65 kHz to the total fluctuation energy is shown in Figure 3 as obtained from integration of power spectra. The streamwise variation of low frequency fluctuations exhibits two characteristic peaks centered near $x/\delta = -0.8$ and $x/\delta \approx 0.1$ which account for approximately 36% and 45%, respectively, of the local total fluctuation energy. In contrast, the higher frequency fluctuations increase continuously through the interaction and peak near $x/\delta \approx 0.45$ which closely follows the pressure fluctuation intensity variation shown previously in Figure 1b. Note that the first low frequency peak in Figure 3 is due to localized shock

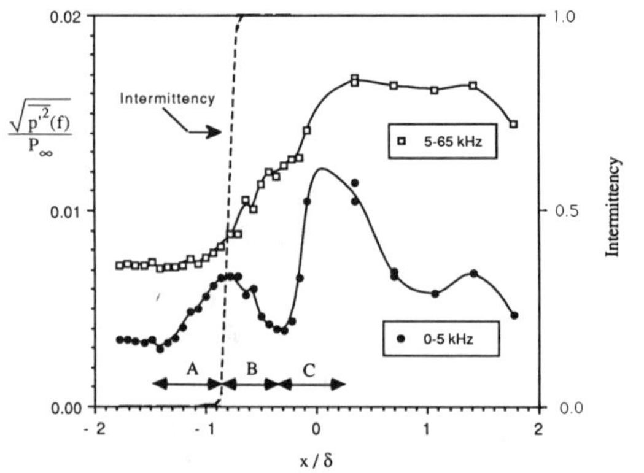

Figure 3. Intermittency and Bandwidth Dependent Intensity

front oscillation. The second low frequency peak which develops at the foot of the compression wedge is associated with an intermittent small scale separation bubble (which would be expected to exhibit characteristic frequency scaling similar to the shock oscillation). Figure 3 also presents the measured shock intermittency function. The intermittency was obtained via a modified form of the two threshold method of Dolling and Dussuage[4] and is described in reference 3. Note that the resulting intermittency function is extremely steep indicating appreciable shock oscillation is quite localized and that it is clearly associated with the first peak in the streamwise variation of low frequency fluctuation intensity.

In order to further characterize the spectral power transfer that occurs across the shock interaction, Figure 4a,b presents the relative spectral power transfer

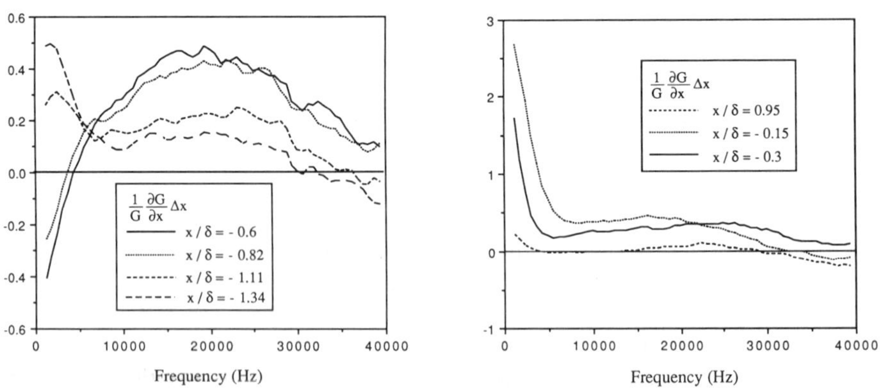

Figure 4a, b. Relative Spectral Power Transfer

$1/G\ (\partial G/\partial x)\Delta x$ as obtained from sequential power spectra. Representative curves covering regions A, B, and C of Figure 3 are presented. Just upstream of the shock (region A) measurements indicate the growth of low frequency fluctuations with f<5kHz. It is of interest to note that such frequencies are characteristic of shock oscillation. In region B,which is immediately downstream of the shock, these low frequency fluctuations lose energy as evidenced by negative $1/G\ (\partial G/\partial x)\ \Delta x$ while higher frequency fluctuations over a wide band exhibit substantial positive power transfer. The strongest positive spectral power transfer occurs near 20-25 kHz. As the wedge apex is approached (region C) very strong positive power transfer to low frequency fluctuations once again occurs within the band f<5kHz. This is associated with an intermittent separation bubble near the foot of the wedge. Downstream of the interaction on the wedge face $1/G\ (\partial G/\partial x)\Delta x \to 0$ as the flow relaxes to a pseudo equilibrium condition.

IV. Digital Estimation of Nonlinear Transfer Functions and Wave Coupling Coefficients

A temporal Fourier transform of the equation governing the fluctuation velocity field of an incompressible flow yields the familiar *wave coupling equation* which indicates that the spatial change in the velocity fluctuation spectrum occurs due to linear and quadratically nonlinear mechanisms. The source terms for the fluctuating pressure field are due both to interaction of velocity fluctuations normal to the wall with the mean shear as well as terms quadratic in the turbulent velocity fluctuations which represent a self-interaction of the turbulent field. Hence, since we view wall pressure fluctuations in the compressible turbulent boundary layer as a manifestation of the local fluctuating velocity field, it seems appropriate to similarly model the *local* spatial change of the pressure fluctuation spectrum as a nonlinear system consisting of both a linear and a quadratically nonlinear input-output relation. That is we assume,

$$\frac{\partial \phi(f,x)}{\partial x} = \Lambda_f^L(f,x)\ \phi(f,x) + \frac{1}{2}\sum_{f_1,f_2} \Lambda_f^Q\ (f_1,f_2)\ \phi(f_1,x)\ \phi(f_2,x) \qquad (2)$$

where x is the streamwise spatial coordinate, $f=f_1+f_2$, f_1, f_2 represent temporal frequencies, $\phi(f,x)$ represents the complex Fourier coefficient of the local fluctuating pressure $p'(t,x)$, $\Lambda_f^L(f,x)$ is a linear coupling coefficient which describes the change of a spectral component due to linear mechanisms and $\Lambda_f^Q(f_1,f_2)$ is a quadratic nonlinear wave coupling coefficient. The linear coupling coefficient $\Lambda_f^L(f,x)=(\gamma_f + i\ \overline{k_f}\ (f)\ )$ where γ_f represents the linear spatial growth

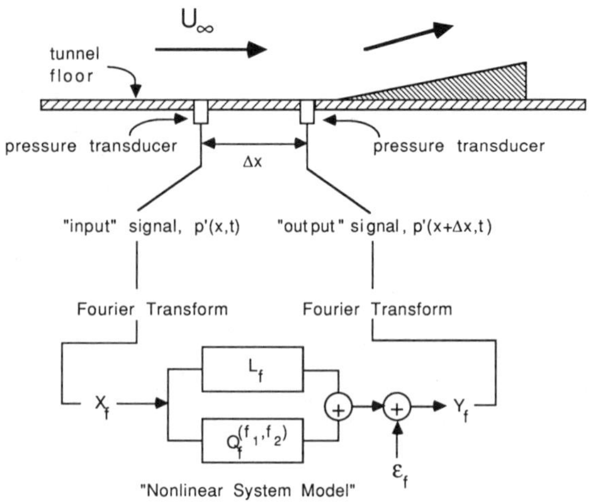

Figure 5. Measurement and Data Analysis Technique

rate of pressure fluctuations and $\overline{k_f}$ (f) is the dispersion relation. The local Fourier coefficient of the fluctuating pressure p'(x,t) may be represented in terms of its amplitude and phase where it is assumed that the amplitude is slowly varying in comparison with phase. Hence,

$$\phi(f,x) = |\phi(f,x)| \exp [i\theta(f,x)] \quad (3)$$

Following Ritz et al[5], the spatial change in the fluctuation pressure spectrum is approximated by a finite difference approach which along with equation (2) yields an expression for $\phi(f,x+\Delta x)$ in terms of $\phi(f,x)$ where the change is due to linear and quadratically nonlinear mechanisms. Hence the local spectral dynamics associated with the boundary layer pressure field will be modelled as a nonlinear system consisting of both linear and nonlinear system elements as shown in Figure 5 which also schematically presents the transducer positioning and signal analysis process utilized for the experiment. Of obvious importance in application of this technique is the question of whether it is actually possible to make a fluctuating pressure measurement that is effectively "local". From flat plate turbulent boundary layer space-time correlation coefficient functions at M_∞=1.5, it was found that the broadband convective velocity for the flow is $U_{cb}=0.77U_\infty$. A comparison was made to broadband convective velocity measurements using hot-wires as obtained in supersonic boundary layers at different y/δ locations by Spina et al[6] which showed that the convective velocity determined from wall pressure fluctuations is most closely associated with those hot wire measurements taken for $y/\delta<0.15$. This result suggests that the wall pressure signals are dominated by contributions from flow structures in the near wall region of boundary layer (i.e $y/\delta<0.15$). Further, when integral time macroscales Λ_T are computed from auto-correlations of the pressure signals, it is found that the product $\Lambda_T U_{cb} \sim d_t$. Hence, although in principle the pressure sensed by the transducer is represented by a volume integral over all fluctuations occurring above the transducer, this result shows that the transducer is most sensitive to fluctuations in the near wall region of the boundary layer directly above the transducer. This result justifies the assumption that the pressure fluctuation data can be treated as effectively *local.*

The "output" complex Fourier coefficient $\phi(f,x+\Delta x)$ will be denoted Y_f and is assumed to be related to the "input" Fourier coefficient $\phi(f,x)$ which is denoted X_f via a Volterra relation

$$Y_f = L_f X_f + \frac{1}{2} \sum_{f_1,f_2} Q_f^{(f_1,f_2)} X_{f_1} X_{f_2} + \varepsilon_f \tag{4}$$

where L_f represents the linear transfer function, and $Q_f^{(f_1,f_2)}$ represents the quadratic nonlinear transfer function. The term ε_f represents noise inherent in the measurement which is unrelated to the input Fourier component. From equations (2), (3) and (4) we obtain

$$L_f = \frac{\Lambda_f^L(f,x)\Delta x + 1 - i[\theta(f,x+\Delta x)-\theta(f,x)]}{e^{-i[\theta(f,x+\Delta x)-\theta(f,x)]}} \tag{5}$$

and

$$Q_f^{(f_1,f_2)} = \frac{\Lambda_f^Q(f_1,f_2)\Delta x}{e^{-i[\theta(f,x+\Delta x)-\theta(f,x)]}} \tag{6}$$

while the phase term in the denominator of equations (5) and (6) is estimated from the cross-spectrum.

Although the computation of the transfer function is quite simple for Gaussian input signals, the method utilized in the experiments reported here is an implementation of the more general multivariate linear regression analysis technique described recently by Kim and Powers[7] in which all fourth order spectral moments of the form $< X_{f_1} X_{f_2} X^*_{f'_1} X^*_{f'_2} >$ where $f = f_1+f_2 = f'_1+f'_2$ are fully retained in the computation (see also Thomas and Chu[8]).

If the wave coupling equation (2) is multiplied by the complex conjugate of the Fourier coefficient $\phi^*(f)$ one may obtain a *wave kinetic equation* for the spatial change in spectral power which is given by

$$\frac{\partial G(f)}{\partial x} = 2\gamma_f\, G_f(f) + \sum_{f_1,f_2} \mathrm{Real}\,[\Lambda_f^Q(f_1,f_2) <\phi(f_1)\,\phi(f_2)\,\phi^*(f)>] \quad (7)$$

where $G(f) = <\phi(f)\phi^*(f)>$. Following Ritz et al[5] the *nonlinear power transfer function* $T_f(f_1,f_2)$ is defined as

$$T_f(f_1,f_2) = \mathrm{Real}\,[\Lambda_f^Q(f_1,f_2) <\phi(f_1)\,\phi(f_2)\,\phi^*(f)>] \qquad (8)$$

In order to avoid a 2π phase ambiguity between each transducer pair, the sampling frequency was set to 80 kHz corresponding to a Nyquist frequency $f_{Nyquist}$= 40 kHz. The blocksize was set to N= 128 points *per channel* which allowed the computation of the nonlinear transfer function to a frequency resolution Δf=625 Hz. Since the spectra of Figure 2 are broadband there is no significant spectral structure that is lost due to frequency resolution limitations. It is also extremely important that the solution of the linear and nonlinear transfer functions is based on accurate estimates of the fourth order spectral moments involved. It was found that the spectral moments changed very little (<2%) when ensemble averaged over more than 2000 realizations. Thus for each streamwise location investigated, the estimation of the nonlinear transfer function and associated nonlinear spectral dynamics was based upon 2000 blocks of 128 time-series points per channel.

V. Experimental Results

A. Nonlinear Coupling Coefficient and Power Transfer Function Results

The relative power transfer presented in Figure 4 documents the spectral redistribution which occurs through the shock interaction but does not address the relative importance of linear and nonlinear mechanisms. In this section the nonlinear aspects of the spectral energy redistribution through the shock interaction will be characterized by both the magnitude of the nonlinear coupling coefficient $|\Lambda_f^Q(f_1, f_2)|$ and the nonlinear power transfer function $T_f(f_1, f_2)$. Due to symmetry conditions, these quantities are presented graphically in the two-dimensional domain (f_1, f_2) as shown in Figures 6 and 7. As reported in detail by Thomas et al[3] that domain may be divided into two primary regions. Region I involves only sum interactions of form $f_1+f_2 \to f_3$ where $f_3>f_1,f_2$. Conversely, region II involves only difference interactions. Region II may be further subdivided into region IIa and IIb. Region IIb involves difference interactions of form f_1-f_2->f_3 where $f_3<f_1,f_2$. Difference interactions in region IIa produce a mode of intermediate frequency.

It is important to note that $|\Lambda_f^Q(f_1, f_2)|$ represents an inherent characteristic of the flow that provides a measure of the local efficiency of nonlinear three-wave interactions. Figure 6 presents the magnitude of the nonlinear coupling coefficient as obtained at three representative locations within the shock/turbulent boundary layer interaction. Figures 6a and 6b present the $|\Lambda_f^Q(f_1,f_2)|$ as obtained upstream of the compression wedge at locations x/δ=-1.34 and -0.44 (i.e. in regions A and B of Figure 3, respectively). These results show that type I sum and type IIb difference interactions are the most efficient interactions in regions A and B. The occurrence of the type I sum interactions that involve $f_1 \geq 15$kHz and $f_2 \geq 5$kHz favor the growth of high frequency fluctuations with $f \geq 20$kHz and such interactions are likely associated with the continuous growth in spectral content for these frequencies as shown in Figure 3. Difference interactions which channel energy into a low frequency band of approximately 5kHz width centered near 2.5kHz are found to exhibit elevated values of $|\Lambda_f^Q(f_1,f_2)|$ and are observed to become most efficient as the wedge is approached. Note that the nonlinear coupling coefficient does not necessarily indicate that such interactions actually do occur. Rather, Figures 6a and 6b indicate that in regions A and B the character of the local flow is such that these two types of interactions would be most efficient should resonant phase coupling occur between the interacting spectral modes f_1, f_2. Figure 6c shows that at x/δ=-0.15 , $|\Lambda_f^Q(f_1, f_2)|$ indicates the flow is most efficient for only type IIb difference interactions between modes of similar frequency $f_1 \sim |-f_2|$ over a wide frequency band which serve to channel energy into a lower frequency fluctuation band centered near 2.5kHz.

Measurement of the nonlinear power transfer function $T_f(f_1, f_2)$ allows the actual energy transfer due to nonlinear mechanisms to be examined. In the contour plots of $T_f(f_1, f_2)$, positive values are associated with nonlinear power transfer in the usual sense of the interaction: f_1, f_2 -> f_3 (where -> is indicative of the direction of energy transfer). In contrast, negative values of $T_f(f_1, f_2)$ are indicative of energy transfer away from the wave that would normally be produced by the given interaction: f_1, f_2 <- f_3 (i.e. toward the other two interacting modes). The power transfer function is sensitive to both the magnitude and phase difference of the coupling coefficient and to the degree of resonant phase coupling as measured by the auto-bispectrum. In order to facilitate comparison of the nonlinear power transfer function results with the

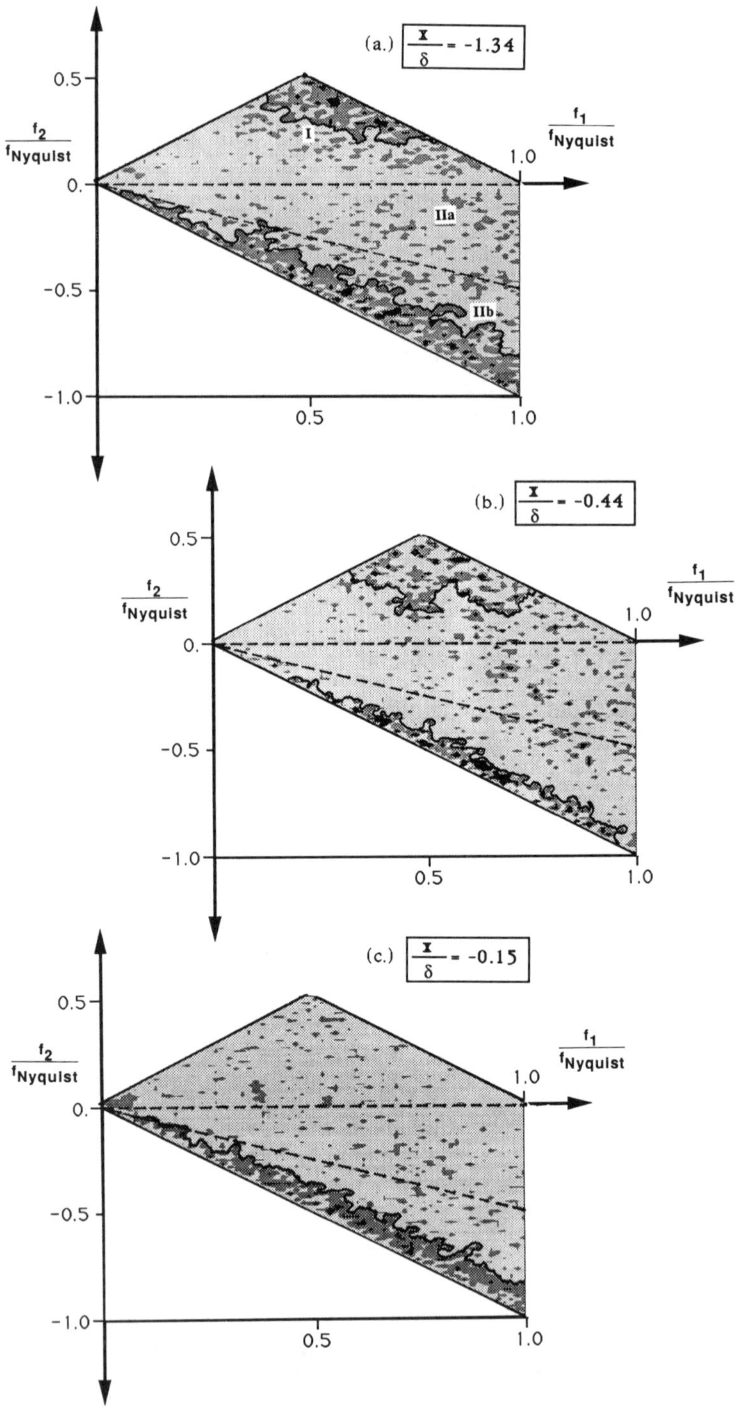

Figure 6. Nonlinear Wave Coupling Coefficients

associated $|\Lambda_f^Q(f_1, f_2)|$, Figure 7 presents the measured nonlinear power transfer function $T_f(f_1, f_2)$ as obtained at the same measurement locations. Figure 7a presents $T_f(f_1, f_2)$ as obtained at $x/\delta=-1.34$. In this contour plot, both positive and negative peaks of $T_f(f_1,f_2)$ are observed in region IIb. Peak positive values of $T_f(f_1, f_2)$ are associated with difference interactions involving modes $f_1<10$kHz and $|-f_2|< 10$kHz which serve to transfer energy into a fairly narrow frequency band centered near 2.5kHz. In contrast, peak negative values of $T_f(f_1, f_2)$ involving modes $f_1>20$kHz and $|-f_2|>10$kHz indicate the transfer of energy from large scale, low frequency fluctuations into higher frequency, fined-grained turbulent fluctuations. It is significant that the frequency band associated with significant power transfer is centered near 2.5kHz which is characteristic of the shock oscillation frequency (i.e. $f \sim 0.1U_\infty/\delta$). This suggests that the shock plays a key role in the two different energy redistribution processes indicated in the type IIb region. The same type IIb difference interaction at location $x/\delta= -1.34$ is also observed to be the most efficient interaction in the corresponding $|\Lambda_f^Q(f_1, f_2)|$ measurement. Figure 7b shows $T_f(f_1, f_2)$ as obtained upstream of the wedge at location $x/\delta= -0.44$. Peak negative values shown in Figure 7b continue to support the production of the high frequency, fine-grained turbulent fluctuations at the expense of low frequency fluctuations. The result of this energy transfer is apparent from inspection of Figures 3 and 4. Consistent with the negative values of power transfer associated with difference interactions is the observation of positive values of $T_f(f_1, f_2)$ in region I which is associated with sum interactions between frequency pairs with $f_1>5$kHz and $f_2\sim2$kHz to channel energy to higher frequencies. In both cases the energy flow is away from the low frequency modes centered near 2.5kHz and is toward associated higher frequency interacting modes. Near the wedge corner at $x/\delta= -0.15$ the character of $T_f(f_1, f_2)$ is very different as shown in Figure 7c. Note that positive values of $T_f(f_1, f_2)$ now characterize the difference interactions of region IIb in direct contrast with Figure 7b. This indicates an "inverse cascade" of energy from fine-grained turbulence into large-scale, comparatively low frequency motions via difference interactions.

B. Relative Quadratic Power Transfer

Since the flow is turbulent the contribution of any particular wave triad to the nonlinear spectral energy transfer will be quite small and this is quite apparent from examination of $T_f(f_1,f_2)$ in Figure 7. Equation (7) suggests, however, that significant nonlinear power transfer can occur via a summation over the contribution from all resonant wave pairs. This power transfer will depend upon the nonlinear wave coupling coefficient (which may be viewed as an efficiency for interaction) and the resonant three-wave coupling expressed through the auto-bispectrum. Figures 8a, b present the measured relative

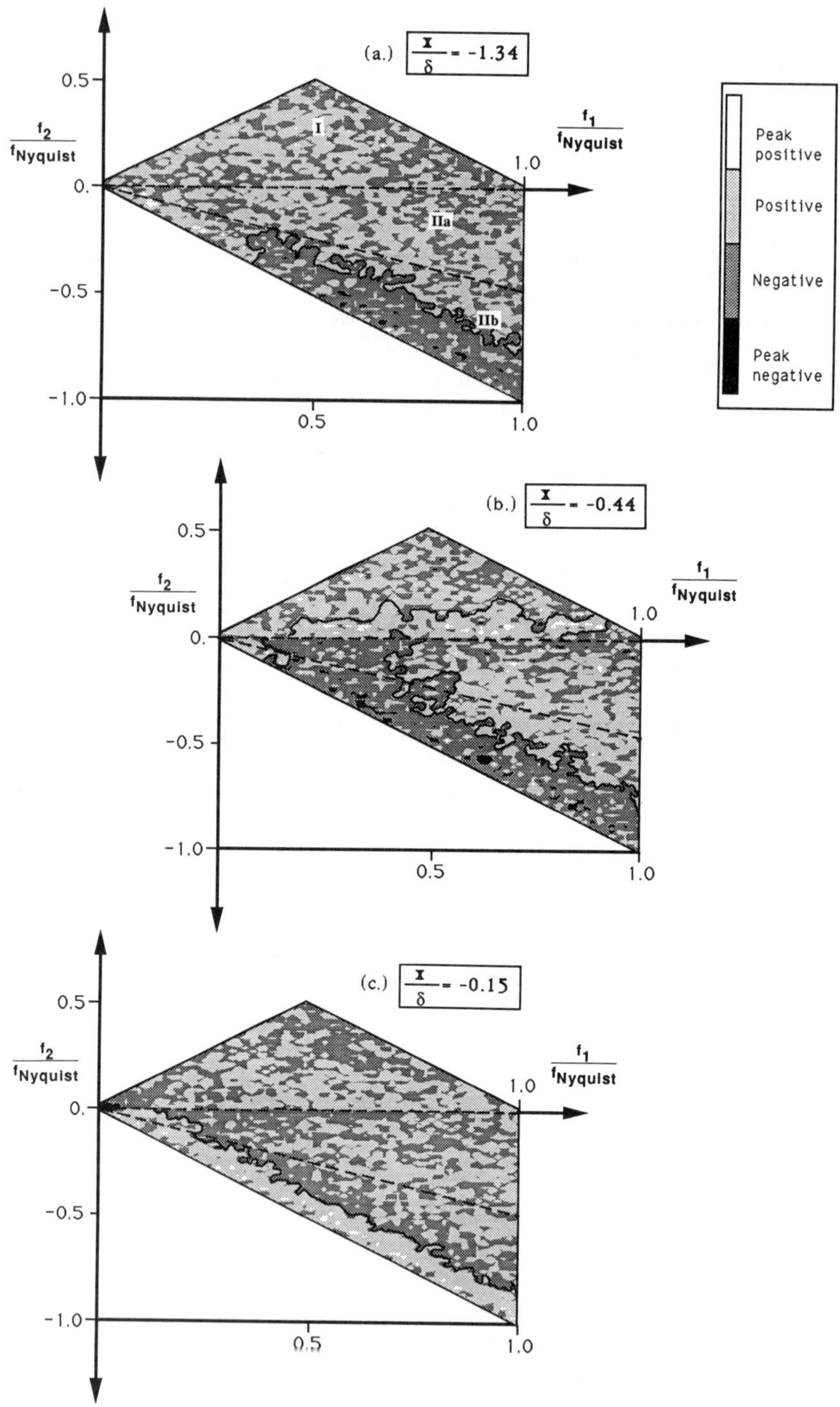

Figure 7. Nonlinear Power Transfer Function

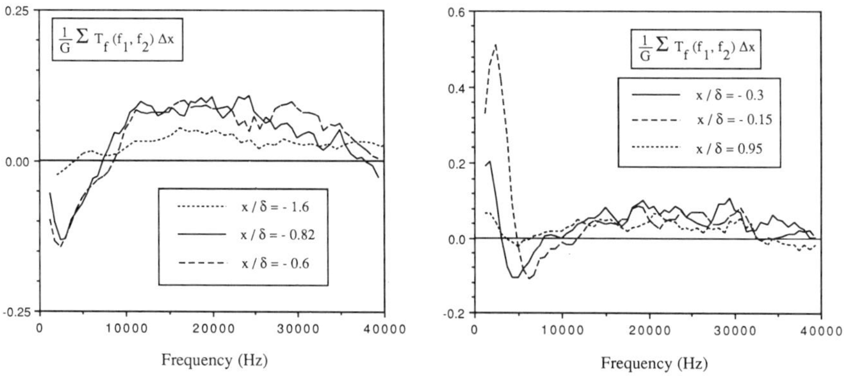

Figure 8a, b. Relative Quadratic Power Transfer

quaratic power transfer, $\Sigma T_f(f_1,f_2) / G$, as obtained at representative locations throughout the shock/boundary layer interaction. At the first location shown which is just upstream of the interaction ($x/\delta = -1.6$) the level of nonlinear power transfer is observed to be small for all frequencies. However, just downstream of the nominal shock position (i.e. the 50% intermittency location) the quadratic power transfer results are characterized by a loss of spectral power for frequencies f< 10k Hz and a concurrent growth of high frequency fluctuations over a wide frequency range f > 10 kHz with maximum growth occurring near 20 kHz. Near the foot of the wedge strong positive nonlinear power transfer to low frequencies occurs with maximum positive transfer occurring near 2.5 kHz. On the back side of the wedge near $x/\delta = 0.95$ the nonlinear power transfer is again found to be smaller for all frequencies. In general then, these nonlinear results directly parallel the spectral power transfer shown in Figure 4 with the exception of the initial energy transfer to low frequencies f<5k Hz which occurs upstream of the shock (i.e. region A). Measurements of the relative linear power transfer in the vicinity of the shock confirm that the initial power transfer to low frequencies as shown in Figure 4 occurs primarily by a linear mechanism. This is perhaps expected due to the sudden perturbation imposed upon the flow by the system of compression waves at the foot of the shock. As an example, Figure 9 presents the relative linear power transfer, $2\,\gamma_f\,\Delta x$, obtained at $x/\delta = -1.11$. As expected, it exhibits the strongest positive power transfer for f<5kHz. It should be noted, however, that the quadratic power transfer measurements indicate that immediately downstream of the shock a significant portion of the spectral redistribution is due to nonlinear wave interactions that transfer this energy from low to high frequencies. In region B for example, it is found that approximately 25% of the power transfer to high frequency fluctuations is due to quadratic transfer. Larger percentages are associated with the nonlinear transfer of energy into low frequencies near the wedge apex.

How well is the nonlinear system shown in Figure 5 capable of modelling the

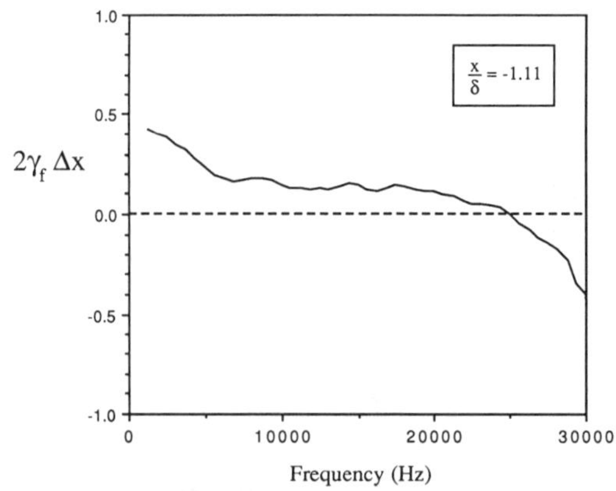

Figure 9. Relative Linear Power Transfer

spectral transfer that characterizes the fluctuating pressure in the shock/boundary layer interaction? One method of addressing this question involves a comparison between the relative spectral power transfer $1/G\ \partial G/\partial x\ \Delta x$ as estimated directly from sequential power spectra with that inferred from the system model via the relation

$$\frac{1}{G}\frac{\partial G(f)}{\partial x}\Delta x = 2\gamma_f\ \Delta x + 1/G\ \{\ \sum_{f_1,f_2} T_f\,(f_1,f_2)\,\}\ \Delta x \qquad (9)$$

As an example Figure 10 presents such a comparison at x/δ =-0.82. This result from within region B may be considered typical of those obtained at other spatial locations in the interaction. It is apparent from inspection of Figure 9 that the agreement between the modelled and true spectral power transfer is excellent at lower frequencies (f<15 kHz). At higher frequencies the trends exhibited by both curves are similar but the agreement is not as favorable. The comparison is quite poor for frequencies approaching the Nyquist frequency for the measurement. There are two primary reasons for this. First, regarding the nonlinear power transfer, as the Nyquist frequency of the measurement is approached there are fewer terms in the summation $\Sigma T_f(f_1,f_2)$. This may also be noted in Figure 8 where it is observed that the relative quadratic power transfer approaches zero at 40 kHz. Hence the error associated with the quadratic power transfer increases at high frequency. Regarding the linear power transfer, the difficulty in measuring spatial growth rates at high frequency is well known. The accurate estimate of γ_f is especially difficult here due to the limited spatial resolution of the pressure transducers. However, it may be concluded from comparisons like that shown in Figure 10 that the nonlinear system model used in the experiments is quite capable of realistically modelling the spectral energy redistribution through the shock and allowing the relative role of linear and nonlinear mechanisms to be established.

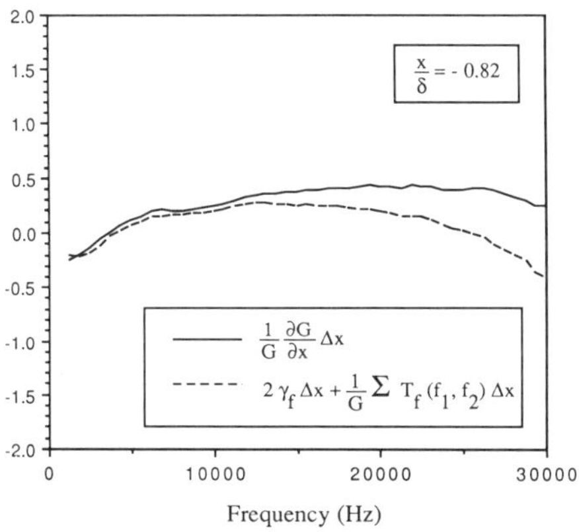

Figure 10. Comparison of Measured and True Spectral Power Transfer

VI. Conclusions

The technique used in this study represents a powerful new tool for the experimental investigation of the spectral energy redistribution which characterizes shock wave/turbulent boundary layer interactions. The magnitude of the nonlinear coupling coefficient provides a direct measure of the local efficiency of nonlinear three-wave coupling. The related nonlinear power transfer function provides a graphic and unambiguous view of the direction of nonlinear spectral energy transfer. Examination of $2\,\gamma_f\,\Delta x$ and $1/G\,\Sigma\,T_f\,(f_1,f_2)\,\Delta x$ enables one to assess the role of linear and nonlinear mechanisms on a local basis throughout the interaction.

It is found that linear mechanisms generally dominate throughout the interaction and at most locations investigated account for approximately 75% of the local spectral power transfer. Obviously, however, significant power transfer also occurs via nonlinear wave coupling. The turbulence interaction with the system of compression waves at the foot of the shock initially gives rise to energy transfer via linear mechanisms to a wide range of frequencies but an especially large contribution is made within a 5kHz bandwidth centered near 2.5 kHz. This frequency is characteristic of the shock oscillation frequency. Analysis of the shock oscillation statistics for the interaction considered here showed that the mean shock oscillation frequency was approximately 4 kHz but that the probability density function had a skewness of 2.6 so that the most probable oscillation frequency was much lower. Downstream of the shock front, but upstream of the wedge apex significant nonlinear spectral power transfer occurs which serves to transfer this excess energy within the low frequency band to higher frequency fine-grained turbulence. This would appear to be a manifestation of turbulent stress amplification due to shock oscillation. This nonlinear transfer is superimposed upon the linear amplification of high frequency fluctuations which is likely associated with direct amplification of vorticity and/or the conversion of incoming entropic modes. As the compression ramp corner is approached, the strongest nonlinear activity is observed and is found to be associated with type IIb nonlinear difference interactions which are found to be most efficient. These function to transfer energy from high frequency wave pairs $f_1,f_2 > 10$kHz to form low frequency modes within a 5kHz band centered near $f \approx 2.5$kHz. This suggests an "inverse" nonlinear energy cascade from small scales to large which is associated with the motions within an intermittent separation bubble near the foot of the wedge. In contrast to the situation upstream of the wedge, the most efficient interactions downstream of the interaction are found to be type I sum interactions. The nonlinear power transfer function demonstrates that these channel the excess energy contained at frequencies $f<5$kHz into high frequency turbulence $f>20$kHz. The flow is found to quickly relax to a pseudo-equilibrium condition and near $x/\delta = 1$ a reduced level of nonlinear power transfer is observed.

Acknowledgements

This work was supported by the Office of Naval Research under Grant No.N00014-88K-0670. This support is gratefully acknowledged.

References

1. Rubesin, M. W., "Turbulence Modeling for Aerodynamic Flows," AIAA Paper 89-0606, 1989.

2. Anyiwo, J. C. and Bushnell, D. M. "Turbulence Amplification in Shock Wave/Boundary Layer Interaction," AIAA Journal, 20, 7, pp. 893-899, 1972.

3. Thomas, F. O., Putnam C. M., & Chu, H. C., "Measurement of the Nonlinear Spectral Dynamics Characterizing a Shock Wave/Turbulent Boundary Layer Interaction" AIAA paper 91-0653, 1991.

4. Dolling, D. S., and Dussauge, J. P., "Fluctuating Wall-Pressure Measurements," in AGARDograph No. 315, ed. E. Reshotko, 1989.

5. Ritz, Ch. P., Powers, E. J., & Bengtson, R. D.,1989, "Experimental Measurement of Three-Wave Coupling and Energy Cascading," Phys. Fluids B, 1, 1, pp. 153-163.

6. Spina, E. F., Donovan, J. F. and Smits, A. J., "Convection Velocity in a Supersonic Turbulent Boundary Layer," Proceedings of the Twelfth Symposium on Turbulence, University of Missouri, Rolla, 1990.

7. Kim, I. K., & Powers, E. J.,1988, "A Digital Method of Modeling Quadratically Nonlinear Systems with a General Random Input," IEEE Transactions, Acoustics, Speech and Signal Processing, 36, 11, pp. 1758-1769.

8. Thomas, F. O. and Chu, H. C.,"Measurement of Nonlinear Wave Coupling and Energy Transfer in Planar Jet Shear Layer Transition," AIAA paper 91-0314, 1991.

Analysis of Blade Unsteady Boundary Layers and Wakes

M. Barnett[1]
J. M. Verdon[1]

ABSTRACT A recently developed nonlinear, compressible viscous-layer analysis and code has been extended and coupled to a linearized inviscid aerodynamic analysis to predict high Reynolds number subsonic unsteady flows through isolated, two-dimensional blade rows. This analysis can be applied to predict the viscous-layer response that arises from imposed inviscid conditions at the blade and wake surfaces. As part of this effort, a similarity analysis has been developed to determine the flow in the vicinity of a moving leading-edge stagnation point. This similarity solution is used to initiate a full unsteady viscous-layer calculation along the blade and wake surfaces. In addition, the viscous wake analysis has been extended to account for possible jumps in the flow variables across vortex-sheet unsteady wakes. At present only a weak or sequential coupling of the inviscid and viscous-layer solutions has been considered. The viscous-layer analysis has been applied to an unstaggered flat-plate cascade and a turbine cascade subjected to acoustic excitations coming from upstream or downstream and traveling toward the blade row in the axial direction. Numerical results presented for the flat-plate cascade indicate that the viscous layer responds linearly, for the most part, to acoustic excitations from upstream, but significant nonlinear response components occur for downstream excitations at high temporal frequency and/or high amplitude, which travel against the mainstream velocity. Results have been obtained for the turbine to demonstrate the present weak inviscid/viscid interaction solution capability for a realistic cascade configuration; they indicate the need for the development of a strong, or simultaneous, coupling of the inviscid and viscous-layer analyses to provide a more useful unsteady viscous aerodynamic analysis for realistic cascade configurations.

[1] United Technologies Research Center, MS 129-20, 411 Silver Lane, East Hartford, CT 06108-1049.

1.1 Introduction

The unsteady aerodynamic analyses intended for turbomachinery aeroelastic and aeroacoustic predictions must be applicable over a wide range of operating conditions and to unsteady excitations at moderate through high reduced frequencies. In addition, because of the large number of controlling parameters involved, there is a stringent requirement for computational efficiency, if the analysis is to be used successfully in the blade design process. To satisfy this latter requirement a number of restrictive assumptions must be introduced into the development of an appropriate unsteady aerodynamic model.

The theoretical analyses that have been developed to predict the aeroelastic and aeroacoustic behavior of turbomachinery blading, i.e., the onset of blade flutter, the amplitudes of forced blade vibration and the sound pressure levels upstream and downstream of the blade row, have, for the most part, been based on the following geometric and aerodynamic assumptions. The blades of an isolated, two-dimensional cascade are usually considered with the effects of neighboring structures being represented via prescribed nonuniform flow conditions at inlet and exit. In addition, the Reynolds number is assumed to be sufficiently high (i.e., $\mathrm{Re} \rightarrow \infty$) so that the displacement of the inviscid flow by viscous layers has a negligible impact on the unsteady pressure field. Finally, the unsteady excitations are assumed to be sufficiently small so that a linearized treatment of the unsteady inviscid flow is justified.

The unsteady flows of practical interest usually occur at high, but finite Reynolds number, so that viscous effects can have an impact on the unsteady pressure response. Provided that large-scale flow separation does not occur, the overall flow field can be divided conceptually into "inner" viscous or dissipative regions, consisting of thin layers that lie along the blade surfaces (boundary layers) and extend downstream from the blade trailing edges (wakes), and an "outer" inviscid region. Solutions to the complete flow problem can then be determined by an iterative process involving successive solutions of the inviscid and viscid equations. If the inviscid/viscid interaction is "weak", then a hierarchical procedure is used: at each step of the iteration, the inviscid and viscid equations are solved sequentially with the pressure being determined by the inviscid flow. However, in most flows, strong inviscid/viscid interactions occur due, for example, to boundary-layer separations, shock/boundary-layer interactions and trailing-edge/near-wake interactions, and the pressure must be determined by solving the inviscid and viscous-layer equations using a non-hierarchical approach in which the two solutions are obtained "simultaneously".

To date, there has been no effort to couple a viscous-layer analysis to the steady and linearized unsteady inviscid aerodynamic analyses that have been developed for turbomachinery aeroelastic and aeroacoustic applications. As a step toward this goal, a fully nonlinear <u>uns</u>teady <u>vis</u>cous-layer analysis (UNSVIS), which can also be applied to predict steady flows, was developed in Ref. [1]. This approach, rather than one in which the results of separate nonlinear steady and linearized unsteady viscous analyses are superposed, will allow an assessment to be made of the relative importance of nonlinear unsteady effects in viscous regions. In the UNSVIS analysis, the flow in the viscous layer is described by Prandtl's equations, with algebraic models used to account for the effects of transition and turbulence. These equations are solved in terms of Levy-Lees type variables using a finite-difference technique in which solutions are advanced in time and in the streamwise direction. Numerical solutions are determined by marching implicitly, first in time and then in the streamwise direction, over several periods of unsteady excitation from an initial steady solution, and from an approximate time-dependent, upstream flow solution. Numerical results were presented in [1] for flat-plate airfoils and for the turbine cascades studied experimentally by Dring *et al* [2],[3]. These were found to be in good agreement with previous analytical and numerical solutions for flat-plate airfoils and with the experimental measurements for the turbine cascades.

In the present study, the nonlinear viscous-layer analysis of Ref. [1] (UNSVIS) has been extended and coupled to the linearized inviscid flow analysis (LINFLO), described in Refs. [4], [5] and [6], to provide a weak inviscid/viscid interaction solution capability for unsteady cascade flows. The values of the inviscid flow variables at moving blade and wake surfaces, i.e., at the edge of viscous layers, needed for an unsteady viscous-layer calculation, are provided by LINFLO. In particular, a similarity analysis has been developed to predict unsteady viscous compressible flow in the vicinity of a moving leading-edge stagnation point. This analysis provides the "initial" upstream information needed to advance or march the viscous-layer calculation downstream along the blade and wake surfaces. In addition, the wake analysis used previously in UNSVIS has been extended so that the changes or jumps in the inviscid quantities that can occur across vortex-sheet unsteady wakes could be properly accommodated.

The nonlinear viscous-layer unsteady aerodynamic analysis is described in this paper, and demonstrated via application to two cascade configurations. This analysis has been applied to study the viscous-layer response of an unstaggered flat-plate cascade and a turbine cascade to external pressure excitations. The flat-plate results indicate that the viscous-layer responses to upstream pressure excitations are essentially linear over a broad range of excitation amplitudes and temporal frequencies, but significant nonlinear responses occur for downstream pressure excitations at high amplitude or high temporal frequency. In addition, the coupled LINFLO/UNSVIS analysis has been applied to a turbine cascade subjected to a pressure excitation from upstream to demonstrate the current weak inviscid/viscid interaction solution capability on a realistic cascade configuration.

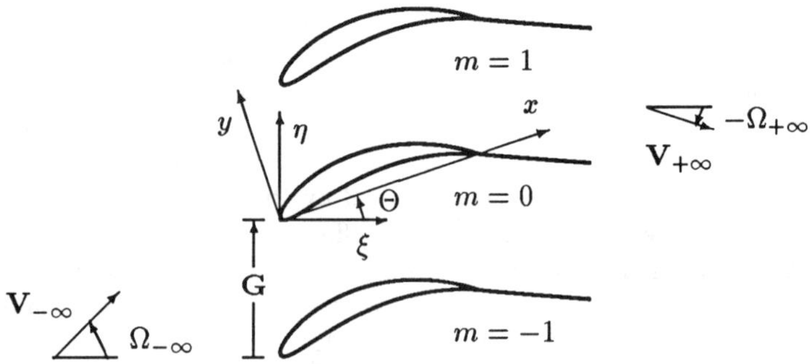

FIGURE 1.1. Two-dimensional compressor cascade.

1.2 Physical Problem

We consider time-dependent, high Reynolds number (Re) flow, with negligible body forces, of a perfect gas with constant specific heats and constant Prandtl number (Pr) through a two-dimensional cascade, such as the one shown in Figure 1.1. The unsteady fluctuations in the flow arise from one or more of the following sources: blade motions, upstream entropic and vortical disturbances, and upstream and/or downstream acoustic disturbances which carry energy toward the blade row. These excitations are assumed to be of small amplitude, periodic in time, and periodic in the cascade "circumferential" direction. For the present study only acoustic excitations are considered and the blades are assumed to be fixed.

The positions of the blade chord lines coincide with the line segments $\eta = \xi \tan\Theta + mG$, $0 \leq \xi \leq \cos\Theta$, $m = 0, \pm 1, \pm 2, \ldots$, where ξ and η are coordinates in the axial-flow and cascade directions, respectively, m is a blade number index, Θ is the cascade stagger angle, and $\mathbf{G}$ is the cascade gap vector which is directed along the η-axis with magnitude equal to the blade spacing. Since the unsteady aerodynamic excitations are periodic in η and t, it is often useful to describe them via Fourier series representations. In this case the complex amplitudes, fundamental frequencies and fundamental circumferential wave numbers of the various excitations are the prescribed quantities.

In the present discussion all physical variables are dimensionless. Lengths are scaled with respect to blade chord, time with respect to the ratio of blade chord to upstream freestream flow speed, density and viscosity with respect to their upstream freestream values, respectively, velocity with respect to the upstream freestream flow speed, normal and shear stresses with respect to the product of the upstream freestream density and the square of the upstream freestream speed, temperature with respect to the upstream freestream speed divided by the fluid specific heat at constant pressure, and entropy with respect to the fluid specific heat at constant pressure. The Reynolds number is defined using the upstream freestream density, velocity and molecular viscosity, with the reference length given by the blade chord.

1.3 Mathematical Model

The field equations that govern the fluid motion follow from the mass, momentum and energy conservation laws and the thermodynamic relations for a perfect gas. After ensemble averaging the resulting equations, equations for the (statistical) mean values of the flow variables are determined. These contain turbulent correlations, which must be estimated empirically. As a consequence of the high Reynolds number assumption, separate sets of equations can be constructed to describe the flow in the "outer" inviscid region and in the "inner" viscous-layer regions. The flow in the inviscid region is governed by a coupled set of nonlinear differential equations (i.e., the Euler equations) along with jump conditions at shocks and at vortex-sheet unsteady wakes. The attached flow assumption requires that the inviscid unsteady flow must be tangential to the moving blade surfaces, and the uniform flow conditions at inlet and exit, the entropic and vortical fluctuations at inlet, and the static pressure disturbances at inlet and exit that carry energy towards the blade row must be specified. The flows in the viscous layers are governed by Prandtl's equations and are subject to no-slip conditions at the moving blade surfaces. In addition, the streamwise velocity and the thermodynamic properties of the fluid at the edges of the viscous layers must be matched to the corresponding inviscid quantities at the blade surfaces and along the reference wake streamline.

1.3.1 The Flow in the Inviscid Region

A detailed description of the inviscid analysis (LINFLO), to which the present viscous analysis is coupled, can be found in Ref. [6]. The following discussion briefly summarizes those points needed to place the viscous analysis, which is the central focus of this paper, in its proper context.

Unsteady Perturbations of a Potential Mean Flow

For small amplitude unsteady excitations, assumed to be $\mathcal{O}(\epsilon) \ll 1$, the time-dependent flow can be regarded as a small perturbation of an underlying nonlinear mean or steady background flow. Thus, for example, we can express the time-dependent velocity in the form

$$\tilde{\mathbf{V}}(\mathbf{x},t) = \mathbf{V}(\mathbf{x}) + \tilde{\mathbf{v}}(\mathbf{x},t) + \cdots , \tag{1.1}$$

where $\mathbf{V}(\mathbf{x})$ is the local mean velocity and $\tilde{\mathbf{v}}(\mathbf{x},t)$ is the first-order (in ϵ) unsteady velocity.

The equations that govern the steady and first-order unsteady flow properties are determined by substituting the foregoing expansions into the full time-dependent, nonlinear governing equations, collecting terms of like power in ϵ and neglecting terms of $\mathcal{O}(\epsilon^2)$ or higher. This procedure provides a nonlinear boundary-value problem for a steady background flow and a linear variable-coefficient problem for the first-order unsteady flow in which the coefficients depend on the underlying steady flow. The first-order or linearized unsteady fluctuations caused by a periodic harmonic unsteady excitation will be harmonic in time, e.g.,

$$\tilde{\mathbf{v}}(\mathbf{x},t) = \mathrm{Real}\{\mathbf{v}(\mathbf{x})\exp(i\omega t)\} . \tag{1.2}$$

In addition, the steady and (for an excitation at circumferential wave number $\kappa_\eta = \sigma G^{-1}$) the linearized unsteady flow properties will satisfy the following circumferential conditions, e.g.,

$$\mathbf{V}(\mathbf{x}+m\mathbf{G}) = \mathbf{V}(\mathbf{x}) \quad \text{and} \quad \mathbf{v}(\mathbf{x}+m\mathbf{G}) = \mathbf{v}(\mathbf{x})\exp(im\sigma) . \tag{1.3}$$

Thus, as described in Ref. [6], solutions to time-independent nonlinear steady and linearized unsteady flow problems are required only over a single extended blade-passage region of the cascade. In addition, since analytic far-field solutions can be determined [7], the numerical solution domain can be restricted further to a single extended blade-passage region of finite extent in the axial direction.

1.3.2 The Flow in the Viscous Layer

Governing Equations

The viscous region consists of thin boundary layers that lie along the upper and lower surfaces of each blade and thin wakes which extend downstream from the blade trailing edges, as indicated in Figure 1.2. For the general case of nonzero blade motions, we assume the motions to be of sufficiently small amplitude, i.e., on the order of the viscous-layer displacement thickness ($\tilde{\bar{\delta}}$), that the system of field equations that governs the flow in a viscous layer is invariant under a transformation from space- to blade-fixed coordinates. Thus, in terms of the blade-fixed τ, n coordinates (see Fig. 1.2) we can write

$$\frac{\partial \tilde{\bar{\rho}}}{\partial t} + \frac{\partial(\tilde{\bar{\rho}}\tilde{U})}{\partial \tau} + \frac{\partial(\tilde{\bar{\rho}}\tilde{V})}{\partial n} = 0 , \tag{1.4}$$

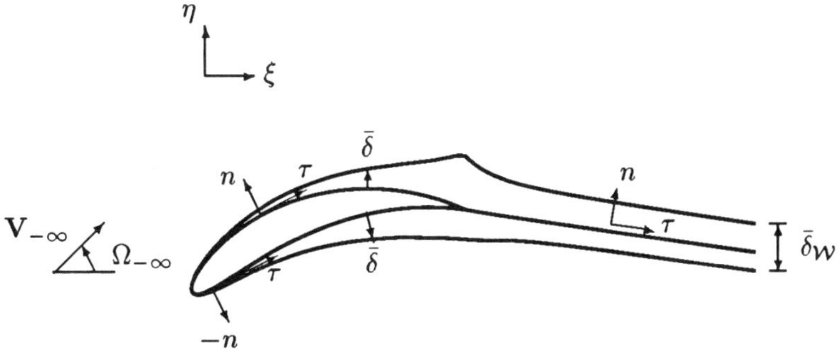

FIGURE 1.2. High Reynolds number steady flow over a blade surface.

$$\tilde{\bar{\rho}}\frac{\tilde{D}\tilde{U}}{Dt} = -\frac{\partial \tilde{P}}{\partial \tau} + \frac{1}{Re}\frac{\partial}{\partial n}\left(\tilde{\bar{\mu}}\frac{\partial \tilde{U}}{\partial n} - \tilde{\bar{\rho}}\overline{u'v'}\right) \tag{1.5}$$

and

$$\tilde{\bar{\rho}}\frac{\tilde{D}\tilde{H}}{Dt} = \frac{\partial \tilde{P}}{\partial t} + \frac{1}{Re}\frac{\partial}{\partial n}\left[\tilde{\bar{\mu}}\left(1 - \frac{1}{Pr}\right)\tilde{U}\frac{\partial \tilde{U}}{\partial n} + \frac{\tilde{\bar{\mu}}}{Pr}\frac{\partial \tilde{H}}{\partial n} - \tilde{\bar{\rho}}\overline{h'v'}\right] . \tag{1.6}$$

Here, $\tilde{U}$ and $\tilde{V}$ are the velocity components, relative to the blade-fixed frame, in the positive τ- and n-directions, respectively, and $\tilde{\bar{\mu}}$ is the molecular viscosity of the fluid. The terms that account for the acceleration of the blade-fixed frame are assumed to be small and have therefore been neglected in reducing the Navier-Stokes equations to viscous-layer equations.

In equations (1.4)–(1.6) the coordinates τ and n measure distances along and normal to, respectively, a blade surface or a reference wake streamline, and $\tilde{D}/Dt \equiv \partial/\partial t + \tilde{\mathbf{V}}\cdot\boldsymbol{\nabla}$ is the convective derivative operator in the blade-fixed frame. The reference wake streamline emanates from a blade trailing edge and is entirely contained within the viscous wake. A distinction should be made between the independent variables that describe the flows in the upper and lower surface boundary layers and in the wakes, e.g., by attaching subscripts to τ and n. However, as a convenience, we are neglecting to make this distinction explicitly.

The symbols $\tilde{\bar{\rho}}$, $\tilde{U}$, $\tilde{V}$, $\tilde{P}$, $\tilde{H}$, and $\tilde{\bar{\mu}}$ in Eqs. (1.4)—(1.6) refer to ensemble (or Reynolds) averaged values of the fluid dynamic variables, where the tilde denotes the instantaneous value of an unsteady quantity; u', v' and h' are the values associated with random turbulent fluctuations, and the overbar indicates a turbulent correlation, which must be determined empirically. As a consequence of the high Re and small-amplitude approximations, the pressure in the thin viscous layer is a function only of τ and t, and $\tilde{H} = \tilde{T} + \tilde{U}^2/2$. Since the interaction between the flows in the viscous layer and the external inviscid stream is assumed here to be weak, the pressure and the flow properties at the edge of the viscous layer are determined by the inviscid values of these variables at the blade and wake surfaces.

In addition to the foregoing conservation equations, the equation of state for a perfect gas, a law relating the molecular viscosity to the temperature (e.g., Sutherland's Law is used here) and equations relating the turbulent correlations $\overline{u'v'}$ and $\overline{h'v'}$ to the ensemble-averaged flow quantities, i.e.,

$$\tilde{\bar{\rho}}\overline{u'v'} = -\epsilon \frac{\partial \tilde{U}}{\partial n} \quad \text{and} \quad \tilde{\bar{\rho}}\overline{h'v'} = -\epsilon_{\tilde{H}} \frac{\partial \tilde{H}}{\partial n} = -\frac{\epsilon}{Pr_T} \frac{\partial \tilde{H}}{\partial n} , \tag{1.7}$$

are also required. The turbulent correlations $\overline{u'v'}$ and $\overline{h'v'}$ have been related to mean flow gradients, using Prandtl's mixing-length hypothesis. Here ϵ and $\epsilon_{\tilde{H}}$ are the eddy viscosity and eddy diffusivity, respectively, and play roles similar to their molecular counterparts. The eddy diffusivity is related to the molecular diffusivity through the introduction of the turbulent Prandtl number, Pr_T.

The models used here and in Ref. [1] to simulate the effects of transition and turbulence on the flow in the viscous layer are the algebraic eddy-viscosity model proposed by Cebeci and Smith [8] and the wake turbulence model proposed by Chang *et al* [9]. Also, since flows in turbomachines are known to be characterized by high freestream turbulence levels, a simple modification developed by Yuhas [10] has been incorporated into the turbulence model to account for the effects of freestream turbulence on the viscous layer. Although the location of the onset of transition could be determined from an empirical model or using an analytical technique, the points where transition starts and ends (i.e., where the flow becomes fully turbulent) are specified in the present study. The streamwise intermittency distribution between those points, γ_{tr}, where $0 < \gamma_{tr} < 1$, is obtained from the transition length correlation proposed by Dhawan and Narashima [11]. These models are easy to implement, and are known to be reasonably accurate for steady flows with mild pressure gradients; details of the models as implemented here are provided in Ref. [6]. Their ability to accurately represent transition and turbulence in unsteady flows is not known, however; therefore, results must be interpreted with some caution.

Initial and Boundary Conditions

The foregoing system of field equations is parabolic in time and in the streamwise direction. Therefore, the streamwise component of the velocity and the total enthalpy must be known for all time at some upstream streamwise location, and throughout the entire solution domain at some initial time. Also, conditions on the fluid properties at the edge(s) of the viscous layer must be specified, i.e.,

$$\tilde{U} \to \tilde{U}_e(\tau, t) \quad \text{and} \quad \tilde{H} \to \tilde{H}_e(\tau, t) \quad \text{for} \quad n \to \pm\infty , \tag{1.8}$$

where the limits $+\infty$ and $-\infty$ refer to the edges of the upper (+) and lower (−) surface boundary layers and the upper and lower edges of a wake, a no-slip condition and either a prescribed temperature or heat flux condition at a solid blade surface must be imposed, i.e.,

$$\tilde{U} = \tilde{V} = 0 \quad \text{and} \quad \tilde{H} = \tilde{H}_w(\tau, t) \quad \text{or}$$
$$\frac{\partial \tilde{H}}{\partial n} = \tilde{Q}_w(\tau, t) \quad \text{for} \quad n = 0\ , \quad \tau \leq \tau_{\text{te}}\ ; \tag{1.9}$$

and a condition on the fluid velocity normal to a reference wake streamline, i.e.,

$$\tilde{V} = 0 \quad \text{for} \quad n = 0\ , \quad \tau > \tau_{\text{te}}\ , \tag{1.10}$$

must be enforced. Here the subscripts w and e denote the values of the fluid properties at a solid wall and at the edge of the viscous layer, respectively, and the subscript te refers to the airfoil trailing-edge point. The fluid velocity, $\tilde{U}_e$, and total enthalpy, $\tilde{H}_e$, at the edges of the viscous layers are determined by the inviscid solution along the blade surfaces and reference wake streamlines.

For the calculations reported herein, the condition at the initial time is the viscous-layer solution in the absence of unsteady excitation, i.e., the solution for steady background flow. The calculated transients resulting from this approximation have been found to die out with increasing time. The upstream profiles, required at each time step of the unsteady viscous-layer calculation, are obtained from a similarity analysis of the flow in the leading-edge region, as appropriate for either a flat plate or for an airfoil with a blunt leading edge (i.e., a stagnation point); the latter is discussed in more detail below.

Transformed Equations

For laminar flows, the Levy-Lees transformation (see Blottner [12]) provides independent variables that effectively capture the growth of the viscous layer with increasing streamwise distance. In addition, in steady flows the transformed equations reduce to similarity equations at an airfoil leading edge. These features facilitate the determination of a numerical solution. Thus, an extension of the Levy-Lees transformation, in which the laminar edge viscosity is replaced by an effective turbulent viscosity [13], is applied herein to accommodate unsteady turbulent flows. The new independent and dependent variables are given in Ref. [6], along with the transformed equations and boundary conditions.

Stagnation Region

We seek a similarity solution for the flow in the stagnation region, which can be used to set the instantaneous upstream tangential velocity and total enthalpy profiles required to initiate a full viscous-layer calculation. The present analysis for unsteady, compressible flow can be viewed as an extension of the incompressible analysis developed independently by Rott [14] and Glauert [15]. Their analysis yields an exact solution to the Navier-Stokes equations for incompressible flow. This is not the case for compressible flow, however, because it is necessary to neglect the dissipation term in the energy equation to obtain a similarity solution. Fortunately, frictional heat dissipation is of limited importance for a wide range of practical flows.

The similarity solution is developed by considering two-dimensional compressible flow directed towards an infinite flat plate oriented normal to the stream direction, which undergoes a harmonic motion its own plane. The flow in the viscous layer is determined in terms of spaced-fixed Cartesian coordinates. The resulting stagnation-region solution is then re-expressed in terms of plate-fixed coordinates as a prerequisite to its implementation into the complete unsteady viscous-layer calculation. A detailed derivation of the stagnation-point analysis can be found in Ref. [6].

1.3.3 Numerical Solution Procedure

The transformed viscous-layer equations [6] are solved by using an implicit finite-difference approach to approximate the various partial derivatives and to convert the system of differential equations into a system of algebraic equations. Streamwise and temporal derivatives are approximated by first-order backward differences and normal derivatives are approximated by second-order central differences. The streamwise momentum and energy equations contain nonlinear terms, and therefore, they are linearized to facilitate their solution at each point in space and time. The finite-difference approximation results in a block tridiagonal system of linear algebraic equations at each step, which is solved using a standard block inversion algorithm.

The analyses for surface boundary layers and wakes are almost identical, except for the implementation of different boundary conditions. In addition to replacing the surface no-slip condition with a zero normal velocity condition at the reference wake streamline, the possibility of jumps in the tangential velocity and total enthalpy across a viscous wake must be taken into account. Further details on the reference wake streamline procedure are given in Ref. [6].

The discretized governing equations are solved on a grid that is nonuniform in both spatial directions, and uniform in time. The grid distribution in the streamwise direction can be specified to cluster points near the blade leading and trailing edges, using geometric stretching defined by the relation

$$\Delta\xi_{l+1} = K_\xi \Delta\xi_l \ , \quad l = 1, 2, \ldots, L \ , \tag{1.11}$$

where l is the streamwise mesh point index ($l = 1$ at the blade leading edge and increases with distance downstream), $K_\xi > 1$ from the leading edge of the blade to midchord, and $K_\xi < 1$ from midchord to the trailing edge . The grid is then stretched (i.e., $K_\xi > 1$) aft of the trailing edge, and generally extends between one and two blade chords downstream of the trailing edge.

A geometrically stretched grid is also employed in the normal direction, with clustering near the surface to capture the large velocity gradients that occur near the wall, and to ensure that, for turbulent flow, there are enough points in the near-wall region to adequately resolve the laminar sublayer. Since the resolution requirements are a function of the solution, the normal grid stretching parameter and the spacing at the wall must be determined for each case. The number of points used in the normal direction is chosen to allow the flow over the entire viscous layer to be resolved accurately at all streamwise stations — generally, between 50 and 100 points are sufficient.

The constant value of the time-step used for the temporal discretization is chosen based on the frequency of the imposed unsteady excitation and the number of time-steps prescribed for each period of oscillation, generally between 20 and 50.

1.4 Numerical Results

Response predictions were obtained for two-dimensional blade rows operating at subsonic Mach numbers and subjected to external unsteady aerodynamic excitations to demonstrate the present capabilities of the unsteady viscous-layer analysis, UNSVIS. Specifically, the viscous-layer response of flat-plate and turbine cascades to incident acoustic excitations is studied here. Detailed results for the flat-plate cascade and a brief discussion of the results for the turbine cascade will be given below.

1.4.1 Unstaggered Flat-Plate Cascade

As a benchmark case to demonstrate the unsteady viscous-layer analysis, UNSVIS, we consider unsteady flow through an unstaggered flat-plate cascade. Here, the uniform inlet velocity $\mathbf{V}_{-\infty}$ is aligned with the blade chord lines. Hence, the steady background flow is uniform with velocity, pressure and total enthalpy given by $\mathbf{V} = \mathbf{e}_\xi$, $P = (\gamma M^2)^{-1}$, and $H = \frac{1}{2} + (\gamma - 1)^{-1} M^{-2}$, respectively. The unsteadiness is excited by an incident pressure disturbance, which carries acoustic energy toward the blade row from either far upstream ($-\infty$) or far downstream ($+\infty$). The acoustic excitation travels in the axial (or chordwise) direction, i.e., $\kappa_\eta = \kappa_y = \sigma/G = 0$. The perturbation velocity potential describing this flow is given in Ref. [6]. The axial wave number is related to the temporal frequency (see Ref. [6]) by

$$\kappa_{\xi,\mp\infty} = \frac{\mp\omega M}{1 \pm M} , \qquad (1.12)$$

where ω and $\kappa_\xi = \kappa_x$ are the temporal frequency and axial wave number of the unsteady pressure excitation, respectively. Note that for a given temporal frequency, ω, and Mach number, $M < 1$, $|\kappa_{\xi,+\infty}| > |\kappa_{\xi,-\infty}|$, because a disturbance coming from downstream travels against the freestream flow direction.

The flat-plate cascade produces no response to the pressure excitation considered here; therefore, the entire first-harmonic inviscid flow field can be described analytically [6]. The time-dependent pressure, velocity and total enthalpy in the flow are then given by

$$\tilde{P} = P + \tilde{p} + \cdots = (\gamma M^2)^{-1} + \text{Real}\{p_{I,\mp\infty} \exp[i(\kappa_{\xi,\mp\infty}\xi + \omega t)]\} + \cdots , \qquad (1.13)$$

$$\tilde{\mathbf{V}} = \mathbf{V} + \tilde{\mathbf{v}} + \cdots = [1 \pm M \cdot \text{Real}\{p_{I,\mp\infty} \exp[i(\kappa_{\xi,\mp\infty}\xi + \omega t)]]\}\mathbf{e}_\xi + \cdots \qquad (1.14)$$

and

$$\tilde{H} = H + \tilde{h} + \cdots = H + (1 \pm M)\text{Real}\{p_{I,\mp\infty} \exp[i(\kappa_{\xi,\mp\infty}\xi + \omega t)]\} + \cdots , \qquad (1.15)$$

where $\tilde{p}$, $\tilde{\mathbf{v}}$ and $\tilde{h}$ are the first-harmonic unsteady pressure, velocity and total enthalpy, respectively. The quantity $p_{I,\mp\infty}$ is the complex amplitude of the unsteady pressure excitation.

The inviscid results in Eqs. (1.13) — (1.15) have been used as edge conditions for a series of unsteady viscous-layer calculations. The latter allow us to examine the effects of pressure-excitation amplitude, temporal frequency and axial wave number on the behavior of the viscous-layer displacement thickness, $\tilde{\tilde{\delta}}(\mathbf{x},t)$, surface shear stress $\tilde{\tilde{\tau}}_w = (\text{Re})^{-1}(\tilde{\tilde{\mu}}\partial\tilde{U}/\partial n)_w$ and wake centerline velocity $\tilde{U}_{\mathbb{C}}$. In particular, we can decompose each of these quantities into Fourier series; e.g.,

$$\tilde{\tilde{\delta}}(\mathbf{x},t) = \int_0^\infty [1 - \tilde{\tilde{\rho}}\tilde{U}/(\tilde{\tilde{\rho}}_e\tilde{U}_e)]dn = \sum_{n=-\infty}^{\infty} \delta_n(\mathbf{x})\exp(in\omega t) , \qquad (1.16)$$

where

$$\delta_n(\mathbf{x}) = \frac{\omega}{2\pi}\int_0^{2\pi/\omega} \tilde{\tilde{\delta}}(\mathbf{x},t)\exp(-in\omega t), \quad n = 0, \pm 1, \pm 2, \ \ldots , \qquad (1.17)$$

to examine the behavior of their Fourier components. In this way we can gain insight into the relative importance of nonlinear viscous effects and, therefore, into whether a linearized viscous analysis could be applied to provide meaningful and efficient unsteady viscous-layer solutions.

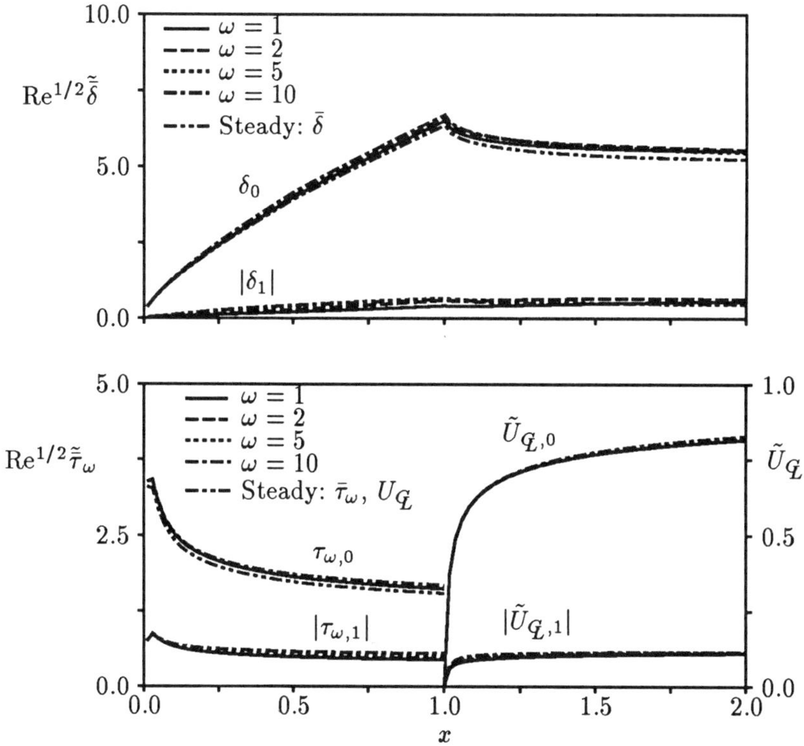

FIGURE 1.3. Fourier amplitudes of $\tilde{\bar{\delta}}$, $\tilde{\bar{\tau}}_\omega$ and $\tilde{U}_{\mathcal{L}}$ for an unstaggered flat-plate cascade ($\Omega = \Theta = 0\,\mathrm{deg}$, $G = 1$, $M = 0.5$ and $\mathrm{Re} = 10^6$) subjected to pressure excitations from upstream with $p_{I,-\infty} = (0.5, 0)$ and $\sigma = 0\,\mathrm{deg}$.

A series of calculations were performed for upstream and downstream acoustic excitations at frequencies, ω, of 1,2,5 and 10 and complex amplitudes, $p_{I,\pm\infty}$, of (0.1,0), (0.3,0), (0.5,0) and (0.75,0). In each case the steady Mach number is 0.5 and the Reynolds number, Re, is 10^6. The viscous-layer calculations were initiated at $(x, t) = (0.01, 0)$, where the laminar flat-plate similarity solution was imposed as the initial condition in x. Also, the flows were assumed to undergo an instantaneous spatial transition from laminar to turbulent at $x = 0.02$ for $t \geq 0$. The surface is assumed to be adiabatic; i.e., $\partial\tilde{H}/\partial n = 0$. A detailed description of the mesh used for the viscous-layer calculations is given in Ref. [6].

The behavior of the zeroth- and first-harmonic components of the viscous parameters is illustrated in Figure 1.3 for upstream acoustic excitations with $p_{I,-\infty} = (0.5, 0)$, $\sigma = 0$, and $\omega = 1, 2, 5$ and 10. Similar results for downstream acoustic excitations are given in Figure 1.4. For the upstream case, the time-mean values and the magnitudes of the first-harmonic viscous-layer response quantities are relatively unaffected by changes in ex-

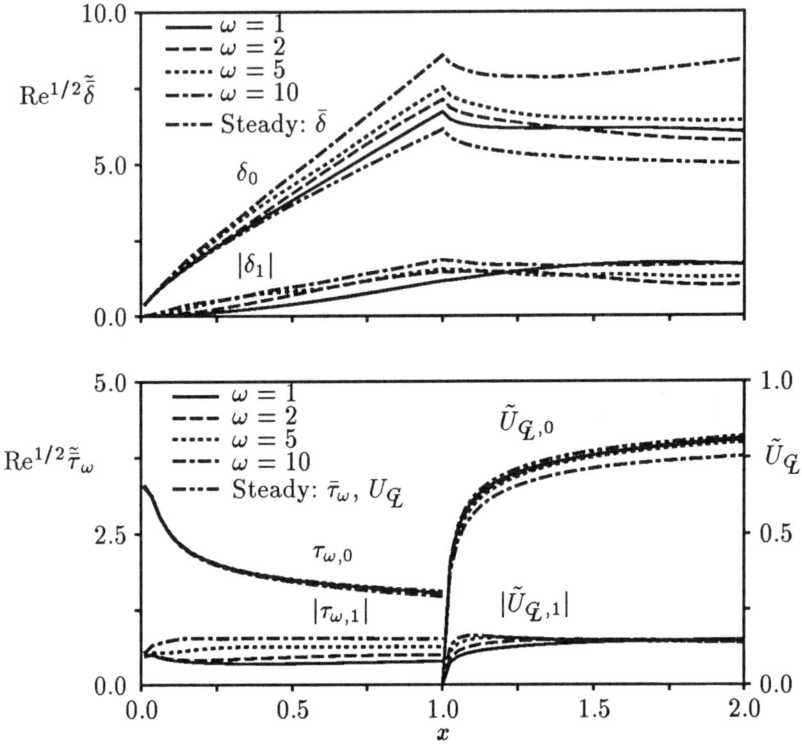

FIGURE 1.4. Fourier amplitudes of $\tilde{\bar{\delta}}$, $\tilde{\bar{\tau}}_\omega$ and $\tilde{U}_{\mathbb{C}}$ for an unstaggered flat-plate cascade ($\Omega = \Theta = 0\,\text{deg}$, $G = 1$, $M = 0.5$ and $\mathrm{Re} = 10^6$) subjected to pressure excitations from downstream with $p_{I,+\infty} = (0.5, 0)$ and $\sigma = 0\,\text{deg}$.

citation frequency. In addition, although not shown here, the first-harmonic displacement thickness is essentially out-of-phase with the first-order velocity fluctuation, $\tilde{u}_e$, at the edge of the viscous layer, and the first-harmonic wall shear stress and wake centerline velocity are essentially in phase with the edge-velocity fluctuation. The corresponding results (Figure 1.4) for the downstream acoustic excitation show that frequency has a greater impact on the temporal mean and the first-harmonic amplitudes of the viscous quantities. In particular, the time-mean displacement thicknesses vary significantly with ω and indicate that nonlinear effects are important at higher frequencies. Although again not shown here, the first-harmonic displacement thicknesses lag the fluctuating component of the viscous-layer edge velocity by phase angles ranging from approximately -140 to -180 deg; the first-harmonic wall shear stresses lead this fluctuating velocity by angles ranging from 0 to 40 deg; and the minimum wake-streamwise velocities are approximately in-phase with the wake-edge velocity fluctuation.

An unexpected result of the calculations for incident pressure disturbances from downstream is the predicted increase in the time-mean of the unsteady displacement thickness with increasing distance along the wake (i.e., as $x \rightarrow 2.0$), particularly evident in the result obtained for $\omega = 10$ (see Fig. 1.4). In an attempt to determine whether this effect is physical or numerical in origin, an unsteady viscous solution was calculated using a grid with twice as many uniformly distributed streamwise points. It was found that, although slightly less pronounced, this behavior was also present in the fine grid calculation and appears to be a grid independent feature of the solution. In addition, similar behavior was observed in the results of a series of laminar calculations carried out in an effort to understand the source of the increasing displacement thickness. Thus, the far-wake displacement thickness growth (as $x \rightarrow 2$) cannot be attributed to the turbulence model used in the present viscous analysis.

As a final illustration, we consider the response of the viscous layer to different amplitudes of acoustic excitation. In particular, numerical results for the displacement thickness are presented in Figures 1.5 and 1.6 for excitations at $\omega = 5$, $\sigma = 0$ and $p_{I,\mp\infty} = (0.1, 0)$, (0.3,0), (0.5,0) and (0.75,0). For brevity, the results for the quantities $\tilde{\bar{\tau}}_\omega$ and $\tilde{U}_{\text{℄}}$ are not shown here; they can be found in Ref. [6]. For disturbances from upstream, the effect of unsteadiness on the time-mean values of $\tilde{\bar{\delta}}$ and $\tilde{\bar{\tau}}_\omega$ is small, increasing as the disturbance amplitude, $|p_{I,-\infty}|$, is increased. The effect on the time-mean wake centerline velocity is negligible. The results for the acoustic excitations from downstream in Figure 1.6 once again indicate that such disturbances produce much stronger nonlinear viscous-layer responses, particularly in the displacement thickness. For both the upstream and downstream acoustic excitations, the phase angles (not shown) of the first-harmonic viscous quantities relative to the edge velocity fluctuation are almost independent of the excitation amplitude.

The foregoing results indicate that the viscous-layer response parameters $\tilde{\bar{\delta}}$, $\tilde{\bar{\tau}}_\omega$ and $\tilde{U}_{\text{℄}}$ behave essentially in a linear manner for acoustic excitations originating upstream of the blade row. For excitations originating downstream, nonlinear effects can be significant at high temporal frequencies (say $\omega \geq 5$) and/or high excitation amplitudes ($|p_{I,+\infty}| > 0.5$). For $M = 0.5$ the magnitude of the axial wave number for a pressure excitation from downstream is three times that for an excitation from upstream (see Eq. (1.12)). This produces a corresponding ratio in the magnitudes of the pressure gradients that drive the viscous solution, and is perhaps responsible for the nonlinear response of the viscous layer to downstream disturbances. Another factor is the direction of propagation of the acoustic excitation. We have performed a series of calculations to isolate these effects and have found the direction of propagation to be the dominant factor in promoting a nonlinear response in the viscous layer.

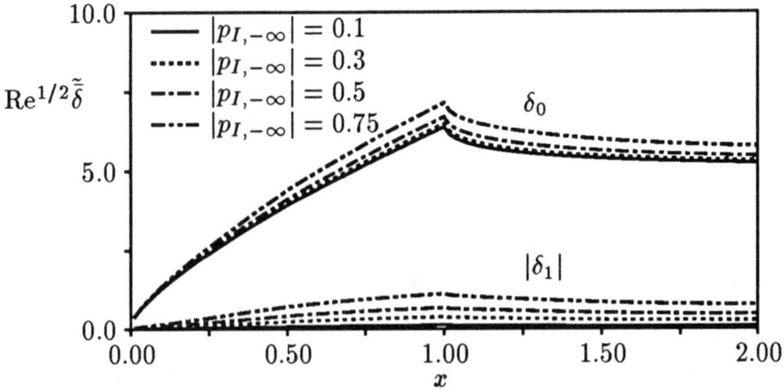

FIGURE 1.5. Fourier amplitudes of $\tilde{\tilde{\delta}}$ vs amplitude, $|p_{I,-\infty}|$, for an acoustic excitation from upstream, with $\omega = 5$ and $\sigma = 0$ deg, interacting with an unstaggered flat-plate cascade ($\Omega = \Theta = 0$ deg, $G = 1$, $M = 0.5$ and $\mathrm{Re} = 10^6$).

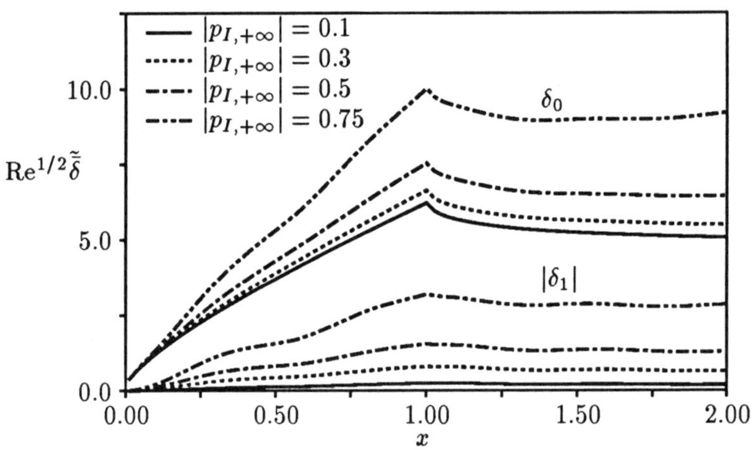

FIGURE 1.6. Fourier amplitudes of $\tilde{\tilde{\delta}}$ vs amplitude, $|p_{I,+\infty}|$, for an acoustic excitation from downstream with, $\omega = 5$ and $\sigma = 0$ deg, interacting with an unstaggered flat-plate cascade ($\Omega = \Theta = 0$ deg, $G = 1$, $M = 0.5$ and $\mathrm{Re} = 10^6$).

1.4.2 Turbine Cascade

In order to demonstrate the coupled LINFLO/UNSVIS weak inviscid/viscid interaction solution capability, we consider an unsteady flow through a representative turbine cascade, namely the fourth standard configuration of Ref. [16]. This flow occurs at a Reynolds number, Re, of 5.0×10^5, an inlet Mach number, $M_{-\infty}$, of 0.19 and an inlet flow angle, $\Omega_{-\infty}$ of 45 deg. The unsteadiness is excited by a pressure disturbance from upstream, characterized by $p_{I,-\infty} = (0.35, 0)$, $\kappa_{\eta,-\infty} = \sigma/G = 0$ and $\omega = 1.0$. For this example the unsteady pressure is essentially in-phase with the upstream acoustic excitation.

Since the turbine blade has a blunt leading edge, the unsteady stagnation-point analysis, described above, was applied at each time step to provide the upstream velocity profiles needed to advance the viscous-layer solutions along the blade suction and pressure surfaces. Upstream velocity profiles calculated using the similarity analysis are shown in Ref. [6]. The viscous-layer calculation for the reference turbine blade and its wake was performed assuming that instantaneous spatial transition from laminar to turbulent flow occurs near the leading edge on both the pressure and suction surfaces of the blade, and remains fixed at this location for $t \geq 0$. A detailed description of the grid used for this calculation can be found in Ref. [6].

Results of the unsteady viscous-layer calculation are shown in Ref. [6] and demonstrate that the unsteady response of the viscous layer is essentially linear for this example, i.e., the temporal mean and the steady viscous solutions are almost identical, and the Fourier amplitudes of the higher ($n > 2$) harmonic components of $\tilde{\tilde{\delta}}$, $\tilde{\tilde{\tau}}_\omega$ and $\tilde{U}_{W,\min}$ are negligible. Although calculations for higher amplitude pressure excitations were attempted, they produced viscous-layer separation just upstream of the blade suction-surface trailing edge, precluding a continuation of the unsteady viscous-layer calculation into the wake.

1.5 Concluding Remarks

An existing unsteady viscous-layer aerodynamic analysis and code has been extended and coupled to a linearized inviscid aerodynamic analysis for use in predicting subsonic unsteady flows through isolated, two-dimensional blade rows. The extensions include the incorporation of a new similarity analysis for the compressible flow in the vicinity of a moving stagnation point, and a treatment which properly accounts for the jumps in the inviscid flow variables across vortex-sheet unsteady wakes. The analysis can be applied to predict the viscous-layer response that arises from imposed inviscid conditions at the blade and wake surfaces. At present only a weak or sequential coupling of the inviscid and viscous-layer solutions has been considered. The development of a strong inviscid/viscid interaction analysis involving a simultaneous coupling of the inviscid and viscous solutions is planned as future work.

The viscous-layer analysis (UNSVIS) has been applied to unstaggered flat-plate and turbine cascades subjected to acoustic excitations coming from upstream or downstream and traveling in the axial direction. The flat-plate example is, perhaps, the simplest unsteady cascade problem that can be analyzed, both because of the geometric simplicity and because the unsteady pressure is nonsingular at the blade leading edges. The numerical results indicate that the viscous layer responds linearly, for the most part, to acoustic excitations from upstream, but significant nonlinear response components occur for downstream excitations at high temporal frequency and/or high amplitude, which travel against the mainstream velocity. The numerical results for the turbine demonstrate the present weak inviscid/viscid interaction solution capability for a realistic cascade configuration.

Because of boundary-layer separation, the range of application of a weak inviscid/viscid interaction analysis is limited. For example, the mean pressure rise produced by a compressor blade row typically causes boundary-layer separation very near the blade trailing edges, thereby precluding a continuation of a direct viscous-layer calculation along the blade and into the wake. Thus, the development of a strong inviscid/viscid interaction analysis will be needed so that the effects of viscous displacement on the unsteady pressure response can be included in aeroelastic and aeroacoustic design calculations. The linear vs nonlinear response of viscous layers to entropic, vortical and blade-motion excitations are other issues that will require further study and clarification.

Acknowledgments: This work was supported by NASA Contract NAS3-25425 with George Stefko serving as NASA Program Manager. The authors would like to thank Drs. Greg Power and Ken Kousen of UTRC for their helpful discussions regarding various aspects of the UNSVIS code, and Ruthann Rudewicz of UTRC for her help in preparing this paper and in running test cases.

References

[1] G. D. Power, J. M. Verdon, and K. A. Kousen. *Analysis of Unsteady Compressible Viscous Layers.* ASME Paper 90–GT–14, 35th International Gas Turbine and Aeroengine Congress and Exposition, Brussels, Belgium, June 11–14 1990. to be published in *Trans. of the ASME.*

[2] R. P. Dring, H. D. Joslyn, L. W. Hardin, and J. H. Wagner. Turbine rotor-stator interaction. *Transactions of the ASME, A: Journal of Engineering for Power*, 104(4):729–742, October 1982.

[3] R. P. Dring, M. F. Blair, H. D. Joslyn, G. D. Power, and J. M. Verdon. *The Effects of Inlet Turbulence and Rotor/Stator Interactions in the Aerodynamics and Heat Transfer of a Large-Scale Rotating Model.* V0l. I — Final Report CR 4079, NASA, June 30 1989.

[4] J. M. Verdon. Unsteady aerodynamics for turbomachinery aeroelastic applications. In D. Nixon, editor, *Unsteady Transonic Aerodynamics, Progress in Astronautics and Aeronautics*, pages 287–347, American

Institute of Aeronautics and Astronautics, Washington, D. C., 1989. Vol. 120.

[5] J. M. Verdon. *Linearized Unsteady Aerodynamics for Turbomachinery Aeroelastic Applications.* Paper 90–2355, AIAA/SAE/ASME/ASEE 26th Joint Propulsion Conference, Orlando, Florida, July 16–18 1990.

[6] J. M. Verdon, M. Barnett, K. C. Hall, and T. C. Ayer. *Development of Unsteady Aerodynamic Analyses for Turbomachinery Aeroelastic and Aeroacoustic Applications.* NASA CR, NASA Lewis Research Center, Cleveland, Ohio, 1991. (UTRC Report R91–957907–3), prepared under Contract NAS3–25425, to be published.

[7] J. M. Verdon. The unsteady flow in the far field of an isolated blade row. *Journal of Fluids and Structures*, 3(2):123–149, March 1989.

[8] T. Cebeci and A. M. O. Smith. *Analysis of Turbulent Boundary Layers*, pages 211–239. Academic Press, New York, 1974.

[9] K. C. Chang, M. N. Bui, T. Cebeci, and J. H. Whitelaw. The calculation of turbulent wakes. *AIAA Journal*, 24:200–201, February 1986.

[10] L. J. Yuhas. *An Optimization Technique for the Development of a Two-Dimensional Turbulent Boundary Layer Model.* Master's thesis, Lehigh University, 1981.

[11] S. Dhawan and R. Narashima. Some properties of boundary-layer flow during the transition from laminar to turbulent motion. *Journal of Fluid Mechanics*, 3, 1958.

[12] F. G. Blottner. Finite difference method of solution of the boundary-layer equations. *AIAA Journal*, 18:193–205, February 1970.

[13] M. J. Werle and J. M. Verdon. *Viscid/Inviscid Interaction Analysis for Symmetric Trailing Edges.* Report R79–914493–5, UTRC, East Hartford, CT, January 1980. Prepared under Contract N00019–78–C–0604 for the Naval Air Systems Command.

[14] N. Rott. Unsteady viscous flow in the vicinity of a stagnation point. *Quarterly Journal of Applied Mathematics*, 13:444–451, 1956.

[15] M. B. Glauert. The laminar boundary layer on oscillating plates and cylinders. *Journal of Fluid Mechanics*, 1:97–110, 1956.

[16] T. H. Fransson and P. Suter. *Two-Dimensional and Quasi Three-Dimensional Experimental Standard Configurations for Aeroelastic Investigations in Turbomachine-Cascades.* Report LTA–TM–83–2, Ecole Polytechnique Federale de Lausanne, Lausanne, Switzerland, 1983.

Active Control of Centrifugal Compressor Rotating Stall

Patrick B. Lawless[1]
Sanford Fleeter[1]

Abstract

A mathematical model is developed to analyze the suppression of rotating stall in a low speed centrifugal compressor, thereby addressing the important need for centrifugal compressor rotating stall and surge control. In this model, the precursor to instability is a weak rotating potential velocity perturbation that eventually develops into a finite disturbance. To suppress the growth of this potential disturbance, a rotating control vortical velocity disturbance is introduced into the impeller inlet flow. The effectiveness of this control is analyzed by matching the perturbation pressure in the compressor inlet and exit flow fields with a model for the unsteady behavior of the compressor. The model predicts that favorable extensions of the compressor operating range are possible with such a control system.

Nomenclature

$\mathbf{i}$ = unit vector, rotor-relative
$\mathbf{j}$ = unit vector, fixed reference
n = wave number
p_t = total pressure
p = static pressure
r = radial coordinate
s = distance along mean flow path
V = absolute flow velocity
W = relative flow velocity
z = axial coordinate

β = control phase
γ = mean path direction cosine
$\overline{\eta}_c$ = isentropic efficiency
κ = specific heat ratio
$\Delta\overline{\phi}\%$ = change in $\overline{\phi}|_{stall}$ due to control
$\overline{\psi} = \dfrac{\overline{p}_5 - \overline{p}_{t3}}{\rho U^2}$
ω = disturbance frequency
ω_I = impeller angular velocity

[1]School of Mechanical Engineering, Purdue University, Chaffee Hall, West Lafayette, IN 47907.

$\xi = \phi_v' \rho U^2 / p'|_3$

$\zeta = \phi_v' / \phi_p'$

superscripts:

- $*$ rotor-relative coordinate
- $'$ perturbation value
- $-$ mean, axisymmetric

subscripts:

- *asy* asymmetric
- D diffuser
- I impeller
- *m* mean flow direction
- *p* potential
- *r* radial direction
- *u* unsteady
- *v* vortical

Introduction

On the performance map of a compression system, the surge line separates the regions of stable and unstable compressor operation. The term "surge line" is somewhat misleading since surge is only one of the possible phenomena that result when this boundary is reached. In general, the types of instabilities found in turbomachines can be categorized as surge and rotating stall. Surge takes the form of a low frequency, global oscillation of the mass flow through the compression system with total flow reversal often occurring over a portion of the surge cycle. Surge is a phenomena of the entire compression system, consisting of the compressor and the system into which it discharges. Rotating stall, in contrast, is an instability local to the compressor itself, and is characterized by a circumferentially nonuniform mass deficit which propagates around the compressor annulus at a fraction of wheel speed.

Active control schemes for the prevention of rotating stall and surge are of interest due to the potential for achieving increased operating range without suffering significant performance penalties. To date, research into the active control of rotating stall has been centered around axial compressors. Experimental investigations into stall inception in axial compressors by McDougall et al.[1] and by Garnier et al.[2] have provided evidence that rotating stall can result from the growth of a modal inlet velocity wave which exists prior to any finite stall cell formation. An analysis by Epstein et al.[3] concerned the suppression of the small amplitude modal waves such as those detected by McDougall. Here, a suggestion was made for an active control system that employed vortical disturbances introduced upstream of the rotor inlet to adjust the aerodynamic damping of the system. The results from an initial analysis indicated this approach could be used to delay the onset of stall and allow operation in the previously unstable region of the compressor characteristic. A control system based on this concept has been reported by Paduano et al.[4] to have significantly reduced the stalling mass flow rate for a low speed, single stage axial compressor.

Thus, the focus of research on the control of rotating stall has been axial flow compressors, thereby excluding an important class of turbomachines. Namely, centrifugal compressors, which are also highly susceptible to rotating stall. However, due to the features unique to centrifugal compressors the axial flow results, both models and experiments, are not directly applicable. The flow field in the centrifugal compressor is also more complicated than its axial equivalent, with separated flow generally occurring in the rotor and a complex mixing region between rotor exit and diffuser entrance. Nevertheless, there is evidence to suggest that modal waves play a significant part in centrifugal compressor stability. For example, an inviscid analysis by Moore[5] for a centrifugal compressor with vaneless diffuser predicted the occurrence of waves of similar modal shape and wave speeds as those observed in finite rotating stall patterns. Also, several researchers have made progress in active surge control techniques for centrifugal compressors. [6],[7]

This paper is directed at the investigation of rotating stall control in a centrifugal compressor. A mathematical model will be developed to analyze the suppression of modal instabilities in a low speed centrifugal compressor by introducing an unsteady vortical velocity field into the impeller inlet. This will be accomplished by significantly extending the axial flow model of reference 3 to the centrifugal case. The precursor to instability is a weak rotating potential velocity perturbation in the inlet flow field that eventually develops into a finite disturbance. To suppress the growth of this potential disturbance, a rotating control vortical velocity disturbance is introduced into the impeller inlet flow. The effectiveness of this control is analyzed by matching the perturbation pressure in the compressor inlet and exit flow fields with a model for the unsteady behavior of the compressor.

Mathematical Model

Centrifugal Compressor Geometry

A diagram of a low speed centrifugal compressor is given in Figure 1. This geometry is based on a research compressor at Purdue University which has demonstrated a tendency to enter rotating stall near the peak of the pressure rise versus flow rate characteristic. The inlet flow to the compressor is axial, entering the annular inlet duct at Position 1. At Position 2 the flow passes an array of control vorticity generators, which are discussed in more detail below. The flow then enters the impeller at Position 3 and exits into the vaneless space at Position 4. The flow proceeds through the diffuser vanes, exiting into a parallel walled vaneless region before discharging into a large exit plenum at Position 5.

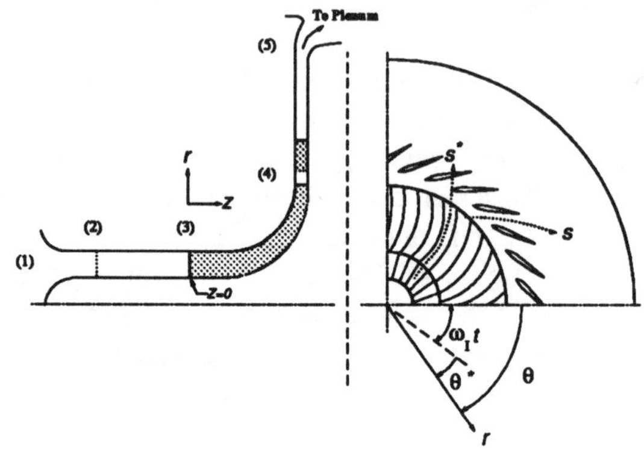

Figure 1. Compressor geometry and coordinate system.

Proposed Control Scheme

Rotating stall is taken to be the overt culmination of a phenomena which manifests itself in its early stages as a weak, circumferentially distorted potential wave. Hence, the occurrence of such a weak distortion serves as a precursor to a fully developed rotating stall condition. The goal of the control system is to detect and attenuate this potential wave before the compressor experiences any performance degradation. Wave detection is accomplished with an array of circumferentially distributed sensors mounted near the impeller inlet. A spatial Fourier transform of these signals is analyzed to detect the presence of a dominant circumferential harmonic. Responding to this information, the control vorticity generators are activated so as to introduce additional damping into the system and suppress the growth of the instability. In practical application these vorticity generators could take the form of oscillating guide vanes, air injection jets, or other types of wake producing devices.

Model Assumptions

The flow is considered to be incompressible, with the centrifugal compressor inlet flow inviscid and two-dimensional. The flow though the impeller and diffuser passages is characterized by a representative mean-flow streamline. Both the impeller and diffuser are assumed to be of high solidity, allowing a constant flow angle assumption to be imposed. The analysis is restricted to the initial stages of instability, where the flow distortions can be modelled as linear perturbations from the mean flow. Therefore the unsteady, asymmetric variations in the flow properties are assumed to be small as compared to the time averaged, axisymmetric values.

$$|\vec{V}'| \ll \overline{V}$$

$$p' \ll \overline{p}$$

The performance of the compressor under steady, axisymmetric operating conditions is defined by the characteristic $\overline{\psi}(\overline{\phi})$ and isentropic efficiency curves $\overline{\eta}_c(\overline{\phi})$ as shown in Figure 2. These curves represent the behavior of the compressor in the absence of rotating stall and surge. The portion of the curve beyond the stall point, represented by the shaded region in Figure 2, in practice is difficult to determine experimentally. However, convincing arguments have been made that the characteristic curve in this region is well represented by an extension of the smooth cubic polynomial curve generated from the stable flow operating points [8].

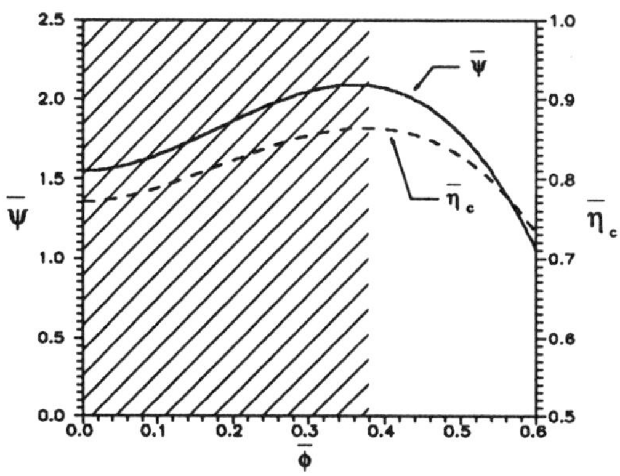

Figure 2. Assumed steady, axisymmetric characteristic and efficiency curves.

Inlet Flow Field

Upstream Potential Precursor

Before the introduction of the corrective waveform from the vorticity generators, the inlet flow is characterized by a neutrally stable weak potential disturbance represented by a disturbance potential Φ, superimposed on an axial mean inlet velocity of magnitude $\overline{V}$. The existence of this potential disturbance represents the stalling point of the compressor in that the wave will progress into rotating stall if the system damping is reduced. Thus, the expression for the velocity in the inlet flow can be represented as

$$\vec{\mathbf{V}}(z,\theta,t) = \overline{V}\mathbf{i}_z + \nabla\Phi(z,\theta,t)$$

where $\Phi \to 0$ as $z \to -\infty$.

Under the assumptions stated above the disturbance velocity potential Φ must satisfy Laplace's equation $\nabla^2\Phi = 0$. The solutions of this equation for the perturbation potential are assumed to be harmonic in time, taking the form

$$\Phi = \sum_{n=-\infty}^{\infty} B_n U e^{-in\theta} e^{i\omega t} e^{\frac{|n|z}{\overline{r}}}$$

where B_n are undetermined complex constants and U is the impeller speed at the mean inlet radius $\overline{r}$.

In this linear model, superposition applies. Thus, for convenience the summation is dropped and the various circumferential harmonics are considered independently. In addition, defining the perturbation velocity as a deviation from the mean, the $n = 0$ contribution to the summation is zero and is excluded. The potential velocity perturbations are then

$$V_{zp}' = \frac{\partial\Phi}{\partial z} = \frac{|n|}{\overline{r}} B U e^{-in\theta} e^{i\omega t} e^{\frac{|n|z}{\overline{r}}}$$

$$V_{\theta p}' = \frac{1}{\overline{r}}\frac{\partial\Phi}{\partial\theta} = -\frac{in}{|n|} V_{zp}' \tag{1}$$

where the subscript p indicates the velocity perturbation is of potential origin.

The total pressure perturbation in the region must also satisfy Laplace's equation, $\nabla^2 p_{tp}' = 0$, and takes the form

$$p_{tp}' = C e^{-in\theta} e^{i\omega t} e^{\frac{|n|z}{\bar{r}}}$$

where C is an undetermined complex constant.

The linearized axial momentum equation is used to relate this total pressure perturbation to the velocity perturbation, resulting in the final expression for the inlet total pressure perturbation due to the inlet potential precursor wave.

$$p_{tp}' = -\rho \frac{i\omega\bar{r}}{|n|} V_{zp}'$$

Control Vortical Waveform

The control velocity wave introduced into the impeller inlet is assumed to be simply convected, solenoidal, and purely rotational. Defined in this manner, the wave does not interact with the static pressure field in the inlet [9]. The parameter that is chosen to characterize the control waveform is the axial part of the velocity perturbation, which for convenience is expressed in a similar form to the corresponding potential perturbation

$$V_{zv}' = \frac{|n|}{\bar{r}} D U e^{-in\theta} e^{i\omega\left(t - \frac{z}{\bar{V}}\right)} \tag{2}$$

where the subscript v denotes the vortical nature of the perturbation and D is a complex constant.

The solenoidal condition determines the relationship between the axial velocity perturbation and the corresponding circumferential velocity perturbation

$$V_{\theta v}' = -\frac{\omega\bar{r}}{\bar{V}n} V_{zv}' = -\frac{k}{\bar{\phi}n} V_{zv}' \tag{3}$$

where $\bar{\phi} = \bar{V}/U$ is the mean inlet flow coefficient and $k = (\omega\bar{r})/U$ is the reduced frequency of the disturbance.

The rotational contribution to the total pressure perturbation in the inlet can then be expressed.

$$p_{tv}' = \rho\bar{V} V_{zv}'$$

Impeller Inlet Field

Combining the expressions for the potential precursor and control rotational flow contributions leads to the final expression for the scaled total pressure perturbation at the impeller inlet ($z=0$).

$$\frac{p'_t}{\rho U^2}\Big|_3 = \bar{\Phi}\phi_v' - \frac{ik}{|n|}\phi_p' \tag{4}$$

where ϕ_v' and ϕ_p' are the perturbation inlet flow coefficients.

Exit Boundary Condition

At the diffuser exit, the flow enters a large plenum of constant static pressure. An abrupt expansion is assumed, resulting in the loss of all remaining dynamic pressure. The resulting boundary condition states that the static pressure perturbation vanishes at this point.

$$p_5' = 0 \tag{5}$$

Unsteady Centrifugal Compressor Characteristic

The behavior of the centrifugal compressor under the influence of a small unsteady inlet flow distortion is modelled by introducing a correction to the mean flow pressure rise predicted by the compressor characteristic $\bar{\psi}$. The new characteristic takes the form

$$\psi = \bar{\psi} + \psi' = \frac{\bar{p}_5 - \bar{p}_{t3}}{\rho U^2} + \frac{p_5' - p_{t3}'}{\rho U^2}$$

The perturbation characteristic ψ' is subdivided into the following terms.

$$\psi' = \psi'|_{\Delta\bar{V}} + \frac{1}{\rho U^2}\{\Delta p'|_{\Delta w} + \Delta p'|_{u.,asy.,I} + \Delta p'|_{u.,asy.,D}\} \tag{6}$$

The first term of this expression represents the correction due to the axial velocity perturbation neglecting any unsteady, asymmetric effects and can be represented as

$$\psi'|_{\Delta\overline{V}} = \frac{\partial\overline{\psi}}{\partial\overline{\phi}}\phi' \tag{7}$$

where $\phi' = \phi_v' + \phi_p'$ is the net perturbation flow coefficient at the inlet.

The second term represents the pressure change due to a loss or gain of work done on the fluid because of the circumferential velocity perturbation in the inlet. This term is written as

$$\Delta p'|_{\Delta w} = \frac{\partial\Delta\overline{p}}{\partial\overline{w}} w'$$

where $w' = -V_{\theta 3}'U$.

Introducing the mean flow isentropic efficiency for the compressor and noting that the contribution of $V_{\theta 3}'$ to the inlet dynamic pressure is second order, the pressure change due to this small inlet prewhirl can be expressed as

$$\frac{\Delta p'|_{\Delta w}}{\rho U^2} = -\overline{\eta}_c \frac{V_{\theta 3}'}{U}\left\{1 + \frac{\rho\overline{V}_5{}^2}{2\overline{p}_{t3}} + \frac{\rho U^2\overline{\psi}}{\overline{p}_{t3}}\right\}^{\frac{1}{\kappa}} \tag{8}$$

From Equations 1 and 2, the total circumferential velocity perturbation at the impeller inlet can be related to the axial perturbations and, by defining a diffusion factor $\overline{D}_D|_5$ such that

$$\overline{V}_5 = \overline{D}_D|_5\overline{V}$$

Equation 8 can be expressed in terms of the mean and perturbation inlet flow coefficients.

$$\frac{\Delta p'|_{\Delta w}}{\rho U^2} = \overline{\eta}_c\left\{1 + \frac{\rho U^2}{\overline{p}_{t3}}\left(\frac{\overline{D}_D^2|_5\overline{\phi}^2}{2} + \overline{\psi}\right)\right\}^{\frac{1}{\kappa}}\left\{\frac{in}{|n|}\phi_p' + \frac{k}{\overline{\phi}n}\phi_v'\right\} \tag{9}$$

The last two terms in Equation 6 represent the pressure rise due to unsteady, asymmetric effects in the impeller and diffuser respectively. The basis for these corrections is that the unsteady, asymmetric pressure rise is balanced by the acceleration of fluid in the compressor. It is recognized that even when operating with an axisymmetric inlet flow, the separated flow in the impeller and jet-wake mixing in the vaneless space result in asymmetric flow in the compressor. It is suggested here, however, that distortion occurring in these regions would be on the scale of the impeller blade or diffuser vane spacing, while the perturbations of interest are of circumferential scale. Hence, while higher harmonic distortions may be generated, the essential character of the lower-order harmonic waves of interest here should be preserved. The pressure correction for the impeller is considered first, followed by that of the diffuser.

Impeller Correction

In a reference frame fixed to the impeller

$$-\frac{1}{\rho}\nabla^{*}p|_{u.,asy.,\mathrm{I}}=\frac{W_{\theta}}{r}\frac{\partial\overline{\mathbf{W}}}{\partial\theta^{*}}+\frac{\partial\overline{\mathbf{W}}}{\partial t} \tag{10}$$

where $\overline{\mathbf{W}}$ is the relative flow velocity and $\theta^{*}=\theta-\omega_{\mathrm{I}}t$ is the impeller-relative circumferential coordinate.

The pressure change across the impeller due to the distortion is calculated by integrating the above equation. To facilitate this, a mean velocity and flow path are defined such that

$$\overline{\mathbf{W}}_{m}=V_{r}\,\mathbf{i}_{r}+W_{\theta^{*}}\,\mathbf{i}_{\theta^{*}}+V_{z}\,\mathbf{i}_{z}$$

$$\mathbf{i}_{m}=\cos\gamma_{1\mathrm{I}}\,\mathbf{i}_{r}+\cos\gamma_{2\mathrm{I}}\,\mathbf{i}_{\theta^{*}}+\cos\gamma_{3\mathrm{I}}\,\mathbf{i}_{z}$$

Note that for the impeller $W_{\theta}=V_{\theta}-\omega_{\mathrm{I}}\,r$.

Assuming that the mean flow path is not altered by the unsteady flow perturbations, Equation 10 can be integrated to give an expression for the pressure change.

$$\frac{\Delta p}{\rho}|_{u.,asy.,\mathrm{I}}=-\int_{s^{*}}\left(\frac{W_{\theta}}{r}\frac{\partial W_{m}}{\partial\theta^{*}}+\frac{\partial W_{m}}{\partial t}\right)ds^{*}$$

Linearization results in the following integral equation.

$$\frac{\Delta p'}{\rho}|_{u.,asy.,\mathrm{I}}=-\int_{s^{*}}\left(\frac{\overline{W}_{\theta}}{r}\frac{\partial W_{m}'}{\partial\theta^{*}}+\frac{\partial W_{m}'}{\partial t}\right)ds^{*} \tag{11}$$

It is necessary to transform this integral into a function of the inlet waveforms. This is accomplished by defining a relative diffusion factor $\overline{D}_{\mathrm{I}}$.

$$\overline{W}_{m}(s^{*})=\overline{D}_{\mathrm{I}}\overline{W}_{m}|_{3}$$

and

$$W_{m}'(s^{*})=\overline{D}_{\mathrm{I}}W_{m}'|_{3}$$

Two assumptions are implicit in these expressions. The first is that the perturbation velocity diffusion is identical to that of the mean flow. The second is that the solenoidal nature of the vortical perturbation is lost at the impeller inlet and, from continuity considerations, the waveform at the impeller exit is in phase with that at the inlet.

Substituting these expressions into Equation 11 leads to

$$\frac{\Delta p'}{\rho}\Big|_{u.,asy.,I} = -\ \overline{W}_m \frac{\partial W_m'}{\partial \theta^*}\Big|_3 \int_3^4 \frac{\cos\gamma_{2I}}{r}\, \overline{D}_I^{\,2}\, ds^* \ -\ \frac{\partial W_m'}{\partial t}\Big|_3 \int_3^4 \overline{D}_I\, ds^* \quad (12)$$

At this point it is useful to relate the derivatives of the mean flow perturbations to the inlet absolute velocities. An analysis of the inlet velocity triangle results in the following expressions.

At impeller inlet

$$\overline{W}_m \frac{\partial W_m'}{\partial \theta^*} = \overline{V} \frac{\partial V_z'}{\partial \theta^*} - U \frac{\partial V_\theta'}{\partial \theta^*} \quad (13)$$

and

$$\frac{\partial W_m'}{\partial t} = P \frac{\partial V_z'}{\partial t} - S \frac{\partial V_\theta'}{\partial t} \quad (14)$$

where $P \equiv \dfrac{\overline{V}}{\overline{W}_m} = \dfrac{\overline{\phi}}{\sqrt{\overline{\phi}^2+1}}$ and $S \equiv \dfrac{U}{\overline{W}_m} = \dfrac{1}{\sqrt{\overline{\phi}^2+1}}$.

These expressions, along with Equations 1 and 3 are substituted into Equation 12 and, after dividing by U^2 and moving to a stationary reference frame give the final expression for the unsteady, asymmetric pressure correction across the impeller.

$$\begin{aligned}\frac{\Delta p'}{\rho U^2}\Big|_{u.,asy.,I} = &-\ \Gamma_{2I}\left\{\frac{in}{|n|}\frac{\partial \phi_p'}{\partial \theta} + \frac{k}{\overline{\phi} n}\frac{\partial \phi_v'}{\partial \theta} + \overline{\phi}\frac{\partial \phi'}{\partial \theta}\right\} \\ &-\ \frac{\overline{r}\Gamma_{1I}}{U}\left\{P\frac{\partial \phi'}{\partial t} + S\left(\frac{in}{|n|}\frac{\partial \phi_p'}{\partial t} + \frac{k}{\overline{\phi} n}\frac{\partial \phi_v'}{\partial t}\right)\right\} \\ &-\ \Gamma_{1I}\left\{P\frac{\partial \phi'}{\partial \theta} + S\left(\frac{in}{|n|}\frac{\partial \phi_p'}{\partial \theta} + \frac{k}{\overline{\phi} n}\frac{\partial \phi_v'}{\partial \theta}\right)\right\} \end{aligned} \quad (15)$$

where $\Gamma_{2I} = \displaystyle\int_3^4 \frac{\cos\gamma_{2I}}{r}\, \overline{D}_I^{\,2}\, ds^*$ and $\Gamma_{1I} = \dfrac{1}{\overline{r}}\displaystyle\int_3^4 \overline{D}_I\, ds^*$.

Diffuser Correction

Proceeding in a similar manner for the diffuser, a mean velocity and flow path are defined in the stationary reference frame.

$$\vec{\mathbf{V}}_m = V_r\,\mathbf{j}_r + V_\theta\,\mathbf{j}_\theta + V_z\,\mathbf{j}_z$$

$$\mathbf{j}_m = \cos\gamma_{1D}\,\mathbf{j}_r + \cos\gamma_{2D}\,\mathbf{j}_\theta + \cos\gamma_{3D}\,\mathbf{j}_z$$

A diffusion factor is defined such that

$$\overline{V}_m(s) = \overline{D}_D \overline{V}|_3$$

and

$$V_m'(s) = \overline{D}_D V_z'|_3$$

The pressure correction for the diffuser thus becomes

$$\frac{\Delta p}{\rho U^2}\Big|_{u.,asy.,D} = -\,\Gamma_{2D}\,\overline{\phi}\,\frac{\partial\phi'}{\partial\theta} - \frac{\overline{r}\,\Gamma_{1D}}{U}\frac{\partial\phi'}{\partial t} \tag{16}$$

where $\Gamma_{2D} = \int_4^5 \frac{\cos\gamma_{2D}}{r}\overline{D}_D{}^2\,ds$ and $\Gamma_{1D} = \frac{1}{\overline{r}}\int_4^5 \overline{D}_D\,ds$.

Final Expression for the Unsteady Characteristic

The final expression for ψ' is given by the summation of the terms represented in Equation 6. For a fixed compressor geometry specified by the Γ diffusion parameters, and for a steady flow performance specified by $\overline{\psi}(\overline{\phi})$ and $\overline{\eta}_c(\overline{\phi})$, this unsteady asymmetric characteristic may be represented as

$$\psi' = \psi'(\overline{\phi}, k, \phi_v', \phi_p')$$

The Neutral Stability Condition

The unsteady, asymmetric compressor characteristic ψ' is used to match the inlet total pressure perturbation described by Equation 4 and the extinction condition on the static pressure perturbation given by Equation 5.

$$\bar{\phi}\phi_v' - \frac{ik}{|n|}\phi_p' + \psi'(\bar{\phi}, k, \phi_v', \phi_p') = 0 \tag{17}$$

For a specified unsteady waveform at the inlet, represented by ϕ_v' and ϕ_p', this complex equation can be solved for $\bar{\phi}$ and k. These values represent the mean inlet flow coefficient and reduced frequency at which a neutrally stable potential velocity disturbance of wavenumber n would exist in the compressor inlet. As such, they form a stability boundary for the growth of the potential wave and define the stall point of the compressor. At flow coefficients greater than the critical value of $\bar{\phi}$, denoted by $\bar{\phi}_{stall}$, the potential wave will have a negative growth rate. As the mass flow is reduced to values lower than $\bar{\phi}_{stall}$, the growth rate will change sign and the wave will develop into a stall condition.

Effectiveness of the Control

At this point, it is useful to express the potential perturbation flow coefficient ϕ'_p in terms of a complex amplitude A in the following manner.

$$\phi_p' = B\frac{|n|}{\bar{r}}e^{-in\theta}e^{i\omega t} \equiv Ae^{-in\theta}e^{i\omega t}$$

where $A \equiv \frac{|n|}{\bar{r}}B$.

The complex quantity A represents the phase and magnitude of the potential axial velocity perturbation, scaled with U, at the impeller inlet. A complex quantity ζ is defined such that

$$\zeta \equiv \frac{\phi_v'}{\phi_p'} = |\zeta|e^{i\beta}$$

The parameter ζ is then the complex control gain for a measured input ϕ'_p and a controller response $\phi_v' = \zeta\phi'_p$. To obtain the optimum value of ζ, the expressions

$$\phi_p' = Ae^{-in\theta}e^{i\omega t}$$

$$\phi_v' = \zeta Ae^{-in\theta}e^{i\omega t}$$

$$\phi' = \phi_p' + \phi_v' = (A + \zeta A)e^{-in\theta}e^{i\omega t}$$

are substituted into Equation 17, along with the individual terms of the unsteady characteristic ψ'. The resulting expression takes the form of

$$\bar{\phi}\zeta - \frac{ik}{|n|} + \frac{\partial\bar{\psi}}{\partial\bar{\phi}}(1+\zeta) + \bar{\eta}_c\left[1+\frac{\rho U^2}{\bar{p}_{t3}}\left(\frac{\bar{D}_D^2|_5\bar{\phi}^2}{2}+\bar{\psi}\right)\right]^{\frac{1}{\kappa}}\left[\frac{in}{|n|}+\frac{k}{\bar{\phi}n}\zeta\right]$$

$$- \Gamma_{2I}\left\{|n| - \frac{ik}{\bar{\phi}}\zeta - in\bar{\phi}(1+\zeta)\right\} - \Gamma_{1I}\left\{iP(k-n)(1+\zeta) - \frac{Snk}{|n|} + \frac{ik^2}{\bar{\phi}n}\zeta + S|n| - \frac{iSk}{\bar{\phi}}\zeta\right\}$$

$$+ (\Gamma_{2D}\, in\bar{\phi} - \Gamma_{1D}\, ik)(1+\zeta) = 0 \qquad (18)$$

Equation 18 is solved to determine the stall point $\bar{\phi}_{stall}$ and corresponding reduced frequency k as a function of the magnitude of ζ and its phase angle β. Comparison of this value to the stall point when no control waveform exists ($\zeta = 0$) then provides a measure of the effectiveness of the control.

Results and Discussion

The control model is investigated by estimating the Γ diffusion parameters for the compressor geometry given in Figure 1. For the purposes of numerical evaluation, the steady characteristic and efficiency curves in Figure 2 are represented by a cubic polynomial function. To characterize the damping achieved by the vortical waveform, the control effectiveness $\Delta\bar{\phi}\%$ is defined as the difference between the mean flow coefficient at neutral stability $\bar{\phi}_{stall}$ for a given controller gain magnitude $|\zeta|$ and phase β and that for the no-control case ($\zeta = 0$).

Figures 3 and 4 show the results obtained for the first two circumferential harmonics, $n = 1$ and $n = 2$, using the baseline geometry. The results present the percentage change in mean flow coefficient at neutral stability $\Delta\bar{\phi}\%$ that results from the control introduction as a function of controller phase β and gain magnitude $|\zeta|$. Negative values of $\Delta\bar{\phi}\%$ represent a favorable shift in the stall point, increasing the stable operating range of the centrifugal compressor. The results indicate that a vortical control waveform amplitude of 40% of the potential perturbation amplitude can achieve reductions in the stall point $\bar{\phi}_{stall}$ on the order of 12% for the $n = 1$ wave and 21% for the $n = 2$ wave. It can also be seen that if the phase angle is not kept near the optimum value, the controller will readily destabilize the compressor. If the compressor mass flow is decreased from a value in the stable region of the characteristic, the first harmonic becomes unstable at $\overline{\phi}_{stall} = 0.36$. If this wave is successfully suppressed the second harmonic will become unstable at $\overline{\phi}_{stall} = 0.34$ and the controller will have to suppress this wave as well.

The final set of results, presented in Figure 5, provide a more practical representation of the results presented in Figure 3. In this case, the precursor to the instability is assumed to be detected with static pressure sensors in the inlet. Since it is this pressure wave, rather that the actual potential inlet flow coefficient which is measured, a comparison of the control waveform magnitude and phase relative to the pressure wave is shown. In the inlet flow field, the static pressure perturbation can be related to the axial velocity perturbation through the linear momentum equation, giving

$$\frac{p'}{\rho U^2} = -\left(\overline{\phi} + \frac{ik}{|n|}\right)\phi_p'$$

The curve represents the effect of the control vortical waveform to scaled pressure perturbation ratio $\xi = \phi_v'\rho U^2/p'|_3$ when the compressor is operating at the no-control stall point $\overline{\phi} = .36$.

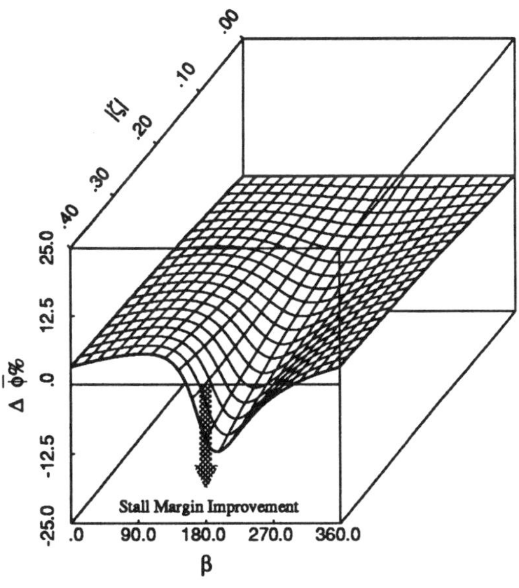

Figure 3. Change in mean flow coefficient at stall as a function of control gain magnitude and phase angle, n=1.

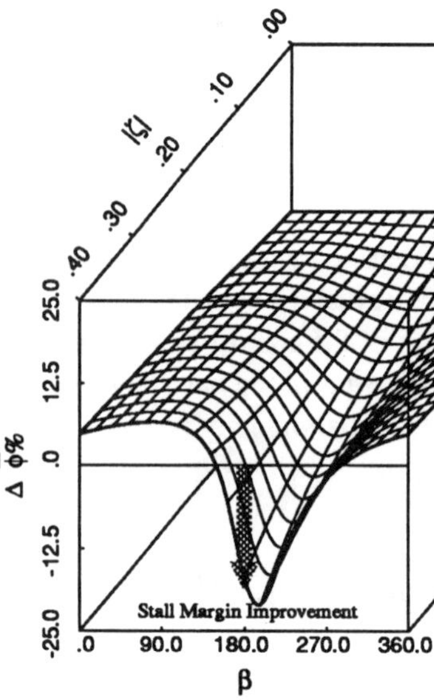

Figure 4. Change in mean flow coefficient at stall as a function of control gain magnitude and phase angle, n=2.

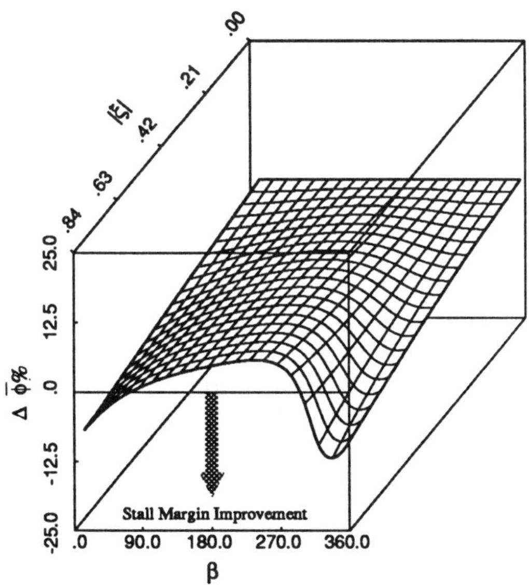

Figure 5. Change in mean flow coefficient at stall as a function of control to static pressure perturbation ratio, n=1.

Summary and Conclusions

A mathematical model has been developed to analyze the suppression of instabilities in centrifugal compressors by the introduction of an unsteady vortical velocity field into the impeller inlet. In this model, the precursor to the instability is a weak rotating potential velocity perturbation in the inlet flow field that eventually develops into a finite disturbance. To suppress the growth of this potential disturbance, a rotating control vortical velocity disturbance is introduced into the impeller inlet flow. The effectiveness of this control is analyzed by matching the perturbation pressure in the compressor inlet and exit flow fields with a model for the unsteady behavior of the compressor.

The trends observed by applying this model to the Purdue Low Speed Research Centrifugal Compressor indicate that the use of an upstream vortical velocity perturbation to suppress instability has merit for the centrifugal compressor with vaned diffuser. It is also clear that any control system will be able to dramatically destabilize the system should it react to a disturbance with an improper phase angle.

Acknowledgements

Support of this research by the Army Aviation Systems Command, NASA Lewis Research Center, Lawrence F. Schumann technical monitor, is most gratefully acknowledged.

References

1. McDougall, N.M., Cumpsty, N.A., and Hynes, T.P., "Stall Inception in Axial Compressors," *ASME Journal of Turbomachinery*, Vol. 112, January, 1990, pp. 116-125.
2. Garnier, V.H., Epstein, A.H., Greitzer, E.M., "Rotating Stall Anticipation and Initiation in Axial Compressors," *ASME Paper 90-GT-156*, 1990.
3. Epstein, A.H., Ffowcs Williams, J.E., and Greitzer, E.M., "Active Suppression of Aerodynamic Instabilities in Turbomachines," *AIAA Journal of Propulsion and Power*, Vol. 5, No. 2, March-April 1989, pp. 204-211.
4. Paduano, J., Epstein, A.H., Valavani, L., Longley, J.P., Greitzer, E.M., and Guenette, G.R., "Active Control of Rotating Stall in a Low Speed Axial Compressor," submitted to 1991 ASME Gas Turbine Conference.
5. Moore, F.K., "Weak Rotating Flow Disturbances in a Centrifugal Compressor with a Vaneless Diffuser," *ASME Journal of Turbomachinery*, Vol. 111, October 1989, pp. 442-449.
6. Pinsley, J.E., Guenette, G.R., Epstein, A.H., and Greitzer, E.M., "Active Stabilization of Centrifugal Compressor Surge," *ASME Paper 90-GT-123*, 1990.
7. Ffowcs Williams, J.E., and Graham, W.R., "An Engine Demonstration of Active Surge Control," *ASME Paper 90-GT-224*, 1990.
8. Koff, S.G., and Greitzer, E.M., "Stalled Flow Performance for Axial Compressors I: Axisymmetric Characteristic," *ASME Paper 84-GT-93*, 1984.
9. Goldstein, M.E., "Unsteady Vortical and Entropic Distortions of Potential Flows Round Arbitrary Obstacles," *Journal of Fluid Mechanics*, Vol. 89, 1978, p433-468.
10. Greitzer, E. M., "The Stability of Pumping Systems," *ASME Journal of Fluids Engineering*, Vol. 103, June 1981, pp. 193-242.

Computational Investigation of Airfoil Stall Flutter

J.D. Clarkson[1]
J.A. Ekaterinaris[2]
M.F. Platzer[3]

ABSTRACT

A fully implicit factorized two–dimensional Navier–Stokes flow solver has been developed and applied to the problem of predicting subsonic airfoil flutter in the light stall regime. The inviscid fluxes are evaluated with a central difference ADI scheme and fourth and second order numerical dissipation is used to obtain oscillation–free solutions. The performance of algebraic and one–equation turbulence models in predicting separated flow is explored for computing subsonic high Reynolds number steady flow and unsteady flows over an oscillating NACA 0012 airfoil. Comparisons of the computed results with available experimental data indicate that even though the lift response is fairly well predicted, the computation of the pitching moment hysteresis loops is very sensitive to turbulence modeling. Results computed with several current models are in good agreement whenever the steady stall angle is exceeded only slightly. However, they fail to capture the vortex shedding process leading to the onset of stall flutter.

I. INTRODUCTION

The phenomenon of unsteady flow separation on airfoils which execute a time–dependent motion is usually referred to as the dynamic stall phenomenon. It occurs on aircraft wings, helicopter and propeller blades, on wind turbines and in turbomachinery bladings. Therefore, its understanding and prediction has wide–ranging practical importance. A review of dynamic stall progress in analysis and prediction has recently been given by Carr[1]. As pointed out by McCroskey and Pucci[2] it is useful to distinguish between deep and light stall, the former occurring during the large–amplitude pitching excursions of the helicopter rotor blade whereas dynamic stall on propeller blades usually begins with an incipient torsional motion near the static stall point. Carta and Lorber[3] recently conducted an experimental study of the aerodynamics of incipient torsional stall flutter and found that small–amplitude oscillations near stall may be extremely unstable. The development of "rational" stall flutter analyses based on

[1]Lt, U.S. Navy, Graduate Student, Naval Postgraduate School, Dept. of Aeronautics and Astronautics, Monterey, CA 93943.
[2]Adjunct Research Professor, Navy-NASA Joint Institute of Aeronautics, NASA Ames Research Center, MS 258-1, Moffett Field, CA 94035.
[3]Professor of Aeronautics and Astronautics, Naval Postgraduate School, Department of Aeronautics and Astronautics, Monterey, CA 93943.

solutions of the viscous flow equations in contrast to semi–empirical methods is still in its infancy. Sisto[4] presented a review of the state–of–the–art of turbomachinery stall flutter and, most recently, presented an approach which combines the inviscid vortex method with unsteady boundary layer calculations[5]. Jang et al[6] developed a viscous– inviscid interaction approach for the computation of unsteady blade boundary layers and the onset of flow separation. Such approaches have the advantage of relative computational efficiency, but more work is required to establish their range of validity. A second "rational", albeit computationally more extensive, approach is offered by the numerical solution of the unsteady Navier–Stokes equations. Recent progress by Visbal[7,8], Ekaterinaris[9], Currier and Fung[10], Dindar et al[11], Grohsmeyer et al[12], Rumsey and Anderson[13] in the computation of dynamic stall using the compressible Navier–Stokes equations suggests to explore the practicality of using this approach to predict torsional stall flutter. The problem of small amplitude incipient torsional stall flutter appears to be more suitable for such calculations because the vertical extent of the viscous zone tends to remain of the order of the airfoil thickness. On the other hand, as already noted by Dindar et al[11], the problem of proper turbulence modeling of the unsteady vortical flows generated during stall flutter introduces major uncertainties. Therefore, in this paper calculations are presented for small–amplitude pitch oscillations of a NACA 0012 airfoil near static stall in a low speed high Reynolds number flow using the compressible Navier–Stokes equations in combination with various turbulence models. This approach and the results and conclusions are described in the following sections.

II. GOVERNING EQUATIONS

The general form of the conservation equations for an inertial reference frame[14] is used and the finite difference discretization is employed. All quantities are discretized at every node in physical space computational mesh with partial derivatives approximated by finite differences. In order to facilitate the numerical implementation for complex geometry domains, the physical space (x, z) points are mapped to the computational domain (ξ, ζ) points resulting from the following generalized coordinate transformation $x = x(\xi, \zeta, t), z = z(\xi, \zeta, t)$. The differential form of the two–dimensional conservation law form for an arbitrary coordinate system ξ, ζ when the thin–layer approximation[15] is applied can be written as

$$\partial_t \hat{\mathbf{Q}} + \partial_\xi \hat{\mathbf{F}} + \partial_\zeta \hat{\mathbf{G}} = Re^{-1} \partial_\zeta \hat{\mathbf{S}} \tag{1}$$

where $\hat{\mathbf{Q}}, \hat{\mathbf{F}}, \hat{\mathbf{G}}$ and $\hat{\mathbf{S}}$ are

$$\hat{\mathbf{Q}} = \frac{1}{J}\begin{pmatrix} \rho \\ \rho u \\ \rho w \\ e \end{pmatrix}, \quad \hat{\mathbf{F}} = \frac{1}{J}\begin{pmatrix} \rho U \\ \rho u U + \xi_x p \\ \rho w U + \xi_z p \\ (e+p)U - \xi_t p \end{pmatrix}, \quad \hat{\mathbf{G}} = \frac{1}{J}\begin{pmatrix} \rho W \\ \rho u W + \zeta_x p \\ \rho w W + \zeta_z p \\ (e+p)W - \zeta_t p \end{pmatrix}$$

$$\hat{\mathbf{S}} = \frac{1}{J}\begin{pmatrix} 0 \\ \mu m_1 u_\zeta + (\mu/3) m_2 \zeta_x \\ \mu m_1 w_\zeta + (\mu/3) m_2 \zeta_z \\ \mu m_1 m_3 + (\mu/3) m_2 m_4 \end{pmatrix}$$

Here

$$\begin{aligned} m_1 &= \zeta_x^2 + \zeta_z^2 \\ m_2 &= \zeta_x u_\zeta + \zeta_z w_\zeta \\ m_3 &= \left(\frac{1}{2}\right)\frac{\partial}{\partial \zeta}(u^2 + w^2)/2 + \frac{1}{Pr(\gamma - 1)}\left(\frac{\partial a^2}{\partial \zeta}\right) \\ m_4 &= \zeta_x u + \zeta_z w \end{aligned}$$

and U and W are the contravariant velocity components given by

$$\begin{aligned} U &= u\xi_x + w\xi_z + \xi_t \\ W &= u\zeta_x + w\zeta_z + \zeta_t \end{aligned}$$

In the above equations all geometrical dimensions are normalized with the airfoil chord length c; ρ is the density normalized with the free–stream density ρ_∞; u and w are the Cartesian velocity components of the physical domain normalized with the free–stream speed of sound a_∞; e is the total energy per unit volume normalized with $\rho_\infty a_\infty^2$; Pr is the Prandtl number; and K is the thermal conductivity. The pressure is related to density and total energy through the equation of state for an ideal gas, $p = (\gamma - 1)\left[e - \rho(u^2 + w^2)/2\right]$.

III. THE NUMERICAL SCHEME

The following scheme is used for the time integration

$$\begin{aligned} \Delta\hat{\mathbf{Q}}^n = &-\frac{\alpha \Delta t}{1+\beta}\frac{\partial}{\partial t}\Delta\hat{\mathbf{Q}}^n + \frac{\Delta t}{1+\beta}\frac{\partial}{\partial t}\hat{\mathbf{Q}}^n + \\ &\frac{\beta}{1+\beta}\frac{\partial}{\partial t}\Delta\hat{\mathbf{Q}}^{n-1} + O\left[\left(\alpha - \frac{1}{2} - \beta\right)\Delta t + \Delta t^3\right] \end{aligned} \qquad (2)$$

here $\Delta\hat{\mathbf{Q}}^n = \hat{\mathbf{Q}}^{n+1} - \hat{\mathbf{Q}}^n$, and with different choices of the coefficients α and β different time integration schemes are obtained. For $\alpha = 1$, $\beta = 0$ the implicit Euler scheme is obtained, $\alpha = .5$, $\beta = 0$ yield the trapezoidal rule, and with $\alpha = 1$, $\beta = .5$ a three point backward scheme is obtained. After linearization and space discretization with central differences the integration is performed by the following Beam– Warming[16], factorized, implicit, iterative numerical algorithm

$$\begin{aligned} &\left\{I + \frac{\Delta t}{2}\left[\delta_\xi \hat{\mathbf{A}}_{i,k}^n + (D_{impl})_\xi\right]\right\}^p \\ &\cdot\left\{I + \frac{\Delta t}{2}\left[\delta_\zeta \hat{\mathbf{B}}_{i,k}^n - Re^{-1}\delta_\zeta \hat{\mathbf{M}}_{i,k}^n + (D_{impl})_\zeta\right]\right\}^p \Delta\hat{Q}_{i,k}^p = \\ &-\Delta t\left\{\hat{Q}_{i,k}^p - \hat{Q}_{i,k}^n + \left[\delta_\xi \hat{F}_{i,k}^n + \delta_\zeta \hat{G}_{i,k}^n\right] - Re^{-1}\delta_\zeta S_{i,k}^n + \epsilon_{expl} D_{i,k}^n\right\} \end{aligned} \qquad (3)$$

Here δ is the central difference operator, $\hat{F}_{i,k}$, $\hat{G}_{i,k}$ are the inviscid fluxes and and $\hat{S}_{i,k}$ viscous flux. In Eq. (3) the quantities $\hat{A}, \hat{B}$ are the inviscid flux Jacobians and $\hat{M}$ is the viscous flux Jacobian matrices obtained from the linearization. A Jameson–type[17,18] blended second and fourth order numerical dissipation based on the computed pressure field is used to suppress high frequency numerical oscillations. For subsonic shock free solutions only the fourth order dissipation is used, while for transonic solutions the second order dissipation is activated in the vicinity of shocks where the pressure has steep gradients. In addition, an implicit fourth–order smoothing is used on the left hand side for numerical stability. Both the implicit and explicit dissipation are scaled by the spectral radius and with the time step. Scaling with the time step makes the steady–state independent of time–step. However, since the added dissipation terms modify the original partial differential equation the dissipation coefficients should be kept as small as possible. Elimination of the error introduced by the linearization and approximate factorization may be obtained by performing Newton subiterations to convergence within each time step. In Eq. (3) the approximation to $\tilde{Q}^{n+1}$ at each subiteration is the quantity $\tilde{Q}^p$. When $p \geq 2$, during a given level of subiteration to convergence, $\tilde{Q}^p = \tilde{Q}^{n+1}$, but when $p = 1$ and no subiterations are performed, then $\tilde{Q}^p = \tilde{Q}^n$, and $\tilde{Q}^{p+1} = \tilde{Q}^{n+1}$. By subiterating to convergence, linearization and factorization errors can be eliminated during the iteration process, because the left hand side of Eq. (3) can be driven to zero at each time step.

IV. BOUNDARY CONDITIONS

All flows were computed at subsonic free stream speeds. For subsonic inflow outflow boundaries the flow variables are evaluated using zero order Riemann invariant extrapolation. At the inflow boundary there are one incoming and three outgoing characteristics. Therefore, three variables, the density, the normal velocity w_1, and the pressure are specified and the fourth variable the axial velocity u_1 is extrapolated from the interior. The inflow boundary conditions are given by

$$\begin{aligned} \rho_1 &= \left(\frac{a_1^2}{\gamma s_1}\right)^{(1/\gamma-1)} \quad , \quad s_1 = \left(\frac{p_\infty}{\rho_\infty^\gamma}\right), \quad a_1 = \frac{(\gamma-1)}{4}\left(R_1^+ - R_2^-\right) \\ u_1 &= (R_1^+ + R_2^-)/2 \\ w_1 &= w_\infty \\ p_1 &= \left(\frac{\rho_1 a_1^2}{\gamma}\right) \end{aligned} \tag{4}$$

where R_1^+, R_2^- are the incoming and outgoing Riemann invariants given by

$$R_1^+ = u_\infty + 2a_\infty/(\gamma-1), \quad R_2^- = u_2 - 2a_2/(\gamma-1)$$

At the outflow boundary there are one incoming and three outgoing characteristics and only one quantity, the pressure, is specified while the others are extrapolated from the interior. For the density and normal velocity simple first–order extrapolation is used, and the axial outflow velocity is obtained from the zero order outgoing Riemann invariant. The outflow boundary conditions are given by

$$\begin{aligned} \rho_1 &= \rho_2 \\ u_1 &= R_1^+ - 2a_1/(\gamma - 1), \quad a_1 = \sqrt{\gamma p_1/\rho_1} \\ w_1 &= w_2 \\ p_1 &= p_\infty \end{aligned} \tag{5}$$

On the body surface the nonslip condition is applied for the velocities and the density and the pressure are obtained from the interior by simple extrapolation. For C–type grids used in this study averaging of the flow variables at the wake cut is used.

V. TURBULENCE MODELS

The turbulence models used for the computation of unsteady separated flows over airfoils undergoing large amplitude oscillatory motion are presented in this section. Only the most important features of these models necessary for the interpretation of the results are discussed here. Further details of the turbulence models can be found in the original papers[15,19,20] and references therein. Two algebraic eddy viscosity models, namely, the well known Baldwin–Lomax algebraic eddy viscosity model[15] and the algebraic RNG–based model[19] are implemented for closure of the compressible Reynolds–averaged equations. In addition, the Johnson–King[20] model based on the solution of an ordinary differential equation for the maximum shear stress, which gave very good predictions of shock separated flows, is also implemented for the computation of unsteady separated flows. The Johnson–King model also gave better preditions[11] of unsteady flow compared to the Baldwin–Lomax and Cebeci–Smith[21] models.

Baldwin–Lomax model

The Baldwin–Lomax model is a two–layer, inner and outer eddy viscosity model for the computation of separated flows. This model is based on the Cebeci–Smith[21] turbulence model formulation and includes modifications that allow computation of the eddy viscosity only from the mean flowfield and do not require to determine the boundary layer edge. A crossover distance from the wall y_c for each streamwise station is defined for which the values of the inner and outer eddy viscosities are the same. For the presentation of the turbulence models the normal to the body is referred to as y. The inner eddy viscosity for $y \leq y_c$ is given by

$$(\mu_t)_{inner} = \rho l^2 |\omega| \qquad y \leq y_c \tag{6}$$

where $|\omega|$ is the vorticity magnitude, $l = ky\left[1 - e^{-y^+/A^+}\right]$, and $y^+ = \rho_w u_t y/\mu_w$. In the outer region the eddy viscosity is given by

$$(\mu_t)_{outer} = \kappa\rho C_{CP} F(y)_{wake} F(y)_{Kleb} \qquad y_c \leq y \tag{7}$$

where

$$F(y)_{wake} = min\left\{y_{max}F(y)_{max} \quad , \quad C_{WK} y_{max} u_{diff}^2 / F(y)_{max}\right\}$$

and

$$F(y) = y|\omega|\left[1 - e^{-y^+/A^+}\right]$$

The exponential term is set equal to zero for the wakes and $F(y)_{max} = max F(y)$ and y_{max} is the value of y where $F(y)_{max}$ occurs. The Klebanoff inermittency factor is given by $F(y)_{Kleb} = \left[1 + 5.5\left(C_{Kleb}/y_{max}\right)^6\right]^{-1}$. All the constants for this model are given in Ref. 15. The advantage of this model is that it models completely the turbulent flow over an airfoil, is easy to implement, does not degrade the convergence of the solution, and requires minimal computational effort.

Algebraic RNG–based model

Recently an algebraic eddy viscosity, as well as a two–equation $\kappa - \epsilon$ model based on the Renormalization Group (RNG) Theory of turbulence[19] were proposed for the closure of the Reynolds averaged Navier–Stokes equations. The algebraic model, although free from uncertainties related to the experimental determination of empirical modeling constants, still requires specification of an integral length–scale of turbulence which reduces the generality of the model. Here the integral scale is assumed proportional to the distance from the wall and the eddy viscosity for the RNG–based algebraic turbulence model is obtained as in Ref. 22 from the following formula

$$\nu = \nu_l\left[1. + H\left\{\frac{\hat{a}}{\nu_l^3}\phi\left(\frac{1}{y} + \frac{1}{0.225\delta}\right)^{-4} - C_c\right\}\right]^{\frac{1}{3}} \tag{8}$$

where $\nu = \nu_t + \nu_l$, H is the Heaviside step function and ϕ is the dissipation function $\phi = \tau_{ij}\left(\partial u_i/\partial x_j\right)$. The parameter $\hat{a} = 0.0192$ or $\hat{a} = 0.0256$, depending whether the Von Karman constant is taken equal to the $\kappa = 0.372$ as is predicted by the RNG theory, or to the standard value $\kappa = 0.4$, and the parameter $C_c = 75$. The RNG–based turbulence model is applied only for the suction surface separated flow region while the pressure side and the wake regions are computed with the Baldwin–Lomax model.

Johnson–King model

The Johnson–King model[20] takes into account the convection and diffusion effects on the Reynolds shear stress $\overline{-u'w'}$ in the streamwise direction. The eddy viscosity is given by

$$\nu_t = \nu_{t_o}\left[1 - exp(-\frac{\nu_{t_i}}{\nu_{t_o}})\right] \tag{9}$$

where ν_{t_i}, ν_{t_o} describe the eddy viscosity variation in the inner and outer part of the boundary layer. The inner eddy viscosity is computed as

$$\begin{aligned} \nu_{t_i} &= D^2 \kappa y \sqrt{(\overline{-u'w'})_{max}} \\ D^2 &= 1 - e^{-(y/A^+)} \end{aligned} \tag{10}$$

where the constant $A^+ = 15$. The outer eddy viscosity is given by

$$\nu_{t_o} = \sigma(x) \cdot [0.168 U_e \delta^* \gamma] \tag{11}$$

where γ is the Klebanoff intermittency function $\gamma = \left[1 + 5.5(\frac{y}{\delta})^6\right]^{-1}$ and $\sigma(x)$ is the solution of an ordinary differential equation which describes the development of $\overline{-u'w'}|_{max}$ along the path of the maximum shear stress. This model accounts for the effects of convection and diffusion on the Reynolds shear stress development through the solution of the following ordinary differential equation

$$\frac{dg}{dx} = \frac{a_1}{2\bar{u}_m L_m}\left\{\left(1 - \frac{g}{g_{eq}}\right) + \frac{C_{dif} L_m}{a_1 \delta\left[0.7 - (\frac{y}{\delta})_m\right]}|1 - \sqrt{\frac{\nu_{t_o}}{\nu_{t_o,eq}}}|\right\} \tag{12}$$

here C_{dif} and a_1 are modeling constants $\bar{u}_m$ is the maximum average mean velocity and $g = \left[\overline{-u'w'}_{max}\right]^{-1/2}, g_{eq} = \left[(\overline{-u'w'})_{max,eq}\right]^{-1/2}$ where L_m is the dissipation length evaluated as

$$\begin{aligned} L_m &= 0.40 y & y_m/\delta \leq 0.225 \\ L_m &= 0.09\delta & y_m/\delta \geq 0.225 \end{aligned} \tag{13}$$

the equilibrium shear stress g_{eq} in Eq. 13 is determined from the following equilibrium eddy viscosity distribution

$$\begin{aligned} \nu_{t,eq} &= \nu_{t_o,eq}\left[1 - exp(-\frac{\nu_{t_i,eq}}{\nu_{t_o,eq}})\right] \\ \nu_{t_i,eq} &= D^2 \kappa y \sqrt{(\overline{-u'w'})_{m,eq}} \\ \nu_{t_o,eq} &= 0.0168 U_e \delta^* \gamma \end{aligned} \tag{14}$$

An implicit Euler method is used for the numerical solution of Eq. 12, and the maximum shear stress at each iteration level is updated as follows

$$\sigma(x)^{n+1} = \sigma(x)^n \frac{\nu_{t_o}^{n+1}}{\nu_{t_o}^n} \tag{15}$$

Solutions with the Johnson–King turbulence model were obtained as follows. First a convergent solution using the Baldwin–Lomax turbulence model for the entire flowfield was obtained. Then the Johnson–King model was applied only for the upper part of the airfoil. To initiate the solution $\sigma(x)$ in Eq. 11 is set equal to unity and it is allowed to change according to Eq. 15 until the final solution is obtained. Note that the Johnson–King model reduces to the Cebeci–Smith[21] model when $\sigma(x)$ is identically one.

VI. RESULTS AND DISSCUSSION

The above described numerical approach was applied to the computation of steady and unsteady subsonic airfoil flows at Reynolds numbers greater than one million. Therefore the assumption was made that fully turbulent flow conditions can be assumed and thus no boundary layer transition calculations need to be incorporated. At first the procedure was used for the computation of steady flow over the NACA 0012 airfoil. Figure 1a shows a comparison of the computed lift curve with the experimental data. It can be seen that the computed maximum lift angle and hence the static stall angle is 13.5 degrees. This agrees quite well with the experimentally measured stall angle of McCroskey et al[23]. The computed lift and pitching moment are in fair agreement with the experiment for small angles of attack, but start to deviate at larger angles. However, much better agreement was obtained with the measurements of Harris[24]. Detailed comparisons of computed steady flows at various flow conditions with experimental data can be found in Ref. 25. Further comparisons for NACA 0012 airfoils undergoing a ramp–type pitch motion are given in reference 12. The computations presented here were performed on a 161×64 point grid. Grid refinement studies were also performed with a 243×91 point grid which provided finer resolution of the upper surface flow. Only small deviations were found between the finer grid and the coarser grid solutions. Further validations, not shown here, can be found in Ref. 25. The time accuracy of the numerical method was investigated for various time steps. It was found that the maximum allowable nondimensional time step is $\Delta t = 0.015 - 0.02$, corresponding to a maximum Courant number of about 2500. For these time steps it was found that Newton subiteration does not improve the accuracy of the unsteady solutions, even though Newton subiteration accelerates the convergence to the steady–state. Next, the code was applied to sinusoidal airfoil oscillations about the quarter chord point. These time periodic solutions were obtained for the second cycle because the results obtained for the third cycle were indistinguishable from the second cycle. Figures 2a and 2b show a comparison of the computed lift and moment hysteresis loops with the Baldwin–Lomax turbulence model for an oscillation of 5 degrees amplitude about a mean angle of attack of 9 degrees with McCroskey's experimental data for a Mach number of 0.3, a Reynolds number of 4 million, and a reduced frequency $k = \omega c/2U_\infty = 0.2$. It is seen that there is a substantial difference between the computed and the experimental lift hysteresis loops. This difference becomes even more pronounced when the computed and experimental moment hysteresis loops are compared with each other. The measured loop consists of two subloops, a clockwise "unstable" subloop and a counterclockwise "stable" subloop. As shown previously by Carta and Niebanck[26], clockwise pitching moment loops represent negative aerodynamic damping and therefore cause oscillations of a free airfoil to grow in amplitude, while counterclockwise loops cause such oscillations to decay. Hence torsional flutter will occur as soon as the area of the clockwise loop exceeds that of the counterclockwise loop. It is seen that the Navier–Stokes calculations in combination with the Baldwin–Lomax turbulence model fail to predict the destabilizing clockwise pitching moment loop. For this reason the sensitivity of the computed hysteresis loops to different turbulence models was studied next. Figure 3 shows the computed lift and pitching moment loops for oscillation about a slightly higher mean angle of 10 degrees at an amplitude of 5.5 degrees. The Mach and Reynolds numbers are again 0.3 and 4 million, and the reduced frequency k = 0.1. It can be seen that the Baldwin–Lomax, Johnson–King, and the RNG turbulence models produce significantly different hysteresis loops. The RNG model produces relatively good agreement with the measured lift hysteresis loop but fails again to predict the destabilizing moment subloop. These calculations were then repeated for oscillations about a mean angle of 11 degrees at an amplitude of 5 degrees, the other parameters left unchanged. Figure 4 shows the computed lift and moment hysteresis loops. As before, the calculations fail to predict the expected destabilizing moment subloop. The reason for this failure can be better appreciated by visualizing the separated flow structures computed with these turbulence models. Figures 5 and 6 show the computed flowfield with particle traces, even though instantaneous velocities or unsteady streaklines are more representative of the unsteady flowfield structure. Figure 5 corresponds to the instant when the airfoil oscillates through an incidence of 15 degrees during the downstroke. Figures 6 show the same information at 14 degrees during the downstroke. Different turbulence models produce substantially different recirculatory flow patterns on the upper surface near the trailing edge. The recirculatory flow region grows and then vanishes again during the course of oscillation. Even though the structure at the trailing edge resembles a vortex it does not shed from the trailing edge. However, as emphasized by Carr[27], the occurrence of the destabilizing moment loop appears to be closely associated with the vortex shedding from the trailing edge, an event which occurs soon after the static angle is substantially exceeded during part of the cycle. Although the static stall angle is significantly exceeded in the present calculations the numerical procedure along with the turbulence modeling used in the calculations appears to be unable to produce the correct vortex shedding process. Finally, figures 7a and b show the computed lift and pitching moment loops for oscillations about a mean angle of 13 degrees with an amplitude of 2.5 degrees, at a Reynolds number of 2 million, for three values of reduced frequency, using the Baldwin–Lomax model. Again, it can be seen that the moment loops indicate torsional stability.

SUMMARY

The preceding application of a fully implicit two–dimensional Navier–Stokes solver to the problem of light stall flutter of a NACA 0012 airfoil in compressible subsonic flow shows that the modeling of the recirculatory flow region with current turbulence models fails to capture the essential physics which governs the onset of stall flutter. Therefore, further detailed studies of improved numerical schemes and turbulence models as well as viscous/inviscid interaction approaches are required to improve the prediction of the unsteady flow separation and vortex shedding phenomenon.

Acknowledgment This investigation was supported by the Naval Postgraduate School, the Naval Air Systems Command, and the Naval Air Development Center.

REFERENCES

[1] Carr, L.W., "Progress in Analysis and Prediction of Dynamic Stall," *Journal of Aircraft*, Vol.25, No.1, pp. 6–17, 1988.

[2] McCroskey, W.J. and Pucci, S.L., "Viscous–Inviscid Interaction on Oscillating Airfoils in Subsonic Flow," *AIAA Journal*, Vol. 20, No.2, pp. 167–174, 1982.

[3] Carta, F.O. and Lorber, P.F., "Experimental Study of the Aerodynamics of Incipient Torsional Stall Flutter," *Journal of Propulsion*, Vol. 3, No.2, pp. 164–170, 1987.

[4] Sisto, F., "Stall Flutter", chapter 7 in AGARD Manual on Aeroelasticity in Axial–Flow Turbomachines, Vol.1, Unsteady Turbomachinery Aerodynamics, AGARD–AG–298, March 1987.

[5] Abdel–Rahim, A., Sisto, F., and Thangam, S., "Computational Study of Stall Flutter in Linear Cascades," ASME/IGTI 36th Gas Turbine Conference, Orlando, FL, June 3–6, 1991

[6] Jang, H.H., Ekaterinaris, J.A., Platzer, M.F., and Cebeci, T., "Essential Ingredients for the Computation of Steady and Unsteady Blade Boundary layers," ASME Paper 90–GT–160, June 1990.

[7] Visbal, M.R., "Effect of Compressibility on Dynamic Stall of a Pitching Airfoil," AIAA Paper 88–0132, January 1988.

[8] Visbal, M.R., "On the Formation and Control of the Dynamic Stall Vortex on a Pitching Airfoil," AIAA Paper 91–0006, January 1991.

[9] Ekaterinaris, J.A., "Compressible Studies on Dynamic Stall," AIAA Paper 89–0024, January 1989.

[10] Currier, J. and Fung, K.Y., "An Analysis of the Onset of Dynamic Stall," AIAA paper 91–0003, January 1991.
[11] Dindar, M., Kaynak, and U. Fujii, K., "Nonequilibrium Turbulence Modeling Studies on Light Dynamic Stall of a NACA 0012 Airfoil," AIAA paper 90–3040, August 1990.
[12] Grohsmeyer, S.P., Ekaterinaris, J.A., and Platzer, M.F., "Numerical Investigation of the Effect of Leading–Edge Geometry on Dynamic Stall of Airfoils," AIAA Paper 91–1798, June 1991.
[13] Rumsey, C.L. and Anderson, W.K., "Some Numerical and Physical Aspects of Unsteady Navier–Stokes Computations over Airfoils Using Dynamic meshes," AIAA Paper 88–0329, January 1988.
[14] Vinocour, M., "An Analysis of Finite–Difference and Finite–Volume Formulation of Conservation Laws," *Journal of Computational Physics,*" Vol. 81, 1989, pp. 1–52.
[15] Baldwin B. S., and Lomax, H., "Thin Layer Approximation and Algebraic Model for Separated Turbulent Flows," AIAA Paper 78–257, Jan 1978.
[16] Beam, R. and Warming R. F., "An implicit Finite–Difference Algorithm for Hyperbolic Systems in Conservation Law Form," *Journal of Computational Physics,*" Vol. 22, 1976, pp. 87–110.
[17] Jameson, A., Schmidt, W., and Turkel, E., "Numerical Solution of the Euler Equations by Finite Volume Methods Using Runge–Kutta Time–Stepping Schemes," AIAA Paper 81–1259, 1981.
[18] Pulliam, T. H., "Artificial Dissipation Models for the Euler Equations," *AIAA J.*, Vol 24, Dec. 1986, pp. 1931–1940.
[19] Yakhot, V. and Orszag, S. A., "Renormalization Group Analysis of Turbulence. I Basic Theory," *Journal of Scientific Computing,* Vol. 1, 1986.
[20] Johnson, D. A. and King, L. S., "A Mathematically Simple Turbulence Closure Model for Attached and Separated Turbulent Boundary Layers," *AIAA Journal,* Vol. 23, No. 11, Nov. 1985, pp. 1684–1692.
[21] Cebeci, T. and Smith, A. M. O., "Analysis of Turbulent Boundary Layers," *Academic Press*, NY, 1974.
[22] Martinelli, L. and Yakhot, V., "RNG–Based Turbulence Transport Approximations with Applications to Transonic Flows," AIAA Paper 89–1950, 1989.
[23] McCroskey, W.J., McAlister, K.W., Carr, L.W., Pucci, S.L., and Lambert, D., "An Experimental Study of Dynamic Stall on Advanced Airfoil Sections", USA AVRADCOM TR 82–A–8, 1982.
[24] Harris, C. D.,"Two–Dimensional Aerodynamic Characteristics of the NACA 0012 Airfoil in the Langley 8 Foot Transonic Pressure Tunnel," NASA TM 81927, 1981.
[25] Clarkson, J.D., MSAE Thesis, Naval Postgraduate School, September 1991.
[26] Carta, F.O. and Niebanck, C.F., "Prediction of Rotor Instability at

High Forward Speeds, Vol. III, Stall Flutter", USAAVLABS Tech. Rept. 68–18C, February 1969.
[27] Carr, L.W., Private Communication, September 1991.

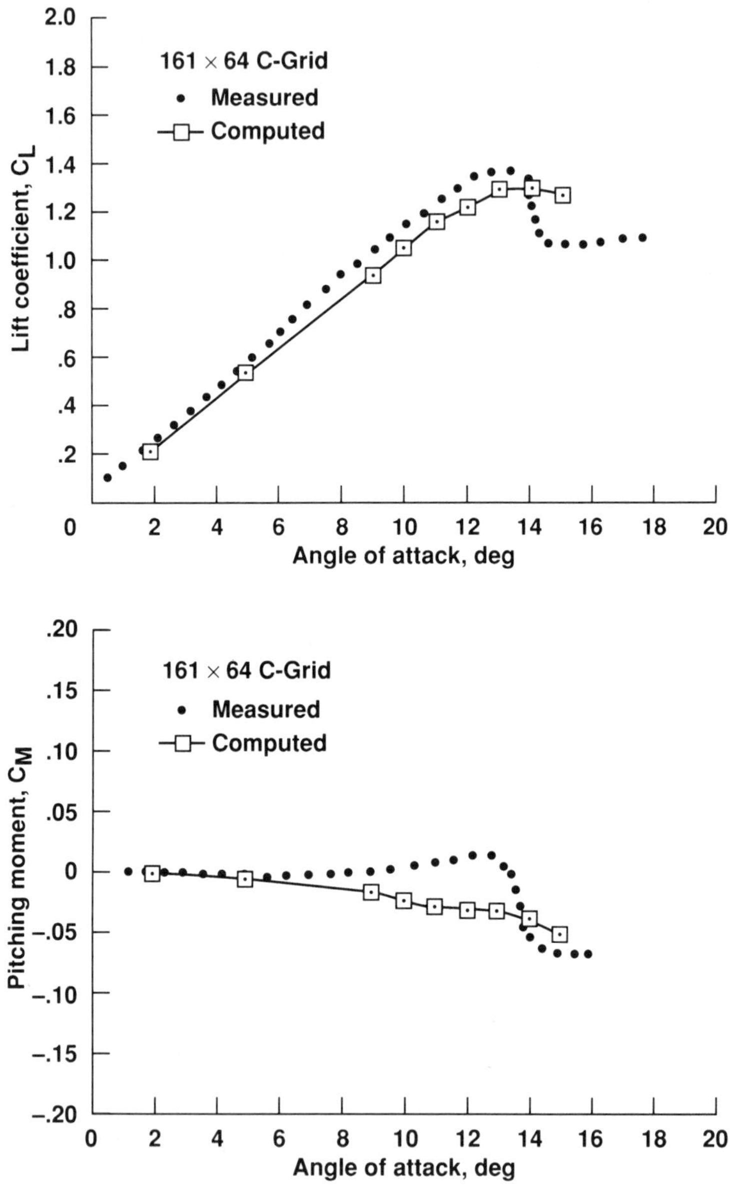

Figure 1. Computed steady–state and measured lift and pitching moment curves: $M_\infty = 0.3, Re = 4 \times 10^6$

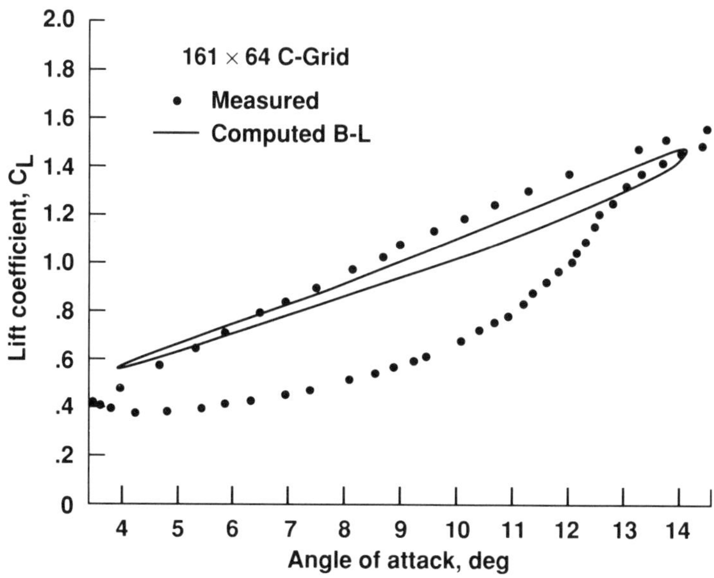

Figure 2a. Computed and measured unsteady lift hysteresis loop, frequency k = 0.20: $M_\infty = 0.3, \alpha = 9^o + 5^o sin(\omega t), Re = 4 \times 10^6$

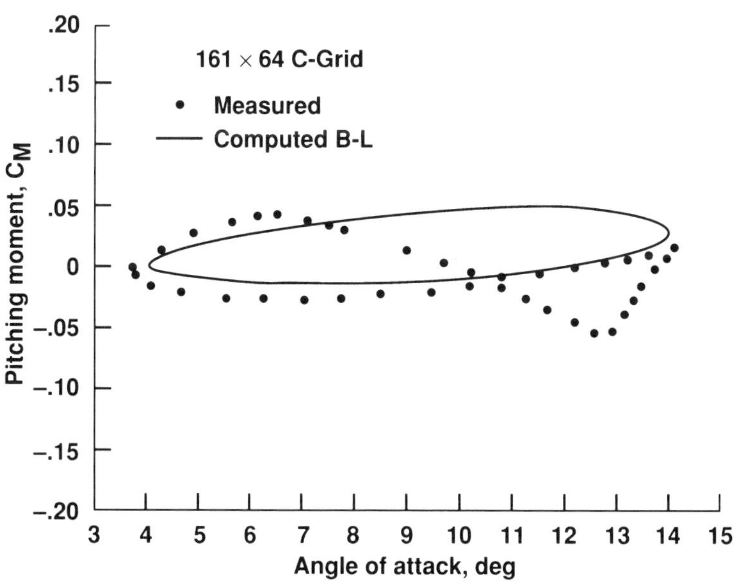

Figure 2b. Computed and measured unsteady pitching moment hysteresis loop, frequency k = 0.20: $M_\infty = 0.3, \alpha = 9^o + 5^o sin(\omega t), Re = 4 \times 10^6$

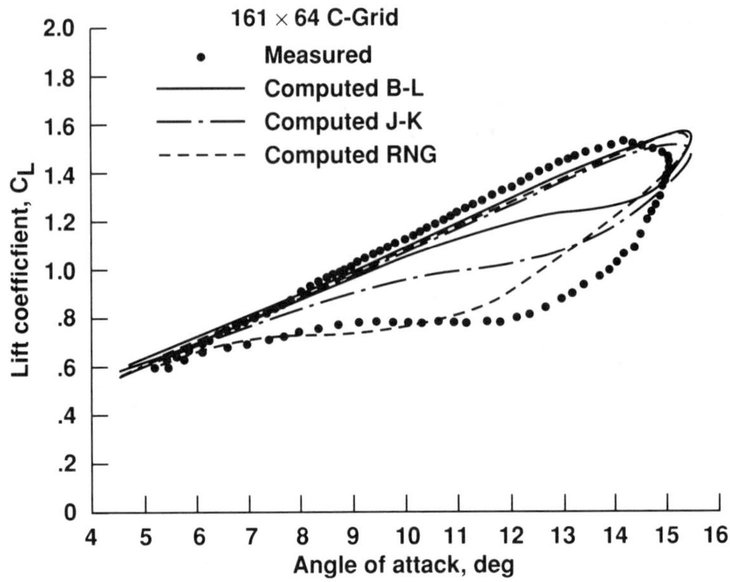

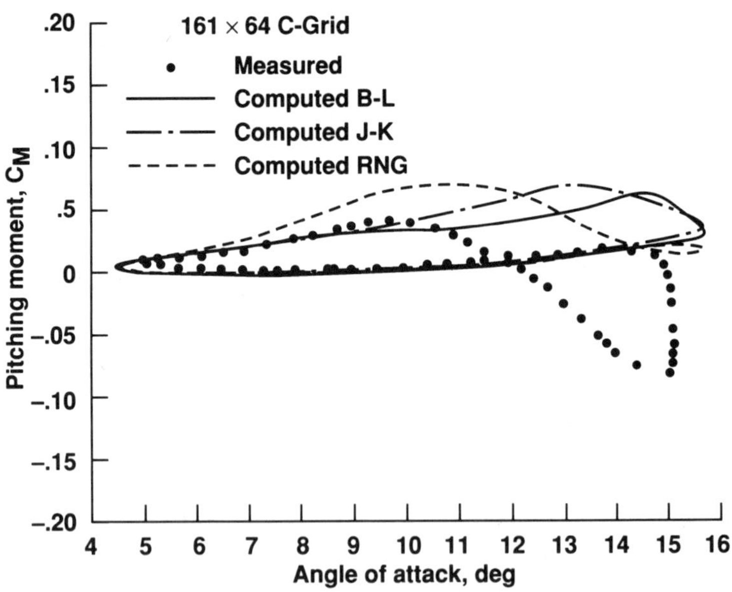

Figure 3. Computed and measured unsteady lift and pitching moment hysteresis loops, frequency k = 0.1: $M_\infty = 0.3, \alpha = 10^o + 5.5^o sin(\omega t), Re = 4 \times 10^6$

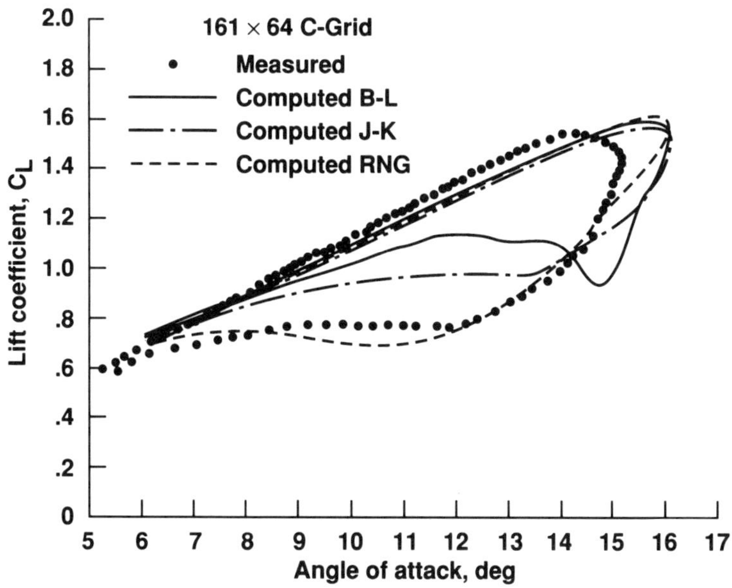

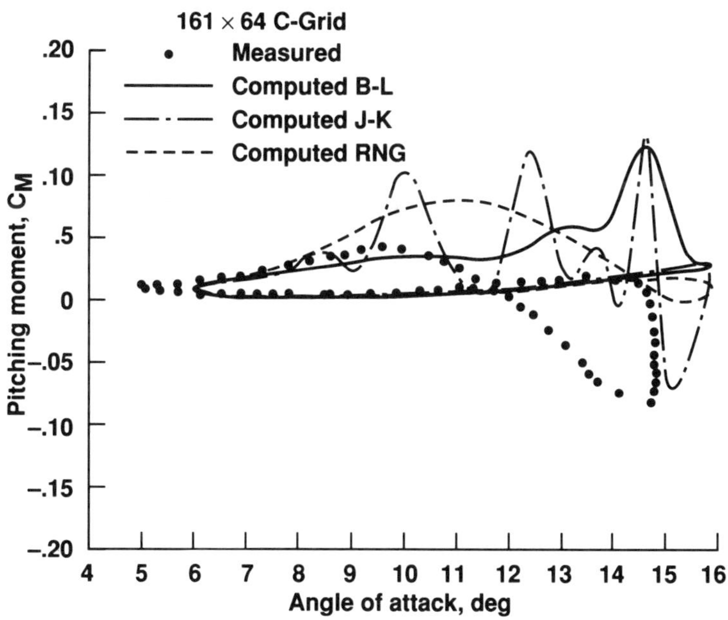

Figure 4. Computed and measured unsteady lift and pitching moment hysteresis loops, frequency k = 0.10: $M_\infty = 0.3, \alpha = 11^o + 5^o sin(\omega t), Re = 4 \times 10^6$

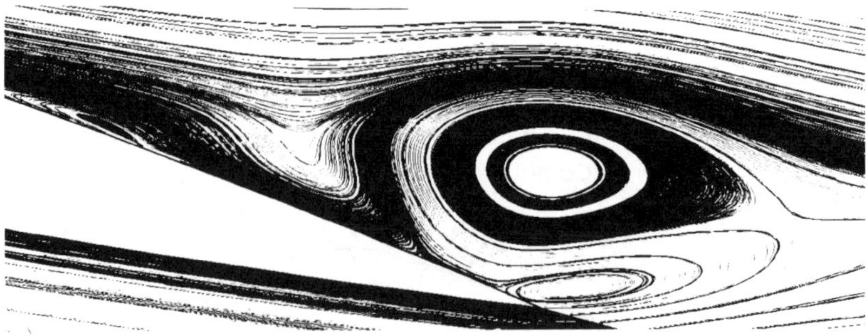

(a) Baldwin-Lomax

(b) Johnson-King

(c) RNG

Figure 5. Computed flowfield at the trailing edge using different turbulence models, frequency k = 0.10:
$M_\infty = 0.3, \alpha = 11^o + 5^o sin(\omega t), Re = 4 \times 10^6$;
$\alpha = 15^o$ downstroke (phase $\omega t = 0.7\pi$)

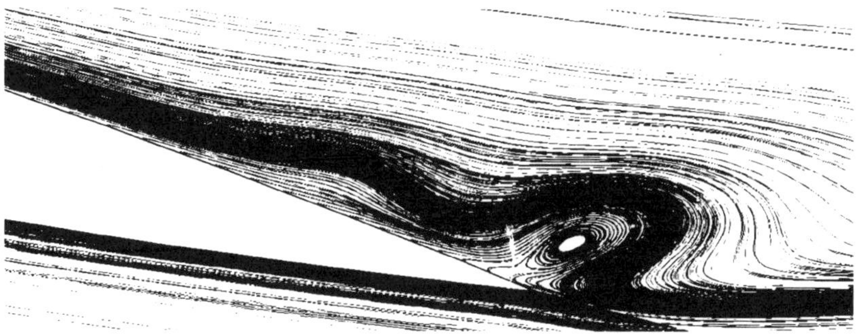

(a) Baldwin-Lomax

(b) Johnson-King

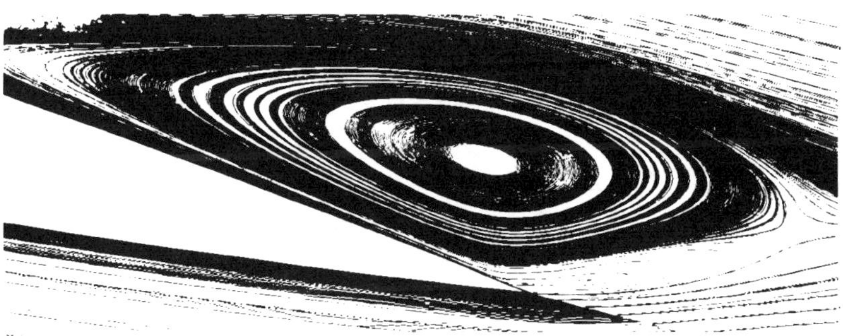

(c) RNG

Figure 6. Computed flowfield at the trailing edge using different turbulence models, frequency k = 0.10:
$M_\infty = 0.3, \alpha = 11^o + 5^o sin(\omega t), Re = 4 \times 10^6$;
$\alpha = 14^o$ downstroke (phase $\omega t = 0.795\pi$)

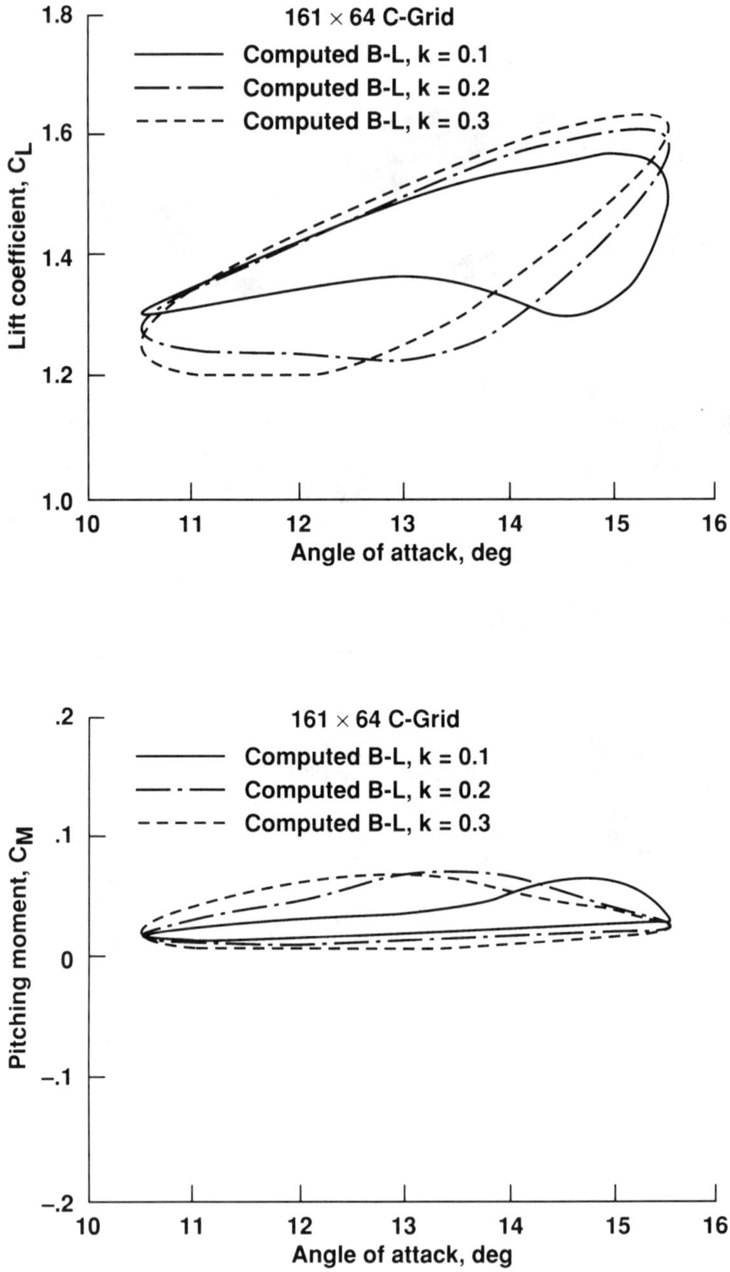

Figure 7. Effect of frequency on lift and pitching moment: $M_\infty = 0.3, \alpha = 13^o + 2.5^o sin(\omega t), Re = 2 \times 10^6$

Compressor Blade Gust Response to Attached and Separated Flow Forcing Functions

Kuk H. Kim[1]
Sanford Fleeter[1]

ABSTRACT

A series of experiments are performed to investigate the fundamental flow forcing function phenomena generating different blade row gust responses, in particular attached and separated flow forcing functions generated with NACA 0024 airfoils. Two NACA 0024 airfoils are installed in an extensively instrumented axial flow compressor inlet to generate the periodic 2-E unsteady aerodynamic forcing functions to the first stage rotor. These forcing functions are measured with a rotating cross hot-wire, with the resulting rotor blade row unsteady aerodynamic gust response measured with dynamic pressure transducers embedded in the rotor blade over a range of steady loading levels. Appropriate rotor blade gust response unsteady aerodynamic data are then correlated with appropriate predictions. These experiments clearly show that the forcing function generator fluid dynamics is significant with regard to the resulting unsteady aerodynamic gust response of the downstream airfoil row.

[1]School of Mechanical Engineering, Purdue University, Chaffee Hall, West Lafayette, IN 47907.

NOMENCLATURE

$\overline{C_l}$	steady lift coefficient
$C_{p,ps}$	pressure surface complex unsteady pressure coefficient
$C_{p,ss}$	suction surface complex unsteady pressure coefficient
ΔC_p, $C_{\Delta p}$	complex unsteady pressure difference coefficient
$\overline{C_{pi}}$	steady pressure coefficient at i^{th} chord position
$\overline{C_{p,p}}$	pressure surface steady pressure coefficient
$\overline{C_{p,s}}$	suction surface steady pressure coefficient
$\Delta\overline{C_p}$	steady pressure difference coefficient
i	rotor relative flow incidence angle
$\bar{i}$	mean rotor relative flow incidence angle
p'_1	1st harmonic of Fourier decomposed pressure
$\overline{P_{exit}}$	static pressure at rotor exit
u	streamwise gust component
u^+	1st harmonic streamwise gust component
U	flow velocity or wheel speed
v	transverse gust component
v^+	1st harmonic transverse gust component
$\overline{V_x}$	mean axial velocity
W	instantaneous rotor relative velocity
$\overline{W}$	mean rotor relative velocity
α	absolute flow angle
$\overline{\alpha}$	mean absolute flow angle
β	rotor relative flow angle
$\overline{\beta}$	mean rotor relative flow angle

INTRODUCTION

The spatial flow nonuniformities generated by inlet guide vanes, stators vanes and struts are periodic temporal variations to downstream rotor blades. These spatial flow nonuniformities in the stationary frame of reference are sources of periodic excitation to the rotor blades. When the frequency of these periodic flow nonuniformities coincides with a blade natural frequency, fatigue failure of the rotor blade may result, thereby compromising the durability of the engine.

The prediction of the flow induced vibratory response of a blade row first requires a definition of the unsteady aerodynamic forcing function in terms of its harmonics, with each harmonic independently considered. Thus, even though forcing functions may be generated by a wide variety of fundamentally different sources, the forcing functions are assumed to be equivalent if their harmonics are the same, i.e., the first harmonic response to forcing functions generated by fundamentally different sources are considered to be equivalent if the first harmonics of the forcing functions are the same. The unsteady aerodynamic response of the blade row to each forcing function harmonic is then assumed to be comprised of two components: the disturbance being swept past the nonresponding airfoils, termed the gust unsteady aerodynamic and the airfoil vibratory response to this disturbance.

Early treatments of unsteady flow due to periodic gusts were developed in the linear approximation wherein the mean flow is assumed to be uniform. In this approximation, the steady and unsteady flow field s are completely uncoupled from one another, with the upstream generated periodic gust aerodynamic forcing functions convected with the uniform mean flow. Semi-analytical unsteady aerodynamic analyses based on this linear model have been developed for cascades in subsonic and supersonic flows [1 2,3,4]. Such models are currently being extended to consider unsteady flows linearized about a nonuniform mean flow, with the gust interacting with the mean flow [5,6,7].

A number of experiments have been directed at the verification of such mathematical models and the determination of their applicability and limitations. Of particular interest herein are the experiments which investigated the unsteady aerodynamic response of a research compressor rotor performed by Manwaring and Fleeter [8]. The first harmonic gust response of the 1st stage rotor generated by two equivalent 2-per-revolution (2-E) unsteady forcing functions were measured: (1) an inlet flow distortion and (2) the wakes behind flat plate airfoils. The inlet distortion and wake first harmonic unsteady aerodynamic forcing functions were fundamentally equivalent, having the same gust characteristics including the ratio of the streamwise-to-normal gust components u^+/v^+ and reduced frequency values.

The first harmonic gust response of the 1st stage rotor generated by these two different but equivalent forcing functions is presented in Figure 1. Clearly, the first harmonic gust response unsteady aerodynamics, shown in the form of the chordwise variation of the complex unsteady pressure difference across the rotor blade chord, are dependent on the particular forcing function. However, because the two first harmonic forcing functions are equivalent in terms of classical unsteady aerodynamic theory, these results can not be predicted.

In this paper, this interesting result is further investigated. This is accomplished through a series of experiments performed in the extensively instrumented Purdue Axial Flow Research Compressor. Two NACA 0024 airfoils are installed in the compressor inlet to generate both an an attached flow and separated flow periodic 2-E unsteady aerodynamic forcing functions to the first stage rotor. These forcing functions are measured with a rotating cross hot-wire, with the resulting rotor blade row unsteady aerodynamic gust response measured with dynamic pressure transducers embedded in the rotor blade over a range of steady loading levels. Appropriate rotor blade gust response unsteady aerodynamic data are then correlated with predictions from the subsonic unsteady aerodynamic flat plate cascade analysis of Smith [9].

RESEARCH COMPRESSOR

The Purdue Axial Flow Research Compressor, depicted schematically in Figure 2, models the fundamental turbomachinery unsteady aerodynamic multistage interaction phenomena which include the incidence angle, the velocity and pressure variations, the aerodynamic forcing function waveforms, the reduced frequency, and the unsteady blade row interactions. The compressor is driven by a 15 HP DC electric motor at a speed of 2,250 RPM. Each identical stage contains 43 rotor blades and 31 stator vanes having a British C4 airfoil profile, with the first stage rotor inlet flow field established by an inlet guide vane (IGV) row of 36 airfoils. The 2-E unsteady aerodynamic forcing functions are generated by two NACA 0024 profile airfoils installed 180° apart in the compressor inlet. The overall compressor and airfoil characteristics are defined in Table 1.

The compressor aerodynamic performance is determined utilizing a 48 port Scanivalve system, thermocouples and a venturi orifice to measure the required pressures, temperatures and flow rate, respectively. The Scanivalve transducer is calibrated each time data are acquired, thus automatically compensating for zero and span shifts of the transducer output. A 95% confidence interval, root-mean-square error analysis of 20 samples is performed for each steady data measurement.

INSTRUMENTATION

Both steady and unsteady rotor blade row data are obtained. The steady data quantify the rotor row mean inlet flowfield and the resulting rotor blade midspan steady loading distribution. The unsteady data define the periodic aerodynamic forcing function and the resulting midspan blade surface periodic unsteady pressure distributions.

The inlet flow field, both steady and unsteady, is measured with a rotating cross hot-wire probe. The inlet flow field of the 1st stage rotor row is measured in the rotating frame of reference by mounting the hotwire on the rotor drum 18.8% chord upstream and 65% blade spacing from a rotor blade. The hotwire is oriented for maximum sensitivity, achieved when the 0° flow incidence corresponds to the position at which the flow angle is 45° to both wires. The hotwire is calibrated for velocities from 21.3 m/s to 62.5 m/s and ±40° flow angle variations. The uncertainties in the velocity and the flow angle measurements were determined to be 5% and ± 0.5°. Centrifugal loading effects on the rotating hot-wire sensor resistances and thus the responses were found to be negligible.

The detailed steady aerodynamic loading on the rotor blade surfaces is measured with a chordwise distribution of 20 midspan static pressure taps, 10 on each surface. The static pressure at the rotor exit plane, measured with a rotor drum static tap, is used as the blade surface static pressure reference. These static pressure measurements are made using a rotor based 48 port constant speed drive Scanivalve system located in the rotor drum.

The measurement of the midspan rotor blade surface unsteady pressures is accomplished with 20 ultra-miniature, high response transducers embedded in the rotor blades at the same chordwise locations as the static pressure taps. To minimize the possibility of flow disturbances associated with the inability of the transducer diaphragm to exactly maintain the surface curvature of the blade, a reverse mounting technique is utilized. The embedded dynamic transducers are both statically and dynamically calibrated. The static calibrations show good linearity and no discernible hysteresis. The dynamic calibrations demonstrate that the frequency response, in terms of gain attenuation and phase shift, are not affected by the reverse mounting technique. The maximum error in gain and phase angle were determined to be 0.60 dB and 1.5°.

The rotor-based static pressure Scanivalve transducer, rotating cross hot-wire probe and 20 blade surface dynamic pressure transducers are interfaced to the stationary frame-of-reference through a 40 channel slip ring assembly. On-board signal conditioning of the transducer output signals is performed to maintain a good signal-to-noise ratio through the slip rings. The remaining 17 channels of the slip-ring assembly are used to provide excitation to the transducers and on/off switching to the Scanivalve DC motor.

DATA ACQUISITION AND ANALYSIS

Steady Pressure Data

The rotor blade surface static pressure data are defined by a root-mean-square error analysis of 20 samples with a 95% confidence interval. The airfoil surface static pressures are presented in terms of a nondimensional steady pressure coefficient, with the steady lift coefficient calculated by integrating the differential steady pressure coefficient across the rotor blade chord.

$$\overline{C}_{p,i} = \frac{\overline{P}_i - \overline{P}_{exit}}{\rho \overline{W}^2} \quad (1)$$

Since the blade surface and the reference static pressures are measured at different radii, a correction derived by Manwaring and Fleeter [8] is applied to the exit steady pressure to account for centrifugal effects

Periodic Data

The periodic data of interest are the harmonic components of the aerodynamic forcing function to the first stage rotor blade row together with the resulting rotor blade surface unsteady pressures and unsteady pressure differences. These are determined by defining a digitized ensemble averaged periodic unsteady aerodynamic data set consisting of the rotating cross hot-wire probe and blade surface dynamic pressure transducer signals at each steady operating point. In particular, these time-variant signals are digitized with a high speed A-D system at a rate of 100 kHz and then ensemble averaged.

The key to this averaging technique is the ability to sample data at a preset time, accomplished by an optical encoder mounted on the rotor shaft. The microsecond range step voltage signal from the encoder is the data initiation time reference and triggers the high speed A-D multiplexer system. To significantly reduce the random fluctuations superimposed on the periodic signals of interest, 200 averages are used. A Fast Fourier Transform (FFT) algorithm is then applied to these ensemble averaged signals to determine the harmonic components of the unsteady aerodynamic forcing function and the resulting rotor blade surface harmonic unsteady pressures and pressure differences.

Forcing Function

The 2-E unsteady aerodynamic forcing functions are generated by two NACA 0024 profile airfoils installed 180° apart in the compressor inlet. Forcing functions categorized as attached flow and fully separated flow, are generated by varying the angle of attack (AOA) of the two NACA 0024 profile airfoils from 5° to 20°. Note that since the airfoils are set at a fixed angle of attack, the values of the resulting streamwise-transverse gust ratio depend on the compressor steady loading level.

The forcing function to the 1st stage rotor, the unsteady rotor inlet flow field, is measured with the rotating cross hot-wire probe which quantifies the relative velocity and flow angle. To the rotor, the flow from the upstream NACA 0024 airfoils appear as deficits in the rotor relative inlet velocity W and fluctuations in the rotor relative inlet flow angle b. Thus, the total flow consists of freestream and wake regions, with the instantaneous value of W increased in the wake region and decreased in the freestream. The total rotor inlet relative velocity gust $\overrightarrow{\Delta W}$ is the vector difference between the instantaneous and mean relative velocity $\vec{W}$. It has two components, u and v, parallel and normal to the mean flow direction. The gust and its components are depicted in the velocity diagram of Figure 3.

The fundamental frequency of interest is the 2-E forcing function frequency. Thus, an harmonic analysis is utilized in the data analysis, accomplished by taking the Fast Fourier Transforms (FFT) of both the time variant rotor inlet flow field and the resulting unsteady aerodynamic response of the 1st stage rotor row, with only the components at the fundamental frequency or its harmonics analyzed. Figure 4 shows the streamwise and transverse gust components and their FFT, with the u and v harmonics denoted by u^+ and v^+ and nondimensionalized by the mean rotor relative velocity. This Fourier transformed inlet flow, defined by u^+ and v^+, is the unsteady aerodynamic forcing function to the downstream rotor row.

Unsteady Pressure Data

The rotor blade pressure and suction surface unsteady pressure data are analyzed to determine the harmonics of the chordwise distribution of the unsteady pressure coefficient.

$$C_{p,i} = \frac{p_i^+}{\rho \overline{W}^2 v^+} \qquad (2)$$

The unsteady differential pressure coefficient is computed by subtracting the unsteady pressure coefficient on the suction surface from that on the pressure surface.

RESULTS

A series of experiments are performed in the Purdue Axial Flow Research Compressor to investigate gust generated rotor blade unsteady aerodynamic response. The effect of inlet flow forcing function type, i.e., generated by an attached flow or a fully separated flow, and rotor steady loading on the gust response of the first stage rotor blade are studied. For each forcing function flow type, four steady loading conditions as characterized by the rotor relative mean flow incidence angles are studied, i = -3°, 0°, 3°, and 6°.

Surface Steady Pressure Distributions

The chordwise rotor blade surface steady pressure distributions for the two forcing function flow types are presented in Figure 5 for the low and high steady rotor steady loading levels, i = -3° and 6°. These data are compared with predictions from an incompressible, inviscid small camber airfoil cascade analysis developed by Henderson and Fleeter [10]. The rotor blade surface steady pressure distribution is a function of the steady loading level but independent of the forcing function.

In the pressure surface leading edge region, the steady pressure coefficient increases with loading. It then increases until approximately 30% chord where it becomes constant, independent of the loading. In contrast, steady loading affects the suction surface over the entire chord, with the suction surface steady pressure coefficient a strong function of the steady loading level. Note that there is no evidence of suction surface steady flow separation . With regard to the differential steady pressure coefficient, the leading edge value is negative at negative values of the relative mean flow incidence, approximately zero at 0° incidence, increasing with increased loading. These data exhibit relatively good correlation with the predictions, accomplished by utilizing the mean rotor relative velocity measured with the rotor blades adjacent to the hotwire removed to determine the value of the mean rotor relative velocity $\overline{W}$ for the nondimensionalization of the steady pressure coefficient.

Attached & Separated Flow Generated Forcing Functions

The Fourier decomposition of the attached flow forcing function generated by the 5° AOA NACA 0024 airfoils is presented in Figure 6. The dominant u^+ amplitude occurs at the 2/rev fundamental frequency, with the higher order harmonic amplitudes generally smaller and decreasing with increased frequency. The Fourier decomposed v^+ amplitudes of the attached inlet flow forcing function are approximately equal and small for all frequencies. Note that not all of the u^+ and v^+ harmonics are of the same magnitude nor are they distinct. Also the characteristics of the u^+ spectrum resemble that of a harmonic forcing function having a high fundamental frequency component with lower amplitude higher harmonics.

The Fourier decomposition of the fully separated inlet flow forcing function generated by the 20° AOA NACA 0024 airfoils is also shown in Figure 6. Although the magnitude of u^+ at the fundamental frequency is large, it no longer dominates the u^+ spectrum. The higher harmonic amplitudes are much larger relative to the fundamental frequency value than that of the attached flow forcing function, i.e., higher harmonics are a significant part of the complete forcing function. However, the higher harmonic amplitudes continue to decrease with increased frequency, as per the attached flow generated forcing function. Similarly, the higher harmonics of v^+ are increased in magnitude relative to the fundamental harmonic and larger in value when the forcing function is separated. No particular harmonic of v^+ dominates the forcing function. However the forcing function generated by the fully separated flow has a frequency at which a minimum v^+ occurs. Thus, the separated flow forcing function more closely resembles an impulse function due to these increases in the higher harmonic amplitudes.

Forcing Function Generator Fluid Dynamic Effects on Gust Response

Pressure Surface

Figures 7 and 8 show the change in the pressure surface gust response when the flow field generating the forcing function is changed from an attached flow to a fully separated flow, accomplished by changing the NACA 0024 airfoil AOA from 5° to 20°. Altering the forcing function generator from an attached flow to a separated flow has a noticeable effect on the pressure surface gust response. With the attached flow generated forcing function, the magnitude and phase of the pressure surface gust response are clearly a function of the steady loading level. As the AOA increases to 20°, the pressure surface response becomes much less dependent on the steady loading level. At 20° AOA, the magnitude still smoothly decreases with chord. However, the level of the magnitude response is increased. Also, the 20° AOA phase response is now independent of the loading level, with a smooth phase increase with chord.

Suction Surface

Figures 9 and 10 show the suction surface gust response resulting from the different forcing function generator flow fields. Altering the forcing function generating flow field from an attached flow to a separated flow also has an effect on the suction surface gust response. With the attached flow forcing function, the response magnitude decreases sharply in the leading edge region, attaining a relative minimum value near 25% chord and then increasing slightly before decreasing to an absolute minimum near 65% chord. In the trailing edge region, there is a relatively large increase and a subsequent decrease. Also, these magnitude data are a function of the steady loading level, particularly in the airfoil leading and trailing edge regions. As the AOA is increased to 20°, the overall trends of these magnitude response data are unchanged. However, the chordwise distribution of the magnitude data becomes smoother, i.e., the magnitude data decrease more rapidly in the leading edge region, attaining a somewhat constant level in the 25% to 50% chord region, then decreasing to a minimum near 60% chord. The rapid increase and decrease in the magnitude near the trailing edge is still evident.

The suction surface gust response phase of the airfoil is also a function of the forcing function generator flow field. At 5° AOA attached flow, the unsteady pressure phase response is constant with chord at the highest steady loading of 6° mean flow incidence. As the loading is changed from this level, there is a large phase decrease in the aft chord region, with the phase remaining constant over the front of the airfoil. Also, the phase is a function of the steady loading level. Increasing the forcing function generator AOA to 20° has a clear effect on the phase response. Namely, in the front chord region, the phase response becomes independent of loading. Also, the large phase change near the trailing edge is suppressed as the AOA is increased. Instead, the phase decreases from the leading to trailing edge more gradually as the AOA is increased and the flow generating the forcing function flow changes to a separated flow field.

Unsteady Pressure Difference

As expected based on the individual airfoil surface data, altering the forcing function generator from an attached flow to one at the onset of separation to a separated flow has a significant effect on the pressure difference gust response, Figures 11 and 12. The pressure difference magnitude data generally decrease with chord, with the decrease in the leading edge region becoming larger as the forcing function generator flow field changes to a separated flow. Note that this general trend is in agreement with the linear theory prediction. However, the degree of correlation with the linear theory prediction is dependent on the forcing function generator fluid dynamics. Namely, at an AOA of 5°, the pressure difference magnitude data are typically increased in value as compared to the prediction.

Increasing the forcing function generator AOA to 20° results in these magnitude data decreasing in value with respect to the prediction.

The interesting but unexplained result of Manwaring and Fleeter [8] can now be attributed to the attached and separated flow nature of the forcing functions utilized in their experiments. Namely, the inlet distortion forcing function was generated by an attached flow forcing function. In contrast , the airfoil wake forcing function was generated by a flat plate at an angle of attack, thereby resulting in a separated flow generated forcing function.

SUMMARY AND CONCLUSIONS

A series of experiments were performed to investigate the fundamental flow forcing function phenomena generating different blade row gust responses, in particular attached and separated flow forcing functions generated with NACA 0024 airfoils. Two NACA 0024 airfoils are installed in the compressor inlet to generate attached and separated flow periodic 2-E unsteady aerodynamic forcing functions to the first stage rotor. These forcing functions were measured with a rotating cross hot-wire, with the resulting rotor blade row unsteady aerodynamic gust response measured with dynamic pressure transducers embedded in the rotor blade over a range of steady loading levels. Appropriate rotor blade gust response unsteady aerodynamic data were then correlated with appropriate predictions.

These experiments revealed the new result that the forcing function generator fluid dynamics has a significant impact on the resulting unsteady aerodynamic gust response of a downstream airfoil row. Namely, attached and separated flow generated forcing functions result in large and important differences in the resulting unsteady aerodynamic gust response of the downstream airfoil row and the correlation of these data with linear theory predictions. Rotor blade steady loading also significantly affected this data-theory correlation

These differences in the detailed forcing function fluid dynamics explain the significant differences in Manwaring and Fleeter's [8] fundamentally equivalent 2-E inlet distortion and wake generated gust response data. Namely, their inlet distortion forcing function was generated by an attached flow whereas the wake forcing function was generated by a separated flow. Hence the differences in the resulting rotor blade row gust unsteady aerodynamic response.

ACKNOWLEDGEMENTS

Research sponsored by the Air Force Office of Scientific Research (AFSC) under Contract F49620-88-C-0022. The United States Government is authorized to reproduce and distribute reprints for governmental purposes notwithstanding any copyright notation hereon.

REFERENCES

1. Whitehead, D.S., "Classical Two-Dimensional Methods," AGARDograph No. 298, AGARD Manual on Aeroelasticity in Axial Flow Turbomachines, Volume 1: Unsteady Turbomachinery Aerodynamics, pp.3.1-3.30.

2. Fleeter, S., "The Fluctuation Lift and Moment Coefficients for Cascaded Airfoils in a Nonuniform Compressible Flow," AIAA Journal of Aircraft, Vol. 10, No. 2, February 1973, pp. 93-98.

3. Adamczyk, J.J. and Goldstein, M.E., "Unsteady Flow in a Supersonic Cascade with Subsonic Leading Edge Locus," AIAA Journal, Vol. 16, No. 12, 1978, pp. 1248-1254.

4. Verdon, J.M. and Usab, W.J., "Application of a Linearized Unsteady Aerodynamic Analysis to Standard Cascade Configurations," ASME Paper 90-GT-11, June 1990.

5. Verdon, J.M. and Hall, K.C., "Development of a Linearized Unsteady Aerodynamic Analysis for Cascade Gust Response Predictions," NASA Report 4308, July 1990.

6. Scott, J.S. and Atassi, H.M., "Numerical Solutions of the Linearized Euler Equations for Unsteady Vortical Flows Around Lifting Airfoils, AIAA Paper 90-, 1990

7. Fang, J., "Compressible Flows With Vortical Disturbances Around Cascades of Airfoils," Ph.D. Thesis, University of Notre Dame, April 1991

8. Manwaring, S.R. and Fleeter, S., "Forcing Function Effects on Rotor Periodic Aerodynamic Response," ASME Journal of Turbomachinery, Vol. 113, No. 2, April 1991, pp. 312-319.

9. Smith, S.N. "Discrete Frequency Sound Generation in Axial Flow Turbomachines," Aeronautical Research Council R&M 3709, Great Britain, March 1972.

10 Henderson, G. and Fleeter, S., "Forcing Function and Steady Loading Effects on Unsteady Aerodynamic Gust Response," NASA-CR, 1991.

Table 1. Overall airfoil and compressor characteristics

	ROTOR	STATOR	IGV
Airfoil type	C4	C4	C4
Number of Airfoils	43	31	36
Chord, C (mm)	30	30	30
Solidity, C/S	1.14	1.09	0.96
Camber, θ	28.0	27.7	36.9
Stagger Angle, γ	36.0	-36.0	21.0
Inlet Metal Angle, β_1	50.0	30.0	0.0
Aspect Ratio	2.0	2.0	2.0
Thickness/Chord (%)	10.0	10.0	10.0
Flow Rate (kg/s)		2.03	
Design Axial Velocity (m/s)		24.4	
Design Rotational Speed (RPM)		2250	
Number of Stages		3	
Design Stage Pressure Ratio		1.0	
Inlet Tip Diameter (mm)		420	
Hub/Tip Radius Ratio		0.714	
Stage Efficiency (%)		85	

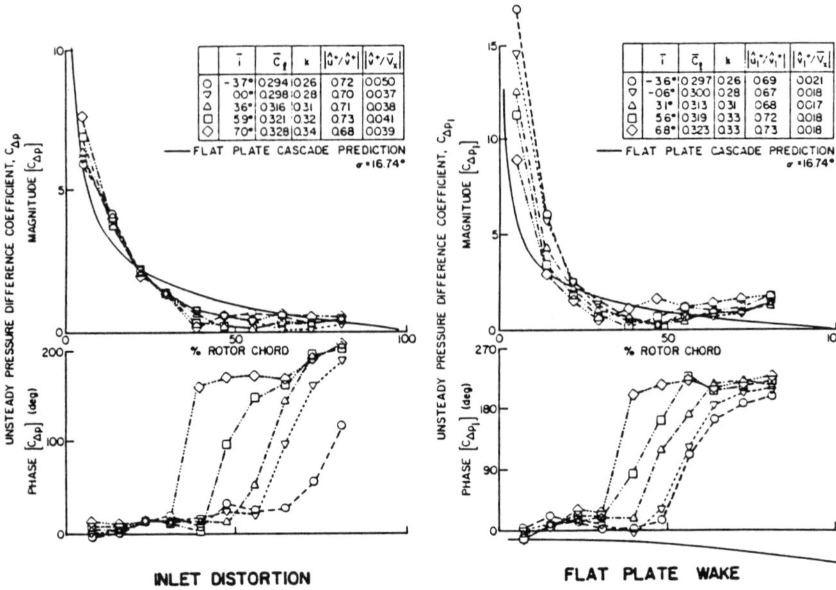

Figure 1. Inlet distortion and wake generated gust response [8]

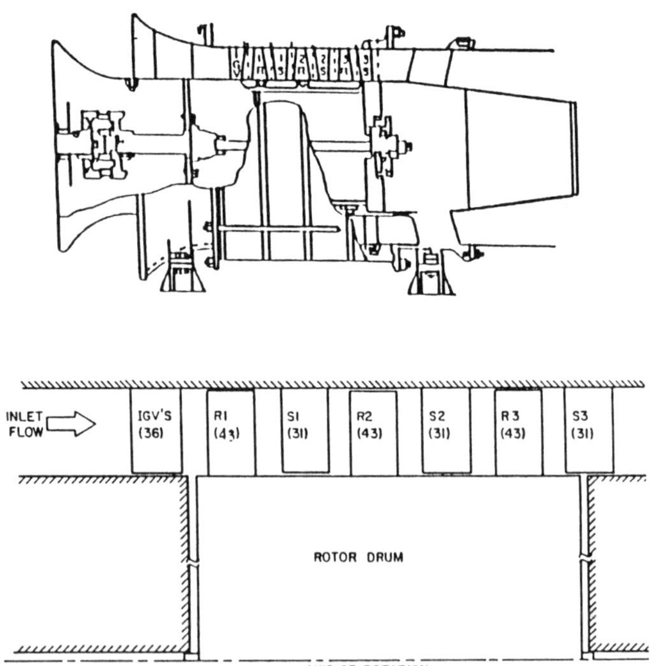

Figure 2. Research compressor schematic

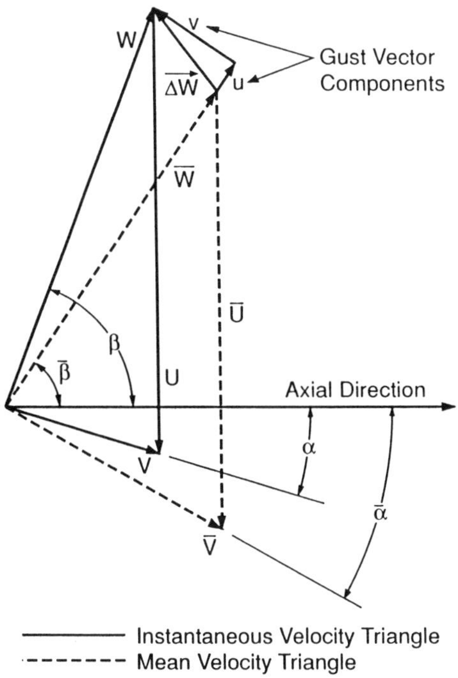

Figure 3. Decomposition of rotor inlet velocity

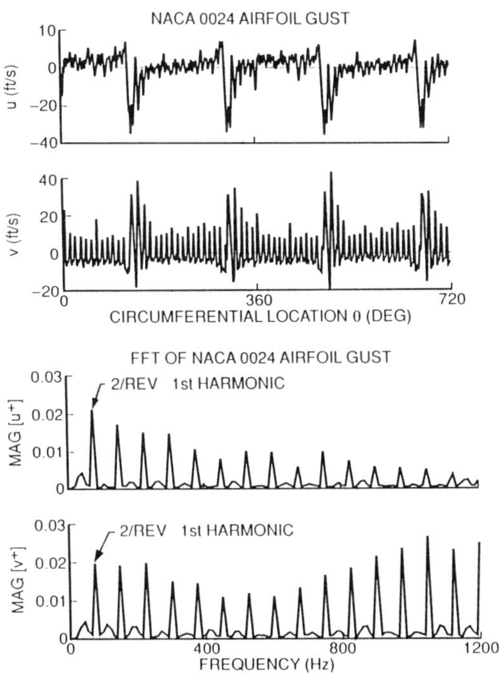

Figure 4. Forcing function and streamwise and normal gust component FFT

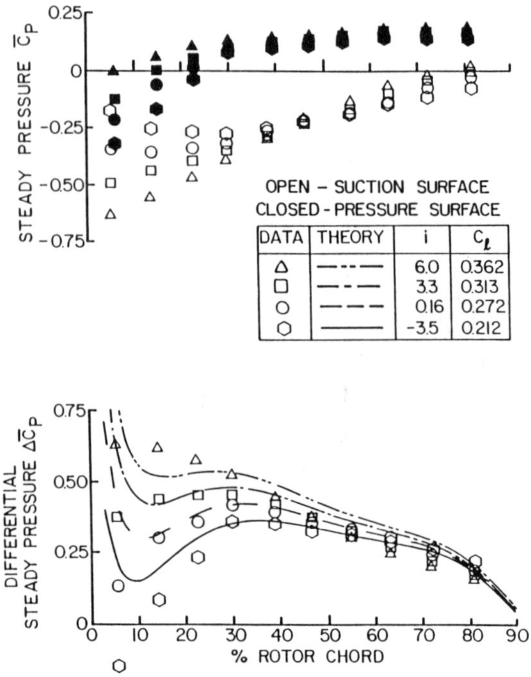

Figure 5. Loading effect on rotor blade surface steady pressures

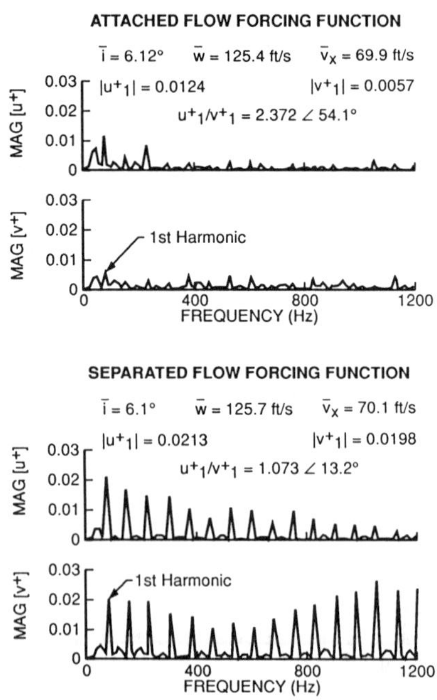

Figure 6. FFT of attached and separated flow generated forcing functions

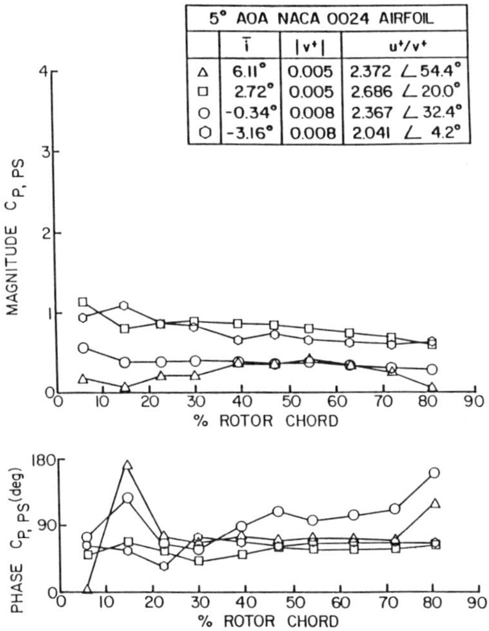

Figure 7. Pressure surface attached flow generated gust response

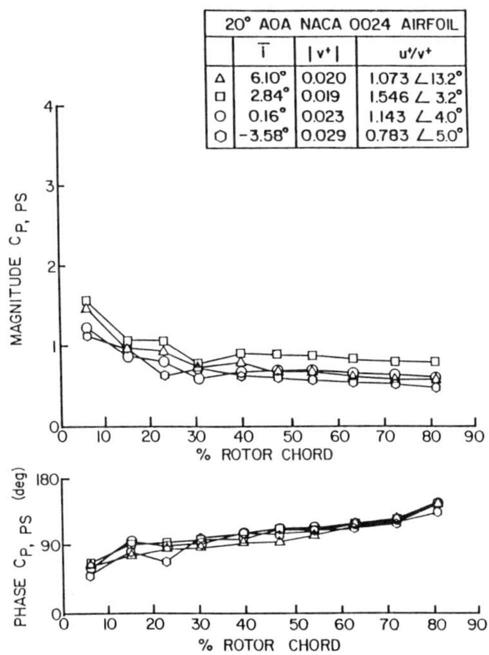

Figure 8. Pressure surface separated flow generated gust response

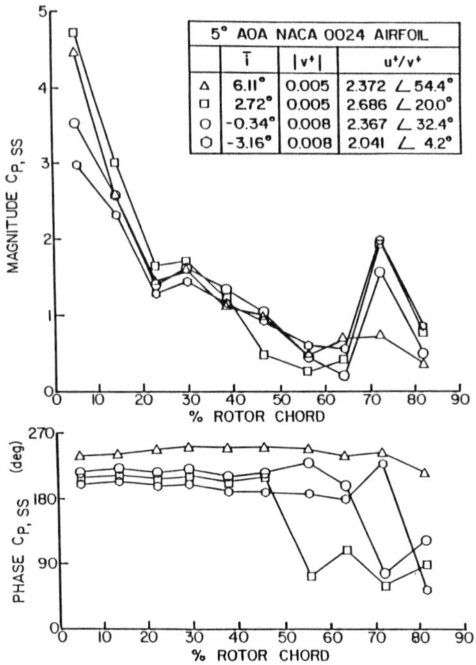

Figure 9. Suction surface attached flow generated gust response

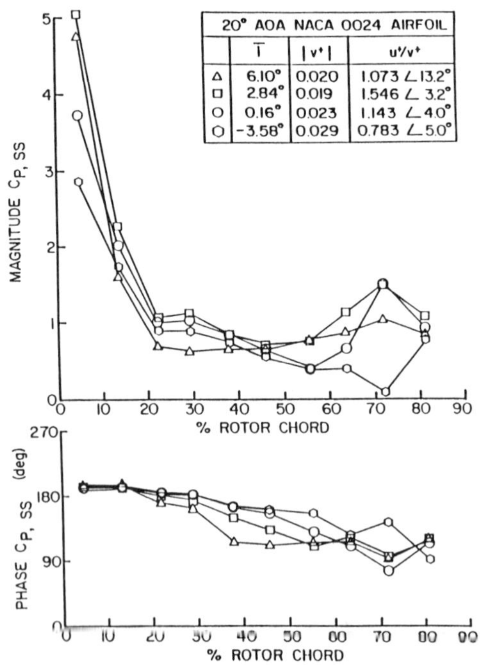

Figure 10. Suction surface separated flow generated gust response

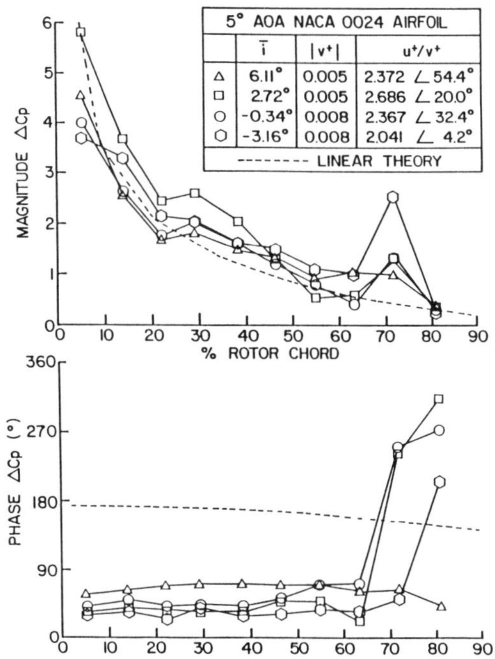

Figure 11. Pressure difference attached flow generated gust response

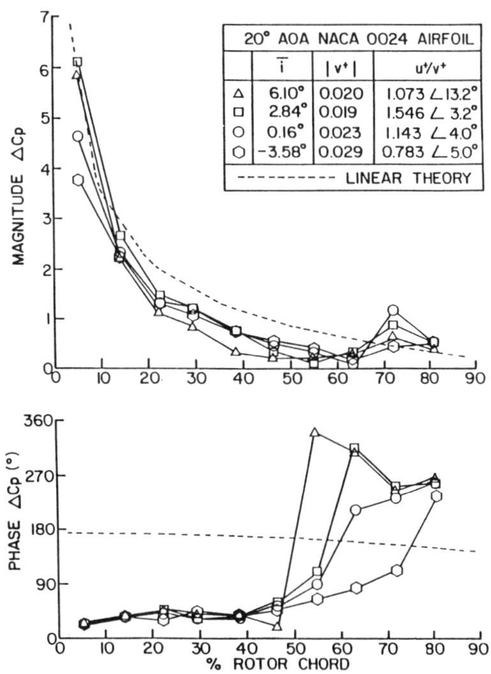

Figure 12. Pressure difference separated flow generated gust response

V. Unsteady Flow Measurements

Experimental Methods and Results in Unsteady Aerodynamics and Aeroelasticity

H.E. Gallus[1]

Abstract

In spite of the great progress in the field of computational fluid dynamics in the last decade, experimental research is still a major development tool in the design process of modern turbomachines. There are a number of different over-all objectives, which include the analysis of data to validate and calibrate advanced numerical calculation schemes, the improvement of the physical understanding of the flow phenomena, and investigations of new concepts to direct theoretical approaches modelling the flow effects.

The unsteady flow in turbomachines is of renewed interest. The traditional working fields on unsteady flows are concerned with the investigation of the aerodynamic stability of machines, the aeroelastic stability of cascade and single stage flows, the noise generation, and the influence of unsteady effects on the effiency. Advances in the experimental methods enabled the experimentalists to investigate the time-dependent, three-dimensional flow in turbomachine stages. A large number of experimental results in this field was published, which underline the necessity to include the unsteady effects in the design process to achieve further improvements of machine performance and reliability.

In this paper, the state of the art concerning unsteady measurements in a turbomachine environment is reviewed. Different techniques as LDV and L2F systems, tracer methods, hot-wire anemometry and subminiature pressure transducers are presented emphasis given to the dominating problems of stator-rotor interaction. An attempt is made to give an overview of the applicability, accuracy and specific limitations of the methods. Special emphasis is laid on the discussion of future needs and development directions.

---

[1]Institut fur Strahlantriebe und Turboarbeitsmaschinen,
RWTH , D-5100 Aachen, Germany.

Introduction

This review paper on experimental methods and results in "Unsteady Aerodynamics and Aeroelasticity" in the field of axial-flow-turbomachines can only be considered as a modest attempt to contribute to the continuity of the former five Symposia where - in view of the exploding progress of computational methods - the experimental part of this science was acknowledged as the other tool necessary for the development of an improved theory.

A detailed and extensive survey of the techniques for unsteady flow and flutter measurement is included in the AGARD Manual on "Aeroelasticity in Axial-Flow Turbomachines", Volume 1 "Unsteady Turbomachinery Aerodynamics" edited by M.F. Platzer and F.O. Carta in 1987.

The state of the art of the experimental techniques is reviewed there in four chapters as follows:

- Unsteady Aerodynamic Measurements in Flutter Research
  by S. Fleeter and R.L. Jay,
- Unsteady Aerodynamic Measurements in Forced Vibration Research
  by R.L. Jay and S. Fleeter,
- Unsteady Fan Blade Flutter through Linear Cascade Aerodynamic Testing
  by E. Szechenyi,
- Unsteady Aerodynamic Measurements on Rotors
  by H.E. Gallus

Thus, it can not be the task of this review paper to repeat the contents of the above-mentioned contributions.

Nevertheless, in the recent years an enormous variety of pertaining experimental work has been published in the journals. Most of these contributions make use of already well-known techniques, at least in the measuring principle, but the field of applications has more and more extended from measurements on single airfoils or 2D-cascades to those taking into account the flow conditions of the environment in the real machine. These requirements gave rise to remarkable modifications of the various sensors, transmitters and data acquisition techniques with emphasis given to the increase of high-frequency response capability and data reduction facing the hardly imaginable amount of data of the three-dimensional viscous and unsteady flow field due to real machine conditions.

In this connection, it is worthwhile to read again the evaluation and conclusion remarks in the Proceedings of the former Symposia and of relevant special meetings, e.g. the AGARD-Conference Proceedings No. 468 "Unsteady Aerodynamic Phenomena in Turbomachines" (1989, Luxembourg). From this view back at the early development one can resume that on the way to progress in the theory of unsteady flow around single airfoils and through rectangular cascades computational and experimental methods kept pace with each other. With increasing efforts and success in establishing CFD-methods, agreement was achieved to provide measured data from various 2D-cascade investigations to list them as "test cases" or "standard configurations" for the validation of computational methods (T. Fransson; A. Bölcs, 1984, 1987). The "Workshop on Aeroelasticity in Turbomachines: Comparison of Experimental and Theoretical Cascade Results" turned out to be very useful as well as successful as a forum of both CFD-people and experimentalists. Until recently, the CFD-methods for predicting unsteady flow were restricted to inviscid flow. Viscosity effects such as boundary layer behavior and separation in unsteady flow, shock/boundary layer interaction, boundary layer transition, inhomogeneous inlet flow fields due to wake and vortex flow from upstream bade rows could not yet be included.

Thus, the comparison between the measured data and the computational results of the standard configurations showed less agreement with higher Mach-numbers, higher reduced frequencies and shock-waves being present in the blade passage.

In the meantime, CFD-methods are under way to predict the 3D unsteady transonic flow field including viscosity and unsteady boundary layer behavior. Moreover, the computation is extended to include the flow through the adjacent upstream and downstream blade rows, thus simulating the flow field due to blade row interaction.

From this rapid development of the computational methods it may be expected that the above mentioned differences between former inviscid flow-calculations and measured data of the 2D-cascade standard configurations will be reduced in the near future. Since these cascade tests and the time-saving 2D prediction methods are powerful tools for the practical use, they will remain a necessary compliment to investigations in rotating machines and should be accomplished to a satisfying level.

In order to include the blade-row interaction into the prediction methods for the resulting time-dependent spanwise aerodynamic load distribution on the blading, measurements of the unsteady flow behavior in stages and multi-stage machines are urgently needed. The benefit of a much better physical understanding of the complex unsteady real flow through machines will improve not only the aero-elastic design of a turbomachine, but -as is well known - many other required properties, e.g. higher efficiency, sufficient range of operation, increased compactness etc.

By this general engagement for a better understanding of the unsteady blade-row interaction in stages and multi-stage machines, the experimental methods and analysis have been pushed intensively, and not only by aeroelastic requirements. Within the last two decades the tools necessary for unsteady flow measurements in turbomachines have been developed to a state of maturity that has encouraged many research groups in industries and universities to improve the understanding and the theory of the unsteady viscous 3D-flow in stages and multi-stage machines by highly sophisticated experimental analyses.

Therefore, the author of this paper will concentrate his review on recently achieved progress in the unsteady aerodynamics of axial-flow machines. There might be a need of this review as the original publications are spread in so many various journals. In favor of the pertaining experimental work done by many research groups it should also be encouraging to find an optimistic answer on a question raised very often in recent time, and also in the Proceedings-Review of the 1989 AGARD Conference on "Unsteady Aerodynamic Phenomena in Turbomachines". There, this question apparently arises considering the stimulating reminder in the following statement: "The large numerical Codes for viscid or inviscid, unsteady flows in turbomachinery will be more and more developed, although the experimental work for validation of these codes is lagging behind them".

This warning statement should enhance the experimental activities in spite of the hopeful level of the state of the art that has already been achieved until now.

Simulation of Real Machine Flow Environments in Low or High Speed Rotating Rigs

The physics of the flow interaction in stationary vanes and rotating blade rows can be studied by arranging them similarly as in the real machine, but with the advantage to make use of scale model rotors and stages and of special rig constructions providing easy access for all the instrumentation needed to identify and quantify the events in spatial unsteady flow fields.

Variations in gust amplitudes and reduced frequencies can be realized quite easily. Using various numbers of rotor and stator blades permits the investigation of inter-blade phasing and reduced frequency effects on both the blading of rotor and stator airfoils. Rotating rigs also allow the experimental analysis of spanwise effects which are of increasing influence for large aspect ratio twisted bladings with varying inlet and exit flow fields along the blade span including tip leakage and secondary flow effects due to viscous flow. Another very important advantage of rotating rigs can be seen in the possible study of periodical flow separation and stall effects which can lead to high aerodynamic blade loads. Such studies are necessarily to be performed in order to improve the mathematical modelling of the complex three-dimensional, viscous, unsteady flows occurring in the real turbomachine, and to obtain the realistic rotor and stator blade surface load distribution. These studies also require measurements of the unsteady flow on the rotor and in the rotating frame of reference. Although very expensive, these measurements are necessary for both the development of satisfactory flow models and the validation of computational methods.

In many cases, disagreements between measured and computed data are due to omitting major features of the real machine flow physics in the computational simulation. From this point of view, quite a number of additional targets of aerodynamic measurements in rotors and stages become important. They refer to the whole range of parameters influencing the steady and unsteady spatial viscous flow field around rotor and stator blades. Special attention is to be paid to well-known physical interaction events and to their consequences as listed below:

- Blade row interactions (potential and viscous flow interaction)
- Rotor and stator blade wake transport and decay
- Vortex transport and cutting
- Behavior of boundary layers in unsteady flow (transition, separation, rotating stall)
- Endwall influences and secondary flow effects (channel vortices etc., tip clearance vortex, corner stall)
- Shock-boundary layer interaction
- Turbulence effects
- Radial mixing
- Unsteady aerodynamic blade load distribution
- Blade vibration

Most of these phenomena influence each other intensively according to machine and blade design and operating data. Thus, Fig. 1 demonstrates the dependency of the spanwise unsteady blade load distribution on the above-mentioned phenomena. This distribution can be introduced then as time-depending forcing function into the equation of motion of the elastic blade system.

The main objectives of such very detailed and expensive unsteady flow field measurements on rotating rigs are - at first - a better understanding of the physics of the complex interaction phenomena for improving the flow models and then in - particular -providing a data base for the validation of advanced CFD-methods for flow field prediction.

Measuring Techniques for Unsteady Flow Field Measurements in Rotating Rigs and Real Axial-Flow Machines

For the study of the flow interaction in stator/rotor combinations such as stages or multistage machines and for establishing a data base for the validation of CFD-methods it is very important to make sure that each one of the above mentioned flow phenomena can be detected by at least one available measuring technique at as many measuring stations as necessary for a sufficiently fine determination of the 3D-flow field. In order to minimize errors due to uncertainties of the applied individual measuring technique, it is very useful to provide corresponding data by other techniques. It is well known that measurements of the unsteady flow data need a highly sophisticated data acquisition and evaluation technique. Redundant measuring techniques are preferably used to cross-check the results and increase the reliability of the measured data.

Appropriate techniques for unsteady flow measurements have been developed with increasing success for about 20 years . They have been

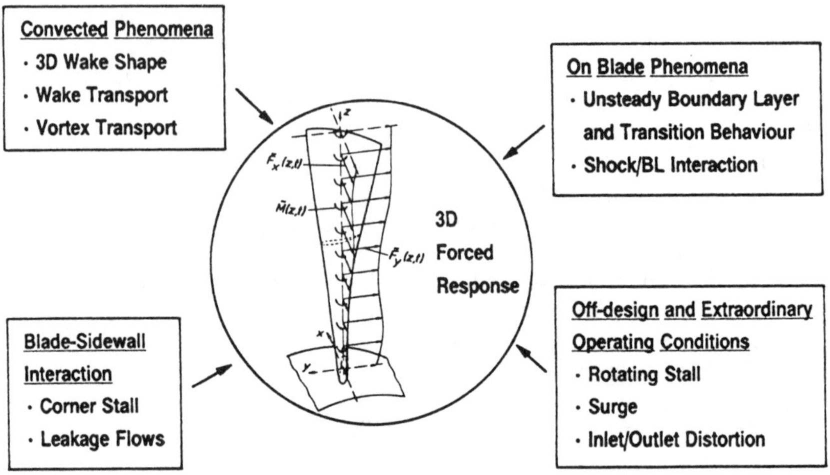

Fig. 1: Forced Response through 3D Unsteady Flow Phenomena in Turbomachines

reported on in former review-papers and, therefore, will not be presented here once more (Platzer, M.F. and F.O. Carta, 1987; Fleeter, S. and R.L. Jay, 1978; Szechnenyi, E., 1987; Gallus, H.E. 1987).

Instead of that, the author would like to demonstrate in the following how these techniques are applied in rotating rigs and machines, and to provide an impression of their capacity and efficiency for the experimental and theoretical analysis of all the complex phenomena connected with blade row interaction and real viscous flow effects. This review will be given by presenting selected examples from recent literature.

Instrumentation for Measurements in Rotating Rigs

Fig 2. displays the instrumentation of an axial-flow turbine stage with adjacent stator downstream the rotor (Zeschky, J. and H.E. Gallus, 1991; Zeschky, J., 1991). This scheme can be considered as a representative example for the experimental investigation of the before mentioned rotor/stator interaction phenomena.

Fig. 2 provides a schematic description of the flow-field data to be measured, the measuring techniques being applied, and the arrangement of measuring stations inclusively the necessary traversing mechanisms for detecting the data in the whole flow field.

Pneumatic pressure Five-hole probes are used here to detect the time averaged data of the local flow velocity as well as pitch and yaw angles α and β. Within the flow angle of about 20 degrees against the axis of the center-hole the time-averaged value of the total pressure can be detected

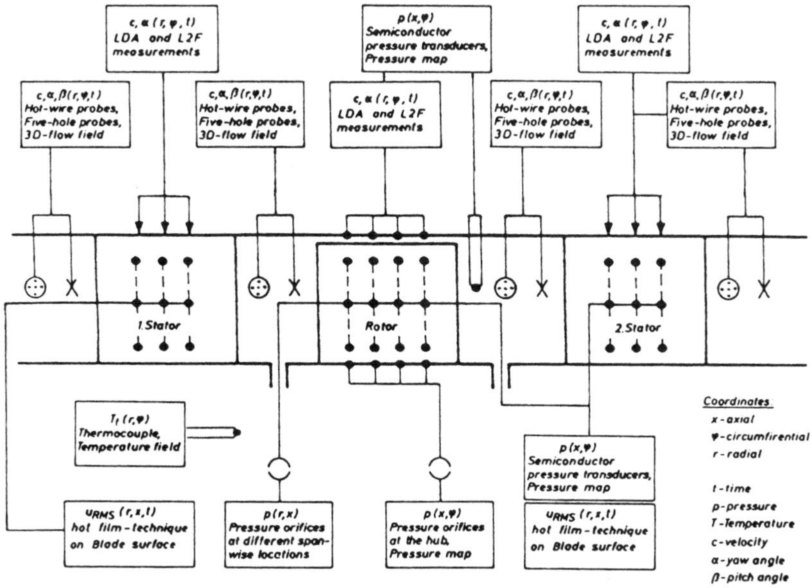

Fig. 2: Instrumentation of an Axial Turbine

locally. By radial and circumferential traversing the total pressure distributions can be measured in planes normal to the machine axis and in various axial stations within the axial gaps between the blade rows. The comparison of the pitchwise and spanwise averaged total pressures between two axial measuring stations delivers the total pressure losses, e.g. for determining the efficiency of each blade row.

Hot Wire Techniques

The 3D unsteady flow field can be detected by use of 3D hot-wire probes. Due to their high frequency response, they also permit measurements of the local pitchwise and spanwise flow-data downstream of the rotors. The data measured in the stationary frame of reference can be recalculated to the relative flow field in the rotating field of reference as is illustrated in Figs. 3a and 3b. The data was measured in a plane 12 per cent of the axial blade width downstream of the rotor blade trailing edge. The presentation of the secondary flow field in the rotating frame of reference shows the physics of the flow events in the rotor passages even more impressively. The two passage vortices as well as the tip clearance vortex are quantitatively determined in both their position and extent. In order to analyze the rotor flow due to stator/rotor interaction, one has to compare with the inlet flow field upstream of the rotor. Therefore, in Fig. 4 the secondary flow field at stator exit is shown in the stationary frame of reference. Additionally, in Fig. 5 the strong fluctuations of the rotor inlet conditions in the rotating frame become evident in a comparison with the pitchwise averaged data. In the channel vortex regions and close to the sidewalls the relative inlet flow angle fluctuates locally between ± 20 degrees, thus leading to strong local incidence and dynamic load changes.

Fig. 6 presents the rotor exit velocity field in the rotating frame of reference measured by the 3-sensor hot-wire probe 12 per cent axial blade width downstream of rotor trailing edge. Compared with the stator exit flow, the centers of the low velocity regions have moved towards midspan.

With respect to changing operating conditions, the following Fig. 7 reveals considerable changes in the relative rotor exit velocity distributions for nominal, decreased and increased loading. Due to the low aspect ratio of the rotor blades (0.917) it becomes obvious, that the centers of the low velocity regions are nearly growing together at midspan with higher loadings. For comparison, Fig. 8 demonstrates this behavior in the secondary flow field plots for increased and decreased loadings.

Fig. 9 shows the influence of the measuring position of the stationary 3-sensor hot-wire probe downstream of the rotor in relation to the circumferential position of the upstream stator vanes. This Figure represents the secondary flow field at rotor exit measured over two blade spacings by the stationary triple-sensor hot-wire probe at four different circumferential positions relative to the stator vane trailing edge. Since the numbers of stator vanes and rotor blades were different, the cross flow fields differ over the two measured spacings for each position. The wakes and channel vortices from the stator vanes in front of the rotor are nearly stationary with respect to a fixed probe position. Thus, the measured flow field differs for the four probe positions as can be recognized in Fig. 9. The shaded zones in these secondary flow charts mark those regions in which the local degree of turbulence is higher than 10 per cent outside the rotor blade wakes and more than 15 per cent inside of them. These shaded regions mark - in agreement with results from LDV-measurements - the stator wakes after their transport through the rotor. They are almost locally fixed, whereas the rotor has moved during the space of time.

There is obviously a certain coincidence of higher flow turbulence in the shaded regions with the increased cross flow towards the blade suction side. This seems to correspond with the "negative jet model" well known from literature.

The relative rotor exit velocity field at the four rotor/stator positions is plotted in Fig. 10. Each of these distributions corresponds with the cross flow distributions in the left one of the two blade spacings plotted in Fig. 9. It can be concluded from Fig. 10, that the flow field at midspan is only slightly influenced by the stator wakes at this operating point.

The above presented selection of data achieved by the application of a 3-Sensor hot wire probe in the axial spacings of the blade rows demonstrates the capacity and efficiency of this technique in subsonic flows. For measurements close to the side-walls, they have to be replaced by other methods or probes, e.g. x-wire probes.

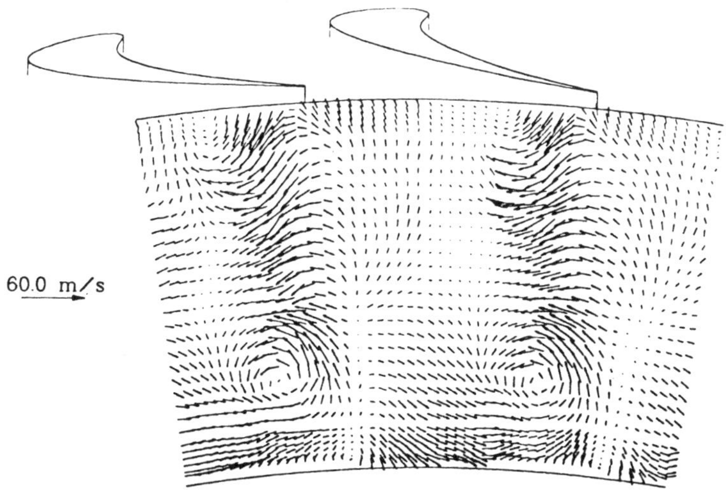

Fig. 3a: Secondary Flow Field in the Rotor Exit Plane x/b_p = 112 per cent
Absolute Frame

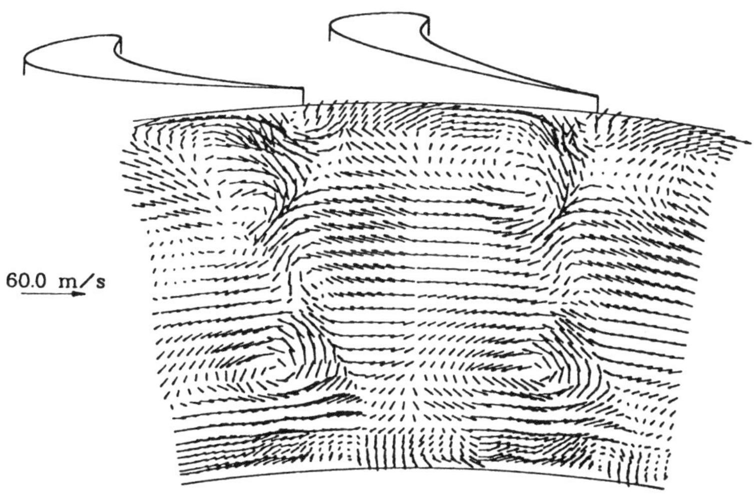

Fig. 3b: Secondary Flow Field in the Rotor Exit Plane x/b_p = 112 per cent
Relative Frame

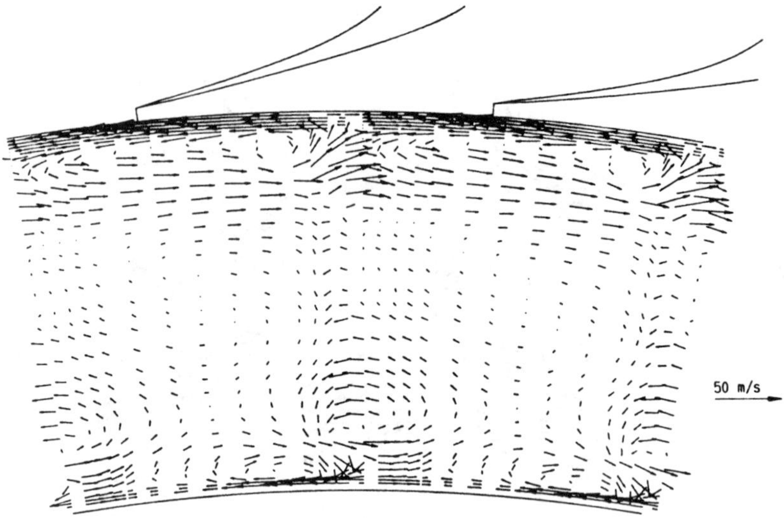

Fig. 4: Secondary Flow Field at Stator Exit

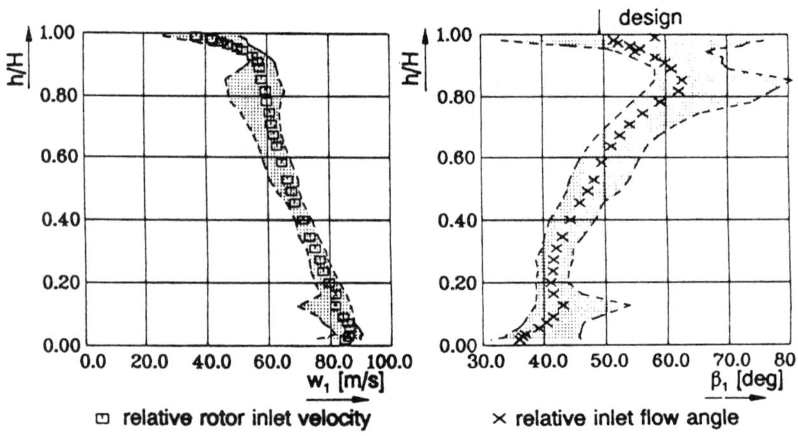

Fig. 5: Rotor Inlet Conditions, Fluctuations Shaded

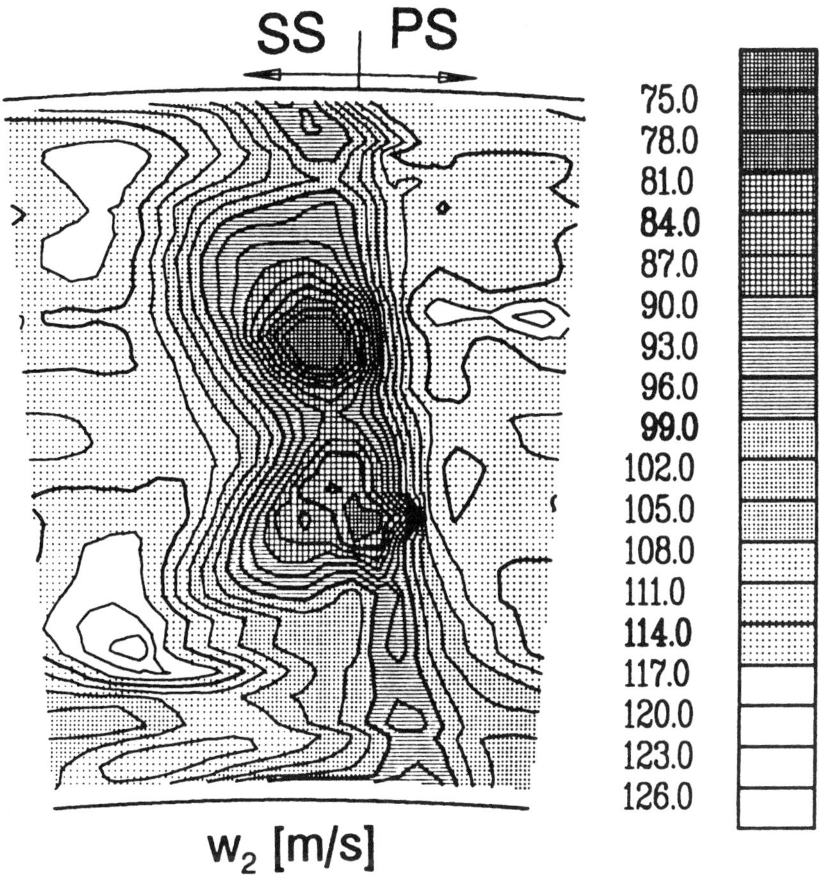

Fig. 6: Rotor Exit Velocity, Relative Frame x/b_p = 112 per cent

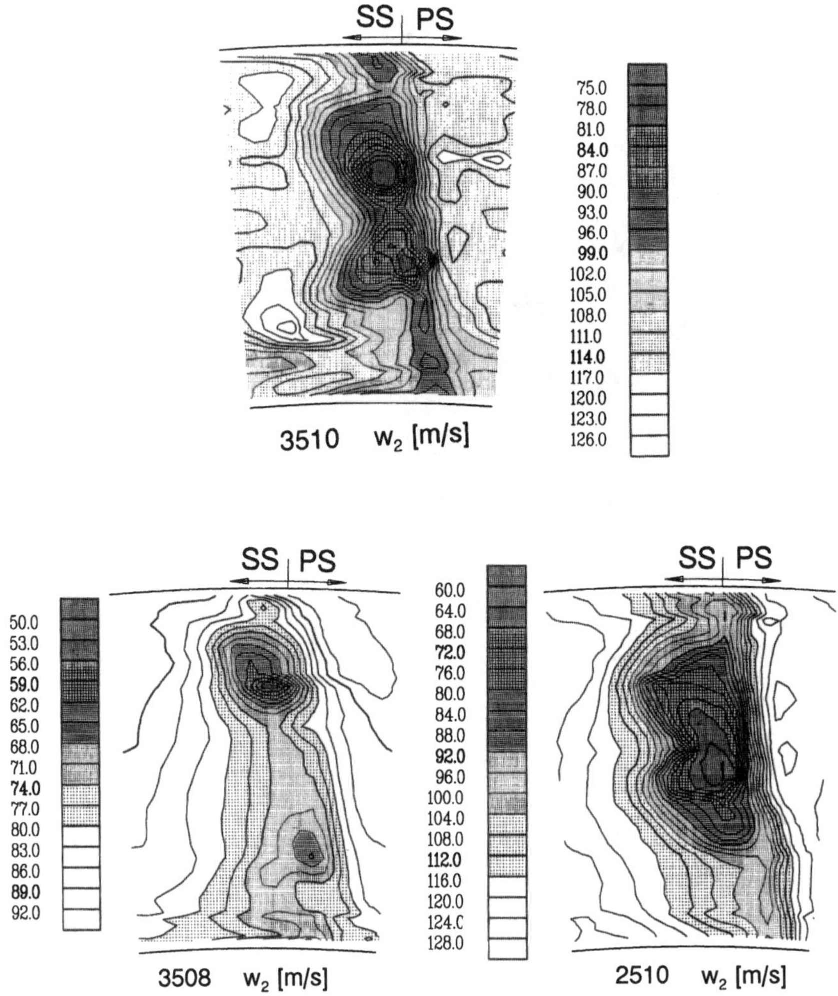

Fig. 7: Relative Rotor Exit Velocity with Nominal, Decreased and Increased Loading

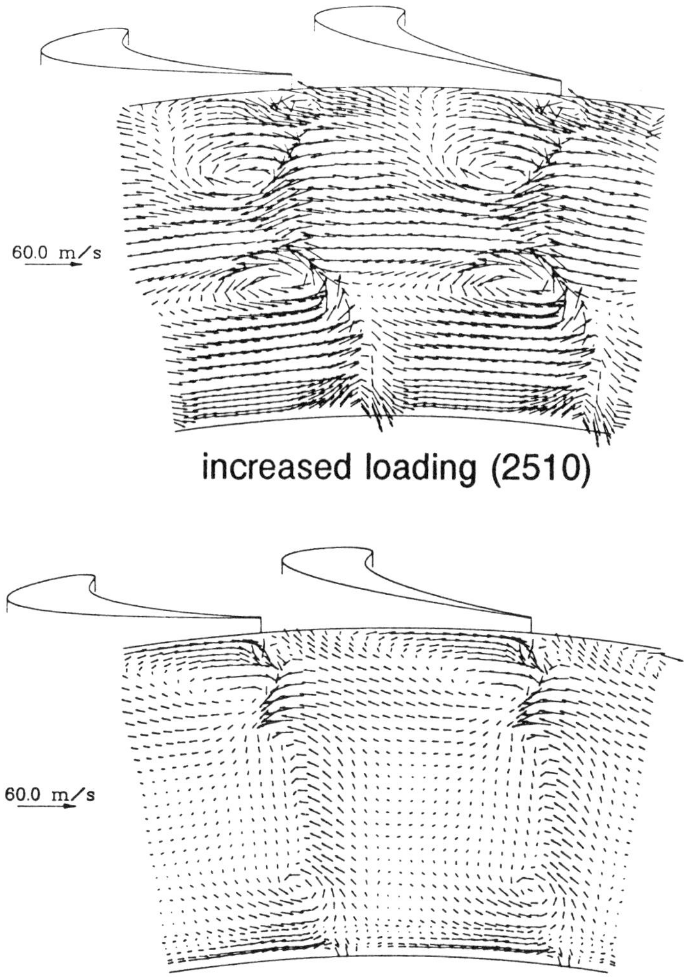

Fig. 8: Secondary Flow Field at Rotor Exit Relative Frame

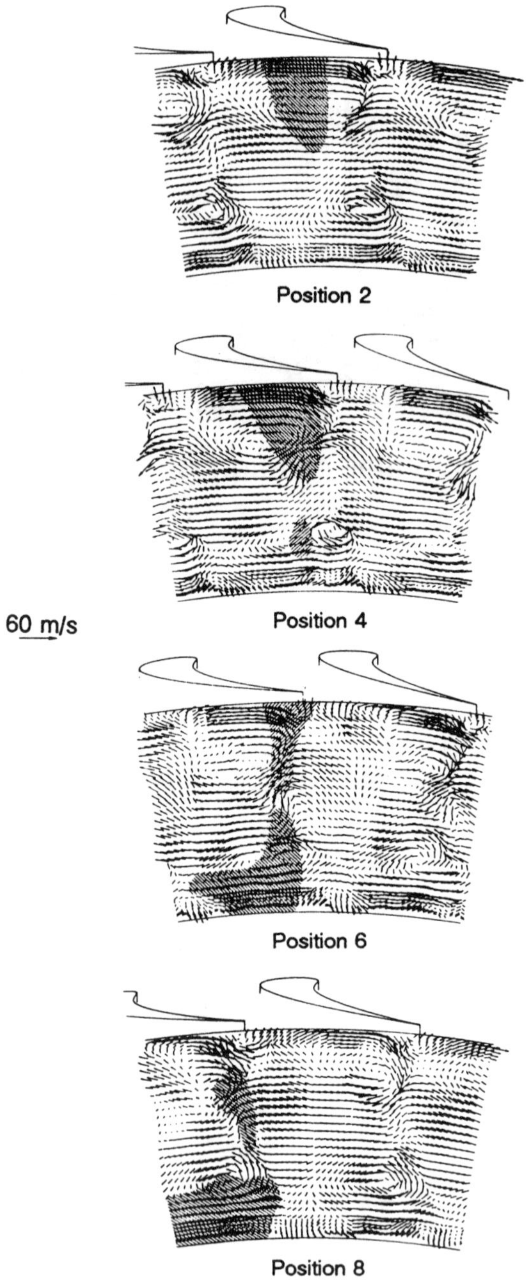

Fig. 9: Secondary Flow Field at Rotor Exit for 4 Rotor/Stator Positions
Relative Frame

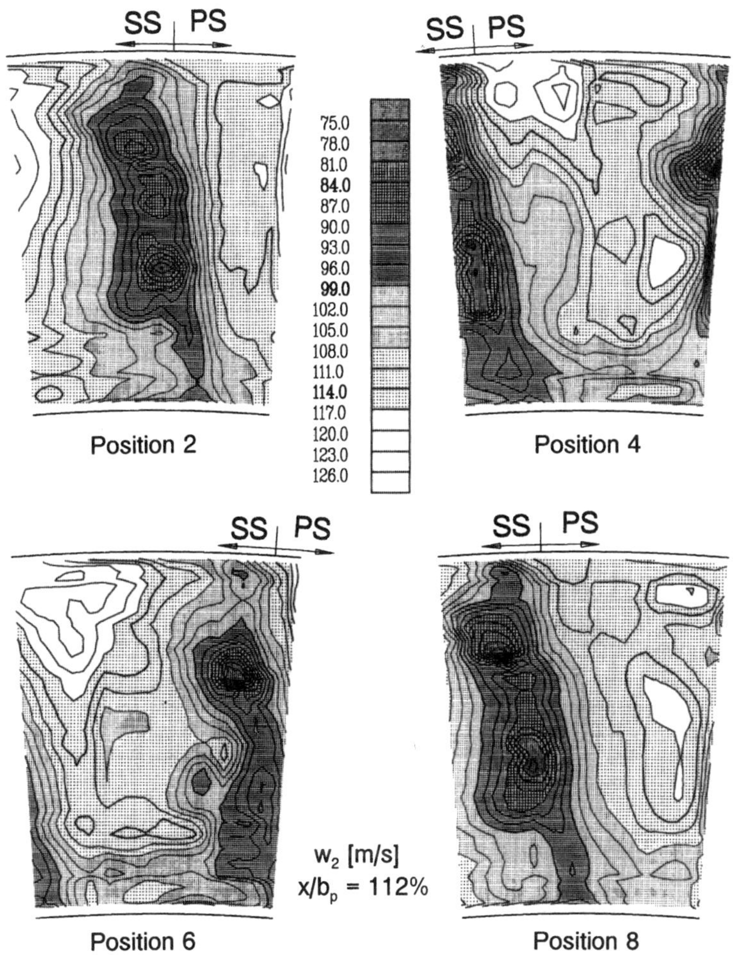

Fig. 10: Relative Rotor Exit Velocity at Four Rotor/Stator Positions

Returning to the instrumentation scheme in Fig. 2, 3-sensor hot-wire probes can also be used to measure the 3D unsteady flow field in the interior of stator vane channels apart from wall and vane surfaces. In the latter regions x-wire probes or 1-wire probes can be used.

An example of triple-sensor hot wire measurements in and downstream of a compressor stator vane passage was published by Ch. Poensgen and H.E. Gallus, 1990. Fig. 11 shows the cascade geometry at midspan and the locations of measurement. The inlet flow angle to the stator was provided by variable inlet guide vanes. Their wakes could be mixed out downstream to the stator cascade. In order to stay compatible with former investigations a rotor with cylindrical spokes was used. The aim of these investigations was to determine the influence of unsteady wake flow on the flow field inside the stator passage. Using cylindrical spokes instead of rotor blade profiles the wake parameters, i.e., wake width, wake depth, and reduced frequency could be kept constant with changing incidence.

In Fig. 12 the decay characteristics of the unsteady flow data (velocity, turbulence, turbulent stress) are plotted together with the cross flow field for the measuring locations 1, 3, 6, and 10. The stator loading was moderate due to an incidence angle of only 2 degrees. The decay characteristics shown in Fig. 12, change dramatically with an increase of incidence to 7 degrees due to increased corner stall effects, as it is depicted and explained in the original paper.

LDV- and L2F-Techniques

For unsteady flow field measurements inside the rotor blade passages LDV- and L2F- techniques have proved to provide sufficient and reliable data. In Vol. 1 of the AGARD-Manual on "Aeroelasticity in Axial-Flow Turbomachines" the author of this review reported on the development of these techniques and also on their application to rotor-flow measurements. Therefore, in the following, only some more examples of application and measured results will be presented with respect to stator/rotor interaction.

According to the scheme of instrumentation of an axial-flow turbine (Fig. 2), these LDV/L2F-techniques shall deliver the rotor-blade passage flow field. For example, the LDV-measurement arrangement is illustrated in Fig.13 (J. Zeschky, 1991).

In the following, some results of these measurements will be presented. Figs. 14 to 17 show the transport of the stator wakes through the rotor passage. The LDV-measurements were carried through over one stator pitch. For eight rotor/stator positions a stop-action sequence was determined which can be understood as "photographs" of the flow field. The rotor turns over 25 per cent of the blade pitch between the positions 2,4,6, and 8. Each figure refers to the same position in the stationary frame of reference, while the rotor blades are passing through the measurement plane from right to left.

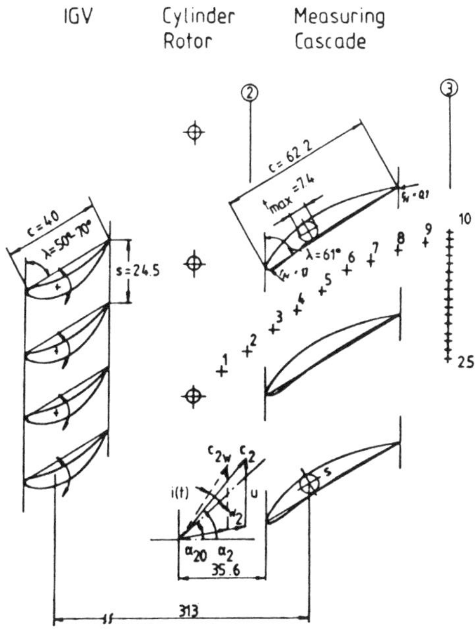

Fig. 11: Cascade Geometry at Midspan and Locations of Measurement

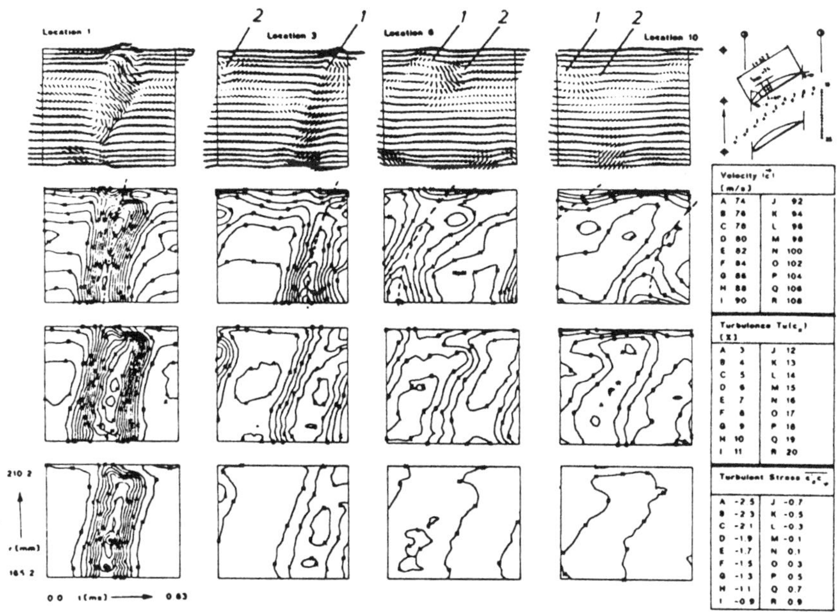

Fig. 12: Decay Characteristics of Unsteady Flow Quantities
(Incidence Angle of 2°)

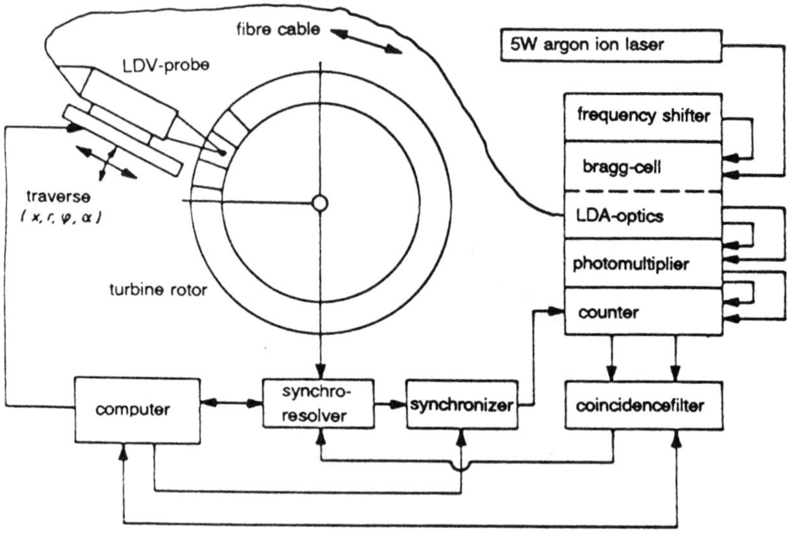

Fig. 13: LDV Measurement Arrangement

Fig. 14 represents the relative flow angle distributions at the four rotor/stator positions at an axial measuring station of 13.9 per cent axial rotor blade width from the leading edge. At rotor position 8, no stator wake is visible in the blade passage. In the tip region, the higher circumferential speed and the lower acceleration cause high flow angles. At rotor position 2 increased flow angles indicating cross flow toward the suction side occur in the corner between pressure side and casing. This is due to the lower velocity and the underturning in the high-loss area of the stator exit flow. At position 4, the stator wake is visible along the entire span, with strong distortions at h/H = 15 per cent and 85 per cent span. At position 6, the flow angles decrease after the wake has passed. Position 8 shows an almost undisturbed flow field.

In Fig. 15 the turbulence distributions for the same rotor/stator positions as above were plotted with the assumption of an isotropic distribution. Since a local passage-averaged velocity could not be calculated, the averaged rotor exit velocity was used as a reference. Corresponding to Fig. 14, the turbulence level is rather low at rotor/stator position 8. At position 2, turbulent spots associated with the stator loss cores appear, and at position 4, the stator wake can be detected along the whole span. At position 6, the wake has almost passed with diminishing turbulence. From Figs. 14 and 15 it becomes evident, that the stator wakes can clearly be identified in the front part of the passage. At about 15 per cent and 85 per cent span, the passing wake causes strong cross flow components to the suction side with an increase of 10 degrees of the flow angle and an almost two-times higher turbulence level.

Further downstream in the blade passage, increased fluctuations occur when the stator wakes cross the measurement planes. Transport and

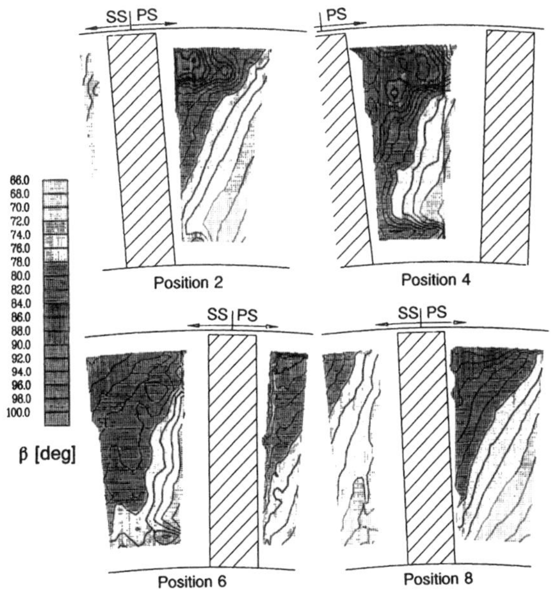

Fig. 14: Relative Flow Angle at 4 Rotor/Stator Positions x/b_p = 13.9 per cent

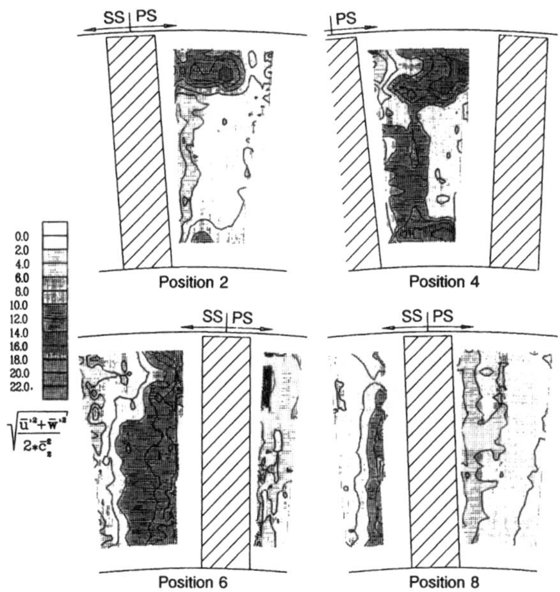

Fig. 15: Turbulence Intensity at 4 Rotor/Stator Positions x/b_p = 13.9 per cent

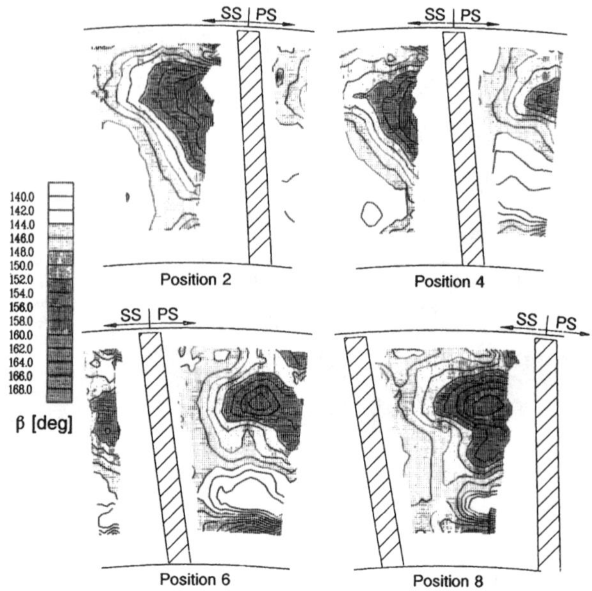

Fig. 16: Relative Flow Angle at 4 Rotor/Stator Positions x/b_p = 94.4 per cent

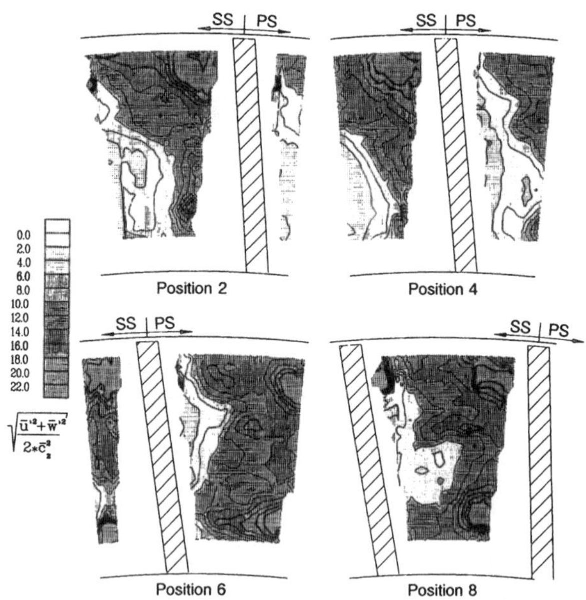

Fig. 17: Turbulence Intensity at 4 Rotor/Stator Positions x/b_p = 94.4 per cent

mixing cause the accumulation of highly turbulent fluid on the suction side and less accentuated edges of the wake. These conditions in the downstream part of the rotor blade passage close to the trailing edge (94.4 per cent axial blade width) are plotted in Figs. 16 and 17.

The relative flow-angle in Fig 16 shows the lowest disturbance at position 4, whereas at positions 6 and 8 intense cross flow appears which becomes weaker at position 2. Nevertheless, the cross flow at about 70 per cent span is present to a considerable extent for all rotor/stator positions. This is explained by the larger extension of the stator wakes in the tip region. As Fig. 17 shows, the turbulence level is generally high at the suction side and in the corner between suction side and casing. The zones of increased turbulence near the hub coincide with a fluctuating passage vortex. In the stator wakes, which act as "negative jets" in the rotor blade passage, the turbulent fluid is transported towards the suction side.

The notation "negative jet" for the transport of low energy wake material to the suction side of the blade channel was introduced by R.X. Meyer (1958). Fig. 18 describes the effect of a slip velocity w_{slip} between the velocity triangles of jet and wake flow at the passage inlet. Within the passage, the low energetic fluid is transported to the blade suction side due to the blade-to-blade pressure gradient in the channel. Using a 2D finite volume method, H.P. Hodson (1985) computed the relative flow field in a turbine rotor, as is impressively shown in Fig. 19. Very recently, J. Zeschky (1991) succeeded in measuring the blade-to-blade velocity field as well as the associated turbulence distribution in a rotor blade channel using the LDV-technique introduced above. Figs. 20a and b display these results for a measurement at 72,7 per cent span. They demonstrate the transport and deformation of the stator wakes on the way through the rotor blade channel at a span position where the highest flow fluctuations occurred. By time-averaging, a mean relative rotor flow field could be calculated for eight rotor positions. The velocity difference vectors between the mean flow field and the locally measured unsteady flow vectors are mapped on the left side of Figs. 20a/b, whereas the turbulence field is represented on the right side. For each rotor position, two wakes can be recognized in the channel simultaneously. At rotor position 4, a stator wake is just entering the rotor blade channel at midspace while the wake, which entered this channel before, can be recognized by the slip velocities towards the blade suction side at about 70 per cent chord. Correspondingly, the turbulence distribution map shows an accumulation of high turbulent fluid along the suction side. In rotor position 6, the cutting of the stator wake by the rotor-blade leading edge is demonstrated, which is associated with high turbulence near the leading edge. This effect was already investigated in detail by A. Binder (1985), who traced it back to the bursting of the vortex and the transformation of its kinetic energy into a turbulent one.

The dominating role of the LDV/L2F-techniques for flow field measurements inside of rotor blade as well as stator vane channels has turned out in numerous publications, e.g.

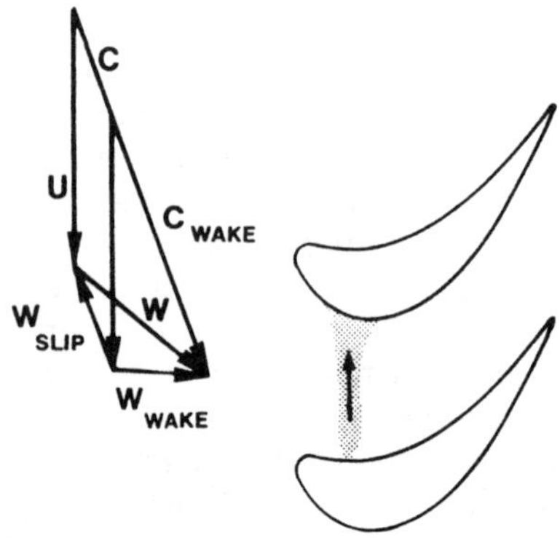

Fig. 18: "Negative Jet" model, Meyer (1958)

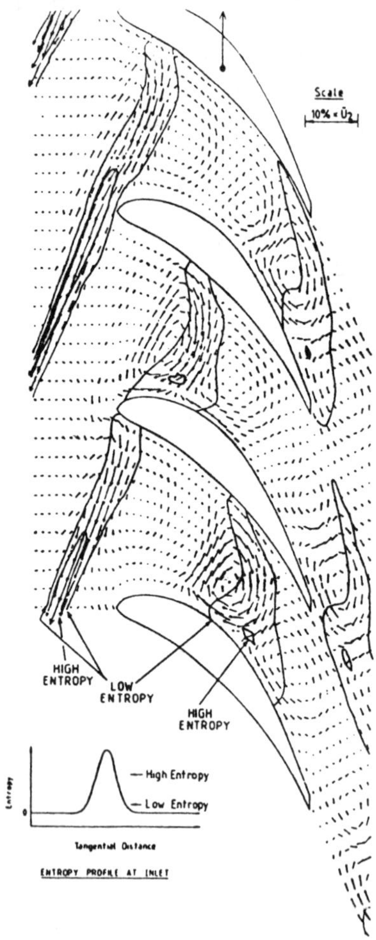

Fig. 19: Computed Relative Rotor Flow Field, Hodson (1985)

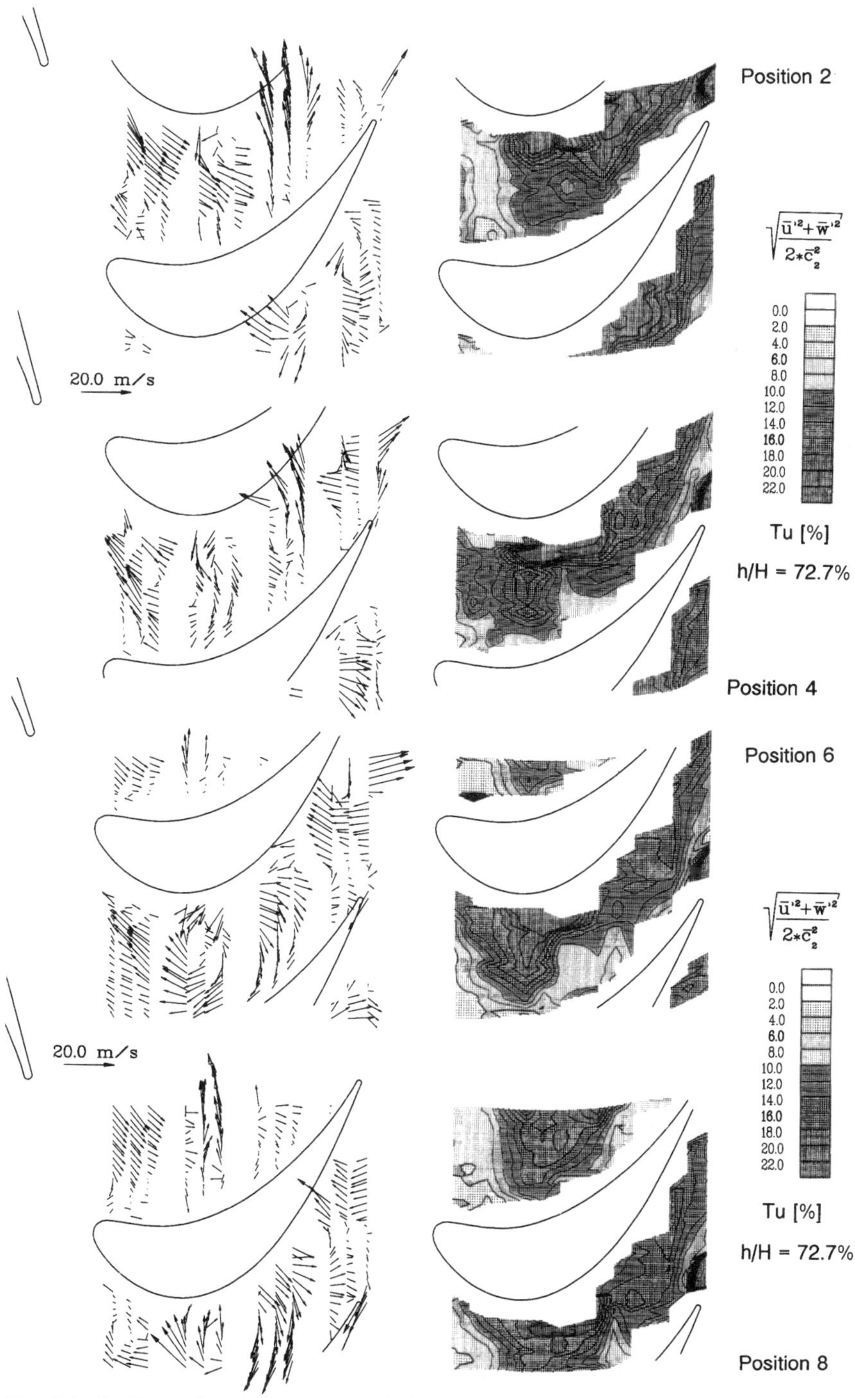

Fig. 20a/b: Secondary Flow and Turbulence Intensity

High Response Pressure Transducers for Unsteady Pressure Field Measurements

This technique is favored for the study of

- time-dependent static pressure distributions along the channel walls at hub and tip as well as on the blade and vane surfaces in order to measure the fluctuating aerodynamic blade loads,
- unsteady flow fields in passages by inserting these miniature transducers in probes.

Many contributions to both fields of application can be found in the literature of the last two decades. A review on the development of those techniques for their use in the stationary as well as rotating frame was presented by the author (H.E. Gallus, 1987). In the meantime, great progress has been achieved by further miniaturization of the sensors, so that they are utilized even more in sophisticated designs of probes for 3D unsteady flow field measurements.

For the application on channel walls, there will be presented some examples in the following. Looking again at Fig. 2, semiconductor pressure transducers are very often used to get detailed information on the unsteady static pressure distribution on the wall in the tip clearance region of the rotor blades. This wall-pressure distribution is extremely informative for the study of the leakage flow in the tip clearance, e.g. J. Zeschky (1991). Figs. 21a/b/c show the static pressure field at the casing above the turbine rotor, measured by a semiconductor pressure transducer at various locations in the wall along the blade channel.

The detailed comparison of the three very different pressure distributions corresponding to the different loading of the rotor blades provides a reliable basis for the study of tip leakage flow and for the validation of computational codes which are still to be improved, as the determination of leakage flow and losses is concerned. In the Fig. 21a, showing the static pressure distribution for an incidence angle near design loading, the dashed line connecting the points of lowest static pressure indicates the estimated location of the tip clearance vortex.

In trans- and supersonic flows, this technique is utilized to detect the shock waves at the rotor-blade tip, as is demonstrated in Fig. 22 according to measurements by R. Larguier (1980).

The measurement of static pressure fluctuations on stator vane and rotor blade surfaces by means of miniaturized high response pressure transducers have become a standard technique to get the time-dependent aerodynamic loads on rigid or oscillating blades. Thus, this technique plays a key role in the field of forced response and aeroelasticity. There has been reported on detailed measurements on rotor-blades, e.g. by R.P. Dring et al. (1982) for a turbine stage, and by H.E. Gallus et al.(1980) for a compressor stage. The voltage signals of the transducers in the rotor blades are to be transmitted to the stationary frame by a slip-ring system or by telemetric systems.

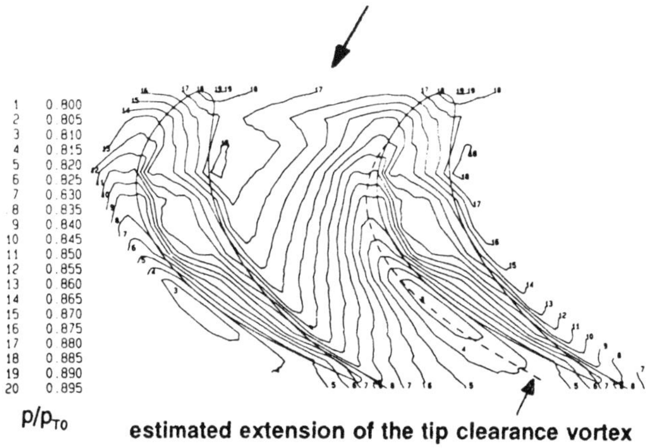

Fig. 21a: Static Pressure Field at the Casing Above the Rotor

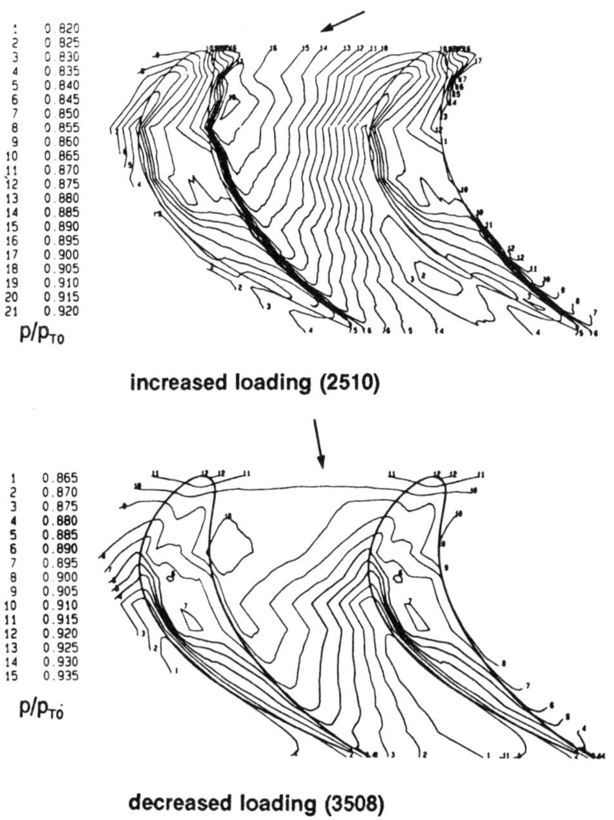

Fig. 21b/c: Unsteady Static Pressure Field at the Casing Above the Rotor

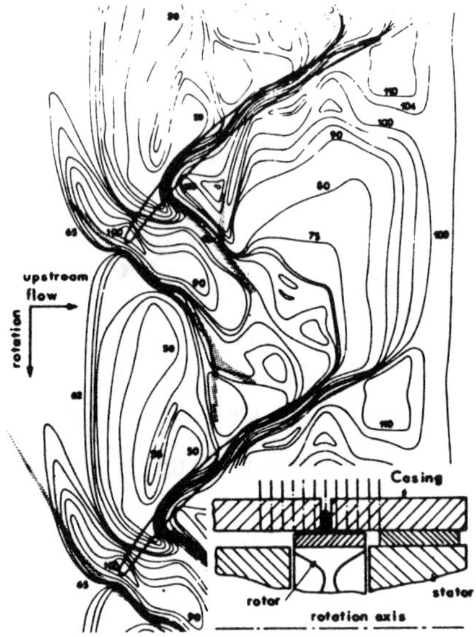

Fig. 22: Chart of Isobars Along the Casing Wall (mbar). Comparison with Shock Waves Observed by Schlieren Photography (Shaded Zones), Larguier (1980)

Surface-mounted Hot Films and Hot Wires for Studies of the Boundary Layer Behavior in Unsteady Flows

The flow unsteadiness due to rotor/stator interaction associated with increasing turbulence, with wake transport and cutting as well as wake impingement on the blade surface has a strong impact on the boundary layer behavior, e.g. boundary layer transition and separation. To achieve an improved knowledge of these complex flow mechanisms is of great importance for studies of loss reduction and stage design, accounting for the viscous and unsteady flow effects in the real machine environment.

The importance of this subject becomes evident from more and more investigations published recently, and also from the fact that this symposium also includes a whole session on "Unsteady Boundary Layer and Transition" and another one on "Unsteady Separated Flows".

To present only one example of boundary layer measurements on a stator vane surface of an axial-flow compressor, Fig. 23 illustrates the use of surface-mounted hot films (H.D. Schulz and H.E. Gallus 1988).

The photo shows the instrumented elongated blade incorporating 31 surface-mounted hot films along the suction side. The inset in this photo is a blowup of the sensor area. The vane elongation permitted the radial traversing of the vane and the sensors on it from hub to tip.

Fig. 24 represents the hot-film real time traces at the blade suction

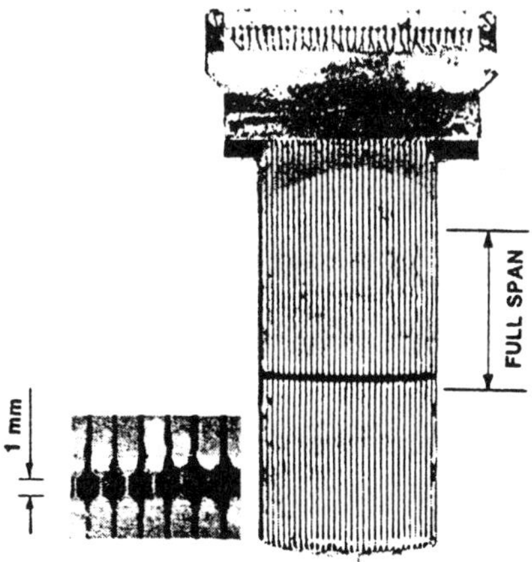

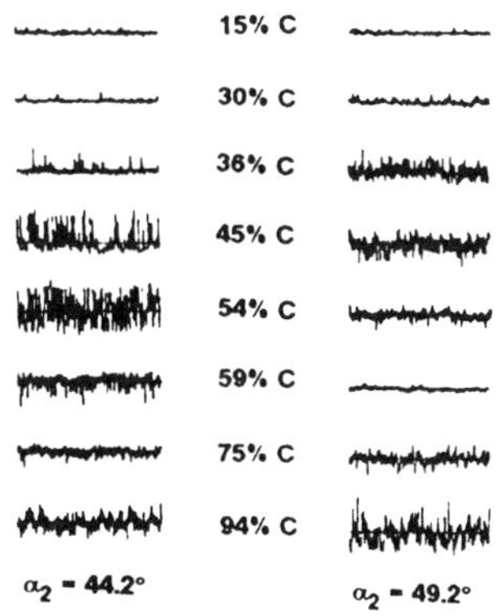

Fig. 23: left side - Instrumented Elongated Blade Incorporating 31 Surface-Mounted Hot-Films Along the Suction Side (the Inset is a Blowup of the Secondary Area)

Fig. 24: right side - Hot-Film Real Time Traces at Blade Suction Surface (Midspan)

surface (midspan) for two different incidence angles at +2° on the left side and +7° on the right one. From top to bottom, the traces reveal the chordwise transition of the boundary layer from laminar flow via a separation bubble to turbulent flow. With increasing incidence, the transition takes place further upstream. Fig. 24 shows the boundary layer development for a stationary flow without a rotor upstream. The variable flow incidence was provided by variable inlet guide vanes far enough upstream the stator cascade to provide a steady inlet flow field.

For the case of the interaction with an upstream rotor further measurements in the suction side boundary layer of the above stator vane were carried through by use of single-wire hot film probes at a very close distance of only 0.15 mm from the blade surface. They were reported by H.D. Schulz et al. in 1990.

Various Additional Measuring Techniques

Flow visualization techniques are very informative even when the results very often have only a qualitative character. Nevertheless, with increasing turbulence and flow velocity most of the methods can no longer deliver accurate results. Schlieren-optical investigations as reported by H.E. Gallus et al. in 1977, for the study of rotor/stator interaction in an axial-flow supersonic compressor stage, may be considered as one of the exceptions.

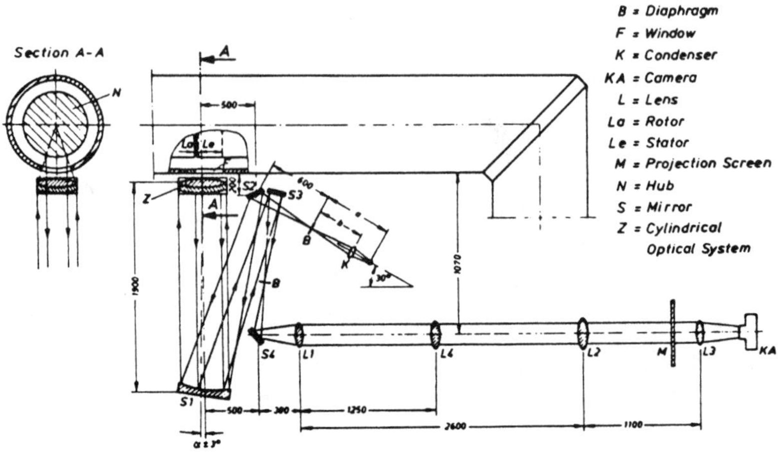

Fig. 25: Schlieren Optical System for Flow Visualization in Supersonic Compressor Stages

Due to a stroboscopic light source that is triggered by a rotor blade, it is possible to achieve a stationary view of the time-dependent flow field. By a very slight frequency shift of the stroboscopic lamp, the unsteady flow field can be studied in slow motion recorded by a video-camera. This Schlieren-optical system is illustrated in Fig. 25. In this arrangement with a cylindric lens system, it can be used only in case of cylindric side walls at the light reflecting hub and the window of glass at the casing. Low blade heights are another restrictive requirement.

For the study of mixing effects, tracer-gas methods have been developed, e.g. the ethylene tracer-gas technique used by D.C. Wisler et al. (1987), to investigate the secondary flow, turbulent diffusion, and mixing in axial-flow compressors.

More information on further techniques and authors can be taken from former review papers mentioned in the introduction of this presentation. The list of References contains further papers on rotor/stator interactions, that have played an important role in the research of these difficult flow mechanisms.

Summary

It has been pointed out in the introduction, that the rapid development of the computational methods recently surprised by first approaches to the prediction of the complex viscous and unsteady flow in stages and multistage axial-flow turbomachines. Facing this progressive development of CFD-methods and the need of their validation by experimental data, the question is open whether the state of the art of experimental techniques is on that level to provide the data, or whether it is lagging behind the development of CFD-codes. Until now, the answer can be neither "yes" nor "no". Reasons for this valuation will be given in the following.

An axial-flow turbine stage has been selected for the demonstration of all the measuring techniques necessary for a widely complete description of the very complex flow fields in stator and rotor due to their interaction. It could be shown by the presentation and discussion of many fields of data, that the measuring techniques for the unsteady flow field analysis of the periodic rotor/stator interaction have reached a very high standard. Many of the interdependent flow phenomena whose physics is not yet sufficiently understood could be measured quantitatively and reasonably interpreted.

Restrictions appeared essentially for flow field measurements close to walls where some methods fail or loose their accuracy owing to sensor-/wall interaction.

Nevertheless, apart from these problems the data that can be provided by the advanced measuring techniques (high response sensors, signal amplification, data acquisition and processing, immediate data reduction and display) represent a powerful base for the validation of advanced computational methods.

Since measurements of the unsteady flow field in stages and multi-stage machines as well as in rotating rigs are extremely expensive, all the data of machine design and operating points should be stored together with the measured data to serve as "Standard Configurations" similar to those of 2D-cascade investigations. With respect to the progressive advances in CFD-methods, no time should be wasted in contributing to such a data base of some very precisely and carefully investigated rigs and machines.

Further fundamental investigations on the unsteady boundary layer behavior under simulated conditions of the real machine flow should be preferred targets for future research, too. This includes also the need of a better understanding of flow separation and rotating stall.

References:

1. AGARD-Conference Proceedings No. 468 "Unsteady Aerodynamic Phenomena in Turbomachines", 1989, Luxembourg.

2. Binder, A., Förster, W. Kruse, H., and Rogge, H.: "An Experimental Investigation into the Effect of Wakes on the Unsteady Turbine Rotor Flow", Trans. ASME, 1985 a, Vol. 107.

3. Binder, A., Rogge, H., Kaurnatschke, W.: "Sekundärströmungen im Laufrad einer Turbine", VDI-Berichte 527.2, 1985 b.

4. Binder, A.: "Turbulence Production due to Secondary Vortex Cutting a Turbine Rotor", J. of Eng. for Gas Turbines and Power, Vol. 107, 1985 c.

5. Bölcs, A. and T.H. Fransson: "Aeroelasticity in Turbomachines - Comparison on Theoretical and Experimental Cascade Results", Communication du Laboratoire de Thermique Appliquée et de Turbomachines, No. 13, Lausanne, EPFL 1986.

6. Dring, R.P., Joslyn, H.D., Harding, L.W. and Wagner, J.H.: "Turbine Rotor-Stator Interaction", ASME-Paper 82-GT-3, 1982.

7. Fleeter, S. and R.L. Jay: "Unsteady Aerodynamic Measurements in Flutter Research", Vol. 1 in AGARD-AG-298, 1987.

8. Gallus, H.E.: "Unsteady Aerodynamic Measurements on Rotors", Vol. 1 in AGARD-AG-298.

9. Gallus, H.E. and S. Servaty, editors: "Unsteady Aerodynamics and Aeroelasticity of Turbomachines and Propellers", Proceedings 4. International Symposium, Aachen 1987, Mitteilung Nr. 88-1, Institut für Strahlantriebe und Turboarbeitsmaschinen, RWTH Aachen, 1988

10. Gallus, H.E., Bohn, D. and Broichhausen, K.D.: "Measurements of Quasi-steady and Unsteady Flow Effects in a Supersonic Compressor Stage", Trans. ASME, J. of Eng. for Power, Vol. 99, 1977.

11. Hodson, H.P.: "Measurements of Wake-Generated Unsteadiness in the Rotor-Passages of Axial-Flow Turbines", J. Power, Vol. 107, 1985.

12. Hodson, H.P. and J.S. Addision: "Wake-Boundary Layer Interactions in an Axial-Flow Turbine Rotor at Off-Design Conditions"., ASME-Paper 88-GT-233, 1988.

13. Jay, R.L. and S. Fleeter: "Unsteady Aerodynamic Measurements in Forced Vibration Research", Vol. 1, AGARD-AG-298, 1987.

14. Larguier, R.: "Experimental Analysis Methods for Unsteady Flows in Turbomachines", ASME Paper No. 80-GT-71, 1980.

15. Mayer, R.X.: "The Effects of Wakes on the Transient Pressure and Velocity Distribution in Turbomachines", Trans. J. of Basic Engineering, 1958

16. Platzer, M.F. and F.O. Carta, editors: AGARD manual on "Aeroelasticity in Axial-Flow Turbomachines", Vol. 1 "Unsteady Turbomachinery Aerodynamics", AGARD-AG-298, 1987.

17. Poensgen, Ch. and H.E. Gallus: "Three-Dimensional Wake Decay Inside of a Compressor Cascade and its Influence on the Downstream Unsteady Flow Field", Part I and II, Trans. J. of. Turbomachinery, Vol. 113, 1991.

18. Schulz, H.D., Gallus, H.E., and Lakshminarayana, B.: "Three-Dimensional Separated Flow Field in the Endwall Region of an Annular Compressor Cascade in the Presence of Rotor-Stator Interaction", Part 1 and 2, J. of Turbomachinery, Vol. 112, 1990.

19. Schulz, H.D. and Gallus, H.E.: "Experimental Investigations of the Three-Dimensional Flow in an Annular Compressor Cascade", ASME J. of Turbomachinery, Vol. 110, 1988.

20. Szechenyi, E.: "Unsteady Fan Blade Flutter through Linear Cascade Aerodynamic Testing", Vol. 1, AGARD-AG-298, 1987.

21. Wisler, D.C., Bauer, R.C. and T.H. Okiishi: "Secondary Flow, Turbulent Diffusion and Mixing in Axial-Flow Compressor", ASME-Paper 87-GT-16, 1987.

22. Yamamoto, A.: "Instantaneous Three-Dimensional Flows due to Rotor-Stator Interaction in a Low-Speed Axial-Flow Turbine", European Conference on Turbomachinery, London 1991, Proceedings of the Institution of Mechanical Engineering, IMechE 1991-3.

23. Zeschky, J. and H.E. Gallus: "Effects of Stator Wakes and Spanwise Nonuniform Inlet Conditions on the Rotor Flow of an Axial Turbine Stage", ASME-Paper 91-GT-93, 1991.

24. Zeschky, J.: "Exprimentelle Untersuchung der dreidimensionalen insationären Rotorströmung einer axialen Kaltluftturbine", Dr. Thesis, RWTH Aachen, 1991.

Three-Dimensional Unsteady Flow in a Single Stage Axial-Flow Turbine and Compressor

H.E. Gallus[1]
C.A. Poensgen[1]
J. Zeschky[1]

Abstract

In this paper the importance of unsteady flow effects is demonstrated for the accelerated turbine and decelerated compressor flow. For the understanding of the entire core flow in turbomachines, it is necessary to investigate the unsteady three-dimensional flow field inside the blade passages. Here the flow is mainly disturbed by periodic unsteadiness due to rotor-stator interaction. In order to determine the unsteady flow quantities by means of the experiment, two problems have to be solved. First it is necessary to investigate the unsteady flow field downstream of a rotating cascade. In a second step the influence of the periodically disturbed inlet flow on the passage flow inside a downstream located blade row has to be resolved. In this paper experimental methods are presented to attack the above described problems. It is demonstrated how different experimental methods and techniques of measurements are employed and supplement each other. The combination of the obtained results lead to a very close description of the flow structures. The importance of the unsteady flow effects will be discussed while presenting the results of tested configurations.

Detailed investigations in a single stage turbine and a single stage compressor were carried out. Using multi-sensor hot-wire probes the unsteady flow field was recorded in the axial gaps and inside the stator blade passages. Additional measurements with subminiature pressure transducers on blade and endwall surfaces as well as in total pressure probes give provided information of the unsteady pressure fields. Inside

[1]Institut fur Strahlantriebe und Turboarbeitsmaschinen,
RWTH , D-5100 Aachen, Germany.

the turbine rotor blade passage LDV-measurements were carried out. The high spatial and temporal resolution provides distinct information of the flow phenomena and also demonstrates the progress of the data acquisition systems and measurement techniques.

Introduction

The unsteady rotor-stator interaction is of renewed interest in the current turbomachinery research. This is partly due to the progress in developing high-response measurement techniques and data acquisition systems. On the other hand, the first advanced numerical methods for the calculations of the unsteady 3d-flow through multi-stage machines are presented, Dawes (1990), Rao and Delaney (1990).

Most of the present analysis and design methods in turbomachinery are based on steady aerodynamics. The unsteady flow associated with blade row interaction has a major influence on the flow field, boundary layers, turbulence intensities, separation, vibration, heat transfer and noise. The demand for higher performance in aircraft engine design results in a close spacing of the blade rows, lower aspect ratios and a reduced number of blades. This requires more sophisticated design methods, including the possibility to calibrate numerical schemes and their turbulence modeling. There is a need to get a better understanding of the three-dimensional unsteady flow in a blade row including the influence of rotor-stator interactions.

Besides the potential interaction the cutting and transport of wakes through a downstream blade row is a major source for the unsteadiness. Meyer (1958) presented an early model of the wake convection as a negative jet in the downstream blade row. Kerrebrock and Mikolajczak (1970) used the model to explain the temperature redistribution by compressor rotor wakes passing through a stator. Very detailed measurements of Adachi and Murakami (1979) in a cascade with an upstream wake generator confirmed the model further. Additionally, the incoming stator wakes have a strong impact on the rotor profile boundary layers. Hodson (1983) reported a 50 per cent increase of the profile losses compared to cascade with steady inflow. This is explained by the early transition of the suction side boundary layer, which is likely to occur, when the turbulent wake impinges on the blade surface. Unsteady transition and the oscillation of the boundary layer between turbulent and laminar state was also found by Evans (1978). Addison and Hodson (1991) presented a model of the development of unsteady transitional boundary layers. The influence of wake disturbed inlet flow on the overall performance of a compressor stator row was investigated in detail by Schulz (1990). Although the profile losses increase, he found an overall

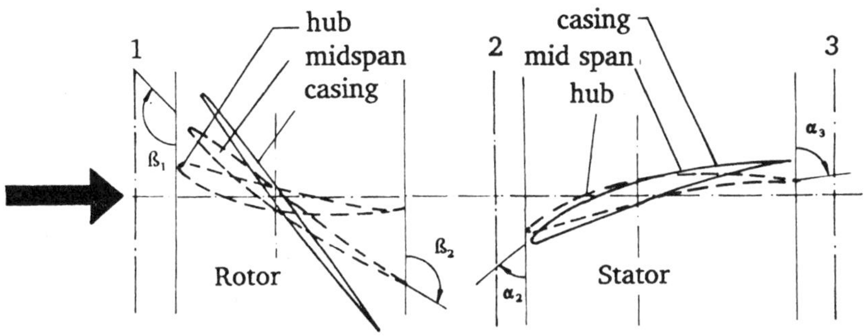

Fig. 1: Cross Section though the Blading of the Axial Flow Compressor Stage

loss reduction due to intense mixing effects which reduce the size of separated flow regions.

In recent time detailed experimental investigations of rotor-stator and rotor-rotor interactions were carried out by Binder et al. (1989) and Schröder (1989) in a five stage turbine. Arndt (1991) showed that rotor-rotor interaction in a multistage turbine causes a strong amplitude modulation of turbulence waves in the circumferential direction.

The objective of the present study is to investigate the characteristic flow parameters which govern the unsteady flow field downstream of a compressor rotor and inside the passage of a turbine rotor. Of particular interest is to show the application of different measurement techniques, which supplement each other and thus provide a detailed picture of the flow field.

Experimental facilities

Compressor

Fig. 1 displays a cross section of the single stage axial-flow compressor. The compressor stage represents an inlet stage of a multistage industrial compressor with a low hub to tip ratio of 0.37. More detailed information about further geometrical data can be found by Poensgen (1991). The rotor is driven by an electric direct current motor with speed control. A downstream located throttle valve controls the mass flow through the machine. The measurements were taken in the axial gap between the rotor and the stator, in front of the rotor and 16 per cent of chord downstream of the stator.

Turbine

The experimental work was carried out in a single-stage axial turbine with untwisted blades (Fig. 2). In the stator, the Traupel profile described by Utz (1972) was used, the rotor consists of modified VKI-profiles. A turbocompressor set provided a continuous airflow to the test rig. The total temperature at turbine inlet was set to 308 K ± 0.5 by cooling the air at the compressor outlet, the total pressure could be adjusted by bypassing part of the compressor mass-flow. With a shaft speed variation of less than ± 0.2 per cent during the run, the tests could be repeated with a flow coefficient variation of less than ± 0.7 per cent. The Reynolds numbers were constant with an accuracy of ± 1.0 per cent.

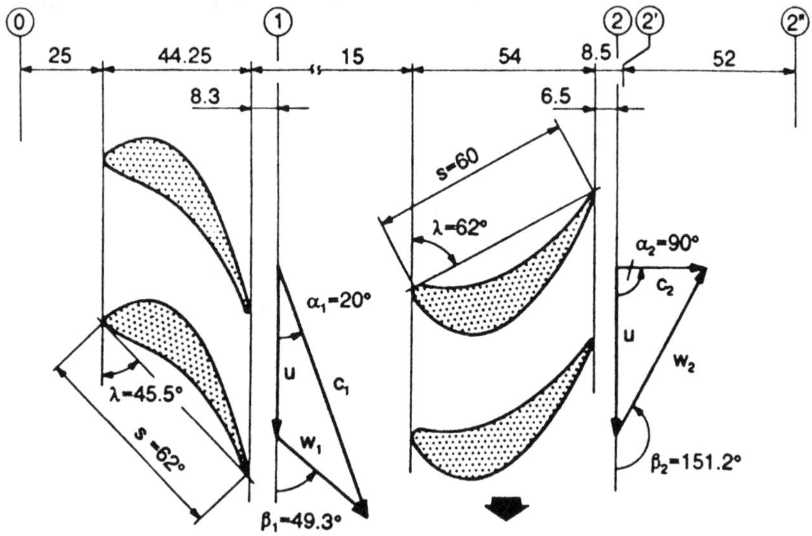

Fig. 2: Cross Section Through the Turbine Stage

Tab. 1: Design Parameters of the Turbine Stage

	stator	rotor
Aspect ratio h/s	0.887	0.917
Pitch (midspan) t	47.6 mm	41.8 mm
Blade number	36	41
Tip clearance	-	0.4 mm
Reynoldsnumber	6.8 10^5	4.9 10^5
Rotational speed	-	3500 rpm
Tip diameter	600 mm	600 mm
Axial vane/blade spacing	15 mm	
Inlet velocity	45 m/s	

Instrumentation

To reduce the design effort and to improve the accuracy of the measurements, only non-rotating probes were used. The real-time data of the high-response probes was correlated to the rotor blade positions using a time scale and the circumferential speed.

Pneumatic probes were used to determine the steady state flow quantities. In the core flow region five-hole probes were radially traversed. Close to the sidewall boundary layers, miniature three-hole probes were taken to get additional information. Except of measurements of the pneumatic pressure distribution of a rotor blade, no measurements were carried out in rotating part of the machines.

Hot-wire probes were a key feature of these investigations. Calibration and data acquisition are described shortly. A more detailed description was presented by Poensgen and Gallus (1990). The unsteady voltages from the hot-wire anemometers were logged with a high-speed (up to 10 MHz) multi-channel data acquisition system. In these experiments, 256 real-time samples were recorded over roughly 4 blade passing periods and averaged over 256 revolutions. For each of the 65000 data points, the velocity vector was calculated based on a polynomial least square fit method. Input for this were the results from a free-jet calibration of the probes, where an accuracy in velocity of 0.5 m/s and in pitch and yaw angle of less than 0.2° was achieved. Phase-locked-averaging (Gostelow (1977)) has been used for the data reduction. Using a square wave signal, the corner frequency of the hot-wires was found to be higher than 20 KHz, which is a conservative estimation. At 50 KHz, the ratio between signal amplitude and probe response is still higher than 0.8, while the blade passing frequency of the turbine rotor was only 2.4 KHz. The recorded real-time signals were filtered by a digital low-pass Blackman filter to eliminate high frequency noise. The Channon factor, which describes the relation between cut-off frequency of the filter process and the sampling frequency is about 0.3.

Two detailed problems concerning hot-wire anemometry had to be solved. First, in the near wake region downstream of turbine rotor and when operating the test compressor in rotating stall, the pitch flow angles are fluctuating within a range, which is larger than the range of the sensor calibration. To overcome this problem, the measurements were taken with different probe pitch position angles. Fig. 3 shows the some typical pitch flow angle data which was taken downstream of the compressor rotor operating in rotating stall. Two measurements of a two cell rotating stall configuration were taken with a triple hot-wire sensor. The sensor was placed at a pitch angle of 55° and 85° respectively. The dashed lines indicate the boundaries of the range of the probe calibration where

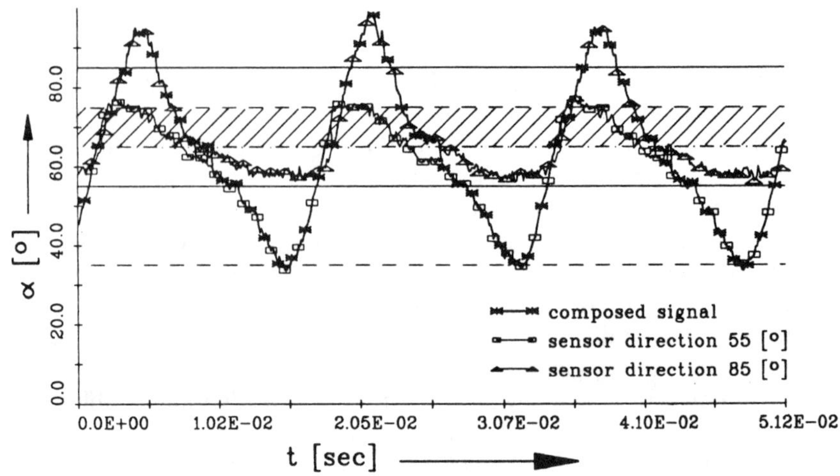

Fig. 3: Example of Unsteady Flow Measurements; The Pitch Angle (α) Fluctuations are Larger than the Calibration Range of the Probe

valid data could be expected. In the overlapping range the data from both positions is smoothed with a linear form function. Secondly, hot-wire measurements close to the side walls were made using conventional x-wire sensors. Thus it was not possible to measure the radial components of the side wall flow. A correction scheme, based on the inviscid Euler equations was derived, in order to correct the measured velocity vector. Fig. 4a shows the tip secondary flow structure downstream of the compressor rotor which is obtained by measurement. Fig. 4b displays the same data after the correction of the radial velocity component.

LDV-measurements were performed at 12 axial locations inside the turbine rotor. The 2d-velocity vector was measured at 22 radial positions, and to include the rotor-stator interaction at 8 circumferential positions. A coincidence filter ensured, that only the burst impulses of particles were evaluated, which lead simultaneously to valid signals in both fringe patterns. In this case the rotor position was detected by a syncro-resolver, allowing an accuracy of 0.08 degrees. The probe head of the LDV-system is connected to the laser-optics by a fibre cable. Therefore only the small probe had to be traversed, thus allowing an accuracy of 0.1 mm in axial and radial direction. The error in the LDV-measurements is estimated to be lower than ± 1 per cent of the velocity. A more detailed description of the real time data acquisition and analysis, including an error estimation for the probe measurements, was given by Zeschky (1991).

The unsteady static pressure field at the casing above the turbine rotor was detected with a miniature semiconductor transducer. The transducer could be inserted at 25 axial locations in the casing. To improve the accuracy, only the AC-voltage of the pressure transducer was recorded and superimposed on the time-mean static pressure measured at the same axial

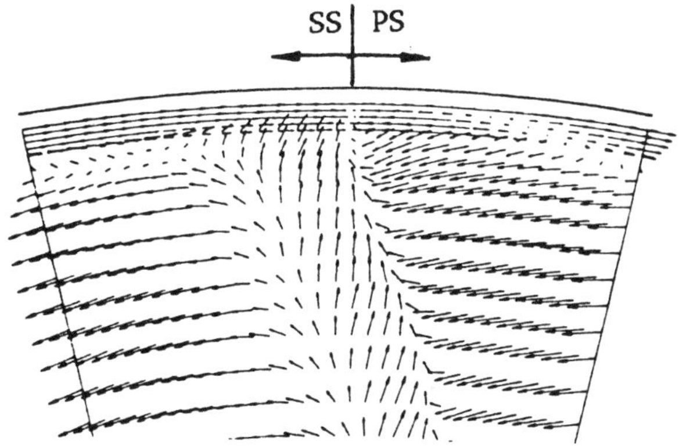

Fig. 4a: Secondary Flow Structure at the Rotor Exit Without Correction of the Radial Flow

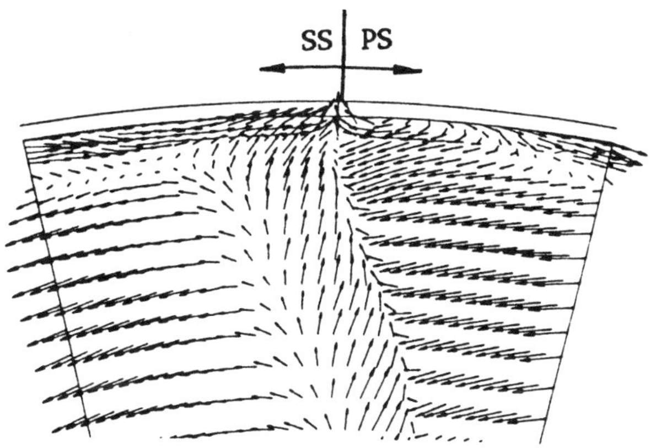

Fig. 4b: Secondary Flow Structure at the Rotor Exit With Correction of the Radial Flow

locations (ee also Fig. 6 and Fig. 12).

Downstream of the turbine and compressor rotor the unsteady total pressure fluctuation was measured with a semiconductor unsteady total pressure probe. A miniature pressure transducer is implemented in the probe head which is formed as a Kiel-Type probe.

Experimental results and discussion

In this part of the paper the results of the different measurement techniques are compared and their contribution to resolve the flow phenomena is discussed.

Compressor

The investigation of the rotor-outlet flow field was the main objective of the investigations on the single stage axial flow compressor. Of particular interest was to observe the development of the unsteady three-dimensional flow with throttling the machine to its stall point. As a question of length, only the data for a moderately throttled operating point $\varphi=0.66$ at 6000 rpm will be discussed here.

The inlet flow with ambient air conditions enters a bellmouth and passes a 9.5 m long duct until it reaches the compressor rotor. As there are no inlet guide vanes the inlet has no swirl. The radial inlet pressure distribution was measured with pneumatic probes (Fig. 5). With throttling the compressor flow, the mass flow decreases and so does the loss produced in the inlet pipe. Thus, the total inlet pressure raises up with throttling the compressor flow.

The rotor outlet flow has been measured with pneumatic probes, x-wire and triple-hot-wire probes and with a semiconductor total pressure probe. Transforming the measured unsteady data into the rotating frame of reference, the quasi-steady 3-d rotor outlet flow field was obtained. The basic flow phenomena can be seen in Fig. 6 which shows the secondary flow vectors. As it is not possible to separate the primary and secondary flow by measurement, the secondary velocity vector is defined as the difference vector from the circumferential averaged mean velocity vector and the local velocity vector.

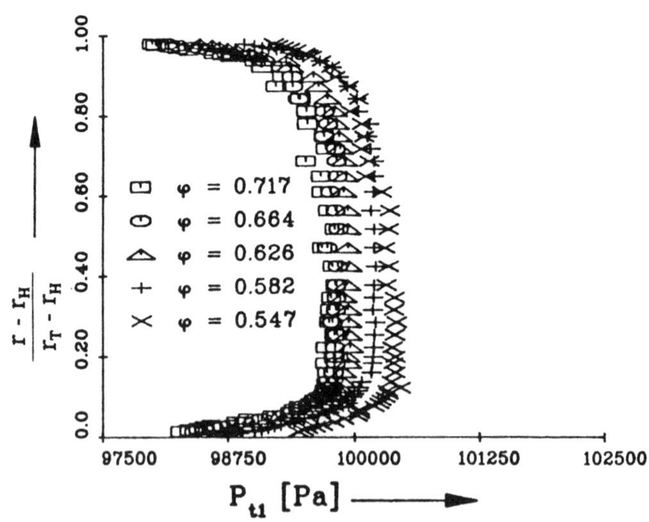

Fig. 5: Total Pressure at Rotor Inlet

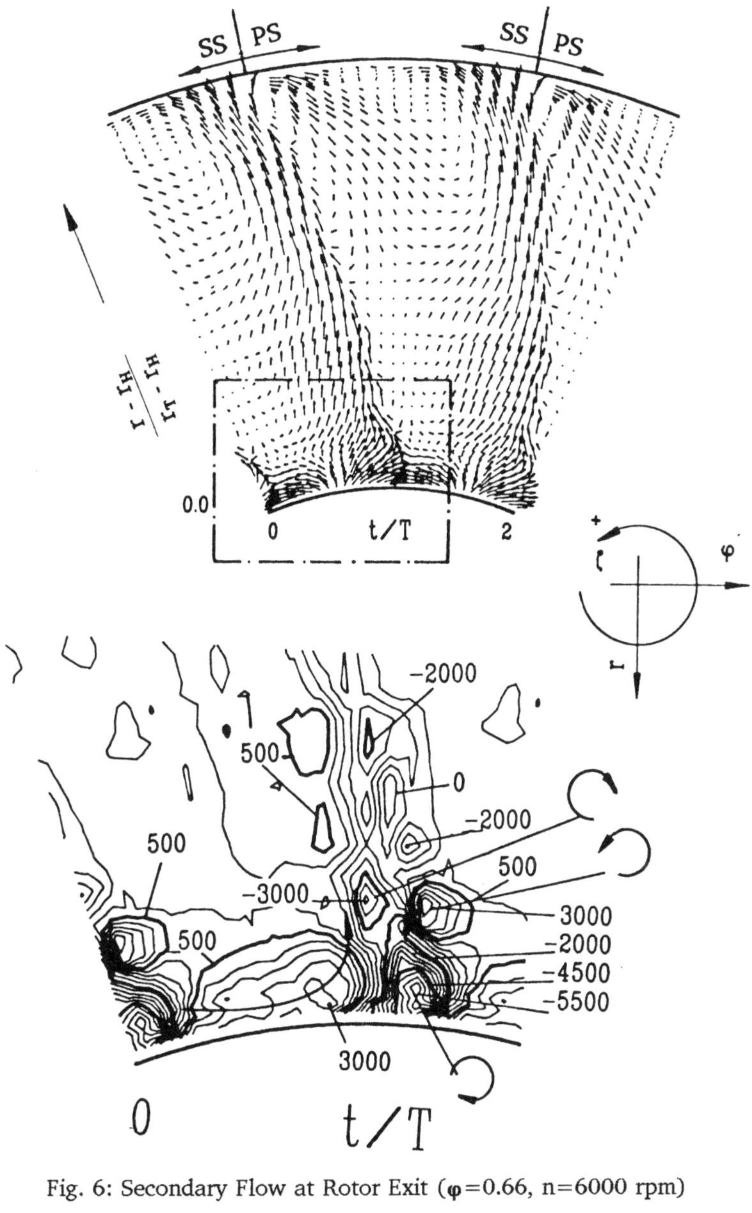

Fig. 6: Secondary Flow at Rotor Exit (φ=0.66, n=6000 rpm)

$$\vec{c}_{sec}(r, \varphi z) = \vec{c}(r, \varphi, z) - \frac{z}{2\pi} \int_0^{\frac{2\pi}{z}} \vec{c}(r, \varphi, z)\, d\varphi$$

Near the hub there exists a strong overturning of the flow due to intense

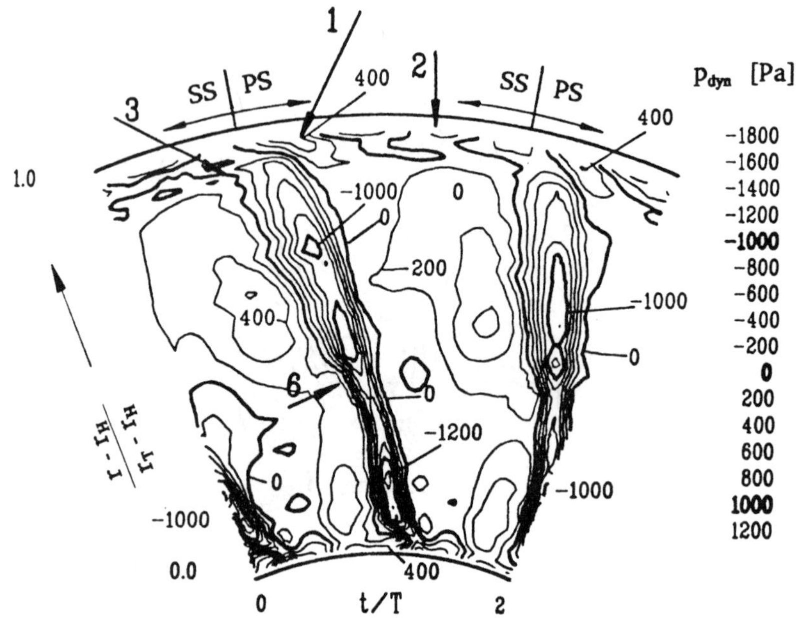

Fig. 7: Periodic Total Pressure Fluctuation at Rotor Exit

secondary flow. This effect cannot be represented by the above given definition. Thus, it is necessary to calculate the vorticity vector perpendicular to the outlet flow plane.

$$\zeta_s = \frac{\partial w_\varphi}{\partial r} - \frac{\partial w_r}{r \partial \varphi}$$

The lower part of Fig. 6 displays the isocontours of the vorticity distribution close to the hub. A significant secondary flow vorticity distribution is visible. In the wake of the corner flow between rotor hub and profile suction side a trailing vortex could be determined.

Inside the rotor wake, the flow has a strong radial component towards the casing due to centrifugation of the blade boundary layers. Because of the thick inlet boundary layers the incidence angle at the rotor tip increases to 12°. A horseshoe vortex is produced at the tip of the rotor (compare no. 1 of Fig. 7 to Fig. 6). This vortex reenergizes the casing boundary layer in the corner between the rotor pressure side and the casing. Due to the sense of rotation, fluid of the core flow is transported towards the casing. High total pressures in this region result from this effect (no. 1 of Fig. 7). At the suction side a small leakage vortex could be detected. Together with the suction side part of the horseshoe vortex the leakage flow vortex forms a sole vortex system. The secondary flow plot indicates that this vortex picks up the centrifuged wake fluid from the blade suction side.

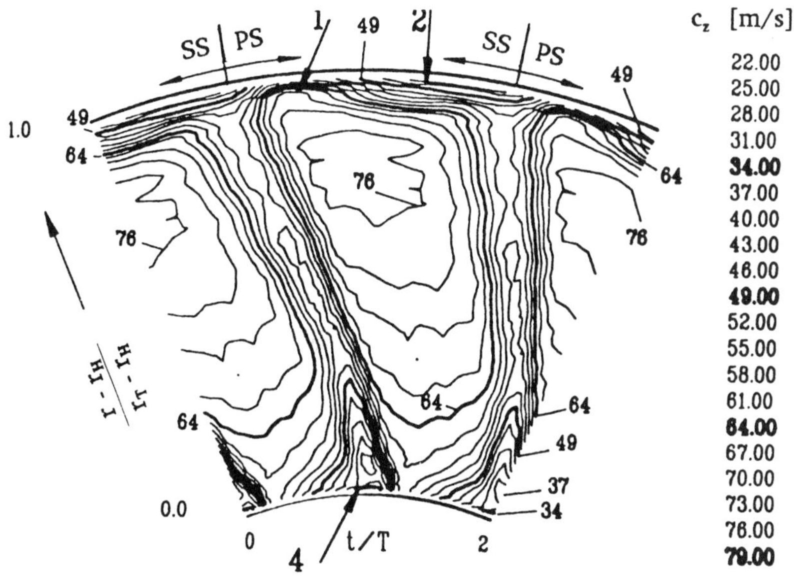

Fig. 8: Axial Velocity at Rotor Exit

Vortex rotation causes this fluid to be mixed with the core flow in the region of half spacing inside the passage flow.

Fig. 8 displays the isocontour lines of the axial flow velocity. Downstream of the corner flow of the rotor hub and the suction side the axial flow isocontours indicate a strong diffusion. In fact, the diffusion factor (Lieblein, 1975) reaches values of about 0.8. But no significant turbulent kinetic energy could be found in this region (no. 5 of Fig. 9). Thus it can be concluded that there exists a strong diffusion of the flow without any separations of the corner flow. This effect may be related on the blade boundary layer centrifugation and the strong secondary flow in the rotor hub passage. The rotor wakes are inclined versus the radius as the rotor blade is twisted. The effect of the flow re-energitization in the corner of the blade pressure side and the casing produces high axial flow rates. The vortex system, which is located at the suction side generates high losses. By this, the axial flow tends to lower values (no. 2 of Fig. 8).

Some details about the structure of the wake flow are displayed in Fig. 10. The time history of the radial flow angle, the relative outlet flow angle and the relative velocity is shown. By definition the radial flow towards the hub corresponds to positive yaw angles. The wake velocity profile is asymmetric in shape. Because of the higher loading of the blade suction side boundary layer, the velocity gradients of the suction side part of the wake are smaller. Due to the centrifugation of the blade boundary layer, a strong radial velocity component develops in the wake flow, which

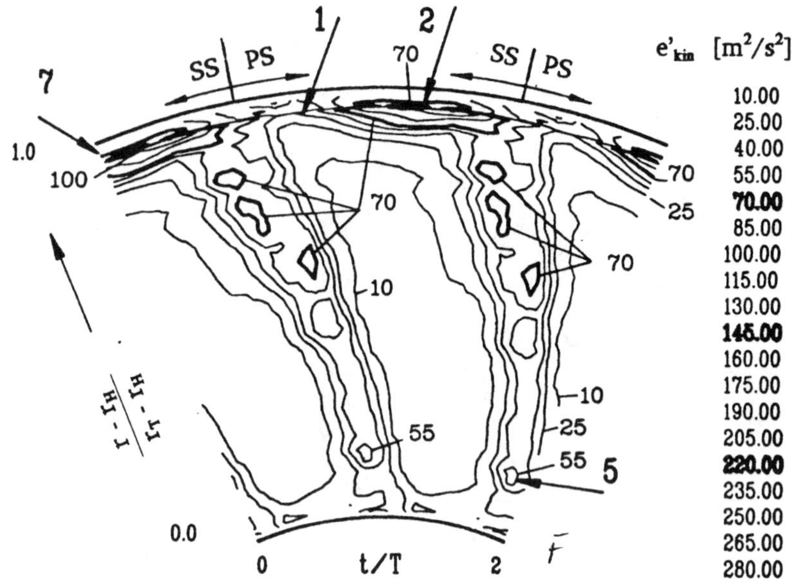

Fig. 9: Turbulent Kinetic Energy at Rotor Exit

raises up to about 15 per cent of the mean relative velocity at midspan. In the wake the pitch angle ($ß_2$) indicates an underturning of the wake flow. This effect is related to the three-dimensional wake decay characteristics. The attached boundary layer flow is overturned compared to the flow at midpassage. This effect generates the blade boundary layer centrifugation. Nevertheless, during the wake decay process, the wake flow keeps its radial momentum. Because there in no net work done on the flow in the axial gaps, the swirl of fluid elements of the wake flow is constant. Thus, moving towards the casing, the circumferential velocity of the wake fluid elements decreases with radius. But the swirl raises from hub to tip which follows from the radial pressure equilibrium desired at the design processes. Thus, at the location of the measurement wake centrifugation generates a significant underturning of the wake flow.

Turbine

The rotor inlet velocity and flow angles in Fig. 11 were obtained by transforming the measured stator exit flow into the relative frame. The shaded area marks the fluctuations when the rotor blades pass the stator wakes. Compared to a midspan incidence of 3 degrees, the loading increases towards the hub and decreases towards the casing. This is due to the higher acceleration in the stator and the lower circumferential speed at the hub.

At 15 and 85 per cent span the incidence is locally reduced. The underturning in the stator passage vortices and the low velocity in two loss cores in the stator wakes at these locations cause a time-averaged decrease of 8 degrees of the incidence angle.
The static pressure distribution at the casing above the rotor is shown in Fig. 12. Because of the near-zero incidence angle at the tip, see Fig. 11, the stagnation point is located at the leading edge. If the direction of the local velocity vector is assumed to be perpendicular to the lines of constant static pressure, Fig. 12 indicates an overturning of the flow at the casing from pressure to suction side. The effect of the tip clearance flow is also visible. The crossflow starts at the leading edge and seems to emerge at the suction side at about 25 per cent axial chord. The extension of the tip clearance vortex is estimated by connecting the locations of the lowest static pressures and is shown as a dashed line in Fig. 12. At the trailing edge it extends over 20 to 25 per cent of the passage.
Fig. 13 compares the circumferentially averaged flow velocity at rotor outlet as detected by hot-wire and pneumatic probes. At locations (2) and (4), where the pitch angle variation is relatively small, good agreement is

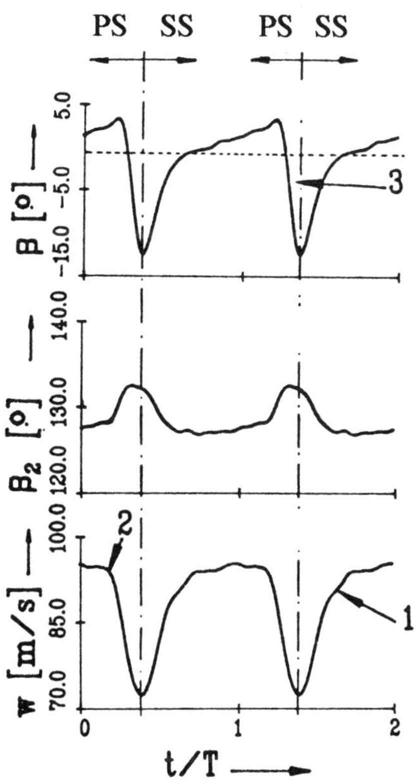

Fig. 10: Wake Profile at 57 per cent Span

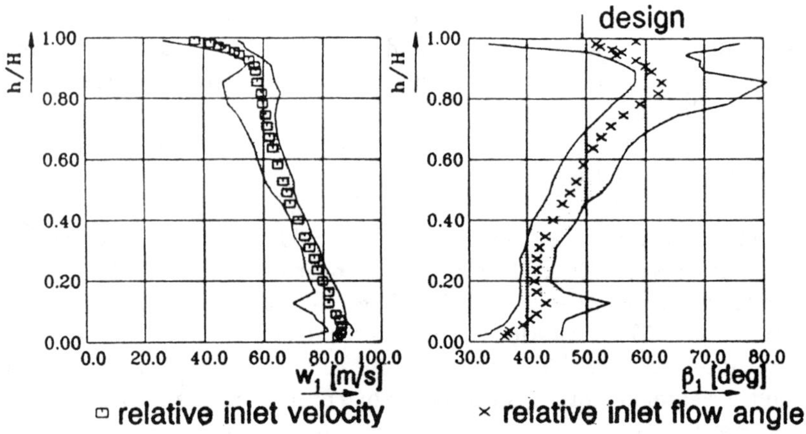

Fig. 11: Rotor Inlet Velocity and Flow Angle in the Relative Frame

reached, whereas at locations (1) and (3) with a pitch angle variation of more than 40° during the passing wake, the differences between steady and unsteady measurements amount to more than 10 per cent. Based on this, the authors felt that the hot-wire technique yields reliable results, confirmed by the good agreement with the five-hole probe and that the steady-state pneumatic probe readings were not reliable in regions of large unsteady fluctuations.

By transforming the hot-wire measurement results into the relative frame, the rotor exit secondary flow field in Fig. 14 was obtained. Downstream of the rotor two passage vortices exist with a strong overturning at 15 and above 85 per cent span. This coincides with the locations of the under-

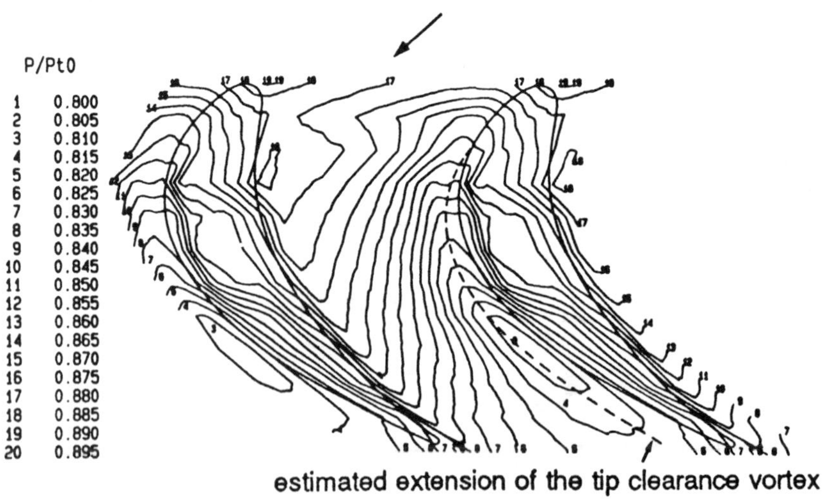

Fig. 12: Static Pressure Field at the Casing above the Turbine Rotor

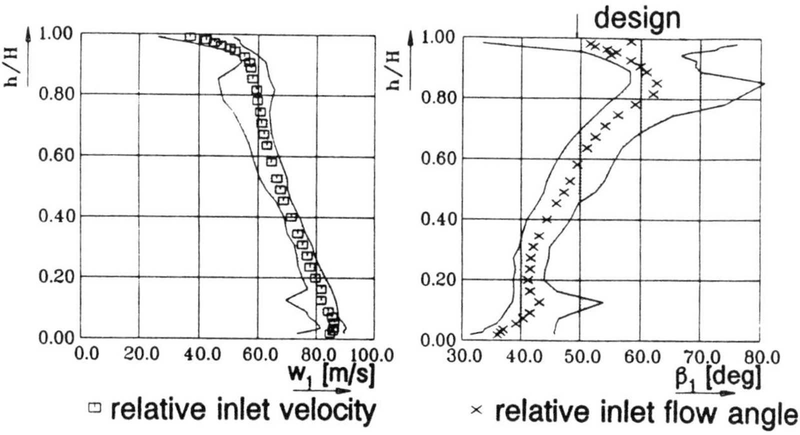

Fig. 13: Comparision of Hot-Wire and Pneumatic Measurements in the Rotor Exit Plane, Circumferentially Averaged

turning and the loss cores in the stator exit flow and partly caused by the locally reduced incidence as was discussed earlier by Zeschky and Gallus (1990). The tip clearance vortex extends to about 20 per cent of the pitch. This is in good agreement with the casing static pressure distribution, Fig. 12. The relatively small tip clearance vortex with a radial extension of only 10 per cent span is due to the small radial gap of 0.4 mm or 0.7 per cent span.

To get information about the flow inside the rotor passages the LDV-technique was applied. Fig. 15 shows the flow field inside the rotor close to the leading edge. For eight rotor/stator positions a stop-action sequence was computed, which can be understood as "photographs" of the flow field. Between the positions 2, 4, 6, 8 the rotor turns over 25 per cent of the blade pitch. Each figure refers to the same position in the absolute frame, with the rotor blades passing through the measurement plane from right to left.

In Fig. 15, the relative flow angles at $x/b_p = 13.9$ per cent are depicted. At rotor position 8, no stator wake is visible in the blade passage. In the tip region, the higher circumferential speed and the lower acceleration cause high flow angles. The blade-to-blade gradient is caused by the different turning and acceleration at pressure and suction side. At rotor position 2, higher flow angles representing crossflow towards the suction side occur in the corner between pressure side and casing. The reason for this are the lower velocity and the underturning in the high-loss area of the stator exit flow. At position 4, the stator wake is visible over the entire span, with strong distortions at h/H = 15 per cent and 85 per cent. The earlier passing of the loss cores and the wake in the tip region is explained by the higher absolute flow angles. At position 6, the flow angles decrease after

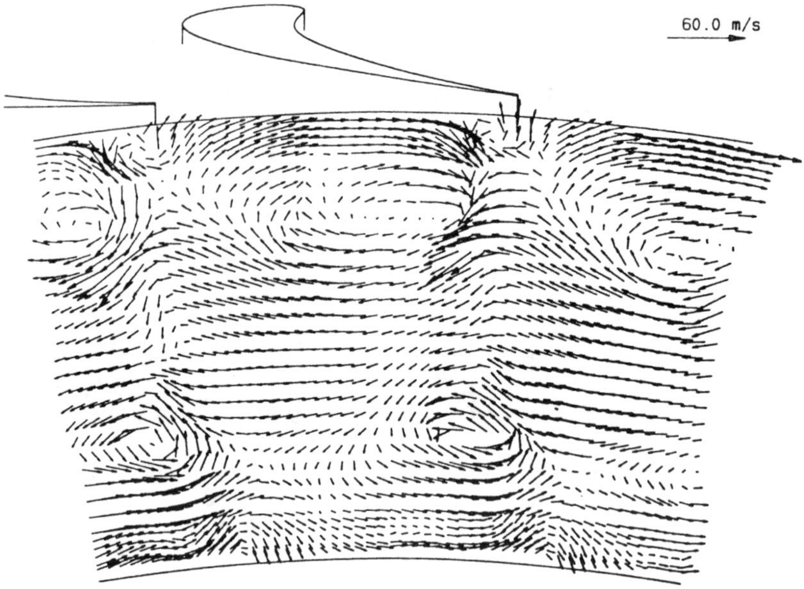

Fig. 14: Rotor Exit Secondary Flow Field in the Relative Frame

the wake has passed. Position 8 shows the undisturbed flow field.

Summary

The experimental analysis of the quasi-steady flow in the rotating frame of reference could be resolved by means of stationary unsteady measurement techniques. Thus, it is possible to get detailed informations about the structure of the rotor flow, as the results from the test compressor have shown. The structure of the wake could be investigated as well as an analysis of the 3-d secondary flow structure was made. Here zones of reenergized flow and transport mechanisms of wake fluid in the upper blade passage were found. Strong secondary flow close the hub of the rotor causes a radial centrifugation of low momentum hub boundary layer fluid. This generates a large overturning of the flow. Therefore the diffusion factors of the flow cross the limits given by Lieblein's diffusion factor without any flow separation.

The flow in the turbine rotor test case is much more complex. The investigations clarified the mechanism of the transport of low momentum fluid, generated by the passage vortices of the upstream located stator, while they are passing the rotor passage. The measurements of the unsteady three-dimensional turbine rotor flow show that the rotor secondary flow is mainly caused by the radial nonuniform stator exit flow and the periodi-

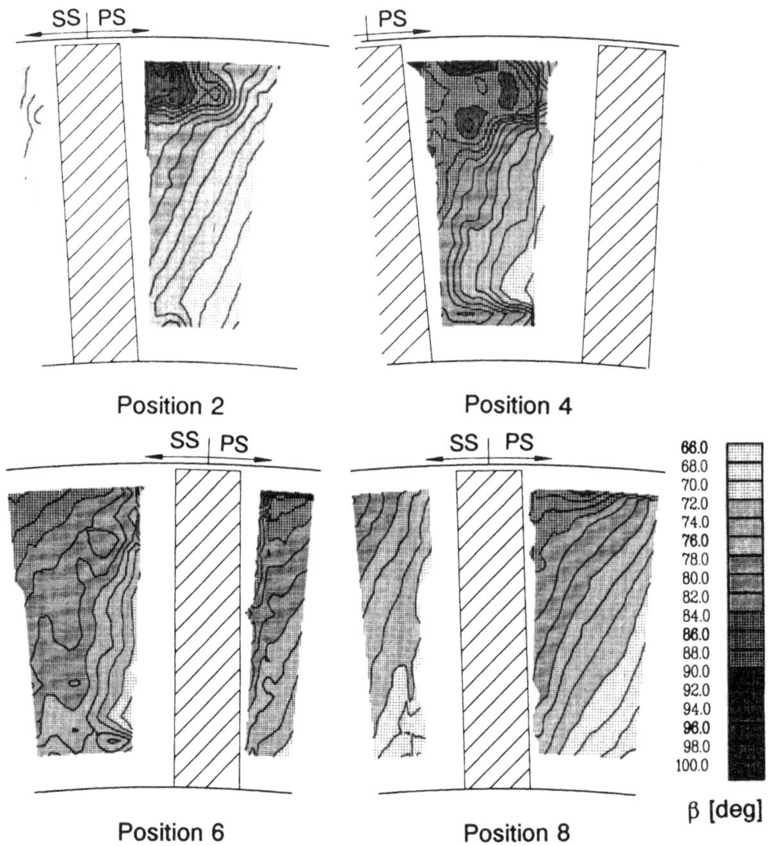

Fig. 15: Flow Field Inside of the Rotor (13.9% Chord from the Leading Edge)

cally unsteady transport of the stator wakes through the rotor passage. The endwall boundary layers at the rotor inlet are very thin, which results from the transport mechanisms and the acceleration in the stator. Therefore they contribute only in a smaller scale to the rotor secondary flow, and the highest overturning in the rotor occurs at 15 and 85 per cent span and not directly at the endwall.

Acknowledgment

This work has been carried out with financial support of the "Forschungsvereinigung Verbrennungskraftmaschinen" and the "Deutsche Forschungsgemeinschaft".

References

Adachi T., Murakamai Y., 1979
"Three-Dimensional Velocity Distribution Between Stator Blades and Unsteady Force on a Blade Due to Passing Wakes"
JSME, Vol. 22, No. 170, Aug. 1979

Addison J.S., Hodson H.P., (1991)
"Modelling of Unsteady Transitional Boundary Layers"
ASME-paper 91-GT-282

Arndt N., 1991
"Blade Row Interaction in a Multistage Low-Pressure Turbine"
Paper of the ASME 91-GT-283

Binder A., Schröder Th., Hourmouziadis J., 1989
"Turbulence Measurements in a Multistage Low-Pressure Turbine"
ASME Journal of Turbomachinery, Vol. 111, pp 153-161

Dawes W.N. 1990
"Towards Improved Throughflow Capability: The Use of 3D Viscous Flow Solvers in a Multistage Environment."
ASME-paper 90-GT-18

Evans R.L., 1979
"Boundary Layer Development on an Axial-Flow Compressor Stator Blade"
ASME Journal of Eng. for Power, Vol. 100, pp. 287 - 293, Apr. 1978

Gostelow J.P., 1977
"A New Approach to the Experimental Study of Turbomachinery Flow Phenomena"
J. Engineering for Power, Vol. 99, pp. 97 - 105

Hodson H.P., 1983
"The Development of Unsteady Boundary Layers on the Rotor of an Axial-Flow Turbine"
AGARD CP 351

Kerrebrock J.L., Mikolajczak A.A., 1970
"Intra-Stator Transport of Rotor Wakes and its Effect on Compressor Performance"
ASME Journal of Engineering for Power, pp. 359 - 368

Lieblein S., 1975
"Experimental Flow in Two-Dimensional Cascade Analysis"
Aerodynamic Design of Axial-Flow Compressors
NASA SP-36, 1975

Meyer R.X., 1958
"The Effects of Wakes on the Transient Pressure and Velocity Distribution in Turbomachines"
Journal of Basic Engineering, pp. 1544 - 1552

Poensgen C.A., Gallus H.E., 1990
"Three-Dimensional Wake Decay Inside of a Compressor Cascade and Its Influence on the Downstream Unsteady Flow Field"
Part 1 and 2
Journal of Turbomachinery, April 1991, Vol. 113

Poensgen C.A., 1991
"Experimentelle Untersuchung der Strömung in einer Unterschall-Axialverdichterstufe bei hoher Drosselung und im Rotating Stall"
Thesis RWTH Aachen

Rao K., Delaney R., 1990
"Investigation of the Unsteady Flow Through a Transonic Turbine Stage"
Jet Propulsion AIAA 90-2408

Schröder Th., 1989
"Measurements with Hot-Film Probes and Surface Mounted hot-Film Gauges in a Multistage Low-Pressure Turbine"
1989 European Propulsion Forum, pp 15.1 - 15.27

Schulz H.D., 1989
"Experimentelle Untersuchung der dreidimensionalen abgelösten Strömung in einem Axialverdichterringgitter."
Thesis RWTH Aachen

Utz C., 1972
"Experimentelle Untersuchung der Strömungsverluste in einer mehrstufigen Axialturbine"
Thesis ETH Zürich

Zeschky J., Gallus H.E., 1991
"Effects of Stator Wakes and Spanwise Nonuniform Inlet Conditions on the Rotor Flow of an Axial Turbine Stage"
ASME-paper 91-GT-93

Zeschky J., 1991
"Experimentelle Untersuchung der dreidimensionalen instationären Rotorströmung einer axialen Kaltluftturbine"
Thesis RWTH Aachen, 1991

Reduced Frequency Effects on Gust Response Unsteady Aerodynamics in Turbomachines

Steven R. Manwaring[1,2]
Sanford Fleeter[1]

ABSTRACT

A series of experiments to investigate rotor unsteady aerodynamic response to high and very high reduced frequency gusts, including the effects of steady aerodynamic loading and detailed forcing function waveform shape are described. These experiments are performed in an extensively instrumented axial flow research compressor. The high reduced frequency forcing function to the first stage rotor blade row is the first harmonic of the wakes generated by the 36 inlet guide vanes (IGV's), with the very high reduced frequency being the second harmonic of these wakes. These forcing functions are each measured in the rotor frame of reference, decomposed into chordwise and transverse velocity components, and then Fourier analyzed. The unsteady aerodynamic gust response is then investigated on the individual surfaces of the first stage rotor blade for two values of the streamwise-to-transverse gust amplitude ratio over a range of compressor steady loadings. Appropriate data are correlated with classical predictions. The results show that there are large and significant differences between the high and very high reduced frequency gust generated unsteady aerodynamic response of the rotor blade row.

---

[1]Thermal Sciences and Propulsion Center, School of Mechanical Engineering, Purdue University, Chaffee Hall, West Lafayette, IN 47907.
[2]Currently Engineer, Aerodynamics Research Laboratory, General Electric Aircraft Engines, Cincinnati, OH 45215.

NOMENCLATURE

b	Rotor blade semichord
$\overline{C}_p$	Rotor blade steady pressure coefficient
C_p	Rotor blade unsteady pressure coefficient
$C_{\Delta p}$	Rotor blade unsteady pressure difference coefficient
$\bar{i}$	Rotor blade mean incidence angle
k	Reduced frequency = $\omega b / \overline{V}_x$
p	Digitized ensembled averaged unsteady pressure
P_s	Rotor blade surface steady pressure
$\hat{p}$	First harmonic complex unsteady pressure
$\hat{u}^+$	Streamwise gust first harmonic component
$\hat{v}^+$	Transverse gust first harmonic component
$\overline{V}_x$	Mean axial velocity
ΔV	Absolute velocity vector difference from mean value
ΔW	Total unsteady velocity
$\overline{\beta}$	Relative mean flow angle
$\Delta\overline{\beta}$	Relative flow angle difference from mean value
ω	Forcing function frequency

INTRODUCTION

The component design trend towards higher stage loadings and specific flow are being attained through increased tip speed, lower radius ratio, higher camber, lower aspect ratio, fewer stages, with swept blade rows a current design consideration. The resultant fan and compressor blade designs utilize thin, highly twisted, low aspect ratio blading with corresponding high steady state stresses. Also, the mechanical damping is considerably reduced in newer rotor designs, particularly those with integral blade-disk configurations (blisks) and in those without shrouds. These are similar trends in turbine blading, but they are further complicated by increased inlet and cooling temperatures that have resulted in thin-walled, complex cooling passage designs. As a result, advanced blade designs often have significant aerodynamically forced response problems.

The primary mechanism of blade failure is fatigue caused by vibrations at levels exceeding material endurance limits. These vibrations occur when a periodic forcing function, with frequency equal to a natural blade resonant frequency, acts upon a blade row. Because a blade may have as many critical points of high stress as it has natural modes, the designer must determine which particular modes have the greatest potential for aerodynamic excitation.

Treatments of unsteady flow due to periodic gusts or to oscillating airfoils have been developed in the linear approximation wherein the mean flow is assumed to be uniform. In this approximation, the steady and unsteady flow fields are uncoupled from one another. Thus, for a periodic gust, the upstream unsteady aerodynamic forcing function components are convected by the uniform mean flow. Semi-analytical unsteady aerodynamic analyses based on this linear model have been developed for cascades in subsonic and supersonic flows [1-4]. Such models are currently being extended to consider unsteady flows linearized about a nonuniform mean flow, with the gust interacting with the mean flow [5-7].

Forced response unsteady aerodynamic models involve many assumptions. Thus, prior to implementing these into a component design system, it is necessary that the various components be experimentally verified and that the applicability of these components to real blade row designs, i.e., compressors and turbines, be established. A number of experiments have been directed at this model verification in the attached flow regime, both subsonic and supersonic and the determination of their applicability and limitations [8-11].

The reduced frequency k defines the importance of the unsteady flow phenomena. In forced response, the complete range of reduced frequencies is significant, including very high values, in contrast to flutter where the concern is only with unit order or smaller reduced frequency values. Also, much different flow phenomena are important at low than at high reduced frequency values. Hence, it is imperative that both low and high reduced frequency unsteady flow phenomena be considered.

In this paper, rotor unsteady aerodynamic response to high and very high reduced frequency gusts, including the effects of steady aerodynamic loading and detailed forcing function waveform shape are investigated for the first time. This is accomplished utilizing an extensively instrumented axial flow research compressor. The lower reduced frequency forcing function to the first stage rotor blade row is the first harmonic of the wakes generated by the 36 inlet guide vanes (IGV's), with the very high reduced frequency being the second harmonic of these IGV wakes. These forcing functions are each measured in the rotor frame of reference, Fourier analyzed and then decomposed into chordwise and transverse components. The unsteady aerodynamic gust response is then investigated on the individual surfaces of the first stage rotor blade for two values of the streamwise-to-transverse gust amplitude ratio over a range of compressor steady loadings.

RESEARCH COMPRESSOR AND INSTRUMENTATION

The Purdue Axial Flow Research Compressor models the fundamental turbomachinery unsteady aerodynamic multistage interaction phenomena which include the incidence angle, the velocity and pressure variations, the aerodynamic forcing function waveforms, the reduced frequency, and the unsteady blade row interactions. The compressor is driven by a 15 HP DC electric motor at a speed of 2,250 RPM. Each identical stage contains 43 rotor blades and 31 stator vanes having a British C4 airfoil profile, with the first stage rotor inlet flow field established by a variable setting inlet guide vane (IGV) row of 36 airfoils. The overall compressor and airfoil characteristics are defined in Table 1.

The compressor aerodynamic performance is determined utilizing a 48 port Scanivalve system, thermocouples, and a venturi orifice to measure the required pressures, temperatures and flow rate, respectively. The Scanivalve transducer is calibrated each time data are acquired, thus automatically compensating for zero and span shifts of the transducer output. A 95% confidence interval, root-mean square error analysis of 20 samples is performed for each steady data measurement.

Both steady and unsteady rotor blade row data are required. The steady data quantify the rotor row mean inlet flowfield and the resulting rotor blade midspan steady loading distribution. The unsteady data define the periodic aerodynamic forcing function and the resulting midspan blade surface periodic unsteady pressure distributions.

The inlet flow field, both steady and unsteady, is measured with a rotating cross hot-wire probe. Disturbances in the stationary frame-of-reference, i.e., the IGV wakes, are the unsteady aerodynamic forcing functions to the first stage rotor row. The rotor periodic unsteady inlet flow field generated by these disturbances is measured with a cross hot-wire mounted in the rotor frame-of-reference. The probe is axially mounted 30% of rotor chord upstream of the rotor leading edge plane. A potential flow field analysis determined this axial location to be such that leading edge potential effects are negligible for all steady loading levels. The probe is angularly aligned to obtain rotor relative velocity and flow angle data. The cross hot-wire probe was calibrated and linearized for velocities from 18.3 m/sec to 53.4 m/sec and ± 35 degrees angular variation, with the accuracy of the velocity magnitude and flow angle were determined to be 4% and ± 1.0 degree, respectively. Centrifugal loading effects on the rotating hot-wire sensor resistances and, thus, the responses, were found to be negligible.

The detailed steady aerodynamic loading on the rotor blade surfaces is measured with a chordwise distribution of 20 midspan static pressure taps, 10 on each surface. The static pressure at the rotor exit plane, measured with a rotor drum static tap, is used as the blade surface static pressure reference. These static pressure measurements are made using a rotor based 48 port constant speed drive Scanivalve system located in the rotor drum.

The measurement of the midspan rotor blade surface unsteady pressures is accomplished with 20 ultra-miniature, high response transducers embedded in the rotor blades at the same chordwise locations as the static pressure taps. To minimize the possibility of flow disturbances associated with the inability of the transducer diaphragm to exactly maintain the surface curvature of the blade, a reverse mounting technique is utilized. The pressure surface of one blade and the suction surface of the adjacent blade are instrumented, with transducers embedded in the nonmeasurement surface and connected to the measurement surface by a static tap. The embedded dynamic transducers are both statically and dynamically calibrated. The static calibrations show good linearity and no discernible hysteresis. The dynamic calibrations demonstrate that the frequency response, in terms of gain attenuation and phase shift, are not affected by the reverse mounting technique. The accuracy of the unsteady pressure measurements, determined from the calibrations, is $\pm 4\%$.

The rotor-based static pressure Scanivalve transducer, rotating cross hot-wire probe and 20 blade surface dynamic pressure transducers are interfaced to the stationary frame-of-reference through a 40 channel slip ring assembly. On-board signal conditioning of the transducer output signals is performed to maintain a good signal-to-noise ratio through the slip rings. The remaining 17 channels of the slip-ring assembly are used to provide excitation to the transducers and on/off switching to the Scanivalve DC motor.

DATA ANALYSIS

Steady Data

The rotor blade surface static pressure data, measured with the rotor-based Scanivalve system, are defined by a root-mean-square error analysis of 20 samples with a 95% confidence interval. The reference for these midspan blade pressure measurements is the static pressure at the exit of the rotor measured on the rotor drum. Thus, the blade surface and the reference static pressures are measured at different radii. Hence, a correction for the resulting difference in the radial acceleration is applied in calculating the blade surface static pressure coefficient defined in Equation 1.

$$\overline{C}_p = \frac{\overline{P}_s - \overline{P}_{exit}}{1/2\ r\ U_t^2} \qquad (1)$$

where U_t is the rotor blade tip speed.

Periodic Data

The periodic data of interest are the harmonic components of the aerodynamic forcing function to the first stage rotor blade row together with the resulting rotor blade surface unsteady pressures and unsteady pressure differences. These are determined by defining a digitized ensemble averaged periodic unsteady aerodynamic data set consisting of the rotating cross hot-wire probe and blade surface dynamic pressure transducer signals at each steady operating point. In particular, these time-variant signals are digitized with a high speed A-D system at a rate of 100 kHz and then ensemble averaged.

The key to this averaging technique is the ability to sample data at a preset time, accomplished by an optical encoder mounted on the rotor shaft. The microsecond range step voltage signal from the encoder is the data initiation time reference and triggers the high speed A-D multiplexer system. To significantly reduce the random fluctuations superimposed on the periodic signals of interest, 200 averages are used. A Fast Fourier Transform (FFT) algorithm is then applied to these ensemble averaged signals to determine the harmonic components of the unsteady aerodynamic forcing function and the resulting rotor blade surface harmonic unsteady pressures and pressure differences.

The unsteady inlet flow field to the rotor row is measured with the rotating cross hot-wire probe which quantifies the relative velocity and flow angle. The velocity triangle relations are then used to determine the unsteady inlet flow field to the rotor, in particular, the streamwise and transverse velocity components, u^+ and v^+, respectively. These are then Fourier decomposed to determine the first and second harmonics of the streamwise and transverse gust components, $\hat{u}^+$ and $\hat{v}^+$.

The various unsteady aerodynamic gust mathematical models reference the gust generated airfoil aerodynamic response to a transverse gust at the leading edge of the airfoil. However, in the experiments described herein, the time-variant data are referenced to the initiation of the data acquisition shaft trigger pulse. Thus, for consistency with the models, the periodic data are further analyzed and referenced to a transverse gust at the leading edge of the first stage rotor blade. This is accomplished by assuming that: (1) the aerodynamic forcing function remains fixed in the stationary reference frame; and (2) the forcing function does not decay from the rotating hot-wire probe axial location to the rotor row leading edge plane.

The rotor blade surface unsteady pressure data, measured with the embedded high response pressure transducers, are analyzed to determine the harmonics of the chordwise distribution of the unsteady pressure coefficient C_p and the unsteady pressure difference coefficient $C_{\Delta p}$. These are defined in Equation 2 and are specified from the Fourier coefficients of the digitized ensemble averaged dynamic pressure transducer signals. Note that introducing $\bar{\beta}$ into the data analysis incorporates the effect of steady loading as characterized by the mean flow incidence angle into the analysis of the data.

$$C_{pi} = \frac{\hat{p}_i}{\rho \bar{V}_x^2 \left(\frac{\hat{v}_i^+}{\bar{V}_x}\right) \bar{b}} \tag{2}$$

$$C_{\Delta pi} = C_{pi,\ \text{pressure}} - C_{pi,\ \text{suction}}$$

where $\hat{v}_i^+$ is the harmonic transverse gust component, $\overline{V}_x$ is the mean axial velocity and $\overline{\beta}$ is the relative mean flow angle.

The final form of the gust generated rotor blade row unsteady aerodynamic data define the chordwise distribution of the harmonic complex unsteady pressure and pressure difference coefficients. Also included as a reference where appropriate are predictions from the transverse gust analysis of Smith [12]. This model analyzes the unsteady aerodynamics generated on a flat plate airfoil cascade at zero incidence by a transverse gust convected with an inviscid, subsonic, compressible flow.

RESULTS

The effects of high reduced frequency on blade row gust unsteady aerodynamic response, including the effect of steady aerodynamic loading, are investigated in an extensively instrumented axial flow research compressor. The rotor unsteady aerodynamic response to high and very high reduced frequency gusts is quantified. The streamwise and transverse gust components u^+ and v^+ are generated by IGV wakes. The high reduced frequency gusts, on the order of 5 to 7, are generated from the first harmonic of the IGV wakes, with the very high reduced frequency corresponding to the wake second harmonic. The effect of the detailed forcing function waveform shape, defined by the streamwise-to-transverse gust amplitude ratio $|u^+/v^+|$ on the periodic gust unsteady aerodynamics of the first stage rotor blade row is quantified, with the steady loading level as parameter.

Forcing Function

Two distinct 36 per revolution forcing functions to the first stage rotor blade row are generated, characterized by nominal streamwise-to-transverse gust amplitude ratios of 0.3 and .45. The harmonic decomposition shows a dominant 36 per rev excitation fundamental harmonic with smaller higher harmonics. As the gust amplitude ratio increases, the transverse harmonic gust amplitudes become smaller while the streamwise harmonic gust amplitudes become larger.

Blade Surface Steady Pressures

The effect of steady aerodynamic loading, characterized by the mean incidence angle, on the rotor blade surface steady pressure coefficient is shown in Figure

1. The level of steady loading only affecting the steady pressure distribution on the pressure surface over the front 40% of the chord. On the suction surface, the steady loading variation has a large effect on the steady pressure distribution over the entire suction surface. Also, these data give no indication of suction surface flow separation. It should be noted that these surface steady pressure distributions are not affected by the characteristics of the periodic unsteady aerodynamic forcing function.

Rotor Row Periodic Aerodynamic Response

The periodic aerodynamic response of the first stage rotor blade row to the IGV wake forcing function first and second harmonics are presented in the form of the chordwise distribution of the complex unsteady airfoil surface pressure and pressure difference coefficients for two nominal streamwise-to-transverse gust amplitude ratios of 0.3 and 0.45, with the steady loading level as parameter.

Figures 2 through 5 show the effect of high and very high reduced frequency values on the first stage rotor pressure surface unsteady gust response for nominal streamwise-to-transverse gust amplitude ratios of 0.3 and 0.45, with the steady loading as a parameter. The data for all steady loading values essentially collapse onto one another. The exception to this is in the quarter chord region, with an increase in the gust streamwise-to-transverse gust amplitude ratio increasing the collapsing of these data. This collapsing of the data is a direct result of the form of the nondimensionalization. Namely, the relative mean flow angle $\bar{\beta}$ is incorporated into all of the normalized unsteady pressure data, per Equation 2. Hence, this collapsing of the pressure surface unsteady pressure response specifies that the effect of loading is characterized by the mean flow incidence angle on this low camber surface. Thus, the gust response on the low camber pressure surface is primarily affected by the mean flow incidence.

The non-collapsing of the data near the quarter chord at the high reduced frequencies is associated with the negative incidence angles for these two steady loading levels. This results in the flow accelerating around the pressure surface leading edge. Thus, on this low camber pressure surface at the high first harmonic reduced frequency value, the interaction between the steady and unsteady flow fields is large, decreasing for the very high second harmonic reduced frequency values.

Figures 6 through 9 present the corresponding results on the high camber suction surface of the first stage rotor blade. There are large differences between the high and very high reduced frequency results, both in the magnitude and phase unsteady pressure data. At these high reduced frequencies, the steady loading primarily affects the unsteady pressure phase on the aft three quarters of the chord. As the mean incidence angle is increased from the low loading level, the chordwise variation of the phase data on the aft part of the surface becomes linear, with the extent of this linear distribution increasing with increasing mean incidence. This linear chordwise distribution indicates the existence of a wave phenomena, with a convective velocity equal to the mean axial velocity through the blade row. This mean axial velocity wave phenomena has been experimentally detected by other researchers, but is yet to be physically explained.

Also, in contrast to the pressure surface, this form of the dimensionless unsteady pressure coefficient does not compress these data with mean flow incidence angle at any reduced frequency. Thus, the magnitude of the unsteady pressure response on the blade suction surface, i.e., the higher camber airfoil surface, is due to the level of steady loading. Also, an increase in the reduced frequency value results in an increased lack of data compression, with an increase in the gust streamwise-to-transverse gust amplitude ratio also increasing the spread in the data - particularly at the very high reduced frequency. That the unsteady pressure data are a function of the steady loading and in particular that these variations can be related to the mean steady pressure loading variations shows that for this high camber suction surface, the mean flow field interacts with the unsteady gust over the entire surface for all reduced frequency values, with this steady-unsteady interaction increasing with reduced frequency value and the gust streamwise-to-transverse gust amplitude ratio.

The resulting gust generated unsteady aerodynamic loading on the rotor blades, the unsteady pressure difference across the airfoil chord, are shown in Figures 10 through 13 together with the corresponding linear theory predictions. The differences between the unsteady pressure difference data and the linear theory predictions is clearly evident, both with regard to the magnitude and the phase. With regard to the magnitude data, increasing the reduced frequency results in a significant decrease in the data-theory correlation, with the data much smaller than the corresponding prediction. Also, the spread in these magnitude data decreases with an increase in the streamwise-to-transverse gust amplitude ratio from 0.3 to 0.45, with this effect more pronounced for the first harmonic reduced frequency data. In fact, although the 0.3 streamwise-to-transverse gust amplitude ratio first harmonic magnitude data exhibit very poor correlation with the prediction and large gradients along the chord, the corresponding 0.45 streamwise-to-transverse gust amplitude ratio data exhibit reasonable correlation with the linear theory prediction.

With regard to the unsteady pressure difference phase data, over the aft 50% to 75% of the blade chord, the linear variation in the phase data noted on the suction surface is evident for all reduced frequency values. Thus, the unsteady pressure loading on the rotor blade is primarily due to the high camber suction surface.

SUMMARY AND CONCLUSIONS

A series of experiments were performed to investigate rotor unsteady aerodynamic response to high and very high reduced frequency gusts, including the effects of steady aerodynamic loading and detailed forcing function waveform shape. These experiments were performed in an extensively instrumented axial flow research compressor. The high reduced frequency forcing function to the first stage rotor blade row was the first harmonic of the wakes generated by the 36 IGV's, with the very high reduced frequency being the second harmonic of these wakes. These forcing functions were each measured in the rotor frame of reference, Fourier analyzed and then decomposed into chordwise and transverse components. The unsteady aerodynamic gust response was then investigated on the individual surfaces of the first stage rotor blade for two values of the streamwise-to-transverse gust amplitude ratio over a range of compressor steady loadings. Appropriate data were correlated with classical predictions.

Analysis of these unsteady aerodynamic data revealed the following.

Pressure Surface

* This low camber surface is primarily affected by the mean flow incidence angle.
* The interaction between the steady and unsteady flow fields at the high first harmonic reduced frequency values is large on this low camber surface, decreasing for the very high second harmonic reduced frequency values.

Suction Surface

* On this high camber surface, the magnitude of the unsteady pressure response is due to the steady loading, with the dimensionless unsteady pressure coefficient not compressing these data with mean flow incidence angle.
* This high camber surface is strongly affected by the interactions between the mean and the unsteady flow field, with this steady-unsteady interaction increasing with reduced frequency and gust streamwise-to-transverse gust amplitude ratio.
* There are large differences between the high and very high reduced frequency results.
* There is an unexplained wave phenomena evident on the high camber suction surface.

Pressure Difference

* The correlation of the unsteady pressure difference data and the linear theory predictions is poor.
* The magnitude data-theory correlation diminishes with increased reduced frequency, with the data much smaller than the corresponding prediction.
* The spread in these magnitude data decreases with an increase in the streamwise-to-transverse gust amplitude ratio, with this effect more pronounced for the first harmonic reduced frequency data.
* Over the aft 50% to 75% of the blade chord, the linear variation in the phase data noted on the suction surface is evident for all reduced frequency values.
* The unsteady pressure loading on the rotor blade is primarily due to the unsteady flow phenomena on the high camber suction surface.

Thus, these experiments show that there are large and significant differences between the high and very high reduced frequency gust generated unsteady aerodynamic response of the rotor blade row, with these effects not predicted by linear theory.

ACKNOWLEDGEMENTS

Research sponsored by the Air Force Office of Scientific Research (AFSC) under Contract F49620-88-C-0022. The United States Government is authorized to reproduce and distribute reprints for governmental purposes notwithstanding any copyright notation hereon.

REFERENCES

1. Whitehead, D.S., "Classical Two-Dimensional Methods," AGARDorgraph No. 298, AGARD Manual on Aerelasticity in Axial Flow Turbomachines, Volume 1: Unsteady Turbomachinery Aerodynamics, pp.3.1-3.30.

2. Fleeter, S., "The Fluctuation Lift and Moment Coefficients for Cascaded Airfoils in a Nonuniform Compressible Flow," AIAA Journal of Aircraft, Vol. 10, No. 2, February 1973, pp. 93-98.

3. Adamczyk, J.J. and Goldstein, M.E., "Unsteady Flow in a Supersonic Cascade with Subsonic Leading Edge Locus," AIAA Journal, Vol. 16, No. 12, 1978, pp. 1248-1254.

4. Verdon, J.M. and Usab, W.J., "Application of a Linearized Unsteady Aerodynamic Analysis to Standard Cascade Configurations," ASME Paper 90-GT-11, June 1990.

5. Verdon, J.M. and Hall, K.C., "Development of a Linearized Unsteady Aerodynamic Analysis for Cascade Gust Response Predictions," NASA Report 4308, July 1990.

6. Scott, J.S. and Atassi, H.M., "Numerical Solutions of the Linearized Euler Equations for Unsteady Vortical Flows Around Lifting Airfoils, AIAA Paper 90-0694, 1990

7. Fang, J., "Compressible Flows With Vortical Disturbances Around Cascades of Airfoils," Ph.D. Thesis, University of Notre Dame, April 1991

8. Fleeter, S., Jay, R.L., and Bennett, W.A., "Wake Induced Time-Variant Aerodynamics Including Rotor-Stator Axial Spacing Effects," ASME Journal of Fluids Engineering, Vol. 103, No. 1, March 1981, pp. 59-66

9. Capece, V.R., and Fleeter, S., "Experimental Investigation of Multistage Interaction Gust Aerodynamics," ASME Journal of Turbomachinery, Volume 111, No.4, October 1989, pp 409-417.

10. Manwaring, S.R. and Fleeter, S., "Inlet Distortion Generated Periodic Aerodynamic Rotor Response," ASME Journal of Turbomachinery, Vol. 112, No. 2, April 1990, pp. 298-307.

11. Buffum, D. and Fleeter, S., "Investigation of Oscillating Cascade Unsteady Aerodynamics by an Experimental Influence Coefficient Technique," AIAA Journal of Propulsion and Power, September-October 1990, pp. 612-660.

12. Smith, S.N. "Discrete Frequency Sound Generation in Axial Flow Turbomachines," Aeronautical Research Council R&M 3709, Great Britain, March 1972.

Table 1. Overall airfoil and compressor characteristics

	ROTOR	STATOR	IGV
Airfoil type	C4	C4	C4
Number of Airfoils	43	31	36
Chord, C (mm)	30	30	30
Solidity, C/S	1.14	1.09	0.96
Camber, θ	28.0	27.7	36.9
Stagger Angle, γ	36.0	-36.0	21.0
Inlet Metal Angle, β_1	50.0	30.0	0.0
Aspect Ratio	2.0	2.0	2.0
Thickness/Chord (%)	10.0	10.0	10.0
Flow Rate (kg/s)		2.03	
Design Axial Velocity (m/s)		24.4	
Design Rotational Speed (RPM)		2250	
Number of Stages		3	
Design Stage Pressure Ratio		1.0	
Inlet Tip Diameter (mm)		420	
Hub/Tip Radius Ratio		0.714	
Stage Efficiency (%)		85	

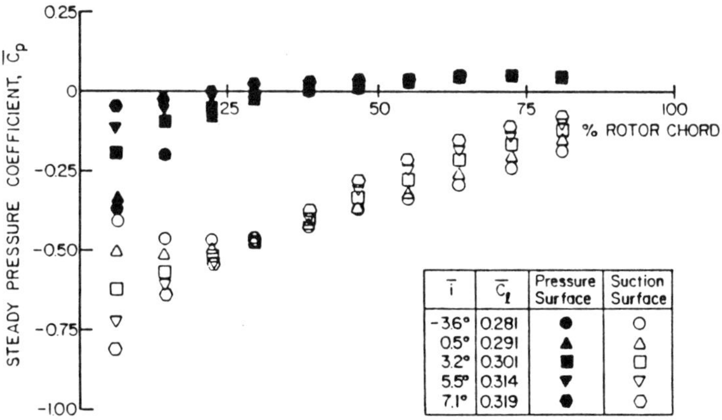

Figure 1. Loading effect on rotor blade surface steady pressures

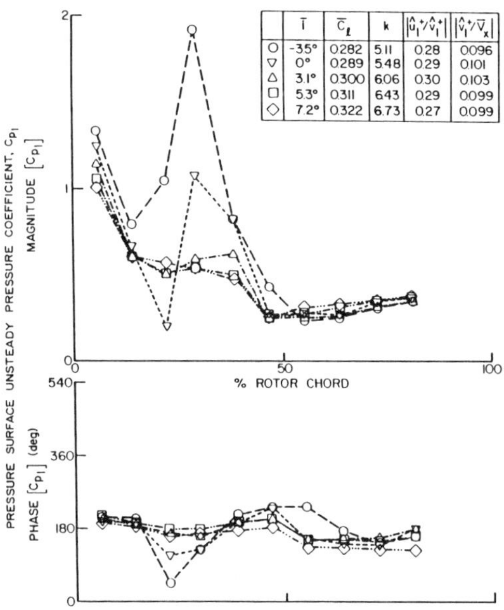

Figure 2. Pressure surface 1st harmonic unsteady gust response, $|u^+/v^+| = 0.30$

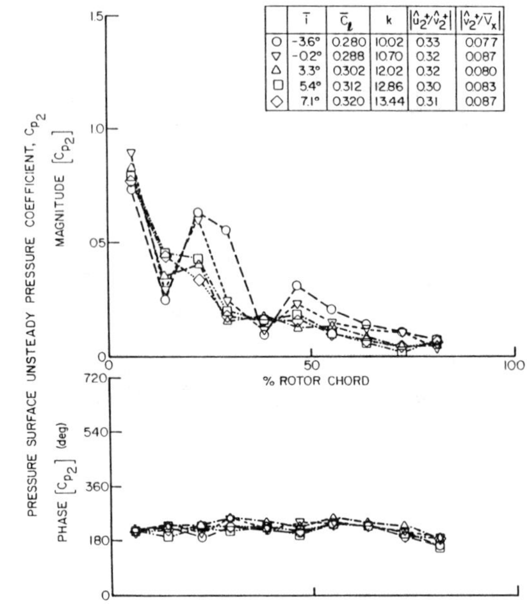

Figure 3. Pressure surface 2nd harmonic unsteady gust response, $|u^+/v^+| = 0.30$

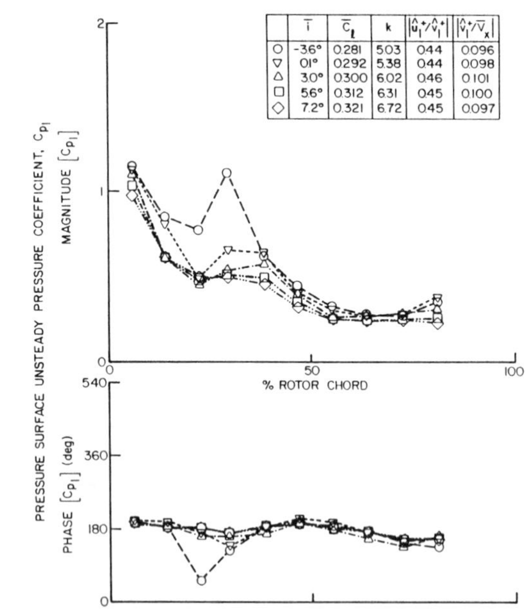

Figure 4. Pressure surface 1st harmonic unsteady gust response, $|u^+/v^+| = 0.4$

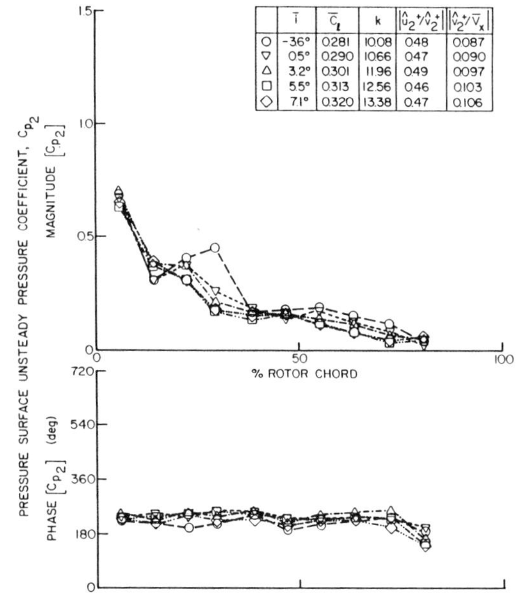

Figure 5. Pressure surface 2nd harmonic unsteady gust response, $|u^+/v^+| = 0.45$

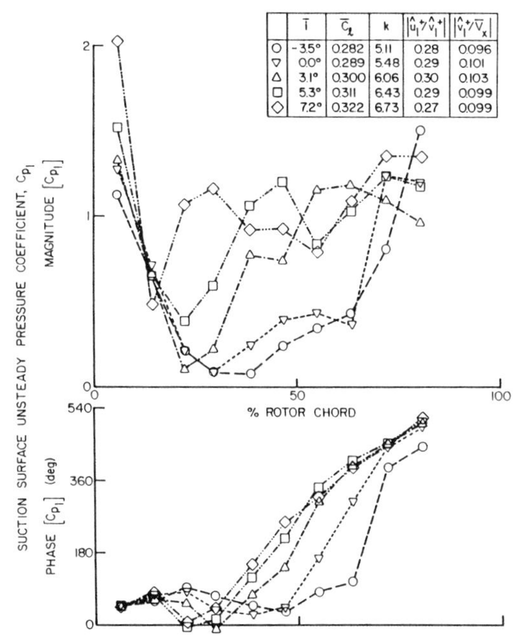

Figure 6. Suction surface 1st harmonic unsteady gust response, $|u^+/v^+| = 0.30$

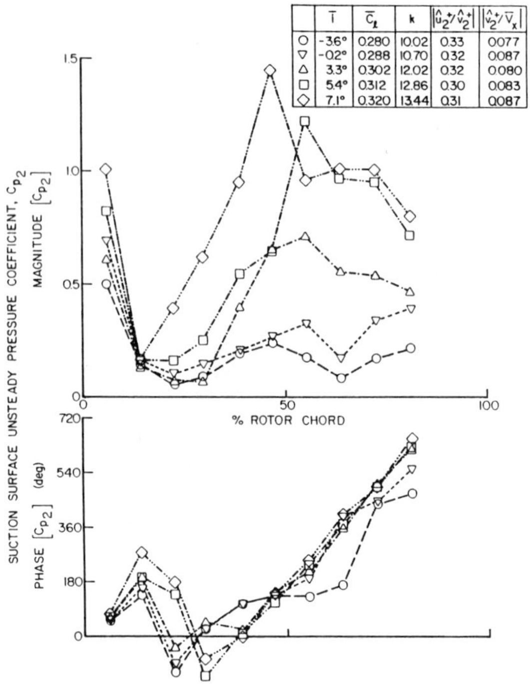

Figure 7. Suction surface 2nd harmonic unsteady gust response, $|u^+/v^+|$ = 0.30

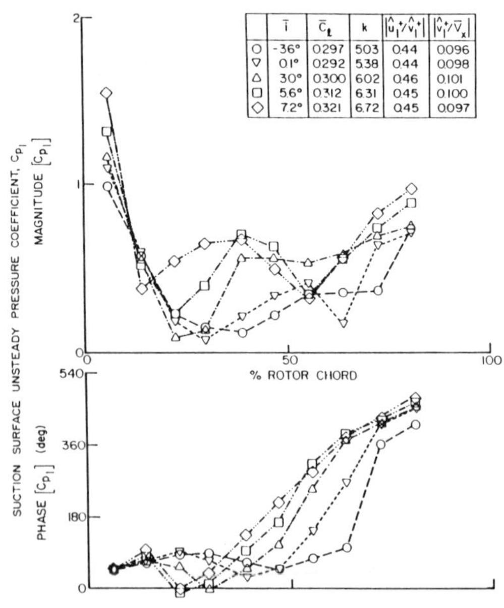

Figure 8. Suction surface 1st harmonic unsteady gust response, $|u^+/v^+|$ = 0.45

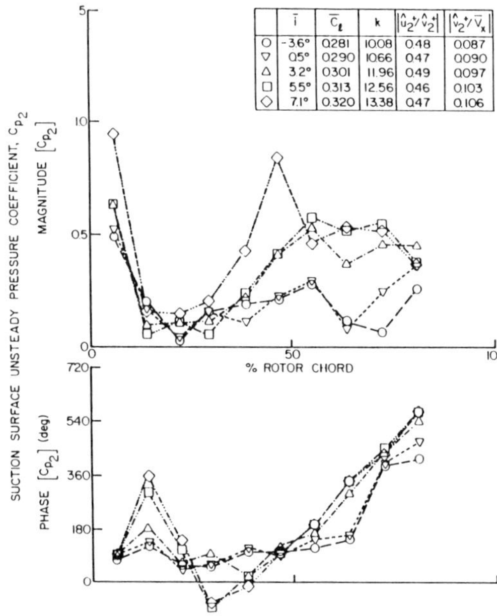

Figure 9. Suction surface 2nd harmonic unsteady gust response, $|u^+/v^+| = 0.45$

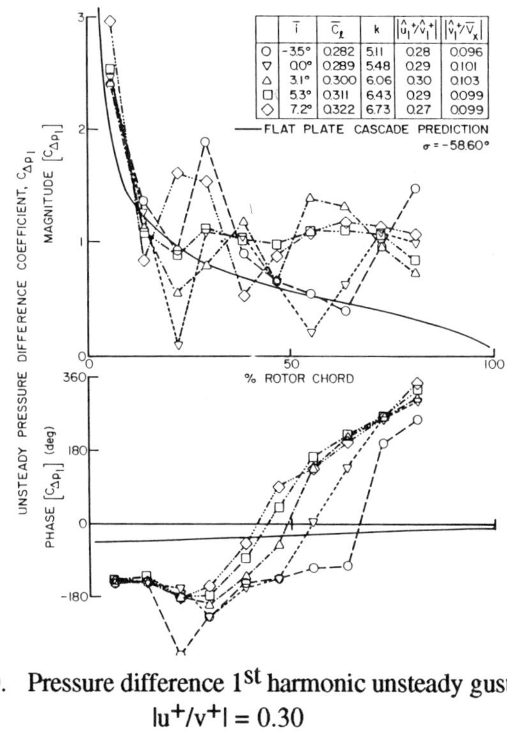

Figure 10. Pressure difference 1st harmonic unsteady gust response, $|u^+/v^+| = 0.30$

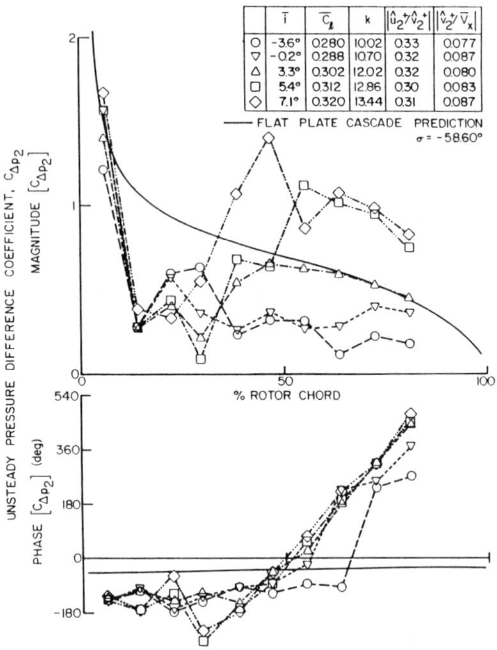

Figure 11. Pressure difference 2nd harmonic unsteady gust response, $|u^+/v^+| = 0.30$

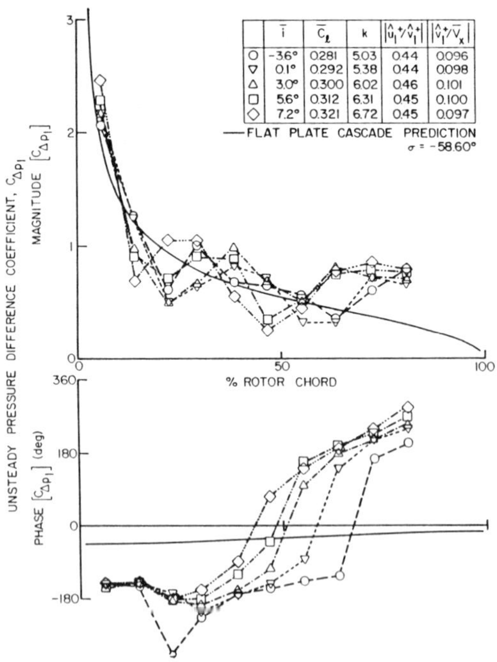

Figure 12. Pressure difference 1st harmonic unsteady gust response, $|u^+/v^+| = 0.45$

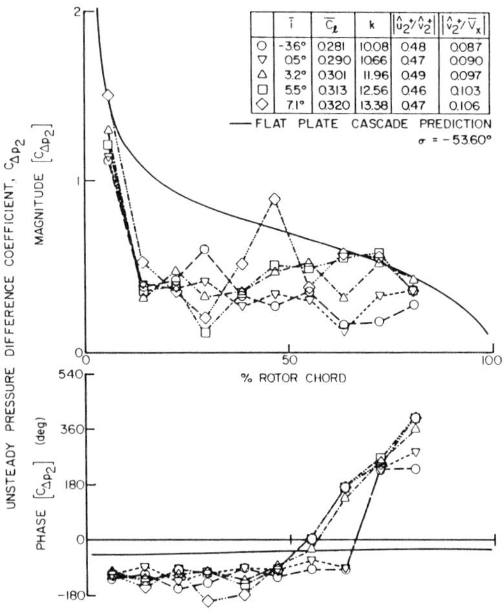

Figure 13. Pressure difference 2nd harmonic unsteady gust response, $|u^+/v^+| = 0.45$

The Response of a Low Solidity Symmetric Airfoil Cascade to Compressor Wake and Linear Theory Gusts

Gregory H. Henderson[1,2]
Sanford Fleeter[1]

ABSTRACT

Linear theory and compressor-type wake generated gusts and the resulting unsteady aerodynamic response of a low solidity airfoil cascade are investigated experimentally by means of a series of experiments performed in the Purdue Annular Cascade Research Facility. The linear theory gust is generated by a rotating row of perforated plates, with the compressor wake forcing functions associated with a rotor blade row. When the forcing function exhibits the characteristics of a linear-theory gust, as is the case for the perforated-plate wake generators, the resulting response on the low solidity downstream stator row is in excellent agreement with the linear-theory models. In contrast, when the forcing function does not exhibit linear-theory gust characteristics, i.e., for the compressor rotor blades, the resulting unsteady aerodynamic response of the downstream stator row is much more complex and correlates poorly with the linear-theory gust predictions.

---

[1]School of Mechanical Engineering, Purdue University, Chaffee Hall, West Lafayette, IN 47907.
[2]Senior Engineer, Advanced Product Development, Cummins Engine Company, Mail Code 50160, Box 3005, Columbus, IN 47202-3005.

NOMENCLATURE

C	stator airfoil chord length
$\overline{C}_p$	steady surface pressure coefficient
$\overline{C}_{\Delta p}$	steady differential pressure coefficient
$\widehat{C}_p$	unsteady surface pressure coefficient
$\widehat{C}_{\Delta p}$	unsteady differential pressure coefficient
$\vec{k}$	gust propagation wave-number vector
k_c	reduced frequency based on chord
$\vec{U}$	absolute velocity
u^+	streamwise gust amplitude
v^+	transverse gust amplitude
w^+	gust amplitude
$\vec{W}$	relative velocity
$\overline{W}_1$	inlet mean-relative velocity
α	angle of attack
β_w	gust amplitude angle
β_w'	gust velocity vector direction
ϕ_w	gust component phase angle

INTRODUCTION

State-of-the-art unsteady aerodynamic analyses are essentially limited to two-dimensional linearized perfect-fluid theory models. In the absolute reference frame, the steady flow past flat plate airfoil cascades is uniform, with a small perturbation unsteady flow resulting from the upstream rotor blade wake flow field resolved into streamwise and transverse components. Advanced analyses which allow the potential field of the downstream airfoils to distort the wake by linearizing about the nonuniform mean flow near the blade row instead of the much simpler upstream uniform mean flow are currently being developed, for example, Goldstein and Atassi [1976], Scott and Atassi [1990], Hall and Verdon [1989] and Fang[1991]. In all of these models, the wake is defined in terms of its harmonics, with each harmonic forcing function modeled as a gust forcing function boundary condition in the prediction of the resulting blade row unsteady aerodynamics. Thus, even though forcing functions can be generated by a wide variety of fundamentally different phenomena, all forcing functions are modeled as being equivalent, with the resulting predicted blade row response independent of the particular wake generator.

A number of experiments have been directed at the verification of unsteady aerodynamic models appropriate for blade row forced response prediction and the determination of their applicability and limitations, for example Frank and Henderson [1980], Fleeter, Jay, and Bennett [1978, 1980, 1981] Gallus, Lambertz, and Wallman [1980] and Capece and Fleeter [1987, 1989]. In these, the unsteady airfoil row response comparisons with linear theory were made without considering the fundamental gust modeling assumptions. However, it has been found experimentally that the harmonic gust response generated by different but equivalent forcing functions are dependent on the particular forcing function , Manwaring and Fleeter [1991] and Kim and Fleeter [1991].

In this paper, linear theory and compressor-type wake generated gusts and the resulting unsteady aerodynamic response of a low solidity airfoil cascade are investigated experimentally by means of a series of experiments performed in the Purdue Annular Cascade Research Facility. The linear theory gust is generated by a rotating row of perforated plates, with the compressor wake forcing functions associated with a rotor blade row, Figure 1.

LINEAR THEORY GUST CHARACTERISTICS

In classical linear perfect-fluid theory, the wake flow field is composed of a uniform mean flow and a superimposed sinusoidal vortical gust $\vec{w}$ propagating according to the wave-number vector $\vec{k}$. The resulting blade-row flow field schematic is presented in Figure 2.

In the stationary reference frame of a downstream airfoil row, the propagation of the harmonic wake velocity field is defined by considering periodicity requirements in the axial-circumferential coordinate system. As a consequence of this periodicity, the downstream mean-relative flow $\vec{W}_2$ and the gust propagation vector $\vec{k}$ are perpendicular.

The stator row flow field is modeled as being compressible and isentropic and is analyzed by superpositioning of rotational and irrotational flow fields which satisfy continuity independently, Goldstein [1978]. The unsteady aerodynamic loads on an airfoil are determined by solving the velocity potential equation using standard techniques subject to the Kutta condition and the solid boundary condition at the airfoil surface. The flow field pressure distribution is obtained by integrating the linearized Bernoulli Equation with the known velocity potential.

Two constraints are imposed by continuity on the downstream rotational forcing function wake flow field. The primary constraint is the requirement that the gust-component phase angle ϕ_W be either 0° or 180°. If this primary constraint is satisfied then the secondary constraint stipulates that the gust-component amplitude vector must be parallel to the downstream mean-relative flow $\vec{w}^{+} \parallel \vec{W}_2$. When the primary constraint is satisfied, the gust amplitude simplifies to $w^+ = \sqrt{u^{+2} + v^{+2}}$, with the periodic velocity vectors parallel over the entire periodic cycle.

Gust profiles are completely defined by specifying $\vec{k}$, w^+, β_W and ϕ_W, where β_W is the gust-amplitude angle. If the primary gust constraint is violated, the gust-amplitude angle β_W is distinct from the velocity vector angle β'_W, the angle of any velocity vector, and w^+ is not equal to $\sqrt{u^{+2} + v^{+2}}$. Note that the velocity vector angle β'_W is the angle of any velocity vector and varies with time and space whereas the gust-amplitude angle β_W is the angle of the maximum gust magnitude or gust amplitude w^+ and is a constant. For a gust that violates the primary constraint, the gust amplitude and gust-amplitude angle are found by maximizing the velocity vector magnitude.

RESEARCH FACILITY

The Purdue Annular Cascade Research Facility is an open-loop draw-through type wind tunnel capable of test section velocities of 70 m/sec (220 ft/sec). The inlet flow, conditioned first by a honeycomb section and then a settling chamber, accelerates into the annular test section via a bellmouth inlet. The test section exit flow is diffused into a large plenum. The 224 kw (300 hp) centrifugal fan located downstream of the plenum draws the air through the facility, with the guide vanes at the fan inlet allowing flow rate adjustment through the facility. The annular test section, Figure 3, houses a rotor independently driven by a 7.5 kw (10 hp) A-C motor controlled by a variable frequency drive to create the desired unsteady flow field together with a downstream stator row.

Figure 4 shows the profile of the symmetric stator airfoils, with the low-solidity stator row comprised of four instrumented airfoils, two for static and two for dynamic data, resulting in a solidity of 0.17. The airfoil chord and span are equal at 0.152 m (6.0 in.). The twelve chordwise tap locations provide highly detailed spatial resolution of the midspan chordwise pressure distributions.

DATA ACQUISITION AND ANALYSIS

The unsteady data defining the gust forcing function are determined from the unsteady velocity and static pressure fluctuations. The rotor-exit flow-field unsteady total pressure is measured with a total pressure pitot tube fitted with an unsteady pressure transducer. A cross hot-wire anemometer located on the downstream stator vane row leading edge line is used to measure the stator cascade inlet flow field.

Static pressure taps and high-response pressure transducers embedded in the stators are used to obtain the detailed airfoil-surface steady and unsteady pressure data. Four instrumented airfoils are required, two to obtain static pressure distributions on each airfoil surface and two to obtain the periodic pressure distributions on each surface. The stator-cascade steady and unsteady velocity fields are measured with the cross hot-wire probe, with these measured wake velocity deficits then decomposed into streamwise and transverse components. The total pressure fluctuations downstream of the rotor are measured with the unsteady total pressure probe, with the static pressure distributions calculated by subtracting the dynamic pressure $\rho_0 \overline{U}_2^2/2$ from the total pressure.

The stator vane row airfoil unsteady surface pressures are measured with PCB Piezotronics model 103A piezoelectric pressure transducers. The transducers have a nominal sensitivity of 0.22 mv/Pa (1500 mv/psi) and a natural frequency of 13 kHz. After the transducers were installed in the airfoil, the dynamic response of each airfoil transducer-passage system was experimentally determined to correct for any passage effects.

The time variant signals are digitized over one rotor revolution using approximately 2000 samples. Ensemble averaging the hot-wire and pressure transducer signals 150 and 100 times, respectively, produces very clean periodic time traces with the random fluctuations averaged away. The Fourier components of the ensemble averaged time traces are numerically determined with Fast Fourier Transform software. The sample frequency is set and the number of samples is adjusted to produce time records of exactly one rotor rotation period to eliminate frequency leakage problems in Fourier Transform analysis.

The stator vane chordwise pressure response data are presented as nondimensional pressure coefficients and are correlated with a steady linear-theory analysis. The unsteady response data are correlated with Smith's [1972] compressible flow analysis. Both of these linear-theory analyses account for the cascade stagger angle γ, solidity C/S, reduced frequency k_C, and interblade phase angle σ.

The steady data are nondimensionalized by calculating a steady surface pressure coefficient

$$\overline{C}_p(x) = \frac{\overline{p}(x) - p_\infty}{\frac{1}{2}\rho_0\overline{U}^2} \tag{1}$$

where $\overline{p}$, is the time-averaged airfoil surface pressure and p_∞ is the free stream pressure measured on the test-section outer diameter just upstream of the stator cascade.

The steady differential pressure coefficient is defined by

$$\overline{C}_{\Delta p} = \overline{C}_{p,l} - \overline{C}_{p,u} \tag{2}$$

where the subscripts l and u refer to the lower and upper airfoil surfaces, respectively.

The unsteady periodic signals are Fourier decomposed and each pressure harmonic nondimensionalized by calculating an unsteady surface pressure coefficient

$$\hat{C}_p(x) = \frac{\hat{p}(x)}{\rho\overline{U}\hat{v}} \tag{3}$$

where $\hat{p}$ and $\hat{v}$ represent the harmonic surface pressure and transverse gust component respectively.

The harmonic differential pressure coefficient is defined by

$$\hat{C}_{\Delta p} = \hat{C}_{p,l} - \hat{C}_{p,u}\,. \tag{4}$$

Direct comparison of the experimental and theoretical steady and unsteady lift coefficients is accomplished by integrating the experimental and theoretical chordwise differential pressure coefficients over the airfoil chord.

$$C_L = \frac{1}{C}\int_{0.025C}^{0.95C} C_{\Delta p}\,dx. \tag{5}$$

The trapezoidal rule is used to evaluate the integral from the discrete differential pressure coefficient data and theoretical predictions. The differential pressure distributions were integrated from the leading edge tap location, 0.025C, to the trailing edge tap location, 0.95C. In a similar manner experimental and theoretical quarter-chord moment coefficients are calculated.

$$C_{M,\,c/4} = \frac{1}{C^2}\int_{0.025C}^{0.95C} C_{\Delta p}(x - 0.25)dx. \tag{6}$$

The complex valued unsteady vane surface pressure data contain both amplitude and phase angle information. These quantities are referenced to the transverse gust sinusoid measured by the cross hot-wire on the airfoil leading edge plane. To obtain pressure coefficients for a single equivalent airfoil, the pressure signals are phase corrected to account for the circumferential locations of the instrumented airfoils.

RESULTS

To investigate linear theory and compressor-type wake generated gusts and the resulting unsteady aerodynamic response of a low solidity airfoil cascade, a series of experiments were conducted in the Purdue Annular Cascade Research Facility with a constant axial velocity corresponding to a Mach number of approximately 0.15.

Forcing Functions

The perforated-plate forcing functions closely resemble linear-theory gusts, as shown in Figure 5 which shows a perforated-plate generated aerodynamic forcing function. The arrows represent the velocity vector fluctuations, with static pressure fluctuations denoted by the solid line. In particular, the periodic gust velocity vectors are parallel to $\vec{W}_2$, the static pressure distribution is very small and the fundamental-harmonic linear gust parameters ϕ_W and β_W very near the linear theory values, with $\phi_W = 182^\circ$ and $\beta_W \approx \overline{\beta}_2$.

The corresponding compressor airfoil-wake forcing function is presented in Figure 6. This forcing function exhibits gust characteristics far from linear-theory gusts. Not only are the wake region velocity vectors misaligned with the downstream mean-relative flow $\overline{\beta}_2$, the free-stream velocity vectors are nonparallel $\phi_W \neq 180^\circ$ and misaligned with the downstream mean-relative flow $\beta_W \neq \overline{\beta}_2$. These compressor blade row forcing functions exhibit also have sizable static pressure fluctuations which are not considered in linear theory and are not present in the perforated-plate forcing functions.

Airfoil Gust Response

Figure 7 presents the airfoil steady chordwise surface and differential pressure coefficients obtained in a steady flow field. The correlation with linear theory is excellent for angles-of-attack less than 15°. At $\overline{\alpha} = 0°$ the surface pressures are coincident resulting in zero differential pressure coefficients. For angles-of-attack greater than 10°, stalling is evident.

The unsteady aerodynamic gust response of the low-solidity symmetric-airfoil cascade to the perforated-plate forcing function is shown in Figures 8 and 9. The upper and lower surfaces correspond to the suction and pressure surfaces for positive mean angles-of-attack.

The lower-surface response magnitude decreases monotonically from the stator leading to trailing edge, whereas the upper-surface response magnitude decreases rapidly over the leading quarter chord then increases slightly, so that a relative minima, a dip, in the data trend is predominant near the quarter chord. The lower-surface response phase angle is relatively flat, with the upper-surface response phase angle more a function of chord location. The upper and lower-surface responses are approximately equal in magnitude and approximately 180° out of phase at the leading edge. At the trailing edge, the Kutta condition is approached as the upper and lower-surface pressure coefficients are again approximately equal in magnitude and are now nearly in phase. The differential-pressure coefficients are in excellent agreement with the theoretical predictions in both magnitude and phase.

The unsteady aerodynamic gust response in terms of the unsteady fundamental-harmonic lift and moment coefficients $\widehat{C}_L$ and $\widehat{C}_{M,\,C/4}$ is presented in Figure 10. The upper two plots present the lift and quarter-chord moment coefficient magnitudes and the lower two plots the lift and quarter-chord moment coefficient phase angles. Also presented are the theoretical predictions from Smith's linearized compressible cascade analysis. With four airfoils comprising the stator cascade, only four interblade phase angles σ are possible. However, the $\sigma = 90°$ and 270° predictions are the same for an unstaggered cascade. The discontinuities in the predictions result from the acoustic resonance condition. The interblade phase angle of each data point is indicated by the location of a tick mark on the symbol. A tick mark at the 12-o'clock, 3-o'clock, 6-o'clock, and 9-o'clock position designates interblade phase angles of 0°, 90°, 180°, and 270°, respectively. Recall that the perforated-plate forcing functions closely match linear theory gusts whereas the compressor blade wake generated gusts do not.

The measured unsteady lift coefficients correlate almost exactly with the theoretical predictions in both the magnitude and phase plots. The quarter-chord unsteady moment coefficient magnitudes are very small and also correlate well with the theoretical predictions. The moment coefficient phase angles deviate from theory when the lift-coefficient phase-angle correlation is used as a basis of judgement. However, the quarter-chord moment-coefficient phase angles are of little consequence because of the very small moment coefficient magnitudes.

The unsteady aerodynamic gust response of the low-solidity symmetric-airfoil cascade to the compressor-wake forcing function is presented in terms of the chordwise unsteady response in Figure 11, 12 and 13. The upper-surface response decreases monotonically from the leading edge to the midchord and then increases slightly. The lower-surface response magnitudes decrease monotonically from the leading edge to the midchord. The lower-surface response phase angles are weak functions of chord position from the stator leading edge to 65% chord. The unsteady differential-pressure-coefficient magnitudes follow the trend of the linear-theory predictions. The phase angles slope slightly downward from the leading edge to 55% chord but are near the linear theory prediction. The variation in the unsteady differential-pressure-coefficient magnitude is due to some rotor-loading dependent characteristic of the forcing function which affects the lower-surface more than the upper-surface by greatly increasing the response magnitude.

The unsteady aerodynamic gust response in terms of the unsteady fundamental-harmonic lift and moment coefficients is presented in Figure 14. The trends of the unsteady aerodynamic coefficients do not following the linear-theory predictions, with the unsteady lift-coefficient magnitude correlation poor. The quarter-chord unsteady moment coefficients, both predicted and measured, are small compared to the unsteady lift coefficient magnitudes. However, the forcing functions are far from linear-theory gusts, i.e. the static pressure fluctuations are high and the linear-gust parameter correlation is poor, which coincides with the poor response correlation.

SUMMARY AND CONCLUSIONS

Linear theory and compressor-type wake generated gusts and the resulting unsteady aerodynamic response of a low solidity airfoil cascade have been investigated experimentally by means of a series of experiments performed in the Purdue Annular Cascade Research Facility. The linear theory gust was generated by a rotating row of perforated plates, with the compressor wake forcing functions associated with a rotor blade row.

The perforated-plate forcing functions closely resemble linear-theory forcing functions, with the static pressure fluctuations small and the periodic velocity vectors parallel to the downstream mean-relative flow angle over the entire periodic cycle. The compressor-wake forcing functions are much more complex, exhibiting characteristics far from linear-theory gusts.

When the forcing function exhibits the characteristics of a linear-theory gust, as was the case for the perforated-plate wake generators, the resulting response on the downstream stator airfoils is in excellent agreement with the linear-theory models. In contrast, when the forcing function do not exhibit linear-theory gust characteristics, i.e., for the airfoil wake generators, the resulting unsteady aerodynamic response of the downstream are much more complex and correlate poorly with the linear-theory gust predictions.

ACKNOWLEDGEMENTS

This research was sponsored, in part, by the NASA Lewis Research Center. Both the financial support and the technical interchanges with Dr. Daniel Hoyniak are most gratefully acknowledged.

REFERENCES

1 Capece, V.R. and Fleeter, S., "Unsteady Aerodynamic Interactions in a Multi-Stage Compressor," *ASME Journal of Turbomachinery*, Vol. 109, No. 3, July 1987, pp. 420-428.

2 Capece, V. R. and Fleeter, S., 1989, "Experimental Investigation of Multistage Interaction Gust Aerodynamics," *ASME Journal of Turbomachinery*, vol. 111, pp. 409-417.

3 Fang, J., "Compressible Flows With Vortical Disturbances Around Cascades of Airfoils," *Ph.D. Thesis, University of Notre Dame*, April 1991

4 Fleeter, S. and Jay, R. L., and Bennett, W. A., 1978, "Rotor Wake Generated Unsteady Aerodynamic Response of A Compressor Stator," *ASME Journal of Engineering for Power,* vol. 100, pp. 664-675.

5 Fleeter, S., Jay, R. L., and Bennett, W. A., 1980, "The Time-Variant Aerodynamic Response of a Stator Row Including the Effects of Airfoil Camber," *ASME Journal of Engineering for Power*, vol. 102, no. 2, pp. 334-343.

6 Fleeter, S., Jay, R. L., and Bennett, W. A., 1981, "Wake Induced Time-Variant Aerodynamics Including Rotor-Stator Axial Spacing Effects," *ASME Journal of Fluids Engineering*, vol 103, no. 1, pp. 59-66.

7 Franke, G. F. and Henderson, R. E., 1980, "Unsteady Stator Response to Upstream Rotor Wakes," *AIAA Journal of Aircraft*, vol. 17, no. 7, pp. 500-506.

8 Gallus, H. E., Lambertz, J. and Wallmann, Th., 1980, "Blade-Row Interaction in an Axial-Flow Subsonic Compressor Stage," *ASME Journal of Engineering For Power,* vol. 102, pp. 169-177.

9 Goldstein, M. E., 1978, "Unsteady Vortical and Entropic Distortions of Potential Flows Round Arbitrary Obstacles," *Journal of Fluid Mechanics*, vol. 89, part 3, pp. 433-468.

10 Goldstein, M. E. and Atassi, H., 1976, "A complete second-order theory for the Unsteady Flow about an Airfoil due to Periodic Gust," *Journal of Fluid Mechanics*, vol. 74, part 4, pp. 741-765.

11 Hall, K.C., and Verdon, J.M., 1989,"Gust Response Analysis for Cascades Operating in Nonuniform Mean Flows," AGARD Conference - Unsteady Aerodynamic Phenomena in Turbomachines, AGARD CPP-468.

12 Kim, K.H. and Fleeter, S., "Compressor Blade Gust Response to Attached & Separated Flow Forcing Functions," *Sixth International Symposium on Unsteady Aerodynamics, Aeroelasticity and Aeroacoustics of Turbomachines and Propellers*, Notre Dame, September 1991.

13 Manwaring, S.R. and Fleeter, S., "Forcing Function Effects on Rotor Periodic Aerodynamic Response," *ASME Journal of Turbomachinery,* Vol. 113, No. 2, April 1991, pp. 312-319.

14 Scott, J.S. and Atassi, H.M., "Numerical Solutions of the Linearized Euler Equations for Unsteady Vortical Flows Around Lifting Airfoils, *AIAA Paper 90-0694*, 1990

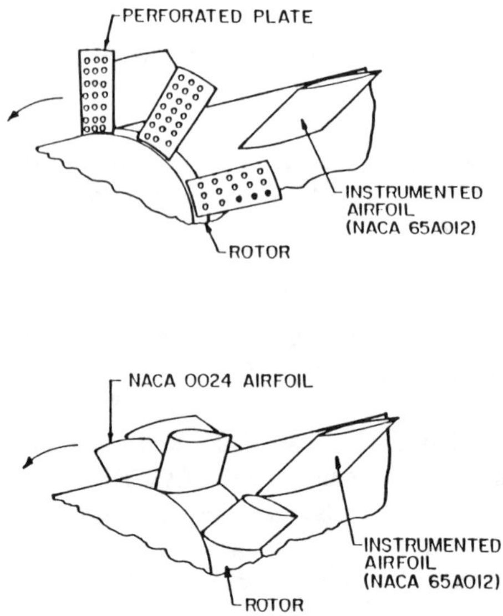

Figure 1. Perforated plate and airfoil gust generators

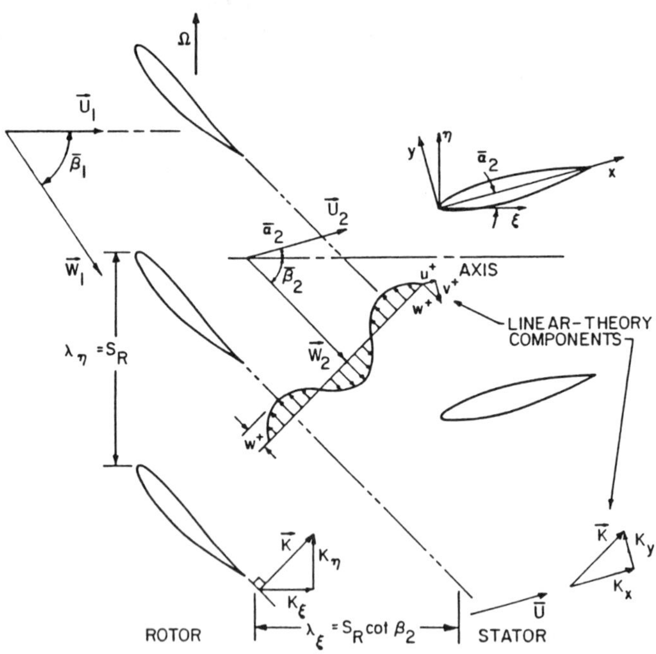

Figure 2. Turbomachine gust propagation

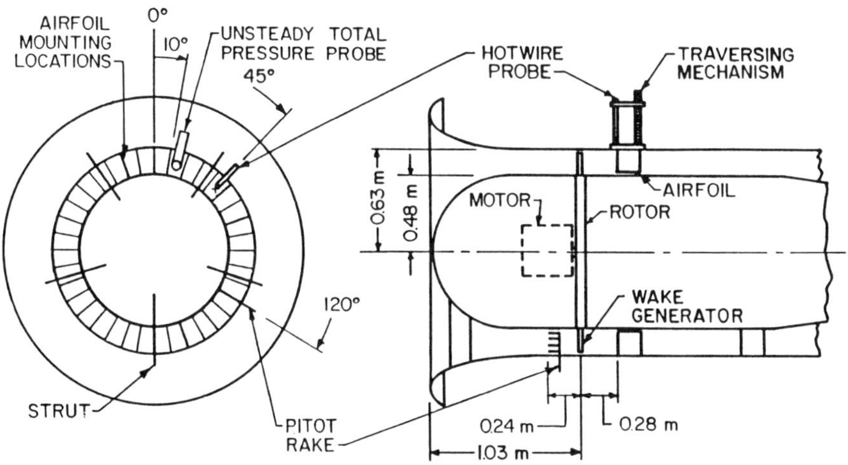

Figure 3. Test section schematic

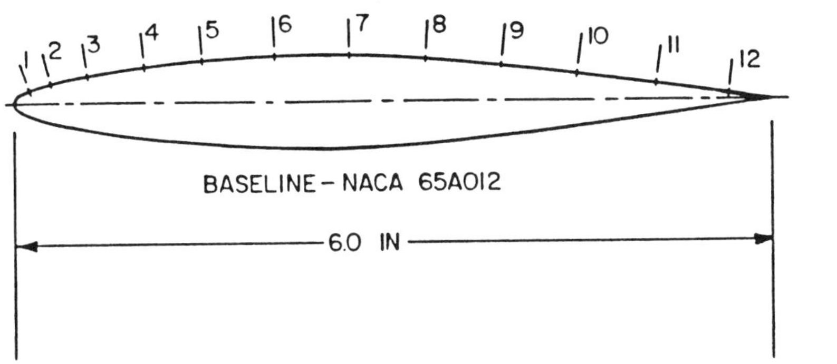

PRESSURE MEASUREMENT LOCATIONS

TAP NO.	1	2	3	4	5	6	7	8	9	10	11	12
% CHORD	2.5	5.0	10.0	17.5	25	35	45	55	65	75	85	95

Figure 4. Instrumented stator airfoil profile

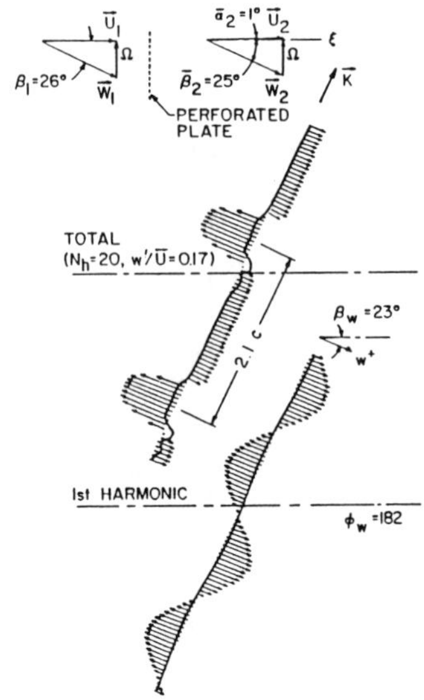

Figure 5. Perforated plate generated forcing function

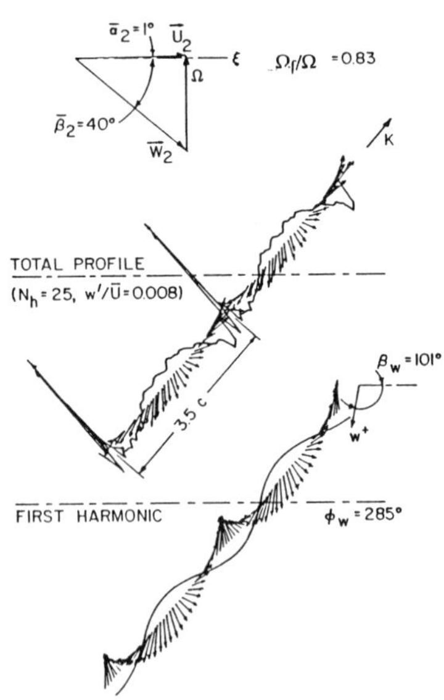

Figure 6. Airfoil generated forcing function

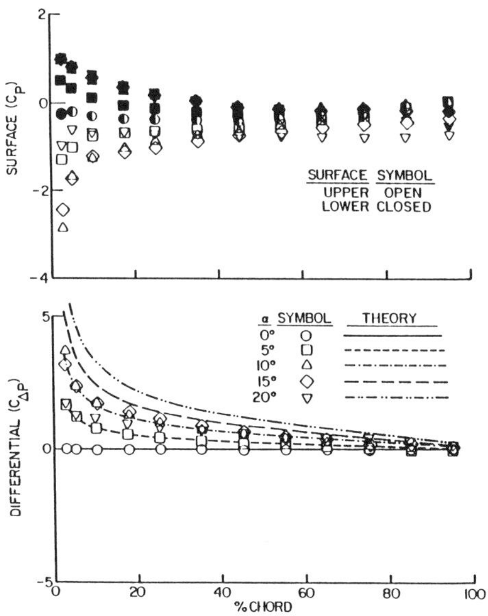

Figure 7. Stator surface steady pressure distributions

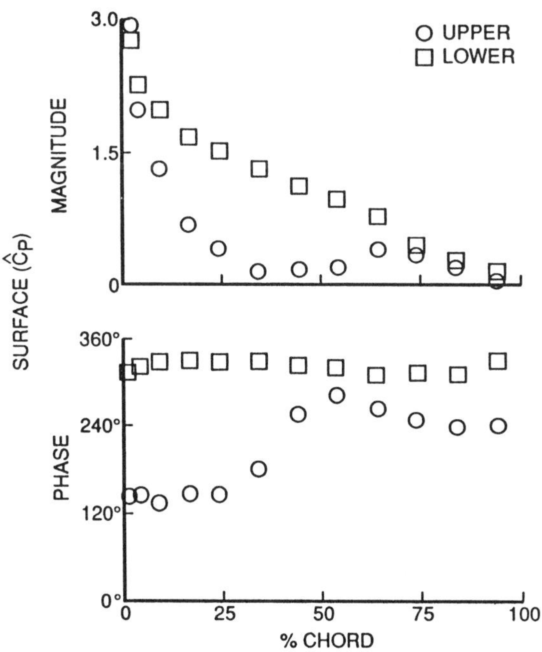

Figure 8. Perforated plate 1st harmonic gust generated pressure response

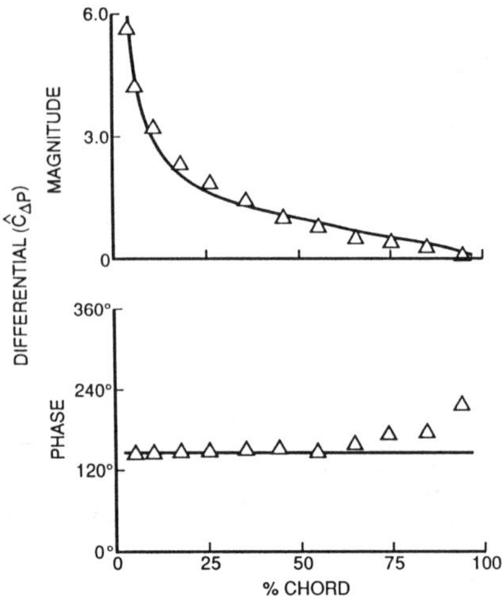

Figure 9. Perforated plate 1st harmonic gust generated pressure difference response

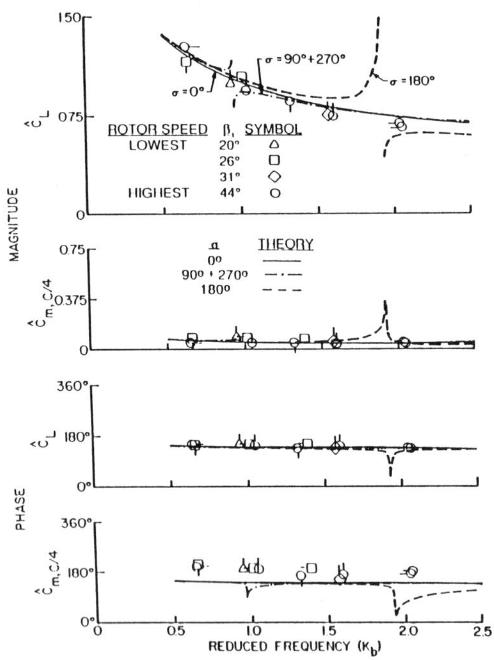

Figure 10. Perforated plate 1st harmonic gust generated lift and moment coefficients

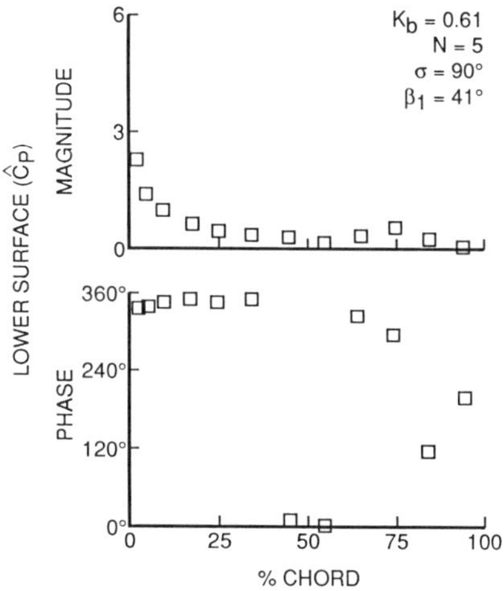

Figure 11. Wake 1st harmonic gust generated pressure response of upper surface

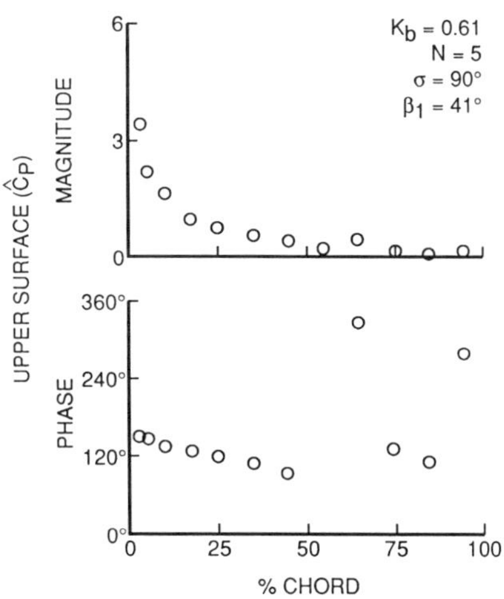

Figure 12. Wake 1st harmonic gust generated pressure response of lower surface

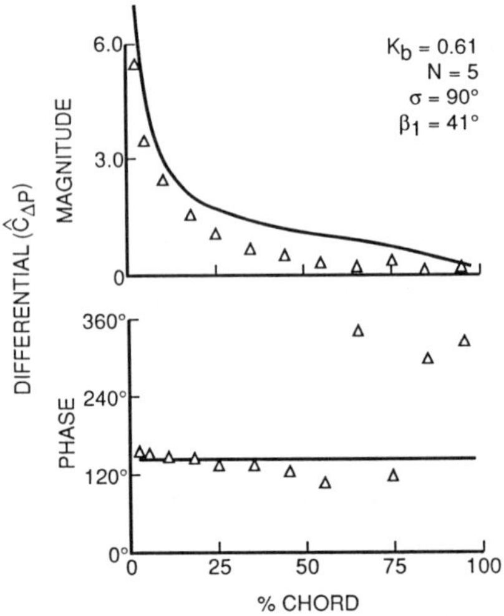

Figure 13. Wake 1st harmonic gust generated pressure difference response

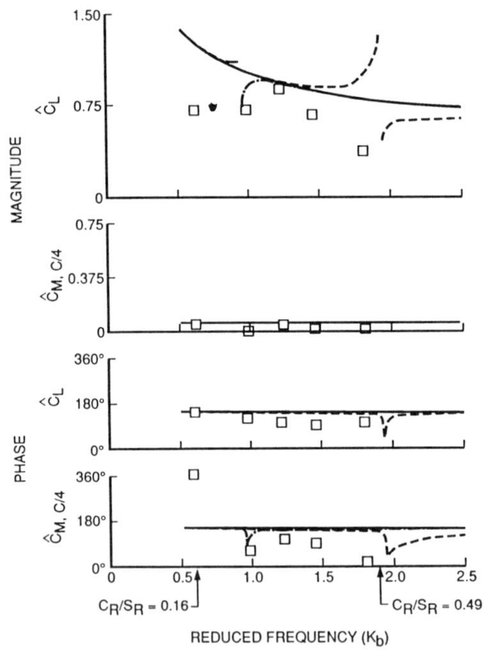

Figure 14. Wake 1st harmonic gust generated lift and moment coefficients

Time-Dependent Pressure Fluctuations on an Oscillating Turbine Cascade at Transonic Off-Design Flow Conditions

A. Bölcs[1]
T.H. Fransson[1,2]
H. Körbächer[1]

ABSTRACT

A cylindrical section of the last stage of a gas turbine blade has been investigated experimentally in an annular non-rotating cascade facility as regards to its steady-state and time-dependent aerodynamic characteristics for off-design inlet and transonic outlet flow conditions at realistic reduced frequencies. The blades vibrate in the first travelling wave bending mode.

The present investigation concentrates upon the detailed unsteady response in the vicinity of an aerodynamic shock wave. Emphasis is also put on using pressure transducers on different blades in order to determine non-periodicities in the flow which may be of large importance when comparing with prediction models.

The time-dependent aerodynamic damping of the cascade, as integrated from the pressure transducers on the blades, shows that the cascade is aerodynamically unstable over a region of interblade phase angles.

The steady-state data indicate a separation bubble on the suction surface until about 30% chord. The length of this region increases with increasing inlet flow angle, but is not largely affected by a change in outlet flow velocity. High unsteady loads are measured in this domain. A second region of high loads is found in the vicinity of the expected impingement of aerodynamic shock waves. The reduction of the data from the travelling wave vibration mode into influence coefficients shows that the unsteady response on the pressure surface is similar for the different steady-state operating conditions. The suction surface data show that the high unsteady loads in the region of a normal shock wave is mostly due to the influence of the reference blade on itself.

[1]Ecole Polytechnique Federale de Lausanne, Laboratoire de thermique appliquee et de turbomachines, Lausanne 1015, Switzerland.
[2]Presently at Royal Institute of Technology, Dept. of Thermal Engineering, Stockholm 100 44, Sweden.

NOMENCLATURE

Symbol	Explanation	Dimension
c	Blade chord length	m
$\tilde{c}_p(x,t)$	Unsteady perturbation pressure coefficient:: $\tilde{c}_p(x,t) = \frac{1}{h} \cdot \frac{\tilde{p}(x,t)}{\bar{p}_{w1} - p_1}$	
$\tilde{c}$	Quasi-steady perturbation pressure coefficient	-
f	Blade vibration frequency	Hz
h	Dimensionless (with chord) bending vibration amplitude	-
k	Reduced frequency: $k = \frac{\omega \cdot c}{2 \cdot u_2}$	-
M	Mach Number	-
p	Pressure	mbar
$\tilde{p}$	Unsteady perturbation pressure	mbar
Re	Reynolds number: $Re = \frac{c \cdot u_1}{\nu_1}$	-
t	Time	sec
t	Blade pitch	m
u	Velocity	m/s
x	Coordinate along chord	m
z	Circumferential position	m
β	Flow angle	deg
β_g	Stagger angle	deg
δ	Blade vibration direction	deg
ν	Kinematic viscosity	m^2/s
ω	Circular frequency of the blades	rad/s
σ	Interblade phase angle σ is positive when blade "n+1" leads blade "n"	deg
ϕ	Phase angle. Positive when disturbance leads blade "n"	deg
Ξ	Aerodynamic damping coefficient	-

Subscripts:

- ic Influence coefficient
- p Pressure
- s Isentropic
- w Stagnation values
- 1 Upstream flow conditions
- 2 Downstream flow conditions
- 1B First bending mode

Superscripts:

n Influence of blade "m" on blade "n"

INTRODUCTION

Vibration-related fatigue failures have been, and remain, a major concern for turbomachine manufacturers and operators world-wide. If such a failure appears, it has major consequences and can become extremely expensive. It is not probable that blade vibration problems will disappear in the next few years while, although the quality of design methods increases, manufacturers constantly push the aerodynamics and materials of blades into higher performances. Modern turbomachine blades are thus constantly at the high end of possibilities of the materials used.

The agreement between theoretical and experimental results in the domain of turbomachine blade aeroelasticity ranges from extremely good to extremely bad, depending on such factors as geometry and flow conditions, but also on experimental procedure and numerical method. Generally it can be concluded that the low subsonic attached flow can today be reasonably well predicted, whereas the transonic flow conditions on realistic profile types show discrepancies between theory and experiments, and that virtually no data or prediction models are available for stalled flow. A large need exists thus for both well documented experimental test cases and for appropriate prediction models. Furthermore, experimental indicate that the unsteady effects from separations or from a shock wave can be of considerable importance. It is thus important to consider in detail unsteady pressure responses at the foot of an oscillating shock wave.

The objectives of the experiments presented here are to obtain time dependant informations about pressure responses at high incidence flow angles and in the vicinity of oscillating shock waves. A comparison with predictions by a numerical unsteady inviscid time-linearized potential flow model should also be performed.

TEST FACILITY AND NUMERICAL MODEL

The tests on the present cascade were performed in the non-rotating annular cascade tunnel at the Swiss Federal Institute of Technology, Lausanne [Bölcs, 1983; Abbassi et al., 1991]. The flow angle in the test section can be regulated from axial to ±75° (measured from axial), and the inlet Mach number varied from 0.3 to 1.6. The flow conditions are measured with aerodynamic probes, pressure taps. Schlieren optics and laser holography are used for flow visualization.

For time-dependent measurements the blades in the test cascade are mounted on elastic springs and are driven into a vibration mode by means of a vibration control system [Kirschner et al., 1980]. Each blade is elastically suspended on a spring-mass system, designed to reproduce the first three eigenfrequencies, as well as the vibration direction of the first bending mode, as determined from the blades in the full-scale turbine, and they are forced into vibration by means of individual electromagnetic exciters, each with its own feed-back loop.

The numerical model employed in the study is based on the linearized unsteady quasi three-dimensional potential flow in turbine and compressor cascades and uses a numerical mesh of triangular finite elements [Whitehead, 1990; Whitehead and Newton, 1985]. It is called "FINSUP". The program calculates both sub- and supersonic steady-state and time-dependent flow conditions, where the unsteadiness is introduced either through wake interactions or through blade vibrations. Most of the predictions were obtained with "FINSUP" in the two-dimensional mode. Only when these calculations did not converge the quasi three-dimensional option was used.

Details about the model-cascade and the nominal flow conditions of the section under investigation are given in Fig. 1 and Table 1. The model blade section is overlapped for about 45% of the chord.

Table 1: Model cascade geometry and nominal flow conditions.

Number of blades:	20	Reynolds no. (Re):	$0.7 \cdot 10^6 \div 1.3 \cdot 10^6$
Blade span:	40 mm	Pressure taps, Suction surface:	31
Chord (c):	77.8 mm	Pressure taps, Pressure surface:	8
Maximum blade thickness:	7.7 mm	Pressure taps, Trailing edge:	14
Camber :	49.2°	Pressure transducers, Suction surface:	19
Hub diameter:	320 mm	Pressure transducers, Pressure surface:	9
Tip diameter:	400 mm	Nominal flow conditions:	
Pitch (midspan)(t):	56.5 mm	M_1:	0.33
t / c (midspan):	0.72	M_2:	1.04
Minimal section (midspan):	28.5mm	β_1:	10.5°
Stagger angle (β_g):	49°	Reduced frequency k_{1B}:	0.1366
Bending mode eigenfrequency (f_{1B}):	205±2Hz		
Bending mode direction (δ_{1B}):	90°		

Eight blades are furnished with pressure taps and five blades are equipped for unsteady pressure measurements with piezoresistive pressure transducers (ENDEVCO Model 8515-50A and KULITE Model LQ-5-080). The ENDEVCO transducers are embedded in the blade and connected to the surface through taps, whereas the KULITE transducers are flush mounted with the surface. Measurement uncertainties for the high-response pressure transducers are expected from the fact that the unsteady pressures are fairly small (i.e. the signals have a relatively high noise level). Although the number of pressure transducers on the blades is fairly high (around 25-30 depending on the test), they are still at discrete locations. It is thus important to realize, especially when comparing the data with numerical predictions, that the measured values

of aerodynamic damping probably give a good idea about the overall tendency of blade stability but that the absolute values are not exact.
Once the results in the travelling wave mode have been obtained, a data reduction into influence coefficients is done [Fransson, 1990; Széchényi and Girault, 1983; Hanamura et al., 1980] .

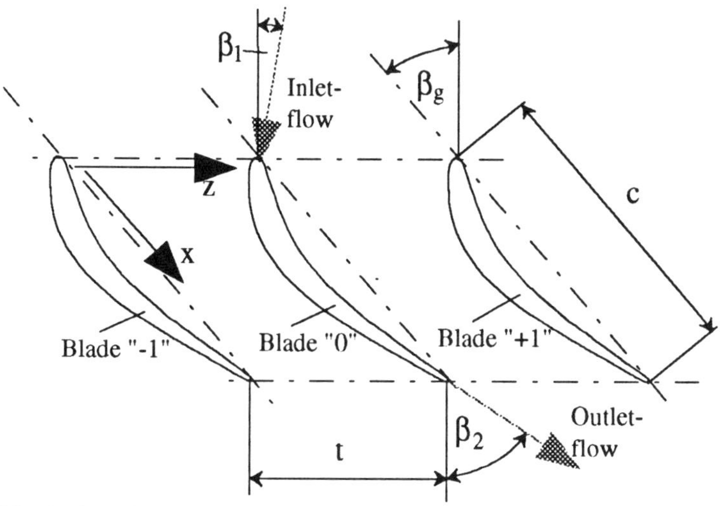

Fig.1: Annular cascade model geometry at midspan

STEADY-STATE RESULTS

As the objectives of the tests were to run the cascade at higher than nominal incidence angles, three inlet angles (β_1~35°, β_1~40° and $_1$~45°) are presented. Different outlet Mach number conditions were regulated, ranging from M_{s2}~0.7 to 1.4.
Before any conclusions about differences between different steady state and time dependant operating conditions can be drawn, the periodicity and repetitivity of the experimental data must be assessed, especially for flow with shock waves. From these tests it is concluded that the steady state results can be repeated with a high degree of accuracy. Small differences are only noticed in the exact position of the shock wave around x/c~0.84.
Maximal and minimal values of 11 circumferential positions have been brought together in Fig. 2. The pressure surface data show a very good periodicity from x/c~0.2 downwards. The two first pressure tappings in the leading edge region show at this high incidence flow conditions a non-periodicity. This leads to the conclusion that the inlet flow angle is not completely constant around the circumference of the test facility.

Further downstream on the suction surface, up to the shock position, the circumferential periodicity is again good. The shock position is however not identical on all blades and for all circumferential positions. The variation in the location of the shock wave impingement on the suction surface is estimated to be about 5% of the blade chord. This information has to be kept in mind when analysing the time-dependant pressure-data.

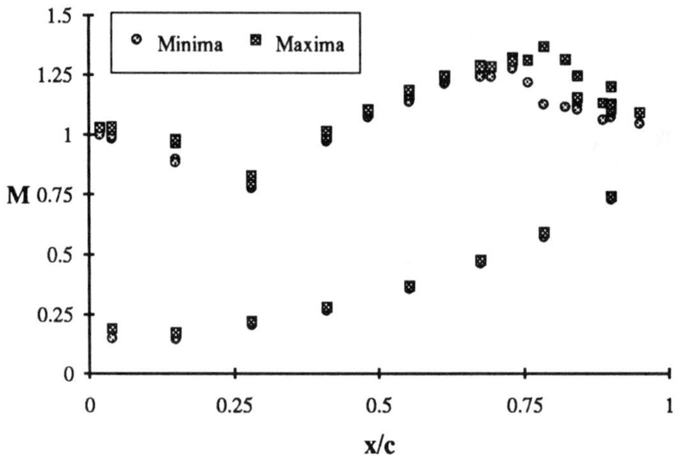

Fig. 2: Periodicity of steady-state Mach number distribution

Comparison at Different Steady-State Operating Conditions

From the blade surface Mach number distributions as well as from surface paint flow visualization, it can be concluded that a separation bubble exists from the leading edge to approximately 30% chord on the suction surface, for the three off-design inlet flow angles presented. From this point onwards the suction surface flow is accelerated as for the nominal inlet flow angle (Fig. 3a). As the outlet flow velocity increases from subsonic to supersonic the blade surface Mach number increases only slightly (for a given inlet flow angle) on the pressure surface. The Mach number distribution on the suction surface shows a small increase in the leading edge separation zone between M_{S2}~ 0.8 -> 0.9, but no further significant increase between M_{S2}~ 0.9 -> 1.4. This holds for all three inlet flow angles, whereas the zone with high Mach numbers in the leading edge region increases as expected, as the inlet flow angle is changed from β_1~33° to 45°. At the highest flow angle, the time averaged pressure is constant in the separation bubble at least up to x/c~0.15.

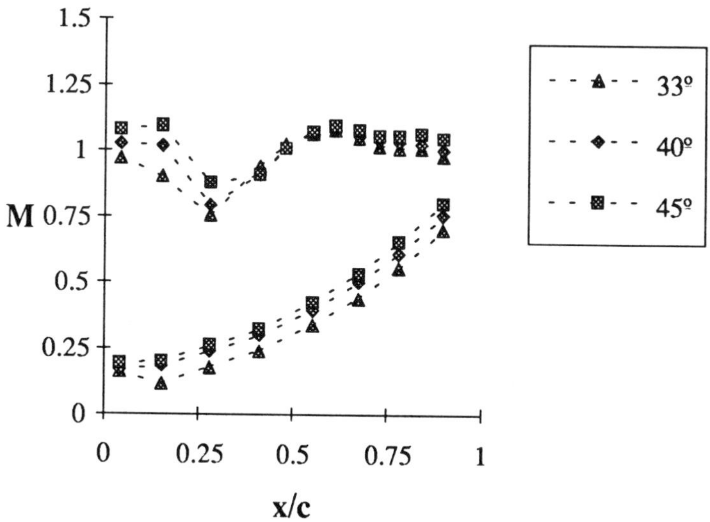

Fig 3a: Steady-state blade surface Mach number distribution for three inlet flow angles at M_{2S}~0.95

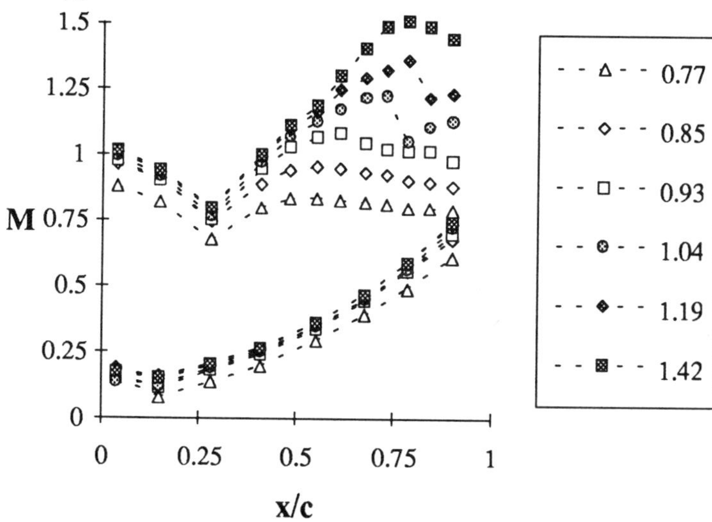

Fig 3b: Steady-state blade surface Mach number distribution for 6 outlet Mach numbers at β1~33°

The blade surface Mach number distribution indicates a supersonic zone for outlet Mach numbers larger than ~0.9 (Fig. 3b) As the flow is not yet choked, the suction surface Mach number still increases with increasing outlet flow velocity. The expansion downstream of the separation bubble is however similar, for a given outlet flow velocity, for the different inlet flow angles. The supersonic zone is finished with a recompression that starts at x/c~0.6 for M_{S2}~ 0.93 and around x/c~0.8 at M_{S2}~ 1.19. The Mach number downstream of this zone is fairly constant.

The steady-state blade surface Mach number distribution of the experimental tests were calculated with "FINSUP". For these "predictions" the experimental steady-state inlet Mach number and flow angle as well as the downstream Mach number where used as guide-lines, but no attempt was made to get the blade surface distribution to agree as well as possible with the data. It is to be expected that a certain amount of three-dimensionality is to be found in the experiments at the off-design flow conditions discussed here. A better agreement between experiment and prediction could certainly be found if this would be considered in detail in the calculations.

It can be concluded that the computed Mach number distribution on the pressure surface agrees well with the data for the lowest inlet flow angle. At the other inlet flow angles a fairly large difference exists between the data and the two-dimensional prediction. The viscous separation bubble on the suction surface is not, as expected, captured by the inviscid code. At β_1~34° and M_2~1.04 the pressure surface still shows a good agreement, whereas the shock on the suction surface is not well captured by "FINSUP". The trends of the prediction towards the suction surface trailing edge agree however (Fig. 3c).

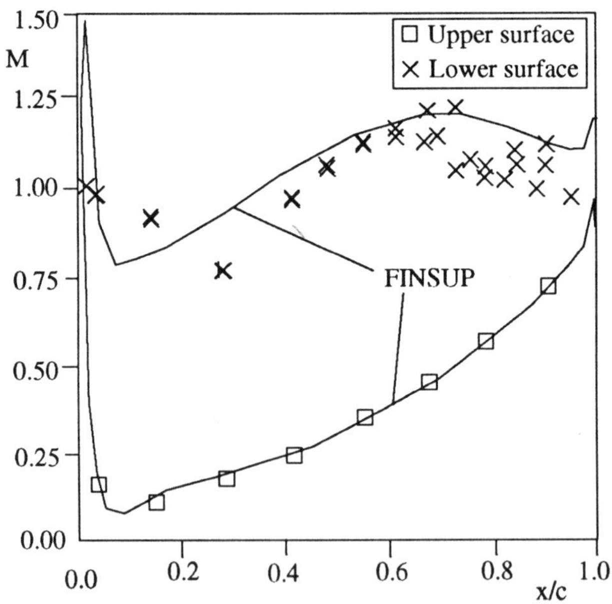

Fig 3c: Comparison of steady state data with FINSUP calulation (β_1~34°; M_2~1.04)

TIME-DEPENDENT RESULTS

For each of the steady-state operating conditions between 10 and 20 interblade phase angles where established.
It is well known that shock waves in the transonic flow regime are highly unstable and very sensitive to small changes in the steady-state and unsteady operating conditions. It is thus necessary to estimate, for each cascade tested and especially for unsteady flow, the uncertainties that may arise from slightly different steady-state flow conditions and, thus, the representativeness of the time-dependent results concerning the studied cascade. It is also important to validate the periodicity, not only the steady-state but also the time-dependent, of the flow and to show any differences that may arise between different blades. Finally, such checks of unsteady repetitiveness and periodicity are important when comparisons should be made with numerical results.
The unsteady repetitivity of the tests is excellent at subsonic flow conditions as well as for supersonic flow conditions even in the vicinity of the shock wave (Fig. 4). Both the amplitudes and the phase angles are highly repetitive. The circumferential periodicity is, as for the steady state flow, good for subsonic flow conditions, whereas some differences are to be expected from the shock wave influence.

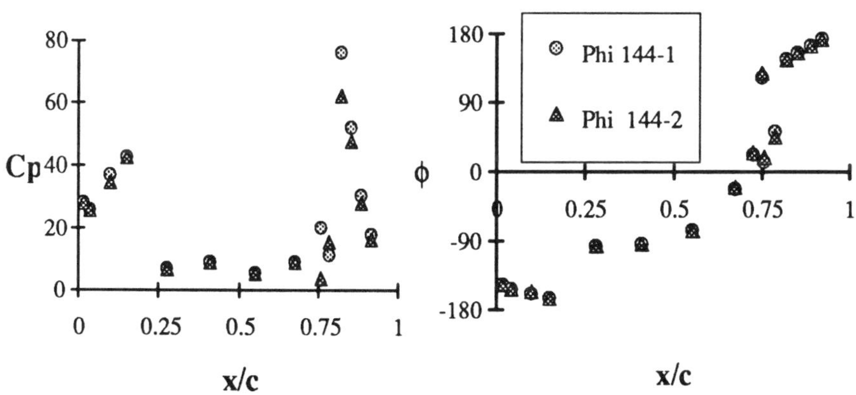

Fig. 4: Amplitude and phase angle repetitivity of suction surface at supersonic outlet flow conditions ($\beta 1=33°$, $M_{2S}=1.19$, $\sigma=144°$)

The representation for the magnitude of the pressure coefficients (Fig. 5) shows a tendency of a slight displacement of the shock position over the circumferential direction of the test facility, and the accompanying large variations in the unsteady load. The figure is interpreted to show that the shock wave is displaced slightly upstream with increasing z/t, from being situated around x/c=0.821 at z/t~6. A small change in steady-state shock wave position has thus a significant influence on the measured unsteady pressures.

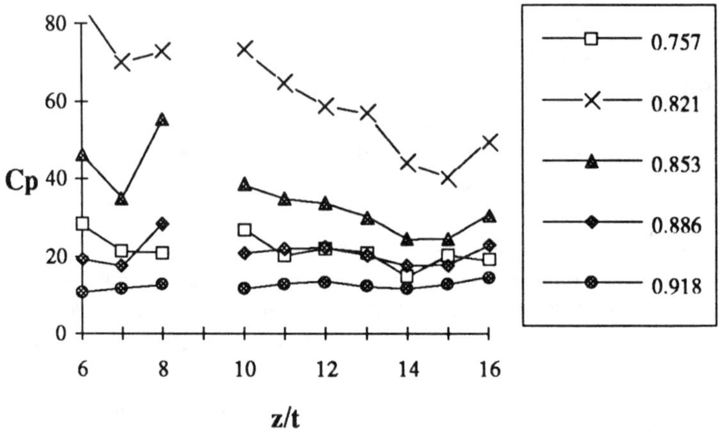

Fig. 5: Time dependent circumferential periodicity of suction surface phase-angles for five transducers positions x/c on one blade ($\beta 1=34°$, $M_{2S}=1.19$, $\sigma=180°$)

It is concluded that valid comparisons between different steady-state and time-dependant operating conditions can be performed, as long as the blades are always in the same circumferential position in the test facility. Comparisons with theory can also be made, but it must be kept in mind that the full extent of shock influences are not always captured experimentally with a discrete number of transducers. Comparisons between different blades must also be carefully evaluated.

Experimental Time-Dependent Aerodynamic Damping and Blade Surface Pressures

The aerodynamic damping integrated from the unsteady pressures for the three different inlet flow angles are shown in Fig. 6. It is seen that the aerodynamic damping indicates a zone around 270° to 340° interblade phase angles where instabilities may occur. The largest positive damping is situated around 180° interblade phase angle.
The transducers on the pressure surface indicate generally a tendency of a slight increase in magnitude from the leading edge towards the trailing edge, especially at 180° interblade phase angle (Figs. 8). It is especially interesting to note that the different inlet flow angles give similar unsteady responses for a given interblade phase angle, on the pressure surface for all studied outlet flow

velocities, both for the amplitude and the phase angle. This is to be expected as the pressure surface Mach number distributions are similar for the presently studied sample cases, and it has also been noted for other cascade tests both for turbines [Fransson, 1990] and compressors [Széchényi and Cafarelli, 1989].
The suction surface transducers give relatively high amplitudes for the 20% foremost part of the blade at β_1=35° and up to about 30% for β_1=45°, with a larger amplitude at the lower inlet flow angle (Figs.7;8). It is noted that the confidence interval is somewhat larger inside than outside the separated flow region, as expected. It is also seen that the first transducer downstream of the presumed separation bubble shows a significantly lower pressure amplitude than the ones in the separation zone.
On the forward part of the suction surface (up to x/c~0.4) the steady-state flow conditions are essentially identical for the subsonic and transonic sample cases studied, for a given inlet flow angle. For different flow conditions the time-dependent pressures are also in close agreement with each other for this part of the blade. As the outlet Mach number increases from subsonic to transonic the time-dependent response on the blade suction surface increases considerably downstream of the 60% chordwise position. This position corresponds to the impingement of a shock wave on the suction surface. This increase is noted on all blades and for all interblade phase angles, although the shock seems to be positioned somewhat more upstream on blade "1" than on blade "2".

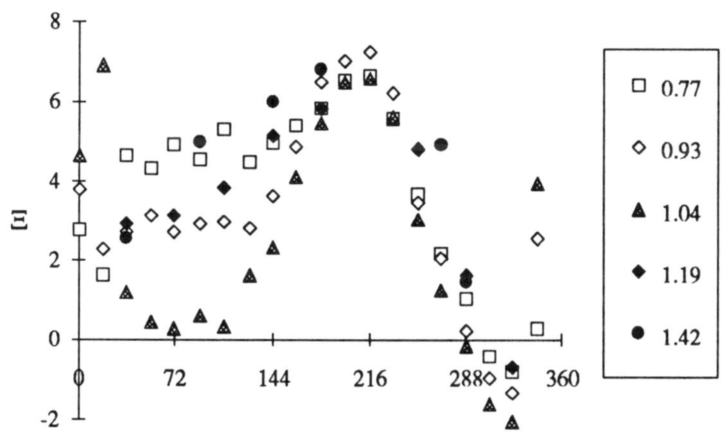

Fig.6: Aerodynamic damping versus interblade phase angle, at different outlet Mach numbers M_{2s} (β1=34°)

To study the effects on the pressure coefficient phase angles we will concentrate upon interblade phase angles around 180° (close to the maximum damping in Fig. 6). It is noted that the pressure surface phase angles are largely independent of the outlet flow velocity and the inlet flow angle (Fig.8).

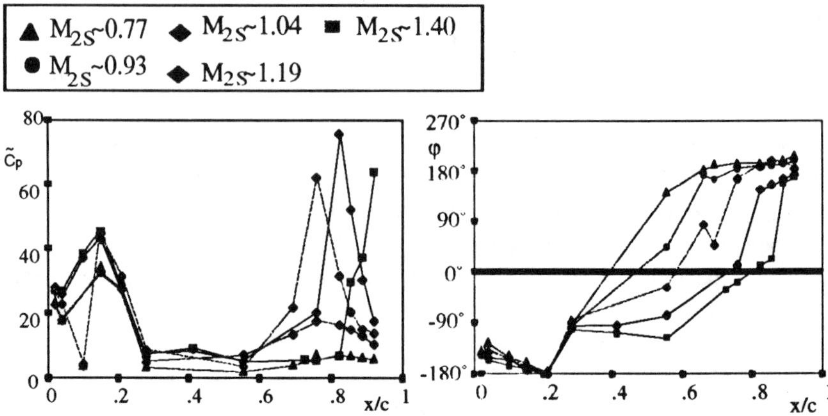

Fig. 7: Time-dependent blade surface pressure coefficients for different outlet Mach numbers M_{2S} ($\beta 1$=34°, σ= 144°)

The suction surface phase angle starts for interblade phase angles around 180° around ϕ_p~-135° at the leading edge. For subsonic outlet flow conditions this value is more or less constant +/-45° along the whole suction surface, whereas a considerable phase shift is measured for transonic and supersonic outlet flow conditions. The chordwise position of the main phase angle increase is seen to be shifted towards the rear for increasing outlet Mach numbers (Figs. 7) and to be positioned somewhat upstream of the maximum unsteady pressure amplitude in the vicinity of the shock wave. It is concluded that the shock wave interference is responsible for this phase angle shift. It is also noted that the trend of the unsteady pressure coefficient on the pressure surface is well predicted by "FINSUP". The predicted magnitude is mostly a little bit larger than the measured one. It is noticed that both the prediction and the data gives a stabilizing effect of the pressure surface (i. e. pressure phase angles between 0° and 180°) for most interblade phase angles.

On the suction surface the magnitude and trend of the pressure coefficient amplitude correlate well between data and prediction for subsonic flow conditions. Large values in the leading edge region are noted from both the data and the prediction although the data are higher for the off-design inlet flow angles studied (apart from just at the leading edge where the calculation shows a high peak). At transonic outlet flow conditions the prediction indicates a small increase in magnitude in the neighbourhood of the expected shock position, but this is not as large as seen from the data. It is obviously expected that a refinement of the computational mesh and the introduction of a stream sheet contraction in the calculation will improve the agreement, but the theoretical model is inviscid and should not be expected to predict separation bubbles and viscous shock effects.

The predicted phase angle of the pressure coefficient on the suction surface does in general not correlate with the data as well as on the pressure surface. The values in the leading and trailing edge regions correlate, with a predicted smooth connection in between.

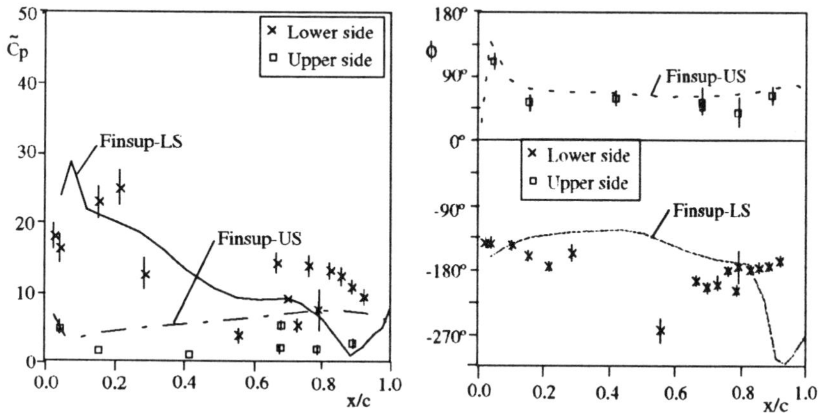

a): (M_{2S}= 0.95, β_1=40°)

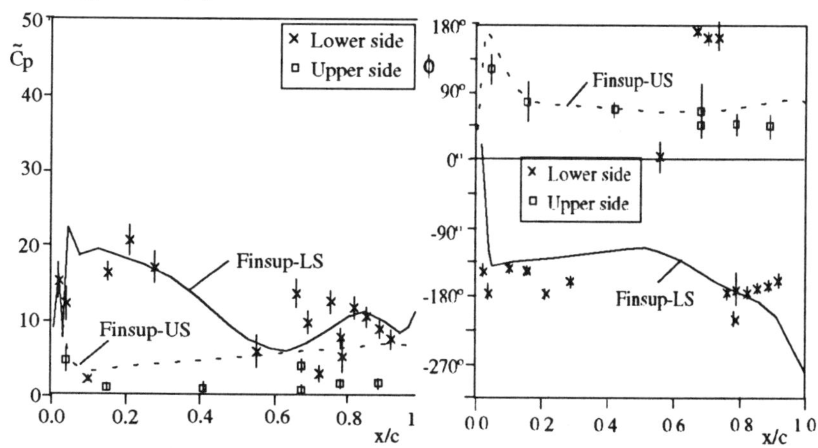

b): (M_{2S}= 0.99, β_1=45°)

Fig. 8: Time-dependent blade surface pressure coefficient with increasing inlet flow angle (M_{2S}~0.95, σ=144°)

Influence Coefficients

A way to reduce the large amount of time dependent data (several interblade phase angles per steady-state operating condition) is to decompose the results from the travelling wave mode into influence coefficients for a single blade excitation. In the present case only the influences from blades "-1","0" and "+1" will be analysed as the blades further away have been experimentally shown to yield a smaller, although not entirely negligible, influence [Bölcs et al, 1989].
The pressure surface pressure influence coefficient amplitude (Figs.9) show similar trends for the prediction and the data with a significant increase in amplitude towards the trailing edge for all outlet flow velocities. The eigen-influence is overpredicted whereas the influence of blade "+1" on blade "0" is similar from the prediction and experiment.
The pressure surface phase angle influences are shown for two outlet Mach numbers in Figs. 10. The agreement between the prediction and the experiment is excellent, especially for the choked cases, both for the influence of blade "0" and "+1". As with two other turbine cascades [Bölcs et al., 1991; Fransson 1990] it is noted that the cascade effects are quasi-steady for realistic reduced frequencies for the first bending vibration mode of gas turbine blades of the presented type, as the phase angle differences between blades "0" and "1" in Figs. 10 are around 180°. The eigen-influence phase angle lies around 90° along the whole blade pressure surface. This, together with the fact that the eigen-influence of the pressure coefficient amplitude is larger than the influence from blade "+1", gives the stable behaviour of the pressure surface for all interblade phase angles. It also indicates that the time-dependent pressure eigen-influence on the whole pressure surface is in phase with the blade vibration velocity, which in turn means for the present operating conditions a more or less "quasi-steady" eigen-influence phenomena for the pressure surface phase angle.
On the blade suction surface the tendency of the amplitude influence coefficients are, for subsonic outlet flow conditions, the same for the data and the prediction, with a higher value in the data towards the trailing edge region, especially with increasing outlet flow velocity (Figs. 11). In this region the data indicate a considerably larger influence from blade "0" than from blade "-1", whereas the prediction gives values of the same order. Note that already at M_{2S} ~0.93 a small supersonic zone is measured on the suction surface (Fig.3).
In these figures the importance of the shock wave on the unsteady loads is clearly noticed. Both the eigeninfluence and the influence from blade "-1" is large in the vicinity of the shock wave (Fig. 11b). The data indicate a large correlation with the reference blade itself and a somewhat smaller with blade "-1".
A large increase in the peak of the unsteady pressure coefficient is noticed between the tests at M_{2S}~0.93 and 1.19, and it is clearly seen that the chordwise

location of the maximum unsteady pressure is displaced towards the trailing edge as the outlet flow velocity increases.

Also the high unsteady loads in the separated zone are clearly demonstrated in this representation. The maximum unsteady pressure in the separated zone stays approximately the same for all outlet flow velocities. As with the shock wave, the data are highly correlated with both blades "0" and "-1". Neither of the unsteady effects in the separated zone or at the foot of the shock wave are expected to be captured by the inviscid prediction model.

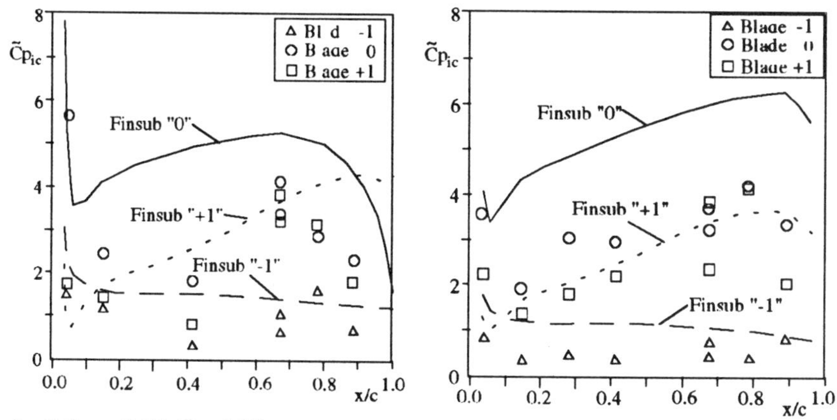

a): (M_{2S}= 0.77, β_1=34°) b): (M_{2S}= 1.42, β_1=33°)

Fig. 9: Amplitude influence coefficients of unsteady blade surface pressure coefficient on the blade pressure surface.

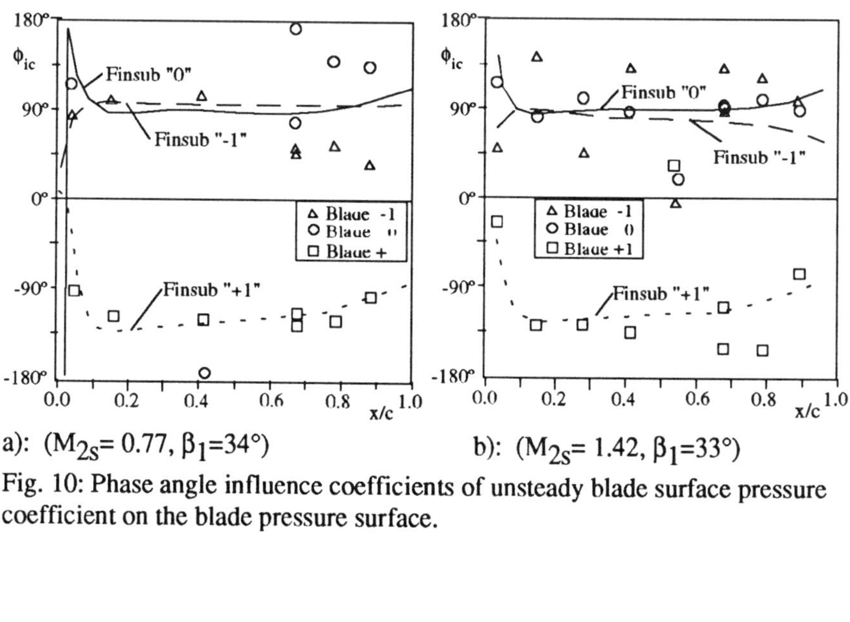

a): (M_{2S}= 0.77, β_1=34°) b): (M_{2S}= 1.42, β_1=33°)

Fig. 10: Phase angle influence coefficients of unsteady blade surface pressure coefficient on the blade pressure surface.

The suction surface phase angle influence is shown in Fig. 12. At all outlet flow velocities the eigen-influence of the pressure transducers lies around 180° in the leading edge region. The tendencies of the prediction and the data are similar for the subsonic flow velocities, but a difference is developed towards the trailing edge. As for the pressure surface the difference in phase between blades "-1" and "0" is around 180° in the forward part of the blade, with only slight changes in the trailing edge region. It is interesting to note that the inviscid code very accurately predicts the time-dependant phase angles although not the amplitudes (Figs. 11), in the separated region close to the leading edge.

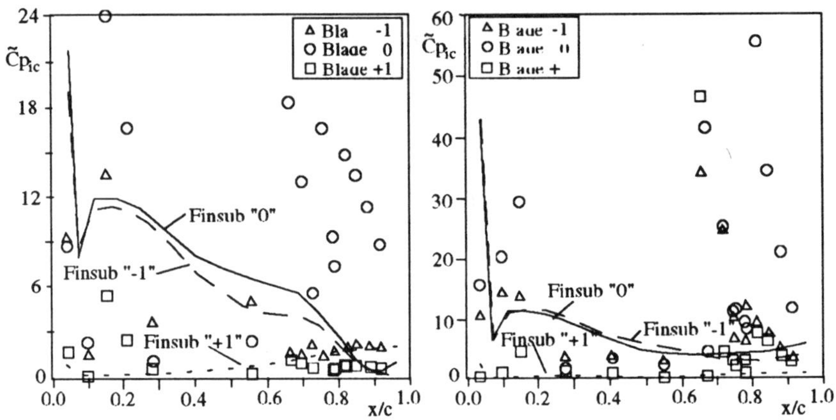

a): (M_{2s}= 0.92, β_1=34°) b): (M_{2s}= 1.19, β_1=33°)

Fig. 11: Amplitude influence coefficients of unsteady blade surface pressure coefficient on the blade suction surface.

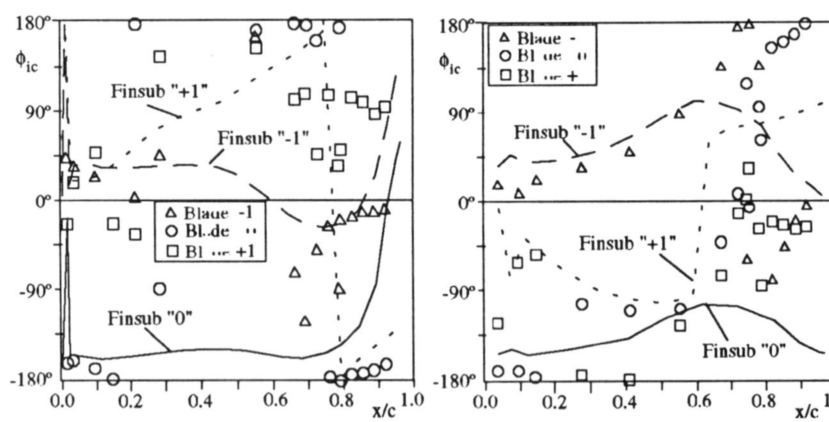

a): (M_{2s}= 0.92, β_1=34°) b): (M_{2s}= 1.19, β_1=33°)

Fig. 12: Phase angle influence coefficients of unsteady blade surface pressure coefficient on the blade suction surface.

For the supersonic outlet flow velocities a tendency of a significant phase shift exists for the eigen-influence in the region where the shock waves are expected to impinge on the blade suction surface (Figs. 12). This shift is almost 360° and close to the trailing edge the predicted and measured values are again close together. There is, however, a considerable discrepancy in the tendency of the curves in the shock region.
A tendency of a similar phase shift is also found in the influence of blade "-1" on blade "0" (Figs. 12). An excellent agreement between the prediction and the data is again found in the forward part of the blade, whereas the trends of the curves differ further downstream.

CONCLUSIONS

The presented experiments at off-design inlet flow angles and high outlet flow velocities have shown that the pressure surface of the cascade is both eigen-damped and mostly damped for all interblade phase angles. It has thus no significant influence on an eventual change of stability of the cascade. The agreement between the data and the prediction is in general good, especially as regards to the phase angle whereas the influence coefficient amplitudes are often overpredicted.
On the suction surface high time-dependent loads are measured in the separated zone close to the leading edge region, up towards 30% chord (depending on inlet flow angle). The predicted and measured pressure phase angles correlate well in this region. Further downstream, from about 60% chord, the measurements indicate a considerably higher unsteady pressure amplitude than the predictions for transonic and supersonic outlet flow conditions.
From the experimental data and the predicted results the following conclusions are drawn, believed to be generally valid for cascades of the type studied:
· Shock waves can significantly influence the unsteady load on the profile, both as regards to magnitude and phase.
· High unsteady loads are present in leading edge separation bubbles.
· The experiments indicate a phase shift in the time-dependent blade surface pressures at the location of the expected shock wave position.
· The unsteady cascade effects are largely quasi-steady, both as regards to the pressure and suction surfaces.
· The inviscid prediction model employed in the study accurately predicts the phase angles on the pressure surface, but overpredicts the amplitudes. The suction surface predictions do not agree so well with the data for supersonic outlet flow conditions.
Finally, it is concluded that small changes in downstream steady state flow conditions can significantly alter the position of high unsteady loads in supersonic

flow. It is thus generally important that the steady-state and unsteady periodicities are well known when comparisons with prediction models should be made.

ACKNOWLEDGEMENTS

The experimental data used for this study were obtained at the "Laboratoire de thermique appliquée et de turbomachines" at the EPF-Lausanne during a project sponsored by Asea Brown Boveri, Baden, Switzerland, and the Swiss Research Fund for Scientific Encouragement (project No. 1684.1) in collaboration with Rolls Royce Ltd, Derby, United Kingdom. The project managers are Dr. K. Vogeler and Dr. A. Cargill, respectively. Dr. D. Schläfli and Mr. P. Ott were helpful in preparing the test cascade and the data acquisition and reduction programs. Our thanks also go to Mr. G. Rossini and Mr. J.-P. Mottier who keep the test facility and instrumentation up to date.
The views expressed are the authors' alone and do not necessarily correspond with the views of the above mentioned companies.

REFERENCES

Bölcs, A.; Cargill, A.; Fransson, T.H.; Sudhoo, A.; Vogeler, K.; 1991: "Time dependant predictions and analysis of turbine cascade data in the transonic flow region", Proceedings of the International Symposium on Aeroelasticity in Turbomachines and Propellers, Notre Dame, USA, 1991.

Bölcs, A.; 1983: "A Test Facility for the Investigation of Steady and Unsteady Transonic Flows in Annular Cascades", ASME Paper 83-GT-34, 1983.

Bölcs, A; Fransson, T. H.; Schläfli, D.; 1989: "Aerodynamic Superposition Principle in Vibrating Turbine Cascades" AGARD, 74th Specialists' Meeting of the Propulsion and Energetics Panel on Unsteady Aerodynamic Phenomena in Turbomachines, Luxembourg, August 28-September 1, 1989.

Fransson, T. H.; 1990: "Analysis of Experimental Time-Dependent Blade Surface Pressures from an Oscillating Turbine Cascade with the Influence-Coefficient Technique", ASME Paper 90-GT-225, 1990.

Hanamura, Y.; Tanaka, H.; Yamaguchi, Y.; 1980: "A Simplified Method to Measure Unsteady Forces Acting on the Vibrating Blades in a Cascade", Bulletin of the JSME, Vol. 23, No. 180, 1980.

Kirschner A.; Fosco B.; Muller E.; 1980: "Control of vibration in aeroelastic cascade experiments."
Communication de L'ITA/ EPFL, No. 10, pp. 285-295,1980.

Széchényi, E.; Cafarelli, I.; 1989: "Le flottement des aubes de compresseur: Une certaine compréhension grâce à des essais en soufflerie", Revue Française de Mécanique No. 1, pp. 73-87, 1989.
Széchényi, E.; Girault, J. P.; 1983: "An Aeroelastic Testing Technique in a Straight Cascade Wind Tunnel", Proceedings of the Symposium "Measuring Techniques in Transonic and Supersonic Cascades and Turbomachines", Aachen, West Germany, pp. 4.1-4.8, 1983.
Whitehead, D. S.; 1990: "A Finite Element Solution of Unsteady Two-Dimensional Flow in Cascades."
International Journal for Numerical Methods in Fluids, Vol. 10, pp. 10-34, 1990.
Whitehead, D. S.; Newton, S. G.; 1985: "A Finite Element Method for the Solution of Two-Dimensional Transonic Flows in Cascades."
International Journal for Numerical Methods in Fluids, Vo. 5, pp. 115-132, 1985.

Linear Oscillating Cascade Unsteady Aerodynamic Experiments

Daniel H. Buffum[1]
Sanford Fleeter[2]

ABSTRACT

Experiments in a linear oscillating cascade reveal that the wind tunnel walls enclosing the cascade have, in some cases, a detrimental effect on the oscillating cascade aerodynamics. In a subsonic flow field, biconvex airfoils are driven simultaneously in harmonic, torsion-mode oscillations for a range of interblade phase angle values. It is found that the cascade dynamic periodicity - the airfoil-to-airfoil variation in the unsteady surface pressure coefficient, C_p - is good for some values of interblade phase angle but poor for others. Calculations based upon linearized unsteady aerodynamic theory indicate that pressure waves reflected from the wind tunnel walls are responsible for the cases where there is poor periodicity in C_p. However, the periodicity in the unsteady pressure difference coefficient ΔC_p is generally good, apparently due to cancellation of acoustic modes imposed by the wind tunnel walls. The unsteady pressure difference coefficients are correlated with linearized analysis predictions.

NOMENCLATURE

C	airfoil chord
C_p	unsteady pressure coefficient, $p_1/(\rho_{in}V_{in}^2\alpha_1)$
$\overline{C}_p$	steady pressure coefficient, $(p_{in}-p_0)/(\rho_{in}V_{in}^2)$
$\hat{C}_p^n$	unsteady pressure influence coefficient for the nth airfoil
k	reduced frequency, $\omega C/V_{in}$

[1]NASA Lewis Research Center, MS 77-6, 21000 Brookpark Road, Cleveland, OH 44135.
[2]Thermal Sciences and Propulsion Center, School of Mechanical Engineering, Purdue University, Chaffee Hall, West Lafayette, IN 47907.

M_{in}	cascade inlet Mach number
$\overline{p}_{in}$	mean inlet static pressure
p_j	j^{th} harmonic of airfoil surface static pressure
S	airfoil spacing
V_{in}	inlet velocity
α_j	j^{th} harmonic of incidence angle
β	interblade phase angle (positive when airfoil n+1 leads airfoil n)
γ	stagger angle
ΔC_p	unsteady pressure difference coefficient
ρ_{in}	inlet density
ω	airfoil oscillation frequency

INTRODUCTION

Both linear and annular cascades have been used to investigate the aerodynamics of airfoils driven in controlled, harmonic, traveling-wave mode oscillations. Certain advantages and disadvantages are associated with these two types of facilities. In particular, a highly two-dimensional flow field may be obtained in a linear cascade while, in an annular cascade, undesirable three-dimensional effects may be problematic. However, while the annular cascade appears to be inherently dynamically periodic, i.e., the temporal fluctuations in the flow field vary from passage to passage according to the interblade phase angle, periodicity is less certain in the linear cascade due to the finite extent of the cascade and the boundaries introduced by the wind tunnel walls. The potential for wall interference is indicated by a number of papers devoted to the effects of wind tunnel walls on a single oscillating airfoil, for example, Runyan et al. (1955).

Carta (1983) investigated the dynamic periodicity of a linear oscillating cascade in a low subsonic flow field. His cascade consisted of 11 NACA 65-series airfoils staggered at 30 degrees with 1.33 solidity. Airfoil surface unsteady pressures were measured at reduced frequency values ranging from 0.14 to 0.30 over a wide range of interblade phase angles with an inlet Mach number of $M \approx 0.18$. Good dynamic periodicity was generally found except for in-phase oscillations, where circumferential gradients in the phase of the unsteady pressure coefficient were found. Carta attributed this to an acoustic resonance condition in the cascade.

In the present study, the dynamic periodicity of an oscillating, linear cascade is investigated for a subsonic flow field but with a significantly higher inlet Mach number, M=0.55, and also a much larger reduced frequency, k=0.64. The cascade solidity, 0.65, was chosen to be representative of an advanced propeller model which fluttered during wind tunnel performance tests (Mehmed, et al., 1982). The torsion mode oscillating cascade aerodynamics are determined for a range of interblade phase angles. To help determine the validity of the data, an investigation is made into the steady state and dynamic periodicity of the cascade. Insight into the effect of the wind tunnel walls on the cascade unsteady aerodynamics is gained from the theoretical acoustic wave generation properties of an oscillating cascade. Then the unsteady airfoil surface pressure data are correlated with the predictions of the linearized subsonic oscillating cascade analysis of Whitehead (1987).

OSCILLATING CASCADE FACILITY

The NASA Lewis Transonic Oscillating Cascade, Figure 1, combines a linear cascade wind tunnel capable of inlet flow approaching Mach one with a high-speed airfoil drive system which imparts torsion-mode oscillations to the cascaded airfoils at specified interblade phase angles and realistic high reduced frequency values.

Air drawn from the atmosphere passes through a smooth contraction inlet section into a constant area rectangular test section of 9.78 cm span which measures 58.6 cm along the stagger line. Upstream of the test section, suction is applied through perforated walls to reduce the boundary layer thickness. Except for these small perforated regions, the wind tunnel walls present smooth, solid boundaries to the flow. Tailboards are used to adjust the cascade exit region static pressure and also form bleed scoops which further reduce upper and lower wall boundary layer effects. Downstream of the test section, the air is expanded through a diffuser into an exhaust header. The exhaust system, part of a central air facility at NASA Lewis, maintains a 30 kPa pressure downstream of the flow control valves. The cascade inlet and the airfoil angles may be adjusted over a wide range of incidence and stagger angle combinations.

A high-speed mechanism drives the airfoils in controlled torsional oscillations. Stainless steel barrel cams, each having a six cycle sinusoidal groove machined into its periphery, are mounted on a common rotating shaft driven by a 74.6 kW electric motor. A cam follower assembly, consisting of a titanium alloy connecting arm with a stainless steel button on one end, is joined on the other end to an airfoil trunnion. The button fits into the cam groove, thus coupling the airfoil to the camshaft. Lubrication of the cam/follower assembly is provided by an oil bath. The amplitude of the torsional airfoil motion is 1.2 degrees as

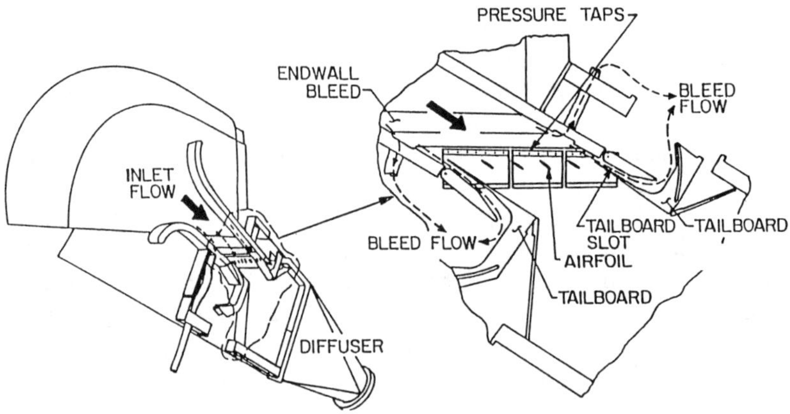

Figure 1. Oscillating cascade facility

dictated by the cam and follower geometry. The drive system is configured for oscillations at a chosen interblade phase angle by fixing the cams at the required relative positions on the shaft. A reduced frequency of 0.64 is achieved at 0.55 inlet Mach number with an oscillation frequency of 245 Hz.

AIRFOILS AND INSTRUMENTATION

Figure 2 illustrates the cascade geometry nomenclature. Details of the geometry are summarized in Table 1. Four uncambered, 7.6% thick biconvex airfoils with rounded leading and trailing edges are used to create a low solidity (C/S = 0.65) cascade. The stagger angle is 45 degrees and the airfoils oscillate about the midchord.

Airfoils instrumented with static pressure taps are used to measure the airfoil surface steady pressure distributions. There are sixteen chordwise measurement locations, with a higher density in the leading edge region used to capture the higher gradients there. Rows of sidewall static pressure taps located upstream and downstream of the cascaded airfoils are used to determine the mean inlet and exit pressures.

Flush-mounted high-frequency-response miniature pressure transducers are used to measure the unsteady surface pressures on the oscillating airfoils. Two airfoils are instrumented, each having six transducers mounted symmetrically about the midchord on one airfoil surface. These transducers, having active sensor diameters of 1.3% of the airfoil chord, are epoxied into milled slots and

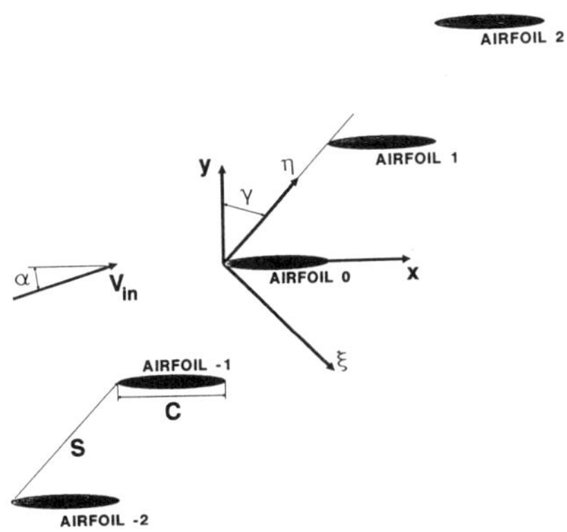

Figure 2. Schematic of cascade geometry

Table 1. Airfoil and cascade geometry

Airfoil	
Shape	symmetric biconvex
Surface radii of curvature	27.4 cm
Leading and trailing edge radii of curvature	0.025 cm
Chord, C	7.62 cm
Maximum thickness / chord	0.076
Elastic axis	midchord
Cascade	
Number of airfoils	4
Airfoil spacing, S	11.72 cm
Solidity, C/S	0.65
Stagger angle, γ	45 degrees
Mean flow incidence angle, α_0	2 degrees
Amplitude of oscillation, α_1	1.2 degrees

potted in room-temperature-vulcanizing rubber for isolation from airfoil strain. A thin coating of rubber is also used to fair the transducer surface into the airfoil contour.

From static and dynamic calibrations, the pressure transducers were found to be highly linear in response over the frequency and pressure ranges of interest. However, the pressure transducers may produce undesirable signals as a consequence of the airfoil motion. This effect was quantified by oscillating the instrumented airfoils under no-flow conditions. The response of each transducer was found to be a linear function of the airfoil acceleration magnitude, implying that the unsteady pressure, which varies with the airfoil velocity magnitude, is dominated by the acceleration response. Thus calibration data were obtained to correct the oscillating airfoil pressure data for acceleration effects.

The instantaneous position of the reference oscillating airfoil is determined by a capacitance-type proximity sensor which produces a voltage proportional to the air gap between it and an adjacent object. This sensor is positioned to face a six cycle sinusoidally-shaped cam mounted on the airfoil drive camshaft so as to be in phase with the reference airfoil motion.

DATA ACQUISITION AND ANALYSIS

Conventional instrumentation is used to quantify the steady flow field. An average of the upstream sidewall static pressures along with the atmospheric pressure (total) are used to calculate the inlet Mach number. Steady flow airfoil surface static pressures are calculated from an average of approximately 100 samples. The steady pressure coefficient is defined in Eq. 1.

$$\bar{C}_p(x,y) = \frac{p_{in} - p_0(x,y)}{\rho_{in} V_{in}^2} \qquad (1)$$

p_{in} is the mean inlet static pressure, p_0 is the time-average airfoil surface static pressure, and ρ_{in} and V_{in} are the inlet values of density and velocity.

Unsteady signals are recorded on magnetic tape for post-experiment processing. During tape playback, the signals are simultaneously digitized at rates sufficient to capture at least three harmonics of the oscillation frequency, with 32,768 samples taken per channel. Each channel of data is divided into contiguous blocks, each block typically with 2048 samples, and then Fourier decomposed to determine the first harmonic of each block of data. The first harmonic pressure of each block is referenced to the airfoil motion by subtracting from it the phase of the first harmonic motion signal of the same block. Once all of the blocks from a channel are decomposed in this manner, the first harmonic block results are averaged and the complex-valued acceleration response is subtracted vectorally. Statistical analysis of the block results is used to estimate uncertainties for the average first harmonic values.

The motion of the nth airfoil is defined by the change in the incidence angle with time, $\alpha^n(t) = \alpha_0 + \alpha_1 \cos(\omega t + n\beta)$, where α_0 is the mean incidence angle, α_1 is the torsional oscillation amplitude, ω is the frequency and β is the interblade phase angle.

The complex-valued unsteady pressure coefficient is defined as

$$C_p(x,y) = \frac{p_1(x,y)}{\rho_{in} V_{in}^2 \alpha_1}. \quad (2)$$

p_1 is the first harmonic airfoil surface static pressure. The dynamic pressure difference coefficient is the difference between the lower (y^-) and upper surface (y^+) unsteady pressure coefficients:

$$\Delta C_p(x) = C_p(x,y^-) - C_p(x,y^+). \quad (3)$$

RESULTS

For a mean inlet Mach number of 0.55 and 2 degrees mean incidence, the airfoil surface steady pressure coefficient distributions are presented first, followed by dynamic periodicity data and correlation of the unsteady pressure difference coefficient data with linearized analysis predictions. A linear analysis is then used to ascertain the effects of the wind tunnel walls on the cascade unsteady aerodynamics.

Steady State Aerodynamics

For a linear cascade to be a valid simulation of a turbomachine blade row, the cascade must exhibit good passage-to-passage periodicity for the steady flow field. Thus airfoil surface steady pressure distributions are obtained for multiple passages in the low solidity cascade. In Figure 3, steady pressure coefficient data are presented for the center cascade passage and the two adjacent passages. Good cascade periodicity is apparent, with the only significant differences found at the leading edge of the airfoil upper surface.

The upper surface pressure coefficient distribution peaks near the leading edge and the pressure difference tends toward zero near the trailing edge. There is negligible loading beyond 50% of chord. Using the method of Kline and McClintock (1953), a 95% confidence interval of ±0.003 is estimated for these pressure coefficients. The mean exit region static pressure divided by the inlet total pressure was 0.8251.

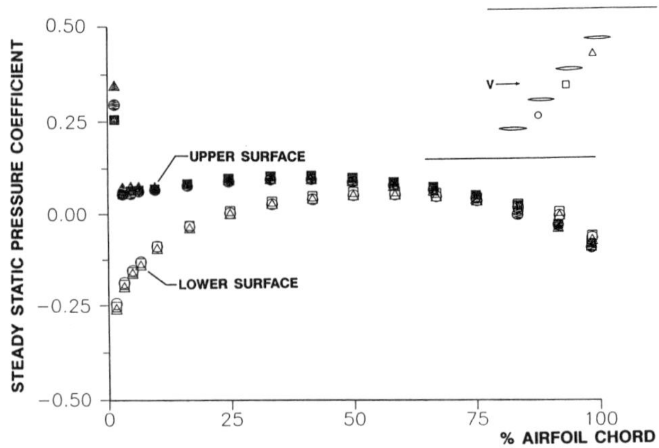

Figure 3. Airfoil surface steady pressure coefficient distributions

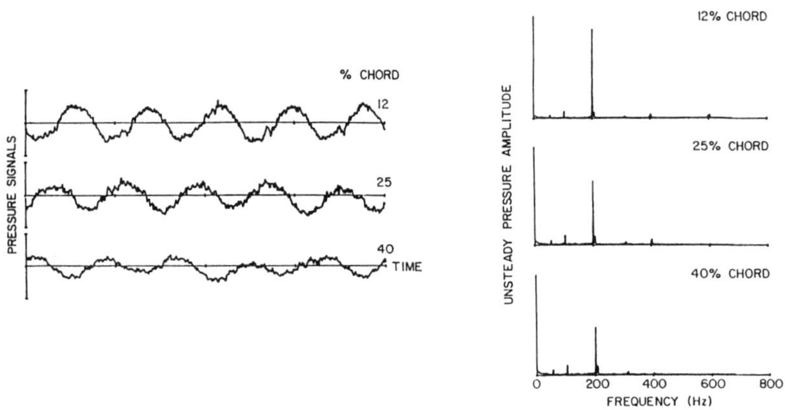

Figure 4. Unsteady pressure signals and spectra

Unsteady Aerodynamics

Airfoil surface unsteady pressure distributions are obtained for interblade phase angle values of 0, 45, -45, 90, -90 and 180 degrees at a reduced frequency of 0.64. For this cascade geometry, acoustic resonances are calculated to occur at interblade phase angles of -23.3 and 59.1 degrees. Sample time-dependent pressure signals presented in Figure 4 along with the resulting pressure spectra illustrate the dominance of the first harmonic component as typically found in the signals. 95% confidence intervals of ± 5% in magnitude and ±4 degrees in phase are calculated for the mean value of the unsteady pressure coefficients.

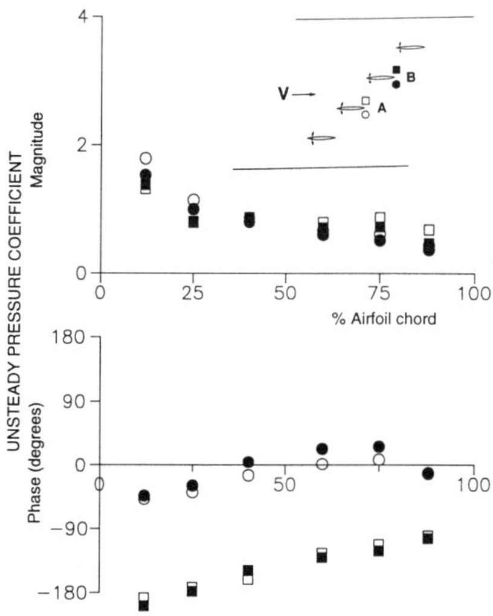

Figure 5. Unsteady pressure distributions, β=-45 degrees

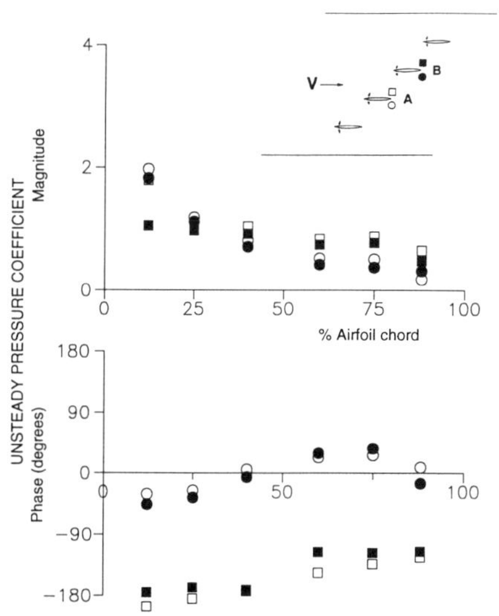

Figure 6. Unsteady pressure distributions, β=-90 degrees

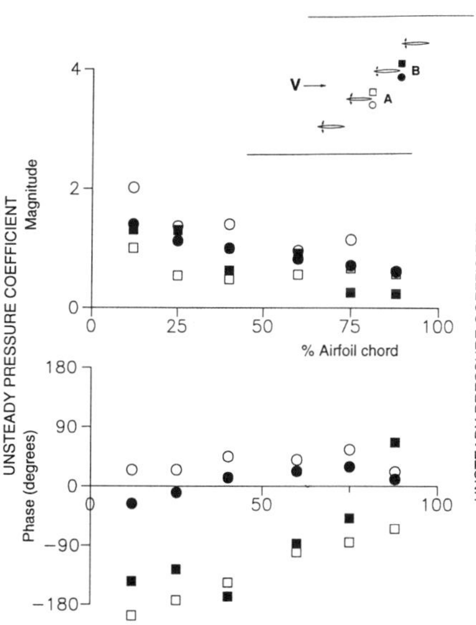

Figure 7. Unsteady pressure distributions, β=0

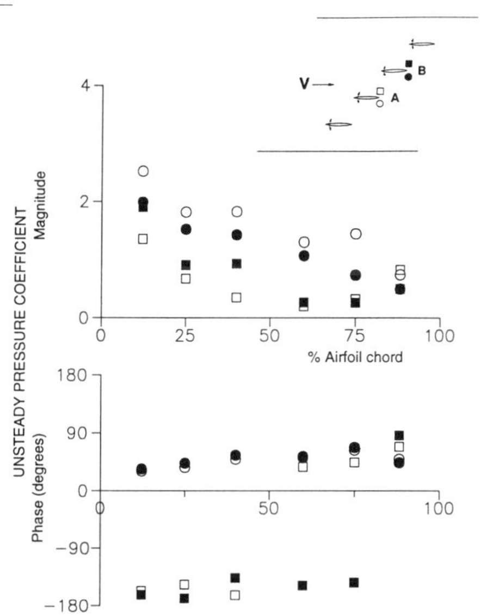

Figure 8. Unsteady pressure distributions, β=45 degrees

Dynamic Periodicity

Airfoil surface unsteady pressure data were obtained for the two airfoil positions on either side of the center passage of the cascade. First harmonic unsteady pressure coefficient data for these two airfoil positions, A and B as shown in the schematic of Figure 5, are presented in Figures 5 through 10. For only two values of β, -45 and -90 degrees (Figures 5 and 6, respectively), is the periodicity good. For the remaining values of β, 0, 45, 90 and 180 degrees (Figures 7 through 10), the periodicity is generally poor.

The poor dynamic periodicity found at some values of β is attributed to the wind tunnel walls which enclose the cascade, causing undesirable reflections of acoustic waves. One might suspect that the periodicity is related to the predicted cascade wave propagation modes, with only superresonant cascades having significant coupling with the duct acoustics. However, this is not the case: although for both superresonant values of β, 0 and 45 degrees, the periodicity is poor, it is also poor for several subresonant values of β, 90 and 180 degrees.

The propagation directions of far field cascade acoustic waves may be calculated from linearized theory assuming a uniform mean flow (Buffum and Fleeter, 1990). The predicted propagation directions of the mode zero cascade waves (i.e., the tangential wave number m satisfies $mS = \beta + 2\pi n$ with the integer specifying the mode, n, set equal to 0) are shown in Figure 11. For this cascade at the flow conditions under consideration, superresonant waves ar found to occur only for $n = 0$. In addition, when the waves are subresonant, they will have the slowest rate of decay when $n = 0$.

The results shown in Figure 11 give an explanation on a physical basis for the observed relationship between cascade periodicity and interblade phase angle. For example, where the periodicity was found to be good, $\beta = -45$ and -90 degrees, decaying cascade waves are predicted which would reflect off the upper wind tunnel wall and move away from the cascade in the downstream direction. Although there is a diffuser shortly downstream of the cascade which might cause reflections, it had no significant effect on the data (i.e., the periodicity was good), perhaps due to the predicted exponential decay of the waves with distance from the cascade. For $\beta=0$, one wave, directed at the upper wind tunnel wall, is shown to reflect back into the cascade, where it would presumably have a detrimental effect on the periodicity. The other wave propagates downstream - reflection of this wave at the diffuser might cause undesirable effects. Similar comments apply for $\beta = 45$ degrees, although now the two waves are directed at the lower wall and upstream, respectively. Although the waves depicted for $\beta = 90$ and 180 degrees do not interfere with the cascade, just a small change from the predicted propagation direction would result in interference. Overall, this simple model gives a physical explanation for the cascade/wind tunnel wall interaction which is reasonably consistent with the periodicity results.

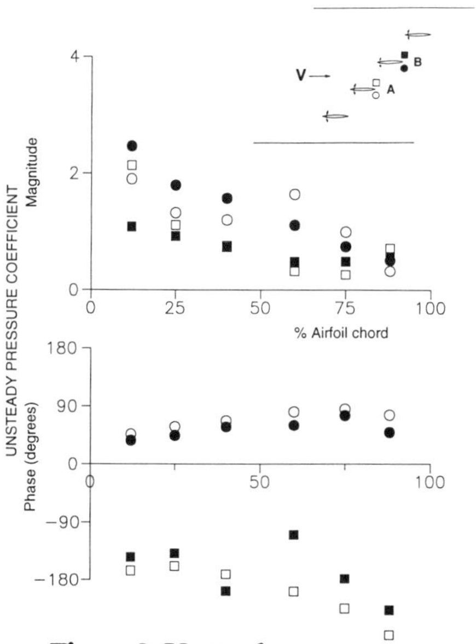

Figure 9. Unsteady pressure distributions, β=90 degrees

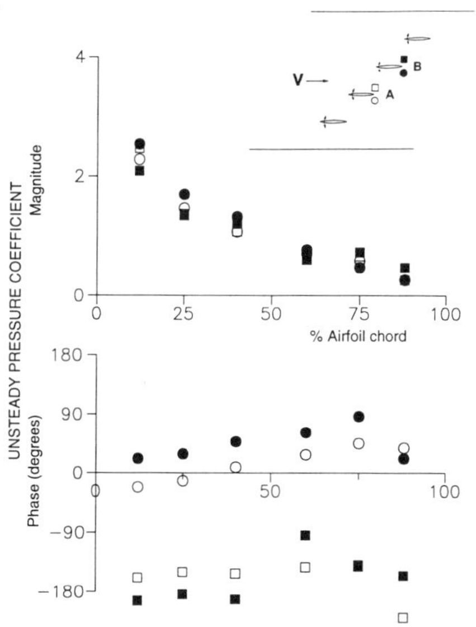

Figure 10. Unsteady pressure distributions, β=180 degrees

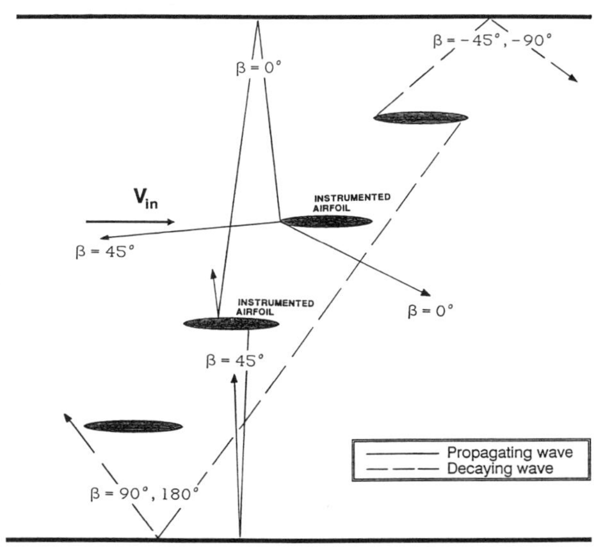

Figure 11. Interaction of pressure disturbances and wind tunnel walls

Correlation with Linearized Analysis

Experimentally-determined dynamic pressure difference coefficient data are correlated with the predictions of a linearized unsteady cascade analysis, LINSUB (Whitehead, 1987). The unsteady pressure difference coefficient, $\Delta C_p(x,\beta)$, calculated by LINSUB for a given chordwise location and interblade phase angle is essentially a Fourier series of the influence coefficients $\Delta \hat{C}_p^n$,

$$\Delta C_p(x,\beta) = \sum_{n=-\infty}^{\infty} \Delta \hat{C}_p^n(x) e^{in\beta}. \tag{4}$$

Inversion of Eq. 4 results in a Fourier integral for the influence coefficients,

$$\Delta \hat{C}_p^n(x) = \frac{1}{2\pi} \int_{-\pi}^{\pi} \Delta C_p(x,\beta) e^{-in\beta} d\beta. \tag{5}$$

After LINSUB results for $\Delta C_p(x,\beta)$ are calculated over the interval $-\pi \leq \beta \leq \pi$, influence coefficients may be calculated by numerical integration of Eq. 5. These influence coefficients are then summed using a truncated form of Eq. 6,

$$\Delta C_p'(x,\beta) = \sum_{n=n_1}^{n_2} \Delta \hat{C}_p^n(x) e^{in\beta}, \tag{6}$$

to predict the unsteady pressure difference coefficient on the reference airfoil (airfoil 0 in Figure 2) for a finite number of airfoils oscillating at a particular value of β. In summing the influence coefficients, the limits of summation are chosen to best simulate the experimental cascade configuration. That is, to predict the unsteady pressure difference coefficient for position A in the cascade, the influence coefficients are summed over the range n_1=-1 to n_2=2, while for position B, the limits of summation are n_1=-2 and n_2=1.

The comparisons presented in Figures 12 and 13 are for the two values of β, -45 and -90 degrees, for which the dynamic periodicity was good. The predictions for these two conditions are in good overall agreement with the data: the predicted magnitudes and phases are in very good agreement with the data except for the β=-45 degrees phases, where the predictions consistently lead the data by a fairly small amount.

Correlations with the ΔC_p data for those values of β where the periodicity was poor are shown in Figures 14 through 17. For all but one condition, β=90 degrees, the ΔC_p data for the two airfoil positions are in surprisingly good agreement with each other despite the poor periodicity found for C_p. In these cases, the data-prediction correlations are good, too: the correlation is quite good for in-phase oscillations and only small differences are found between the predictions and the data for β=45 and 180 degrees. At these conditions, the

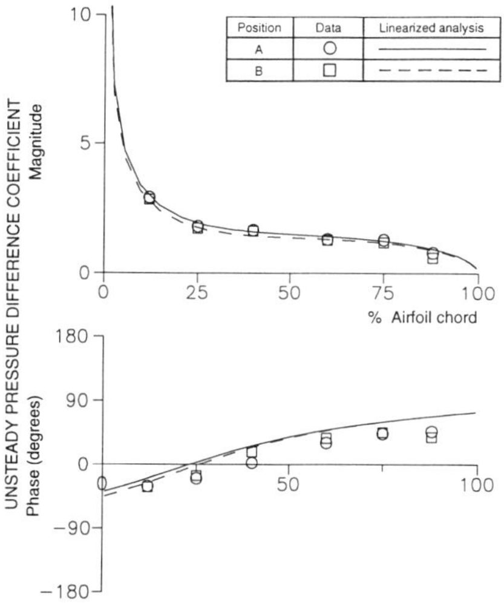

Figure 12. Unsteady pressure difference coefficients, β = −45°

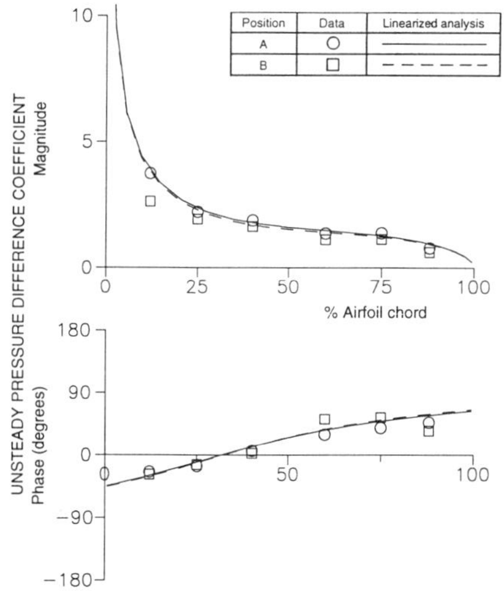

Figure 13. Unsteady pressure difference coefficients, β = −90°

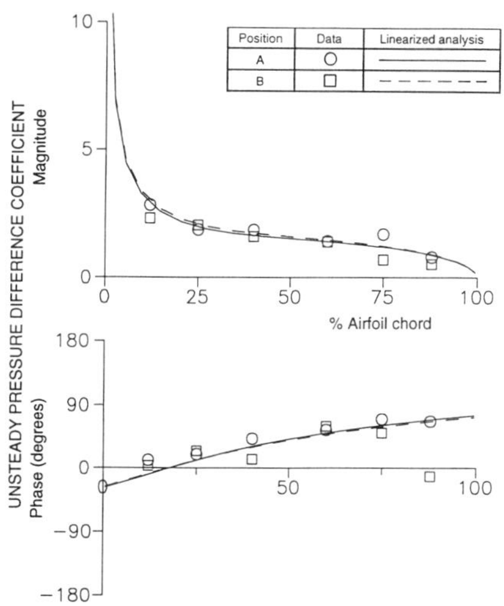

Figure 14. Unsteady pressure difference coefficients, β = 0°

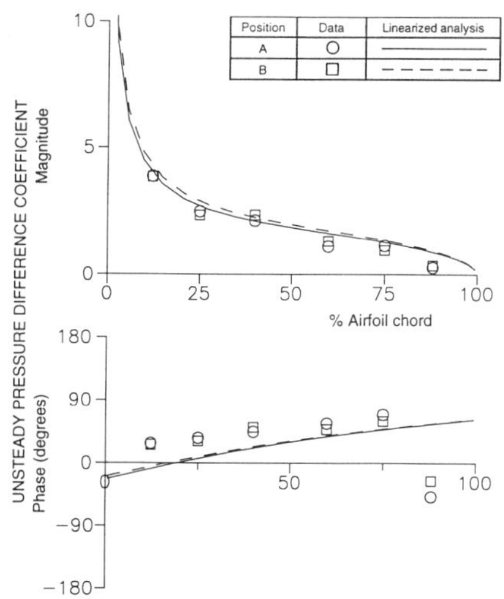

Figure 15. Unsteady pressure difference coefficients, β = 45°

oscillating cascade apparently excites an acoustic mode (or modes) in the wind tunnel, the effect of which destroys the periodicity of C_p but cancels when ΔC_p is calculated, thereby resulting in good ΔC_p periodicity and also good correlation with the predictions.

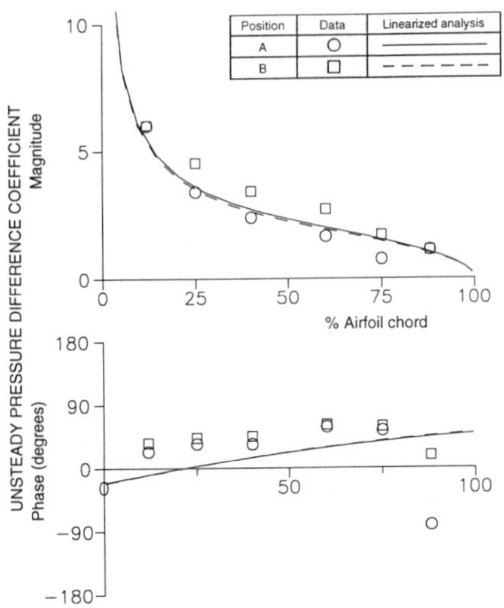

Figure 16. Unsteady pressure difference coefficients, β = 90°

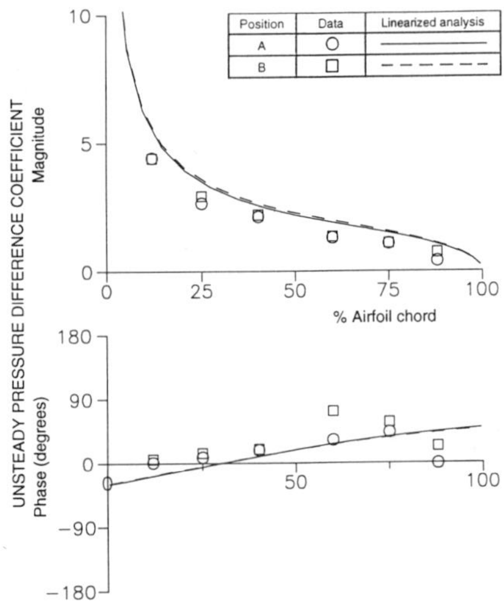

Figure 17. Unsteady pressure difference coefficients, β = 180°

In all of these cases, the differences between the linearized analysis predictions for airfoils A and B are slight. This indicates that the oscillations of airfoils -2 and 2 have only a small influence on the unsteady pressure distribution of the reference airfoil. Only airfoils -1 and 1 have significant influences on the predicted variations in ΔC_p with β.

SUMMARY AND CONCLUSIONS

Steady and unsteady airfoil surface pressure distributions have been obtained in a low solidity linear cascade oscillating at reduced frequencies of 0.64 with 0.55 inlet Mach number. The airfoils were driven in simultaneous torsion-mode oscillations for a range of interblade phase angles. Steady and unsteady airfoil surface pressure distributions were measured, the latter using flush-mounted miniature pressure transducers. Discrete Fourier analysis techniques were used to analyze the unsteady pressure data and determine the first harmonic components. Periodicity of the cascade was investigated by measuring the steady and unsteady pressures on the airfoil surfaces defining the two passages at the center of the cascade. The unsteady pressure difference coefficient data were compared to flat plate cascade predictions.

It was found that the cascade dynamic periodicity for the unsteady pressure coefficient was good for some values of interblade phase angle but poor for others despite good steady flow periodicity. Predicted propagation directions for cascade acoustic waves were found to give a reasonable physical explanation for the cascade/wind tunnel wall interactions and the resulting problems with cascade dynamic periodicity. However, periodicity of the unsteady pressure difference coefficient data was usually good, even for conditions with poor periodicity. In those cases, cancellation of wind tunnel acoustic mode effects apparently occurred when the pressure difference was calculated. Linearized analysis ΔC_p predictions for a finite number of oscillating airfoils were in good agreement with the data. These predictions also indicated that the dominant effects on the unsteady aerodynamics of this cascade are due to the self-induced unsteadiness and the oscillations of the airfoils adjacent to the reference airfoil.

To make this a reliable facility for the investigation of oscillating cascade aerodynamics at arbitrary interblade phase angles, the effects of the wind tunnel walls must be made negligible. An effort is currently under way to replace the solid tunnel walls in the vicinity of the cascade with acoustically-treated walls as developed to reduce aircraft gas turbine engine noise (Groeneweg and Rice, 1987). The effectiveness of the acoustic treatment will be investigated in part by repeating the experiments reported upon herein.

REFERENCES

Buffum, D.H. and Fleeter, S., 1990, "Aerodynamics of a Linear Oscillating Cascade," *NASA Technical Memorandum 103250.*

Carta, F.O., 1983, "Unsteady Aerodynamics and Gapwise Periodicity of Oscillating Cascaded Airfoils," *Journal of Engineering for Power,* Vol. 105, pp. 565-574.

Groeneweg, J.F. and Rice, E.J., 1987, "Aircraft Turbofan Noise," *Journal of Turbomachinery*, Vol. 109, pp. 130-141.

Kline, S.J. and McClintock, F.A., 1953, "Describing Uncertainties in Single-Sample Experiments," *Mechanical Engineering*, pp. 3-8.

Mehmed, O., Kaza, K.R.V., Lubomski, J.F. and Kielb, R.E., 1982, "Bending-Torsion Flutter of a Highly Swept Advanced Turboprop," *NASA Technical Memorandum 82975.*

Whitehead, D.S., 1987, "Classical Two-Dimensional Methods," AGARD Manual on Aeroelasticity in Axial Flow Turbomachines, Volume 1: Unsteady Turbomachinery Aerodynamics, *AGARDograph No. 298*, pp. 3-1 - 3-30.

Runyan, H.L., Woolston, D.S. and Rainey, A.G., 1955, "Theoretical and Experimental Investigation of the Effect of Tunnel Walls on the Forces on an Oscillating Airfoil in Two-Dimensional Subsonic Compressible Flow," *NACA Report 1262*.

ACKNOWLEDGEMENTS

This investigation was supported by the Turbomachinery Technology Branch, Propulsion Systems Division of the NASA Lewis Research Center. The experiments were performed with the assistance of R. Olsey, S. Orahoske and V. Verhoff. Helpful comments on the contents of this paper were provided by L. Bober, D. Hoyniak and D. Huff.

Experimental Investigation of the Unsteady Pressure Field on a Vibrating Blade in Steady and Unsteady Flow

C.A. Poensgen[1]
H.E. Gallus[1]

Abstract

For the prediction of flutter in turbomachine bladings a numerical program has been developed to solve the two dimensional unsteady Euler equations. In order to get a data bank to check the numerical code, flutter experiments have been carried out. Because of the necessity to predict the self excited flutter of a blade as well as the interaction of unsteady blade forces due to blade motion and incoming wakes, measurements of the unsteady pressure field on a blade surface were made in steady and in unsteady inlet flow. Thus, the interaction between incoming wakes and blade motion on the unsteady blade pressure field could be determined.

The experimental apparatus is build up in a way that the boundary conditions of the experiment can be simulated by the numeric program. In order to reduce the costs of the experiment, only one blade of an annular cascade was mounted on a spring and forced into vibration (influence wave mode). The test facility is able to carry an annular compressor stator row or an annular turbine vane row. Results were obtained for both cascades, but only the data for the compressor test cases will be addressed in this paper. Optionally, a wake generating rotor with uncambered airfoils can be mounted upstream.

Nomenclature

A	amplitude of bending	[mm]
A	amplitude of torsion	[°]
c	absolute velocity	[m/s]
f	vibration frequency	[Hz]
k	reduced frequency	

[1]Institut fur Strahlantriebe und Turboarbeitsmaschinen, RWTH , D-5100 Aachen, Germany.

u	speed of rotation	[m/s]
w	relative velocity	[m/s]
z	number of blades	
α	flow angle	[°]
Ψ	phase angle between the rotor wake and the stator blade motion	[°]
Ma	inlet Machnumber	

Subscripts

R	rotor
S	stator
w	wake
2	inlet condition of the stator
3	outlet condition of the stator 35% chord downstream

All symbols not marked here are compatible to the nomenclature used in the "Workshop of Aeroelasticity" (Fransson and Suter (1983))

Introduction

The prediction of flutter of turbomachine bladings is a very difficult task in the research of turbomachinery. Though, the flow is three-dimensional, unsteady and viscous, most of the numerical prediction methods are restricted to two-dimensional unsteady inviscid flow models (Fransson (1986), Verdon (1987) and (1990), Hall and Crawley (1987), Kau (1990)). In recent time numeric program systems came up solving the unsteady three-dimensional Euler equations (Fourmaux and Le Meur (1987), Kirtely et. al. (1990), Ni and Sharma (1990)). In the early years of turbomachinery research, progress of the understanding on the nature of flutter could only be achieved by experimental work connected to simplified analytical flutter models. The rapid increase in computational power led to the development of numeric programs, solving the nonlinear two- and three-dimensional Euler equations. Thus, the prediction of the flutter boundaries for a given blade row geometry may become possible in the near future. By this, the role of the experiment changed in the last decade. As experimental work is expensive and tied to a specific type of blading, experimental flutter investigations are connected to the development of numeric flutter prediction programs in order to get benchmark database. Thus, the numeric programs can be checked against experimental data.

The assumption of inviscid flow around a blade is only valid at operating points near design. Stall flutter, the behavior of aerodynamic stability while throttling a compressor, and choke flutter are of major practical interest. But the experimental results of flutter investigations at off-design are influenced by viscous effects. Besides that, the intensities of secondary flows increase. Thus, it is necessary to quantify the impact of the viscous effects and the influence of three-dimensional flow on the obtained data.

Test Facility and Test Program

The unsteady flutter investigations were carried out in an annular compressor cascade described in detail by Schulz and Gallus (1988) and Poensgen (1990). The major problem of doing flutter experiments in annular cascades is the achievement of sufficiently large oscillation amplitudes at high frequencies. Because of the low aspect ratio, the tested blades cannot be run into vibration with large amplitudes. Therefore, these blades can be mounted on springs. Significant displacement amplitudes can only be obtained, when the blade-spring systems are forced into vibration near their first eigenfrequency by electrodynamic or mechanical shakers. Nevertheless, the output power of electrodynamic shakers is limited. The maximum dynamic displacement amplitude of a blade mounted on a spring, is proportional to the square of the excitation frequency. Thus, it is much more convenient to concentrate the excitation energy on a single blade-spring system. As the numerical code for predicting the flutter boundaries for a given profile should be able to take into account different interblade phase angles, the calculation procedure cannot be restricted to a single blade. The number of blades, which are necessary for the numeric calculation depends on the periodic boundary condition in the circumferential direction. Therefore, it should be possible to do a numeric calculation for the specific constellation of the here described

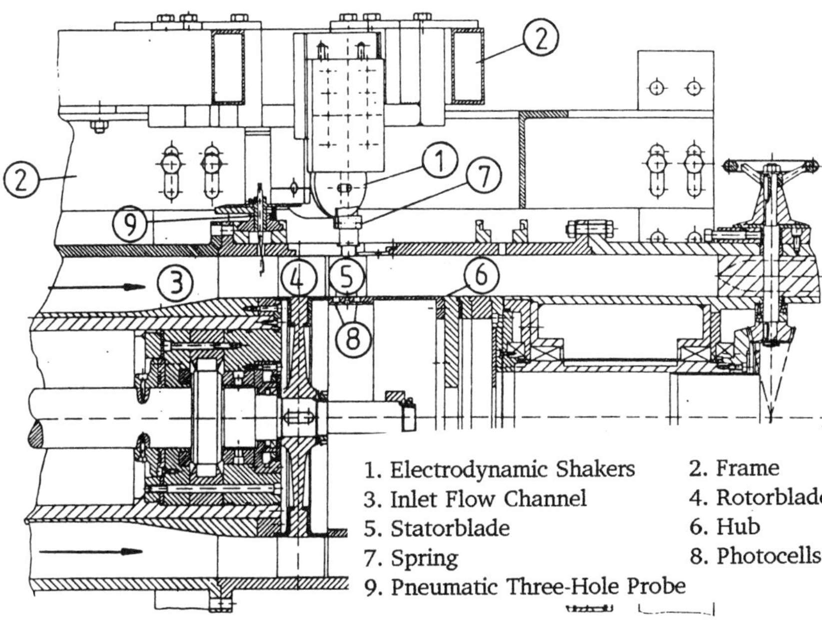

Fig. 1: Cross Section of the Annular Cascade

Fig. 2: Shakers working on a Blade-Spring System

experiment. A single blade is vibrating in an annular cascade while all the other blades are fixed. This method to attack the problem is confirmed by Whitehead (1984) in his concluding comments to the UATP Conference at Cambridge.

The advantages of this experimental setup are lower investment costs and the blade-spring system can be fixed outside of the casing. Outside, there is enough room to place strong shakers (Fig. 1). Fig. 2 shows two shakers fastened on a frame outside of the casing, working on a torsion cross spring.

The experimental investigations were carried out in steady and in unsteady flow. In order to synchronize the wake passing frequency of the rotor blades with the vibrational frequency of the test blade, a rotor with uncambered profiles was manufactured. The rotor is driven by a direct current electromotor. The rotor airfoils are staggered, so that the rotor blade passing frequency is equal to the vibration frequency of the stator test blade, and the steady inlet conditions to the stator are the comparable to the test cases without rotor. At midspan, no work on the fluid is done by the rotor. Table 1 summarizes further geometrical parameters.

The experiments were carried out within two steps:

1. The unsteady pressure field around a single vibrating compressor blade was measured in steady flow. The points of operation were set according the parameters, which can be changed independently in the annular cascade. The mass flow was controlled by a bypass valve. Variations of the Machnumber from 0.18 to 0.43 are possible. Nevertheless, above inlet Machnumbers of 0.3 the static inlet pressure rose to much, so that the leakage flows throughout the opening of the casing, where the spring is put through, became too high.

Table 1: Geometrical parameters of the blading

Outer Diameter	428 mm
Inner Diameter	321 mm
Design parameters of the stator	
Radial gap at the casing	0.55 mm
Radial gap at the hub of the test blade	0.2 mm
Radial gap at the hub of the fixed	0.0 mm
blades	61°
Stagger angle	62.2 mm
Length of chord	0.78
Solidity	0.229
Design inlet Machnumber	42.0°
Design inlet flow angle	24°
Number of stator blades	
Design parameters of the rotor	78 mm
Length of chord	0.628
Solidity	52°
Stagger angle of the rotor blades	12/18
Number of rotor blades	NACA
Bladeprofile	0008
Axial spacing	3.5 mm

The influence of the incidence angle on the unsteady pressure distribution could be varied within a wide range. Through the adjustable inlet guide vanes, 313 mm upstream of the rotor, the incidence angle was set from -2° to +12°. The output energy of the vibration shakers was controlled by an oscillation amplifier. Confident data of unsteady pressures were obtained at bending amplitudes above 0.15 mm and torsion amplitudes of 0.25 degree, respectively. As the photocells, detecting the blade motion, are operating in a linear range of 1.8 mm peak to peak amplitude, the vibrational displacement of the blades could be varied up to amplitudes of 0.75 mm for the 150 Hz bending mode. Amplitudes of 0.65° were achieved for the 370 Hz torsion mode. It was possible to perform a parametric study on the influence of the incidence angle, the oscillation amplitude and, in a smaller range, of the Machnumber on the resulting unsteady pressure distribution.

2. The interaction of wake disturbed inlet flow and blade motion on the unsteady pressure distribution was the final goal of the investigation. As the vibration frequency of the blade is bound to the resonance frequency of the blade-spring system, the rotor wake passing frequency had to be synchronized to the blade motion. The variation of the phase angle between the rotor wake and the motion of the stator wake gives the possibility to get an amplification of both influences at a certain fixed phase angle. After a phase shift of 180° wake and blade motion will work contrary to each other so that the resulting

pressure fluctuations will be minimized. The measurements were carried out for 8 different phase angles. In order to determine the influence of the rotor wake on the resulting pressure distribution around a fixed stator blade, a measurement was taken where the amplifiers for the electrodynamic shakers were switched off.

The velocity profile of the incoming wakes was measured using 3-d hot-wire anemometry. The flowfield downstream of the rotor was resolved in 18 spanwise points.

Table 2 summarizes the matrix of the experiments, which were run with a vibrating single stator blade. In the test cases with rotor the phase angles between the rotor wake and the stator blade motion was set from 0° to 315° in steps of 45°

Table 2: Test Program (Remark: Ben.=Bending; Tor.=Torsion)

f [Hz]	Mode	Ma	α_2	k	A [mm] or [°]	z_R
155	Ben.	0.218	40.0°	0.405	0.26 0.51 0.63	0
155	Ben.	0.224	44.5°	0.394	0.33 0.55 0.70	0
155	Ben.	0.224	49.0°	0.394	0.34 0.50 0.74	0
155	Ben.	0.230	54.0°	0.383	0.38 0.53 0.76	0
230	Ben.	0.186	40.0°	0.708	0.10 0.21 0.40 0.45	0
230	Ben.	0.196	44.5°	0.673	0.10 0.20 0.37 0.42	0
230	Ben.	0.196	49.0°	0.671	0.10 0.29 0.36 0.41	0
						0
230	Ben.	0.270	40.0°	0.488	0.17 0.27 0.33 0.38	0
230	Ben.	0.275	44.5°	0.477	0.17 0.26 0.33 0.39	0
230	Ben.	0.282	49.0°	0.466	0.17 0.26 0.33 0.39	0
230	Ben.	0.281	54.0°	0.468	0.17 0.26 0.32 0.39	0
370	Tor.	0.231	40.0°	0.922	0.30 0.40 0.51 0.62	0
370	Tor.	0.235	44.5°	0.900	0.28 0.40 0.52 0.64	0
370	Tor.	0.234	49.0°	0.902	0.29 0.40 0.52 0.64	0
370	Tor.	0.235	54.0°	0.897	0.30 0.41 0.53 0.65	0
370	Tor.	0.279	40.0°	0.757	0.41 0.56 0.65	0
370	Tor.	0.284	44.5°	0.744	0.37 0.55 0.65	0
370	Tor.	0.279	49.0°	0.757	0.28 0.52 0.65	0
370	Tor.	0.286	54.0°	0.739	0.30 0.41 0.54 0.66	0
155	Ben.	0.238	44.5°	0.370	0.35 0.54	12
230	Ben.	0.238	44.5°	0.549	0.25 0.34	18

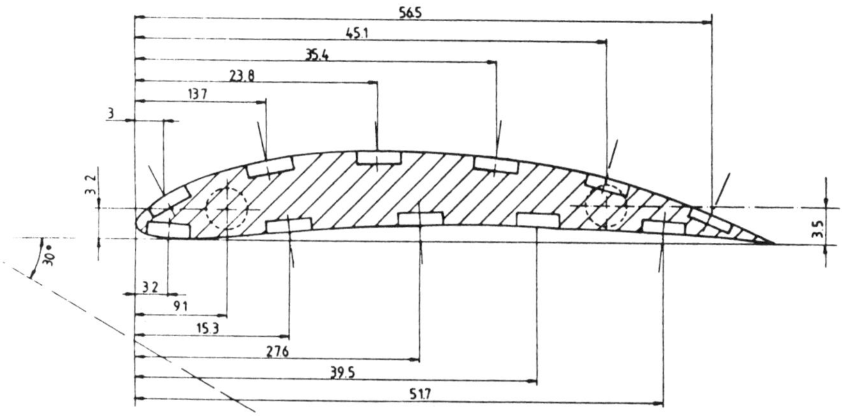

Fig. 3: Locations of the Semi-Conductor Pressure Transducers

Data Acquisition and Data Reduction

The unsteady pressure distribution around the stator blade was measured using flat pressure semiconductors (Kulite LQ-80-10-M). Six pressure sensors were placed on the blade suction side compared to five sensors on the pressure side. Fig. 3 shows the mid span section of the compressor blade and the accurate locations of the unsteady pressure transducers. The pressure sensors were statically calibrated. As the sensitive area of the transducers is on the blade surface, and as the oscillation frequency is lower than 400 Hz, there is no need to perform a dynamic pressure calibration. Mounted on an oscillating surface, pressure transducers are sensitive to acceleration. The influence of the blade acceleration can be calibrated. The calibration function was found to be linear. As the blade motion is measured, the exact value of the periodic acceleration is known. Through a complex subtraction in the Fourier domain, the components of the acceleration can be filtered out. Then, the inverse Fourier transformation leads to the time-dependent pressure signals. Fig. 4 displays the complete data reduction process.

The blade motion was detected by photooptical position detectors, which are mounted in the hub. Two photovoltaic elements, having main sensitivities in the circumferential and the axial direction, give the output signals from which the complex amplitudes of blade heaving and blade pitch can be calculated. Two infrared light emitting diodes were placed inside the blade at the hub (see Fig. 5). The emitted light was focused by a small lens. The sensitivity of the photovoltaic position decoders is linear within two perpendicular directions in a range of ± 1.5 mm. More details about this position detecting system can be found by Knauf (1985).

The operating point was measured by a three-hole pressure probe, located at midspan 82 mm upstream of the stator leading edge. Thus, the inlet flow angle, inlet Machnumber and total inlet pressure were determined.

An eight channel transient recorder was used for the data recording.

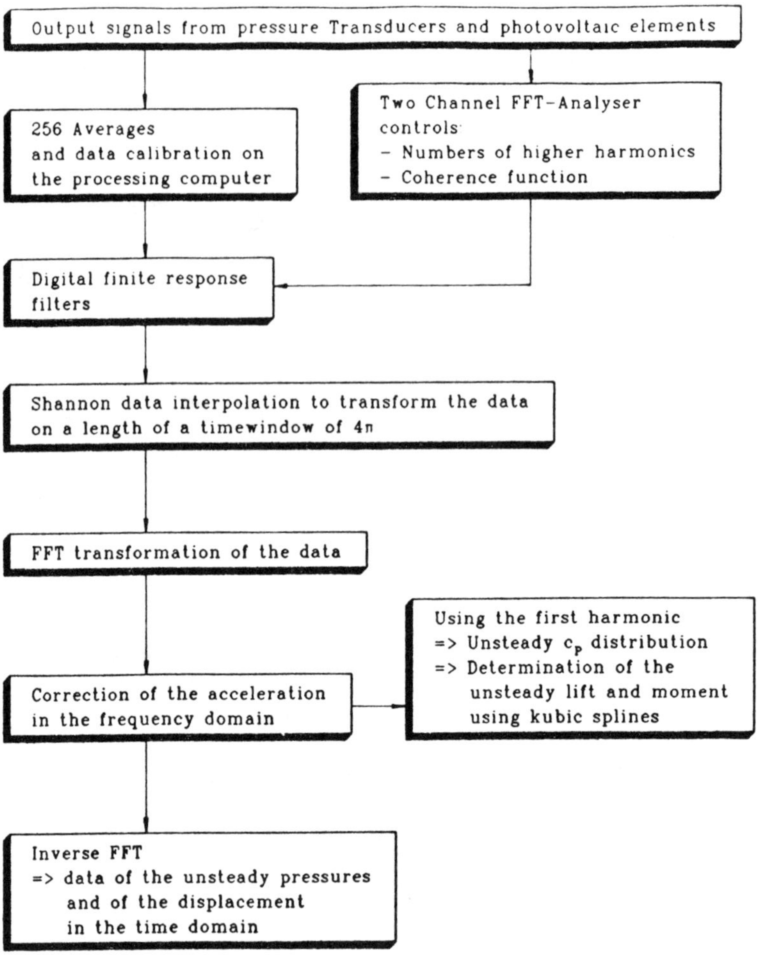

Fig. 4: Flow-Chart of the Data Reduction Process

Each pressure sensor and each photovoltaic element are connected to separate amplifiers. Their output plugs were connected via a multiplexer to the transient recorder. The sampling frequency was set so that 256 digital samples contained two periods of the blade oscillation. Using the full data storage of the transient recorder, 256 time windows were recorded. The trigger signal was taken from the power amplifier of the electrodynamic shakers. Based on the time-average procedure the data was reduced by the processing computer and calibrated. Thus, the unsteady pressure distribution and the blade motion could be controlled during the test.

A two channel FFT-Analyzer was used to control ON-LINE the data quality during the test run. The coherence function between the blade pressure signals and the blade motion in circumferential direction was

Fig. 5: Instrumented Test Bed With the Photooptical Detection System

computed and analyzed. By this, the number of necessary averages and the signal to noise ratio were determined. Disturbing periodical signals could be detected and observed.

In unsteady flow the blade vibration and the rotor blade passing frequency have to be synchronized. Fig. 6 shows a schematic sketch of the data acquisition and vibrational control apparatus. The rotor speed is set according to the desired oscillation frequency. A photosensitive trigger-element mounted in the casing detected the reflection of a passing rotor blade. Its signal was amplified and connected to a PLL (Phase Lock Loop) digital function generator. Here the TTL signal was converted into a sinusoidal signal, whose amplitude and phase could be shifted. This signal is the controlling input signal for the power amplifier of the electrodynamic vibrators. The corresponding TTL signal was connected to the transient recorder and used as trigger signal for the data acquisition. This control loop gave the guarantee that the phase shift between blade motion and rotor wake was kept constant for the time of measurement. The phase shift between these both signals can be changed due to the digital function generator over more then one period. The measurements of the blade motion and unsteady pressures were taken for eight different phase angles between the rotor blade position and the blade motion. In a final test the electric shakers were switched off. So the blade was only excited by the wake. As the resulting amplitudes are less than 1/100 mm the influence of the blade motion on the unsteady pressure field can be neglected. Thus, this data represents the impact of the wake flow.

The velocity profiles of the rotor wakes were detected by a stationary three-sensor hot-wire probe. Each sensor was connected to a DISA M55 anemometer. Their output signals were recorded by the transient recorder.

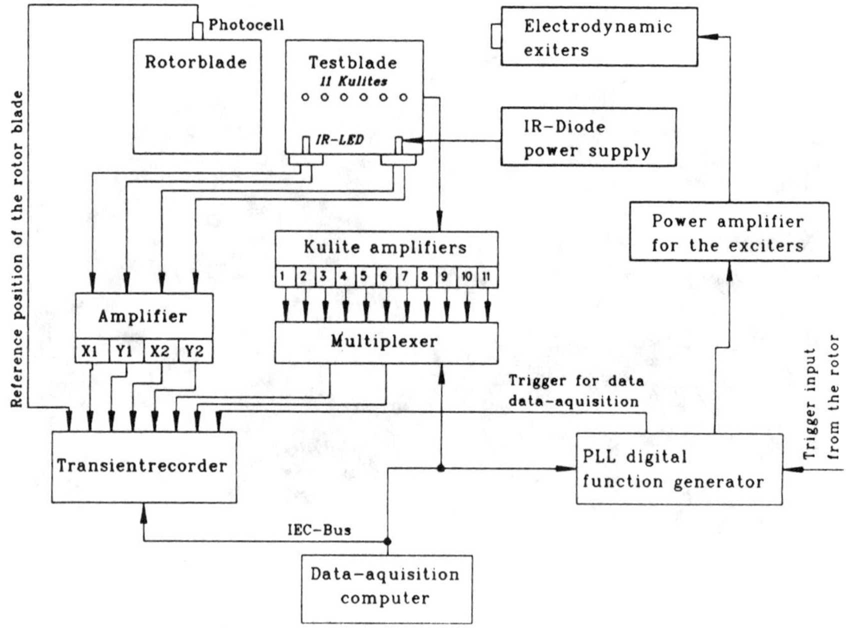

Fig. 6: Scheme of the Data Acquisition

More details of the data acquisition and data reduction procedure on multi-sensor hot-wire anemometry are reported by the authors (1989).

Experimental Results

Because of length, not all results (see table 2) are shown in this paper. Detailed information is available on request from the authors. Fig. 7 shows the time-dependent pressure and blade motion signals for $Ma_2=0.224$, $f=155$ Hz and $\alpha_2=44.2°$. The pressure amplitudes detected from the first transducers at the pressure and at the suction side are of nearly equal magnitude. Their phase difference is about 180°. With chord the amplitudes decay in a quasi exponential way. The signals of the blade motion are nearly sinusoidal. The amplitudes of the measured higher harmonics are of second order small. Thus, for the correction of the influence of the blade acceleration on the output signals of the pressure transducers only the base frequency has to be taken into account.

If the blade motion is of pure heaving or pure pitching, the unsteady c_p distribution can be evaluated from the measured data. This is the case for all tests at 155 Hz and 230 Hz bending mode. Here, the signals of the blade motion do not yield confident data of the torsion amplitudes as they are smaller than 0.05°. The influence of the variation of the blade amplitude at constant vibration frequency and constant inlet Machnumber is

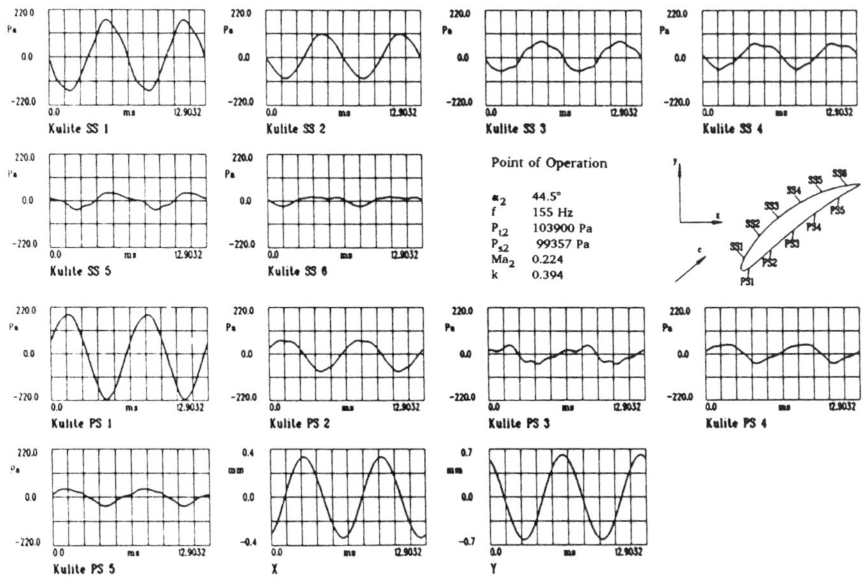

Fig. 7: Unsteady Pressure Distribution along Chord for the Bending Mode

shown in Fig. 8. First confident results of unsteady pressure signals (controlled by the use of the coherence function) were obtained at a blade heaving amplitude of 0.3 mm. The maximum measured was 0.7 mm. However, the unsteady pressure coefficients remain constant, while the blade heaving amplitude is increased. Thus, the assumption of linearity between the amplitude of the blade displacement and the resulting unsteady pressure field around the blade is verified very good by this experiment.

Fig. 9 presents the development of the unsteady pressure coefficients for two different inlet Machnumbers. The oscillation amplitude, frequency, and the inlet flow angle were kept constant. Higher Machnumbers result in an increase of the unsteady pressure coefficients. In both test cases, the resulting aerodynamic forces were damping the blade motion.

A comment should be made on the development of the phase angle along the suction side. At the location of the sensor SS5 there is a strong delay of the pressure signal. This effect cannot be explained by uncertainties of the measurement technique, as the measured amplitudes are small, but it can also be seen at other points of operation. At an inlet flow angle of 40° (-2° incidence) there exists a laminar separation bubble due the boundary layer transition (Schulz (1989)). The sensor SS5 is placed at a point, where the separated flow reattaches. The sequence of the unsteady pressure coefficients in Fig. 10 show the same effect. As the transition bubble moves upstream with increasing inlet flow angle, and as the transition length becomes shorter, the influence of the transition separation bubble tends to be smaller. A similar effect has been observed by the authors (1990) on the suction side of an oscillating turbine blade.

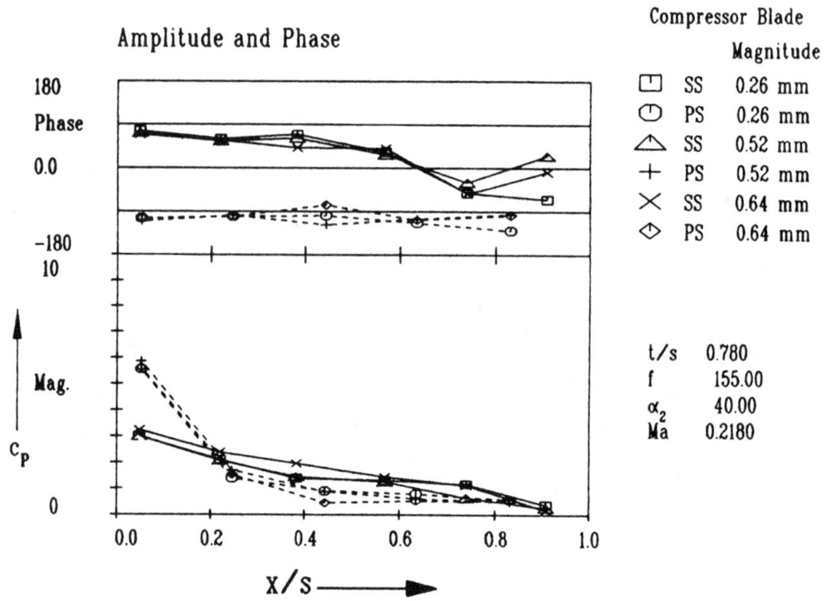

Fig. 8: Dimensionless Unsteady Pressure Distribution Along Chord with Variation of the Bending Amplitude

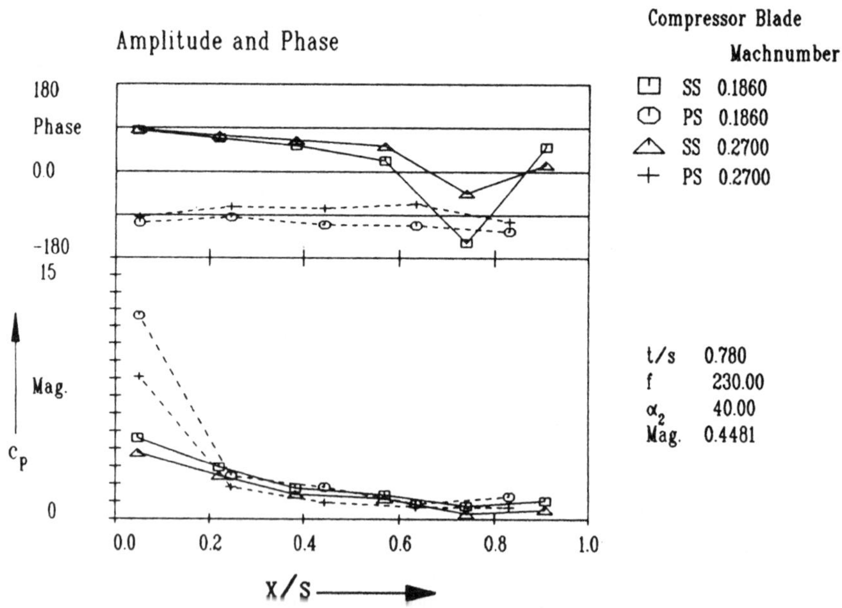

Fig. 9: Dimensionless Unsteady Pressure Distribution Along Chord with Variation of the Inlet Machnumber

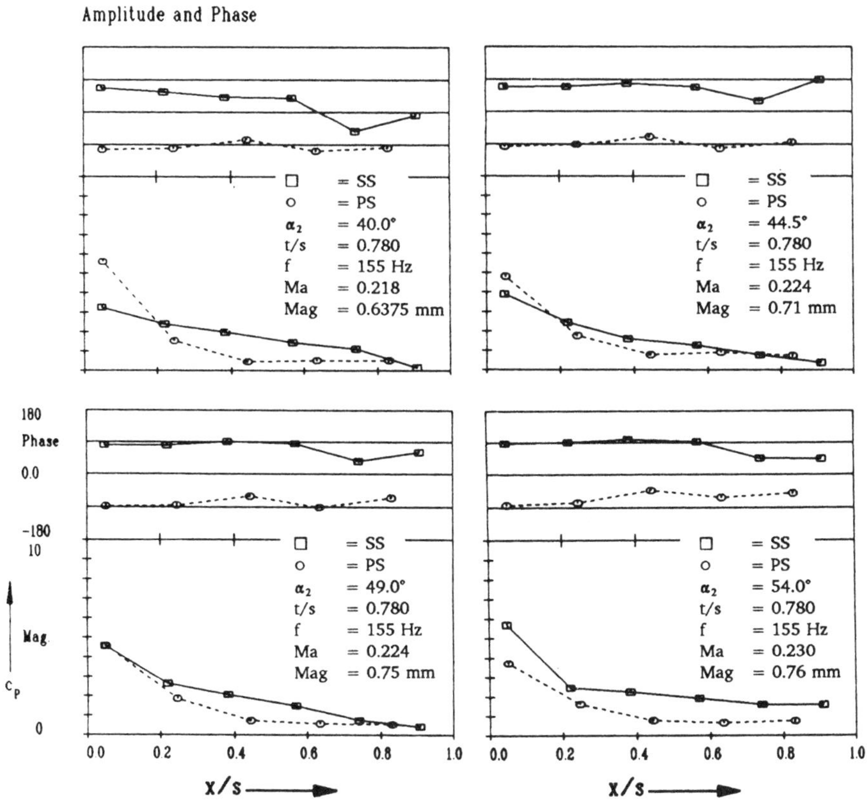

Fig. 10: Dimensionless Unsteady Pressure Distributions Along Chord with Variation of the Inlet Flow Angle

Table. 3: Coefficients of the unsteady aerodynamic force and moment with varying the inlet flow angle (Ma_2=0.23, f=155 Hz, A=0.76 [mm])

Inlet flow angle	Re(F)	Im(F)	Re(M)	Im(M)
40.0°	0.739	-0.345	1.545	-0.837
44.5°	0.479	-0.014	-0.469	-0.117
49.0°	0.463	0.407	1.731	-0.099
54.0°	0.714	0.987	10.983	0.500

The influence of the inlet flow angle at constant bending amplitude and constant inlet Machnumber is displayed in the series of Fig. 10. The unsteady pressure distribution was measured at four different inlet flow angles from 40° up to 54°. For these data, the three dimensional flow of the tested cascade was investigated in detail by Schulz (1989). Increasing

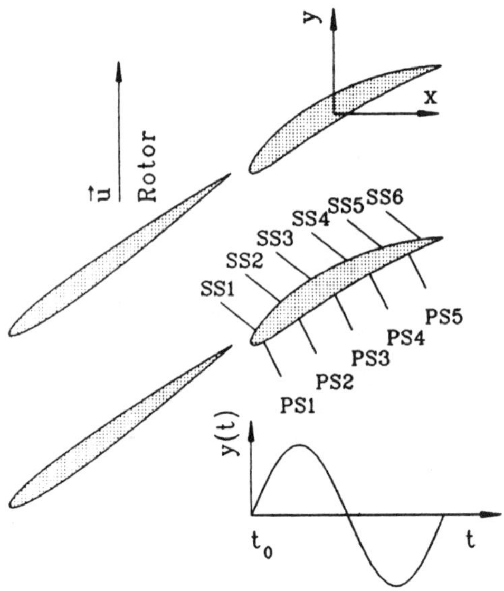

Fig. 11: Definition of the Phase Angle Between Rotor Wake and Blade Motion

the inlet flow angle will lead to an increase of the unsteady pressure peak at the first location of measurement at the suction side, while the magnitude on the pressure side becomes smaller. At an inlet flow angle of 40° (-2° incidence) the maximum pressure amplitude occurs near the leading edge at the pressure side. Here the magnitude is twice as high as at the suction side. The integration of the unsteady forces and moments show, that with increasing incidence flow angle, the aerodynamic damping becomes more and more unstable (Table 3).

The investigations of Schulz (1988) have shown that there exists a hub corner stall at the suction side of the stator blade. At inlet flow angles above 52° the hub corner separation reaches midspan. By this, the sensors SS5 and SS6 at the suction side are covered by the separation at an inlet flow angle of 54°, as the real-time data showed. But, there is no discontinuity in the distribution of the unsteady pressure coefficients. The pressure amplitudes are twice as high compared to those obtained at an inlet flow angle of 40°. The development of the phase angles shows a delay of the unsteady pressure signals. It should be marked out that the aerodynamic instability of the flow is not generated by large flow separations. The major impact on the integrated value of the unsteady force comes from the pressures round the blade leading edge (SS1-SS2 and PS1-PS2). Compared to the pressure amplitudes measured near the leading edge, the pressure fluctuations measured at the locations SS3-SS6 and PS3-PS5 are of minor magnitude. As the flow is attached at all instances of time and as there exists no full span blade separation, the aerodynamic instability is not connected to stall flutter.

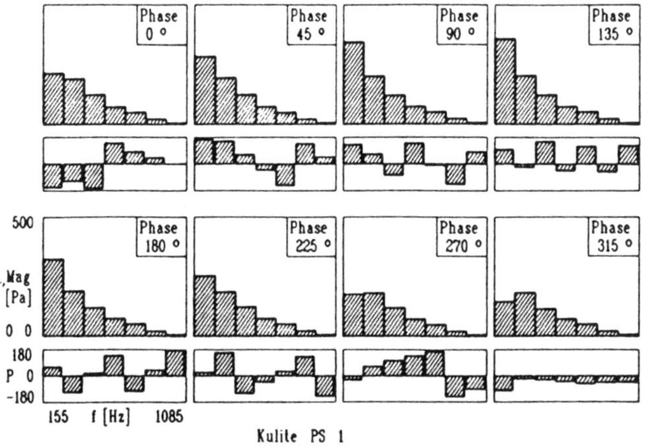

Fig. 12: First Seven Harmonics of the Pressure Signal at the Location PS1

In a next step the unsteady pressure field around a vibrating blade in unsteady flow was investigated. In order to attack this problem, the measurements were carried out at constant inlet flow conditions, but the phase angle between the vibrating blade and the rotor blade position was shifted. The phase angle of 0° in the here described investigation is defined as follows:

The phase angle of zero degree is at the instance of time, when the rotor trailing edge at midspan and the stator leading edge have the same circumferential position. At this instance of time the stator blade motion is zero and becomes positive (see Fig. 11). A phase difference of 90° between stator blade motion and rotor wake represents a delay of the rotor wake of a quarter of a period. The phase angle between the rotor position and the stator blade motion will be labeled as "rotor-stator phase angle".

The results discussed further in this paper are obtained at an inlet flow angle of 44.5° and an inlet Machnumber of 0.238. The stator blade was excited at 155 Hz and 12 rotorblades were acting. Thus, the ratio of stator to rotor blades was 2:1. The major influence of the incoming wakes occurs near the leading edge. The unsteady fluctuations decay very fast. Fig. 12 shows the first seven harmonics of the pressure signal of the pressure sensors SS1, SS3 and PS1. Each sequence of eight frequency plots represents the pressure signals at different rotor-stator phase angles. The pressure amplitudes at oscillation frequency show a sweeping characteristic with the variation of the rotor-stator phase angle. The corresponding phase angles of this pressure signals decrease by an amount of nearly 45°.

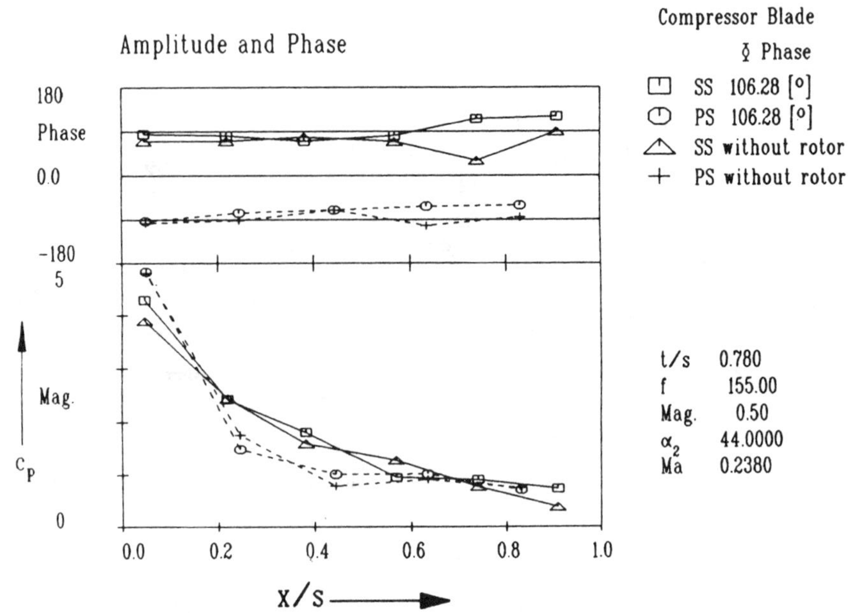

Fig. 13: Comparision of the Unsteady Pressure Distribution Due to the Blade Motion Obtained with and without Rotor

The amplitudes of the higher harmonics do not show this sweeping effect. Thus, it may be concluded that the unsteady pressure field of an oscillating stator blade in unsteady inlet flow, is a linear superposition of the unsteady pressures generated by the rotor wakes and the unsteady pressure distribution due to the blade motion. Because of this presumption the unsteady pressure field around the non vibrating stator blade was determined. By correcting the rotor-stator phase angle, the measured unsteady time-dependent pressures from the test case with a fixed stator blade were subtracted from the corresponding pressures, which were measured on a vibrating blade in unsteady flow. Doing this, a fictive pressure distribution is obtained, whereby the influence of the rotor blade wakes is eliminated. This pressure distribution can be compared to those, obtained on an oscillating blade with steady inlet flow. Fig. 13 shows the comparison of the unsteady pressure distributions, which are acquired by direct flutter measurements and the pressure distribution derived from the measurements in unsteady inlet flow. The good agreement between the compared pressure distributions confirms the assumption that in subsonic flow the unsteady pressure distribution around a vibrating blade results from a linear superposition of the unsteady pressure produced by the incoming wakes and those caused by the blade oscillation.

The velocity profile of the rotor wake has been measured using 3-d hot-wire anemometry. For this test case the rotor was equipped with 12 blades. The velocity defect in the wake is 12 m/s and the wake covers

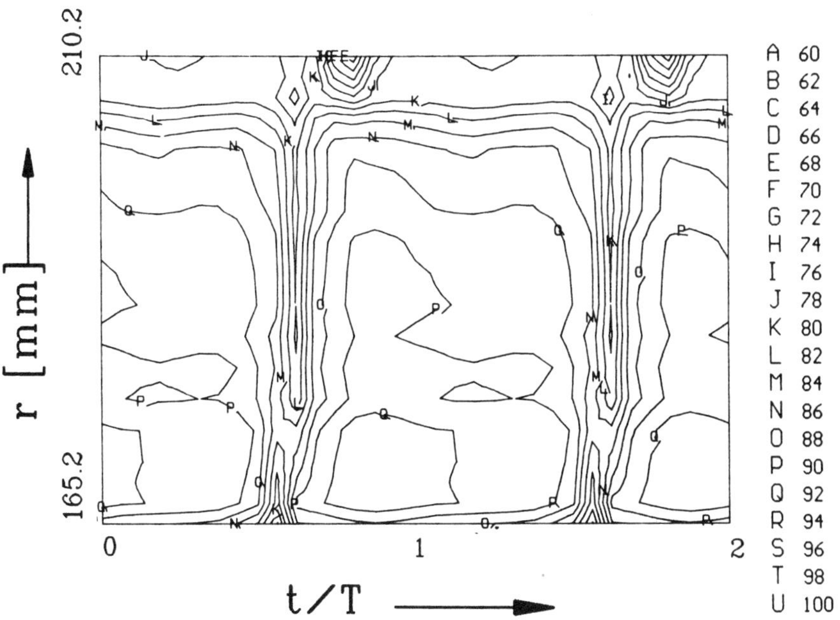

Fig. 14: Velocity Countours Downstream of the Rotor at Stator Mid-Passage

about 20% of the spacing of the rotor blades. As the direction of the relative velocity is close to the absolute velocity, there are only small fluctuations of the inlet flow angle due the wake. The impact of three-dimensional wake flow on the measured pressure distributions can be neglected. Fig. 14 displays the velocity contours downstream of the rotor. At the tip a leakage vortex is visible. Inside of the wake, there exists a radial inward flow, but the radial components do not reach 4 m/s. Further informations about the three-dimensional wake decay characteristics inside and downstream of the stator is given by the authors (1990).

Conclusions

The parametric study of the unsteady pressure distribution around a single vibrating blade has shown that the presumption of linearity between the blade amplitude and then unsteady pressure distribution is sufficient. Higher unsteady pressure coefficients were obtained with increasing Machnumber. As expected in the here investigated low subsonic flows, no changes of the aerodynamic damping characteristics were found. Increasing the inlet flow angle from -2° incidence to 12° incidence has a major effect on the damping of the aerodynamic forces. While the aerodynamic forces damp a motion at negative and low positive incidence angles, an aerodynamic instability will develop at high inlet flow angles. The corner separation has only a minor influence.

Viscous influences on the data could not be avoided. Flow separations like separation bubbles due to boundary layer transition and large flow separation like the corner stall did not affect the pressure amplitudes significantly, but a delay in the pressure signals was found.

In unsteady flow, the unsteady blade pressure distribution due to the blade motion and the incoming wakes was found to be a linear superposition of both effects. As the axial spacing between the rotor and the stator is small, potential effects cannot be neglected.

References

Fourmaux A., Le Meur A. (1987)
"Computation of Unsteady Phenomena in Transonic Turbines and Compressors"
UAATP-Symposium, Aachen, Sept. 1987

Fransson T.H., Suter P. (1983)
"Two-Dimensional and Quasi Three-Dimensional Experimental Standard Configurations for Aeroelastic Investigations in Turbomachine-Cascades"
École Polytechnique Fédérale de Lausanne, Report LTA-TM-83-2

Fransson T.H. (1986)
"Numerical Investigation of Unsteady Subsonic Compressible Flows Through an Oscillating Cascade"
Communication du Laboratoire de Thermique Appliquée et de Turbomachines
EPFL Lausanne

Hall K.C., Crawley E.F. (1987)
"Calculation of Unsteady Flows in Turbomachinery Using the Linearized Euler Equations"
UAATP-Symposium, Aachen, Sept. 1987

Kirtely K.R., Beach T.A., Adamczyk J.J. (1990)
"Numerical Analysis of Secondary Flow in a Two-Stage Turbine"
AIAA-90-2356

Knauf W.B. (1985)
"Experimentelle Untersuchungen zur Ermittelung der Aerodynamischen Schaufeldämpfung an einem schwingenden Axialverdichtergitter"
Disseration RWTH Aachen, 1985

Ni Ron-Ho, Sharma Om (1990)
"Using 3D Euler Flow Simulations to Asses Effects of Periodic Unsteady Flow Through Turbines"
AIAA 90-2357

Poensgen C.A. (1989)
"Ein Verfahren zur Vermessung der instationären dreidimensionalen Geschwindigkeitsvektoren in Turbomaschinen"
Internal Technical Report No. 89-01, RWTH Aachen

Poensgen C.A. (1990)
"Experimentelle Untersuchung des instationären Druckfeldes um eine schwingende Schaufel"
Internal Technical Report No. 90-01, RWTH Aachen

Poensgen C.A., Gallus H.E. (1990)
"Three Dimensional Unsteady Flow Field Inside and Downstream of an Annular Compressor Cascade"
Part I: Unsteady Wake Decay
Part II: Downstream Flow Field
Journal of Turbomachinery, Vol. 113, April 1991

Schulz H.D., Gallus H.E. (1988)
"Experimental Investigation of the Three-Dimensional Flow in an Annular Compressor Cascade"
Journal of Turbomachinery, Vol. 110

Schulz H.D., Gallus H.E., Lakshminarayana B. (1989)
"Three- Dimensional Separated Flow Field in the Endwall Region of an Annular Compressor Cascade in the Presence of Rotor-Stator Interaction"
Part I : Quasi Steady Flow Field and Comparison with Steady- State Data
Part II : Unsteady Flow and Pressure Field

Schulz H.D. (1989)
"Experimentelle Untersuchung der dreidimensionalen abgelösten Strömung in einem Axialverdichterringgitter"
Dissertation RWTH Aachen, 1989

Verdon J.M. (1987)
"The Unsteady Flow in the Far Field of an Isolated Blade row"
Report R87-957333-1, April 1987

Verdon J.M. (1990)
"Linearized Unsteady Aerodynamics for Turbomachinery Aeroelastic Applications"
AIAA-90-2355

Whitehead D.S. (1984)
"Concluding Report and Comments"
Symposium on "Unsteady Aerodynamics of Turbomachines and Propellers"
Cambridge England, September 1984

Test Facility for Flutter Investigations with Variable Frequency of the Vibrating Blade

Th. Steer[1]

ABSTRACT Within this paper there is presented the setup of a new straight test rig, which overcomes some restrictions concerning the determination of test parameters. Furthermore a new concept of measurement is presented.

In several test rigs the blades are fixed be means of a spring-damping element in order to achieve the oscillation of the blades. The osciallation is realized by mechanic or magnetic excitation. The restriction is that the frequency of oscillation has to be near the natural frequency. In order to alter the natural frequency there are commonly used supplemental masses. At the new test rig there is selected another approach. The blades are not fixed at the test-rig but are beared by air-bearings. An axial magnetic bearing is used to supply the forces for excitation. As the magnetic bearing simulates the formerly used spring-damping-element. the natural frequency can be choosen freely by adjusting the parameters of the magnetic bearing. This yields an independent setting of the parameters reduced frequency and Ma-number. Furthermore it should be possible to measure directly the aerodynamic damping coefficient by variing the damping characteristic of the magnetic bearing.

Up to now there were maintained many measurements dealing with the pressure distribution on the blade surfaces in order to achieve some knowledge about the flutter boundaries of blades.

The new test rig allows another approach. The main theme of interest will be the oscillation of downstream velocity and angle of the fluid flow. It will be investigated whether there are dependencies between the oscillation of the downstream characteristica on one hand and the unsteady lift coefficients calculated by the traditional means of integrating over the dynamic blade surface pressure on the other hand. It is obvious that the dynamic pressure distribution is measured and validated, too.

The paper describes the experimental setup, the measurement equipment and the boundary conditions concerning the wind channel.

[1] Institute for Thermal Power Plants, Technical University of Munich, Arcisstrasse 21, 8000 Munich 2, Germany.

1.1 Introduction

Since the first investigation of *Ch. Bellenot and J. Lalive d' Epinay* [Bd50] concerning self-excited flutter of a 2-D blade row there have been numerous investigations on flutter. Several test rigs were built up. They were straight channels or annular rings, with 2- or 3-dimensional experiments.

Although Bellenot and Lalive d'Epinay and some of their successors maintained self-excited flutter it is quite ticklish to realize self-excited flutter in a test rig. The vibration occurs at the natural frequency of the vibrating blades and therefore the amplitudes become quite high. Such it is easily possible to destroy the blades.

To avoid the danger of blade destruction in most experiments the blades were forced to vibrate. The vibration was controlled by some external electronic or mechanic system. It was tried to reach the flutter conditions from the save side. Within these experiments there were realized especially measurements of the static and dynamic pressure distribution on the vibrating blades. It was found that there are different criteria for the onset of blade flutter, e.g. potential flutter, stall flutter, shock-induced flutter. The maesured data were used to calculate dynamic lift coefficients as an integral value and as a function of the chord length. These values were used to calculate an aerodynamic damping coefficient as the overall interesting value to determine whether blade flutter will occur or not.

A second approach was the numeric calculation of the forced vibration by various methods. Calculations need boundary conditions especially at the trailing edge of the blades in order to maintain numeric stability. As this implies a numeric modeling of the downstream behaviour at the trailing edge, there is a lack in the numeric prediction of the behaviour at the trailing edge. As the modeling of the trailing edge has great influence onto the numeric results it is fairly possible that already little inaccuracy in the numeric model can result in false predictions of the flow field and the pressure distribution.

1.2 Existing Test Rigs - Capabilities of the New Test Rig

Fig. 1.1 shows the dependencies between the reduced frequency k and the Ma-number of the standard configurations. The reduced frequency is calculated by formula 1.1 (the nomenclature is taken from [BF86], c^* is the velocity of sound).

$$k = \frac{c\omega}{2v_{ref}} = \frac{c\omega}{2c^* Ma_{ref}} \tag{1.1}$$

It is important that there is a hyperbolic dependence between the Ma-number and the reduced frequency. In order to maintain an independent

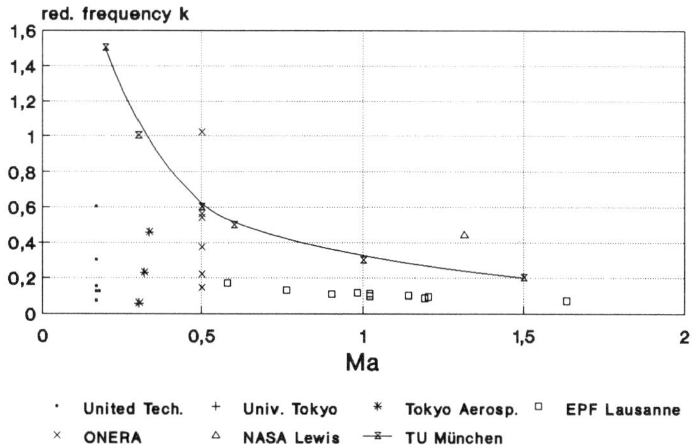

FIGURE 1.1. Standard Configurations 1 - 7 in the k-Ma-plane

setting of the reduced frequency k and the Ma-number it is therefore necessary to realize another degree of freedom.

Usually the supplemental degree of freedom is the frequency of vibration. Much more critical is the variation of the velocity of sound for reasons of condensation of humid air at higher Ma-numbers up to the problem of freezing.

The figure shows all aeroelastic test cases of the 7 experimental standard configurations. Some regions of interest can be determined in fig. 1.1. On one hand there is the region of shock-induced flutter at $Ma \approx 1$. Within this region one can find e.g. the test rig of the EPFL (standard configurations 4 and 6). The entries of the test rig of the EPFL show evidently the dependence between the reduced frequency and the Ma-number, the hyperbolic characteristic can easily be seen.

On the other hand there is the region of potential flutter. To achieve potential flutter it must be realized that the time for a single oscillation and the time a paricle needs to cross the blade row are of the same size. This region lies at high reduced frequencies. *Hashimoto* [Has84] reported of some investigations at high reduced frequencies at the meeting in Cambridge.

At all investigations done up to now the measurement of the dynamic pressure distribution has been taken. The pressure distribution is taken to calculate dynamic lift coefficients as a function of the chordwise coordinate and also as an integral value of the pressure and suction surface. Furthermore these values are combined to calculate an aeroelastic damping coefficient for the blade as the value of interest to indicate the possible onset of blade flutter.

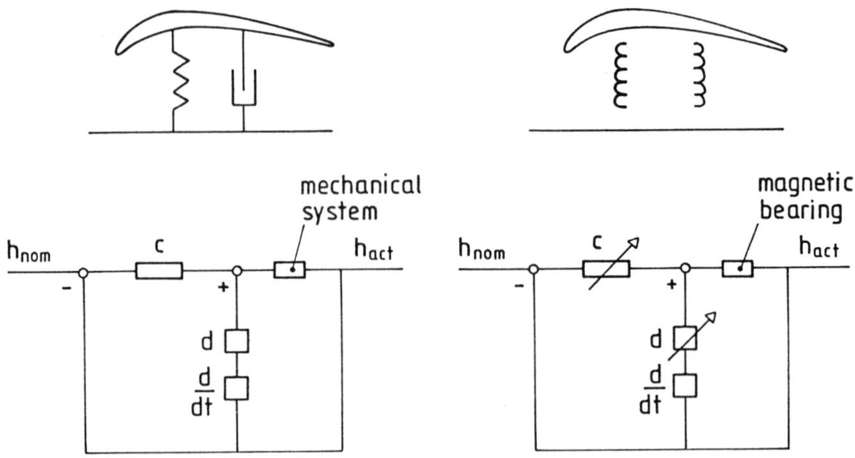

FIGURE 1.2. The magnetic bearing - Simulator - Automatic control

As the problem of flutter in general is not totally resolved today it seems interesting to investigate the behaviour of the dynamic flow characteristic around the vibrating blades. Therefore a new test rig was designed at the Technical University of Munich. This test rig allows experiments with a variable frequency of the vibrating blade. The possible range is shown in the k-Ma-plane. The main target of investigation will be the downstream behaviour. It is obvious that measurements of the dynamic pressure distribution are taken, too.

1.3 Mechanism for Blade Excitation - The Axial Magnetic Bearing

1.3.1 Functional Details

Several solutions are possible to force a blade to vibrate.

One approach is to use a spring-damping element. Supplemental excitation can be realized by some magnetic equipment. If the natural frequency should be variable it is necessary to use supplemental masses, located at some kind of intertia bar. By variing the masses the natural frequency can be altered due to formula 1.2. With the exception of the excitation unit such a system is mechanically controlled, stiffness and damping coefficients are fixed.

$$f = \frac{1}{2\pi}\sqrt{\frac{c}{m}} \qquad (1.2)$$

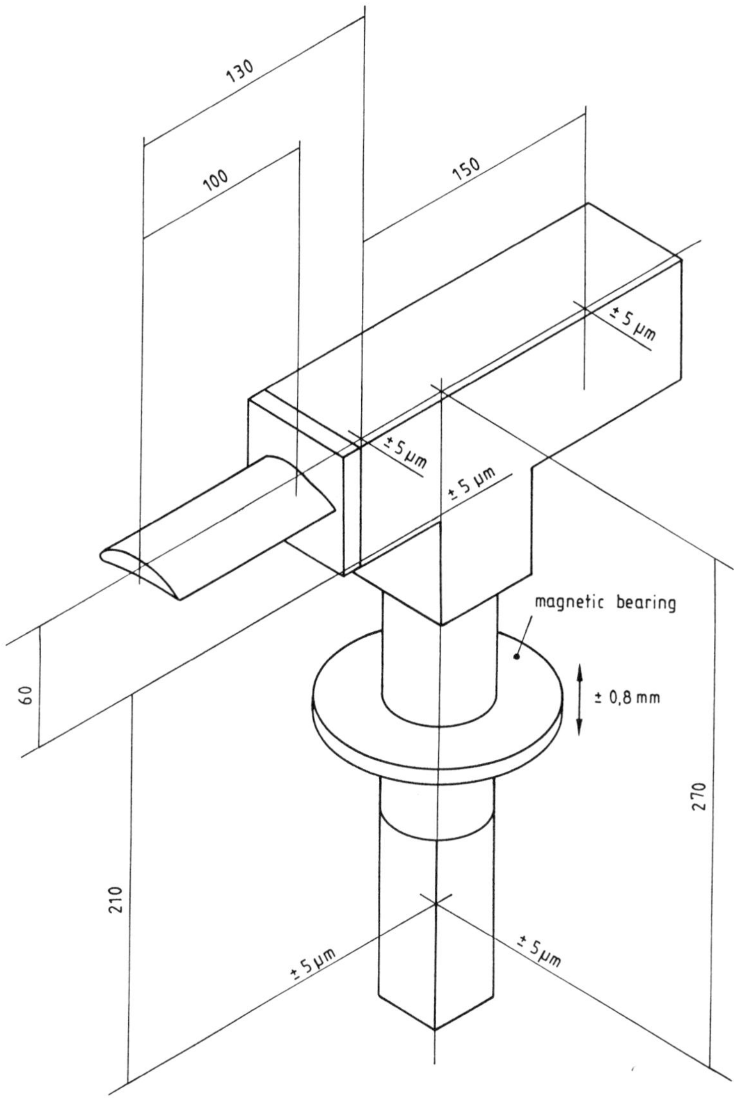

FIGURE 1.3. Functional design of the excitation unit

Such a solution is implemented at the *ONERA* [SF81].

Another possibility is the usage of electro-dynamic shakers as well for the excitation as for the fixing. A variable frequency is quite simple to obtain by altering the frequency of the shaker.

A setup of that kind was used by *Hashimoto* [Has84]. His experimental setup used shakers and a phase-shifter to realize the excitation, but there was no closed loop for the control of the blade position. He reached frequencies up to 100 Hz but only velocities up to 35 m/s. The reduced frequency range was up to 4 at a velocity of 7 m/s and below 0,9 for 35 m/s, a maximum shake force of 20 N and an amplitude up to 0,5 mm.

The new test rig maintains excitations by use of an axial magnetic bearing (fig. 1.2). This implies vibration only in bending mode. The gap between the traction disk and the coils is 0,8 mm to each side in the middle. Therefore an amplitude of 0,5 mm can be realized easily in the linear region of the magnetic bearing. The maximum force at 100 Hz is at least 1000 N. The magnetic bearing is restricted to frequencies below 1000 Hz.

The advantage of a magnetic bearing against the spring-damping element and the electro-dynamic shaker is that the stiffness and damping coefficients can be altered in a wide range. The dynamic position of the blade can be governed by an automatic control system. The automatic control system needs just 2 potentiometers to select stiffness and damping values and one entry for the blade position. Such two experimental configurations are possible. The first case is controlled vibration by adjusting *maximum stiffness and zero damping by sinusoidal input position of the blade.* The second case is *adjusting stiffness and damping to certain values and constant middle position of the blade.* By variing the damping coefficient it should be possible to measure directly the aerodynamic damping coefficient.

1.3.2 Details of Design

Within the magnetic bearing there is no fixing of the blade. The fixing must be done otherwise. As it is necessary to avoid secondary motions the blade vibration unit must be beared with gaps as small as possible. Therefore air bearings with a gap of typically 5 μm are used (fig. 1.3). With a distance between the bearings from 150 mm to 270 mm the bearing gap restricts a maximum secondary vibration to a maximum amplitude possible of 5 μm in every direction, that is 1 % of the amplitude of the vibration Taking into account the stiffness of the air bearing at a value of about 80 N/μm the real amplitudes of the secondary motion will be even smaller. The mass of the vibrated parts is 6 kg.

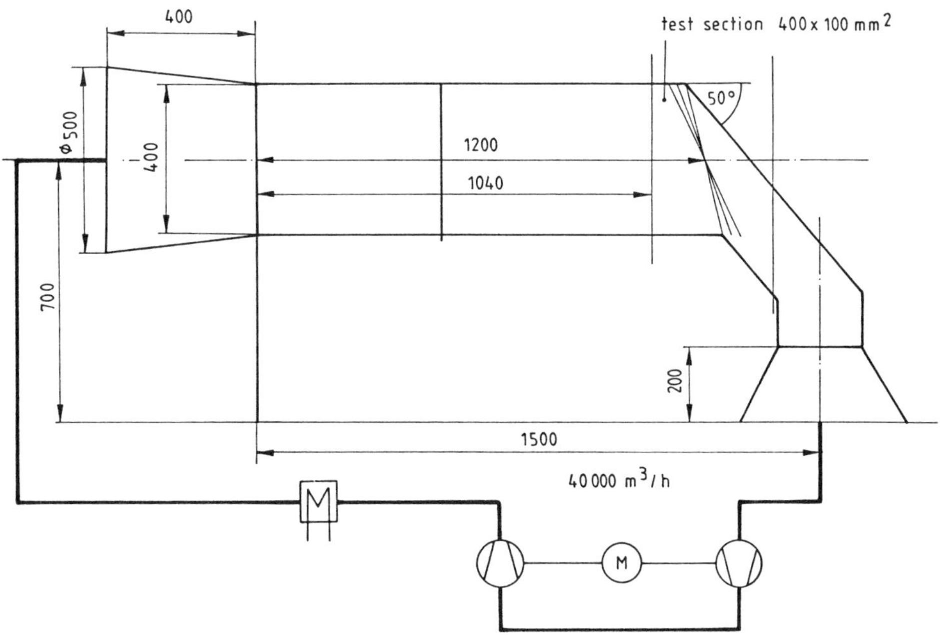

FIGURE 1.4. Geometric conditions for the design of the test section

1.4 The Wind Channel

The wind channel used for the experiments is a closed channel with an integrated cooler with water as the cooling medium. Regulation of the water mass flow keeps the air temperature at constant values. For the air supply of the test rig there is used a 2-stage compressor with included measurement of mass- and volume flow by diaphragm, thermoelement and pressure transducers. The maximum volume flow is about 40.000 m^3/h at a maximum pressure ratio of 3. The maximum power consumption is about 1,7 MW. The compressor is used for experiments in annular cascades and rotating turbines as well as in the straight wind channel.

The dimensions of the test section are shown in fig. 1.4. The duct upstream is orientated horicontally. At the inlet of the test section there is a nozzle with a diameter of 550 mm at the inlet and a square of 400 × 100 mm^2 at the outlet. The following flange contains a grid to straighten the fluid flow. The great diameter of the pipes avoids large pressure losses due to high velocities. The flange is followed by a calmation section of 1000 mm. This leads into the section where the grid is arranged. The mean deflection angle of the grid is 50 °, so the channel also is deflected by 50 ° downwards. As the downstream duct is vertically after the test section there is a second deflection of 40 °. A diffuser adapts the small dimensions of the test

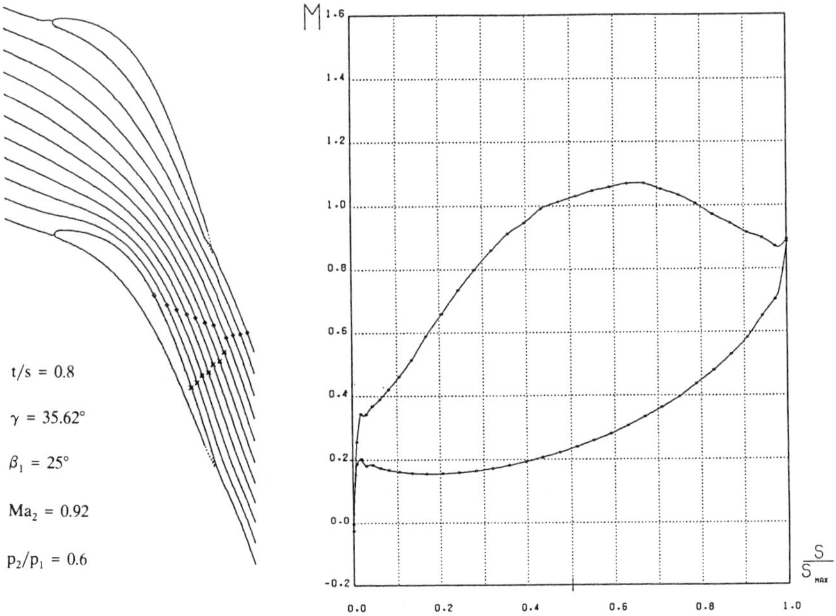

FIGURE 1.5. Ma-number distribution and streamlines of the ABB-blade

section onto the broad dimensions of the downstream duct. To aviod flow separations there are inserted flow straighteners into the flanges.

1.5 The Blade Characteristics - Design of Test Section

1.5.1 The Blade

The new final stage of the ABB steam turbine [BW91] shows transsonic conditions at midspan. The cut taken for the investigations in the 2-D channel is cut H, the 8th cut out of 14 (fig. 1.5). Although the real fluid flow is conical the cut is taken cylindrical.

The Ma-number upstream and downstream are both subsonic. At the suction side there is a small region with supersonic flow conditions in the trailing part of the blade. The natural frequency is about 112 Hz. The angle of attack chosen for the experiments is 20 ° but shall be varied to 15 ° and 25 ° . The stagger angle ist 35,6 ° . The pressure ratio at design conditions is 0,6. The pitch to chord ratio is greater than 1.0 in the real machine. As numeric calculations showed already beginning flow separation in 2-D calculation of the cylindrical cut for the nominal 3-D pitch to chord ratio the ratio was reduced to 0,8 for the experiments in 2-D.

1.5.2 The Grid

The test grid consists of 8 blades (fig. 1.6). Two of them are positioned at the top and the bottom boundary of the channel and are supplementally used for the termination of the test grid. The deflection of the fluid flow induces a lift force in opposition to the gravity. Such greater lift reduces the gravity the magnetic bearing has to compensate, leaving more force for the excitation of the blade.

Schlaefli [Sch89] showed that dynamic lift coefficients can be calculated by a superposition of complex influence coefficients between the blades. Furtheron he showed that for the determination of these coefficients it is sufficient to take into account two adjacent blades at each side.

Such it is possible, and realized, to take measurements with one single blade vibrated. The experimental data are taken at every blade subsequently. In that way already for a long time measurements are taken at the *ONERA* [SF81]. At the new test rig it is planned to use the same scheme, that means positioning the blade with the measuring equipment at various positions in the grid and repeating the measurement for all positions.

To avoid the influence of the boundary layer it is cut off at the walls of the channel right before the beginning of the grid. At one side wall there is located a window to control optically the behaviour of the vibrated blade and the fluid flow, too.

The periodicity condition for the blades is maintained by flaps in the upper and lower boundary section and by tailboards behind the final blades used as boundary of the test grid.

The angle of attack can be altered by using different adapters between the plate where the grid is fixed at the test section and the test section itself.

Furthermore it is planned to take measurements at different directions of the bending oscillation. To maintain this function there are two different positions to which the vibrating unit can be fixed. One provides a vibration in the direction of minimal stiffness of the blade, the other a vibration in the direction of the grid.

1.6 Data Acquisition and Evaluation

1.6.1 Data Acquisition

The data aquisition will be controled by a HP Series 300 computer. The measurement of the volume flow and the mass flow is realized - as mentioned above - by diaphragms, thermoelements and pressure transducers within the wind channel.

The mean values of the upstream and downstream flow will be taken from pressure tappings and thermocouples upstream and downstream of the test rig.

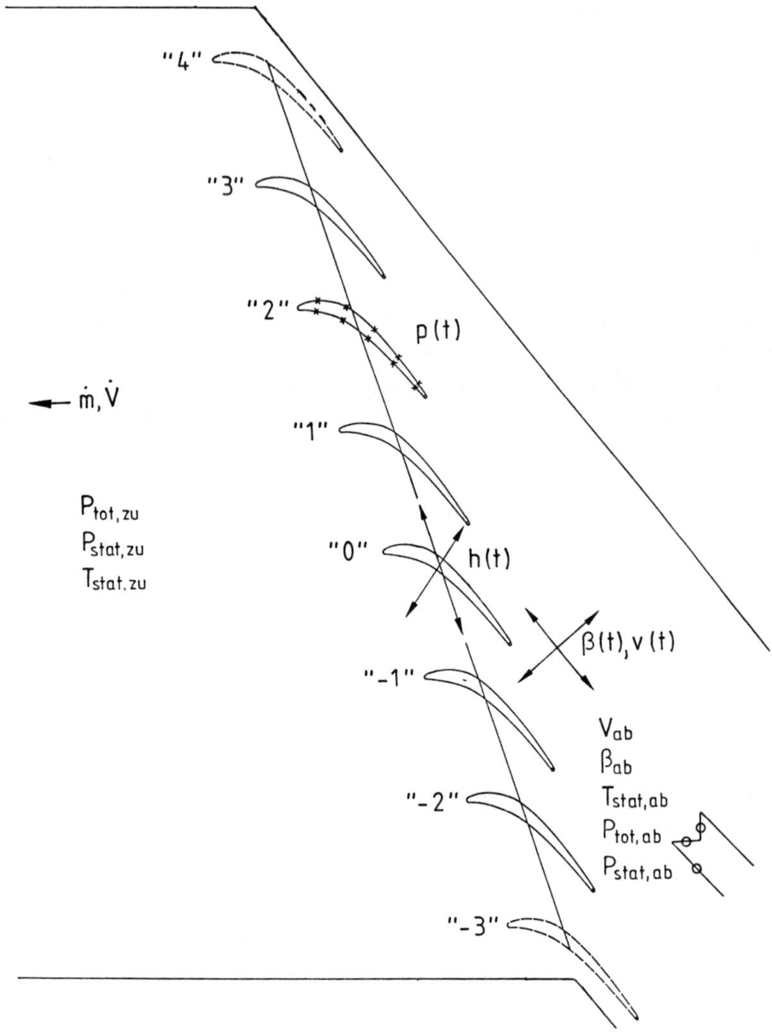

FIGURE 1.6. Design of the grid

Downstream of the test grid there is located a split film anemometer to measure the value and angle of the downstream conditions. As the split film wil be run at CTA conditions it is useable at high frequencies. The probe is traversable in both the grid direction and the mean flow direction. Such the flow field can be measured for every blade and at various distancies from the trailing edge. It will be used on one hand to maintain the periodicity condition of the fluid flow and on the other hand for the measurement of the dynamic downstream conditions.

The dynamic blade position is taken directly from the blade excitation and control unit.

The dynamic pressure transducers are KULITE transducers. As the blade thickness is only 1,5 mm in the trailing section type LQ-47 transducers are used. The thickness of these transducers is 0.64 mm; the signal transport is maintained by enamelled wire. Two blades will be instrumented with 5 transducers at each side, such in total 20 transducers are installed.

To change the position of an instrumented blade the compressor has to be shut down and the test section must be opened.

To minimize errors due to inexact reproduction of the general flow conditions one of the instrumented blades will remain at the position of the vibrating blade all the time. The dynamic measurement of the vibrated blade will be taken as supplemental reference for the general flow conditions. The second instrumented blade will be positioned at the different locations to take measurement for every fixed blade.

Each turn of the compressor will include various up- and downstrem conditions and vibrating frequencies. Measurements can be made only for a single blade at a time. Nevertheless the measurement of the velocity behind the trailing edge will be done for every blade. Such the dynamic downstream conditions will provide another reference to minimize the errors due to slightly different general flow conditions.

The dynamic data are stored in a 20-channel data memory. Such it is not possible to keep the data of all pressure tappings of the vibrating blade when that blade is taken just for reference, as already 3 channels are used for the measurement of the downstream flow and the blade position.

1.6.2 Data Evaluation

To secure fairly constant experimental conditions the general flow will be controled online in volume and mass flow including pressure and temperature conditions.

The dynamic data will be treated immediately after each experiment. Beginning with a fourier analysis of the blade vibration the dynamic pressure and velocity data will be reduced to the coefficients of the vibration at the blade frequency (fig. 1.7).

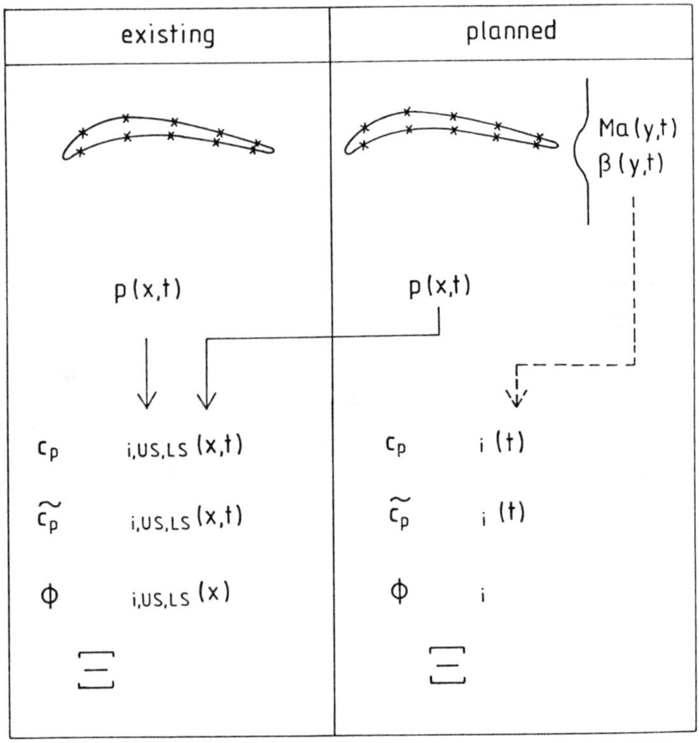

FIGURE 1.7. Evaluation of measured data

Classical Method

The pressure distribution will be used to calculate dynamic pressure coefficients as a function of the chordwise coordinate. Vectorial integration over the blade surface leads to the integral pressure coefficient and furthermore to the dynamic lift coefficient. This procedure is well-known.

Interpreting Downstream Velocities

Numeric calculations often fail for reasons of not totally correct formulations of the boundary conditions. This problem increases especially at numerically singular points. A typical example for such a singular point is the trailing edge of a blade. The diameter there is very small thus the numeric coordinate system usually has a sharp edge which leads to quickly changing values for the transformation from physical to numerical grid and viceversa. To keep the numeric solution stable it is necessary to implement some kind of boundary condition which controls the behaviour of the singular point. The effect is that the flow field puts its characteristic behaviour onto the boundary. Especially at the trailing edge the physical situation is the other way round. The sharp edge can impose e.g. flow separations which cannot be handled by the numeric formulation of the boundary condition as reason and effect are changed.

It is a basic idea of this test rig to correlate real downstream behaviour near the trailing edge with the well-known pressure distributions. Numeric methods then can update their boundary conditions at the trailing edge with the obtained velocity data.

A second idea is the transformation of the velocity data into vortices leaving the blade at the trailing edge. This theory is not yet fully developed and will be presented later, together with results obtained from the new test rig.

1.7 Conclusion

The basic ideas and principles concerning the setup and the design of the new test rig at the Technical University of Munich were presented. The main theme of interest is to find or validate influence coefficients for oscillating cascades by means of analyzing the dynamic fluid flow. For reasons of compatibility there are taken the traditional measurements of the dynamic pressure distribution, too.

References

[Bd50] Ch. Bellenot and J. Lalive d'Epinay. Selbsterregte Schaufelschwingungen in Turbomaschinen. *Brown Boveri Mitteilungen*, 368ff, Oktober 1950.

[BF86] A. Boelcs and T.H. Fransson. *Aeroelasticity in Turbomachines - Comparison of theoretical and experimental cascade results.* L'Ecole Polytechnique Federale de Lausanne - L'Institut de Thermique Appliquee, 1986.

[BW91] J. Buetikofer and U. Wieland. Moderne Niederdruck-Dampfturnine. *VGB Kraftwerkstechnik*, 71(4):341ff, 1991.

[Has84] M. Hashimoto. An experimental study on the aerodynamic damping forces acting on cascade blades in a bending motion. In *Unsteady Aerodynamics in Turbomachines and Propellers*, page 459ff, 1984.

[Sch89] D. Schlaefli. *Experimentelle Untersuchung der instationaeren Stroemung in oszil/-lierenden Ring/-gittern.* L'Ecole Polytechnique Federale de Lausanne - L'Institut de Thermique Appliquee, 1989.

[SF81] E. Szechenyi and R. Finas. Aeroelastic testing in a straight cascade wind-tunnel. In *Aeroelasticity in Turbomachines*, page 143ff, 1981.

VI. Aeroacoustics

Aeroacoustics in Turbomachines and Propellers — Future Research Needs

Philip R. Gliebe[1]

1.1 Introduction

The aeroacoustic design and development of turbomachinery components in aircraft engine propulsion systems are important steps in the total propulsion system development process. Engines for subsonic transport aircraft are becoming increasingly more complex and sophisticated as manufacturers strive to provide products which are more fuel-efficient, lighter in weight, more reliable and maintainable, and produce less noise, all-the-while being competitively priced.

The design of an aircraft engine to be quiet is, by nature, a "necessary evil", because it adds no performance value to the product. To the contrary, current state-of-the-art methods for reducing engine noise often add weight, reduce performance and increase complexity of the propulsion system. On the other hand, certification of an aircraft requires meeting prevailing noise regulations, such as the U.S. FAR36 Stage 3 rules, and more airline operators are requesting noise level guarantees which are significantly more stringent than these rules, in order to meet tougher local airport noise requirements. It is felt that the requirements and customer demands for lower aircraft noise levels will only become more stringent in the next 10 to 20 years.

The currently-used methods and procedures for designing engines that meet or beat prevailing noise standards make full use of the current understanding of noise generation and suppression processes in modern bypass turbofan engines. The prospects for achieving significant noise reduction beyond today's modern turbofan levels using existing knowledge and methods is felt to be pretty small, without imposing a significant penalty to the propulsion system weight, size, performance and cost. Figure 1 illustrates the noise level trends that have prevailed since the advent of jet propulsion for commercial service. The trend shows that further noise reductions are

[1]GE Aircraft Engines, 1 Neumann Way, M/D A304, Cincinnati, Ohio 45215

going to be more difficult to achieve, and will require new, innovative technology advances to achieve them. The amount of reductions needed will depend on how much more stringent future noise rules become.

As new engines are being designed with higher bypass ratios, turbomachinery noise (fan, compressor, turbine) becomes more important, and jet exhaust mixing noise contributes less and less to the total engine noise signature. This is illustrated qualitatively in figure 2. The GE-designed Unducted Fan Engine noise signature was totally controlled by fan noise, having a bypass ratio of about 36:1. New ducted-fan engines with bypass ratios greater than 10:1 are being studied which will also have little or no jet noise contribution. Hence, progress in noise reduction of future commercial propulsion systems is dependent on a better understanding of turbomachinery noise, and on novel methods for reducing it which do not carry the penalties associated with today's noise reduction methods.

This paper reviews the current understanding of turbomachinery noise generation mechanisms and the currently-favored procedures and methods for reducing and suppressing noise. An assessment is given as to where the current problems lie in achieving "cost-effective" noise reduction for turbomachinery components of a typical modern, high-bypass turbofan engine. An overview of the unsteady flow processes in turbomachinery stages which produce noise is given, and thoughts on where quantitative knowledge of these processes may be lacking are offered. Finally, recommendations are made for research and development avenues which could provide potential noise reductions beyond that currently achievable with today's technology.

1.2 Noise Generation Mechanisms Overview

Since the 1960's, it has been recognized that unsteady flow processes are responsible for many of the noise-generating mechanisms in turbomachinery, and in aircraft engine turbines, compressors and fans in particular. Early progress in identifying and quantifying unsteady flow-related noise generation, e.g., Tyler and Sofrin [1] and Lowson [2] dealt with compressor noise, a significant noise source for early turbojets and low-bypass engines. The fundamental physics of noise generation uncovered in these early investigations have relevance to modern high-bypass turbofan engine turbomachinery components as well.

Much work has been done in developing analytical models and experimental information on noise generation mechanisms in a turbofan propulsion system. Figure 3 illustrates several of the primary noise sources in a typical engine, with more specific identification given to the fan assembly sources. For a modern, high-bypass turbofan engine, the primary turbomachinery sources to consider are the fan, the low-pressure or "booster" compressor, and the low pressure drive turbine.

Circumferential non-uniformities in the fan inflow, caused by geometric asymmetry (droop), by inlet angle-of-attack, and by inflow turbulence and vortex ingestion, produce excess fan noise. This is primarily a result of the fan blades experiencing fluctuations in loading or lift as they rotate through the circumferentially nonuniform flow or "gusts". The degree to which these inflow-induced unsteady lift fluctuations produce noise is of course dependent on how efficiently the resulting pressure fluctuations couple with the inlet duct and exhaust duct to produce propagating acoustic modes [1]. Published analysis and experimental investigations of these fan excess noise mechanisms include those of Hanson on inflow turbulence [3], Mani on inflow distortion and turbulence [4, 5], and Smith et al [6] on inlet geometric asymmetry effects.

Rotor-Stator interactions are a primary source of turbomachinery noise, as pointed out in figure 3. Fan rotor-stator interaction with bypass duct Outlet Guide Vanes (OGV's) as well as with core duct stator vanes are primary sources. Rotor-Stator interactions produced by the first stages of the low pressure compressor (LPC) and by the last stages of the low pressure turbine (LPT) are also potentially significant sources of noise. References [7, 8, 9, 10] document early investigations of rotor-stator (or stator-rotor) interaction modelling and experimental measurements which identify and attempt to quantify the role of unsteady blade row interaction flow phenomena on turbomachinery noise.

Another potentially significant noise generation mechanism involving unsteady flow phenomena is that of unsteady fan blade lift or load fluctuations produced as a result of imposed circumferentially non-uniform pressure fields by downstream support struts, service struts and mounting pylons. This "backpressure distortion-caused" lift fluctuation on the fan blades usually results in additional blade-passing frequency noise, and usually becomes most noticeable at transonic tip speeds. An experimental investigation of this mechanism was published by Ho in reference [11]. This mechanism involves rotor blades rotating through a stationary, fixed, but circumferentially varying, pressure field, in contrast to the "convected gust" interactions associated with inflow distortions and blade or vane wake interactions.

1.3 Current Noise Control and Reduction Methods

Noise control and reduction methods used in modern high-bypass turbofan engines usually are a combination of source control methods and source suppression methods, as illustrated in figure 4. The turbomachinery interaction-caused noise sources are controlled by increasing axial spacings between adjacent blade rows, to reduce the "gust" strengths, and by selecting blade and vane number combinations to produce "cut-off", whereby the highest-energy acoustic modes decay exponentially with distance along the ducts. "Cut-off" as a noise reduction concept is discussed in reference [1], and has been applied to fan rotor-OGV interaction, fan-core stator IGV interaction, LP compressor stage interaction, and LP turbine interaction, to eliminate blade-passing frequency tones at selected operating conditions.

Providing adequate axial spacing between rotor and stator rows in order to minimize interaction effects is not without penalties, because this adds length, and therefore weight, to the engine. It can also have a small, but non-trivial, negative effect on aerodynamic performance. General practice is that axial spacing for fan-OGV interaction is fairly generous (and effective), but LPC and LPT stage spacings tend to be small relative to axial chord, on the order of 10 to 30%.

Acoustic treatment or sound suppression liners are utilized to suppress the turbomachinery noise sources, almost always in the fan inlet and exhaust ducts, and sometimes in the core exhaust duct. Inlet treatment has very little effect on LPC noise because of the distant proximity to the noise sources, so LPC noise is best controlled by proper selection of blade and vane numbers.

1.4 Future High Bypass Engine Noise Trends

As illustrated in figure 2, subsonic transport engines for the 21st century are forecast to have significantly higher bypass ratios, in order to realize the potential fuel consumption savings provided by the improved propulsive efficiency that higher bypass ratio offers. In the extreme, the GE Unducted Fan Engine, with a bypass ratio of 35 to 40:1, offered a potential 20 to 30% fuel savings, compared to today's engines.

Considering ducted fans for the moment, the fan noise, as illustrated qualitatively in figure 2, will dominate the noise signature above bypass ratios of 8:1 or so. For a given thrust size/requirement, increasing bypass ratio implies increasing fan diameter and reducing fan pressure ratio. Increasing fan diameter to accommodate higher bypass ratio therefore means higher engine (and aircraft) weight, thus offsetting some of the fuel savings advantage. It is currently thought that nacelle length, for example, should not scale with diameter, but should remain at the current bypass ratio size, and therefore inlet duct and exhaust duct length/diameter ratios will be smaller. Current liner design performance models therefore suggest less liner noise suppression will be realized.

As the fan diameter becomes larger, the blades and vanes become larger, and the maintenance of "acoustically-desired" axial spacings becomes a more severe weight penalty. Further, the trend toward low blade number, wide-chord fan designs poses a further threat to maintaining large spacing/chord ratios. Larger diameters, lower blade numbers, and lower tip speeds all produce lower blade-passing frequencies, which require deeper treatment liners to achieve comparable suppression. But large-diameter nacelles need to be thin to provide low drag, and so it can be expected that treatment liners may provide significantly less source suppression using current liner design methods.

All of the above suggests that higher bypass ratio engines will require more source noise control by design, rather than depending too heavily on liner suppression, to even maintain current engine noise levels, let alone reduce noise levels significantly below those of current engines.

1.5 Unsteady Flow — Generated Turbomachinery Noise Mechanisms

For the majority of cases, unsteady interactions in turbomachinery components (fans, LP compressors, LP turbines) can be characterized by a process chart like that shown in figure 5. For a given blade row, whether it be a rotor or a stator, we have a source of excitation or gust, such as upstream wakes, inflow distortions and turbulence, and downstream potential field distortions from adjacent blade rows, struts and/or pylons, etc. The "gust" is caused by relative motion of the blade row. The blade row responds to the gust, producing fluctuating (usually periodic) lift and drag forces, which can be resolved into axial and tangential forces. These fluctuating force components then couple to the duct to produce acoustic energy. The efficiency of conversion to acoustic energy is a function of the flow Mach number, normalized frequency, and mode number (number of generated pressure lobes per revolution).

The control of noise generated at the source can be done by controlling the gust amplitude and/or phase, by controlling the response to the gust (amplitude or phase), and by controlling the efficiency of conversion of the unsteady force energy into acoustic energy. "Significant" acoustic energy can be a tiny fraction of the total flow energy, and so we are dealing with being able to quantify and control extremely small energy fluctuations. For example, the total acoustic power of the prototype UDF®* engine configuration was estimated from scale model tests, and is shown in figure 6. It can be seen that the total acoustic power is less than 0.1% of the shaft horsepower of the fan over most of the operating range. See reference [12].

Considering the specific example of rotor-stator interactions, several interaction mechanisms which can produce noise have been identified. Figure 7 illustrates four of these. Mechanism 1 in figure 7 is that of the downstream stator potential (pressure) field interacting with, and causing unsteady lift fluctuations on, the upstream rotor, designated SPF-R interaction. In this case, the rotor, passing through the stator potential field pressure fluctu-

---

*UDF® is a registered trademark of General Electric Company, U.S.A.

ations , undergoes unsteady lift fluctuations, and is the source of noise. The stator could be replaced by struts and/or pylons, and the physical mechanism would be the same.

Mechanism 2 in figure 7 is that of the upstream rotor potential (pressure) field passing by the stator row, causing unsteady lift fluctuations on the stator vanes. In this case, the stator is the source of noise. This mechanism is designated RPF-S.

Mechanism 3 of figure 7 is a consequence of mechanism 1. Because of the lift fluctuations on the rotor caused by the rotor passing through the stator potential field, the rotor blades shed vorticity. This unsteady vortex shedding convects downstream, producing unsteady gusts over the stator vanes. The stator vanes undergo fluctuations in lift, and generate noise. This mechanism is designated RV-S.

Mechanism 4 shown in figure 7 is that of viscous wakes shed from the rotor blades convecting downstream and producing fluctuating lift on the stator vanes. This mechanism, designated RW-S, is the one which has received the most attention in the literature, and is widely believed to be the most important.

The first two mechanisms (potential field interactions) are thought to be small for spacing-to-chord ratios of 0.5 or greater. However, in multistage compressors and turbines, axial spacings are quite often less than half of the "gust-making" blade row chord, and therefore this mechanism can be significant. In fact, it is possible that this mechanism can even dominate viscous wake-generated noise for spacings of 20% or less. The potential field amplitudes are a function of blade (or vane) loading, and to some extent blade thickness , and their axial propagation upstream and downstream depend on flow Mach number. Very little is known about the potential field interactions of high-camber, high-loading, high Mach number turbine blades and vanes, for example, and the relative importance of viscous wake interactions vs. potential field interactions in generating noise.

The third mechanism, RV-S, is the result of another interaction, SPF-S, and therefore tends to be an order-of-magnitude smaller in amplitude than its parent interaction. The fourth mechanism, RW-S, is the dominant interaction mechanism for fan stages. The mechanism also embodies rotor tip-vortex interaction with a downstream stator. For small tip clearances, this interaction is thought to be small compared to the viscous wake interactions. For the unducted fan, however, it is a dominant noise-generation mechanism. See references [12] and [14] for experimental study results on ducted and unducted fan interaction mechanisms, respectively.

Experimental studies of rotor-stator interaction sensitivity to axial spacing and vane number, reported in reference [12], are summarized in figure 8. Here, the first three fan tone harmonic acoustic power levels are shown as a function of axial spacing normalized by rotor tip chord. Increasing spacing beyond 1.5-2.0 tip chords yields little additional noise reduction. Also, the "cut-off" vane number trends show little effect of spacing on the fundamental blade-passage frequency tone, which suggests that this tone is due to some other source such as inflow distortions.

An example of pressure-field induced unsteady response is shown in figure 9, taken from reference [12]. Here, the influence of downstream strut potential field distortions on rotor unsteady lift was measured using dynamic pressure transducers mounted on the rotor blade surfaces. Again, the rotor lift response decays as the rotor-to-pylon spacing increases, leveling out beyond about 5 to 6 strut widths.

1.6 Unsteady Flow-Generated Propeller Noise Mechanisms

The previous sections discussed noise generation mechanisms in ducted turbomachinery. Much of what has been said there applies in principle to unducted turbomachines, e.g., propellers and unducted fans. The following paragraphs discuss some of the unique features of unducted rotors, and the focus will be on counter-rotating unducted rotors in particular, as that is where the bulk of GE research and development on unducted fans has been focused.

A cross section sketch of the first prototype Unducted fan engine is shown in figure 10. The primary mechanisms for noise generation on an unducted, counter-rotating fan are shown in figure 11. As can be seen, the interaction mechanisms are similar to those of a ducted fan (figure 3), and include inflow distortion interactions , viscous wake and potential field interactions, and tip vortex interactions.

A distinctive feature of the unducted fan is that the preferential selection of propagating and non-propagating modes, i.e., duct acoustic coupling, no longer applies, and so the concept of cut-off cannot be employed. Further, no duct means no acoustic liners, so all noise control has to be at the source.

A scale model Propulsion simulator was designed, built and tested extensively to provide design data and test design concepts for the counter-rotating fan. A picture of the wind-tunnel version is shown in figure 12. This vehicle was used to test and gather data on over 100 different fan configurations, which included many different blade shapes, axial spacing variations and installation effects studies. The aerodynamic performance characteristics, blade aeroelastic behavior, and acoustic characteristics were extensively evaluated during the various test programs. A special vertical-mount version of this propulsion simulator was designed and built for the GE Aircraft Engines Anechoic Free-jet Wind Tunnel Facility (Cell 41), as illustrated in figure 13.

As has been discussed extensively in the published literature (see references [15, 16, 17] , for example), counter-rotating propellers (or unducted fans) produce rotor self-noise, a result of thickness and steady loading, similar to that of a conventional, single-rotation propeller. In addition, rotor-rotor interaction, caused by forward rotor wakes, tip vortices and potential field exciting the aft rotor, and by the aft rotor potential field exciting the forward rotor, contributes significantly to the total noise signature.

Both rotor self noise and unsteady interaction noise are functions of rotor blade numbers, rotor disk loading (e.g., thrust or shaft horsepower per unit annulus area), rotor-to-rotor spacing, blade planform shape, and planform area relative to annulus area. Some typical noise sensitivity trends obtained from analytical parametric studies and model tests are shown in figure 14. The sensitivity to axial spacing between rotors is a good indicator of the total contribution of unsteady interaction mechanisms to the total noise.

A unique feature of the scale model propulsion simulator for the unducted fan was the ability to independently control forward and aft rotor speeds, and test different blade numbers on the forward and aft rotors. Because of this flexibility, each rotor has its own blade-passing frequency and harmonics. In addition, the unsteady interaction tones occur at the sums of integer multiples of each rotor blade-passage harmonic. Hence a narrowband spectrum of the farfield noise can show the contributions of each rotor self-noise component and of each interaction mode component as individual tones, as illustrated by the example in figure 15.

By analyzing measured narrowband noise spectra from different unducted fan configuration tests, we were able to sort out the relative contributions of unsteady interaction-generated noise and "steady" self-noise from each rotor. Some examples of this decomposition are shown in figures 16, 17, 18 and 19. These figures display the mean-square sound pressure sums, expressed in decibels, for the "steady" rotor self-noise, and the unsteady interaction tones, separately. Figures 16a, 16b and 16c show the unsteady tone sums and "steady" rotor self-noise tone sums for an unducted fan with 9 forward rotor blades and 8 aft rotor blades, in the forward quadrant, plane-of-rotation, and aft quadrant, respectively. The tone sum trends with forward-rotor tip speed are shown.

The results shown in figures 16a-c show how unsteady interactions are dominant at low tip speeds, and have greater dominance in the forward and aft quadrants away from the plane of rotation. In the plane of rotation, the rotor self-noise becomes controlling above about 750 ft/sec tip speed.

Figures 17a, b and c show the tone sum directivity patterns for the same results of figure 16 at low (610 ft/sec), intermediate (750 ft/sec) and high 880 ft/sec) tip speeds, respectively. These results show how the unsteady interaction noise has a much flatter directivity pattern than the rotor self-noise, and how the rotor self-noise increases much more rapidly with tip speed than does the unsteady interaction noise.

From the parametric trend study results (figure 14), and the source decomposition results such as those of figures 16 and 17, it was estimated that significant noise reductions could be obtained by increasing blade numbers and axial spacing between rotors. Figure 18 shows experimental results obtained for a configuration with the blade numbers increased from (8+8)-blades to (11+9)-blades and axial spacing increased from 0.17 diameters to 0.24 diameters. The results are shown for plane-of-rotation noise levels, and are plotted vs. thrust, since increasing blade numbers produces comparable thrust at a lower tip speed. These results demonstrated a 2 to 7 dB reduction in tone sum levels for both rotor self-noise and unsteady interaction noise.

As illustrated in figure 11, another important unsteady interaction mechanism identified is that of the pylon wake convecting into the rotor(s). Again by measuring narrowband spectra (figure 15) with and without a mounting pylon in place, and using a configuration with a differential blade number, the effect of this interaction mechanism on both rotor self-noise and rotor-to-rotor interaction noise can be deduced.

Figure 19 shows the results of just such a decomposition analysis for the (11+9)-blade configuration of figure 18. From this figure it can be seen that pylon-rotor interaction (which occurs at the same frequencies as rotor self-noise) has its greatest effect at the lowest thrust (and hence lowest tip speed), and adds to the steady loading and thickness components of the rotor self-noise sums. It has no appreciable effect on unsteady interaction noise. The deduced pylon-rotor interaction component trend is also shown in this figure, and it tends to level off as thrust is increased (although the accuracy of subtracting two approximately equal decibel levels can be questioned for the highest thrust data points). It was also observed, although not obvious from figure 19, that the pylon wake-rotor unsteady interaction generated noise was primarily a forward rotor blade-passage frequency effect. No appreciable increase in higher harmonics or aft rotor blade-passage harmonics was observed. This suggests that the forward rotor acts as a filter to smooth out the pylon wake before its interaction with the second rotor.

In summary, the relative contributions of various unsteady interaction mechanisms to the tone noise of counter-rotating propellers has been quantified through far-field narrowband acoustic measurements, using configurations and operating conditions which displace the interaction tone frequencies relative to the rotor self-noise frequencies. This diagnostic decomposition process is useful in providing noise reduction design guidance, validating analytical models, and providing insight into the relative contributions of the various mechanisms to the total noise signature. Much of this insight can hopefully be transferred to ducted fan applications.

1.7 Recommendations for Future Research

It has been made clear that unsteady flow interactions in turbomachinery play a major role in the noise produced by aircraft engines. Fan, LP compressor, and LP turbine tones are primarily a result of these interactions.

For ducted rotors operating at supersonic tip speeds, and unducted rotors operating at any tip speed, rotor self-noise is also a major contributor.

The unsteady interactions which produce noise can be described in term of a process made up of building blocks or modules. These are:

1. Gust description
2. Blade row response description
3. Unsteady Force energy acoustic efficiency description
    (a) Ducted
    (b) Unducted

Each of the above building blocks need more accurate, more general descriptions, in the form of better analytical models, broader experimental data bases, and/or numerical simulations using Computational Fluid Dynamics (CFD) techniques.

As suggested in a previous section, the end result (noise) is a very small fraction of the total system energy. Further, the ingredients (gust, response, acoustic coupling) all deal with quantification of small variations of large values. Hence accurate modelling of each of the component ingredients and assembling them in a delicate balance is required in order to arrive at an accurate prediction of the final outcome.

Some specific recommendations for future research include the following:

1. Develop better models and descriptions of rotor wakes and vortex structure for fans and turbines.
2. Develop and verify better models of fan rotor and stator response to convected gusts — including blade/vane solidity, loading and camber effects.
3. Develop and verify better models of fan rotor response to potential field gusts.
4. Investigate the effects of blade vibratory motion on noise radiation.
5. Extend current lift response models to account for 3-dimensional effects — see reference [19].
6. Develop diagnostic mechanism decomposition techniques for ducted fans similar to those described herein for unducted (counter-rotating) fans.
7. Investigate the unsteady interactions caused by random/turbulent sources.

Very little has been quantified on the generation of broadband noise. It has been estimated that if one were to completely remove all fan tones from a typical turbofan noise spectrum, the effective perceived noise level metric (EPNL) would be reduced by only 2 EPNdB. Broadband noise reduction is the key to further progress in quieting aircraft engines.

REFERENCES

[1] Tyler, J.M., and Sofrin, T.G., "Axial Flow Compressor Noise Studies", *SAE Journal*, Vol. 70, pp. 309–332 (1962).

[2] Lowson, M.V., "Theoretical Analysis of Compressor Noise", *Journal of Acoustical Society of America*, Vol. 47, No. 1, 1970, pp. 371–385.

[3] Hanson, D.B., "Spectrum of Rotor Noise Caused by Atmospheric Turbulence", *Journal of Acoustical Society of America*, Vol. 56, No. 1, July 1974, pp. 110–126.

[4] Mani, R., "Noise Due to Interaction of Inlet Turbulence with Isolated Stators and Rotors", *Journal of Sound and Vibration*, Vol. 17, No. 2, 1971, pp. 251–260.

[5] Mani, R., "Isolated Rotor Noise Due to Inlet Distortion or Turbulence", NASA Contractor Report No. CR-2479, October 1974.

[6] Smith, E.B., Moore, M.T., and Gliebe, P.R., "Distortion — Rotor Interaction Noise Produced by a Drooped Inlet", AIAA paper No. 80-1050, 6th Aero Acoustics Conference, Hartford, Ct., June 1980.

[7] Hanson, D.B., "Spectrum of Rotor Noise Caused by Inlet Guide Vane Wakes", *Journal of Acoustical Society of America*, Vol. 55, No. 6, June 1974, pp. 1247–1251.

[8] Hanson, D.B., "Unified Analysis of Fan Stator Noise", *Journal of Acoustical Society of America*, Vol. 54, No. 6, 1973, pp. 1571–1591.

[9] Mugridge, B.D., and Morfey, C.L., "Sources of Noise in Axial Flow Fans", *Journal of Acoustical Society of America*, Vol. 51, No. 5, 1972, pp. 1412–1426.

[10] Fleeter, S., "Discrete Frequency Noise Reduction Modeling for Application to Fanjet Engines", *Journal of Acoustical Society of America*, Vol. 68, No. 3, 1980, pp. 957–965.

[11] Ho. P.Y., "The Effect of Vane-Frame Design on Rotor-Stator Interaction Noise", AIAA paper No. 81-2034, 7th Aeroacoustic Conference, Palo Alto, Calif., October 1981.

[12] Kantola, R.A., and Gliebe, P.R., "Effects of Vane/Blade Ratio and Spacing on Fan Noise", AIAA paper No. 81-2033, AIAA 7th Aeroacoustics Conference, Palo Alto, Calif., Oct. 1981.

[13] Janardan, B.A., and Gliebe, P.R., "Acoustic Power Level Comparisons of Model-Scale Counter-rotating Unducted Fans", AIAA paper No. 91-0595, AIAA 29th Aerospace Sciences Meeting, Reno Nevada, January 1991.

[14] Janardan, B.A., and Gliebe, P.R., "Acoustic characteristics of Counter-rotating Unducted Fans from Model Scale Tests", AIAA Journal of Aircraft, Vol. 27, No. 3, March 1990, pp. 268–275.

[15] Hanson, D.B., "Noise of Counter Rotation Propellers", AIAA paper No. 84-2305, AIAA 9th Aeroacoustics Conference, Williamsburg, Va., October 1984.

[16] Parry, A.B., and Crighton, D.G., "Prediction of Counter-Rotation Propeller Noise", AIAA paper No. 89-1141, AIAA 12th Aeroacoustics Conference, San Antonio, Texas, April 1989.

[17] Whitfield, C.E., and Gliebe, P.R., "Predicted vs. Scale Model and Flight Test UDF® Engine Noise", AIAA paper No. 90-3936, AIAA Aeroacoustics Conference, Tallahassee, Fla., October 1990.

[18] Taylor, A.C., and Ng, W.F., "Analytical Prediction of the Unsteady Lift on a Rotor Caused by Downstream Struts", ASME paper No. 87-GT-145, June 1987.

[19] Attassi, H., and Hamad, G., "Sound Generated in a cascade by Three-dimensional disturbances Convected in a Subsonic Flow", AIAA paper No. 81-2046, 7th AIAA Aeroacoustics Conference, Palo Alto, Calif., October 1981.

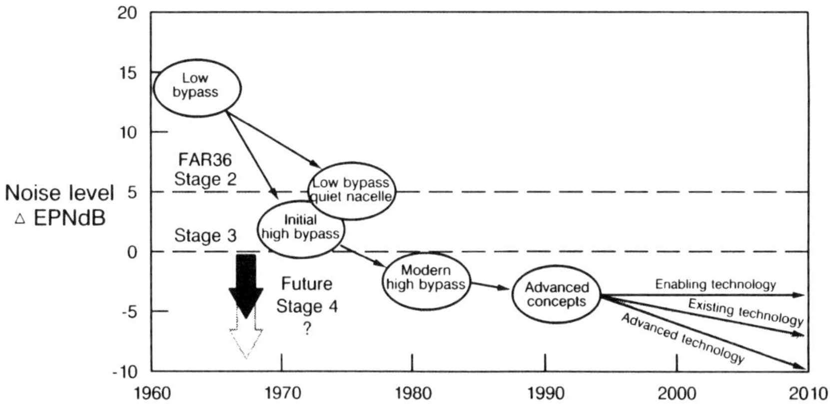

FIGURE 1 - AIRCRAFT NOISE TRENDS

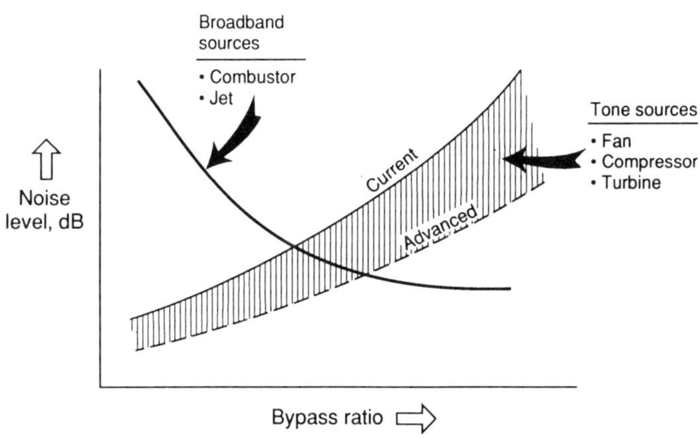

FIGURE 2 - RELATIVE IMPORTANCE OF ENGINE NOISE SOURCES

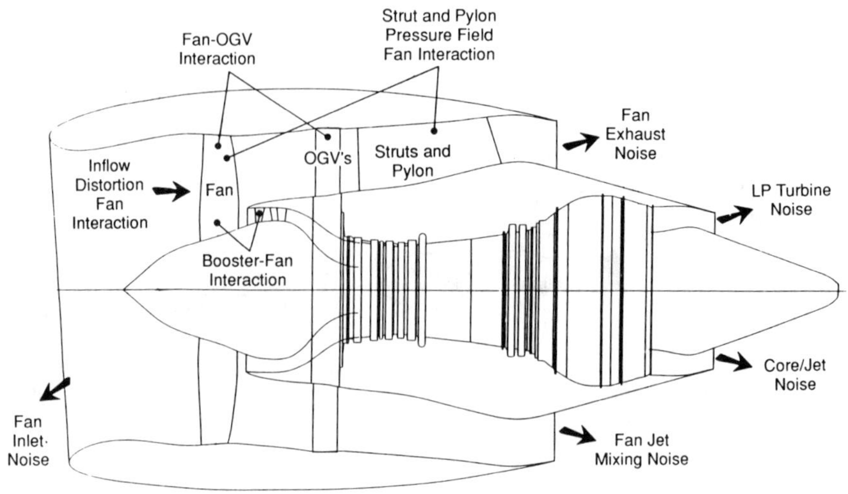

FIGURE 3 - TURBOFAN ENGINE NOISE GENERATION MECHANISMS

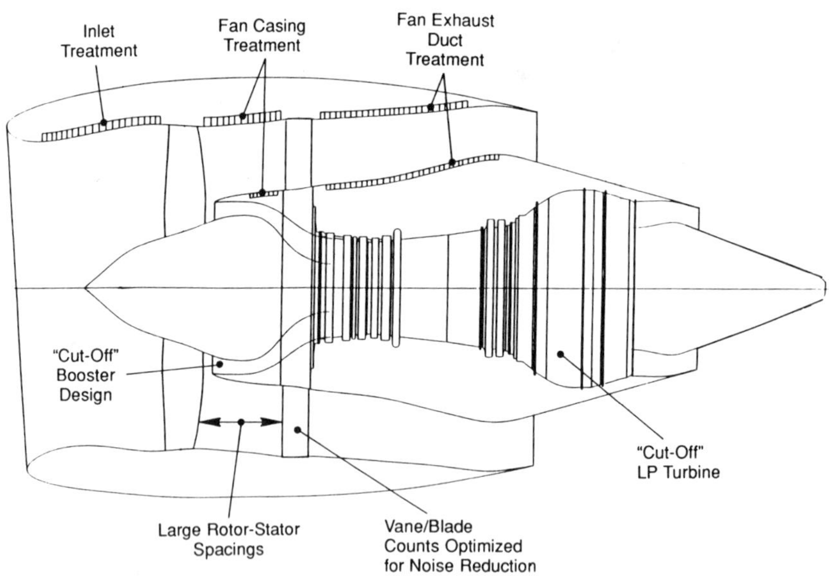

FIGURE 4 - TURBOFAN ENGINE NOISE REDUCTION FEATURES

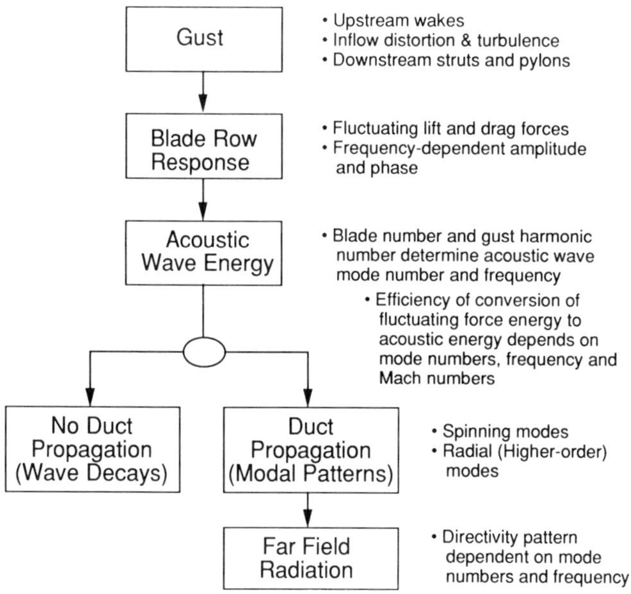

FIGURE 5 - THE TURBOMACHINERY NOISE GENERATION PROCESS

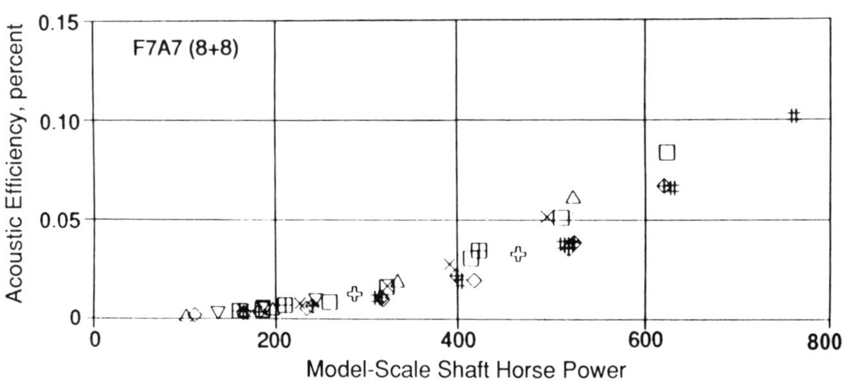

FIGURE 6 - COUNTER-ROTATING UNDUCTED FAN ACOUSTIC EFFICIENCY

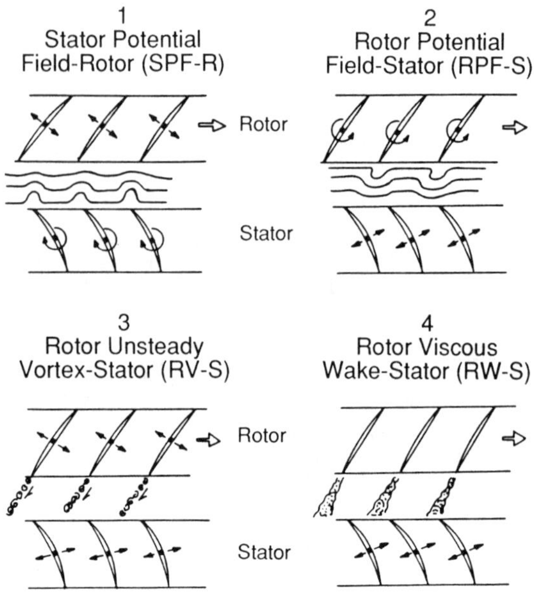

FIGURE 7 - UNSTEADY BLADE AND VANE FORCE MECHANISMS

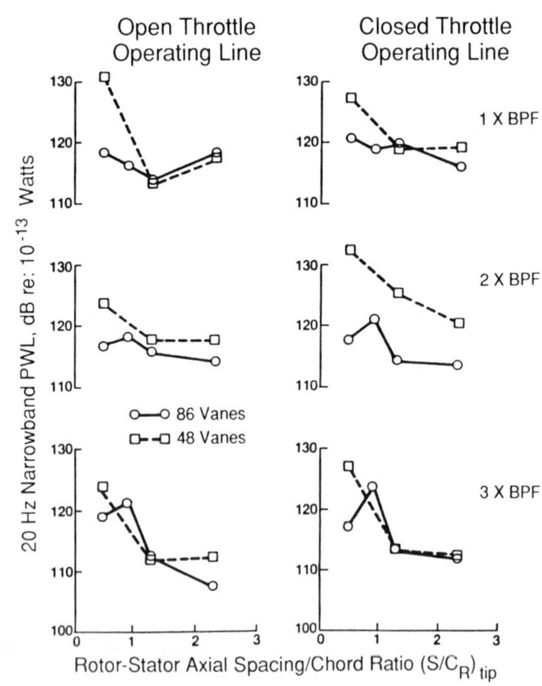

FIGURE 8 - ROTOR-STATOR SPACING EFFECTS ON FAN NOISE

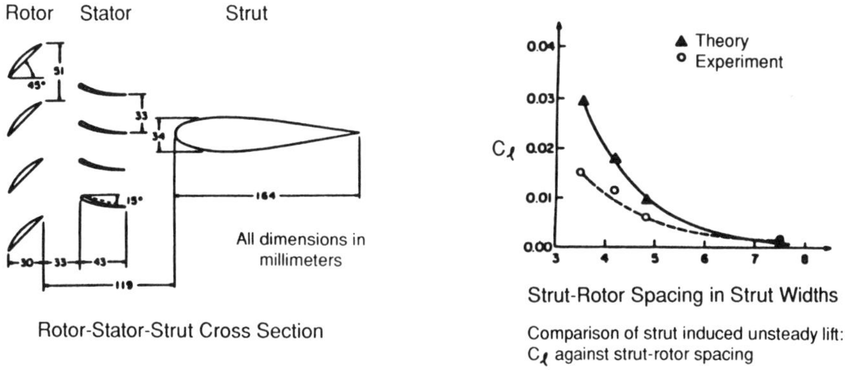

FIGURE 9 - ROTOR UNSTEADY LIFT DUE TO STRUTS

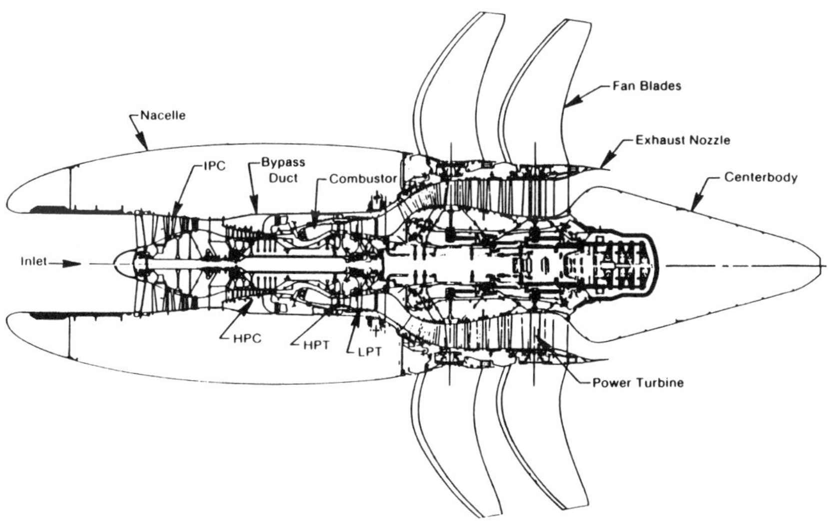

FIGURE 10 - CROSS-SECTION OF UNDUCTED FAN ENGINE

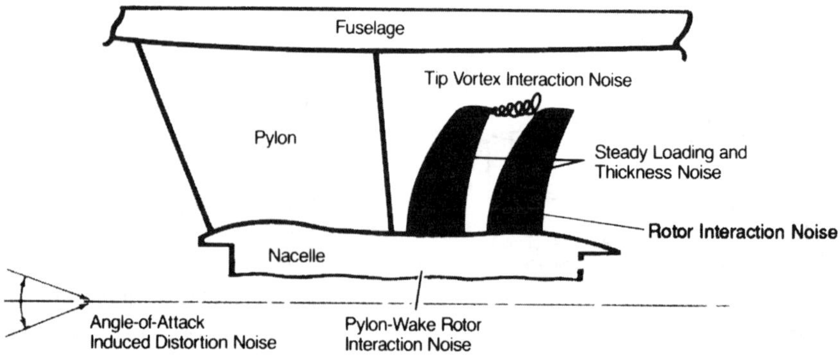

FIGURE 11 - UNDUCTED FAN DOMINANT NOISE GENERATION MECHANISMS

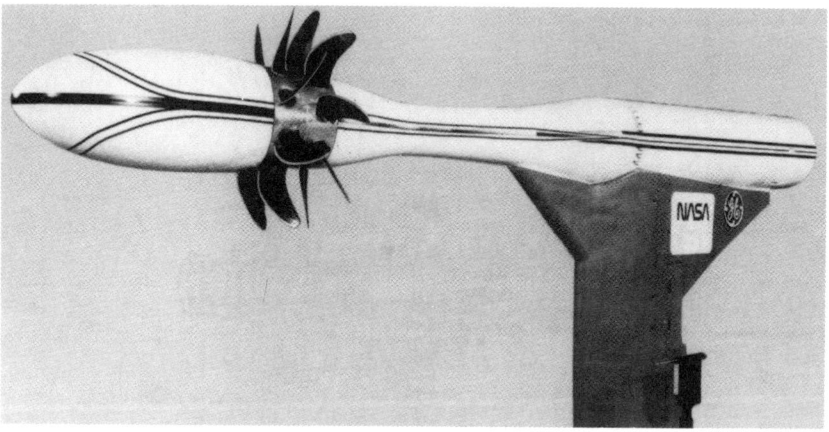

FIGURE 12 - UNDUCTED FAN MODEL PROPULSION SIMULATOR

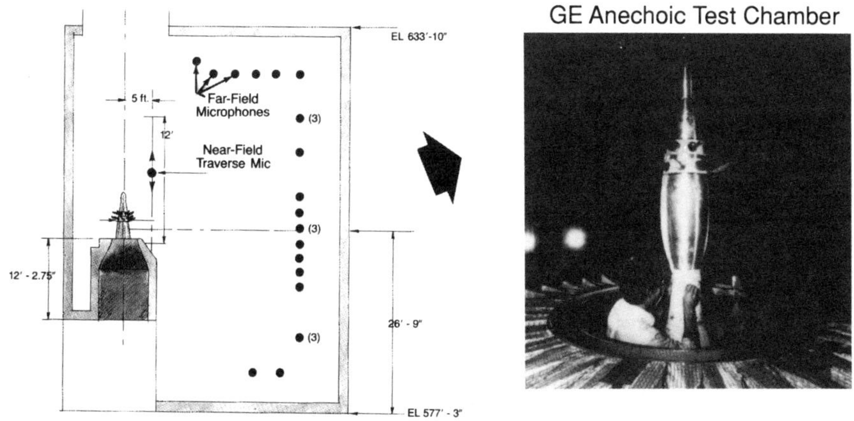

FIGURE 13 - GE ANECHOIC CHAMBER FOR UNDUCTED FAN SIMULATOR ACOUSTIC TESTING

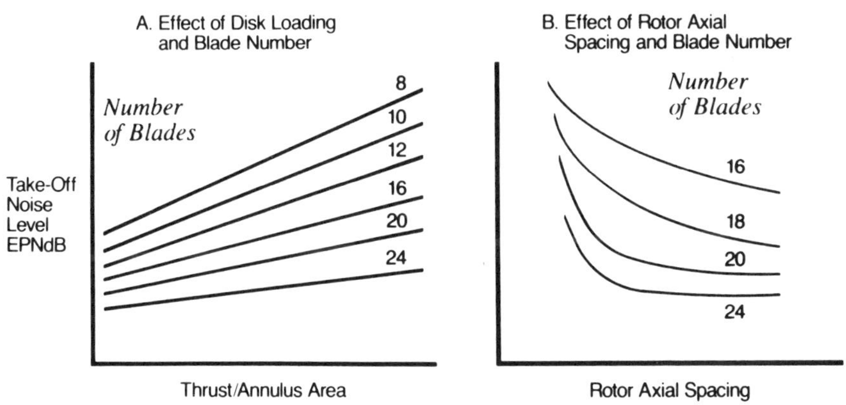

RE 14 - EFFECTS OF VARIOUS DESIGN PARAMETERS ON PERCEIVED UDF® NOISE AT TAKEOFF

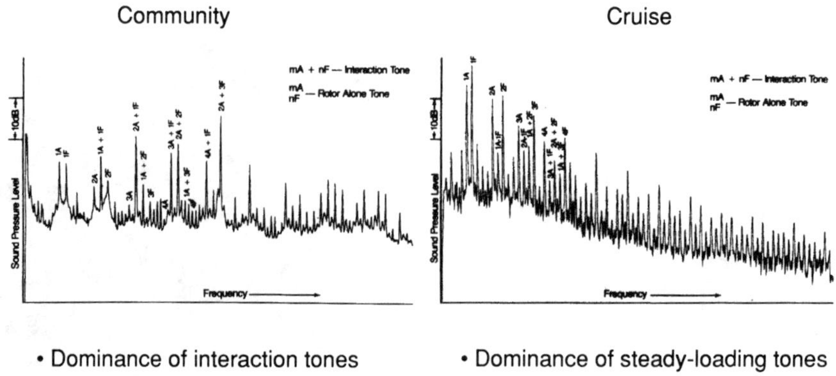

FIGURE 15 - UDF ® ENGINE TONE SOURCE DECOMPOSITION FROM NARROWBAND SPECTRA

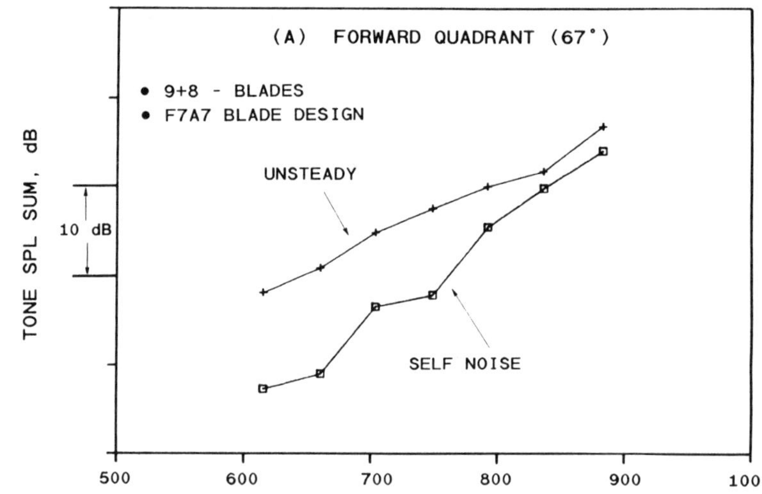

FIGURE 16A - COUNTER-ROTATING UNDUCTED FAN NOISE SOURCE DECOMPOSITION TRENDS WITH TIP SPEED

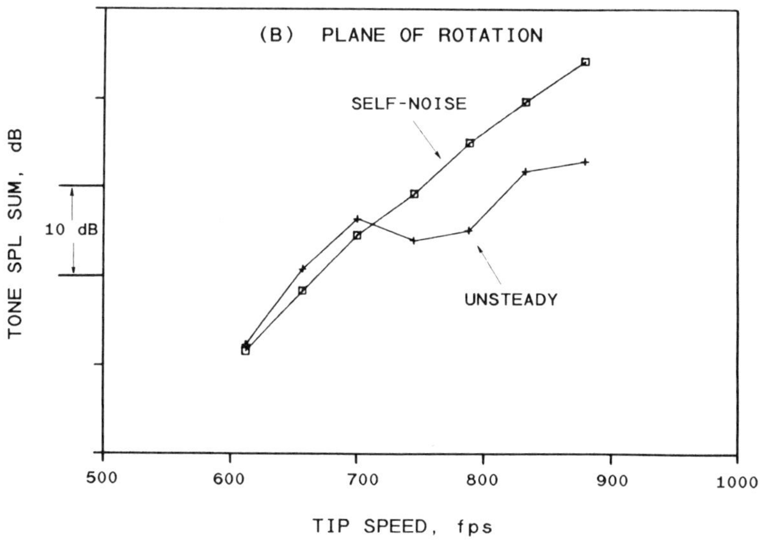

FIGURE 16B - COUNTER-ROTATING UNDUCTED FAN NOISE SOURCE DECOMPOSITION TRENDS WITH TIP SPEED

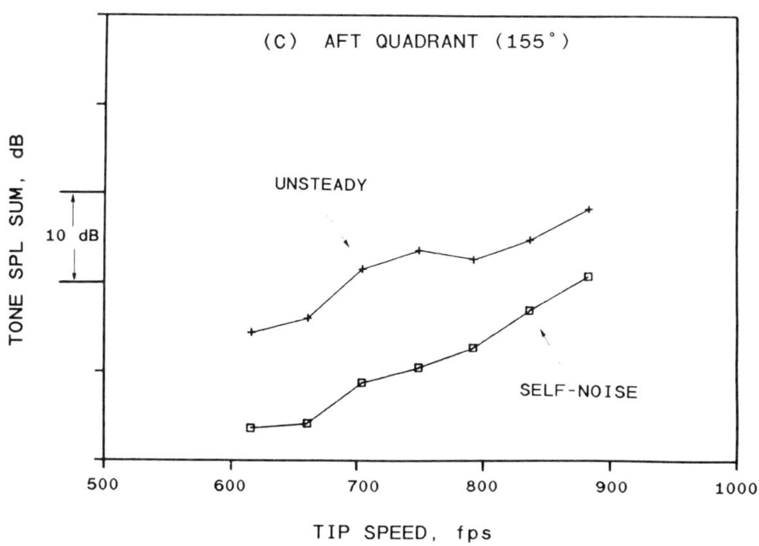

FIGURE 16C - COUNTER-ROTATING UNDUCTED FAN NOISE SOURCE DECOMPOSITION TRENDS WITH TIP SPEED

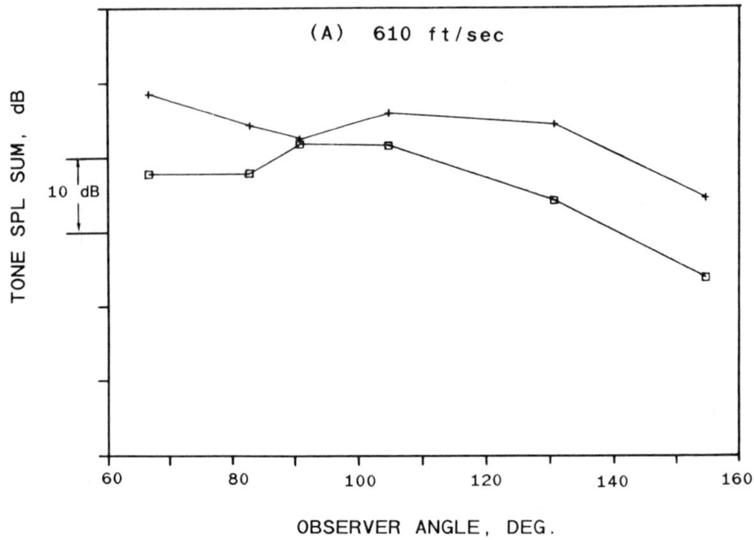

FIGURE 17A - COUNTER-ROTATING UNDUCTED FAN NOISE SOURCE DECOMPOSITION DIRECTIVITY PATTERNS

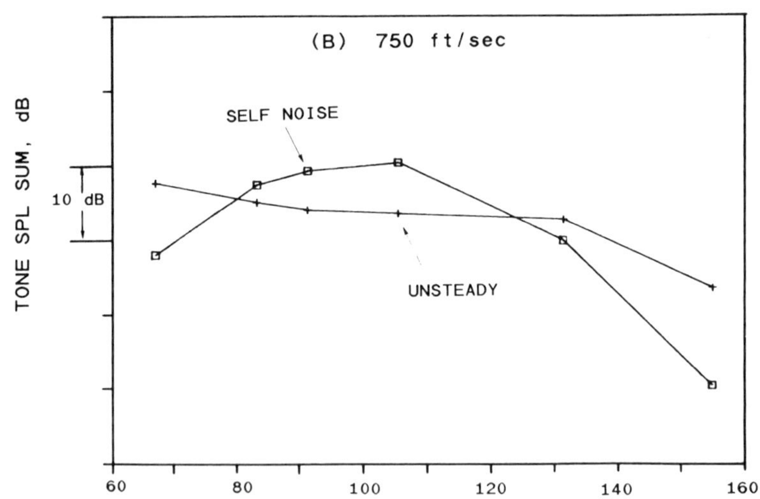

FIGURE 17B - COUNTER-ROTATING UNDUCTED FAN NOISE SOURCE DECOMPOSITION DIRECTIVITY PATTERNS

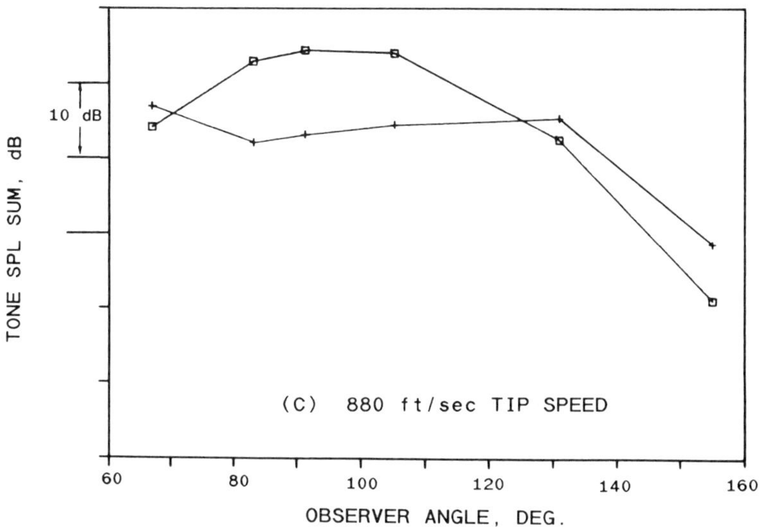

FIGURE 17C - COUNTER-ROTATING UNDUCTED FAN NOISE SOURCE DECOMPOSITION DIRECTIVITY PATTERNS

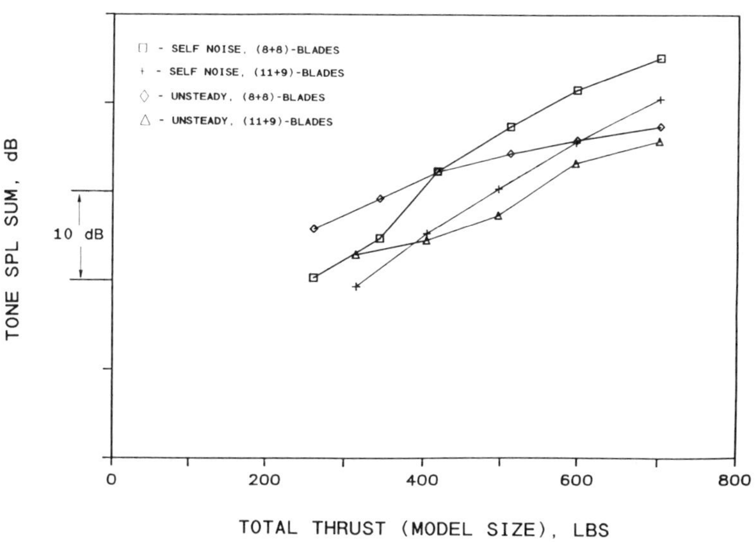

FIGURE 18 - EFFECTS OF INCREASING BLADE NUMBER AND AXIAL SPACING ON UNDUCTED FAN ROTOR SELF NOISE AND UNSTEADY INTERACTION NOISE COMPONENTS

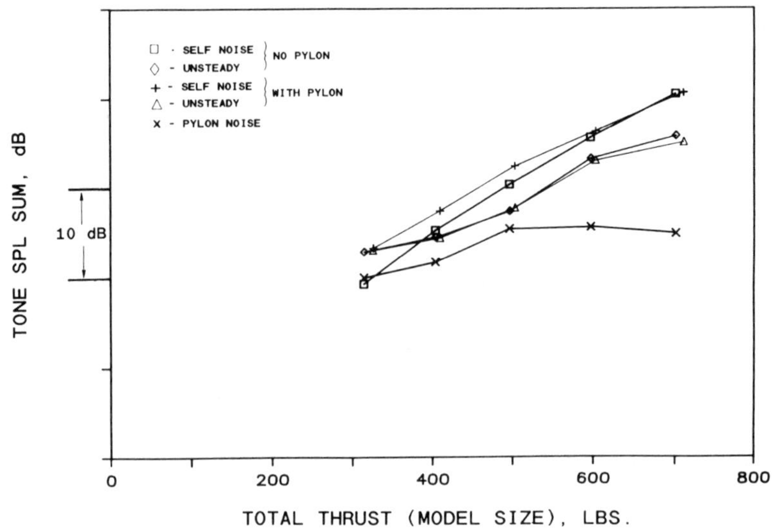

FIGURE 19 - MOUNTING PYLON WAKE EFFECTS ON ROTOR SELF NOISE AND UNSTEADY INTERACTION NOISE

Acoustic Control of Flow Instabilities

J. E. Ffowcs Williams[1]

ABSTRACT The lecture reviews the common origin of sound and instabilities in weak perturbations of a smooth flow. The degree to which sound and instabilities interact, or indeed the degree to which they are separable, depends on their semantic definition. At one level they are merely secondary unsteady effects dependent on the main flow while at the other they display distinctive characteristics of propagation and exponential growth. Even if they exist as separable entities in a pure flow they remain coupled by the inevitable inhomogeneities of realistic geometrical arrangements so that in practice flow instability and sound are actually inseparable.

Sound, being essentially linear, admits to the attenuation brought by superposition of its phase-inverted replica. Anti-sound cancels sound, a principle now being implemented in rapidly growing noise control technology. The same principle implies change in stability whenever its coupled sound field is modified. Flows that are traditionally unstable can be stabilized by active systems. Whether all flows can be stabilized or not might depend on the fundamental constraint that it is impossible to forecast the future. More probably the limit is set by technical complexity and by the need to keep all perturbations within the linear regime where superposition is valid. The scope of the idea will only become clear as definite experiments are completed.

Experiments have progressed from the control of a small combustion system with a single instability mode to the acoustic suppression of violent unsteadiness at aero-engine reheat conditions. A wing section flexibly mounted in a wind tunnel has been brought out of flutter oscillation by switching on a wall-mounted loudspeaker. The fundamental Strouhal frequency in the Karman street wake of a circular cylinder has been suppressed by acoustic feedback and both rotating stall and surge have been avoided at conditions which made those instabilities inevitable without active control. An ONR programme to examine the applicability of these methods to a real gas turbine engine has led to a power increase of some 10 % and to active methods of restoring the stable condition following the onset of surge. The lecture will conclude by reviewing these developments and issues that have arisen

[1] Cambridge University, Engineering Dept., Trumpington Street, CB2 0PZ Cambridge, United Kingdom.

in developing the control methods. In particular it will raise the prospect of aerofoil sections controlled to operate with radically superior characteristics and speculate on that being an element of the performance improvement observed on the test engine.

1.1 Introduction

Sound is an unsteady perturbation of a steady condition, the pure form existing only as an idealization of natural conditions. Sound is organised into waves which propagate through their supporting medium without changing that medium's state. Again, natural conditions are rarely such that this identifying aspect of sound is met exactly, but the recognizable features of practical sound closely resemble those which can be studied and analysed exactly in idealizations where they are unambiguous and precisely defined. What is and what is not sound is largely a matter of opinion until the defining idealization is made, notwithstanding the fact that it is utterly familiar to all who possess the basic senses.

Stability is another universally familiar concept whose precise meaning depends on a definition made on a precise idealization of natural conditions, the definition usually sharing the perturbation and scale-separation aspects common to sound. Unstable flows are noisy and stable conditions are rarely maintained in a noisy environment, making evident the natural coupling of the two phenomena. Whether or not they could ever co-exist without interacting is an issue that requires another idealization, and with it a semantic definition of exactly what is meant by the words 'sound' and 'flow instability'. This lecture concerns the issue of managing flow disturbances using techniques built up in acoustics and it is well to recognise that the two subjects, distinct only in some idealized world, are two sides of the same coin in the practical case where both are possible but neither exists in its pure form.

Stability studies usually concern the issue of whether or not a weak perturbation of a basic flow would naturally grow or die. The techniques exploit the linearity of a weak-enough disturbance and concern the existence or otherwise of the steady state. An unstable steady flow would support an exponentially growing parasitic perturbation of no material consequence to that steady case - but that is very short lived. Such studies are commonly made to enquire into the origin of energetic non-uniformities and have an important place in fluid mechanics even though the developed non-linear unsteady condition is grossly different from the steady flow whose instability brought about its destruction. The natural case may actually be the result of an evolutionary cascade of instabilities of ever-increasing remoteness from the supposed initial condition to which it might have no deterministic connection, yet instability theory provides a strong intellectual discipline for addressing the problem. This lecture presumes the usefulness and relevance of instability mechanics and takes it as axiomatic that were the instabilities of the hypothetical steady case controlled, the resulting flow, steady or otherwise, might be interestingly different and even useful.

Sound fields are enriched by the variety of ways in which different waves interfere both constructively and destructively. The deliberate creation of a phase inverted replica with equipment that monitors a field, filters and reproduces it in anti-phase, is the technique of anti-sound first recognized by Paul Lueg (1936), realized by Olson and May (1953) and is evolving slowly into a technology of noise control, Nelson and Elliot (1991). Linear instability physics obeys exactly the same natural laws as does sound; they are distinguished only by the different idealized forms of the quiescent state for which they are defined. An instability wave is nullified by the presence of its phase-inverted replica, a replica that if artificially manufactured to interfere destructively with an observed instability, would guarantee the permanence of the steady state. Just as anti-sound produces silence so will active control produce fluid motion of unnatural smoothness.

The most common unease at the idea of anti-sound emerges in a perception that some system for broadcasting silence through a loudspeaker is in prospect, clearly impossible and counter to energetic principles. The loudspeaker would in fact be a very carefully constructed array directing into the target region a field that, though phased to destroy a chosen short wave, would fail to suppress any other sound. And the supposed energetic principle does not exist, that being a concept restricted to incoherent sounds. Instability avoidance by wave superposition raises similar intuitive conflict in anyone supposing the power of the control system should be equated to the power of the disturbance that grows from a failure to maintain stability. There is no power-equating principle of that kind. The controlling disturbance is as small as the instability itself, exponentially small when caught in its early infancy.

With that introduction I will now discuss the development of the subject by illustrating specific cases in which theory and experiment have gone hand in hand to prove the validity of the general concept. My view of the subject has developed as more ambitious cases were tackled, the distinctions between different approaches fading as the subject evolved into its broad natural framework, a framework enriched by bringing together the historically separate discipline of control, unsteady fluid mechanics and acoustics.

1.2 Instability Control

A sound field ϕ satisfies the linear wave equation $\Box^2\phi = 0$. A convenient and precise definition of the sound source Q is $\Box^2\phi, \phi$ then being given by the solution of the inhomogeneous wave equation

$$\Box^2\phi = Q \ , \tag{1.1}$$

subject to a causality or radiation condition.

ϕ_a, the anti-sound is the opposite of sound in a chosen region of space, a region defined by $H = 1$; elsewhere $H = 0$.

$$\phi_a \ = \ - \ H\phi \text{ and its source is } - \Box^2(H\phi) \ .$$

This gives a clear prescription for one (of the many) ideal sound suppression schemes. Measure or otherwise determine the unwanted sound ϕ in the region $H = 1$. Construct the source $-\Box^2(H\phi)$ and add to the existing field to produce the mutually destructive fields $(\phi - H\phi)$ that satisfies the inhomogeneous equation

$$\Box^2(\phi - H\phi) \ = \ Q - \Box^2(H\phi) \ . \tag{1.2}$$

The combined field is zero in the controlled region where $H = 1$; the sources of both sound and anti-sound lie outside or on the edge of that region, which may be infinitely large. Acquiring a knowledge of ϕ in time to manufacture its opposite is a fundamental limitation on the scope of the subject which is otherwise limited only by the state of technology.

When instabilities are approached in this light, we have the natural instability ξ, say, a causal solution of the unstable linear operator $L(\xi) = 0$, and require a different causal and static linear device to drive a response $\xi_a = -\xi$ from some measured linear function of ξ. This is added to the natural case to interfere destructively with the instability.

We may say that the ideal linear control device will produce a disturbance ξ_a from a 'source' $\mathcal{L}(\xi_a)$, so that if ξ is 'measured' and the source $\mathcal{L}(-\xi)$ constructed, the device will produce $-\xi$. Any practical implementation is bound to be imperfect and the device's source might have superimposed noise n, making

$$\mathcal{L}(\xi_a) \ = \ -\mathcal{L}(\xi) + n$$

and

$$\xi_a \ = \ \mathcal{L}^{-1}(n - \mathcal{L}(\xi)) \ = \ -\xi + \mathcal{L}^{-1}(n) \ . \tag{1.3}$$

This superimposed on the natural instability results in the finite noise-driven response of the controlled system being,

$$\xi + \xi_a = \mathcal{L}^{-1}(n) , \tag{1.4}$$

which bears no trace of the natural system's characteristics.

This amounts to a rather contrived statement of the fact that if control is exercised by coupling two linear systems together, the response of the coupled system need have nothing to do with the characteristics of the uncontrolled system. The coupled system is completely different and we are intending to make it stable thereby 'controlling' the natural system's instability. When the instability occurs at audio frequency, the control device will have the qualities of anti sound apparatus and the actuators might well be loudspeakers; it is in this sense that the control is acoustic. The instability, and its antidote may not be in the least influenced by compressibility, the essence of sound.

Instabilities cannot remain weak and linear for long; they grow and often change the character of the flow that bore them; laminar flow becomes turbulent. Linear control applies only to the weakly disturbed state but its effect can prevent the energetic consequence of instability. Control at insignificant energy levels avoids breakdown of powerful flows; there is no connection between the effort needed to avoid instability and that which might be needed to fight it in its mature state. The activation requirements are more likely set by noise as Equation (4) indicates.

Of course the early experimental work on this technique concerned the most simple geometrical arrangement one could find. Though Crocco and Cheng (1956) recognised the essential principle it was much later that the necessary apparatus became available to suppress the powerful instabilities that arise in combustion systems. Dines (1984) wrote a thesis on suppressing broad-band combustion noise with anti-sound and demonstrated also the acoustic control of a combustion instability. The Rijke tube is a device where a contained heat source transfers heat to its environment at a rate dependent on the unsteady motion around the source. The motion is unsteady because of the internal sound made by the unsteady release of heat. The loud instability exists when the internal sound stimulates the sound-producing release of heat to produce more acoustic power than is lost from the system by leakage and viscous dissipation. Dines changed the sound around the heat source by activating a nearby loud-speaker driven by a filtered version of the measured sound signal, to discover that the system was stabilized over a wide range of filter characteristics.

Heckl (1985) went into the mechanics of this process in great detail and Dowling et al. (1988) developed the technique far enough to achieve control of the instability known as 'buzz' on aircraft engine re-heat systems. That experiment was probably the first occasion on which really energetic noisy instabilities were avoided using low power acoustical apparatus.

Experiments confirm the idea that active control prevents instability growth. Switching off the controller allows the disturbance to be studied experimentally as it grows, a study previously impossible because the calm weakly disturbed unstable regime could not exist. Experiments also show that switching on the controller again can re-establish calm conditions even though the instabilities had grown well outside their linear range and settled down into their non-linear limiting forms. This was particularly evident in Heckl's (1985) study, and also in Huang's (1988) experiments controlling a wing section undergoing binary flutter in a wind tunnel, the control being effected there with a wall-mounted loud-speaker activated by a filtered signal derived from wing accelerometer measurements. In the non-linear limit cycle the equilibrium level is set by a balance between the energy supplied to the instability and that lost by dissipative effects. The controller alters that balance and therefore leads to a new equilibrium. The controller only has to ensure that the energy supply is less than that dissipated for the cycle to die. The controller must be capable of providing the power differential between supply and demand - it need not match the absolute level of either.

Tilting the energy balance provides a highly robust approach to limit cycle control because it may be effected in a rich variety of ways without a need for the extreme phase accuracy that characterizes destructive interference in the linear regime. Day's (1991) controlled arrest of the periodic surge cycle in an axial compressor system demonstrates this point well as does the Paduano et al. (1991) experimental demonstration that smooth conditions can be recovered by control action on well advanced states of compressor instability.

Steady flow has apparently been imposed by a controller on a regime that naturally supports the unsteady wake structure known as Karman's vortex street. This is reported by Ffowcs Williams and Zhao (1989) evidently unaware of Berger's (1967) earlier demonstration! Flow-excited cavity oscillations provide another example where acoustic feedback is effective in maintaining steady flow at conditions naturally prone to strong oscillations, Sunyach and Ffowcs Williams (1986). There is now no doubt that acoustic control of flow is possible in some interesting regimes, some studied more for their curiosity value than for any practical purpose. But the recent developments are also pointing to the possibility of significant practical benefit, so inevitably the pace of research has quickened. ONR's 'accelerated research initiative on active control' is producing evidence of major controller-induced change in the performance of turbomachinery. So are other studies coming to that view from an entirely different starting point. Fast reacting controllers which sense the onset of compressor instability and move the operating point to a safe regime have long been sought, those safe regimes being ones of lower steady load, c.f. for example Ludwig and Nenni (1980). The need to recover performance by resuming full load as soon as the coast is clear drives towards machinery getting a performance improvement through intermittent operation in the high-load dangerous territory while retreating to safe ground at the first sign of danger.

It would be better still to maintain continuous high load, avoiding trouble by destroying any embryo instability with some auxiliary device activated once the embryo is detected. Day (1991a) has put this approach into practice on an axial compressor in which rotating stall is a recognizable precursor of surge. Day's scheme blows away the stall cell before it triggers off surge, and does it very effectively. At one level his 'monitor trouble and act decisively' type of controller is only subtly different from the linear stabilization device envisaged by Epstein et al. (1986) and realized experimentally by Ffowcs Williams and Huang (1989), Pinsley et al. (1990) and Paduano et al. (1991) on small scale radial and axial compressors. At another level it is far from subtle, for the fully coupled linear stabilizer changes the compressor system to an inherently stable one in which there is no instability to be detected, let alone to be acted on; both the compressor and controller have, in combination, lost their individual identities.

The subject is moving fast and though the applicability of the simple models and experimental test results to real engines is not yet known, a growing body of opinion expect there to be a carry over and one with very important consequences.

Ffowcs Williams and Graham (1990) report engine experiments aimed at implementing Epstein et al.'s (1986) surge control strategy, in essentially the same way as was successfully done on a small scale centrifugal compressor. The remainder of this paper concentrates on gas turbine instabilities and gives some of the results coming from the ONR initiative.

1.3 Engine Surge Control

When a gas turbine surges a massive amount of energy is put into a strong perturbation that frequently causes the flow through the compressor to reverse. Surge is a performance-limiting condition in which the risk of engine damage is high, so operational procedures avoid all conditions where surge is likely.

Enormous effort has gone into understanding surge, part of that effort addressing the possibility that surge arises from the growth of a weak instability. If that is so it should be amenable to control. Certainly a smooth engine flow could not change abruptly, continuity being a basic axiom of fluid mechanics, and Moore and Greitzer's one dimensional modelling of compressor instability does have much in common with the behaviour of real systems. Compressors are also prone to other instabilities of greater geometrical complexity, and again, if they arise as instabilities of a smooth flow, their early life will conform with linear theory and they should be amenable to treatment by low energy active control. That was the point of view taken by Epstein et al. (1986) who demonstrated that if the existing linear models of engine instability are relevant, then engine stability can definitely be maintained by active control. They went on to devise several definite control strategies for avoiding instabilities thought likely candidates for being the embryo stages of rotating compressor stall and engine surge. Experimental verification of that strategy followed in a matter of months, Huang (1988) halving the unstable mass flow range of a compressor system with a loudspeaker actively driven to suppress pressure fluctuations in the compressor delivery duct. An almost identically effective system using a variable position throttle as the control actuator was proved by Pinsley et al. (1990) on a similar turbocharger-based experimental arrangement. In both experiments it was found that not only could the compressor be operated stably well beyond its natural surge line, but that stable operation could be recovered by switching on the controller with the compressor settled into a highly non-linear normally permanent periodic surge cycle. Weak excitation with carefully controlled amplitude frequency and phase was proven to be relevant to the control of compressor surge at model scale. That was expected at small perturbation amplitude but its large amplitude effect was a bonus, apparent in hindsight as the modulation of non-linear modal interaction - but its failure could have been rationalized equally easily! The real life conditions inside a powerful gas turbine are so far away from the weakly disturbed state that modelling small perturbation effects carry little engineering conviction. The ONR programme took the pragmatic view that testing a real engine for sensitivity to the kind of control found effective at model scale would establish quickly the practical relevance of the techniques. It covered also an in-depth study of specific engine-related instabilities, the compressor and combustion systems, notoriously vulnerable to extraneous effects, being the focus for the effort. Early results from that programme were reported by Ffowcs Williams and Graham (1990). They reported results of experiments on a 60 h.p. Rover gas turbine auxiliary power unit.

Control, by feeding the right effect into the engine, depends on the availability of a signal containing information about the effect.

The Rover engine is actually very far removed from the weakly disturbed model, the background noise inside the compressor outlet duct having a wide spectrum and a variance as big as the engine's full operational pressure range, yet coherence between an unsteady air supply into the plenum and the plenum field is high enough for some effective control. Using controlled air injection activated from a compressor delivery sensor array it was possible to make the engine immune to transients that would otherwise inevitably cause surge. And despite the extremely high natural noise level it was possible for that control to bring the engine out of a naturally permanent repetitive cycle of surge.

It was noticed that there tended to be high frequency pressure signals within the engine that indicated the probability of surge, signals with the character expected for rotating stall. That is another instability susceptible to active suppression according to Epstein et al. (1986). A suitably phased 4 sensor 4 actuator method, modulating the compressor flow rate around the compressor inlet disk, was successful in inhibiting stall in the Rover engine to the degree that it delivered 10 % more power than its uncontrolled form could provide. Both these controllers used power proportional to the noise of the system, the actuating amplitude being a negligible fraction of the engine flow.

A most interesting fact to emerge from this study is that operation made possible by the controller in a normally prohibited range, achieves a higher performance level than that described by the mean engine characteristic and has led to the view that the mean condition does not actually exist as an equilibrium state. The engine evidence is that the controlled equilibrium state is more powerful than the mean (but rough) stalled condition and that the mean is not a stable state suitable for small amplitude stability modelling.

I conclude this lecture by noting that the great advances being achieved at M.I.T., Paduano et al. (1991) and at Cambridge University, Day (1991), on compressor test rigs provide evidence in the same direction, small amplitude active control definitely affects major engine instabilities and does so with the power levels, methodology and apparatus of acoustics.

Acknowledgements

Much of the experimental results quoted here comes from the O.N.R. initiative at Topexpress, Ltd., and permission to use that information is gratefully acknowledged as is Dr. Mark Harper's important contribution to that programme.

References

1. Berger, E. (1967). Suppression of vortex shedding and turbulence behind oscillating cylinders. Phys. Fluids 10, S191 - S193.

2. Crocco, L. and Cheng, S. I. (1956). Theory of combustion instability in liquid propellant rocket motors. Interscience.

3. Day, I. J. (1991). Compressor performance during surge. Proc. Tenth Int. Symp. on Airbreathing engines. Nottingham. Published by A.I.A.A.

4. Day, I. J. (1991a). Active Suppression of rotating stall and surge in axial compressors. A.S.M.E. paper 91-GT-87.

5. Dines, P. J. (1984). Active control of flame noise. Ph.D. thesis. Cambridge University, U.K.

6. Dowling, A. P., Bloxsidge, G. J., Hooper, N. and Langhorne, P. J. (1988). Active Control of Reheat Buzz. A.I.A.A. Journal, 26, pp. 783-790.

7. Epstein, A. H., Ffowcs Williams, J. E. and Greitzer, E. M. (1986). Active suppression of aerodynamic instabilities in turbomachines. J. Prop. Power, Vol. 5, No. 2, pp. 204-211.

8. Ffowcs-Williams, J. E. and Zhao, B. C. (1989). The active control of vortex shedding. J. Fluids and Structures, 3, 115-122.

9. Ffowcs Williams, J. E. and Huang, E. Y. (1989). Active stabilization of compressor surge. J. Fluid Mech. 204, pp. 245-262.

10. Ffowcs Williams, J. E. and Graham, W. R. (1990) An engine demonstration of active surge control. Proceedings of the gas turbine and aeroengine Congress and Exposition, June 11-14, Brussels, Belgium.

11. Heckl, M. (1985). Active control of Rijke tube. Ph.D. thesis, Cambridge University.

12. Huang, X. (1988). Active Control of Aerodynamic Instabilities. Ph.D. Thesis, Cambridge University, U.K.

13. Lueg, P. (1936). Process of silencing sound oscillations. U.S. Patent # 2043, 416.

14. Ludwig, G. R. and Nenni, J. P. (1980). Tests on an improved rotating stall control system on a J.85 turbojet engine. A.S.M.E. paper 80-G.T.-17.

15. Nelson, P. A. and Elliot, S. J. (1991). Active noise control; a tutorial review. Proceedings of the Acoustical Society of Japan International Conference on the Active Control of Sound and Vibration. Tokyo.

16. Olson, H. F. and May, E. G.] (1953). Electronic sound absorber. J.A.S.A., 25, 1130-1136.

17. Paduano, J., Epstein, H., Longley, J. P., Valavani, L., Greitzer, E. M. and Guenette, G. R. (1991). Active control of a rotating stall in a low speed axial compressor. Proceedings of the A.S.M.E./I.G.T.I. Conference in Orlando.

18. Pinsley, J. E., Guenette, G. R., Epstein, A. H. and Greitzer, E.M. (1990). Active stabilization of centrifugal compressor surge. A.S.M.E. Gas Turbine Conference, Brussels.

19. Sunyach, M. and Ffowcs Williams, J. E. (1986). Contrôle actif des oscillations dans les cavités excitées par un écoulement. C. R. Acad. Sc. Paris, t. 303, Série 11, No. 12.

High-Speed Helicopter Rotor Noise-Shock Waves as a Potent Source of Sound

F. Farassat [1]
Yung-Jang Lee [1]
H. Tadghighi [2]
R. Holz [2]

ABSTRACT In this paper we discuss the problem of high speed rotor noise prediction. In particular, we propose that from the point of view of the acoustic analogy, shocks around rotating blades are sources of sound. We show that, although for a wing at uniform steady rectilinear motion with shocks the volume quadrupole and shock sources cancel in the far field to the order of $1/r$, this cannot happen for rotating blades. In this case, some cancellation between volume quadrupoles and shock sources occurs, yet the remaining shock noise contribution is still potent. A formula for shock noise prediction is presented based on mapping the deformable shock surface to a time independent region. The resulting equation is similar to Formulation 1A of Langley. Shock noise prediction for a hovering model rotor for which experimental noise data exist is presented. The comparison of measured and predicted acoustic data shows good agreement.

[1] NASA Langley Research Center, Hampton, VA 23665.
[2] McDonnel Douglas Helicopter Company, Mesa, AZ 85201.

1.1 Introduction

Noise prediction of helicopter rotors is an important technical activity because of its commercial and military significance. The issues of community noise, passenger comfort, and detectability have encouraged the research on noise prediction. One can now claim that the noise generating mechanisms of helicopter rotors are understood well. We are approaching the time that the noise of helicopter rotors at various flight conditions can be predicted at the design stage. Modifications to the rotor can then be proposed to control rotor noise. This has become possible because of theoretical research, availability of powerful computers and many superb aerodynamic and acoustic tests for validation and improvement of theoretical models.

In this paper, we discuss the problem of noise generation by shock surfaces around helicopter rotors. The governing equation of shock noise generation obtained by Farassat, Brentner, and Myers [1-3], is based on the quadrupole source term of the Ffowcs Williams - Hawkings (FW-H) equation [4]. This governing equation has two source terms of monopole and dipole types which lie on the moving and deforming shock surface. Although one can argue, on the basis of rank ordering of the poles, that shock surface sources are more important than volume quadrupole sources, such a reasoning can fail because of the possibility of the cancellation of acoustic pressure generated by these sources [5]. We therefore maintain an open mind about the possibility of some cancellation between shock surface and volume quadrupole sources.

A brief derivation of the governing equation of shock noise generation for an observer in the plane containing the rotor and in the far field is given first. This derivation is based on the quadrupole source term of the FW-H equation [4]. It is shown that for the special case of a wing in steady, rectilinear, subsonic flight, the volume quadrupoles and all surface sources, including shock sources, cancel in the far field up to the order of $1/r$. Then two plausibility arguments are presented to show that this cancellation is incomplete in the case of the rotating blades.

The formula for prediction of the noise of a subsonically moving shock is presented next. Finally an example for a hovering model rotor using a recently developed code based on *WOPWOP* is given. This example for which experimental acoustic pressure signature is available serves to highlight the importance of the shock noise in generation of high speed rotor noise.

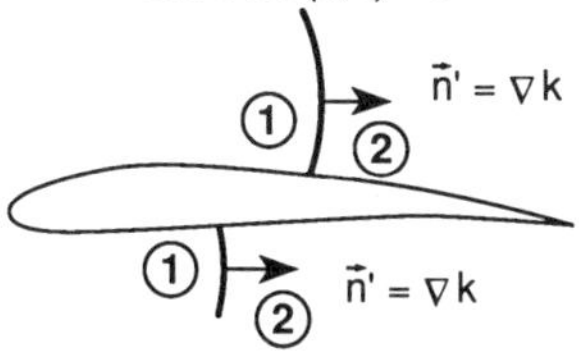

FIGURE 1.1. Definition of shock geometry and regions 1 and 2. The unit normal $\vec{n}'$ points into region 2.

1.2 The Governing Equation

The shock noise must come from the quadrupole source term of the FW-H equation. It was shown in references [1-3] that hidden in this quadrupole source term are surface sources on the blade, wake and shocks. These sources are obtained by taking the generalized derivatives [6,7] of the quadrupole source term of the FW-H equation. Let the blade and shock surfaces be defined by $f(\vec{x},t) = 0$ ($f > 0$ outside the blade) and $k(\vec{x},t) = 0$, respectively, see figure 1.1 1. The local normal vector on the shock surface pointing downstream of the shock is denoted by $\vec{n}'$. The governing equation of the quadrupole noise generation is

$$\Box^2 p'_Q = \frac{\bar{\partial}^2}{\partial x_i\, \partial x_j} [T_{ij} H(f)] \tag{1.1}$$

where T_{ij} is the Lighthill stress tensor and the bar over the derivatives denotes generalized differentiation [6,7].

Taking the derivatives on the right of eq. (1.1) and identifying the source terms on the shock surfaces, we get the governing equation of shock noise generation [8]:

$$\Box^2 p'_Q = \psi(\vec{x},t)\delta(k) + \frac{\partial}{\partial x_i} [q_i\, \delta(k)] \tag{1.2}$$

where

$$\psi(\vec{x},t) = \Delta\left(\frac{\partial T_{ij}}{\partial x_i}\right){n'}_j \tag{1.3}$$

$$q_i(\vec{x},t) = \Delta T_{ij}{n'}_j. \tag{1.4}$$

In these equations $\Delta = [\,]_2 - [\,]_1$ is the jump in a quantity across the shock where the subscripts 1 and 2 refer to the upstream and downstream regions of the shocks, respectively. The parameters $\psi(\vec{x},t)$ and $q_i(\vec{x},t)$ are recognized as monopole and dipole source strengths, respectively. These are obtained from CFD codes which also determine the shock geometry. The shock geometry is in general time dependent which makes the solution of eq. (1.2) somewhat more difficult than for sources on rigid surfaces. We will give the solution of eq. (1.2) in a later section.

1.3 Volume Quadrupole Versus Shock Sources

It can be shown that in eq. (1.1), part of the volume sources cancel part of the blade and wake surfaces for an observer in the $x_1\,x_2$-plane and located in the far field [5]. In fact, the governing equation of noise generation, eq. (1.1), is now approximated as

$$\Box^2 {p'}_Q = \frac{\bar{\partial}^2}{\partial x_\alpha\,\partial x_\beta}\left[T_{\alpha\beta}H(f)\right], \qquad (\alpha,\beta) = 1,2. \tag{1.5}$$

Note that the shock noise source is in this equation. We ask the question: is there further cancellation between the volume quadrupoles and shock surface sources in eq. (1.5)? We can envision a circumstance where this does occur to the order $1/r$ in the far field. The volume source strength in eq. (1.5) is found to be $H(f)\bar{\partial}^2 T_{ij}/\partial x_i\,\partial x_j$ by taking the derivatives in this equation explicitly.

We consider a wing in steady, rectilinear and transonic motion with shocks. We assume that the observer is in the far field and on the x_1-axis along which the wing moves at speed $V < c$. The acoustic pressure ${p'}_Q$ by the collapsing sphere method is:

$$4\pi {p'}_Q = c\int_{-\infty}^{t}\int_{\Omega}\frac{1}{r}\frac{\bar{\partial}^2}{\partial x_\alpha\,\partial x_\beta}\left[T_{\alpha\beta}H(f)\right]\,d\Omega\,d\tau, \qquad (\alpha,\beta) = 1,2, \tag{1.6}$$

where Ω is the sphere $r = c(t-t)$ with center at the observer. In this equation, $r = |\vec{x}-\vec{y}|$, and $(\vec{x},t)$ and $(\vec{y},t)$ are the observer and source space - time variables. If the observer is in the far field on the x_1-axis, the collapsing sphere can be approximated by a plane parallel to $x_2\,x_3$- plane. In addition, only the longitudinal quadru pole term $\overline{\partial}^2\,[T_{11}H(f)]\,/\partial y_1^2$ contributes to the noise at the observer. Let us now set up a new moving coordinate system $\vec{y'}$- frame with axes parallel to those of the $\vec{x}$- frame and origin on the wing. In this frame, the collapsing sphere is seen to move with speed $c-V$ along the y_1 - axis. We can convert the time integration in eq. (1.6) to an integration over y_1' and we no longer have any time dependence

$$4\pi\, p'_Q(\vec{x}) = \frac{1}{r_0\,(1-M)} \int_{-\infty}^{\infty}\int_{-\infty}^{\infty}\int_{-\infty}^{x_1} \frac{\overline{\partial}^2}{\partial {y_1}'^2}\,[T_{11}H(f)]\; dy'_1\, dy'_2\, dy'_3, \tag{1.7}$$

where $r_0 = |x_1|$. We know that $T_{11} \to 0$ as $x_1 \to \infty$ Let $S({y_1}')$ be a plane parallel to ${y_2}'\;{y_3}'$-plane and assume that R is found such that

$$\int_{S(R)} \left|\frac{\partial T_{11}}{\partial {y_1}'}\right|\, dy'_2\, dy'_3 = \epsilon \tag{1.8}$$

where $\epsilon > 0$ is a given small number. Then for $x_1 > R$, we have from eq. (1.7)

$$|p'_Q(\vec{x})| < \frac{\epsilon}{4\pi(1-M)x_1} \tag{1.9}$$

Since ϵ is arbitrary, this means the volume quadru poles and shock sources cancel each other to the order $1/r$ in the far field.

We can find the cause of the above cancellation rather easily. The longitudinal quadrupoles T_{11} are all aligned parallel to the y_1-axis and we were able to convert source time integration into integration along the y_1-axis. Such complete cancellation cannot happen for rotating sources. The major component of the volume quadrupoles is the chordwise component [9-11]. The collapsing sphere intersects these quadrupoles obliquely whenever the blade is not at 90° azimuthal angle which is most of the time. Thus the volume chordwise quadrupoles never align along the y_1- axis. Furthermore, because of blade rotation, the collapsing sphere crosses these quadrupoles faster than when the blade is at 90° azimuthal angle. However, the collapsing sphere crosses the shock when the blade is at about 90° for the peak negative acoustic pressure signature. This means that there is incomplete cancellation between the volume quadrupoles and shock surface sources. We have postulated, subject to further verification, that the shock noise is so dominant that it can account for most of the noise generated by the quadrupole source term of the FW-H equation.

We give yet another plausibility argument about dominance of shock surface noise. Schultz and Splettstoesser have shown that for an observer in the $x_1\,x_2$-plane, the chordwise quadrupole sources can be integrated along vertical lines normal to the rotor disc as shown in figure 1.2 [11]. The

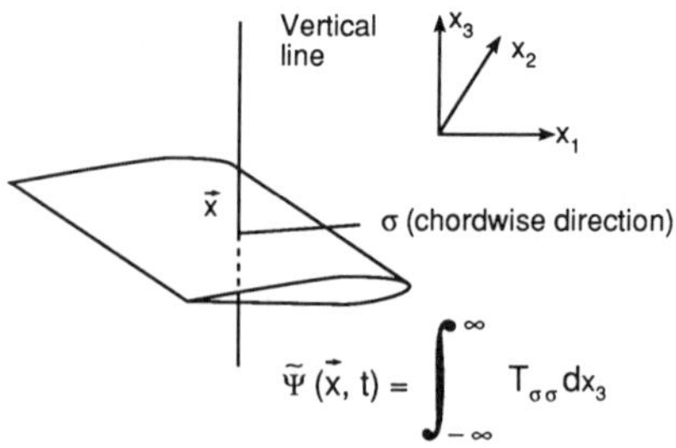

FIGURE 1.2. The calculation of the source strength $\tilde{\Psi}$ in reference 11.

resulting integral can then be assigned to the foot of these vertical lines on the rotor disc as a new source. This means that the governing equation of quadru pole source generation, including the shock sources, for an observer in the far field and in the $x_1 x_2$-plane, is

$$\Box^2 p'_Q = \frac{\overline{\partial}^2}{\partial \sigma^2}\left[\tilde{\Psi} H(h)\right] \tag{1.10}$$

where $h(\vec{x}, t) > 0$ denotes the source region on the rotor disc on which $\tilde{\Psi} \neq 0$. The source strength $\tilde{\Psi}(\vec{x}, t)$ is the line integral along a vertical line on the rotor disc of the longitudinal chordwise quadrupoles. This vertical line intersects the rotor disc at $\vec{x}$ (see figure 1.2). In the above equation, $\frac{\partial}{\partial \sigma}$ is the directional derivative in the chordwise direction.

From the construction of $\tilde{\Psi}$, it is obvious that $\tilde{\Psi}$ has a jump along a curve γ which is where the shock surface intersects the blade (called the shock trace here). When the generalized derivatives in eq. (1.10) are taken, the jumps in $\tilde{\Psi}$ and $\frac{\partial}{\partial \sigma}\tilde{\Psi}$ produce two line sources along γ. These line sources are, in general, powerful noise generators as demonstrated in Formulation 3 of Langley [12] in which line sources on the leading and trailing edges of the blade as well as the shock trace γ appear. This is one more plausibility argument in favor of dominance of shock noise as compared to pure quadrupole noise.

1.4 The Shock Noise Formulation and an Example

Farassat and Tadghighi [8] have given the solution of eq. (1.2) for a subsonically moving shock. This solution is based on mapping the shock surface onto a time independent computational domain Ω with coordinate variables (u^1, u^2). The Jacobian of transformation $\sqrt{g_{(2)}}$ is, therefore, time dependent. We will present this solution here. We define the following symbols first:

$$\dot{q}_r = \hat{r}_i \frac{\partial q_i}{\partial \tau} \tag{1.11}$$

$$\dot{\sigma} = \frac{1}{\sqrt{g_{(2)}}} \frac{\partial}{\partial \tau} \sqrt{g_{(2)}} \tag{1.12}$$

$$\dot{M}_r = \hat{r}_i \frac{\partial M_i}{\partial \tau} \tag{1.13}$$

where $M_i = v_i/c$ is the Mach vector of the point on the shock with coordinates (u^1, u^2) and $M^2 = M_i \cdot M_i$. The solution of eq. (1.2) is given by

$$\begin{aligned} 4\pi\, p'(\vec{x}, t) = & \int_\Omega \left[\frac{\Psi(\vec{y}, \tau)\sqrt{g_{(2)}}}{r(1 - M_r)} \right]_{\tau^*} du^1\, du^2 \\ - & \frac{1}{c} \int_\Omega \left\{ \frac{\left[(\dot{q}_r + q_r \dot{\sigma})(1 - M_r) + q_r \dot{M}_r\right] \sqrt{g_{(2)}}}{r(1 - M_r)^3} \right\}_{\tau^*} du^1\, du^2 \\ - & \int_\Omega \left\{ \frac{\left[(q_r(1 - M^2) - q_i M_i (1 - M_r)\right] \sqrt{g_{(2)}}}{r^2(1 - M_r)^3} \right\}_{\tau^*} du^1\, du^2 \end{aligned} \tag{1.14}$$

Here τ^* is the emission time of the point on the shock with coordinates (u^1, u^2). This result is denoted Formulation 1S of Langley.

An efficient algorithm for implementation of this equation in *WOPWOP* was devised because of the similarity of eq. (1.14) to Formulation 1A of Langley for thickness and loading noise prediction [13]. The new code which is known as *WOPWOP-S* is in the research stage [14]. Some numerical results for a model hovering rotor and a rotor in forward flight were presented in reference 14. Below we present the results for the hovering rotor for the observer in the rotor plane.

The blade, the operating conditions and the observer position are given below:

blade number = 2
blade radius = 1.042 m (3.428 ft)
blade chord = 0.0762 m (0.25 ft)
blade planform = rectangular
blade twist at the tip = 1.5°
rpm = 2741
$M_{tip} = 0.88$
inplane observer position = 3.127 m
(10.26 ft) from rotor center

In this case, there was only a single shock on the upper surface of the blade and no delocalization was observed. Delocalization is the phenomenon where the shock surfaces around the blade extend far beyond the blade tip. Experimental acoustic data in the form of the acoustic pressure signatures were available for this rotor [15].

The monopole and dipole source strength and the shock geometry were obtained from an unsteady full potential aerodynamic code using a shock capturing technique. This code known as RFS2 [16], produced some numerical problems discussed in reference 14. These problems will be investigated in the future by using more recent sophisticated CFD codes. Figure 1.3 shows the acoustic pressure signatures for the monopole and dipole contributions of the shock surface to the shock noise for the observer in the rotor plane. As expected [8], the dipole contributes much less than the monopole to the shock noise. The monopole signature is composed of a single negative pulse and is symmetric with respect to the time t_0 of the occurrence of the negative peak. The dipole signature is asymmetric with respect to t_0. This time t_0 is also close to the time of occurrence of the negative peak of the thickness noise. It is somewhat surprising that the monopole signature has a relatively large negative mean value which is not usually measured by AC- coupled microphones.

Figure 1.4 shows the pressure signatures of thickness, loading, and shock noise for the observer in the rotor plane. The loading noise is very small compared to the thickness and shock noise. It is seen that the thickness and shock noise (the sum of the monopole and dipole contributions of the shock surface) are of the same order of magnitude. This demonstrates the claim of Farassat, Brentner, Myers, and Tadghighi [1-3,8] that shocks around rotating rotors are potent noise generators. Figure 1.5 shows the comparison of the predicted acoustic pressure signature (thickness + loading + shock noise) with the measured signature [15] for the observer in the plane. Note that since the experimental signature in reference [15] is less than one blade passage period, the predicted signature is shown for the same duration. The agreement is quite good. Indeed, only the slight steepening seen

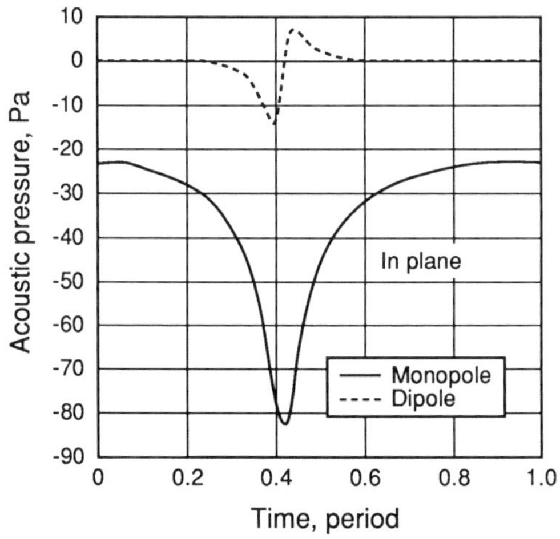

FIGURE 1.3. The acoustic pressure signatures of the monopole and dipole components of the shock noise for a hovering rotor. Observer in rotor plane.

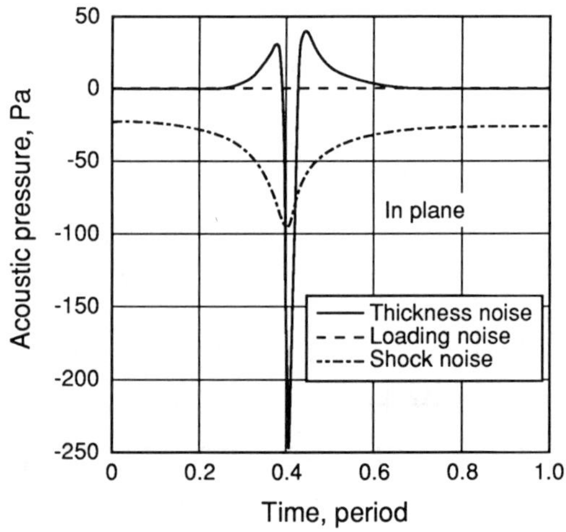

FIGURE 1.4. The relative magnitudes of thickness, loading, and shock noise for a hovering rotor. Observer in rotor plane.

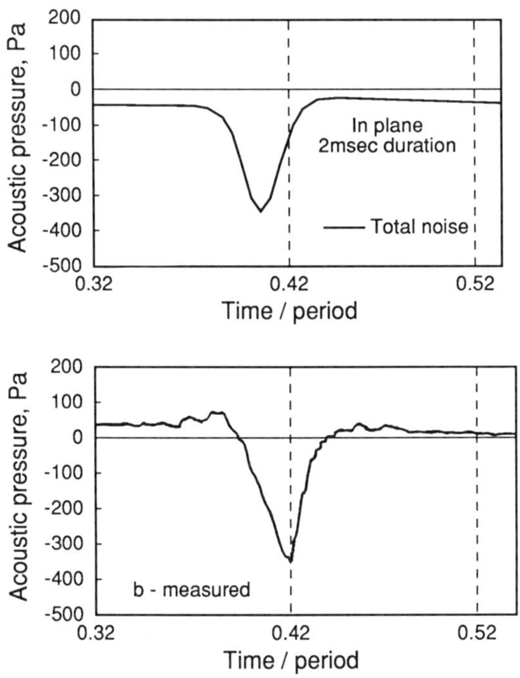

FIGURE 1.5. Comparison of measured and predicted acoustic pressure signatures for a hovering rotor. Observer in rotor plane. Note that the time origin of the measured and predicted waveforms do not correspond.

in the measured signature is not predicted well while the negative peak value and signature width are predicted remarkably well. The shock noise has removed the first positive peak and reduced the second positive peak of the thickness noise. From this calculation, one may conclude that since in our present scheme the volume quadrupoles and blade surface sources (coming from quadrupole term of FW-H eq.) are left out, the net effect of these additional sources is essentially steepening of the acoustic waveforms. The comparison with measured data suggests that the cancellation between the shock source and volume quadrupoles is not great. The complete understanding of the cancellation or even reinforcement of the volume and surface sources requires further numerical work.

1.5 Concluding Remarks

In this paper, we have discussed the possibility of shock surfaces on rotors as potent sources of noise. The governing equation of shock noise generation is derived from the quadrupole source term of the FW-H equation. It is argued here that although there may be some cancellation between pre quadrupole noise and shock noise, the geometry of radiation favors dominance of shock noise. Two plausibility arguments are put forward here. The solution of the wave equation of shock noise generation is given in this paper based on mapping of the shock surface to a time independent computational domain. This equation is very similar to Formulation 1A of Langley. An algorithm for implementing the shock noise generation formula in *WOPWOP* has been devised and incorporated in this code. Numerical results for a hovering model rotor for shock noise is presented here. These indicate the importance of shock noise in the rotor plane in increasing the negative peak of the acoustic pressure signature and widening the width of the main pulse of the signature. One can, therefore, conclude on the basis of theoretical and numerical analysis that shocks on rotors should be viewed as powerful sources of sound for high speed helicopters.

References

1. F. Farassat and K. S. Brentner, "The Uses and Abuses of the Acoustic Analogy in Helicopter Rotor Noise Prediction", Journal of the AHS, 33(1), 1988, pp 29-36.

2. F. Farassat, "Quadrupole Source in Prediction of the Noise of Rotating Blades - A New Source Description", AIAA-87-2675, 1987.

3. F. Farassat and M. K. Myers, "An Analysis of the Quad rupole Noise of High Speed Rotating Blades", Computational Acoustics, Vol. 2, D. Lee, A. Cakmak, and R. Vichnevetsky (Eds.), Elsevier Science Publishers, 1990, pp 227-240.

4. J. E. Ffowcs Williams and D. L. Hawkings, "Sound Generation by Turbulence and Surfaces in Arbitrary Motion", Philosophical Transactions of the Royal Society, A264, 1969, pp 321-344.

5. F. Farassat and K. S. Brentner, "The Influence of Quad rupole Sources in the Boundary Layer and Wake of a Blade on Helicopter Rotor Noise", paper presented at the International Technical Specialists Meeting on Rotorcraft Acoustics and Rotor Fluid Dynamics, sponsored by AHS and RAeS, October 15-17, 1991 Philadelphia, PA.

6. I. M. Gel'fand and G. E. Shilov, Generalized Functions, (Vol. 1) Properties and Operations, Academic Press, 1964.

7. R. P. Kanwal, Generalized Functions - Theory and Technique, Academic Press, 1983.

8. F. Farassat and H. Tadghighi, "Can Shock Waves on Helicopter Rotors Generate Noise? A Study of the Quadrupole Source", paper presented at the 46th Annual Forum and Technology Display of the AHS, May, 1990, Washington D.C.

9. D. B. Hanson and M. R. Fink, "The Importance of Quad rupole Sources in Prediction of Transonic Tip Speed Propeller Noise", Journal of Sound and Vibration, 62(1), 1979, 19-38.

10. F. H. Schmitz and Y. H. Yu, "Theoretical Modeling of High Speed Impulsive Noise", Journal of the AHS, 24(1), 1979, 10-19.

11. K. J. Schultz and W. R. Splettstoesser, "Prediction of Helicopter Rotor Impulsive Noise Using Measured Blade Pressures", paper presented at the 43rd Annual Forum and Technology Display of the AHS, May 1987, St. Louis, MO.

12. F. Farassat, "Theoretical Analysis of Linearized Acoustics and Aerodynamics of Advanced Supersonic Propellers", AGARD CP- 366, 1984, Aerodynamics and Acoustics of Propellers.

13. K. S. Brentner, "Prediction of Helicopter Rotor Discrete Frequency Noise - A Computer Program Incorporating Realistic Blade Motions and Advanced Acoustic Formulation", NASA TM-87721, 1986.

14. H. Tadghighi, R. Holz, F. Farassat and Y.-J. Lee, "Development of a Shock Noise Prediction Code for High-Speed Helicopters - The Subsonically Moving Shock", paper presented at the 47th Annual Forum and Technology Display of the AHS, May 1991, Phoenix, Arizona.

15. D. A. Boxwell, Y. H. Yu and F. H. Schmitz, "Hovering Impulsive Noise - Some Measured and Calculated Results", Vertica, Vol. 3(1), 1979, pp 35-45.

16. L. N. Sankar and D. Prichard, "Solution of Transonic Flow Past Rotor Blades Using Conservative Full Potential Equation", AIAA-85- 5012, 1985.

Unsteady Flowfield of a Propfan at Takeoff Conditions

M. Nallasamy[1]
J.F. Groeneweg[2]

ABSTRACT

The unsteady flowfield of a propfan operating at takeoff conditions with angular inflow is examined by solving the three-dimensional Euler equations. The operating conditions considered are: Mach number = 0.31, advance ratio = 1.6 and inflow angle to the propfan =8.3°. The predicted results clearly show the cyclic variations of the blade power and thrust coefficients due to angular inflow. The flow changes from blade passage to passage are illustrated in terms of static pressure contours. The predicted blade surface pressure waveforms are compared with flight measurements. The predictions at the inboard radial station, r/R=0.68, show reasonable agreement with flight data. At the outboard radial station, r/R = 0.95, where the interactions of the tip vortex, the tip-region flow and the blade wake appear to result in a complex nonlinear measured response, the prediction shows poor agreement.

[1]Sverdrup Technology, Inc., 2001 Aerospace Parkway, Brook Park, OH 44142.
[2]NASA Lewis Research Center, 21000 Brookpark Road, MS 7-6, Cleveland, OH 44135.

INTRODUCTION

The advanced propfan design incorporates thin, highly loaded, highly swept blades to achieve superior performance at high speed over the conventional straight bladed propellers. These blades can produce a complex flowfield with leading edge/tip vortices and or shock waves depending on the operating conditions. An understanding of the propfan flow features at the design and off-design operating conditions is crucial to the improvement of the design methodologies for future propfan designs. To aid this understanding, wind tunnel and flight tests have been conducted on a large-scale 9-ft (2.74 m) diameter SR7L propfan.

Steady and unsteady blade surface measurements were carried out on a two-bladed configuration in a transonic wind tunnel in Modane, France [1,2]. Flight tests of the eight bladed design configuration were done in two sets. In the first set called the Propfan Test Assessment (PTA) program detailed acoustic and structural measurements were taken [3]. In the second set called the PTA follow-on test detailed unsteady blade surface pressure measurements were done [4]. This is the first in-flight detailed blade surface pressure measurement on any advanced propfan. Thus we have an unique data base which may be used to understand propfan aerodynamic characteristics, validate prediction methods and to improve design methodologies.

For the flight tests, the propfan was installed on the left wing of a modified, instrumented Gulfstream II testbed aircraft. A nacelle tilt arrangement was used to vary the inflow angle to the propfan. The unsteady blade surface pressures were measured on a specially designed instrumented blade using 30 pressure transducers. The suction surface had 20 pressure transducers distributed over three radial stations (r/R = 0.68, 0.86 and 0.95, where r is the radial distance and R is the blade tip radius) while the pressure surface had 10 transducers distributed over two radial stations (r/R=0.68 and 0.95).

The unsteady blade surface pressures were measured for three nacelle tilt angles, $-3°$, $-1°$ (tilt down) and $2°$ (tilt up). The effective inflow angle α to the propfan, however, depends on the airplane angle of attack, nacelle tilt and upwash angle of the flow into the propfan. The propfan was tested over a wide range of inflow angles at takeoff and cruise operating conditions. The unsteady flowfield of a propfan at cruise operating conditions was examined by Nallasamy and Groeneweg [5]. In the present paper attention is directed to the unsteady flowfield of the propfan at takeoff conditions. The flow features are examined by obtaining the unsteady three-dimensional Euler solution. The predicted blade surface pressure waveforms are compared with flight measurements. This perhaps is the first time that such a comparison of predicted blade response with in-flight measurements has become feasible.

NUMERICAL SOLUTION

The unsteady three-dimensional Euler equations governing the inviscid flow through a propfan are solved employing a solution procedure developed by Whitfield et al. [6,7]. In this procedure, the equations in conservative differential form are transformed from a Cartesian reference frame to a time dependent body fitted curvilinear reference frame. The transformed equations are discretized employing a finite volume technique. An approximate Riemann solver is used for block interface flux definitions. A lower-upper (LU) implicit numerical scheme which possesses apparent unconditional stability is used to solve the discretized equations. The flowfield is represented by multiblock composite grids to limit the core memory requirements.

COMPUTATIONAL GRID

The configuration considered is that of the eight-bladed SR7L flight test. The direction of rotation of the propfan and the axes of reference are shown in Fig. 1. An H-grid is employed to represent the flowfield. Each blade passage is described by 107 by 41 by 13 (axial by radial by circumferencial) grid points and each passage is divided into three blocks with 10734135 grid points in each. Thus 24 blocks of grid describe the entire flowfield (8 blade passages) with 456,248 nodal points. Each blade surface is represented by 49 by 27 (chordwise by spanwise) grid points with higher resolution near the leading and trailing edges, the hub and the tip. This grid resolution was found sufficient to capture the leading edge vortex for the $M = 0.2$ and $J = 0.88$ case in [8]. Figures 2(a) and (b) show the bladewise surface and spanwise surface views of the grid, respectively. Figure 2(c) shows the distribution of the grid points on the blade surface and around it. The far-field boundary is three blade radii from the blade tip, the inflow plane is two blade radii upstream of the spinner and the exit plane is two blade radii behind the blade. These boundary locations have been found to be adequate to produce accurate results [9,10].

RESULTS AND DISCUSSION

The solution presented here is for the takeoff condition, Mach number, $M = 0.31$ and advance ratio, $J = 1.6$ (run = 40 and id = 3A in ref. 4). For this case, the nacelle tilt was $-1°$ and the airplane angle of attack was $5.4°$. In addition to these angles, the upwash angle at the propfan is needed to determine the effective inflow angle to the propfan. In the absence of a simple computational procedure to determine the upwash angle, an experimental correlation obtained by Heidelberg and Woodward [11] from their SR7A model propfan test in the NASA Lewis 9- by 15-ft wind tunnel, was used to estimate the inflow angle to the propfan. They first measured the pressure response of the blades as a function of inflow angle for the propfan alone configuration. Then the wing was installed downstream of the propfan (as in the flight test, tractor configuration) and blade pressure response was measured over a range of wing angles of attack and nacelle tilt. It is assumed that the local inflow angle and the propfan angle of attack are the same in the propfan alone case. Then by matching the measured first harmonic response of the blades with wing installation to that of propfan alone configuration, they found the equivalent inflow angle as a function of wing angle of attack (Fig. 13 in [11]). This correlation was obtained for takeoff conditions, Mach number = 0.2 and advance ratio 0.88. The same correlation is used here to obtain the effective inflow angle to the propfan. For an airplane (wing) angle of attack of 5.4° and a nacelle tilt (droop) angle of –1° considered here, a value of 8.3° is obtained for the inflow angle from that correlation. Thus, the solution presented and discussed here was obtained with an inflow angle of 8.3° to the propfan axis (Fig. 1).

From an impulse start, the unsteady Euler solutions are obtained for three complete revolutions of the propeller to obtain a reasonably accurate solution. By the third revolution, the results have stabilized as indicated by the periodic variation of per blade power coefficient (Fig. 3(a)) during the second and third revolutions of the propfan. The figure shows the variation of the single blade power coefficient with azimuth angle for four blades, when started at ϕ_s = 0, 90, 180 and 270° respectively, for three complete revolutions of the propfan. The expected sinusoidal variation of the blade loading due to angular inflow is clearly observed in each case. The initial transient loading at the start of the run, however, is quite different for each blade and depends on the azimuthal position of the blade. Irrespective of the magnitude of this transient, the power coefficients become nearly sinusoidal at the end of the first revolution. The amplitude of the stabilished power coefficient during the third revolution varies +61 and –53 percent about the mean. The difference between the minimum and maximum levels about the mean results from the higher positive inflow angle that has been considered here, as a look at the velocity triangles would show. For low inflow angles the variation about the mean in the positive and negative directions has been found to be nearly the same [5,8]. Even though the inflow angle is higher than that in [8], the amplitude variation is small due the low Mach number of the flow considered here.

To see the harmonic content of the loading, the power coefficient for a single blade in the third cycle is Fourier transformed to yield CP = ao + ai cos wt + bi sin wt, where ao, ai, and bi are the Fourier coefficients and t is the time. The predicted mean blade power coefficient, ao, is 0.127 and is the same as that for the steady flow (α= 0). The mean power coefficient measured (per blade) in the flight test is 0.126 for the propfan operating conditions considered here. The loading spectrum is shown in Fig. 3(b). The loading is dominated by the first harmonic, but contributions do come from higher harmonics. The first harmonic loading lags the blade motion by 3.7°. This is much smaller than the 17.5° lag obtained for the takeoff conditions (M = 0.2 and J = 0.88, $\alpha = 3°$) of the two-bladed configuration considered in [8]. It is due to the fact that the blade rotational frequency in the present case is only about half that of the case in [8], thus resulting in a low reduced frequency and flow unsteadiness. In addition, the two-bladed case showed the formation of a leading edge vortex which may have been partly responsible for the higher lag.

Figure 4(a) shows the spanwise variation of the elemental power coefficient for four azimuthal locations of the blade. Also shown in the figure is the curve for the steady flow ($\alpha = 0$). It is seen that at $\phi = 0$ and 90° azimuthal locations the blade loading is lower than the steady levels whereas at $\phi = 180$ and 270° it is higher than the steady value. The magnitude of the cyclic variation of the loading depends on the spanwise location. The elemental thrust coefficient variations for the same four azimuthal locations and steady flow are shown in Fig. 4(b). It is seen that near the hub region, a small portion of the blade experiences negative thrust even in steady flow due to off-design operation of the propfan. The variations at different azimuth angles are similar to those of elemental power coefficient: at $\phi = 0$ and 90° the values of the elemental thrust coefficient are less than the steady value while at $\phi = 180$ and 270° the values are greater than the steady value. It is seen that in either case (Figs. 4(a) or (b)), the maximum deviation (in absolute value) from the steady value occurs at the radial station where elemental power/thrust coefficient attains a maximum, as would be expected from angle of attack variations.

The detailed flowfield information in the blade passages are presented in the form of static pressure contours at the spanwise station, r/R = 0.68, in Fig. 5. In a steady flow (zero degree inflow angle) the static pressure contours (shown on the top of the figure) in all the eight blade passages will be identical. But for an inflow angle of 8.3° changes in the flow occur from passage to passage. In the highly loaded portion of the revolution, the leading edge region near the pressure surface experiences higher pressure levels as indicated by the additional contours and levels (see blade passages marked 180-225, 225-270 and 270-315°). Of course, due to the low Mach number of the flow in the present case no passage shock forms as in the cruise case [3].

Next, we present detailed comparisons of the predicted blade pressure waveforms with flight data for two radial stations r/R = 0.68 and r/R = 0.95. In these waveform comparisons, 0° corresponds to the (top-dead-center) vertical direction for the aircraft installation, as in the presentation of flight data in [4]. First of all, it should be mentioned that during the flight tests no steady pressure measurements were made. However, some steady pressure data were obtained during the unsteady pressure test by retaining the DC component of the pressure signal. These data do not indicate the formation of a leading edge vortex for the operating conditions, $M = 0.31$ and $J = 1.6$, chosen here. Consequently, the measured pressure waveforms on the suction surface at the outboard radial station do not show (see the discussion below for Fig. 8) the double hump nature observed in Modan tests [2] for $M = 0.2$ and $J = 0.88$.

Figure 6 shows the unsteady blade surface pressure as a function of azimuth angle for six transducer locations at $r/R = 0.68$, on the suction side. The measured and predicted blade responses indicate that the response is the largest at the transducer near the leading edge ($x/c = 0.05$, where x is the axial distance and c is the blade chord) and the response reduces gradually towards the trailing edge. The predicted phases of the waveform are in close agreement with the data. However, the amplitudes are overpredicted, the maximum overprediction occurring at the transducer near the leading edge. Similar agreements of the predicted waveforms with wind tunnel data at cruise conditions were also found in [12].

The pressure waveforms for the four transducer locations on the pressure side of the blade are shown in Fig. 7. Here again, the phases of the predicted waveforms agree quite well with flight measurements while the amplitudes are overpredicted. Thus the pressure waveforms for transducer locations at the inboard radial station on suction and pressure surfaces show reasonable agreement with flight data.

Figure 8 shows the pressure waveforms on the suction surface at $r/R = 0.95$ for the five transducer locations. Even at this radial station, the prediction indicates the largest response to occur near the leading edge and to reduce gradually towards the trailing edge. However, the flight measurements show that the highest response occurs downstream of the leading edge at $x/c = 0.25$. At this and other transducer locations downstream ($x/c = 0.42$, 0.58, and 0.92) we observe that the magnitudes are underpredicted and the measured pressure waveforms show a relative phase lag. The measured response at $x/c = 0.25$ is higher than the predicted one by as much as 100 percent at an azimuthal location of 300° which is in the lightly loaded (blade retreating) region of the revolution. Such a large response can only result from the movement of the tip vortex during the revolution and/or strong interactions between the tip vortex and tip-region flow which are not adequately resolved in the present solution. The phase difference between the waveforms may result from two things: (1) due to the interactions mentioned above and (2) the wing installation (or installation effect) in the flight tests as compared to propfan alone configuration of the computation. Heidelberg and Woodward [11] compare the measured pressure waveforms with and without wing for one transducer location, $r/R = 0.75$ and $x/c = 0.1$ for $\alpha = 10°$. They find that the waveforms are identical except for a small relative phase lag of 11° in the wing case. In the present case, the measured (wing case) waveforms also show a phase lag relative to the predicted (no wing) ones, which depends on the transducer location. The transducer near the leading edge ($x/c = 0.08$) show a small lead whereas all the remaining transducers show a phase lag ranging from 8 to 30°. More detailed investigation is necessary to understand the complex interactions in the tip region and the installation effects.

It should, however, be noted that when data indicated the formation of a leading edge vortex as in [2], the blade pressure response on the suction surface was double humped at all the transducer locations on the chord (at r/R = 0.91). The phases of the predicted wave forms were uncorrelated with measurements [8]. In the present case no such double hump response was measured, again perhaps indicating that no leading edge vortex forms at this operating condition.

Figure 9 shows the pressure waveforms for the six transducer locations on the pressure side of the blade at r/R = 0.95. The measured waveform at x/c = 0.58 is erratic for this run (=40) as noted in [4] and is not shown here. At the transducer locations x/c = 0.08, 0.25, 0.42 and 0.75 the predicted peak amplitudes agree reasonably well with flight data. At all the four locations the measured waveforms show a relative phase lag which again may be the result of wing/ no wing configurations and/or the tip-region flow being not resolved well in the predictions. The response at the transducer location x/c = 0.92 is small and the measured waveform is nearly 180° out of phase with the predicted one and with the wave forms at other four transducer locations. The transducer at x/c = 0.92 is not listed as erratic or inoperable. It is not clear as to the kind of blade wake-tip region interaction that will produce such a phase variation.

CONCLUDING REMARKS

Numerical solution of the flow through a propfan operating at takeoff conditions with angular inflow has been obtained. The results show that the blade power and thrust coefficients undergo cyclic variations due to angular inflow. The predicted blade surface pressure waveforms were compared with flight measurements. At the inboard radial station, r/R = 0.68, the predicted pressure waveforms are in reasonable agreement with flight data. However, at the outboard radial station r/R = 0.95 the agreement between the measurement and prediction is poor. The discrepancies between the prediction and data at this radial station seem to arise due to inadequate flow resolution in the blade tip region. The installation effects may also contribute to the observed phase differences between the measured and predicted pressure waveforms.

REFERENCES

1. Bushnell, P.; Measurement of the Steady Surface Pressure Distribution on a Single Rotation Large-Scale Advanced Propfan Blade at Mach Numbers from 0.03 to 0.78. NASA–CR–182124, 1988.
2. Bushnell, P.; Gruber,M.; and Parzych, D.; Measurement of Unsteady Blade Surface Pressure on a Single Rotation Large Scale Advanced Propfan with Angular and

Wake inflow at Mach Numbers from 0.02 to 0.70. NASA–CR–182123, 1988.

3. Little, B.H.; Poland, D.T.; Bartel, H.W.; Withers, C.C.; and Brown, P.C.; Propfan Test Assessment (PTA) Final Project Report, NASAS–CR–185138, 1989.
4. Parzych, D.; Boyd, L.; Meissner, W. and Wyrostek, A.; In Flight Measurement of Steady and Unsteady Blade Surface Pressure of a Single Rotation Large Scale advanced Propfan Installed on the PTA Aircraft, NASA–CR to be published, 1991.
5. Nallasamy, M; and Groeneweg, J.F.; Unsteady Euler Analysis of The flow Field of a Propfan at an Angle of Attack, AIAA–90–0339, also NASA–TM–102426, 1990.
6. Whitfield, D.L.; Swafford, I.W.; Janus, J.M.; Mulac, R.A.; and Belk, D.M.; Three Dimensional Unsteady Euler Solutions for Propfans and Counter-Rotating Propfans, AIAA paper 87–1197, June 1987.
7. Janus, J.M.; and Whitfield, D.L.; A Simple Time Accurate Turbomachinery Algorithm with Numerical Solutions of Uneven Blade Count Configuration, AIAA paper 89–0206, Jan. 1989.
8. Nallasamy, M.; and Groeneweg, J.F.; Unsteady Blade Surface Pressures on Large-Scale Advanced Propeller: Prediction and Data., AIAA paper 90–2402, also NASA–TM–103218, July 1990.
9. Nalllasamy, M.; Clark, B.J.; and Groeneweg, J.F.; Euler Analysis of the Three-Dimensional Flow Field of a High Speed Propeller: Boundary Condition Effects, Journal of Turbomachinery, Vol. 109, pp. 3332-339, 1987.
10. Nallasamy, M.; Yamamoto, O.; Warsi, S.; and Bober, L.J.; Large-Scale Advanced Propeller Blade Pressure Distributions: Prediction and Data, Journal of Propulsion and Power, Vol. 7, pp. 452-461, 1991.
11. Heidelberg, L.J.; and Woodward, R.P.; Advanced Turboprop Wing Installation Effects Measured by Unsteady Blade Pressure and Noise, AIAA paper 87–2719, also NASA–TM–100200, 1987.
12. Heidelberg, L. J.; and Nallasamy, M.; Unsteady Blade Pressure Measurements for the SR-7A Propeller at Cruise Conditions, AIAA paper 90–4022, Oct. 1990, also NASA–TM–103606.

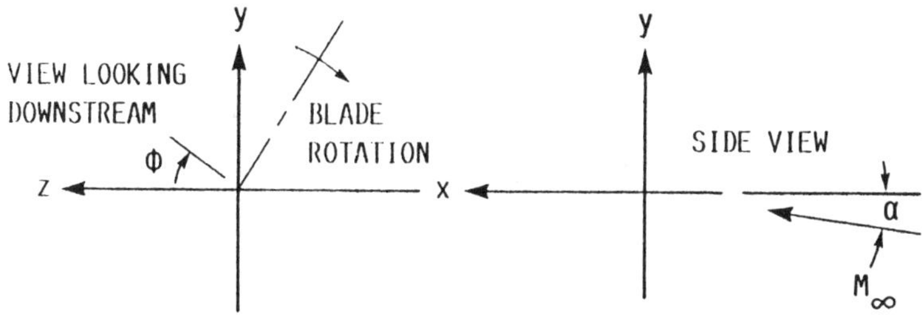

Figure 1. – Axes of reference.

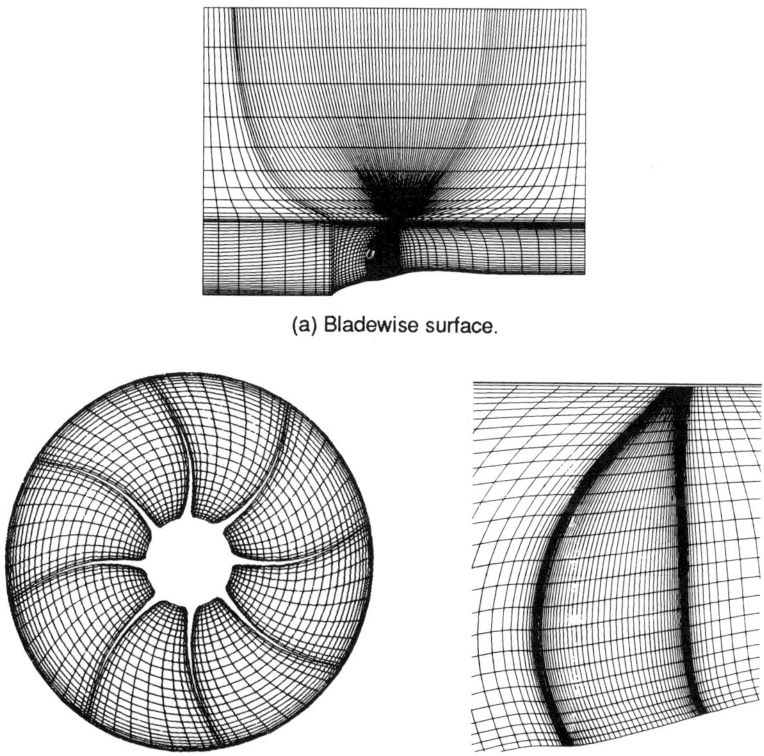

(a) Bladewise surface.

(b) Spanwise surface.

(c) Distribution of grid points on and near the blade surface.

Figure 2. – Computational grid.

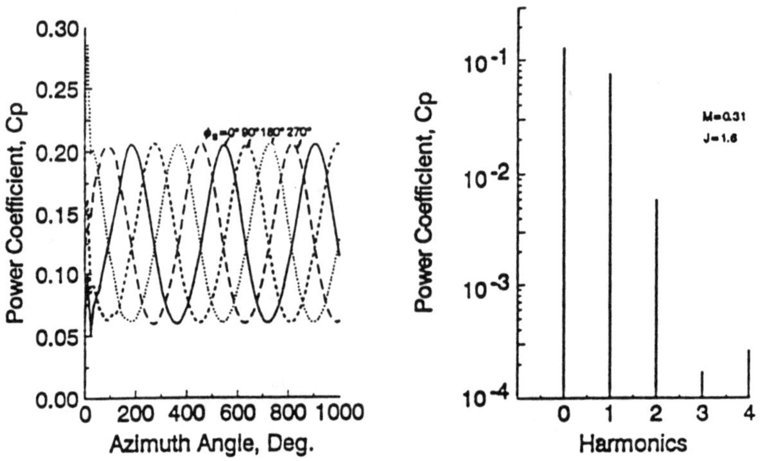

(a) Power

(b) Loading

Figure 3. – Power per blade variation.

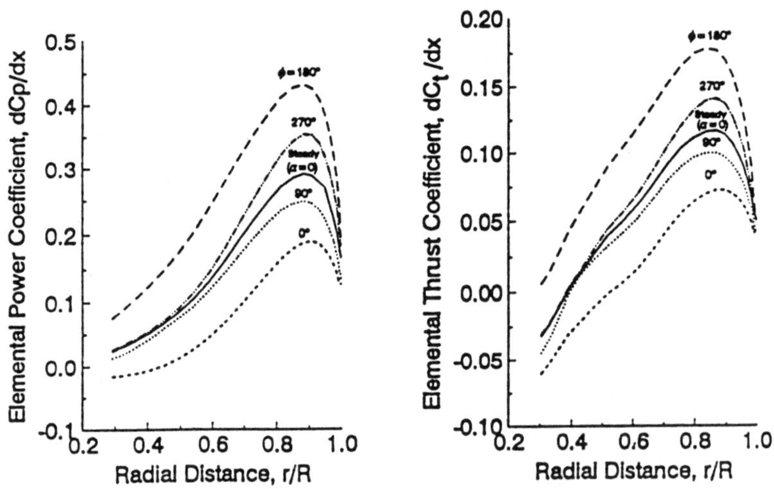

(a) Power.

(b) Thrust.

Figure 4. – Elemental Power and thrust coefficient.

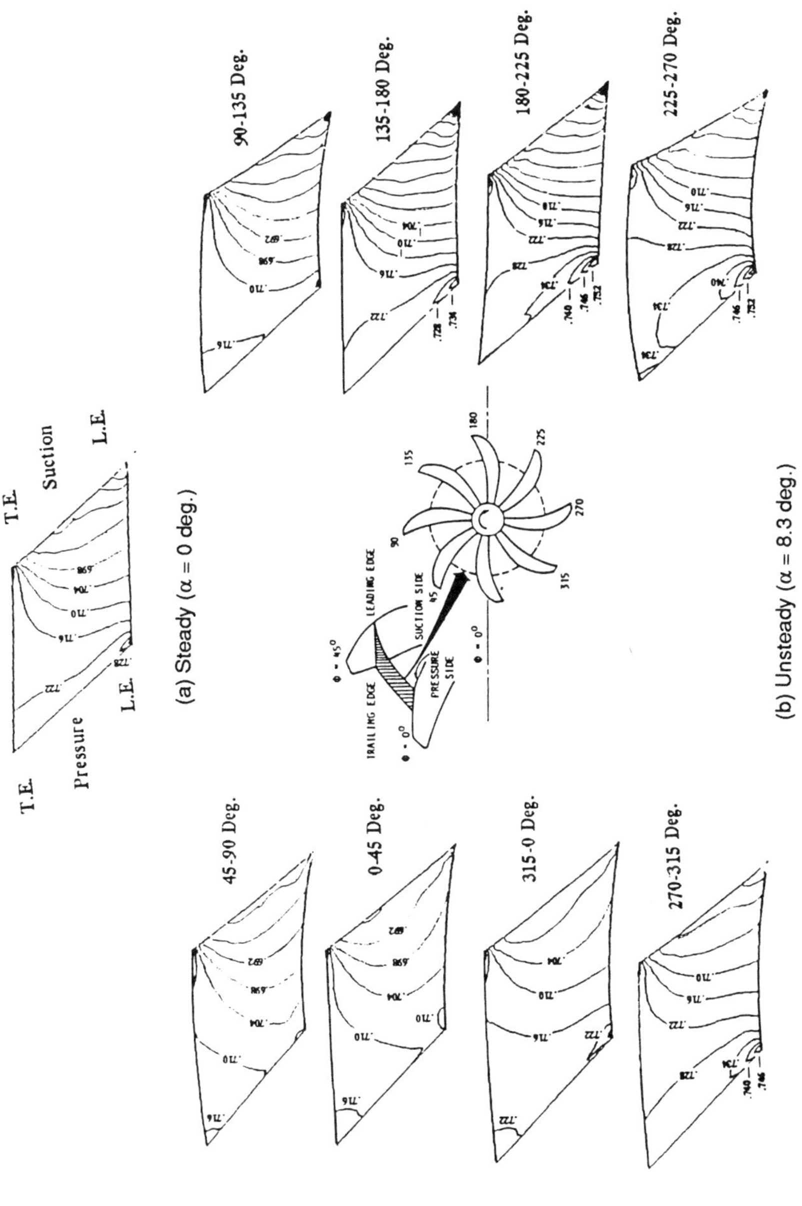

(a) Steady (α = 0 deg.)

(b) Unsteady (α = 8.3 deg.)

Figure 5. – Instantaneous static pressure contours in the blade passages at r/R=0.68. M=0.31, J=1.6.

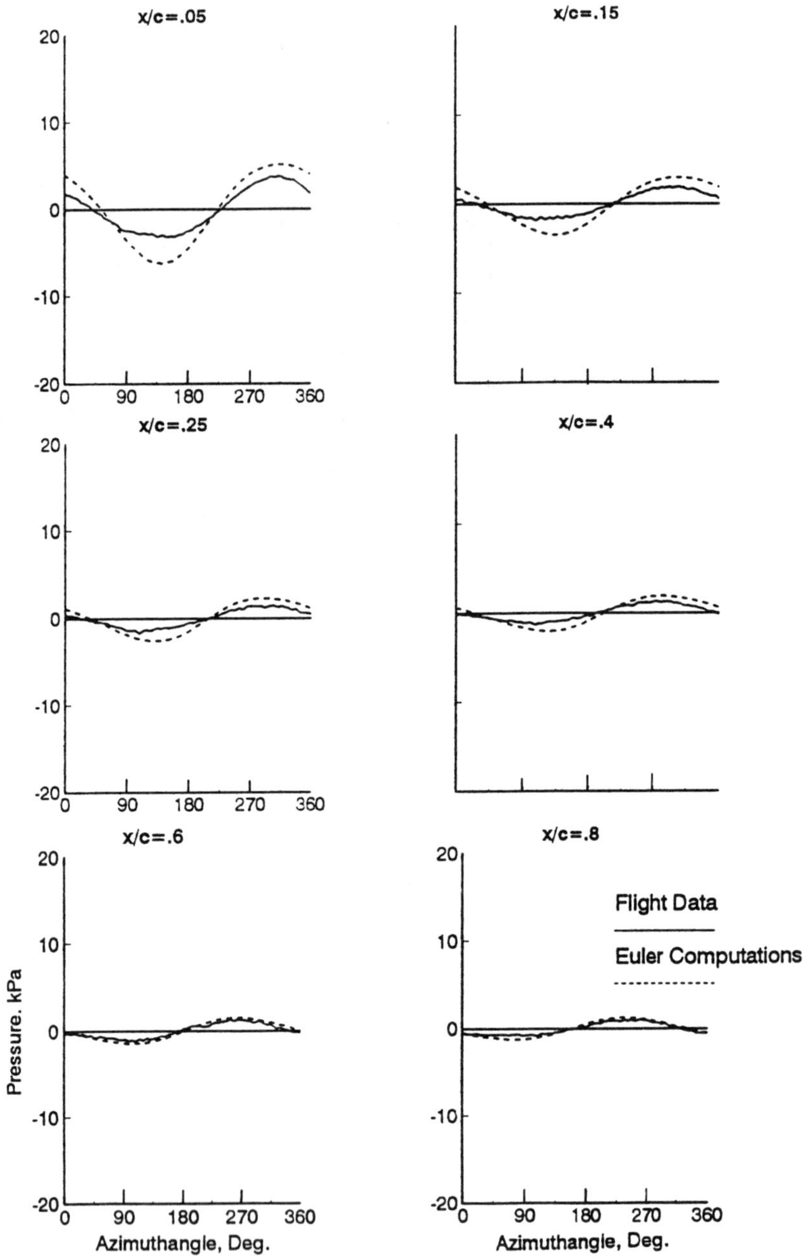

Figure 6. – Predicted and measured presure waveforms at r/R=0.68 on the suction surface. M=0.31, J=1.6, α =8.3°.

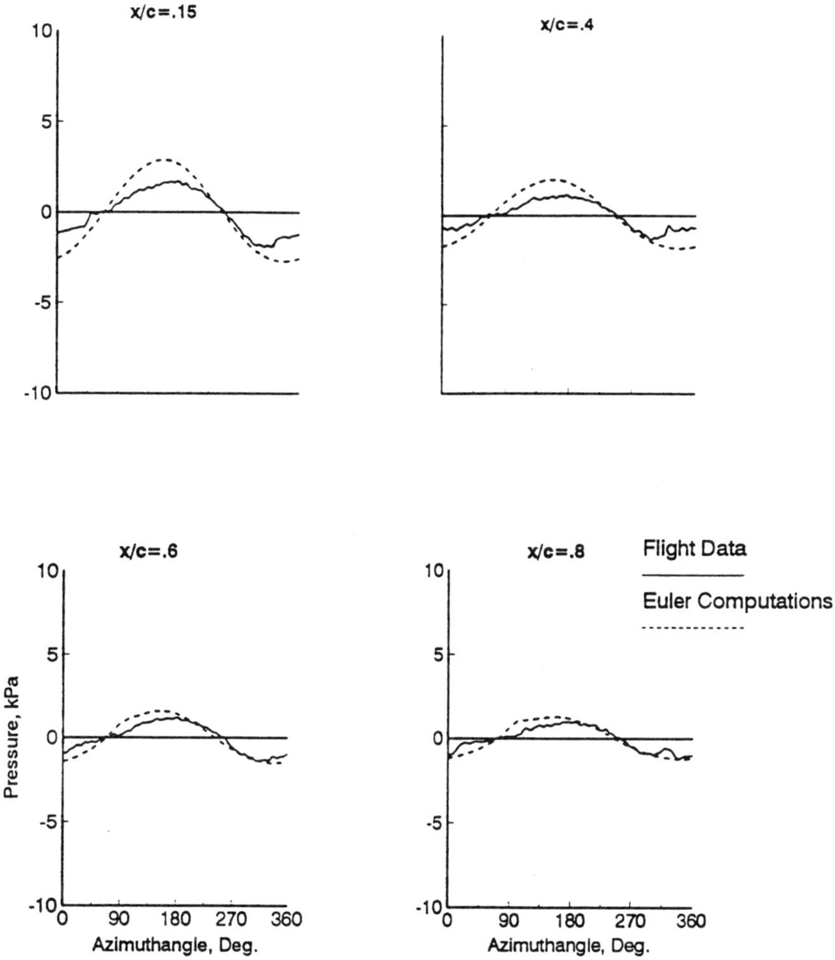

Figure 7. – Predicted and measured presure waveforms at r/R=0.68 on the presure surface. M=0.31, J=1.6, α =8.3°.

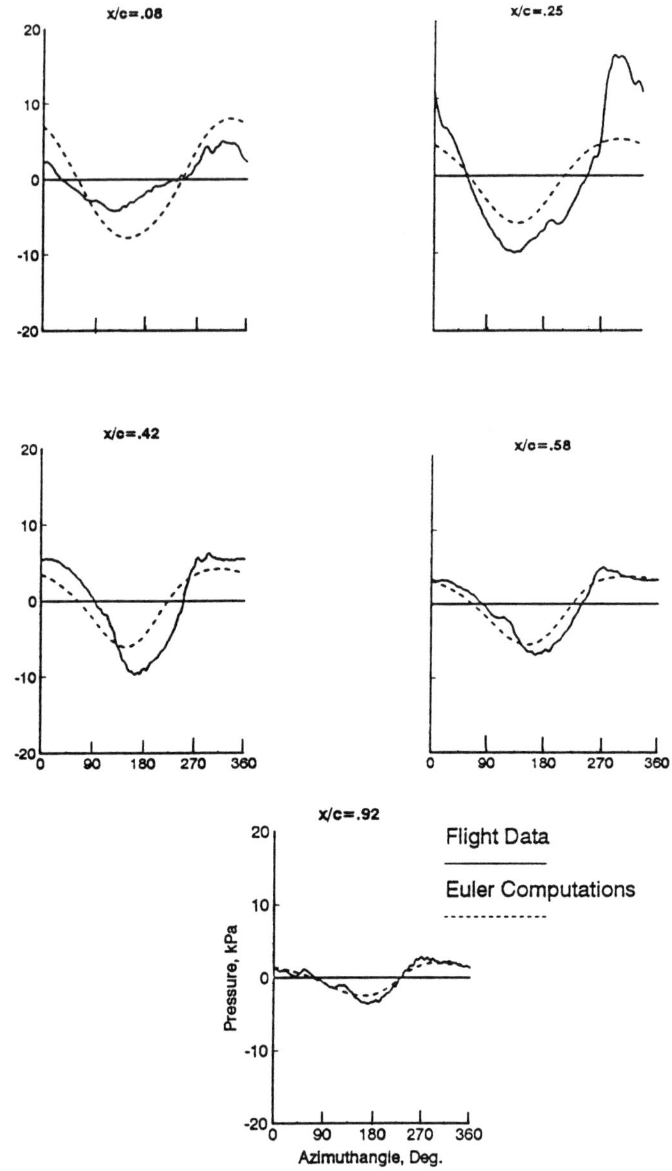

Figure 8. – Predicted and measured presure waveforms at r/R =0.95 on the suction surface. M=0.31, J=1.6, α =8.3°.

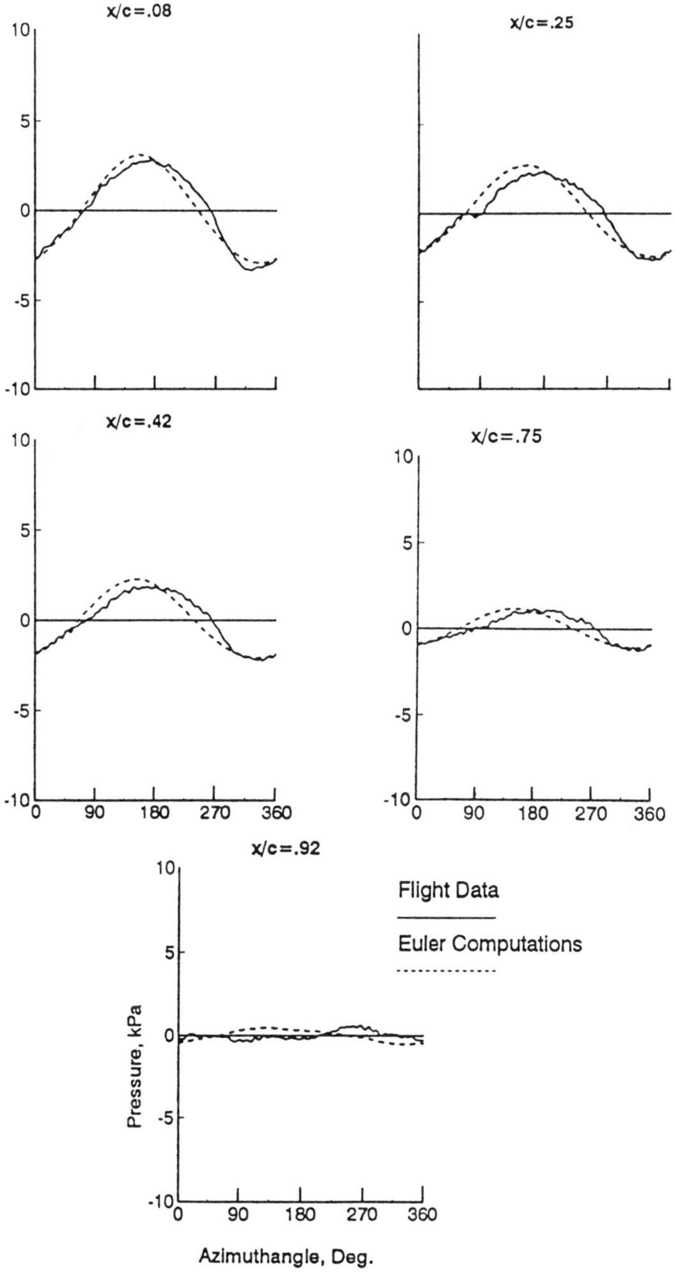

Figure 9. – Predicted and measured presure waveforms at r/R =0.95 on the presure surface. M=0.31, J=1.6, α =8.3°.

Prediction of Noise Field of a Propfan at Angle of Attack

Edmane Envia[1]

ABSTRACT A method for predicting the noise field of a propfan operating at angle of attack to the oncoming flow is presented. The method takes advantage of the high-blade-count of the advanced propeller designs to provide an accurate and efficient formula for predicting their noise field. The formula, which is written in terms of the Airy function and its derivative, provides a very attractive alternative to the use of numerical integration. A preliminary comparison shows rather favorable agreement between the predictions from the present method and the experimental data.

1.1 Introduction

The need to limit both in-flight cabin noise and community noise has prompted the development of methods for accurate prediction of acoustic performance of propfans. A particular aim of these methods has been the prediction of the noise field of a propfan operating at angle of attack to the oncoming flow. Experimental evidence [1] suggest that the noise radiation characteristics of a propfan at angle of attack are significantly different from those of a propfan at zero angle of attack. These differences stem from the changes which occur both in the aerodynamics and aeroacoustics of a propeller whose axis is inclined with respect to the direction of oncoming flow.

One of the most important of these changes is the azimuthal variation of aerodynamic loading experienced by a propeller at angle of attack. Measurements [2], as well as numerical simulations [3], have shown that the loading variation or "unsteadiness" can be quite substantial over one period of revolution. Physically, the modulation occurs because propeller blades experience periodic variations in the local flow incidence. Acoustically, the loading unsteadiness creates a noise field which is axially non-symmetric. Methods which predict this effect have been in use for some time (see Hanson [4], for example).

Another feature of a propeller operating at angle of attack, which is ab-

[1]Sverdrup Technology, Inc., 2001 Aerospace Parkway, Brook Park, OH 44142

sent at zero angle of attack case, is the cross-flow convective phase effect. This effect, first pointed out by Mani [5], may be explained as follows. For an inclined propeller, the oncoming flow has a nonzero velocity component in the plane of rotation. As a result, an acoustic source point fixed to the surface of the propeller "sees" a medium whose convective velocity varies periodically. This means that the convective phase of the sound radiated from this point is also periodically modulated. Krejsa [6], recently developed a prediction scheme which accounts for this effect by assuming that propeller sources are acoustically compact in the chordwise direction.

In this paper, a method which incorporates both the loading unsteadiness and cross-flow phase effects in a straightforward manner, without the simplifying assumptions such as chordwise compactness, is presented. This method, which is based on a moving-medium variant of the Ffowcs Williams and Hawkings [7] equation, relies on a frequency-domain formulation similar to that employed by Hawkings and Lawson [8] or Hanson [9], however, it differs from them in that the resulting Fourier integrals are evaluated asymptotically in the "large-blade-count limit" rather than in the far- or near-field limits. The advantage of this approach lies in the fact that it provides a uniform representation for the sound field in both the near- and far-fields while, at the same time, it affords the flexibility of using realistic blade geometries and aerodynamic loading distributions without their associated computational penalties.

1.2 Analysis

The starting point for the analysis is the moving-medium solution of the Ffowcs Williams and Hawkings equation for the acoustic pressure, $p(\mathbf{x}, t)$, as given by Goldstein [10];

$$p(\mathbf{x},t) = -\int_{-\infty}^{+\infty} \int_{S(\tau)} \rho_0 v_n \frac{D_0 G}{D\tau}\, ds(\mathbf{y})\, d\tau + \int_{-\infty}^{+\infty} \int_{S(\tau)} f n_i \frac{\partial G}{\partial y_i}\, ds(\mathbf{y})\, d\tau$$
$$+ \int_{-\infty}^{+\infty} \int_{V(\tau)} T_{ij} \frac{\partial^2 G}{\partial y_i \partial y_j}\, d\mathbf{y}\, d\tau, \qquad (1a)$$

$$\frac{D_0}{D\tau} = \frac{\partial}{\partial \tau} + U_{0i} \frac{\partial}{\partial y_i}, \qquad (1b)$$

where ρ_0 is the ambient density, v_n the normal component of the surface relative velocity, f the amplitude of the aerodynamic loading, and T_{ij} the Lighthill stress tensor. $S(\tau)$ and $V(\tau)$ represent the propeller blade surfaces and the volume surrounding the blades, respectively. n_i's are components of the outward unit normal to S. It should be noted that the convention for the direction of the unit normal is opposite that of Goldstein's. The medium convection velocity is denoted by U_0. $\mathbf{y}$ and $\mathbf{x}$ are the source and observer coordinates, respectively. τ denotes the source time and t the observer (i.e., retarded) time. The three terms in Eq. (1a) represent contributions from the thickness (monopole), loading (dipole) and stress (quadrupole) sources of noise. In this paper, however, the contribution from the last term will be ignored. The extension of the analysis to include quadrupole sources will be discussed elsewhere.

$G = G(\mathbf{x}, t/\mathbf{y}, \tau)$ in Eq. (1a) denotes the free-space moving-medium Green's function which may be written in the following form;

$$G = \frac{1}{4\pi\kappa R}\,\delta\left(t - \tau - g_c R/C_0\right), \tag{2a}$$

$$g_c(\tau) = \frac{1}{\beta_0^2}\left(\kappa - M_{0_R}\right), \qquad \kappa(\tau) = \left(M_{0_R}^2 + \beta_0^2\right)^{1/2},$$

$$M_{0_R}(\tau) = M_{0i}\, e_i, \qquad e_i(\tau) = \frac{(x_i - y_i)}{R}, \qquad \beta_0^2 = 1 - M_{0i}^2,$$

$$M_{0i} = \frac{U_{0i}}{C_0}, \qquad R(\tau) = |\mathbf{x} - \mathbf{y}(\tau)|. \tag{2b}$$

Here M_{0_R} is the component of the medium convection Mach number in the radiation direction, e_i's the components of the unit vector in the radiation direction, R the length of the radiation vector, and C_0 the medium speed of sound. Parameters g_c and κ represent the effects of medium convection on the retarded time and the spherical spreading rate, respectively. The explicit dependence of various parameters on the source time, τ, is indicated where necessary.

In order to simplify the description of the source motion, the 1-axis is chosen to coincide with the propeller shaft (see Fig. (1)). As a result the motion of the sources is confined to the transverse planes described by the 2– and 3–axes. In this coordinate system, the medium convection Mach number can be written as:

$$M_{01} = M_0 \cos\alpha, \qquad M_{02} = 0, \qquad M_{03} = M_0 \sin\alpha, \tag{3}$$

where α is the propeller angle of attack with respect to the oncoming flow as shown in Fig. (1).

Given the blade geometry and aerodynamic loading, the acoustic pressure, $p(\mathbf{x}, t)$, may be computed from Eq. (1a) for any observer location. This, of course, is the so called time-domain approach (see Farassat [11], for example) which entails solving transcendental equations for the source time, τ. Alternatively, one could use a frequency-domain approach which involves expanding the pressure field, $p(\mathbf{x}, t)$, in terms of its complex Fourier harmonic components, $p_k(\mathbf{x})$, i.e.,

$$p(\mathbf{x},t) = \sum_{k=-\infty}^{+\infty} p_k(\mathbf{x})\, e^{-ik\Omega t}, \tag{4a}$$

with the individual harmonic components given by:

$$p_k(\mathbf{x}) = \frac{\Omega}{2\pi}\int_0^{2\pi/\Omega} e^{ik\Omega t}\left[\int_{-\infty}^{+\infty}\int_{S(\tau)} \rho_0 v_n \frac{D_0 G}{Dt}\, ds(\mathbf{y})\, d\tau\right] dt -$$

$$\frac{\Omega}{2\pi}\int_0^{2\pi/\Omega} e^{ik\Omega t}\left[\int_{-\infty}^{+\infty}\int_{S(\tau)} f n_i \frac{\partial G}{\partial x_i}\, ds(\mathbf{y})\, d\tau\right] dt, \tag{4b}$$

where Ω is the propeller angular speed. In writing Eq. (4b), the derivatives of G with respect to the source coordinates from Eq. (1a) have been replaced by its derivatives with respect to the observer coordinates through the use of the simple relations:

$$\frac{\partial G}{\partial y_i} = -\frac{\partial G}{\partial x_i}, \qquad \frac{D_0 G}{D\tau} = -\frac{D_0 G}{Dt}. \tag{5}$$

Using the usual phase relationship arguments it can be shown that for B identical blades only p_k's for which $k = mB$ (where m is an integer) contribute to the infinite sum in Eq. (4a). This contribution is simply B times the contribution of a single blade. Therefore, from here on B will explicitly appear in the expressions for p_{mB} and, correspondingly, S will denote the surface of a single blade.

To further simplify Eq. (4b), the spatial derivatives, $\partial/\partial x_i$, of G may be rewritten in terms of the temporal derivative, $\partial/\partial t$, through the use of the chain rule;

$$\frac{\partial G}{\partial x_i} = -\left[\frac{1}{C_0}\frac{\partial (g_c R)}{\partial x_i}\frac{\partial}{\partial t} + \frac{1}{R}\frac{\partial R}{\partial x_i} + \frac{1}{\kappa}\frac{\partial \kappa}{\partial x_i}\right] G. \tag{6}$$

The temporal derivatives can now be removed using integration by parts and the integral over t, which involves a delta function, can then be easily evaluated to yield,

$$p_{mB}(\mathbf{x}) = \int_0^{2\pi/\Omega}\left[\int_{S(\tau)} Q_{\text{monopole}}(\mathbf{x},\mathbf{y},\tau)\, e^{imB\Omega(\tau+g_c R/C_0)}\, ds(\mathbf{y})\, d\tau\right] +$$

$$\int_0^{2\pi/\Omega}\left[\int_{S(\tau)} Q_{\text{dipole}}(\mathbf{x},\mathbf{y},\tau)\, e^{imB\Omega(\tau+g_c R/C_0)}\, ds(\mathbf{y})\, d\tau\right], \tag{7}$$

where Q_{monopole} and Q_{dipole} represent the expressions for the "amplitudes" of the two acoustic source. Note that the limits on the integration over τ

have been changed to reflect the fact that an interval of size $2\pi/\Omega$ in t is mapped exactly into an interval of size $2\pi/\Omega$ in τ. The expressions for the monopole and dipole source amplitudes are given by:

$$Q_{\text{monopole}} = -\frac{B\Omega\rho_0 v_n(\tau)}{8\pi^2}\left[\frac{imB\Omega}{\kappa R} - \frac{M_{0_i}}{\kappa^2 R}(e_i - g_c M_{0_i}) + \right.$$

$$\left.\frac{M_{0_i}}{\kappa^3 R^2}\left(M_{0_R}\, M_{0_i} + \beta_0^2 e_i\right)\right], \quad (8a)$$

$$Q_{\text{dipole}} = \frac{B\Omega f(\tau)}{8\pi^2}\left[-\frac{imB\Omega n_i}{\kappa^2 R}(e_i - g_c M_{0_i}) + \right.$$

$$\left.\frac{n_i}{\kappa^3 R^2}\left(M_{0_R}\, M_{0_i} + \beta_0^2 e_i\right)\right]. \quad (8b)$$

Note that, for a propeller operating at an angle of attack, v_n and f depend on τ. n_i and e_i also depend on τ even if the propeller is operating at zero angle of attack. Note also that, the definition of the source amplitude here differs from the standard one in that it includes the fall-off with distance R.

Once source amplitude distributions Q_{monopole} and Q_{dipole} are estimated, it remains to carry out the integrations in Eq. (7). Ordinarily, in the context of a frequency-domain analysis, the integral over τ is computed first. Since, for general geometries and source amplitudes, this integral is not tractable analytically, it is usually computed asymptotically for, say, near- or far-field observer locations for which the integrand may be simplified significantly. The result is given in terms of the appropriate Bessel functions. The remaining surface integral is then carried out analytically or numerically depending on how faithfully the blade geometry is to be represented.

There is, however, a less restrictive, and at the same time, more accurate way of computing $p_{mB}(\mathbf{x})$. This is accomplished by evaluating the τ integral asymptotically for large-blade-count, B (or more precisely mB), using a modification to the standard steepest descent and saddle point methods. The advantage of this approach lies in the fact that it does not require a near- or far-field approximation and, therefore, is applicable to general geometries and source amplitudes. The details of the derivation will be presented for the zero angle of attack case first. The results will then be used to derive an extension to the angle of attack cases most often encountered in practice. For the sake of brevity, the formula will be given in terms of a generic source amplitude Q. The resulting expression is therefore applicable to both the monopole and dipole sources.

1.2.1 Zero Angle of Attack Case ($\alpha = 0$)

Since, it is desirable to represent the blade geometry as accurately as possible, the surface integrals cannot be carried out analytically and must, instead, be performed using a suitable quadrature scheme. For practical considerations, it is more convenient to evaluate the surface integral first. To do so, we begin by dividing the blade surface into N_p small panels. If the typical panel size is sufficiently small the integrand may be assumed to be constant over the panel and, thus, the surface integral may be approximated by some appropriate "mean value" of the integrand times the panel surface area ΔS. If the mean value is chosen to be the value of the integrand at the geometric center of the panel, designated $\mathbf{y_S}$, the error in this approximation is $O(L^2)$ for a given N_p, where $L = |\mathbf{y} - \mathbf{y_S}|_{\max}$. Of course, the error could be made arbitrarily small by choosing a large enough N_p. Therefore, $P_{mB}(\mathbf{x})$ may be written as:

$$p_{mB}(\mathbf{x}) \simeq \sum_{n_p=1}^{N_p} \Delta S_{n_p}\, e^{-imB\Psi_c}\, I_{n_p}, \tag{9a}$$

$$I_{n_p} = \int_0^{2\pi} Q(\theta)\, e^{mB\Phi(\theta)}\, d\theta, \tag{9b}$$

$$\Phi(\theta) = i\left(\theta + a\sqrt{1 - b\cos\theta}\right), \tag{9c}$$

$$a = \frac{M_{tip}}{\beta_{0_1}\sqrt{\chi^2 + r^2 + r_s^2}}, \qquad b = \frac{2rr_s}{\chi^2 + r^2 + r_s^2}, \tag{9d}$$

$$\Psi_C = \frac{M_{tip} M_{0_1} \chi}{\beta_{0_1}} + \phi_s - \phi,$$

$$\chi^2 = \frac{1}{\beta_{0_1}^2}(x_1 - y_{s_1})^2, \qquad \beta_{0_1}^2 = 1 - M_{0_1}^2, \tag{9e}$$

where for brevity (x_1, x_2, x_3) and $(y_{s_1}, y_{s_2}, y_{s_3})$ have been replaced by their cylindrical polar counterparts (x_1, r, ϕ) and (y_{s_1}, r_s, ϕ_s), respectively. Furthermore, the integration variable τ is replaced by a new variable $\theta = \Omega\tau + \phi_s - \phi$. M_{tip} is the tip rotational Mach number (i.e., $R_{tip}\Omega/C_0$), ΔS_{n_p} the surface area of the n_pth panel and Ψ_C is a convective phase factor representing the collection of phase terms which do not depend on τ. $Q(\theta)$ is the source amplitude function written in terms of the variable θ. In order to simplify the notation, the explicit dependence of various variables on $\mathbf{y_S}$, the centroid of the n_pth panel, is suppressed. $\Phi(\theta)$ is the canonical phase function for a propeller operating at zero angle of attack, with a and b representing a combination of geometric, convective and kinematic factors. In deriving the above expressions, both the observer and source spatial coordinates are non-dimensionalized with respect to the propeller tip radius R_{tip}. (Note that, $M_{01} = M_0$ for zero angle of attack case.)

For a typical propfan $B = 8$, thus, even at the blade-passing-frequency (BPF), i.e., $m = 1$, the integrand in Eq. (9b) is highly oscillatory. Of course, one way to tackle the problem is to integrate the θ-integral using a suitable quadrature scheme. That was indeed the approach taken in Nallasamy et al. [12]. Unfortunately, accurate computation of higher harmonics (i.e., $m \geq 2$) requires an ever increasing quadrature resolution to capture the oscillatory nature of the integrand, and that may prove to be computationally expensive. It turns out, however, that it is possible to evaluate the integral I_{n_p} rather accurately using the steepest descent and saddle point methods. Naturally, in that case, the variable θ must be allowed to be complex and the integration path (i.e., $[0, 2\pi]$) be deformed into an appropriate contour.

Since in most practical applications of interest, propfans operate at a supersonic relative tip Mach number (i.e., $M_{(rel)tip} = (M_{tip}^2 + M_{01}^2)^{1/2}$), in the subsequent analysis it is assumed that $M_{(rel)tip} > 1$. Replacing θ with the variable, $\nu = \theta + i\sigma$, leads to a phase function, $\Phi(\nu)$, which is now also complex. It is fairly straightforward to find the saddle points of $\Phi(\nu)$ (i.e., $\Phi'(\nu) = 0$, where ($'$) denotes differentiation with respect to ν) and the appropriate steepest descent contours. The easiest way to find the saddle points is to define an auxiliary variable $\xi = \cos\nu$, and rewrite $\Phi'(\nu) = 0$ in terms of ξ;

$$(ab/2)^2\,\xi^2 \;- b\xi + 1 - (ab/2)^2 = 0, \tag{10}$$

where a and b were defined in Eq. (9d). Eq. (10) is clearly quadratic in ξ and may readily be solved. Thus, for given observer and source locations, $\Phi(\nu)$ has, in general, two simple saddle points in the interval $[0, 2\pi]$ which depending on whether the component of the source relative Mach number in the direction of the observer, i.e., $M_{(rel)_i}e_i$, is subsonic or supersonic, have different forms. The two are a complex conjugate pair if $M_{(rel)_i}e_i < 1$, and are real if $M_{(rel)_i}e_i > 1$. When $M_{(rel)_i}e_i = 1$ (i.e., the "sonic condition"), these two saddle points merge to give rise to a single second order saddle point. For a subsonic source, it turns out, that only one of the two saddle points lies on the appropriate steepest descent path and, hence, only that saddle point contributes to the integral. It should be noted that, the integrand also has an infinite number of branch points, coinciding with those of $\Phi(\nu)$, of which only four lie in the region of interest. A judicious choice of the branch cuts guarantees that the contributions to the contour integral along the cuts in the ν-plane is exponentially small compared with the contributions from the neighborhood of the saddle points. The location of the appropriate saddle points, branch cuts and steepest descent contours for typical source locations are shown in Fig. (2). Note that, the choice of the steepest descent path depends on the type of the source under consideration. Also shown in this Figure, are the auxiliary (descent) contours needed for deforming the original contour, $[0, 2\pi]$, into the steepest descent contour(s). Due to the periodicity of the integrand in θ, the contributions from these contours cancel each other out exactly.

Naturally, the asymptotic structure of I_{n_p} for a sonic source point is different from that for a non-sonic one. In fact, it is fairly easy to show that the asymptotic expansion is given in terms of the inverse fractional 1/3-powers of parameter mB for sonic sources, and in terms of the inverse fractional 1/2-powers of mB for non-sonic sources. Since, there is no simple way of constructing a composite expansion from these two expansions, to allow for a smooth transition through the sonic condition, they are not very convenient to use. Fortunately, employing a theorem due to Chester, Friedman and Ursell [13], it is possible to develop a uniform expansion. The details of the method may be found in Bleistein and Handelsman [14] among others. The basic idea is, however, rather simple. The phase function $\Phi(\nu)$ is mapped conformally ($\nu \to \zeta$) into a much simpler function (i.e., a cubic polynomial in this case) which displays the relevant features of the original phase function. The region of interest in the complex ν-plane (i.e., the region containing the saddle points and steepest descent contours) is correspondingly mapped into a region in the complex ζ-plane. With the introduction of a few necessary parameters, the standard steepest descent and saddle point methods could then be applied to develop the required uniform expansion. The key definitions and parameters along with the final result are summarized below.

The cubic is given by:

$$\Phi(\nu) = -\left(\frac{\zeta^3}{3} - \gamma^2\zeta\right) + \mu, \tag{11a}$$

$$\mu = \frac{1}{2}\left[\Phi(\nu^+) + \Phi(\nu^-)\right], \qquad \gamma^3 = \frac{3}{4}\left[\Phi(\nu^+) - \Phi(\nu^-)\right], \tag{11b}$$

where ν^+ and ν^- denote the locations of the saddle points of Φ in the complex ν-plane and μ and γ are parameters defining the conformal map. Note that, γ, as given above, can take on three possible values or branches. The aforementioned theorem guarantees that one of the branches defines the desired conformal map (see Ref. [14] for details). That branch turns out to be the one for which γ^2 is purely real for the problem at hand. With these parameters determined, the map may be constructed and the uniform asymptotic expansion carried out. In principle, the expansion of the integral I_{n_p} could be carried out to an arbitrary order in the parameter mB, however, it turns out that, for most applications, the first term provides a very reasonable approximation for even the BPF component (i.e., $m = 1$). Therefore, in the subsequent analysis only the first term is considered.

After a fair amount of algebra, the final result can be written as the following formula:

$$I_{n_p} \simeq 2\pi i e^{mB\mu}\left\{d_0\frac{A_i[(mB)^{2/3}\gamma^2]}{(mB)^{1/3}} + d_1\frac{A_i'[(mB)^{2/3}\gamma^2]}{(mB)^{2/3}}\right\}, \tag{12a}$$

$$d_0 = \frac{\Gamma_0(\gamma) + \Gamma_0(-\gamma)}{2}, \qquad d_1 = \frac{\Gamma_0(\gamma) - \Gamma_0(-\gamma)}{2\gamma}, \tag{12b}$$

$$\Gamma_0(\zeta) = Q(\nu(\zeta))\,\frac{d\nu}{d\zeta}, \qquad \frac{d\nu}{d\zeta} = \frac{\gamma^2 - \zeta^2}{\Phi'(\nu(\zeta))}, \tag{12c}$$

where A_i and A_i' are the Airy function and its derivative, respectively, and d_0 and d_1 are coefficients in the asymptotic expansion. It is worth mentioning that, $\pm\gamma$ turn out to be the locations of the saddle points $\nu^{\pm}$ in the ζ-plane. The Airy function and its derivative provide a smooth transition from the subsonic portion of the blade to the supersonic portion. For $\gamma = 0$ (i.e., sonic sources) A_i and A_i' are $O(1)$ and, consequently, I_{n_p} is proportional to the inverse fractional 1/3-powers of mB. For $\gamma \neq 0$ (i.e., subsonic/supersonic sources), and large mB, A_i is $O((mB)^{-1/6})$ and A_i' is $O((mB)^{1/6})$ and, consequently, I_{n_p} is proportional to the inverse 1/2-power of mB as expected. Upon inserting for I_{n_p} in Eq. (9a) from Eq. (12a) and adding the contributions from all the N_p panels, the complex Fourier harmonic component, $p_{mB}(\mathbf{x})$, can be calculated.

At this point it should be pointed out that the analysis just described differs from that presented by Parry and Crighton [15] in that they use asymptotic forms of the Bessel functions in conjunction with the two-variable stationary phase method to asymptotically evaluate the surface integrals in Eq. (7). The key difference between the two analyses lies in their differing points of view. While, the approach described in Ref. [15] seeks to take advantage of the interference between the acoustic pressure disturbances arriving at the observer location emitted from the various sources distributed over the blade surface, the present method is built upon the idea that there is "self-interference" between the acoustic pressure signals emitted by the same source over one period of its revolution. From a practical stand point, the goal in Ref. [15] is to circumvent the time-consuming computations required in a full scale analysis by developing simple approximations to the noise field of a many-bladed propeller in order to gain insight into the importance of various physical parameters which influence the noise radiation characteristics of a propeller. The aim of the current approach is to provide an efficient way of carrying out a full scale analysis which, at the same time, is also amenable to systematic parametric studies of the sort envisioned in Ref. [15].

In order to assess the accuracy of the formula given by Eq. (12a), its predictions for a test problem were compared with those obtained through direct numerical integration of the θ integral in Eq. (9b). The test case chosen was an SR7 propfan operating in cruise conditions. An estimate for the aerodynamic loading distribution on the propfan blades, found by Nallasamy and Groeneweg [16] using an Euler CFD code developed by Whitfield et al. [17], was used along with the detailed geometric description of the blade planform as input to the acoustic calculations. The results are summarized in Figs. (3) and (4).

In Fig.(3), sideline directivity of the BPF sound pressure level (SPL) is shown for two representative sets of observer locations; a near-field set and a far-field set. These correspond to two different sideline distances; one at $x_2 = 1.5R_{tip}$ and the other at $x_2 = 5R_{tip}$. In each case, the observer axial location was varied between $x_1 = -R_{tip}$ (i.e., forward of the plane of rotation) and $x_1 = +2R_{tip}$ (i.e., aft of the plane of rotation). The solid lines represent the asymptotic estimates of the SPLs, and the dotted lines their numerically computed values. As it can be seen, both in the near- and far-fields, the agreement between the numerical results and the asymptotic estimates is quite good, with the maximum error of about $1.5dB$ occurring at $0.5R_{tip}$ aft of the plane of rotation for the near-field observer.

Due to the nature of the asymptotic expansion, the maximum error in approximation occurs for the lowest value of the mB parameter, which is $m = 1$ corresponding to the BPF. As m increases, the error in the asymptotic approximations diminishes rapidly. Therefore, it is no surprise that the agreement between the estimates obtained from the asymptotic formula and the numerically computed results improves with increasing harmonic order. This result is highlighted in Fig. (4), where the waveforms of the radiated sound at the $0.5R_{tip}$ aft location for the two sideline distances mentioned earlier are presented. Note that, at the axial location chosen, the asymptotic estimate of the BPF sound pressure level deviated the most from their numerically computed values (see Fig. (3)). Once again, the agreement is exceptionally good, particularly at the far-field location where the asymptotic and numerical results are virtually indistinguishable from each other. In constructing the waveforms the first ten harmonics were used, which meant that, in carrying out the numerical integration, a progressively finer integration step size had to be utilized. This of course was reflected in an increase in the required computation time. The asymptotic estimates on the other hand, required virtually no additional time as compared with that required to compute the BPF results. This is simply due to the fact that the location of the saddle points and steepest descent paths, and the structure of the conformal map are independent of the parameter mB. It is worth mentioning that the amount of time required to construct the waveforms using the asymptotic method was more than an order of magnitude smaller than that required in the full numerical integration!

Having demonstrated the utility of present approach, we next outline the extension to include the angle of attack effect. Once again, for brevity, only important steps in the analysis are presented.

1.2.2 Non-Zero Angle of Attack Case ($\alpha \neq 0$)

In general, the development of the analysis in this case follows closely that presented in the previous section. However, the details of the analysis itself are somewhat more complicated. For example, the phase function $\Phi(\theta, \alpha)$, (which is now more complicated than that for $\Phi(\theta, 0)$ given by Eq. (9c)), has four saddle points (i.e., $\Phi'(\nu, \alpha) = 0$ is a quartic equation). While, in principle, it is possible to find the roots of a quartic analytically, the expressions involved are rather lengthy. Fortunately, since in most practical applications of interest, the propeller angle of attack is small (typically no more than 5°), it is possible to find analytic approximations to the saddle points. This is done by treating the angle of attack case as a small perturbation to the zero angle of attack case (henceforth called the unperturbed case). The details of this development are given below where for convenience the analysis is carried out in terms of the parameter $M_{03} = M_0 \sin \alpha$, instead of α itself.

We begin by noting that for a small angle of attack, M_{03} is a small parameter, i.e., $|M_{03}| \ll 1$). Using perturbation analysis (see, for example, Bender and Orszag [18]) it can be shown that two of the four roots of $\Phi'(\nu, M_{03}) = 0$, denoted by $\hat{\nu}^{\pm}$, lie near those of the unperturbed problem $\Phi'(\nu, 0) = 0$ (i.e., $\nu^{\pm}$), while the other two scale with the inverse powers of M_{03} and, thus, lie much farther away. Therefore, since the perturbations in the integration contours are also small, the asymptotic behavior of I_{n_p} is controlled by the pair of saddle points $\hat{\nu}^{\pm}$. Of course, they could also merge to give rise to a second order saddle point as in the unperturbed case.

To find $\hat{\nu}^{\pm}$, first the phase function $\Phi(\nu, M_{03})$ is expanded in a Taylor series about $M_{03} = 0$ as follows;

$$\Phi(\nu, M_{03}) = \Phi_0(\nu) + M_{03}\Phi_1(\nu) + \frac{1}{2}M_{03}^2\Phi_2(\nu) + O(M_{03}^3), \tag{13a}$$

where

$$\Phi_0(\nu) = \Phi(\nu, 0), \qquad \Phi_1(\nu) = \left[\frac{\partial\Phi(\nu, M_{03})}{\partial M_{03}}\right]_{M_{03}=0},$$
$$\Phi_2(\nu) = \left[\frac{\partial^2\Phi(\nu, M_{03})}{\partial M_{03}^2}\right]_{M_{03}=0}. \tag{13b}$$

$\Phi_0(\nu)$ is given by Eq. (9c) (with θ replaced by ν), and $\Phi_1(\nu)$ and $\Phi_2(\nu)$ can be easily found by carrying out the differentiations with respect to M_{03}. Since, the salient features of the analysis which follows could be demonstrated without the need for the explicit expressions for $\Phi_1(\nu)$ and $\Phi_2(\nu)$, they are not given here in order to conserve space. The reason for retaining second order terms in Eq. (13a) will become clear shortly.

Next, Eq. (13a) is differentiated with respect to ν and set equal to zero;

$$\Phi'(\nu, M_{03}) \simeq \Phi_0'(\nu) + M_{03}\Phi_1'(\nu) + \frac{1}{2}M_{03}^2\Phi_2'(\nu) = 0, \tag{14}$$

where the higher order terms in M_{03} are neglected to maintain consistency. Rewriting the resulting equation in terms of the auxiliary variable $\xi = \cos\nu$ and employing the notation used in Eq. (10), we find that,

$$(ab/2)^2\xi^2 - b\xi + 1 - (ab/2)^2 + 2M_{03}\left[(1-b\xi)\Phi_1'(\cos^{-1}\xi)\right] + M_{03}^2(1-b\xi)\left[\Phi_1'^2(\cos^{-1}\xi) + \Phi_2'(\cos^{-1}\xi)\right] = 0, \qquad (15)$$

where, as before, only the second order terms are retained. Note that, when $M_{03} = 0$, Eq. (15) reduces to the unperturbed case given by Eq. (10). Therefore, as pointed out earlier, it is reasonable to expect that two of its roots lie near those of the unperturbed case. In order to solve Eq. (15) for the desired roots, $\hat{\xi}^{\pm} = \cos\hat{\nu}^{\pm}$, the functions inside the curly brackets are expanded in a Taylor series about the unperturbed roots $\xi^{\pm} = \cos\nu^{\pm}$. After collecting terms we find that Eq. (15) can be approximated by:

$$\left[(ab/2)^2 + M_{03}H_2^{\pm}\right]\xi^2 - \left[b - 2M_{03}(H_1^{\pm} + \xi^{\pm}H_2^{\pm})\right]\xi + \left[1 - (ab/2)^2 + M_{03}(2H_0^{\pm} - 2\xi^{\pm}H_1^{\pm} + \xi^{\pm^2}H_2^{\pm}) + M_{03}^2K_0^{\pm}\right] = 0, \qquad (16a)$$

where

$$H_0^{\pm} = (1-b\xi^{\pm})\Phi_1'(\cos^{-1}\xi^{\pm}), \qquad (16b)$$

$$H_1^{\pm} = -\left[\frac{1-b\xi^{\pm}}{(1-\xi^{\pm^2})^{1/2}}\Phi_1''(\cos^{-1}\xi^{\pm}) + b\Phi_1'(\cos^{-1}\xi^{\pm})\right], \qquad (16c)$$

$$H_2^{\pm} = \left\{\frac{1-b\xi^{\pm}}{1-\xi^{\pm^2}}\Phi_1'''(\cos^{-1}\xi^{\pm}) - \left[\frac{\xi^{\pm}(1-b\xi^{\pm})}{(1-\xi^{\pm^2})^{3/2}} - \frac{2b}{(1-\xi^{\pm^2})^{1/2}}\right]\Phi_1''(\cos^{-1}\xi^{\pm})\right\}, \qquad (16d)$$

$$K_0^{\pm} = (1-b\xi^{\pm})\left[\Phi_1'^2(\cos^{-1}\xi^{\pm}) + \Phi_2'(\cos^{-1}\xi^{\pm})\right]. \qquad (16e)$$

$H_0^{\pm}$, $H_1^{\pm}$, $H_2^{\pm}$ denote the coefficients of successive terms in the Taylor series expansion of the function inside the first curly bracket in Eq. (15), and $K_0^{\pm}$ denotes the leading term in the Taylor series expansion of the function inside the second curly bracket. Note that, $\cos^{-1}\xi^{\pm} = \nu^{\pm}$ in Eqs. (16a-16e)

The roots of this "new" quadratic can now be easily found. Since, Eq. (16a) has two roots for each of ξ^+ and ξ^-, we choose the root that reduces to ξ^+ or ξ^- in the limit of $M_{03} \to 0$. As expected, $\hat{\xi}^\pm$ turn out to be only slightly different from the unperturbed roots $\xi^\pm$. It is worth mentioning that alternative perturbation methods could be used to find $\hat{\xi}^\pm$. However, the particular approach described above provides a uniform representation of the roots even when the unperturbed part of Eq. (16a) has a single double root (i.e., the sonic condition for the unperturbed problem). Near that condition the accuracy of the approximation drops to $O(M_{03}^{5/2})$. This is why in the original expansion (Eq. (13a)) the second order terms were retained. If only the first order terms were kept, near the sonic condition, the accuracy would be no more than $O(M_{03}^{1/2})$. With the two perturbed saddle points $\hat{\nu}^\pm$ now known, the asymptotic expansion of I_{n_p} can be carried out following the same procedure as was outlined in the previous section; the difference being that $\Phi(\nu, \alpha)$ is now replaced by the three terms on the right hand side of the Eq. (13a), and $\nu^\pm$ are replaced by $\hat{\nu}^\pm$.

Preliminary applications of the angle of attack formula to a few test cases have been rather encouraging. As in the case of zero angle of attack, it provides accurate estimates for the integral I_{n_p}. The utility of the method in solving realistic problems is demonstrated with the following example which involves the prediction of the spectrum of noise of SR7 propfan operating in cruise and at an angle of attack of $\alpha = 1.6^\circ$. The unsteady aerodynamic loading distribution for this case was computed numerically by Nallasamy and Groeneweg [19] and was shown (see Heidelberg and Nallasamy [20]) to be in quite reasonable qualitative , as well as quantitative, agreement with the experimentally measured airloads. Using this aerodynamic loading distribution along with a detailed geometric description of the blade geometry as input, calculations for the spectrum of the noise of the propfan were carried out utilizing the asymptotic formula. The results were then corrected for non-free-field effects (such as scattering) which are not accounted for in the theoretical model. The corrections used, which were found in Spence [21], were 1.0, 3.6 and $5.3dB$ for the first three harmonics, respectively. For higher harmonics a uniform correction of $5.0dB$ was assumed. The experimentally measured spectrum corresponding to this case was presented in Ref. [22], as part of the Propfan Test Assessment (PTA) program (see Poland et al. [23]). The comparison between the measured and asymptotically computed spectra is presented in Fig. (5). In general, the agreement is quite good with appreciable deviations occurring only at higher harmonics where the neglected quadrupole sources probably play an important role.

1.3 Concluding Remarks

An asymptotic approach was presented which allows for accurate and efficient calculation of the noise field of a propfan at angle of attack without the need for simplifying assumptions usually employed in propeller noise theories. A closed form expression, involving the Airy function and its derivative, gives a uniform representation of the propeller noise field characteristics in both the near- and far-fields. The preliminary results shown here indicate that the method provides a useful theoretical tool in propeller noise research.

1.4 Acknowledgements

This work was supported under NASA grant NAS3-25266 and carried out at NASA Lewis Research Center in Cleveland, Ohio. The author wishes to thank Drs. Robert Rubinstein, Ambady Suresh and Paul Giel for many stimulating discussions. The help of Mrs. Tammy Langhals in typing and preparation of the manuscript is also gratefully acknowledged.

1.5 References

1. Woodward, R.P., "Measured Noise of a Scale Model High Speed Propeller at Simulated Takeoff/Approach Conditions," AIAA paper 87-0526, January 1987, also NASA-TM-88920.

2. Heidelberg, L.J. and Clark, B.J., "Preliminary Results of Unsteady Blade Surface Pressure Measurements for the SR-3 Propeller," AIAA paper 86-1893, July 1986, also NASA-TM-87352.

3. Nallasamy, M. and Groeneweg, J.F., "Unsteady Blade Surface Pressure on a Large-Scale Advanced Propeller: Prediction and Data," AIAA paper 90-2402, July 1990.

4. Hanson, D.B., "Unified Aeroacoustic Analysis for Highspeed Turboprop Aerodynamics and Noise. Vol. I: Development of Theory for Blade Loading, Wakes and Noise," NASA-CR-4329, to be published.

5. Mani, R., "The Radiation of Sound from a Propeller at Angle of Attack," NASA-CR-4264, January 1990.

6. Krejsa, E., "Prediction of the Noise from a Propeller at Angle of Attack," AIAA paper 90-5954, October 1990.

7. Ffowcs Williams, J.E. and Hawkings, D.L., "Sound Generation by Turbulence and Surfaces in Arbitrary Motion," Phil. Trans. R. Soc. Lond. Series A, Vol. 264, May 8, 1969, 321-342.

8. Hawkings, D.L. and Lowson, M.V., "Theory of Open Supersonic Rotor Noise," J. Sound Vib. 36(1), 1974, 1-20.

9. Hanson, D.B., "Near Field Frequency Domain Theory for Propeller Noise," AIAA J., Vol. 23, April 1985, 499-504.

10. Goldstein, M.E., AEROACOUSTICS, McGraw-Hill, New York, 1976, 189-192.

11. Farassat, F., "Linear Acoustic Formulae for Calculation of Rotating Blade Noise," AIAA J., Vol. 19, 1981 1122-1130.

12. Nallasamy, M., E. Envia, B.J. Clark, and J.F. Groeneweg, "Near-Field Noise of a Single Rotation Propfan at an Angle of Attack," AIAA paper 90-3953, also NASA-TM-103645, 1990.

13. Chester, C., B. Friedman and F. Ursell, "An Extension of the Method of Steepest Descent," Proc. Camb. Phil. Soc. 53, 1957, 599-611.

14. Bleistein, N. and R.A. Handelsman, ASYMPTOTIC EXPANSIONS OF INTEGRALS, Dover Publications Inc., New York, 1986, 369-374.

15. Parry A.B. and D.G. Crighton, "Asymptotic Theory of Propeller Noise-Part I: Subsonic Single-Rotation Propeller," AIAA J., Vol. 27(9), 1989, 1184-1190.

16. Nallasamy, M. and J.F. Groeneweg, "Prediction of Unsteady Blade Surface Pressures on an Advanced Propeller at an Angle of Attack," J. Aircraft, Vol. 27(9), September 1990, 789- 803.

17. Whitfield, D.L., T.W. Swafford, J.M. Janus, R.A. Mulac, and D.M. Belk, "Three Dimensional Unsteady Euler Solutions for Propfans and Counter Rotating Propfans," AIAA paper 87-1197, June 1987.

18. Bender, C.M. and S.A. Orszag, ADVANCED MATHEMATICAL METHODS FOR SCIENTISTS AND ENGINEERS, McGraw-Hill, New York, 1978, 319-330.

19. Nallasamy, M. and J.F. Groeneweg, "Unsteady Euler Analysis of the Flow Field of a Propfan at an Angle of Attack," AIAA paper 90-0339, October 1990.

20. Heidelberg, L.J. and M. Nallasamy, "Unsteady Blade Pressure Measurements for the SR-7A Propeller at Cruise Conditions," AIAA paper 90-4022, October 1990, also NASA-TM-103606.

21. Spence, P.L., "Progress Report on PTA Noise Prediction," April 5, 1990. 22. Lockheed/NASA-Lewis Propfan Test Assessment Program, Acoustic Flight Research Data, Yearly Status Review - Dec. 1987.

23. Poland, D.T., H.L. Bartel and P.C. Brown, "PTA Flight Test Review," AIAA paper 88-2803, July 1988.

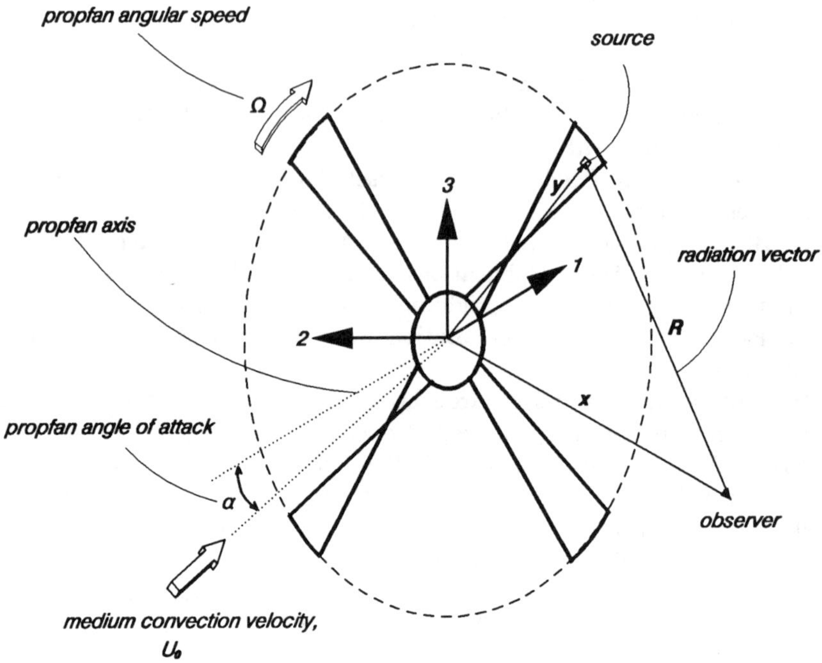

Fig. (1) Definitions of Geometry and Coordinate System.

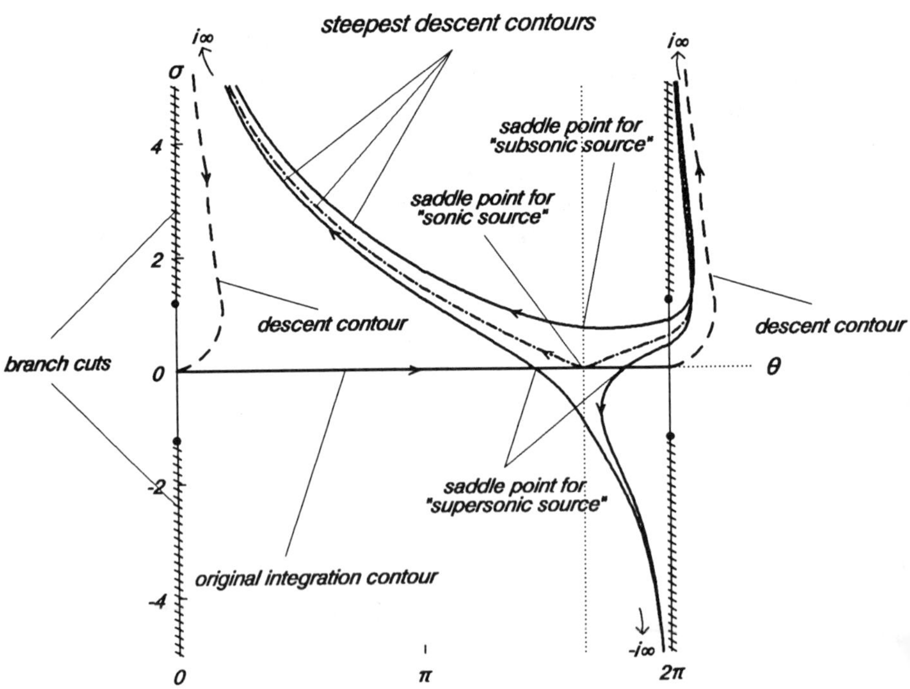

Fig. (2) Saddle Points and Steepest Descent Contours in the Complex ν-Plane.

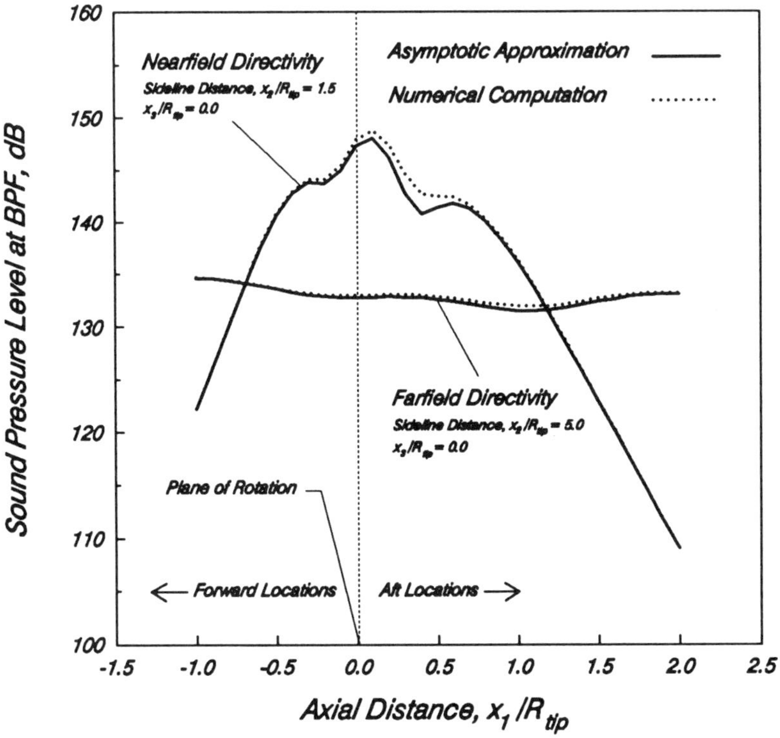

Fig. (3) Comparisons of Predicted Sideline Directivities of SR7 Propfan at Cruise Condition and Zero Angle of Attack ($\alpha = 0$).

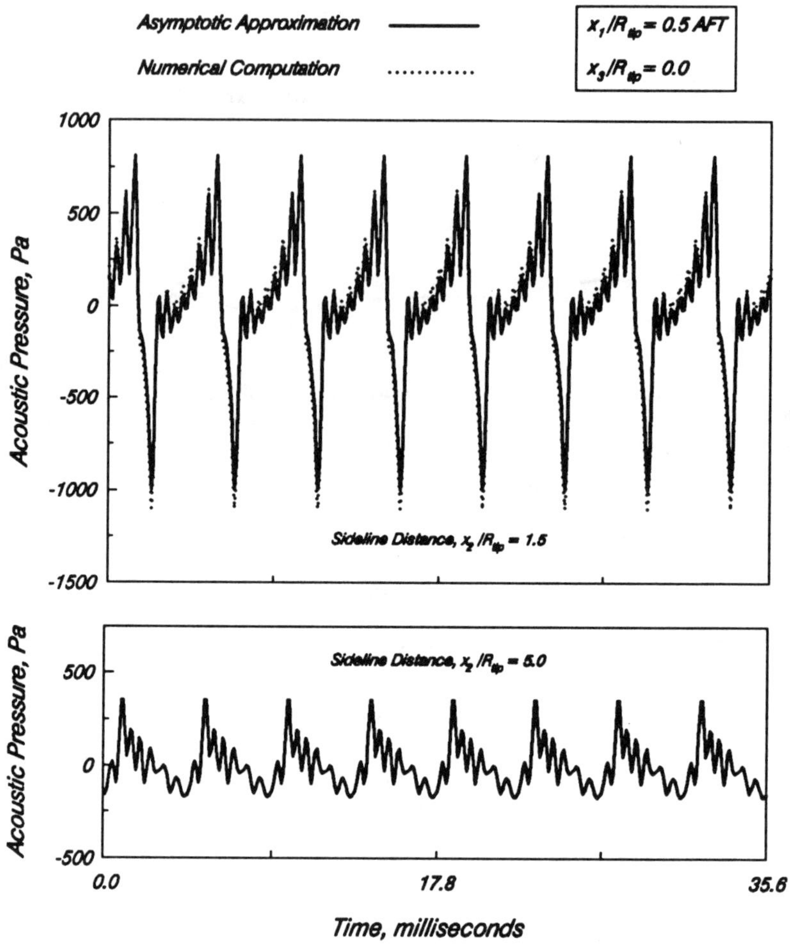

Fig. (4) Comparisons of Predicted Pressure Waveforms for SR7 Propfan at Cruise Conditions and Zero Angle of Attack ($\alpha = 0$).

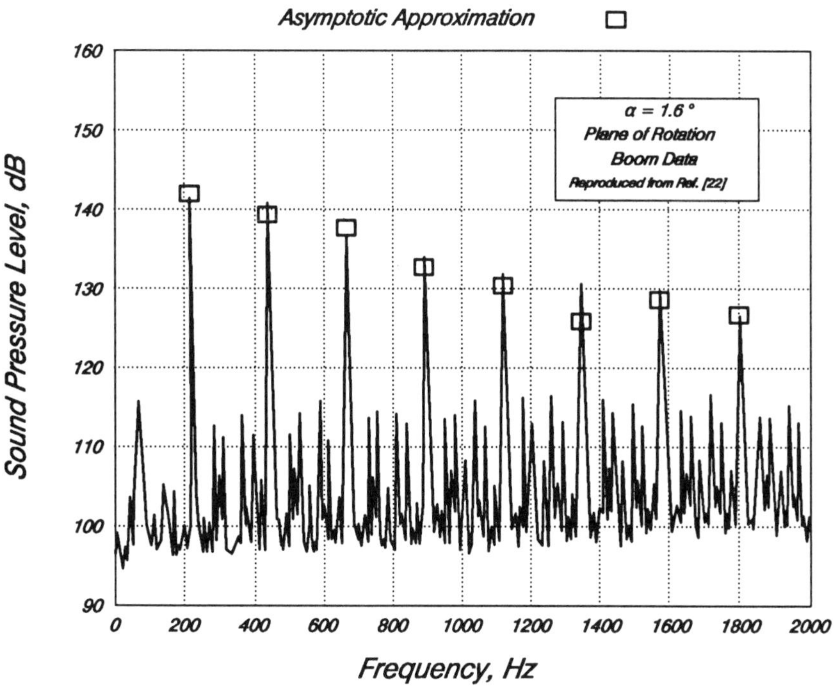

Fig. (5) Comparison of Measured and Predicted Pressure Spectra for SR7 Propfan at Cruise Conditions and Angle of Attack ($\alpha = 1.6°$). Spectra Presented for a Boom Location in the Plane of Rotation of Propfan (see Ref. [22] for details).

The Application of Experimental Data to Blade Wake Interaction Noise Prediction

Stewart Glegg[1]
William J. Devenport[2]

ABSTRACT Blade Wake Interaction noise (BWI) was defined by Brooks et al [3] as the broadband noise generated by the ingestion of turbulent trailing tip vortices by helicopter rotors. This was shown to be the dominant contributor to the subjectively important part of the acoustic spectrum for the approach stage of a helicopter flyover. Glegg [6] developed a prediction method for BWI noise based on the calculated trailing vortex trajectories and estimates of the vortex turbulence based on measurements by Phillips and Graham [11]. These measurements were made on a trailing vortex from a split wing arrangement and did not give the spectrum of the velocity fluctuations. This paper will describe a recent experiment which has been carried out to measure the turbulence associated with a trailing vortex and the application of the results to BWI noise prediction.

1.1 Introduction

Experimental measurements, by Brooks et al [3], have identified the mid-frequency broadband noise from helicopter rotors to be generated by the interaction of the rotor with its own wake. A typical noise spectrum for a rotor in forward flight is shown in figure 1.1, and this demonstrates that, for constant rotor thrust, the broadband noise level is a strong function of the tip path plane angle. As the tip path plane angle is increased the spectral level drops significantly, and this has been shown by Glegg [6] to correlate with the ingestion of the wake into the forward sector of the rotor disc plane. At small values of the tip path plane angle the rotor wake is close to

[1]Center for Acoustics and Vibration, Department of Ocean Engineering, Florida Atlantic University, Boca Raton, Florida 33431

[2]Department of Aerospace and Ocean Engineering, Virginia Polytechnic and State University, Blacksburg, Virginia 24061

the blades, but as the tip path plane angle is increased the wake moves away from the rotor and there is a corresponding reduction in noise level. This noise mechanism was defined by Brooks et al [3] as Blade Wake Interaction (BWI) noise, and has been shown to be the dominant contributor to the subjectively important part of the noise spectrum for the approach stage of a helicopter flyover. A method for predicting BWI noise was developed by Glegg [6] using the calculated trailing vortex trajectories from the rotor blades, and estimates of the vortex turbulence based on measurements by Phillips and Graham [11]. It was found that good agreement could be obtained with the spectral level, and the correct scaling with tip path plane angle was obtained, but the spectral shape was not well predicted. The measurements by Phillips and Graham [11] were made on a trailing vortex from a split wing arrangement and did not give the spectrum of the turbulent velocity fluctuations which define the spectral shape of the radiated noise. Consequently the prediction method estimated the noise spectrum by assuming a Von Karman approximation for the turbulence energy spectrum and estimating the integral lengthscales from similar flows. However to obtain a better understanding of BWI noise more details are required on the features of the turbulent flow in a vortex. This paper will describe an experiment which has been specifically designed to measure this flow using a single wing in a wind tunnel, and will show how the measurements can be incorporated into the BWI noise prediction scheme. Comparisons of the results with rotor noise measurements will also be given.

1.2 Experimental Work

To provide the information about the turbulence structure and velocity spectrum required for BWI noise predictions velocity measurements have been made on the tip vortex shed by a single rectangular NACA 0012 wing.

1.2.1 Apparatus and Instrumentation

The vortex was produced in the Virginia Tech Stability Wind Tunnel. This facility has a test section 1.83×1.83×7.33m, the long axis being in the flow direction. Flow through the empty test section is closely uniform and has a very low turbulence intensity (<0.05%). The wing, 0.20m in chord and 1.22m in span, was mounted vertically in the center of the test section (see figure 1.2), the root being bolted to a turntable assembly flushed into the upper wall of the wind tunnel. The turntable allowed the wing to be rotated to any angle of attack about its quarter chord location. At zero angle of attack the wing leading edge was 0.33m downstream of the test-section entrance. To minimize possible unsteadiness and non uniformity that might result from natural transition, the boundary layer on the wing was tripped. 0.5mm-diameter glass beads were glued to the wing in a random pattern covering a strip extending from the 20% to 40% chord locations with an average density of 200 beads/cm^2. A few measurements were also performed with no trip. Helium bubble visualizations performed prior to the velocity measurements (Devenport and Sharma [4], [5]) showed the vortex generated by this wing to be insensitive to probe interference and relatively stable over a broad range of conditions. Thirty chordlengths downstream of the wing some very low frequency lateral motions were observed but these had a magnitude of less than 3% of the wing chord.

Velocities were measured using single, cross and triple hot-wire probes mounted 0.84m upstream of a traverse gear fixed in the test section (see figure 1.2). All sensors were operated using Dantec 56C01 and 56C17 anemometer and bridge units interfaced through an analogic HSDAS-12 A/D converter to an IBM AT compatible computer. The HSDAS-12 can sample 4 channels simultaneously at rates of up to 100kHz per channel. Raw data was linearized and processed on line using an 18-8 laboratories PL1250 array processor and also stored on optical disc cartridges. All probes were calibrated before and after each traverse. Corrections for temperature drift (typically 1 to 2°C per hour) were applied according to the method of Bearman.

The single hot-wire probe (TSI type 1210T1.5) was used to measure profiles of mean velocity and turbulence intensity and spectra in the wing boundary layer as it left the trailing edge. The cross and triple hot-wire probes (Dantec types 55P51 and 55P91) were used to measure radial profiles of all mean velocity, Reynolds stress and triple product components in the vortex at 20, 25 and 30 chordlengths downstream of the wing. Auto and cross spectra of all velocity components were also measured at representative radial positions. To obtain these data with the cross hot-wire probe it was necessary to rotate it about its axis in steps of 45 degrees. These rotations were accomplished using a computer-controlled stepper motor located at the downstream end of the steel tube, connected to the probe stem by a shaft, sprockets and a timing belt.

The axial and pitch sensitivity of all sensors of the cross and triple- wire probes were measured and taken into account during the data reduction, see Sharma [9]. For cross and triple-wire statistical measurements, between 30000 and 90000 velocity samples (depending on local turbulence level) were taken at each probe position over a total sampling time of about 50 seconds. For spectral measurements, data were collected in 50, 10240-point records at a 30 kHz sampling rate. At selected locations 10240-point records were also measured at a 1 kHz sampling rate.

1.2.2 Results

Figure 1.2 shows the co-ordinate directions (x,y,z) (U,V,W) to be used in presenting results. The origin of the streamwise coordinate 'x' is the leading edge of the wing tip at zero angle of attack. Coordinates 'y' and 'z' are measured from the vortex center. Distances are normalized on the wing chord 'c' and velocities on the nominal free-stream speed U_{ref} measured using a pitot-static probe located in the forward part of the test section. Velocity measurements were performed at Reynolds numbers ($R_{ec} = U_{ref}(c/\nu)$) from 130000 to 530000, angles of attack from 0° to 7.5° and streamwise positions (x/c) from 20 to 30. Because of the volume of results, only a representative sample will be presented and discussed here. Figures 1.3 through 1.7 show velocities and spectra measured with the cross-wire probe during a z-wise traverse through the center of the vortex at x/c = 30, R_{ec} = 400000, with the wing at 5° angle of attack (rotating from the x to the z axis) and the trip attached. This set of conditions will be refered to as the baseline case. In this case the boundary-layers leaving both sides of the wing were fully turbulent. The thicknesses of the suction and pressure-side boundary layers at the trailing edge were 0.052c and 0.037c respectively.

The mean-velocity field of the vortex (figure 1.3) is, at first sight, much as would be expected. Tangential velocities, visible in the V component profile, increase to a peak of $0.25U_{ref}$ at the edge of the vortex core, which appears to have a diameter of about 0.09c. The small core size is circumstantial evidence for the relative stability of the vortex and its insensitivity to probe interference. The U profile shows an axial velocity deficit of about $0.12U_{ref}$ here. Close to the vortex center the normal stress profiles (figure 1.4) are dominated by strong central peaks generated by the small lateral motions of the vortex and the steep V gradient in the core. The flow behavior here is perhaps best illustrated by figure 1.6, which shows autospectra of the velocity components at the center of the core. In this plot the frequency has been normalized as fc/U_{ref} , and the area under each curve is unity. The sharp peaks in these spectra at very high frequencies, $fc/U_{ref} > 50$ are a consequence of electrical noise. Ignoring these there appear to be three distinct regions in the spectra. The first is located around a peak centered at fc/U_{ref}=5. It seems likely that this peak is a result of laminar instability in the vortex core of the type previously observed by Singh and Uberoi [10] . Their velocity measurements made in the core of a vortex from a laminar flow wing, show a regular instability at a frequency of fc/U_{ref}=3.5. We suspect laminar flow in the core of the present vortex because comparisons of spectra measured inside and outside the core show an order of magnitude less energy in the former at frequencies typical of small scale turbulence. The second region, located at very low frequencies ($fc/U_{ref} <$ about 0.3) is a direct consequence of the low-frequency lateral motions observed in flow visualizations. The energy associated with these motions is surprisingly large and the frequencies surprisingly low. Figure 1.6 shows the peak frequency to be at most fc/U_{ref}=0.003, corresponding to a streamwise lengthscale of at least 300 chordlengths. Cross spectra (not shown here) indicate substantial coherence between V and W in this frequency range implying that these long wavelength disturbances have a preferred direction. Given this, the only satisfactory explanation of the lateral motions would seem to be that they are a result of low frequency changes in flow direction through the test section. Changes of only $\pm 0.16^{\circ}$ would be sufficient to account for these motions. The third region lies in the approximate frequency range $fc/U_{ref} = 0.3$ to 1, where the spectra show a strong inflection. Here the implied lengthscales, between 1 and 3 chords, seem too small to be a result of wind tunnel disturbances and there is little or no coherence between V and W fluctuations, and so the motion is isotropic. We suspect that velocity fluctuations here may be a result of random lateral motions of the core generated as it is buffeted by the surrounding wake turbulence. In turbulent boundary layers 'inactive' motion of this type contributes as much as 30% to the normal stresses in the near wall region (see Bradshaw [2]). Without the constraint of a solid interface we might reasonably expect such motion to have a much greater effect.

Figure 1.4 shows a first attempt at reducing the variation of the normal stresses in the core in the absence of wind tunnel disturbances. The time records used to calculate these stresses were high-pass filtered at a frequency fc/U_{ref} of 0.013. They thus exclude a large portion of the energy presumeed to result from wind tunnel effects. Outside the vortex core the effects of vortex motions are much smaller and the true turbulence structure is immediately visible in the normal stress profiles (figure 1.5). This structure, however, is somewhat different from that we had expected. Here the measurements show two, if not three, distinct half turns of the wing wake as it spirals around the vortex. These not only appear as peaks in the normal-stress profiles (figures 1.4 and 1.5) but, on close inspection, as inflections and depressions the W and U profiles (figures 1.3) respectively, centered at $z/c= +0.66$, -0.31 and +0.19. In contrast to the results and discussion of previous workers (in particular Phillips and Graham [11]) there appears to be little or no region surrounding the core where successive turns of the wing wake have merged to form a continuous axisymmetric structure.

To illustrate this point figure 1.7 shows a cross section through the vortex at $x/c=30$. This figure contains contours of u^2/U_{ref}^2 drawn by interpolating y and z profiles measured with the triple-wire probe along a spiral path deduced from the positions of peaks in the profiles. Note that contours have not been drawn in the region $y/c > 0$, $z/c > 0.5$ since this would have involved extrapolation of the data. Apart from the lack of axisymmetry, the most striking feature of figure 1.6 is the continuous and rapid reduction in u^2 with distance along the spiral wake towards the vortex center. (This is despite the fact that turbulence levels in the wake are already very low.) A similar reduction is also visible in all the other stresses (see figures 1.4 and 1.5). This may indicate that the greater circumferential shear experienced by the wing wake towards the vortex center inhibits the development of, or breaks up, large stress-producing turbulent eddies. Figure 1.6 does not show whether the reduction in turbulence stresses continues into the core of the vortex, since the true turbulence levels here are obscured by the small lateral motions of the vortex. Measurements and discussion of the other cases listed in table 1 are presented by Sharma [9]. One relevant conclusion from that work is that turbulence levels in the spiral wake are very strongly dependent on the properties of the wing boundary layer, e.g. removing the wing trip lowers normal-stress levels by 60%. This and the above observations lead us to believe that the wing tip vortex merely serves to dissipate and diffuse the turbulence of the wing wake.

1.3 Modelling of the Turbulence

The turbulent flow measurements in the vortex have identified two different mechanisms for generating the unsteady velocity fluctuations at a fixed point. Close to the vortex center the flow is dominated by the small lateral motions of the vortex core, and at the edge of the vortex the turbulence in the spiral wake is the most important feature. In this section we will consider both of these flows and develop an analytical model to describe their features.

1.3.1 The Small Lateral Motions of the Vortex

To specify the unsteady velocity caused by small lateral motions of the vortex we will consider the vortex center as displaced from the center of the co-ordinate system by $\mathbf{z}(t)$ which is a random function of observer time at a fixed streamwise location. A turbulent eddy located at $\mathbf{x}$ relative to the origin of the co-ordinate system, is displaced from the center of the vortex by $\mathbf{y}$, such that $\mathbf{x} = \mathbf{y} + \mathbf{z}(t)$. The mean velocity field of the vortex is defined as $\mathbf{V}(\mathbf{y})$ and the unsteady turbulence component is $\mathbf{w}(\mathbf{y},\mathrm{t})$ and are a function of the displacement from the vortex center. The velocity measured by a fixed probe will have a mean component $\mathbf{U}(\mathbf{x})$ and an unsteady component $\mathbf{s}(\mathbf{x},t)$, given by

$$\mathbf{U}(\mathbf{x}) + \mathbf{s}(\mathbf{x},t) = \mathbf{V}(\mathbf{x}-\mathbf{z}(t)) + \mathbf{w}(\mathbf{x}-\mathbf{z}(t),t) \tag{1.1}$$

If the displacements of the vortex are small in comparison with the the core radius then we can expand the right hand side of this equation in a Taylor series, to first order, giving

$$\mathbf{U}(\mathbf{x}) + \mathbf{s}(\mathbf{x},t) = \mathbf{V}(\mathbf{x}) + \mathbf{w}(\mathbf{x},t) - (\mathbf{z}(t).\mathbf{div})(\mathbf{V}+\mathbf{w}) \tag{1.2}$$

This shows that the measured mean component $\mathbf{U}(\mathbf{x})$ is, to first order, the same as the mean flow about the vortex. The unsteady component is a combination of the turbulence associated with the vortex and velocity fluctuations induced by the time varying lateral displacements.

We assume that $\mathbf{w}$ and $\mathbf{z} = (0,\alpha,\beta)$ are uncorrelated so that the mean square velocity components $\mathbf{u}^{(2)} = < \mathbf{s}^{(2)} >$ are given by

$$\mathbf{u}^{(2)} = \mathbf{v}^{(2)} + < (\alpha \frac{\partial \mathbf{V}}{\partial y} + \beta \frac{\partial \mathbf{V}}{\partial z})^2 > \quad (1.3)$$

If the vortex displacements can be assumed to be isotropic with an rms displacement ϵ, so that

$$< \alpha^2 > = < \beta^2 > = \epsilon^2 < \alpha\beta > = 0 \quad (1.4)$$

the three components of the normal turbulent stresses $\mathbf{u}^{(2)} = (u^2, v^2, w^2)$ due to the lateral vortex motion can be defined, on $y=0$, as

$$u^2 = \epsilon^2 (\frac{\partial U}{\partial r})^2 \quad (1.5)$$

$$v^2 = \epsilon^2 (\frac{\partial V_\theta}{\partial r})^2 \quad (1.6)$$

$$w^2 = \epsilon^2 (\frac{V_\theta}{r})^2 \quad (1.7)$$

where U is the mean axial velocity and V_θ is the mean azimuthal velocity of the vortex. Schlinker and Amiet [12] and Howe [7] define a model of the azimuthal and axial velocity in a vortex as a function of the radial distance from it's center. This model couples a solid body rotation in the viscous core with a $1/r$ decay in the outer region. The azimuthal velocity is given by:

$$V_\theta(r) = C \frac{v_0 r_0}{r} [1 - \exp(-\alpha (r/r_0)^2)] \quad (1.8)$$

where $\alpha = 1.25643$ and C = $(1+0.5/\alpha) = 1.398$. The maximum azimuthal velocity is v_o and occurs at a radius of r_o. The circulation is given by $\Gamma_o = 2(1+\alpha/2) v_o \, r_o$. The axial velocity deficit model is

$$U(r) = U_{ref} - U_D \exp(-\alpha (r/r_0)^2) \quad (1.9)$$

where U_D is the axial velocity deficit on the centerline. The experimental results suggest that $v_o = 0.25 U_{ref}$, $r_o = 0.045$c and $U_D = 0.12 U_{ref}$. In general this model gives a good fit to the azimuthal velocity in the viscous core but underpredicts the velocity in the outer region, suggesting that the field is not decaying as rapidly as expected from the simple model given here. The width of the axial velocity deficit is also underpredicted. However, in this analysis we are concerned with the velocity induced by lateral motions of the vortex core which will not be affected by errors in the outer region or in the axial velocity deficit. By using equations 1.8 and 1.9 in equations 1.5 - 1.7 we obtain

$$u^2 = \epsilon^2 (U_D \frac{2\alpha r}{r_0^2} \exp(-\alpha (r/r_0)^2))^2 \quad (1.10)$$

$$v^2 = \epsilon^2 (C \frac{v_0 r_0}{r^2} [1 - (1 + 2\alpha (r/r_0)^2) \exp(-\alpha (r/r_0)^2)])^2 \quad (1.11)$$

$$w^2 = \epsilon^2 (C\frac{v_0 r_0}{r^2}[1 - \exp(-\alpha(r/r_0)^2)])^2 \tag{1.12}$$

Notice here that as $'r'$ becomes small equations 1.11 and 1.12 can be approximated as

$$w^2 = v^2 = \epsilon^2 (C\frac{\alpha v_0}{r_0})^2 = 3.08(\frac{\epsilon v_0}{r_0})^2 \tag{1.13}$$

In figure 1.4 we see that $v^2 = w^2$=0.0064U^2_{ref} and since $v_o = 0.24U_{ref}$ and r_o=0.045c, we can compute ϵ as 0.0086c.

Figure 1.8 shows the predicted normal turbulent stresses obtained using this model and the agreement with the measurements in Figure 1.4 is excellent. The only noticeable error is in the predicted value of u^2 close to z=0. However the effects of finite probe size and low turbulence levels at this point may cause the predicted dip to be filled in. The turbulence spectrum of the velocity fluctuations can also be obtained from this model. For example if we consider the spectrum of the velocity components in the z direction, then equation 1.7 implies that

$$G_{ww}(f) = G_{\epsilon\epsilon}(f)(V_\theta/r)^2 \tag{1.14}$$

which indicates that the spectra can be reduced to a universal form by the normalization G_{ww}(f)/w^2=$G_{\epsilon\epsilon}$(f)/ϵ^2. An empirical form for the measured velocity spectra can be obtained using the Von Karman interpolation formula, with a correction for viscous dissipation based on Pao's postulate, and is given by

$$G_{\epsilon\epsilon}(f) = 12.57\epsilon^2 \frac{c}{U_{ref}} \frac{\hat{k}_1^2 \exp(-\frac{3}{2}\left(\frac{\hat{k}_1}{k_d L}\right)^{\frac{4}{3}})}{(1+\hat{k}_1^2)^{\frac{11}{6}}} \tag{1.15}$$

where $\hat{k}_1^2 = \omega L/U_{ref}$ The experimental measurements imply that for the very low frequency lateral motions of the vortex (which are not isotropic) that k_d L = 30 and L/c=50. For the mid frequency range corresponding to fc/U_{ref} greater than 0.3 the lateral motions are isotropic and the spectra can be estimated using k_dL=15 and L/c=0.6 and ϵ = 0.00275c.

1.3.2 THE TURBULENCE IN THE WAKE SPIRAL

If it is assumed that the turbulence in the wake spiral is generated by the blade boundary layers, it may be considered as a distorted form of two dimensional wake turbulence. Wygnanski et al (1986) have shown that the turbulence in the wakes of many different types of body can be scaled on the wake width L_o which can be defined in terms of the momentum thickness Θ of the boundary layer on the blade as

$$L_o = 0.32\Theta\sqrt(x/\Theta + 380) \tag{1.16}$$

One of the unexpected results of this study was the remarkably good scaling which this type of normalization gave to the turbulent spectrum of the velocity fluctuations in the wake spiral. Figure 1.9 shows the collapse of the $G_{ww}(f)$ spectra when they are normalized as $G_{ww}(f)U_{ref}/w^2L_o$ and plotted against $f\,L_o/U_{ref}$. An empirical fit to these spectra can be obtained based on the Von Karman interpolation formula as

$$G_{ww}(f) = w^2 \frac{4\gamma L_0}{3U_{ref}} \frac{(1+4.66\hat{k}_1^2)}{(1+\hat{k}_1^2)^{\frac{11}{6}}} \left(C\exp(-s\hat{k}_1^2)\right) \tag{1.17}$$

where $\hat{k}_1^2 = 4\omega\gamma L/3U_{ref}$ and $s = 0.9248\,(c/\gamma\kappa R_{ec}\Theta)$

where γ is defined as the ratio of the integral lengthscale of the turbulence to the wake width and is determined from the measurements to be γ=0.2. The parameter κ is proportional to the turbulence intensity and is found to be 0.2 and the constant C ensures integral closure and is 1.36. This normalized spectrum is shown in figure 1.9, and gives remarkably good agreement with the measurements in this part of the wake.

1.4 Noise Predictions

Blade wake interaction noise is generated by the unsteady blade loads which result from the passage of the blades through their own wake. To calculate BWI noise we must define the mean properties of the inflow turbulence at each location of the rotor, and use this as an input to an aeroacoustic model, Amiet [1], Glegg [6], to calculate the broadband noise level in the acoustic far field. The turbulence in the wake is assumed to be dominated by the flow near the trailing tip vortices, and so it is first necessary to define the loci of the blade/vortex interactions, Glegg [6], and then to use a model of the turbulence associated with the vortex to calculate the unsteady flow encountered by the blades. The experimental results suggest that there are two dominant features of the unsteady flow which must be considered. The first is the turbulence in the spiral wake, and the second is the unsteady upwash caused by the small lateral displacements of the vortex. In this section we will consider the noise levels generated by each of these mechanisms.

1.4.1 Noise Generated by the Spiral Wake

The turbulence in the spiral wake was found to collapse using a Von Karman energy spectrum which gives a simple formulation for the wavenumber spectrum of the turbulence, and was used in the original BWI noise prediction method, Glegg [6]. However, the measurements show that the turbulence intensities and integral lengthscales are smaller than used in the original model. The modification of the prediction method to incorporate the measured spectra is therefore relatively straightforward, the major difficulty being to estimate the geometry of immersion of the blade in the spiral wake. This will affect both the local turbulence intensity distribution and the spanwise length of the blade which is exposed to turbulent velocity fluctuations, and will be a function of the displacement of the vortex center from the blade. The vortex displacement can be calculated from the wake code, but the functional dependance of the turbulence intensity distribution on this displacement is complex. However it will only affect the intensity of the radiated noise and will have a minor affect on the spectral shape. Consequently a relatively simple model of the blade immersion in the wake is sufficient to give first estimates of the BWI noise spectrum. Figure 1.10 shows the calculations of the radiated noise spectrum using the wake spiral model in comparison with experimental noise measurements given by Brooks et al [3]. It is seen that the levels are of the correct order of magnitude at the high frequencies, but do not predict the low frequencies well. The peak of the broadband noise in the measured spectra occurs at 1500 Hz, while the peak in the predicted spectra occurs at 4500 Hz. This indicates that the turbulence lengthscales in the spiral wake are significantly shorter than required to predict BWI noise, and so the turbulence in the wake spiral is not the dominant contributor to the acoustic source levels.

1.4.2 Unsteady Motions of the Vortex

The blade vortex interactions which cause BWI noise occur when the vortex core is perpendicular to the blade span, and so if the vortex is stationary there are no velocity fluctuations which occur as the blade passes through the vortex with a constant displacement from the vortex center. However if the vortex has a random lateral displacement from its mean position then there will be unsteady velocity fluctuations on the blade, since the velocity distribution will move relative to the blade surface. The upwash will be anti-symetric about the instantaneous spanwise location of the vortex core, and the spatial location of the distribution will be time varying in blade based co-ordinates. The anti-symetric feature of the upwash gives an anti-symetric unsteady load distribution which is acoustically inefficient, since it is equivalent to a quadrupole source mechanism. Also the lateral motions of the vortex are predominantly of low frequency, and so it would appear that this mechanism of generating unsteady loads on the blades is not acoustically important. However, since the spiral wake turbulence does not account for the measured BWI noise spectrum, calculations have also been carried out to evaluate the noise generated by the lateral motions of the vortex. The acoustic field from a blade moving through a vortex with lateral motions the flow can be modelled as described in section 3.1. The wavenumber spectrum of the velocity fluctuations on the blade are defined using the spectrum of the lateral displacements $G_{\epsilon\epsilon}(f)$ and the velocity distribution given by equations 1.2 and 1.8. To obtain the wavenumber spectrum it is necessary to evaluate the Fourier transform of the upwash velocity as function of span, which is difficult using the distribution given by equation 1.8 unless the viscous core is ignored. However since the mean flow about the blade will displace the vortex core from its ideal trajectory, it is unlikely that the blade will split the core. The dominant feature of the upwash will therefore be the outer region of the vortex which has an azimuthal velocity distribution which is inversely proportional to the the distance from the core. An upwash field which has this velocity distribution can be easily integrated to give the wavenumber spectrum required by the noise prediction scheme. The main difficulty therefore comes in estimating the $G_{ee}(f)$ spectrum, which is currently not well understood, since it could be caused by many different mechanisms. However using the spectrum given by equation 1.14 an estimate of the importance of the lateral motions of the vortex can be obtained.

The analysis shows that the acoustic efficiency will have a quadrupole type directionality, given by $z_o\ y_o\ /\ |\mathbf{x}_o|^2$ for an observer at

$$\mathbf{x}_o = (x_o, y_o, z_o) \tag{1.18}$$

in blade based co-ordinates, and will scale with frequency as

$$\omega^4 \exp(-|\omega y_o \Delta / c_o\, |\mathbf{x}_o|) \tag{1.19}$$

where Δ is the displacement of the vortex from the blade in the z direction, and c_o is the speed of sound. This demonstrates that as Δ increases, there will be a change in the spectral shape of the radiated noise. When Δ becomes comparable with the acoustic wavelength the exponential term will reduce the noise levels at high frequencies, ($\Delta >> \lambda$), while not affecting the levels at low frequencies ($\Delta << \lambda$). The minimum distance between the vortex core and the blade is therefore an important parameter and it is assumed that this must be at least one core radius. Using the parameters given in section 1.3 for the mid frequency region of the turbulent spectrum and calculating the wavenumber spectrum as described above enables a prediction to be made of the BWI noise caused by this mechanism. This prediction will depend on estimates of the circulation in the vortex core and the radius of the core, which are difficult to estimate for a helicopter rotor which has a continuously varying angle of attack. Using reasonable estimates of these parameters enables a best fit to the measured noise spectrum to be obtained as shown in figure 1.10. The results compare favorably, but the predictions are significantly less than the measured levels. At the present time the scaling of the mid frequency turbulence spectrum on the flow parameters is not completely defined, and so applying the results to a helicopter rotor configuration can only give crude estimates of the radiated noise levels. However, based on the approaches described above it appears more likely that the mid frequency turbulence spectrum, which is currently considered to be caused by isotropic lateral motions of the vortex, is the main contributor to the BWI noise spectrum.

1.5 Conclusion

This paper has described wind tunnel measurements of the unsteady flow about a trailing tip vortex and how they may be applied to the prediction of helicopter BWI noise. The flow measurements have identified three different frequency regions of the flow. A low frequency region caused by variations of the mean flow in the wind tunnel which causes anisotropic lateral motions of the vortex , and induce large unsteady velocities at a point due to high velocity gradients in the vortex core. A mid frequency region in which the vortex is displaced isotropically from its mean value and a high frequency region which occurs in the spiral wake which is wrapped around the vortex core. The turbulence levels in the spiral wake fall as the core is approached and flow in the core itself appears laminar. Estimates of the noise produced by these different frequency regimes of the turbulence show that it is the mid frequency region of isotropic vortex motions which is the most important mechanism for noise production.

Acknowledgments: This work was supported by NASA grant No 1-1119, and the authors would like to thank Dr. T.F. Brooks for useful discussions during the course of the project. The authors thank Barry Steer for his help in preparing this document using LaTeX.

References

[1] Amiet R., 1975, "Acoustic radiation from an airfoil in a turbulent stream", Journal of Sound and Vibration, Vol 41, pp 407-420

[2] Bradshaw P., 1967, "Inactive motion and pressure fluctuations in turbulent boundary layers", J. Fluid Mechanics, Vol 30, Pt 2, pp 241-258

[3] Brooks, T. F., Marcolini, M. A., and Pope, D. S., February 1987. "Main rotor noise study in DNW." Presented at the AHS Specialist Meeting in Aerodynamics and Aeroacoustics, Arlington, Texas.

[4] Devenport W.J. and Sharma G., 1990, "Flow visualizations of a wing-tip vortex in the presence of a probe", VPI&SU, Report VPI-AOE-177, September.

[5] Devenport W.J. and Sharma G., 1991, "Far-field turbulence structure of the tip vortex shed by a single rectangular wing", 8th Symposium on Turbulent Shear Flows, Munich, Germany, September 9-11.

[6] Glegg S.A L., 1989, "The prediction of blade-wake interaction noise based on a turbulent vortex model", AIAA 89-1134, presented at the 12th Aeroacoustics Conference, San Antonio, Texas.

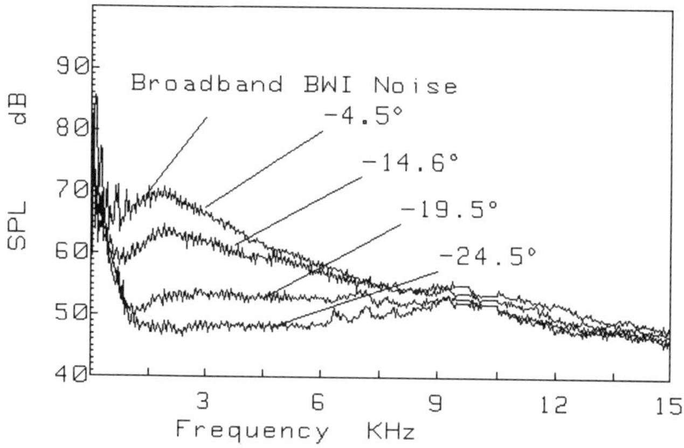

FIGURE 1.1. Measured helicopter noise spectra from a model scale rotor in a wind tunnel showing the broadband BWI noise as a function of the tip path plane angle of the rotor to the flow direction.

[7] Howe M.S., 1988, "Research on Lifting Surface Noise", BBN Report 6890, Bolt Beranek and Newman, Cambridge, Mass.

[8] Mason W H and Marchman J F, 1972, "Farfield structure of an aircraft trailing vortex, including effects of mass injection", NASA CR 62078.

[9] Sharma G., 1991, M.S. thesis, VPI&SU, expected in September, 1991.

[10] Singh P I and Uberoi M S, 1976, "Experiments on vortex stability", Physics of Fluids, vol 19, p 1858.

[11] Phillips W R C and Graham J A H, 1984, "Reynolds stress measurements in a turbulent trailing vortex", Journal of Fluid Mechanics, vol 147, p 353.

[12] Schinkler, R. H., and Amiet, R. K., 1983 "Rotor vortex noise" AIAA Paper 83-0720, Presented at the 8th Aeroacoustics Conference, Atlanta, Georgia.

[13] Wygnanski I., Champagne F.,and Marasli B., 1986, "On the large-scale structures in two-dimensional,small-deficit, turbulent wakes", Journal of Fluid Mechanics vol 168, pp31-71

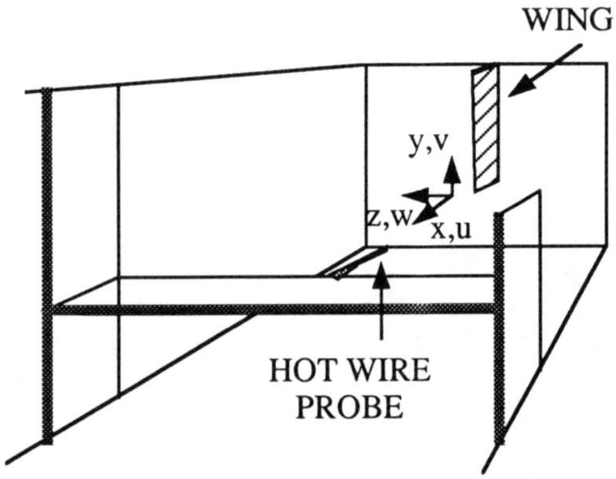

FIGURE 1.2. Schematic of the Stability Wind Tunnel test wing, traverse gear and coordinate system

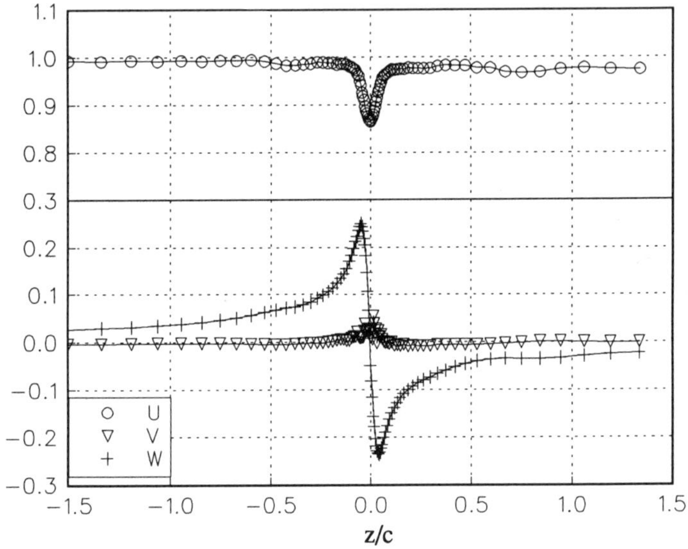

FIGURE 1.3. Velocity profiles measured with the cross-wire probe at x/c=30 normalized on U_{ref}. Baseline case. Mean velocities.

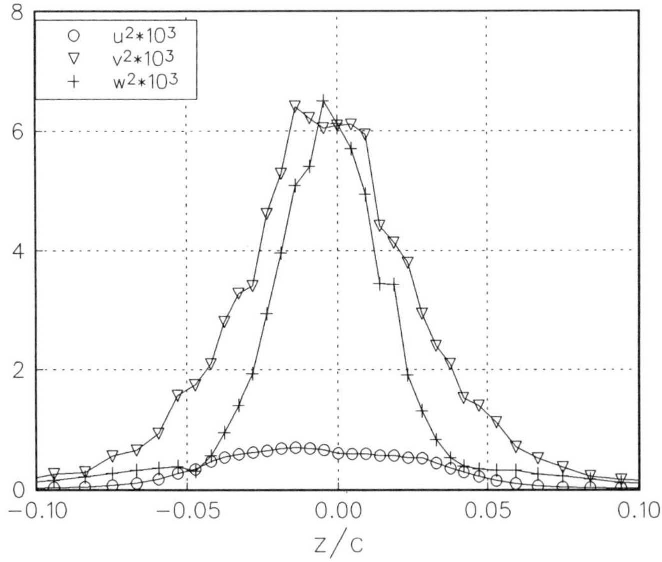

FIGURE 1.4. Velocity profiles measured with the cross-wire probe at x/c=30 normalized on U_{ref}. Baseline case. Filtered normal stress profiles in core.

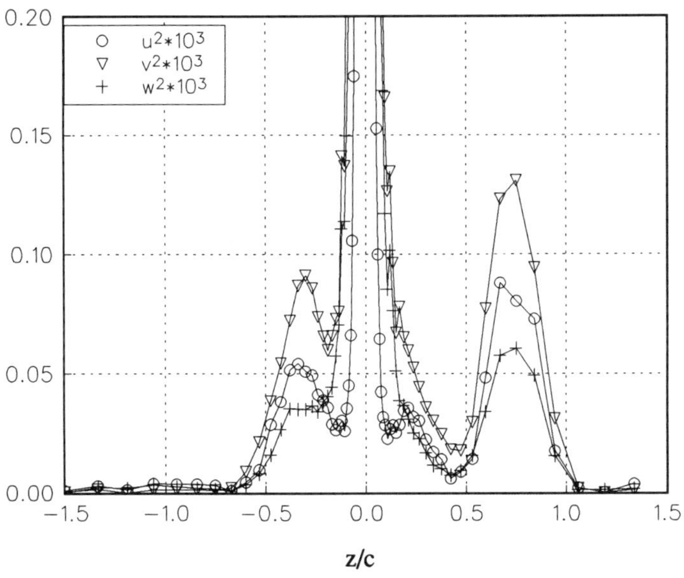

FIGURE 1.5. Velocity profiles measured with the cross-wire probe at x/c=30 normalized on U_{ref}. Baseline case. Normal stress profiles outside core.

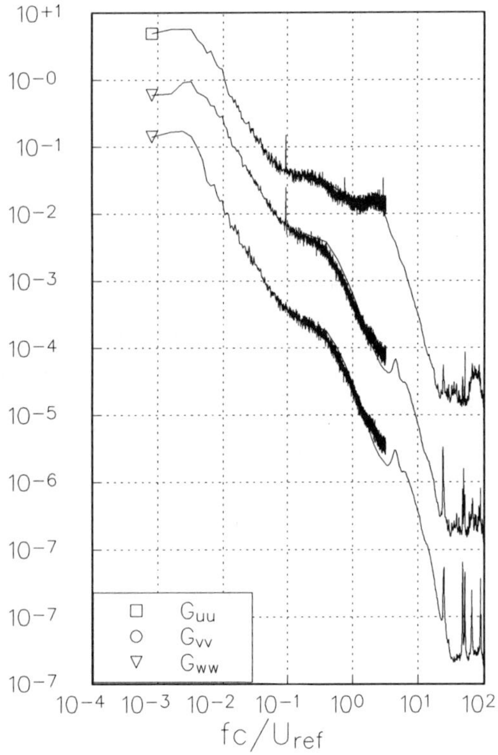

FIGURE 1.6. Autospectra of the normal stresses in the core of the vortex at 30 chords downstream of the blade.

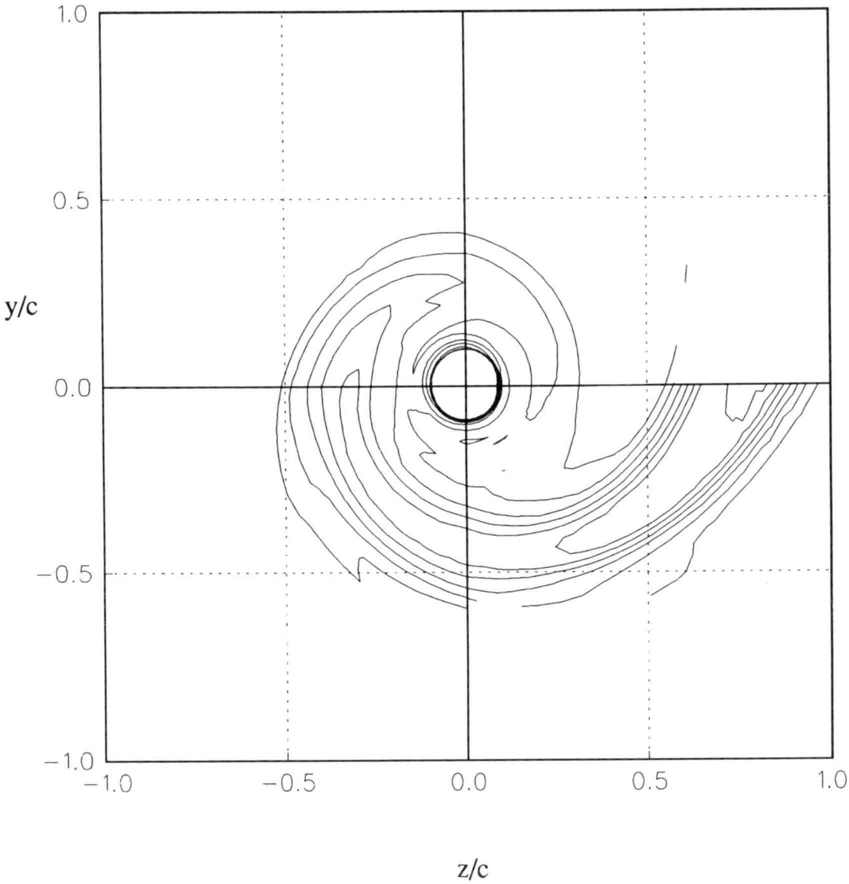

FIGURE 1.7. Contours of $/U^2_{ref} \times 1000$ drawn from triple hot-wire measurements made at x/c=30. Baseline case.

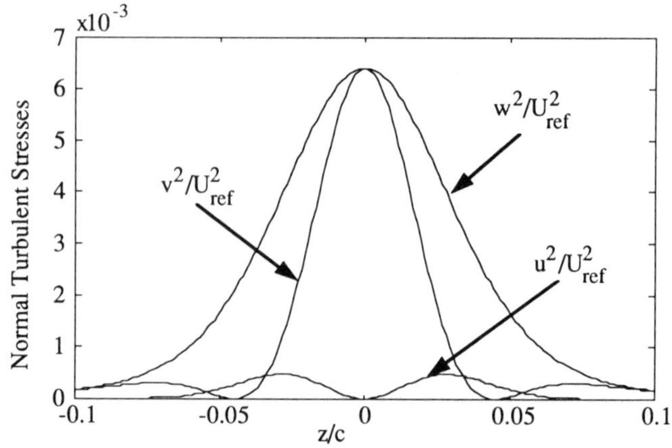

FIGURE 1.8. The predicted normal stresses based on the isotropic lateral vortex motion model.

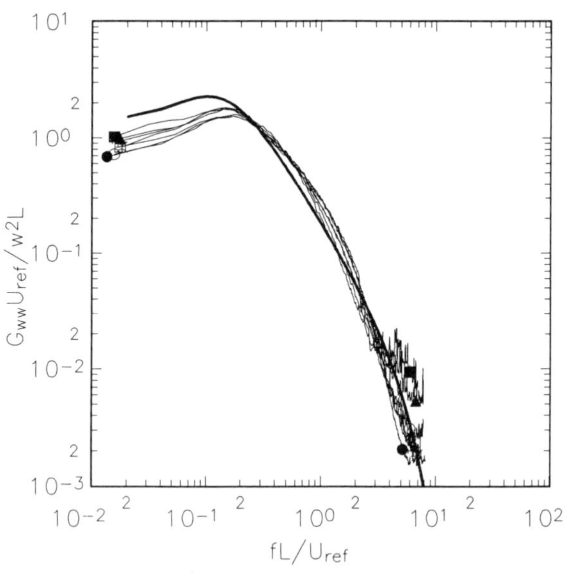

FIGURE 1.9. Autospectra of W-component velocity fluctuations measured in the spiral wake and the estimated spectrum (solid line) using equations 1.10 - 1.12.

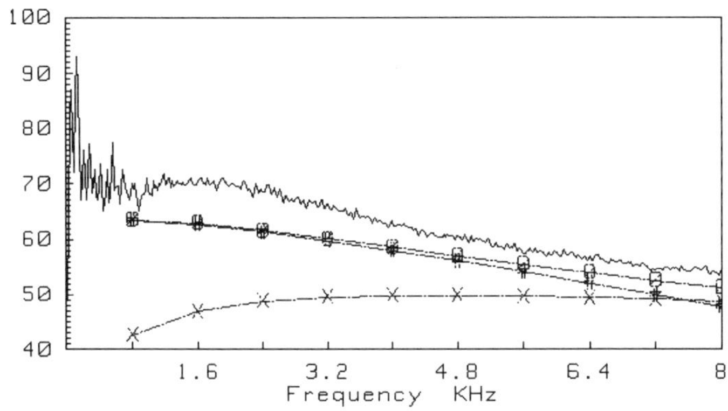

FIGURE 1.10. Comparison between measured and predicted BWI noise levels using the wake model and the lateral vortex motion model. The turbulent wake model lateral vortex motion model total predicted measurements.

Aeroacoustics of Real Blade Cascades

Kenneth A. Kousen[1]
Joseph M. Verdon[1]

ABSTRACT A linearized unsteady aerodynamic analysis is applied to study the aeroacoustic properties of cascades. The analysis allows an assessment of the effects of real blade geometry and mean blade loading on the acoustic response of a cascade to external aerodynamic excitations. The acoustic responses to vortical excitations at multiples of the blade passing frequency for a cascade of fan exit guide vanes are described. The acoustic responses to acoustic excitations at various incidence angles are also described and compared to analogous flat plate results, where qualitative differences are observed.

1.1 Introduction

An analysis of the noise signature of an axial turbomachine reveals large spikes at multiples of the blade passing frequency. While there are many acoustic sources contributing to the overall response spectrum, these discrete tones are generated by the aerodynamic interactions between adjacent blade rows, of which there are two primary types. Potential-flow interactions result from variations in the velocity potential or pressure field associated with the blades of a given row and their effect on the blades of a neighboring row moving at a different rotational speed. This type of interaction is of concern when the axial spacings between neighboring blade rows are small or flow Mach numbers are high. Wake interactions result from the impingement of wakes shed by one or more upstream rows upon the flow through a downstream blade row. This type of interaction can persist over considerable axial distances. Both of these interactions result in the generation of acoustic waves which may propagate away from the blade row unattenuated in amplitude. The propagating acoustic waves may then also interact with other blade rows.

Previous analyses of blade row interaction phenomena have generally

[1] United Technologies Research Center, 411 Silver Lane, East Hartford, CT 06108.

relied on two-dimensional cascade representations of the responding blade row, and either analytical formulations or classical linearized models for the unsteady aerodynamics. Implicit in these models is the assumption that the cascades can be considered to be made up of flat plate blades aligned with the mean flow direction. Representative of these investigations are those by Smith [Smi71], Kaji and Okazaki [KO70b], and Whitehead [Whi70], who applied the classical linearization to analyze the unsteady subsonic flows through cascades, Amiet [Ami71] and Kaji and Okazaki [KO70a], who used actuator disk theory to model acoustic wave/blade row interaction, and Koch [Koc71] and Mani and Horvay [MH70], who applied the Wiener-Hopf technique to the same problem.

The assumption that the cascades are composed of flat plate blades aligned with the incoming steady flow implies that the steady Mach number and flow angle are constants throughout the field. Consequently, the conditions for nondecaying acoustic propagation in both the upstream and the downstream regions are identical. In a cascade composed of realistic blade geometries, the resulting mean blade loading will cause the upstream and downstream steady flowfields and the corresponding propagation conditions to differ.

Linearized unsteady aerodynamic analyses which include the effects of real blade geometry and mean or steady blade loading on the unsteady pressure response have recently become available. Here the unsteady flow is regarded as a small-amplitude perturbation about a fully nonuniform steady flow. The steady flow is governed by a nonlinear inviscid equation set; the unsteady flow, by a system of linear equations with variable coefficients that depend upon the underlying steady flow [Ver89b, Ver90, Whi90]. With the development of these more general linearizations, aeroacoustic phenomena in cascades can now be re-examined in a more comprehensive fashion.

Some preliminary studies involving real blade cascades have recently appeared. Chiang and Fleeter [CF90] evaluated the effects of airfoil and cascade geometry and steady loading on the unsteady lift response in an incompressible flow. Suddhoo and Stow [SS90] examined the effect on the unsteady blade pressure distribution of inlet acoustic disturbances using a linearized potential code. Rangwalla and Rai [RR90] used an unsteady, thin-layer Navier-Stokes analysis to calculate rotor-stator interaction noise for blade rows having rotor/stator ratios of 1:1 and 3:4. The present paper is concerned with examining the effects of real blade geometries on vortical wave/blade row interactions and acoustic wave/blade row interactions.

1.2 The Unsteady Aerodynamic Model

In the following discussion all variables are dimensionless. Lengths have been scaled with respect to blade chord, time with respect to the ratio

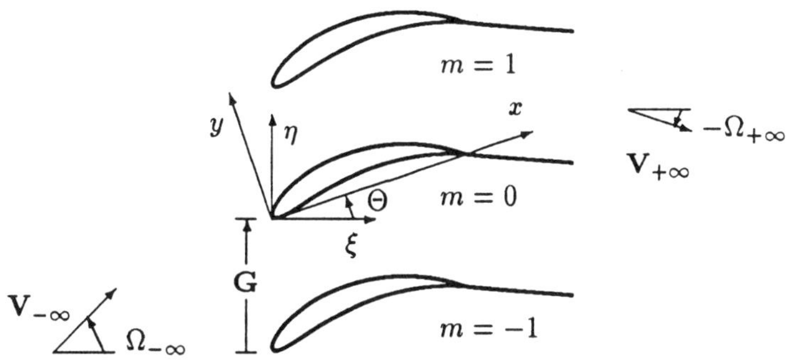

Figure 1. Cascade model.

of blade chord to upstream free-stream speed, density and velocity with respect to the upstream free-stream density and flow speed, respectively, pressure with respect to the product of the upstream free-stream density and the square of the upstream free-stream speed, and entropy with respect to the fluid specific heat at constant pressure. The scalings for the remaining variables can be determined from the equations given below.

We consider time-dependent, adiabatic flow, with negligible body forces, of an inviscid non-heat-conducting, perfect gas through a two-dimensional cascade (see Fig. 1). The mean or steady-state positions of the blade chord lines coincide with the line segments $\eta = \xi \tan\Theta + mG, 0 \leq \xi \leq \cos\Theta, m = 0, \pm 1, \pm 2, \ldots$, where ξ and η are coordinates in the axial and "circumferential" directions, respectively, m is a blade number index, Θ is the cascade stagger angle, and $\mathbf{G}$ is the cascade gap vector which is directed along the η-axis with magnitude equal to the blade spacing. The time-dependent or unsteady fluctuations in the flow can arise from one or more of the following sources: blade motions, upstream entropic and vortical disturbances, and upstream and/or downstream acoustic disturbances which carry energy toward the blade row. These excitations are assumed to be of small amplitude, periodic in time, and periodic in the cascade direction.

We assume that far upstream (say $\xi < \xi_-$) the mean flow is at most a small irrotational perturbation from a uniform free stream, and the unsteady flow remains attached to the blade surfaces. Consequently, the mean or steady background flow is isentropic and irrotational, and thin vortex sheets emanate from the blade trailing edges and extend downstream.

1.2.1 Small-Amplitude Unsteady Disturbances

For small-amplitude unsteady excitations, i.e., $O(\epsilon)$, the time-dependent flow can be regarded as a small perturbation about an underlying nonlinear mean or steady background flow. Thus, for example, the time-dependent fluid velocity $\tilde{\mathbf{V}}(\mathbf{x},t) \approx \mathbf{V}(\mathbf{x}) + \tilde{\mathbf{v}}(\mathbf{x},t)$ where $\mathbf{V}(\mathbf{x})$ is the local mean velocity and $\tilde{\mathbf{v}}(\mathbf{x},t)$ is the first-order (in ϵ) unsteady velocity. The equations that govern the steady and first-order unsteady flow properties are determined by substituting expansions of this form into the full time-dependent governing equations, collecting terms of like power in ϵ and neglecting terms of $\mathcal{O}(\epsilon^2)$ or higher. As a consequence of our assumptions regarding the flow far upstream of the blade row, the steady background flow will be isentropic and irrotational; i.e., $\mathbf{V} = \nabla\Phi$, where Φ is the steady velocity potential. Thus, the field equations that govern the underlying steady flow are

$$\nabla \cdot (\bar{\rho}\nabla\Phi) = 0 \tag{1.1}$$

and

$$(M_{-\infty}A)^2 = \bar{\rho}^{\,\gamma-1} = (\gamma M_{-\infty}^2 P)^{(\gamma-1)/\gamma} = 1 - \frac{\gamma-1}{2} M_{-\infty}^2[(\nabla\Phi)^2 - 1] \,, \tag{1.2}$$

where M, A, $\bar{\rho}$ and P are the local steady Mach number, speed of sound propagation, density, and pressure, respectively, and γ is the specific heat ratio of the fluid. Since, by assumption, the flow remains attached to the blade surfaces, a flow tangency condition applies at such surfaces. In addition, the steady pressure and normal velocity component must be continuous across blade wakes.

The first-order fluid motions that arise from the various modes of unsteady excitation are not coupled. Hence, it is sufficient to develop solution procedures for a single harmonic (in t and η) component of a given excitation. Solutions for an arbitrary excitation can then be obtained by superposition. The first-order or linearized unsteady flow properties caused by a periodic unsteady excitation at temporal frequency ω will be harmonic in time, e.g.,

$$\tilde{\mathbf{v}}(\mathbf{x},t) = \mathrm{Re}\{\mathbf{v}(\mathbf{x})e^{i\omega t}\} \,. \tag{1.3}$$

In addition, the steady and, for an excitation at circumferential wave number $\kappa_\eta = \sigma G^{-1}$, the first-order unsteady properties will satisfy circumferential periodicity conditions, e.g.,

$$\mathbf{V}(\mathbf{x} + m\mathbf{G}) = \mathbf{V}(\mathbf{x}) \quad \text{and} \quad \mathbf{v}(\mathbf{x} + m\mathbf{G}) = \mathbf{v}(\mathbf{x})e^{im\sigma} \,, \tag{1.4}$$

where σ is the interblade phase angle of the unsteady excitation. Thus, solutions to time-independent nonlinear steady and linearized unsteady flow problems are required only over a single extended blade-passage region of the cascade. In addition, since analytic far-field solutions can be determined [Ver89c], the numerical solution domain can be further restricted to a single extended blade-passage region of finite extent in the axial direction.

1.2.2 The Linearized Unsteady Equations

The field equations that govern the first-order unsteady perturbation of a nonlinear isentropic and irrotational steady background flow can be written in the form [HV91]

$$\frac{\bar{D}s}{Dt} = 0 \, , \tag{1.5}$$

$$\frac{\bar{D}}{Dt}(\mathbf{v}_R - s\nabla\Phi/2) + [(\mathbf{v}_R - s\nabla\Phi/2)\cdot\nabla]\nabla\Phi = 0 \tag{1.6}$$

and

$$\frac{\bar{D}}{Dt}(A^{-2}\frac{\bar{D}\phi}{Dt}) - \bar{\rho}^{\,-1}\nabla\cdot(\bar{\rho}\nabla\phi) = \bar{\rho}^{\,-1}\nabla\cdot(\bar{\rho}\mathbf{v}_*) \, . \tag{1.7}$$

Here, the dependent variables $s, \mathbf{v}_R$, and ϕ are the complex amplitudes of the first-order entropy, rotational velocity (i.e., the component of the velocity associated with the fluid vorticity), and velocity potential, respectively, $\bar{D}/Dt = i\omega + \nabla\Phi\cdot\nabla$ is a convective derivative based on the mean-flow velocity, and $\mathbf{v}_* = \nabla\phi_* + \mathbf{v}_R$, where ϕ_* is a convected or pressure-less potential, i.e., $\bar{D}\phi_*/Dt = 0$ and $\mathbf{v}_*\cdot\boldsymbol{n} = 0$ at the mean blade and wake surfaces.

The rotational velocity is divergence-free far upstream of the blade row, i.e., $\nabla\cdot\mathbf{v}_R = 0$ for $\xi < \xi_-$, the unsteady pressure depends only on the unsteady potential ϕ through the relation $p = -\bar{\rho}\bar{D}\phi/Dt$, and the unsteady vorticity is given by $\boldsymbol{\zeta} = \nabla\times\mathbf{v}_R$. In addition to the foregoing equations, the unsteady perturbation is subject to prescribed flow conditions far upstream and far downstream from the blade row, a flow tangency condition at moving blade surfaces, and, since the field equations apply only in continuous regions of the flow, jump conditions based on the fluid-dynamic conservation laws at blade wakes.

The linearized equations for the unsteady entropy, rotational velocity, and velocity potential are coupled only sequentially, therefore the unsteady equations can be solved in order to determine $s, \mathbf{v}_R$, and ϕ.

Far upstream ($\xi < \xi_-$) and far downstream ($\xi > \xi_+$) from the blade row the first-order unsteady field equations can be reduced to constant coefficient equations for which analytical solutions can be determined. For example, it follows after replacing $\nabla\Phi$ by $\mathbf{V}_{-\infty}$ in Eqs. (1.5) and (1.6) that the entropy and rotational velocity fluctuations far upstream ($\xi < \xi_-$) of the blade row must be of the form

$$s(\mathbf{x}) = s_{-\infty}\exp(i\boldsymbol{\kappa}_{-\infty}\cdot\mathbf{x}) \, , \tag{1.8}$$

and

$$\mathbf{v}_R(\mathbf{x}) = \mathbf{v}_{R,-\infty} \exp(i\boldsymbol{\kappa}_{-\infty} \cdot \mathbf{x}) \ . \tag{1.9}$$

where $\mathbf{v}_{R,-\infty} = -i(\boldsymbol{\kappa}_{-\infty} \times \boldsymbol{\zeta}_{-\infty})/|\boldsymbol{\kappa}_{-\infty}|^2$, and $s_{-\infty}$ and $\boldsymbol{\zeta}_{-\infty}$ are the prescribed complex amplitudes (at $\mathbf{x} = 0$) of the incident entropic and vortical gusts, respectively.

The velocity potential fluctuations in the far upstream and far downstream regions depend upon the acoustic excitation as well as the acoustic and vortical response of the blade row to the imposed unsteady excitation [Ver89c]. Thus, we can write

$$\phi(\mathbf{x}) = \phi_E(\mathbf{x}) + \phi_R(\mathbf{x}) \quad \text{for } \xi \lessgtr \xi_{\mp} \ . \tag{1.10}$$

where the potential component ϕ_E accounts for acoustic excitations, i.e., pressure disturbances that either attenuate as they approach the blade row or propagate and carry energy towards or along the blade row, and is given by

$$\phi_E(\mathbf{x}) = \phi_{I,\mp\infty} \exp(-\beta_{\mp\infty}\xi + i\boldsymbol{\kappa}_{\mp\infty} \cdot \mathbf{x}) \ , \tag{1.11}$$

where

$$\phi_{I,\mp\infty} = \bar{\rho}_{\mp\infty}^{-1}\{\beta_{\mp\infty} V_{\mp\infty} \cos\Omega_{\mp\infty} - i[\omega + (\boldsymbol{\kappa}_{\mp\infty} \cdot \mathbf{V}_{\mp\infty})]\}^{-1} p_{I,\mp\infty} \ , \tag{1.12}$$

$p_{I,\mp\infty}$ is prescribed, $\kappa_{\eta,\mp\infty} = \sigma/G$, and $\beta_{\mp\infty}$ and $\kappa_{\xi,\mp\infty}$ depend upon the inlet/exit free-stream conditions, the cascade blade spacing and the temporal frequency (ω) and interblade phase angle (σ) of the unsteady pressure excitation. The potential component ϕ_R is associated with the acoustic and vortical response of the blade row and therefore, must be determined as part of the unsteady solution [Ver89c].

1.2.3 The Linearized Unsteady Flow Solution

Entropy and Rotational Velocity Fluctuations

Closed form solutions can be determined for the entropy and rotational velocity fluctuations in terms of the dependent variable $\mathbf{X} = \mathbf{V}_{-\infty}\Delta + \mathbf{e}_z \times \mathbf{V}_{-\infty}\Psi$, where Δ and Ψ are the drift and stream functions, respectively, of the steady background flow [HV91, Gol78], and $\mathbf{e}_z$ is a unit vector that points out of the page. Since $\bar{D}\mathbf{X}/Dt = \mathbf{V}_{-\infty}$ and $\mathbf{X} \to \mathbf{x}$ as $\xi \to -\infty$, the solution to the entropy transport equation (1.5) which satisfies the upstream condition (1.8) is

$$s(\mathbf{x}) = s_{-\infty} \exp(i\boldsymbol{\kappa}_{-\infty} \cdot \mathbf{X}) \ , \tag{1.13}$$

and the solution to the rotational velocity equation (1.6) which satisfies the far upstream condition (1.9) is given by

$$\mathbf{v}_R(\mathbf{x}) = [\boldsymbol{\nabla}(\mathbf{X} \cdot \mathcal{A}_{-\infty}) + s_{-\infty}\boldsymbol{\nabla}\Phi/2] \exp(i\boldsymbol{\kappa}_{-\infty} \cdot \mathbf{X}), \tag{1.14}$$

where $\mathcal{A}_{-\infty} = \mathbf{v}_{R,-\infty} - s_{-\infty}\mathbf{V}_{-\infty}/2$. Thus, the entropy and rotational velocity fluctuations can be evaluated once the drift and stream functions, and their derivatives, are determined from the solution for the underlying steady flow.

If the steady background flow stagnates at blade leading edges, as it will for a realistic configuration, the rotational velocity will be singular along the mean blade and wake surfaces. This feature stems from the singular behavior of the drift function, i.e., $\Delta \to a_0 \ln n$ as $n \to 0$, where n is the normal distance from a mean blade or wake surface and a_0 is a constant [AG89]. This singular behavior has been removed from the blade and wake surface conditions that determine the unsteady potential in Ref. [HV91], by prescribing a suitable form for the convected potential based on an analysis in Ref. [AG89]. Their choice of ϕ_* ensures that $\mathbf{v}_* \cdot \mathbf{n} = (\mathbf{v}_R + \nabla\phi_*) \cdot \mathbf{n} = 0$ at blade and wake mean positions.

The unsteady potential (ϕ) is determined as a solution of the field equation (1.7) subject to conditions at the mean blade and wake surfaces and in the far field. Flow tangency applies at the blade surfaces, and the fluid pressure and normal velocity must be continuous across blade wakes. Linearized conditions for the flow tangency along the blades and jump conditions across the wakes are given in Ref. [Ver90].

Note that as a consequence of the small unsteady-disturbance approximation, the conditions on the linearized unsteady perturbation at the moving blade and wake surfaces have been imposed at the mean positions of these surfaces (see Ref. [Ver87]), with the mean wake, i.e., the downstream stagnation streamlines, locations being determined from the nonlinear steady solution.

A finite-difference model for solving the boundary value problem for the unsteady potential has been described in [CV81]. Recent extensions to this basic model have been incorporated to treat the unsteady transonic flows excited by prescribed blade motions [UV90] and unsteady subsonic flows excited by entropic and vortical gusts [HV91]. This linearized unsteady flow code, LINFLO, will be applied herein to study the acoustic response of a cascade to vortical and acoustic excitations.

1.2.4 Excitation and Response

For a vortical excitation, the temporal frequency, ω, interblade phase angle, σ, and complex amplitude of the gust vorticity are prescribed quantities. Alternatively, we can prescribe the normal component of the rotational velocity, $\mathbf{v}_{R,-\infty} \cdot \mathbf{e}_N$, in lieu of the vorticity. The component of the gust velocity in the inlet free-stream direction, $\mathbf{v}_{R,-\infty} \cdot \mathbf{e}_T$, is determined by the divergence-free condition, i.e., $i\boldsymbol{\kappa}_{-\infty} \cdot \mathbf{v}_{R,-\infty} \equiv 0$. The wave number $\boldsymbol{\kappa}_{-\infty}$ has a component $\kappa_{\eta,-\infty} = \sigma G^{-1}$ in the cascade- or η-direction and a component $\kappa_T = -\omega V_{-\infty}^{-1} = -\omega$ in the inlet free-stream direction. Therefore,

$$\begin{aligned} \kappa_{-\infty} &= -(\omega \sec \Omega_{-\infty} + \sigma G^{-1} \tan \Omega_{-\infty}) \mathbf{e}_\xi + \sigma G^{-1} \mathbf{e}_\eta \\ &= -\omega \mathbf{e}_T + (\omega \tan \Omega_{-\infty} + \sigma G^{-1} \sec \Omega_{-\infty}) \mathbf{e}_N \ . \end{aligned} \tag{1.15}$$

For acoustic excitations, generally only those that are of propagating type are considered. For subsonic inlet and exit conditions ($M_{\mp\infty} < 1$) the velocity potential corresponding to a propagating acoustic excitation at temporal frequency $\omega > 0$ has the form (1.11) with $\beta_{\mp\infty} = 0$, and

$$\kappa_{\eta,\mp\infty}^{(-)} \leq \kappa_{\eta,\mp\infty} = \sigma G^{-1} \leq \kappa_{\eta,\mp\infty}^{(+)} \ , \tag{1.16}$$

where the circumferential wave numbers at which cut-off occurs, $\kappa_{\eta,\mp\infty}^{(\mp)}$, are given by

$$\kappa_{\eta,\mp\infty}^{(\mp)} = \frac{\omega V_{\mp\infty}^{-1} M_{\mp\infty}}{1 - M_{\mp\infty}^2} \left(M_{\mp\infty} \sin \Omega_{\mp\infty} \pm \sqrt{1 - M_{\mp\infty}^2 \cos^2 \Omega_{\mp\infty}} \right) \ . \tag{1.17}$$

The axial wave number of the propagating acoustic excitation is

$$\kappa_{\xi,\mp\infty} = i(\mp|d_{\mp\infty}| + M_{\mp\infty}^2 \delta_{\mp\infty} \cos \Omega_{\mp\infty}) \ , \tag{1.18}$$

where

$$|d_{\mp\infty}| = |(1 - M_{\mp\infty}^2 \cos^2 \Omega_{\mp\infty})^{-1} \kappa_\eta^2 - M_{\mp\infty}^2 \delta_{\mp\infty}^2|^{1/2} \ , \tag{1.19}$$

and

$$\delta_{\mp\infty} = (\omega V_{\mp\infty}^{-1} + \kappa_\eta \sin \Omega_{\mp\infty}) / (1 - M_{\mp\infty}^2 \cos^2 \Omega_{\mp\infty}) \ . \tag{1.20}$$

Equations (1.15) and (1.18) together with $\kappa_{\eta,-\infty} = \sigma G^{-1}$ determine the wave numbers of vortical or acoustic excitations in terms of the steady flow quantities, the temporal frequency ω, and interblade phase angle σ of the excitation. In studies involving acoustic excitations, however, it is often of interest to be able to specify the angle of incidence as one of the prescribed quantities.

Consider propagating acoustic waves with wave number magnitudes $k = |\boldsymbol{k}_{\mp\infty}|$ defined such that $k_{\xi,\mp\infty} = k \cos \alpha$ and $k_{\eta,\mp\infty} = k \sin \alpha$. In order to satisfy the governing equations of motion in the far field, it is necessary that k also satisfy

$$\begin{aligned} &[1 - M_{\mp\infty}^2 \cos^2(\alpha - \Omega_{\mp\infty})] k^2 \\ &\quad - 2M_{\mp\infty}^2 (\omega V_{\mp\infty}^{-1}) \cos(\alpha - \Omega_{\mp\infty}) k - M_{\mp\infty}^2 (\omega V_{\mp\infty}^{-1})^2 = 0 \end{aligned} \tag{1.21}$$

which comes from the equations for a propagating acoustic excitation, (1.18), (1.19), and (1.20), above. In reaching this equation, however, all sign information distinguishing the excitation axial wave number from the response axial wave number has been lost. Viewed as a quadratic equation for k, this equation has two roots, which are given by

$$k_1 = M(\omega V_{\mp\infty}^{-1})/[1 - M_{\mp\infty}\cos(\alpha - \Omega_{\mp\infty})] \tag{1.22}$$

and

$$k_2 = -M(\omega V_{\mp\infty}^{-1})/[1 + M_{\mp\infty}\cos(\alpha - \Omega_{\mp\infty})] \tag{1.23}$$

Which of these represents an excitation wave number and which a response wave number can be determined by referring to the critical angle of incidence, α_{crit}.

The critical angle of incidence [KO70a] determines whether an input acoustic excitation wave from the boundaries will impact the cascade or not. It is defined by equating the axial velocity of the input acoustic disturbance in the absence of a mean flow with the axial velocity of the mean flow. An acoustic disturbance input from the downstream boundary at an angle α will propagate with an axial velocity equal to $A_{+\infty}\cos\alpha$. The axial component of the steady flow at the downstream boundary is $V_{+\infty}\cos\Omega_{+\infty}$. At the critical angle of incidence, the net component of axial velocity of the acoustic wave is zero, which gives

$$\alpha_{\text{crit}} = \cos^{-1}(M_{+\infty}\cos\Omega_{+\infty}) \tag{1.24}$$

A similar expression can be written for the critical angle of incidence for waves from the upstream boundary. Note that since by definition acoustic waves at the critical angle of incidence have no net axial velocity, there is no transfer of acoustic energy in the axial direction, which can be shown to be an equivalent definition of acoustic resonance [Ver89c].

Any incident acoustic wave input into the field from downstream at an angle outside the range $-\alpha_{\text{crit}} \leq \alpha \leq \alpha_{\text{crit}}$ will not reach the cascade. Also, an input acoustic wave from upstream will reach the cascade only if its incidence angle is greater than α_{crit} and less than $-\alpha_{\text{crit}}$. This implies that for excitations from upstream between the angles α_{crit} and $2\pi - \alpha_{\text{crit}}$, and for excitations from downstream between the angles $\pm\alpha_{\text{crit}}$, the excitation wave number is given by $\kappa_{\mp\infty} = k_1$ above, and the response wave number is given by k_2 above.

1.2.5 Response Magnitude Diagrams

Analytic solutions to the first-order unsteady field equations in the far-field that satisfy the requirements that acoustic response disturbances either attenuate with increasing axial distance from the blade row or propagate carrying energy away from the blade row are given in Ref. [Ver89a]. By matching the far-field analytic solutions to a near field numerical solution, the Fourier amplitudes of the acoustic responses can be determined. A normalized response magnitude will be defined to be the ratio of the magnitudes of an acoustic response disturbance to that of the specified input acoustic excitation. A response magnitude diagram is a plot of normalized response magnitudes of the propagating response disturbances for a set of excitation incidence angles.

This definition of a response magnitude diagram is similar to that for the reflection/transmission diagrams used in several previous flat plate cascade analyses [KO70a, KO70b, MH70]. In the flat plate cascade studies, usually only one unattenuated wave existed in the upstream and downstream flow fields. For an acoustic excitation wave input at the upstream boundary, the propagating response wave in the upstream flow field was designated as a "reflected" wave, while the propagating response wave in the downstream flow field was designated as a "transmitted" wave, and the labels were reversed for an acoustic excitation from the downstream boundary. Since in a more general cascade the number of propagating response waves in the upstream and downstream flowfields can differ, it seems more intuitive to view the entire phenomenon in the context of excitation and response, instead.

We set $a_k = |p_{k,-\infty}|/|p_{I,\mp\infty}|$ for each of the upstream propagating acoustic response waves, and $b_k = |p_{k,+\infty}|/|p_{I,\mp\infty}|$ for each of the downstream propagating acoustic response waves, where $|p_{k,\mp\infty}|$ is the magnitude of the kth spatial Fourier component at the boundary in question, and $|p_{I,\mp\infty}|$ is the magnitude of the incident excitation wave. The a_k's and the b_k's are then the normalized response magnitudes of the propagating waves and can be plotted as in Figure 5, which is discussed in the Results section below.

1.3 Results

1.3.1 Sound Generation due to Vortical Excitation

The interaction of wakes from upstream with a turbomachinery blade row generates an acoustic response. To simulate this process, the present analysis determines the acoustic response of a blade row to prescribed vortical excitations at the inflow boundary. The following results present the acoustic response of a blade row to a prescribed vortical excitation with $\mathbf{v}_{R,-\infty} \cdot \mathbf{e}_N = (1, 0)$. The cascade used is a set of modified NACA 0008 airfoils, with an 8% circular arc camber line and a wedge-shaped trailing edge. The blade spacing G is 0.84 and the cascade stagger angle is 10 degrees. This cascade provides a reasonable model of a modern fan exit guide vane. The prescribed freestream Mach number ($M_{-\infty}$) and flow angle ($\Omega_{-\infty}$) at inlet are 0.55 and 30 degrees, respectively, and the calculated freestream values at exit are $M_{+\infty} = 0.45$ and $\Omega_{+\infty} = -2.39$ degrees. The blade passing frequency (BPF) of the upstream fan rotor occurs at $\omega = 4.106$, and the interblade phase angle at this frequency is $\sigma = -2.513$.

An analysis of the far field acoustic behavior reveals that at BPF, all of the acoustic response waves present in the flow are of decaying type. At twice BPF, however, the $n = 1$ component propagates in both the upstream and downstream flow fields, and at three times BPF, both the $n = 1$ and the $n = 2$ components propagates both upstream and downstream.

Figure 2a shows contours of the real part of the unsteady vorticity field for a vortical excitation at twice BPF. The convection of the vorticity through the blade row as it is sliced by the blades is clear. Contour plots for vortical excitations at BPF and 3 BPF are qualitatively similar.

Figures 2b,2c, and 2d show the corresponding contour plots of the real parts of the unsteady pressure due to the vortical excitations at BPF, 2 BPF, and 3 BPF. Note that in Figure 2b the magnitude of the unsteady pressure drops with increasing axial distance ($|\xi|$) from the blade row. In Figure 2c, a single propagating plane wave upstream of the cascade and a single propagating wave downstream of the cascade is visible. The wave numbers of the propagating response waves indicate that the propagation angle for the upstream response wave is -170.8 degrees and its wavelength is 0.676, while the propagation angle for the downstream response wave is -27.8 degrees and its wavelength is 1.98. The complex amplitudes upstream and downstream propagating waves are $p_{1,\text{up}} = (-0.16, 0.06)$, and $p_{1,\text{dn}} = (0.24, -0.03)$, respectively. The 3 BPF acoustic field is shown in Figure 2d. There are two propagating response waves upstream and downstream in this case. The wave numbers of the upstream waves yield propagation angles of 172.8 degrees and -156.5 degrees, and wavelengths of 0.52 and 0.42, respectively. The wave numbers of the downstream yield propagation angles of 18.7 degrees and -79.0 degrees, and wavelengths of 1.33 and 1.03, respectively. The complex amplitudes of the propagating waves are

$$\begin{aligned} p_{1,\text{up}} &= (0.05, 0.10) \\ p_{2,\text{up}} &= (-0.13, -0.80) \\ p_{1,\text{dn}} &= (0.03, 0.06) \\ p_{2,\text{dn}} &= (0.22, 0.06) \end{aligned}$$

The linear superposition of two acoustic waves of equal magnitude and orthogonal propagation directions would result in a series of circles on the contour map of the unsteady pressure. The magnitudes of the upstream propagating waves at 3 BPF are relatively close to each other, and the angle between them is approximately 30 degrees. It can be seen that the acoustic field in Figure 2d reflects this fact by showing elongated ellipses upstream of the cascade. The downstream flow field is dominated by the $n = 2$ mode, which is much larger in magnitude than the downstream $n = 1$ mode, and the contour plot shows this as well.

To illustrate the differences in the acoustic response that can arise between real blade cascades and flat plate cascades, a cascade of flat plate blades similar to that used for the fan exit guide vanes was analyzed. The

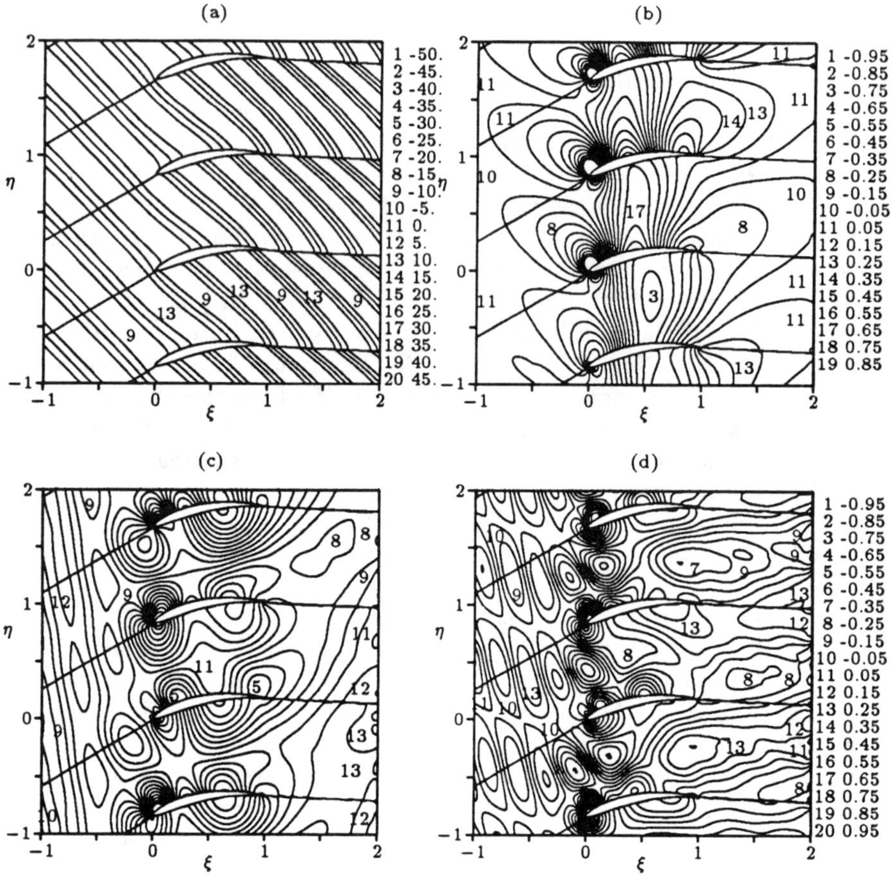

Figure 2. Contour plots of the real part of the vorticity and the real part of the unsteady pressure due to an upstream rotational velocity excitation: (a) vorticity at 2 BPF, (b) pressure at BPF, (c) pressure at 2 BPF, (d) pressure at 3 BPF.

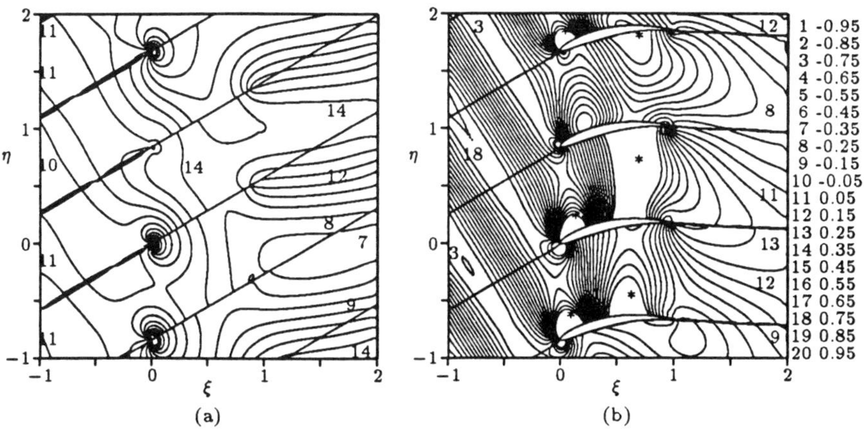

Figure 3. Contour plots of the real part of the unsteady pressure for the case $\sigma = 2.31, \omega = 4.106$. (a) Flat plate cascade, (b) Fan exit guide vane cascade. Regions marked by an asterisk are outside the range of contour levels shown.

flat plate cascade was devised to have the same geometric and upstream freestream flow parameters as the fan exit guide vane, i.e., the gap/chord ratio $G = 0.84$, the stagger angle $\Theta = 30$ degrees, the inlet flow angle $\Omega_{-\infty} = 30$ degrees and the inlet Mach number $M_{-\infty} = 0.55$.

Figure 3 shows contour plots of the real parts of the unsteady pressure for the flat plate and fan exit guide vane cascades subject to an acoustic excitation at an interblade phase angle of 2.31 radians, which corresponds to an incidence angle α of approximately 100 degrees. Figure 3a shows the real part of the acoustic field for the flat plate cascade, and Figure 3b gives the corresponding result for the fan exit guide vane, for an input acoustic excitation from upstream. The upstream steady flows are identical for both cases, and the unsteady flows both show a single upstream travelling acoustic response wave with a wavelength $\lambda = 1.255$ and a propagation angle of $\alpha = 213.3$ degrees relative to the positive ξ axis. The magnitudes of the response waves in the upstream flow differ substantially, however. For the flat plate cascade, $|p_{0,\mathrm{up}} = 0.09$, while for the fan exit guide vanes, $|p_{0,\mathrm{up}} = 0.76$. In the downstream flow field, the steady flows, and the corresponding propagation conditions, differ. For the flat plate cascade, one propagating response wave exists with a wavelength $\lambda = 2.25$, a propagation angle of $\alpha = 279.7$ degrees, and a magnitude $p_{0,\mathrm{dn}} = 0.38$. For the fan exit guide vanes, all of the downstream acoustic response waves are of decaying type.

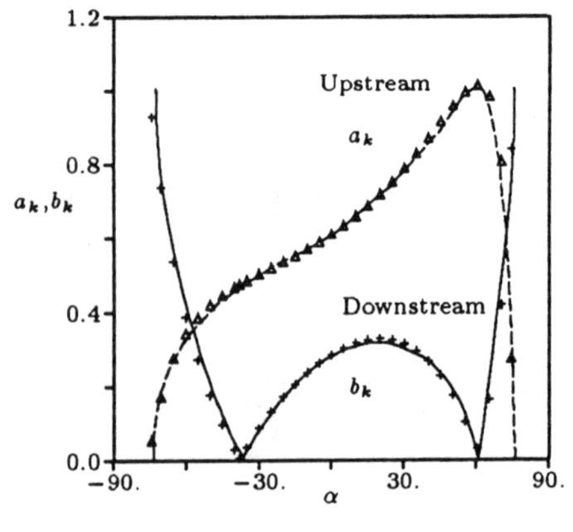

Figure 4. Response magnitudes for acoustic excitations from downstream; the discrete points are the results of the LINFLO analysis and the lines are taken from Smith [Smi71] and Kaji and Okazaki [KO70b].

1.3.2 Acoustic Excitation and Response

To demonstrate the ability of the linearized unsteady aerodynamic analysis LINFLO to reproduce previous results, flows through a cascade of flat plates are now considered. The cascade consists of flat plate blades spaced at unit gap/chord ratio ($G = 1$) and aligned with the inlet flow direction at a stagger angle of 60 degrees. The inlet Mach number is 0.5. Since there is no turning of the flow, the inlet and exit flow angles are both equal to the stagger angle. The unsteady excitations consist of upstream travelling acoustic waves imposed at the exit boundary, over a range of excitation incidence angles between ±74 degrees. In each case, the frequency of the acoustic excitation $\omega = \pi/2$.

For excitations with interblade phase angles σ satisfying $-0.56 < \sigma < 1.47$, a single propagating response wave exists in each of the upstream and downstream flowfields. The normalized response magnitudes for a range of acoustic excitations entering the flow from the downstream boundary are given in Figure 4. In the figure, the discrete data points correspond to the results of the LINFLO calculations, while the solid lines are data taken from Ref. [Smi71] and Ref. [KO70b]. The agreement between the LINFLO results and the solid curves is excellent.

Note that the upstream response coefficients drop abruptly to zero at angles of approximately ±74 degrees. These angles are the critical angles $\pm\alpha_{\text{crit}}$ referred to above. There are also two angles at which the downstream response magnitude goes to zero. One of these occurs at approximately $\alpha =$ 60 degrees, which is the cascade stagger angle. At this angle, the acoustic

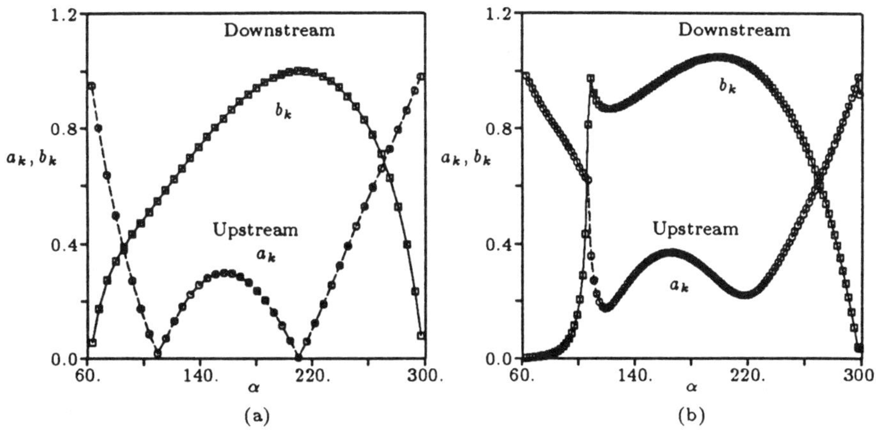

Figure 5. Response magnitude diagrams for acoustic excitations from upstream, $\omega = 4.106$. (a) Flat plate cases, (b) Fan exit guide vane cases.

excitation passes through the cascade without interacting with it. The other angle at which the downstream response goes to zero is $\alpha \approx -37.2$ degrees. At this excitation angle, it is the *response* wave that propagates at 60 degrees, i.e., the wavefronts of the response wave are normal to the steady flow.

Test cases for flat plate cascades responding to acoustic excitations imposed at the inlet boundary were examined as well, but are not shown here. The agreement between the LINFLO results and those of classical linear theory [KO70a, Ami71] was again excellent.

Next, the effect of real blade geometry is examined. The relevant geometries are the cascade of modified NACA 0008 airfoils and the analogous flat plate cascade given in the generation section, above. By keeping the upstream Mach number the same in each case and setting the stagger angle of the flat plates equal to the inlet flow angle of the fan exit guide vanes, the upstream conditions will be identical for both cascades. The frequency of the excitation waves was held fixed at $\omega = 4.106$ for the entire range of incidence angles studied.

A response magnitude diagram for the flat plate cascade interacting with an acoustic wave from upstream is given in Figure 5a. For the flat plate cascade, downstream response magnitude coefficients drop to zero at the critical angles $\alpha = 61.6$ degrees and $\alpha = 360 - 61.6 = 298.4$ degress, corresponding to the critical angle of incidence conditions. There are also two angles at which the upstream coefficients go to zero, given by $\alpha = 210$ degrees, at which point the incoming acoustic wave approaches at the stagger angle, and $\alpha \approx 105.9$ degrees, at which point the response wave propagates at 210 degrees.

The analagous plot for the fan exit guide vanes is shown in Figure 5b. Even though the upstream flow conditions are identical to those for the flat plate cascade, there are significant differences in the acoustic response results. For angles below 107.3 degrees, the far field conditions indicate that there are no propagating waves in the downstream flow field. Consequently, the downstream response coefficients for the incident excitation angles less than this angle are quite small, and would become identically zero as $|\xi| \to \infty$.

Even if the number of propagating acoustic waves is the same for both the flat plate and fan exit guide vane cases, qualitative differences remain in the response magnitude diagrams. Due to the significant turning of the flow by the fan exit guide vanes, however, there are no angles at which the upstream coefficients go to zero. Rather, there are minima in the upstream response curve at approximately $\alpha = 126.7$ degrees and $\alpha = 210$ degrees. The former gives rise to a response wave propagating at an angle of 200.2 degrees, while the latter represents an excitation approximately aligned with the inlet flow angle.

The foregoing example illustrates the significant differences in the overall noise levels predicted for flat plates and real blades. The effects of real blade geometry on the noise level is significant under conditions in which the number of propagating waves is the same in both flow fields, and can be substantial under conditions in which the number of propagating waves differ.

1.4 Conclusions

An existing linearized unsteady aerodynamic analysis capable of providing response predictions for both acoustic and vortical excitations has been used to investigate the effects of real blade geometries on aeroacoustic response phenomena. Acoustic fields representing the generation of acoustic waves by the interaction of rotational velocity or vortical disturbances and blade rows have been shown. The magnitudes of propagating acoustic response waves have also been determined for acoustic excitations at the inlet and exit boundaries for a range of incidence angles. The results have been collected in response magnitude coefficient diagrams, and compared to flat plate cascade flows having identical upstream steady flow conditions. Significant differences in the results have been observed, particularly when the number of propagating acoustic waves in the corresponding downstream flows are different.

References

[AG89] H. M. Atassi and J. Grzedzinski. Unsteady disturbances of streaming motions around bodies. *Journal of Fluid Mechanics*, 209:385–403, December 1989.

[Ami71] R. K. Amiet. Transmission and reflection of sound by a blade row. Paper 71–181, AIAA 9th Aerospace Sciences Meeting, New York, NY, January 1971.

[CF90] H-W. D. Chiang and S. Fleeter. Cascade aeroacoustics including steady loading effects. *Noise Control Engineering Journal*, pages 61–72, March-April 1990.

[CV81] J. R. Caspar and J. M. Verdon. Numerical treatment of unsteady subsonic flow past an oscillating cascade. *AIAA Journal*, 19(12):1531–1539, December 1981.

[Gol78] M. E. Goldstein. Unsteady vortical and entropic distortions of potential flows round arbitrary obstacles. *Journal of Fluid Mechanics*, 89(Part 3):433–468, December 1978.

[HV91] K. C. Hall and J. M. Verdon. Gust response analysis for cascades operating in nonuniform mean flows. *AIAA Journal*, 29(9):1463–1471, September 1991.

[KO70a] S. Kaji and T. Okazaki. Propagation of sound waves through a blade row, i. analysis based on the semi-actuator disk theory. *Journal of Sound and Vibration*, 11(3):339–353, March 1970.

[KO70b] S. Kaji and T. Okazaki. Propagation of sound waves through a blade row, ii. analysis based on the acceleration potential method. *Journal of Sound and Vibration*, 11(3):355–375, March 1970.

[Koc71] W. Koch. On the transmission of sound waves through a blade row. *Journal of Sound and Vibration*, 18(1):111–128, 1971.

[MH70] R. Mani and G. Horvay. Sound transmission through blade rows. *Journal of Sound and Vibration*, 12(1):59–83, May 1970.

[RR90] A. Rangwalla and M. Rai. A kinematical/numerical analysis of rotor-stator interaction noise. Paper 90–0281, AIAA 28th Aerospace Sciences Meeting, Reno, NV, January 1990.

[Smi71] S. N. Smith. Discrete frequency sound generation in axial flow turbomachines. R&M 3709, British Aeronautical Research Council, London, UK, 1971.

[SS90] A. Suddhoo and P. Stow. Simulation of inviscid blade-row interaction using a linearized potential code. Paper 90–1916, AIAA/SAE/ASME/ASEE 26th Joint Propulsion Conference, Orlando, FL, July 1990.

[UV90] Jr. Usab, W. J. and J. M. Verdon. Advances in the numerical analysis of linearized unsteady cascade flows. ASME Paper 90–GT–11, Gas Turbine and Aeroengine Congress and Exposition, Brussels, Belgium, June 11–14 1990. to be published in the *Trans. of the ASME*.

[Ver87] J. M. Verdon. Linearized unsteady aerodynamic theory. In M. F. Platzer and F. O. Carta, editors, *AGARD Manual on Aeroelasticity in Axial-Flow Turbomachines*, chapter II. AGARD, March 1987. Vol. 1, *Unsteady Turbomachinery Aerodynamics*, AGARD–AG–298.

[Ver89a] J. M. Verdon. The unsteady aerodynamic response to arbitrary modes of blade motion. *Journal of Fluids and Structures*, 3(3):255–274, May 1989.

[Ver89b] J. M. Verdon. Unsteady aerodynamics for turbomachinery aeroelastic applications. In D. Nixon, editor, *Unsteady Transonic Aerodynamics, Progress in Astronautics and Aeronautics*, pages 287–347, Washington, D. C., 1989. American Institute of Aeronautics and Astronautics. Vol. 120.

[Ver89c] J. M. Verdon. The unsteady flow in the far field of an isolated blade row. *Journal of Fluids and Structures*, 3(2):123–149, March 1989.

[Ver90] J. M. Verdon. Linearized unsteady aerodynamics for turbomachinery aeroelastic applications. Paper 90–2355, AIAA/SAE/ASME/ASEE 26th Joint Propulsion Conference, Orlando, FL, July 1990.

[Whi70] D. S. Whitehead. Vibration and sound generation in a cascade of flat plates in subsonic flow. Report CUED/A–Turbo/TR 15, Cambridge Univ. Engineering Department, Cambridge, UK, 1970.

[Whi90] D. S. Whitehead. A finite element solution of unsteady two-dimensional flow in cascades. *International Journal for Numerical Methods in Fluids*, 10:13–34, 1990.

High Frequency Gust Interaction with Single Loaded Airfoils in Subsonic Flows

James R. Scott[1]
Hafiz M. Atassi[2]

ABSTRACT This paper presents the high frequency extension of a first-order, finite-difference numerical scheme which the authors have previously developed for solving unsteady, subsonic, periodic vortical flows around isolated airfoils. The method which we have developed is based on the rapid distortion approximation, and fully accounts for the convection and distortion of the upstream vorticity by the nonuniform mean flow around the airfoil. The problem is formulated in the frequency domain, and a transformation is made into computational coordinates using an elliptic coordinate transformation. Numerical solution to the problem is then obtained by solving a frequency dependent matrix equation by a sparse matrix solver. Numerical results are presented for the unsteady lift of various airfoils in one-, two-, and three-dimensional gusts. Comparisons are made with analytical solutions where possible to demonstrate the accuracy of the numerical results.

[1] NASA Lewis Research Center, MS 5-11, 21000 Brookpark Road, Cleveland, OH 44135.
[2] Aerospace and Mechanical Engineering, University of Notre Dame, 380 Fitzpatrick Hall, Notre Dame, IN 46556.

I. INTRODUCTION

Turbomachinery and propeller blades generally operate in an environment with nonuniform upstream flow conditions. Because of the blade rotation, the upstream flow nonuniformities will manifest themselves in the airfoil reference frame as propagating vorticity waves called gusts, even when the upstream nonuniform flow conditions are steady. The vortical waves act to induce a fluctuating pressure field around the airfoil, which in turn leads to unsteady lift and moment forces acting on the blade surface and to the radiation of acoustic waves to the far field. Because of these undesirable effects, along with their potential negative impact on blade performance, there is much interest by aerodynamicsts in being able to predict the unsteady flow around a rotating blade or blade row.

There are a number of reasons, however, why the accurate prediction of such an unsteady flow field turns out to be a difficult task. First, there is a large variation in length scales of the unsteady disturbances that have to be modeled. Depending on the source of the incoming flow nonuniformities, the associated length scales that have to be resolved may vary by one or two orders of magnitude. Second, there is also a large variation in the time scales of the unsteady flow. This makes it both difficult and costly to try to model the unsteady flow with a time-marching approach, since a very small time step must be taken to capture the effects with small time scales. Third, there are strong gradients in the flow around heavily loaded rotating blades which act to strain and distort the upstream vortical disturbances as they are convected past the airfoils. Accurate solution of the unsteady flow field requires that this distortion effect be taken into account. Fourth, because the flow is unsteady, and because of the large variation in scales mentioned above, it is difficult to choose a suitable far field boundary condition. Whatever far field condition is used, it must be a radiation type boundary condition which allows acoustic waves to leave the solution domain, and it must be accurate for the length scales that are being modeled. Otherwise, large errors may result in the numerical solution. Fifth, and finally, it is very difficult to develop a numerical scheme which is accurate enough to model both the unsteady aerodynamics and the aeroacoustics of unsteady vortical flows around airfoils. The acoustic pressure is orders of magnitude smaller than the hydrodynamic pressure, and is very difficult to accurately predict with standard numerical models.

The numerical options which are available to researchers to model these kinds of unsteady flows around rotating blades generally fall into one of two categories. The first of these is the nonlinear approach in which the unsteady Navier-Stokes or Euler equations are solved using a time-marching algorithm. While this approach offers the advantage of being general purpose and widely applicable to a broad range of unsteady flows, it suffers from numerous disadvantages in terms of its practical application. For example, because of the multiple scales problem mentioned above, a very fine grid must be used to capture both the small and large scale effects. The usage of the fine grid, along with its associated small time step, leads to unacceptably long CPU times. Another difficulty is the problem of choosing an accurate and physically correct unsteady far field boundary condition. While progress is being made in the study of far field boundary conditions for unsteady flows governed by nonlinear equations, it is not clear at the present time how well the current methods work for the solution of aeroacoustic problems. Finally, present day algorithms used for solving unsteady flow problems were developed with only unsteady aerodynamics, and not aeroacoustics, in mind. CFD methods for the accurate solution of both unsteady aerodynamics and aeroacoustics problems are just now being developed.

The alternative to solving the full nonlinear equations is to solve their time-linearized counterparts. There are two main drawbacks to this approach. The first is that the unsteady part of the flow must be small compared to the underlying mean flow. The second drawback to the linearized approach is that it should be applied only to flows in which the time scale associated with the mean flow is much less than the time scale associated with the upstream disturbances. This is essentially the "rapid distortion" [1,2] condition, which simply says that the interaction between upstream eddies occurs on a time scale much greater than the time scale of the mean flow, and that the dominant unsteady effects are due to the high speed convection of the eddies past the airfoils. For many flows encountered in applications, these conditions are satisfied, so that a linearized approach is fully justified.

In previous papers [3] - [7] the authors have presented a linearized unsteady aerodynamic analysis for the numerical solution of unsteady flows around single airfoils. The finite-difference, frequency-domain numerical scheme that we have proposed is based on the theoretical work of Goldstein [8] and Atassi and Grzedzinski [9], and offers significant advantages over other numerical approaches. For example, our approach fully accounts for the distortion effects of the nonuniform mean flow around the airfoil on the convected upstream vortical disturbances, whereas previous potential formulations for the gust response problem have neglected mean flow distortion effects [10,11]. Second, in addition to being able to solve the gust response problem, our scheme can be used equally well to solve the oscillating airfoil problem. This means that for a given frequency, both the oscillating airfoil problem and the gust response problem can be solved at the same time for the cost of only one matrix inversion. Third, due to the computational efficiency of our scheme, numerical solutions can be obtained in a couple of minutes per frequency or less on a scientific workstation. Fourth, our formulation is not limited to a specific range of reduced frequencies, but can be applied equally well to both low and high reduced frequencies. Fifth, and finally, the present approach can be coupled with a Green's function formulation to predict the far field noise [12,13]. Because of these features, a single code can be used to determine the unsteady lift and moment on the airfoil due to the gust, plus predict the far field acoustics for the cost of only one calculation.

In the present paper we concentrate on the high frequency gust response problem for single airfoils in subsonic flows. Our previous papers have dealt with unsteady flows with reduced frequencies in the range of 0 to 4 (with normalization by the half chord). As reported previously, our approach works equally well for all reduced frequencies in that range. However, as also reported earlier, the solution times required for a given frequency do increase significantly as the frequency increases, due to the shorter wavelength of the gust and the resultant increase in the number of grid points required to model the unsteady flow field. Because of the longer solution times required for the higher frequencies, our previous work did not consider flows with reduced frequencies much higher than 4. In the recent past, however, significant improvements have been made in the computational efficiency of our numerical scheme to the point that solution times for reduced frequencies even as high as 10 have become reasonable (a few minutes for a symmetric airfoil, and less than an hour for loaded airfoils, on a scientific work station). This has made it possible to extend our approach to include even the very high reduced frequency cases. One of the main conclusions of the present paper is that the linearized unsteady numerical approach that we have developed is essentially not limited in the range of reduced frequencies that can be considered. The present approach can be used for practically any reduced frequency, so long as the computational grid is appropriately constructed for the frequency of interest.

Another of the main conclusions of this paper is that the theoretical formulations of Goldstein [8] and Atassi and Grzedzinski [9] provide an attractive alternative to the full system of linearized unsteady Euler equations for solving unsteady flows. The only limiting assumption in their formulation (as compared to the full linearized Euler equations) is that the mean flow is a potential flow. In view of the significant contribution to the design of aircraft wings and turbomachinery blades by steady potential flow codes, and the major role that potential flow theory has had in aerodynamics, the potential flow assumption for the analysis of unsteady flows is not a serious limitation. Indeed, for many flows of interest in applications, the potential flow approximation is a very good approximation. On the other hand, there are numerous advantages which accrue due to the potential flow assumption. Perhaps the biggest advantage, as first shown by Goldstein, is that the linearized unsteady Euler equations can in this case be reduced to a single partial differential equation for an unsteady potential. This significantly reduces the mathematical complexity of the resulting boundary value problem that must be solved for a given flow. As a result, numerical schemes that are developed to implement this approach are simpler and more efficient than equivalent methods for the full linearized unsteady Euler equations. A second advantage is that the potential flow assumption makes it possible to perform theoretical analyses that would not otherwise be possible. This can be a big advantage when it comes to the development and validation of a numerical code. Finally, the resulting mathematical formulation proposed by Goldstein [8] and Atassi and Grzedzinski [9] has broad application to a large range of unsteady flows. It is valid for incompressible flows, compressible flows, and transonic flows with weak shocks. And as we show in the present paper, there is no practical limitation in the range of reduced frequencies that this formulation can be used for. This is a significant advantage, because it implies that the multiple scales problem faced by time-marching codes will not be a difficulty for the present approach.

Our main purpose in the present paper is to present the high frequency extension of our linearized unsteady aerodynamic numerical scheme which we have developed in previous papers. Since we have presented the details of our approach previously, we will only briefly review the theoretical aerodynamic formulation and numerical approach that we have developed. This will be done in Sections II. and III., respectively. In Section IV., we present and discuss a large variety of numerical results for the gust response problem.

II. LINEARIZED UNSTEADY AERODYNAMIC FORMULATION

Consider an inviscid, compressible flow past a two-dimensional airfoil placed at nonzero incidence to a stream with uniform upstream velocity U_∞ in the x_1 direction. We assume in the present discussion that the fluid is an ideal, non-heat-conducting gas with constant specific heats, and that there are no shocks in the flow. Under the above assumptions there will be a steady potential flow around the airfoil so that we may write

$$\vec{U}_0(\vec{x}) = \vec{\nabla}\Phi_0, \tag{1.1}$$

where 0 subscripts are used to denote steady mean flow quantities.

Let us assume that far upstream an unsteady vortical disturbance is imposed on the flow. The only restriction that we place on the upstream disturbance is that it can be expressed as a generalized Fourier integral so that we may write

$$\vec{u}_\infty(\vec{x} - \vec{i}U_\infty t) = \int_{\vec{k}} \vec{a}(\vec{k}) e^{i\vec{k}\cdot(\vec{x}-\vec{i}U_\infty t)} d\vec{k}. \tag{1.2}$$

In addition, we assume that its length scale l' and characteristic velocity u_∞ are such that the condition

$$\frac{c}{U_\infty} \ll \frac{l'}{u_\infty} \tag{1.3}$$

is satisfied, where c is the airfoil chord length. We thus require that the time scale associated with the mean flow be an order of magnitude less than the time scale associated with the upstream unsteady disturbances. Since we present a linearized mathematical formulation, we may without loss of generality consider a single Fourier component of the incident vortical disturbance, and solve for more general disturbances by superposition. We therefore consider incident vortical gusts of the form

$$\vec{u}_\infty = \vec{a} e^{i\vec{k}\cdot(\vec{X} - \vec{i} U_\infty t)} \tag{1.4}$$

where $\vec{a}$ and $\vec{k}$ must satisfy

$$\vec{a} \cdot \vec{k} = 0 \tag{1.5}$$

to ensure that the continuity equation is satisfied. (See Figure 1.)

In general, the components of $\vec{X}$ in equation 1.4 are not the spatial coordinates, but rather are Lagrangian coordinates of the mean flow. For the case of two-dimensional mean flow the components of $\vec{X}$ are given by

$$X_2 = \frac{\Psi_0}{\rho_\infty U_\infty} \tag{1.6}$$

and

$$X_3 = x_3, \tag{1.7}$$

where Ψ_0 is the stream function of the mean flow and x_3 is the spatial coordinate in the spanwise direction. The component X_1 is defined by

$$X_1 = U_\infty \Delta, \tag{1.8}$$

where Δ is the Lighthill "drift" function [14], which can be expressed in terms of Φ_0 and Ψ_0 as

$$\Delta = \frac{\Phi_0}{{U_\infty}^2} + \int_{-\infty}^{\Phi_0} \left(\frac{1}{{U_0}^2} - \frac{1}{{U_\infty}^2}\right) d\Phi_0, \tag{1.9}$$

where the integration is carried out on $\Psi_0 =$ constant. The difference in Δ between two points on a streamline is the time it takes a mean flow fluid particle to traverse the distance between those two points. Note that for the thin airfoil case in which the mean flow is a uniform parallel flow, the components of $\vec{X}$ reduce precisely to the spatial coordinates.

We assume that the total unsteady flow field can be represented by

$$\vec{U}(\vec{x}, t) = \vec{U}_0(\vec{x}) + \vec{u}(\vec{x}, t) \tag{1.10}$$

$$p(\vec{x}, t) = p_0(\vec{x}) + p'(\vec{x}, t) \tag{1.11}$$

$$\rho(\vec{x}, t) = \rho_0(\vec{x}) + \rho'(\vec{x}, t) \tag{1.12}$$

$$s(\vec{x}, t) = s_0 + s'(\vec{x}, t) \tag{1.13}$$

where the entropy s_0 is constant, and $\vec{u}$, p', ρ', and s' are the unsteady perturbation velocity, pressure, density and entropy, respectively. Quantities with 0 subscripts are the steady mean flow quantities which are assumed to be known. Note that these quantities obey the steady nonlinear equations of motion, so that the linearization of the unsteady flow is about the fully nonlinear mean flow.

Substituting relations 1.10 - 1.13 into the nonlinear Euler equations and neglecting products of small quantities, one obtains the linearized continuity, momentum, and entropy conservation equations

$$\frac{D_0\rho'}{Dt} + \rho'\vec{\nabla}\cdot\vec{U}_0 + \vec{\nabla}\cdot(\rho_0\vec{u}) = 0 \qquad (1.14)$$

$$\rho_0(\frac{D_0\vec{u}}{Dt} + \vec{u}\cdot\vec{\nabla}\vec{U}_0) + \rho'\vec{U}_0\cdot\vec{\nabla}\vec{U}_0 = -\vec{\nabla}p' \qquad (1.15)$$

$$\frac{D_0 s'}{Dt} = 0, \qquad (1.16)$$

where $\frac{D_0}{Dt} = \frac{\partial}{\partial t} + \vec{U}_0\cdot\vec{\nabla}$ is the convective derivative associated with the mean flow.

It is shown in References [8] and [9] that if the unsteady velocity is decomposed into the sum of a known vortical component and an unknown potential component, then the problem for determining the unsteady flow may be reduced to solving a single, non-constant coefficient, inhomogeneous convective wave equation which may be written

$$\frac{D_0}{Dt}(\frac{1}{{c_0}^2}\frac{D_0\phi}{Dt}) - \frac{1}{\rho_0}\vec{\nabla}\cdot(\rho_0\vec{\nabla}\phi) = \frac{1}{\rho_0}\vec{\nabla}\cdot(\rho_0\vec{u}^{(R)}) \qquad (1.17)$$

where

$$\vec{u}(\vec{x},t) = \vec{u}^{(R)} + \vec{\nabla}\phi. \qquad (1.18)$$

The unsteady pressure is related to ϕ through the equation

$$p' = -\rho_0(\vec{x})\frac{D_0\phi}{Dt}. \qquad (1.19)$$

The vortical velocity $\vec{u}^{(R)}$ is a known function of the mean flow Lagrangian coordinates and the upstream vortical disturbances and is given by

$$\vec{u}^{(R)} = [\vec{\nabla}(\vec{a}\cdot\vec{X})]e^{i\vec{k}\cdot(\vec{X}-\vec{i}U_\infty t)} + \vec{\nabla}\tilde{\phi}, \qquad (1.20)$$

where

$$\tilde{\phi} = \frac{i}{k_1}(a_1 + \frac{a_2k_1 - a_1k_2}{1 + ia_0U_\infty k_1}\frac{1 - e^{-ik_2X_2}}{k_2})e^{i\vec{k}\cdot(\vec{X}-\vec{i}U_\infty t)}, \qquad (1.21)$$

and

$$\vec{a} = (a_1, a_2, a_3) \qquad \text{and} \quad a_0 = -(\frac{\partial U_0}{\partial n})_S^{-1}. \qquad (1.22)$$

Here n denotes the direction of the outward unit normal, S denotes the stagnation point near the airfoil leading edge, and $U_0 = |\vec{U}_0|$ is the magnitude of the mean velocity. (See [9] for more details concerning the purpose and derivation of the function ϕ .)

Finally, the potential ϕ must satisfy the boundary conditions

$$\begin{aligned} \vec{\nabla}\phi \cdot \vec{n} &= 0 && \text{airfoil surface} && (1.23)\\ \frac{D_0}{Dt}(\Delta\phi) &= 0 && \text{wake} && (1.24)\\ \Delta[\vec{\nabla}\phi \cdot \vec{n}] &= 0 && \text{wake} && (1.25)\\ \vec{\nabla}\phi \rightarrow &- \vec{\nabla}\tilde{\phi} && \text{as} \quad x_1 \rightarrow -\infty, && (1.26) \end{aligned}$$

where equation 1.23 is the impermeability condition at the airfoil surface, equations 1.24 and 1.25 impose continuity of the pressure and normal velocity across the wake, and equation 1.26 ensures that $\vec{u}(\vec{x},t) \rightarrow \vec{u}_\infty(\vec{x},t)$ as $x_1 \rightarrow -\infty$.

The linearized boundary value problem for the unsteady gust response problem thus consists of the governing equation 1.17 and boundary conditions 1.23 - 1.26 together with the requirement that ϕ is continuous at the airfoil trailing edge.

III. BASIC NUMERICAL APPROACH

In our previous papers we have shown that it is necessary to reformulate the boundary value problem given by equation 1.17 and boundary conditions 1.23-1.26 in order to obtain accurate numerical solutions to the gust response problem. The main difficulty arises due to condition 1.26. From a numerical standpoint, it is far preferable to have a condition which specifies that the gradient of ϕ vanishes in the far field. We therefore reformulate the problem in such a way as to replace ϕ by a function whose gradient vanishes as $r \rightarrow \infty$, where r is the distance from the airfoil center.

To this end, we introduce the potential functions ϕ_1 and ϕ_2, where

$$\phi = \phi_1 - \phi_2 \tag{1.27}$$

and ϕ_2 is a known function which is constructed such that

$$|\phi_2 - \tilde{\phi}| \rightarrow 0 \qquad \text{as} \quad r \rightarrow \infty. \tag{1.28}$$

Equation 1.27 together with conditions 1.26 and 1.28 then show that the new potential function ϕ_1 will satisfy

$$\vec{\nabla}\phi_1 \rightarrow \vec{\nabla}\phi_2 - \vec{\nabla}\tilde{\phi} \rightarrow 0 \qquad \text{as} \quad r \rightarrow \infty. \tag{1.29}$$

The problem may then be reformulated in terms of the unknown potential ϕ_1.

In the present paper we will not concern ourselves with the explicit expression of the potential function ϕ_2. The reader may consult [6] or [7] for details.

In what follows, we assume the following nondimensionalization:

$x_1, x_2, x_3, X_1, X_2, X_3$	by	$\frac{c}{2}$
Φ_0	by	$\frac{c}{2}U_\infty$
Ψ_0	by	$\frac{c}{2}\rho_\infty U_\infty$
U_0, c_0	by	U_∞
ρ_0	by	ρ_∞
p'	by	$\rho_\infty U_\infty a_2$
t, Δ	by	$\frac{c}{2U_\infty}$
ω	by	$\frac{2U_\infty}{c}$
k_1, k_2, k_3	by	$\frac{2}{c}$
$\phi, \tilde{\phi}, \phi_1, \phi_2$	by	$\frac{c}{2}a_2$
$\vec{a}$	by	a_2

The normalized wave number k_1 is the reduced frequency. The authors would also like to point out that we have normalized the unsteady pressure by the normal component of the gust a_2, rather than the gust magnitude $|\vec{a}|$. (We assume throughout the paper that $a_2 > 0$.) With this normalization, the unsteady lift response function for a thin airfoil in a transverse and longitudinal gust becomes identical to that of a thin airfoil in a transverse gust. For an incompressible flow, the unsteady lift response function for both cases would be given by the Sears function. The longitudinal component of the gust will only have an effect on the unsteady lift if the airfoil has a nonzero thickness, camber, or angle of attack. Nondimensionalization of the unsteady pressure by the normal component of the gust will thus better demonstrate the effect of the airfoil geometry on the unsteady lift.

With the above reformulation and nondimensionalization, the present boundary value problem then becomes

$$\frac{D_0}{Dt}\left(\frac{1}{c_0{}^2}\frac{D_0\phi_1}{Dt}\right) - \frac{1}{\rho_0}\vec{\nabla}\cdot(\rho_0\vec{\nabla}\phi_1) = \frac{1}{\rho_0}\vec{\nabla}\cdot(\rho_0\vec{u}^{(R)}) + \frac{D_0}{Dt}\left(\frac{1}{c_0{}^2}\frac{D_0\phi_2}{Dt}\right) - \frac{1}{\rho_0}\vec{\nabla}\cdot(\rho_0\vec{\nabla}\phi_2) \quad (1.30)$$

$$\vec{\nabla}\phi_1\cdot\vec{n} = \vec{\nabla}\phi_2\cdot\vec{n} \qquad \text{airfoil surface} \quad (1.31)$$

$$\frac{D_0}{Dt}[\Delta(\phi_1 - \phi_2)] = 0 \qquad \text{wake} \quad (1.32)$$

$$\Delta[\vec{\nabla}(\phi_1 - \phi_2)\cdot\vec{n}] = 0 \qquad \text{wake} \quad (1.33)$$

$$\vec{\nabla}\phi_1 \to 0 \quad \text{as} \quad x_1 \to -\infty. \tag{1.34}$$

where

$$\tilde{\phi} = \frac{i}{k_1}(a_1 + \frac{a_2 k_1 - a_1 k_2}{1 + i a_0 k_1} \frac{1 - e^{-ik_2 X_2}}{k_2}) e^{i\vec{k}\cdot\vec{X} - ik_1 t} \tag{1.35}$$

$$X_1 = \Delta, \quad X_2 = \Psi_0, \quad X_3 = x_3 \tag{1.36}$$

$$\vec{u}(\vec{x}, t) = \vec{u}^{(R)} + \vec{\nabla}(\phi_1 - \phi_2) \tag{1.37}$$

$$\vec{u}^{(R)} = [\vec{\nabla}(\vec{a} \cdot \vec{X})] e^{i\vec{k}\cdot\vec{X} - ik_1 t} + \vec{\nabla}\tilde{\phi}. \tag{1.38}$$

$$\vec{u}_\infty = \vec{a} e^{i\vec{k}\cdot\vec{X} - ik_1 t} \tag{1.39}$$

The first step in obtaining numerical solutions to equation 1.30 and its associated boundary conditions is to determine the steady potential flow about the airfoil for the given flow conditions. In view of equations 1.35 - 1.39, it is particularly desirable to obtain the steady flow as a function of Φ_0 and Ψ_0, the mean flow potential and stream functions. In previous papers we have shown how this can be done by using a coupled approach which employs the potential flow code FLO36 [15] along with a Gothert's rule approximation. Details may be found in [6] or [7].

Frequency Domain Formulation

The boundary value problem given by equations 1.30 - 1.34 may be transformed into the frequency domain by making the following change of variables:

$$\phi_1 = \varphi e^{-ik_1 t + ik_3 x_3} \tag{1.40}$$

and

$$\varphi = \psi e^{-iK_0 \Phi} \quad \text{where} \quad K_0 = \frac{k_1 M_\infty^2}{\beta_\infty^2} \tag{1.41}$$

Moreover, it is convenient to introduce the new independent variables Φ and Ψ with the transformation

$$\Phi = \Phi_0 \tag{1.42}$$

$$\Psi = \beta_\infty \Psi_0 \tag{1.43}$$

where $\beta_\infty = \sqrt{1 - M_\infty{}^2}$, and M_∞ is the free stream Mach number.

The governing equation then becomes

$$
\begin{aligned}
- & \beta_\infty^2 [\frac{\partial^2\psi}{\partial\Phi^2} + \frac{\partial^2\psi}{\partial\Psi^2} + (\frac{k_1^2 M_\infty^2}{\beta_\infty^4} - \frac{k_3^2}{\beta_\infty^2})\psi] \\
+ & A_1\psi + A_2\frac{\partial\psi}{\partial\Phi} + A_3\frac{\partial\psi}{\partial\Psi} + A_4\frac{\partial^2\psi}{\partial\Phi^2} + A_5\frac{\partial^2\psi}{\partial\Psi^2} \\
= & e^{iK_o\Phi}(S_1 + S_2 + S_3 - S_4) \qquad (1.44)
\end{aligned}
$$

where the coefficients $A_1 ... A_5$ are known functions which depend on the mean flow, and $S_1 ... S_4$ are given by

$$
\begin{aligned}
e^{-ik_1t+ik_3x_3}S_1 &= \frac{\vec{\nabla}\rho_0}{\rho_0}\cdot\{[\vec{\nabla}(\vec{a}\cdot\vec{X})]e^{i\vec{k}\cdot\vec{X}-ik_1t} + \vec{\nabla}\tilde{\phi}\} \\
e^{-ik_1t+ik_3x_3}S_2 &= \vec{\nabla}\cdot\{[\vec{\nabla}(\vec{a}\cdot\vec{X})]e^{i\vec{k}\cdot\vec{X}-ik_1t} + \vec{\nabla}\tilde{\phi}\} \\
e^{-ik_1t+ik_3x_3}S_3 &= \frac{D_0}{Dt}(\frac{1}{c_0{}^2}\frac{D_0\phi_2}{Dt}) \\
e^{-ik_1t+ik_3x_3}S_4 &= \frac{1}{\rho_0}\vec{\nabla}\cdot(\rho_0\vec{\nabla}\phi_2) \qquad (1.45)
\end{aligned}
$$

In the far field both the coefficients $A_1 ... A_5$ and the quantity $(S_1 + S_2 + S_3 - S_4)$ tend to zero, so that the equation reduces to a Helmholtz equation.

Equation 1.44 may be solved by any one of a number of standard numerical methods. The authors' approach has been to use second order accurate finite differences with central differencing for the governing equation and one-sided differencing for the boundary conditions. The resulting finite difference matrix equation which must be solved is not block tridiagonal, but it is sparse and can be efficiently solved with a general purpose, complex variable sparse matrix solver.

As it turns out, however, the independent variables (Φ, Ψ) are not suitable computational coordinates for the gust response problem [3] - [7]. Difficulties arise in obtaining consistently accurate results over a large range of Mach numbers and reduced frequencies. In addition, there are problems with the implementation of far field boundary conditions. Fortunately, these problems can be remedied by the introduction of the elliptic coordinates (η, ξ) through the transformation

$$
\begin{aligned}
\Phi &= a^\star \cos(\pi\eta)\cosh(\pi\xi) \\
\Psi &= a^\star \sin(\pi\eta)\sinh(\pi\xi)
\end{aligned}
\qquad (1.46)
$$

where a^* is a constant which depends on the airfoil steady circulation. Note that in the far field the elliptic coordinates reduce essentially to cylindrical coordinates, and that the $\Phi - \Psi$ plane is mapped into a semi-infinite strip in the $\eta - \xi$ plane.

With this change of variables, the governing equation takes the form

$$\begin{aligned}
- & \beta_\infty^2 [\frac{\partial^2 \psi}{\partial \eta^2} + \frac{\partial^2 \psi}{\partial \xi^2} + J(\eta, \xi)\,(\frac{k_1^2 M_\infty^2}{\beta_\infty^4} - \frac{k_3^2}{\beta_\infty^2})\,\psi] \\
+ & A_1\, J(\eta, \xi)\,\psi + T_1 \frac{\partial \psi}{\partial \xi} + T_2 \frac{\partial \psi}{\partial \eta} \\
+ & T_3 \frac{\partial^2 \psi}{\partial \xi^2} + T_4 \frac{\partial^2 \psi}{\partial \eta^2} + T_5 \frac{\partial^2 \psi}{\partial \eta \partial \xi} \\
= & e^{iK_0 \Phi} (S_1 + S_2 + S_3 - S_4)\, J(\eta, \xi), \qquad (1.47)
\end{aligned}$$

where $J(\eta, \xi)$ is the Jacobian of the transformation 1.46, and $T_1 ... T_5$ are known functions of (η, ξ) which depend upon the mean flow.

The frequency domain boundary value problem then consists of the governing equation 1.47 and boundary conditions 1.31 - 1.34, together with the requirement that the velocity is finite at the trailing edge.

However, we point out that condition 1.34, which requires that $\vec{\nabla}\phi_1 \to 0$ at upstream infinity, cannot be imposed throughout the far field on a boundary at a finite distance from the airfoil. To implement such a condition would impose a reflecting far field boundary condition which can lead to large errors in the numerical solution.

To correctly model the physics of the present unsteady boundary value problem, it is necessary to use a radiation type boundary condition which allows outgoing acoustic waves to leave the solution domain without being reflected back into the computational grid. While this can be done in more than one way, the authors have found that the best approach is to use a Sommerfield radiation condition on the unsteady pressure. This avoids any difficulty which might arise due to the discontinuity of the unsteady velocity in the wake. This approach is simple and easy to implement, and leads to a matrix equation that does not require pivoting. Perhaps most importantly, this condition has been shown through our previous work to be very accurate over a large range of reduced frequencies.

For the present boundary value problem, the Sommerfield radiation condition can be written

$$\left[\frac{\partial}{\partial R} - i\sqrt{(\frac{k_1 M_\infty}{\beta_\infty^2})^2 - (\frac{k_3}{\beta_\infty})^2}\right](\frac{\partial}{\partial \Phi} - i\frac{k_1}{\beta_\infty^2})\psi = 0 \qquad (1.48)$$

where

$$\Phi = R\cos\Theta$$

$$\Psi = R \sin \Theta.$$

This condition is applied for all grid points on the far field boundary.

Grid Determination

As reported previously [3] - [7], it is not possible to use a single grid and obtain accurate solutions to the gust response problem for a large range of reduced frequencies. Rather, the unsteady grid must be determined as a function of both the mean flow Mach number and the reduced frequency.

This requirement is dictated by both the accuracy of the far field boundary condition 1.48, and the need to adequately model the wavelike structure of the numerical solution. The accuracy of the far field boundary condition 1.48 depends on the reduced frequency k_1, the free stream Mach number M_∞, and the distance R to the far field boundary. It can be shown that when the parameter $\frac{k_1 M_\infty}{\beta_\infty^2}$ is large, the accuracy of boundary condition 1.48 is $O\left(1/\sqrt{2\pi \frac{k_1 M_\infty}{\beta_\infty^2} R^{\frac{3}{2}}}\right)$. This shows that the location of the outer boundary of the grid must be determined as a function of k_1 and M_∞. In particular, this shows that for the high frequency case, it is possible to bring in the outer boundary to within a few chord lengths of the airfoil. The other grid requirement is that there should be enough grid points per wavelength to accurately represent the airfoil and wake boundary conditions. It turns out that the distribution of the grid points in the wake is most crucial. As shown in [6] and [7], the location of the j th ξ grid point in the wake should be given by

$$\xi_j = \frac{1}{\pi} \cosh^{-1}\left(j \frac{\pi \beta_\infty^2}{6a^\star k_1} + 1\right), \tag{1.49}$$

except near the airfoil trailing edge where the spacing is uniform for the first few grid points. Numerical experiments have shown that this method of grid determination works equally well for both low and high reduced frequencies.

IV. DISCUSSION OF NUMERICAL RESULTS

In this section of the paper we present a large variety of numerical results for the gust response problem. Figures 2–5 show comparisons between numerical and analytical results for the unsteady lift of a zero thickness airfoil in a periodic gust at different Mach numbers. Figures 6–8 show the effect of airfoil geometry on the unsteady lift of various airfoils in a 1-D, 2-D, and 3-D gust. Finally, in Figure 9 we demonstrate the effect of the gust dimensionality on the unsteady lift for a fixed airfoil geometry.

In Figure 2, we show a comparison between numerical results and the Sears solution for a flat plate airfoil in a transverse gust in incompressible flow. The reduced frequency varies from zero to 10.0. As can be readily seen, the agreement between the numerical and theoretical results is excellent for the full range of frequencies shown.

Figure 3 shows a comparison between numerical and analytical results for the unsteady lift of a flat plate airfoil in an oblique gust in incompressible flow. The spanwise wave number k_3 is .442, and we take $k_2 = k_1$, with $\frac{a_2}{a_1} = -\frac{7}{4}$. The analytical results are obtained from a Possio [16] solver, which solves an integral equation form of the solution using numerical collocation techniques. It should be pointed out that while we refer to these results as "analytical" results, they are not an exact analytical solution as is the Sears solution. There is some numerical error associated with the "analytical" results, particularly as the reduced frequency becomes large. It is evident from the figure that the agreement between the results from the Possio solver and the present numerical predictions is very good. At the highest reduced frequencies, there is a small discrepancy between the two sets of results. It is not possible to say which are more accurate.

Figures 4 and 5 present results analogous to those of Figures 2 and 3, respectively, except now the flow is compressible with a Mach number of .8. As in the incompressible case, the agreement between the numerical and analytical results is in general quite good. As in Figure 3, one cannot conclude which set of results is more accurate.

In Figure 6, we show results that demonstrate the effect of airfoil geometry on the unsteady lift of three different airfoils in a transverse gust. The reduced frequency varies from zero to 6.0, and the mean flow Mach number is .5. Two of the airfoils are 12% thick Joukowski airfoils, one with zero camber and the other with 5% camber, and the third airfoil is a simple flat plate. The incidence angle for all three airfoils is zero degrees, and the steady lift coefficient for the cambered airfoil is .82. By comparing the response functions of the two Joukowski airfoils, it can be readily seen that the mean loading of the cambered airfoil has very little effect on the unsteady lift for a purely transverse gust. This result was also found for incompressible flows by Atassi in a previous paper [17]. On the other hand, in comparing the results for the 12% thick airfoils versus those of the flat plate airfoil, it is evident that airfoil thickness leads to a significant increase in the unsteady lift when the reduced frequency is small, but a decrease in the unsteady lift when the reduced frequency is large.

The results shown in Figures 7 and 8 are analogous to those in Figure 6, except the gust is a transverse and longitudinal gust in Figure 7 and a fully three-dimensional gust in Figure 8. For the results in Figure 7, $k_2 = k_1$ and $-a_1 = a_2 = 1$. For the results in Figure 8, the spanwise wave number k_3 is .4, $k_2 = k_1$, and $\frac{a_2}{a_1} = -\frac{7}{4}$. In the case of the transverse and longitudinal gust, the velocity fluctuations are in the same plane as the airfoil cross-section, and the gust propagates at a 45° angle to the chord line. For the three-dimensional gust, there is a spanwise component to the velocity fluctuations, and the gust propagates in a direction which is oblique to the airfoil chord. In comparing the results shown in Figures 7 and 8 with those shown in Figure 6, it is clear that the effect that the airfoil geometry has on the unsteady lift depends strongly upon the characteristics of the gust. This is apparent in looking at the effects of both airfoil camber and airfoil thickness. For example, in the simple transverse gust case, airfoil camber had almost no effect on the unsteady lift, whereas in the 2-D and 3-D gust cases, airfoil camber has a very strong effect on the unsteady lift for both the low and high reduced frequencies. The exact effect depends to a large extent on the value of the reduced frequency, although one can say that the general trend is for camber to decrease the magnitude of the unsteady lift at the low frequencies and to increase both the magnitude and phase of the unsteady lift at the high reduced frequencies. This trend is noticeably stronger for the 2-D case than the 3-D case. The effects of airfoil thickness, on the other hand, are somewhat different between the 2-D and 3-D cases. In the 2-D case, as in the transverse gust case, airfoil thickness tends to increase the magnitude of the unsteady lift when the reduced frequency is small, but decrease the unsteady lift when the frequency is large. For the 3-D case, however, airfoil thickness does not increase the unsteady lift for the low reduced frequency cases. It does decrease the unsteady lift at the high frequencies, but not as strongly as in the 2-D gust case.

The results presented in Figure 9 are shown to demonstrate the effect that the gust dimensionality has on the unsteady lift response function for a fixed airfoil. The airfoil under consideration is the same 12% thick, 5% cambered Joukowski airfoil that was considered in the last three figures. As before, the mean flow Mach number is .5, the angle of incidence is zero degrees, and the steady lift coefficient is .82. The gust conditions for the two- and three-dimensional results are the same as in Figures 7 and 8, respectively. It is readily apparent from the figure that the dimensionality of the gust has a very strong effect on the unsteady lift for reduced frequencies lower than about 1. In general, the effect of the gust dimensionality becomes stronger as the reduced frequency approaches zero. Amazingly, however, for reduced frequencies above about 1, the three response functions roughly collapse to a single curve, with the exception of a slight difference in phase for the 1-D case. This remarkable result occurs as a direct consequence of normalizing the unsteady lift by the normal component of the gust, rather than by the gust magnitude. By normalizing in this way, it becomes clear that the magnitude of the unsteady lift for this airfoil depends primarily on the value of the reduced frequency and not on the dimensionality of the gust, for reduced frequencies above about 1.

V. CONCLUSION

In this paper we have presented the high frequency extension of the linearized unsteady aerodynamic numerical scheme which we have previously developed. On the basis of the results presented in this and previous papers, a number of conclusions can be made. First, our numerical approach is essentially not limited in the range of reduced frequencies that can be considered. For thin, unloaded airfoils, comparisons made between our numerical results and known analytical solutions have shown that our scheme is very accurate for reduced frequencies as high as 10. The authors are not aware of any other computational fluid dynamics code that can accurately solve the gust response problem for this range of reduced frequencies. Second, the present approach can also be used to predict the far field noise of an airfoil in a vortical gust, including the high frequency cases. Third, our numerical scheme can be used to solve the oscillating airfoil problem as well as the gust response problem. Fourth, and finally, the present numerical approach offers an attractive alternative to solving the full system of linearized Euler equations for unsteady flows around airfoils. Our work to date has shown that the present approach holds much promise for the accurate and efficient solution of unsteady aerodynamic and aeroacoustic problems.

References

[1] Ribner, H. S. and Tucker, M., "Spectrum of Turbulence in a Contracting Stream," *NACA Report No. 1113, 1953.*

[2] Batchelor, G. K. and Proudman, I., "The Effect of Rapid Distortion of a Fluid in Turbulent Motion," *Quart. J. Mech. Appl. Math. 1,* 1954, pp. 83-103.

[3] Atassi, H. M. and Scott, J. R., " Analysis of Nonuniform Subsonic Flows About a Row of Moving Blades," *Proceedings of the Fourth International Symposium on Unsteady Aerodynamics and Aeroelasticity of Turbomachines and Propellers,* H. E. Gallus and S. Servaty, eds., Institute fur Strahlantriebe und Turbomachine, University of Aachen, Federal Republic of Germany, 1988, pp. 39-67.

[4] Scott, J. R. and Atassi, H. M., " Numerical Solution of Periodic Vortical Flows About a Thin Airfoil," AIAA Paper 89-1691, June, 1989.

[5] Scott, J.R. and Atassi, H.M., " Numerical Solutions of the Linearized Euler Equations for Unsteady Vortical Flows Around Lifting Airfoils," AIAA Paper 90-0694, January 1990.

[6] Scott, J.R. and Atassi, H.M., "A Finite-Difference, Frequency-Domain Numerical Scheme for the Solution of the Linearized Unsteady Euler Equations," *Proceedings of the Lewis CFD Symposium on Aeropropulsion,* April 24-26, 1990.

[7] Scott, J.R., "Compressible Flows with Periodic Vortical Disturbances Around Lifting Airfoils," Ph. D. Dissertation, University of Notre Dame, Notre Dame, IN, 1990.

[8] Goldstein, M. E., "Unsteady Vortical and Entropic Distortions of Potential Flows Round Arbitrary Obstacles," *J. Fluid Mech.,* Vol. 89, 1978, pp. 433-468.

[9] Atassi, H. M. and Grzedzinski, J., "Unsteady Disturbances of Streaming Motions Around Bodies," *J. Fluid Mech.,* Vol. 209, Dec. 1989, pp. 385-403.

[10] McCroskey, W.J. and Goorjian, P.M., "Interactions of Airfoils with Gusts and Concentrated Vortices in Unsteady Transonic Flow," AIAA Paper 83-1691, July 1983.

[11] McCroskey, W. J., "The Effects of Gusts on the Fluctuating Airloads of Airfoils in Transonic Flow," *Journal of Aircraft,* Vol. 22, No. 3, March 1985, pp. 236-243.

[12] Atassi, H.M., Subramaniam, S., and Scott, J.R., "Acoustic Radiation from Lifting Airfoils in Compressible Subsonic Flow," AIAA Paper 90-3911, October 1990.

[13] Davis, C.M. and Atassi, H.M., "The Far Field Acoustic Pressure of an Airfoil in Nonuniform Subsonic Flows," NCA-Vol. 11/FED-Vol.130, Flow Noise Modeling, Measurement, and Control, ASME 1991.

[14] Lighthill, M.J., "Drift" *J. Fluid Mech.,* Vol. 1, 1956, pp. 31-53.

[15] Jameson, A. and Caughey, D.A., "A Finite Volume Method for Transonic Potential Flow Calculations," *Proceedings of the AIAA 3rd Computational Fluid Dynamics Conference,* Williamsburg, Va., July 1979, pp. 122-146.

[16] Possio, C., "L'Azione Aerodinamica sul Profilo Oscillante in un Fluido Compressible a Velocita Ipsonora," *L'Aerotecnica XVIII,* 4, 1938.

[17] Atassi, H.M., "The Sears Problem for a Lifting Airfoil Revisited - New Results," *J. Fluid Mech.* Vol. 141, 1984, pp. 109-122.

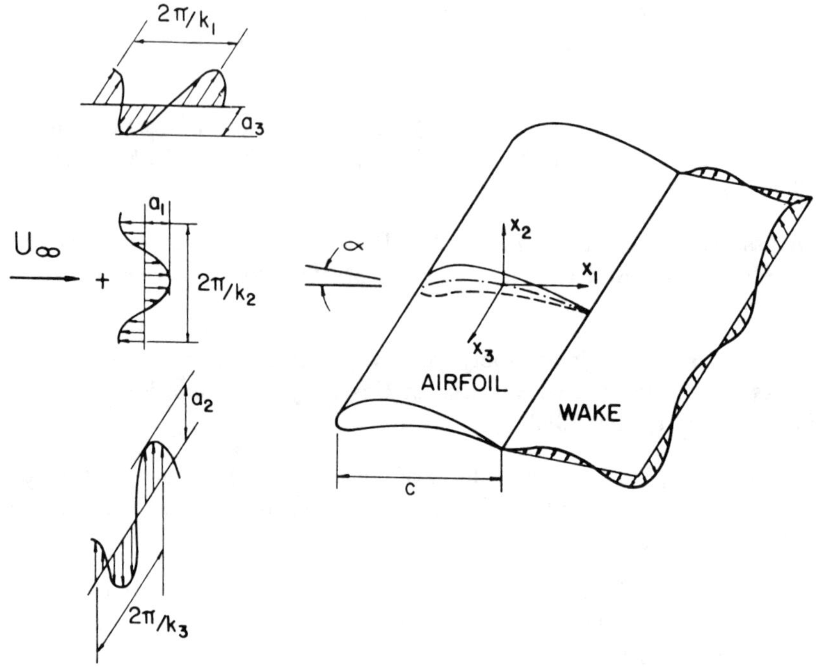

Figure 1. Airfoil in a three-dimensional gust.

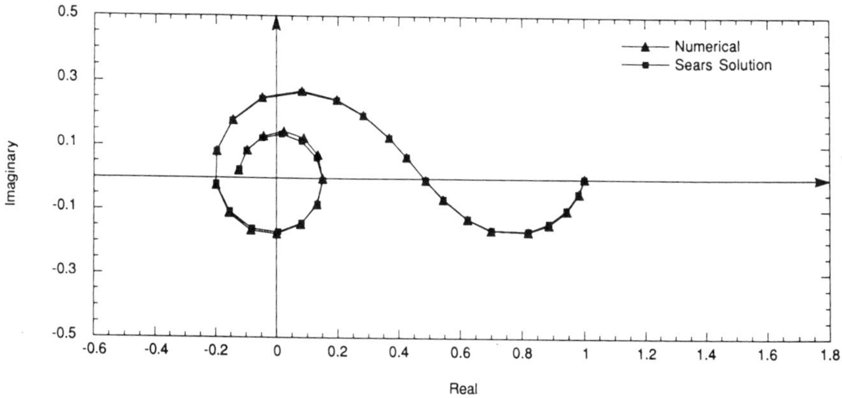

Figure 2. Comparison between the numerically computed unsteady lift and the Sears Solution for a flat plate airfoil in a transverse gust at $M = 0.001$.

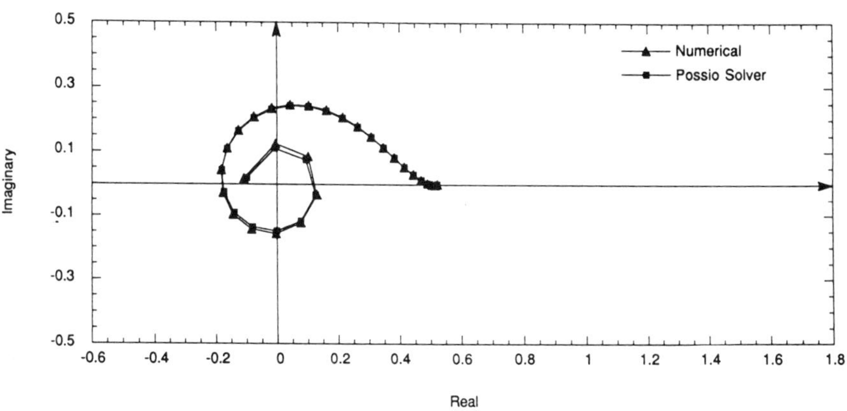

Figure 3. Comparison between the numerically computed unsteady lift and analytical results for a flat plate airfoil in a three-dimensional gust at $M = 0.001$.

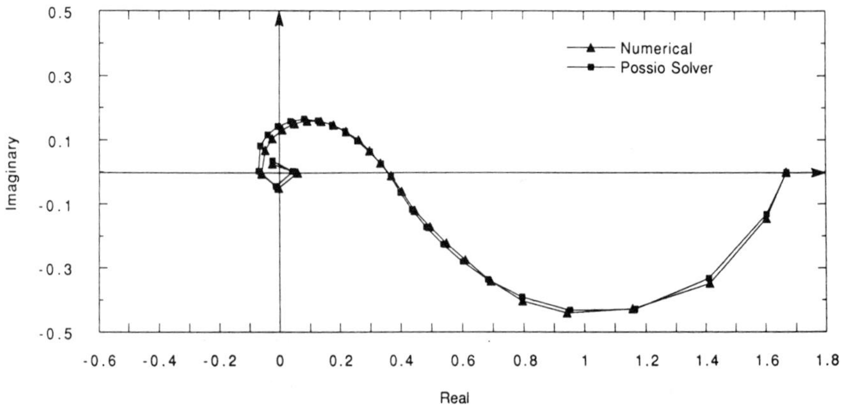

Figure 4. Comparison between the numerically computed unsteady lift and analytical results for a flat plate airfoil in a transverse gust at $M = 0.8$.

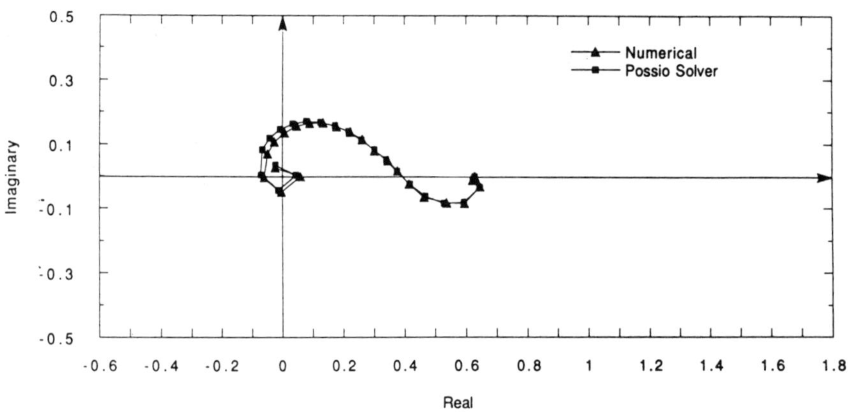

Figure 5. Comparison between the numerically computed unsteady lift and analytical results for a flat plate airfoil in a three-dimensional gust at $M = 0.8$.

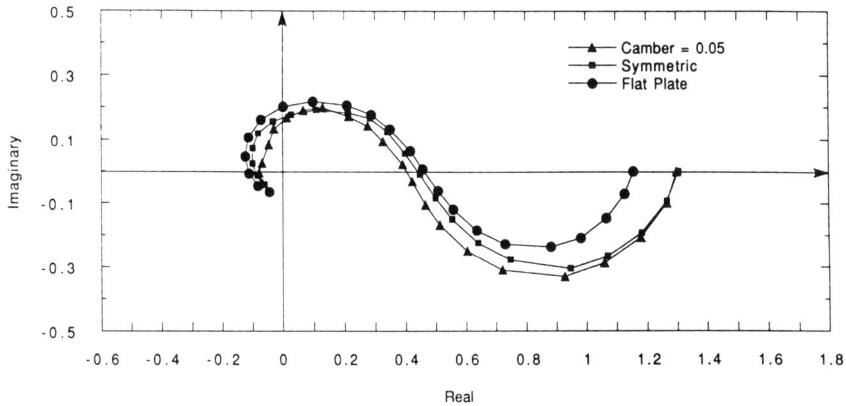

Figure 6. Comparison of the unsteady lift response function for a flat plate airfoil, a 12% thick symmetric Joukowski airfoil, and a 12% thick, 5% cambered Joukowski airfoil in a transverse gust at a free stream Mach number of .5.

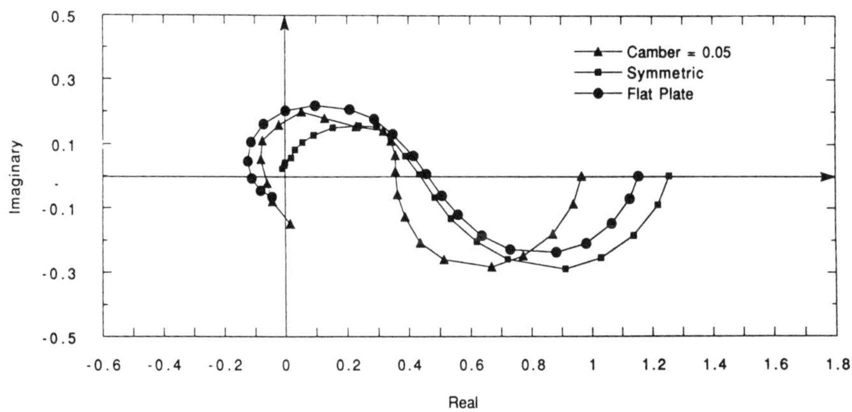

Figure 7. Comparison of the unsteady lift response function for a flat plate airfoil, a 12% thick symmetric Joukowski airfoil, and a 12% thick, 5% cambered Joukowski airfoil in a two-dimensional gust at a free stream Mach number of .5.

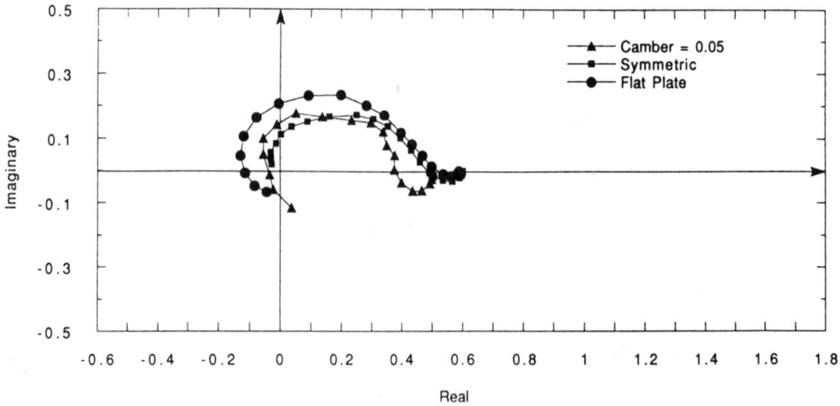

Figure 8. Comparison of the unsteady lift response function for a flat plate airfoil, a 12% thick symmetric Joukowski airfoil, and a 12% thick, 5% cambered Joukowski airfoil in a three-dimensional gust at a free stream Mach number of .5.

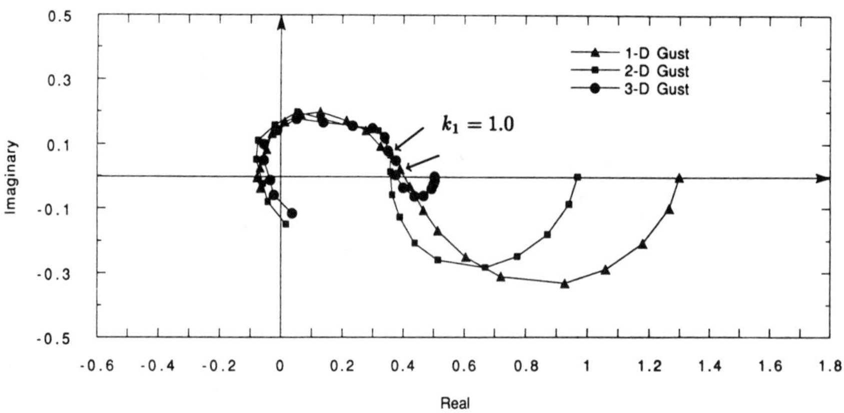

Figure 9. Comparison of the unsteady lift response function for a 12% thick, 5% cambered Joukowski airfoil in a transverse gust, a two-dimensional gust, and a three-dimensional gust at a free stream Mach number of .5.

Influence of Airfoil Shape and Incidence Angle on High-Frequency Gust Interaction Noise

E.J. Kerschen[1]
C.T. Tsai[1]
M.R. Myers[2]

ABSTRACT A theoretical model is developed for noise generated by the interaction of convected vortical and entropic disturbances with a thin airfoil in a subsonic mean flow. The primary objective is to determine how the airfoil thickness, camber and incidence angle affect the unsteady interaction and the resultant sound field. The analysis is based on a linearization of the unsteady motion about the steady, compressible flow past the airfoil. The solution is developed using singular perturbation methods, with the gust and sound wavelengths assumed short compared to the airfoil chord ($k = \omega b/U_\infty >> 1$). The thickness, camber and incidence angle are assumed to be small ($\epsilon << 1$) and the product ϵk is assumed to be O(1). The sound generation is concentrated in the local leading-edge region, which scales on the gust wavelength, with the trailing edge only scattering the leading-edge field. Away from the airfoil edges, there is an outer region in which the solution has a multiple-scales behavior. There are also transition regions which extend over domains of angular width $O(1/\sqrt{k})$ along and behind the airfoil. We find that the influence of airfoil camber and incidence angle on the total far-field sound power is correlated by the parameter $\alpha_{eff}\sqrt{k}$, where α_{eff} is a measure of the strength of the local mean flow around the leading edge. Similarly, the influence of airfoil thickness on the total far-field sound power is correlated by the parameter $\sqrt{S}$ where $S = \omega r_n/U_\infty$ is a Strouhal number based on the airfoil nose radius.

---

[1]Aerospace and Mechanical Engineering, University of Arizona, Tucson, AZ 85721.
[2]Hydrodynamics and Acoustics Branch, U.S. Food and Drug Administration, Rockville, MD 20857.

1 Introduction

The interaction of airfoils with convected disturbances is a significant source of noise in turbofan engines and in counter-rotation propfans. Convected gusts of importance are viscous wakes of upstream airfoils or struts, inflow turbulence, and entropy (or temperature) disturbances for the case of turbines and for some counter-rotation propfan configurations.

In predicting noise generated by the interaction of airfoils with convected gusts, the airfoils involved in the interactions are often modeled as zero-thickness flat plates aligned with a uniform mean flow. However, this approach ignores important physical effects, such as the distortion of the gust by the mean-flow gradients in the vicinity of the airfoil. Ffowcs Williams and Hawkings (1969a, b) first showed that this gust distortion produces a quadrupole source in the acoustic analogy.

The acoustic analogy is formally exact, but its use requires specification of the unsteady velocity fields which appear in the source terms. Since this information is not generally available, heuristic models must be introduced. For acoustically compact interactions, in which the sound wavelength is large compared to the airfoil chord, the results are fairly insensitive to details of the source model. However, for the non-compact case, the details of the source are more crucial and heuristic modeling is less straightforward. The present analysis involves non-compact sources, since the unsteady flow has a length scale which is short compared to the airfoil chord.

For the gust interaction problem, Rapid Distortion Theory (Goldstein, 1978; Kerschen and Myers, 1986; Atassi and Grzedzinski 1989) provides an attractive alternative to the acoustic analogy. In Rapid Distortion Theory, the inviscid flow equations are linearized about the non-uniform, steady flow past the airfoil. Rapid Distortion Theory captures the influence of the mean-flow gradients on the unsteady flow in a systematic approximation to the governing equations.

The first application of Rapid Distortion Theory to airfoil-gust interactions was presented by Goldstein and Atassi (1976), who analyzed the incompressible flow problem. Flow compressibility introduces significant complications into the airfoil-gust interaction problem, and closed-form solutions for general parameter values are not available even for the simplified case of a zero-thickness, flat-plate airfoil at zero angle of attack. In order to obtain approximate, closed-form results for the influence of airfoil shape and incidence angle on sound generated by gust interactions in compressible flow, the present authors have applied singular perturbation techniques to the Rapid Distortion Theory equations. In these analyses, the airfoil camber, thickness and incidence angle are assumed to be small, and the gust frequency is assumed to be large. Kerschen and Myers (1983, 1985, 1987) and Myers and Kerschen (1984, 1986) analyzed the influence of airfoil mean loading on sound generated by high-frequency, airfoil-gust interactions in subsonic mean flows. The case of a zero-thickness, cambered airfoil was considered. Kerschen and Tsai (1989) and Tsai and Kerschen (1990) analyzed the influence of airfoil thickness on sound generated by subsonic, high-frequency airfoil-gust interactions. The present paper presents a summary of the results of these asymptotic theories. Numerical solutions to the Rapid Distortion Theory equations for compressible flow have been presented by Scott and Atassi (1989) and Atassi et al. (1990), and more recent results can be found in other papers in this symposium.

2 Review of Theory

In this section we present a brief discussion of the analysis. A complete description can be found in the papers referenced above. We consider linear, three-dimensional disturbances of a two-dimensional, inviscid, irrotational, compressible mean flow. Convected velocity and entropy disturbances are assumed to be present in the uniform flow far upstream of the airfoil. Since the disturbances are linear, we can represent arbitrary upstream disturbances as a superposition of harmonic components. Thus, the solution is developed for a single harmonic component.

Goldstein (1978) showed that the complete disturbance velocity can be expressed as

$$\mathbf{u} = \nabla G + \mathbf{v} \tag{1}$$

where $\mathbf{v}$ is a known function of the upstream convected gust and contains the additional vorticity generated by the interaction of the upstream velocity and entropy disturbances with the mean-flow gradients. Substituting the expression for $\mathbf{u}$ into the continuity equation produces the following equation for G:

$$\frac{D_0}{Dt}\left(\frac{1}{c_0^2}\frac{D_0 G}{Dt}\right) - \frac{1}{\rho_0}\nabla \bullet (\rho_0 \nabla G) = \frac{1}{\rho_0}\nabla \bullet (\rho_0 \mathbf{v}) \tag{2}$$

On the airfoil surface, the disturbance velocity potential satisfies the boundary condition $\partial G/\partial n = -\mathbf{v} \bullet \mathbf{n}$, where $\mathbf{n}$ is the unit normal vector. The radiation condition applies far from the airfoil. The pressure fluctuations are given by $p = -\rho_0 D_0 G/Dt$.

The above equation is linear, but for a general compressible mean flow the variable coefficients must be determined numerically. To allow the development of approximate closed-form solutions, we assume that the mean flow is a small perturbation to a uniform flow. (The mean-flow perturbation is assumed, however, to be large compared to the size of the unsteady disturbances.) Prandtl-Glauert theory can then be utilized for the mean-flow description. It proves convenient to separate the effects of airfoil camber and incidence angle from the effects of airfoil thickness. Contributions due to airfoil camber and incidence angle are represented by the small parameter α, while contributions due to airfoil thickness are represented by the small parameter ϵ. Removing the harmonic dependence of G on time and the spanwise coordinate, applying a Prandtl-Glauert transformation, introducing the mean-flow velocity potential ϕ and streamfunction ψ as coordinates, and removing an additional phase factor which depends nonlinearly on ϕ, the differential equation for G simplifies to the form (Kerschen and Myers, 1986)

$$L_0(h) + \alpha L_1(h) + \epsilon L_2(h) = So(\phi, \psi) exp(i\Omega) \tag{3}$$

where So is $O(\alpha, \epsilon)$ and second order terms have been neglected. The function h is a modified form of the original disturbance velocity potential G. The operator L_0 is the standard Helmholtz operator,

$$L_0 = \partial^2/\partial\phi^2 + \partial^2/\partial\psi^2 + k^2 w^2 \tag{4}$$

where $w^2 = (M/\beta^2)^2 - (k_z/\beta)^2$, M is the Mach number of the upstream mean flow, $\beta = \sqrt{1-M^2}$, and the gust wavevector far upstream is given by $k(1, k_n, k_z)$ with respect to the coordinates (ϕ,ψ,z). The large parameter k has been extracted in the definition of the wavenumber and the coordinates have been normalized by $U_\infty b$, where b is the airfoil semi-chord. The operators L_1 and L_2 contain variable coefficients, and account for the effects on the acoustic propagation of the mean-flow gradients due to airfoil camber and incidence angle, and airfoil thickness, respectively.

Equation (3) is analyzed by a combination of singular perturbation and complex variable techniques. We assume that the wavelength of the disturbance is small compared to the airfoil chord ($k >> 1$). The various regions involved in the asymptotic analysis are illustrated in Fig. 1. Local regions, whose size scales on the gust wavelength, are present at the leading and trailing edges. Away from the airfoil edges, there is an outer region in which the solution has a multiple-scales behavior. In addition, along the airfoil surface and behind the airfoil there are "transition" regions which extend over an angular domain of $O(1/\sqrt{k})$. In these transition regions, the ray-acoustic outer solution gives way to diffraction-type solutions.

The following physical picture emerges from the analysis. For short wavelength gusts, the primary sound generation is concentrated in the local leading-edge region. In this region the scale of the mean-flow gradients is the same as the wavelength. Sound is generated by interaction of the gust with the airfoil surface, by interaction of the gust with mean-flow gradients, and by scattering effects due to sound propagation through the locally non-uniform mean flow. The sound which is generated in the leading-edge region propagates through the outer region, undergoing O(1) phase shifts due to the mean-flow variations. Two rays propagate along the airfoil upper and lower surfaces, and these rays are scattered by the trailing edge. The far-field sound is a combination of the direct rays from the leading edge and the scattered rays from the trailing edge. We now briefly outline the analysis in the various regions.

First consider the local leading-edge region. Here we utilize coordinates scaled on the disturbance wavelength, and the airfoil chord appears to extend to downstream infinity. In the vicinity of the leading edge, the local behavior of the mean flow is that of a uniform flow past the edge, plus a disturbance flow field which has an inverse square-root singularity at the leading edge. This disturbance flow field has two components. The first component is an asymmetric flow around the leading edge whose strength is measured by the parameter α_{eff}. Physically, α_{eff} is the incidence angle for which a flat-plate airfoil would have the same flow around the leading edge. The value of α_{eff} is determined by the incidence angle of the airfoil and its camber distribution. The second component of the disturbance flow field is a symmetric distribution due to the airfoil thickness. The amplitude of this symmetric disturbance flow in the local leading-edge region depends only on the nose radius r_n of the parabolic leading edge, which we incorporate in a Strouhal number $S = \omega r_n / U_\infty$. (In contrast, the amplitude of the local asymmetric flow component depends on the full camber distribution of the airfoil.)

An additional assumption, $\alpha_{eff}^2 k << 1$ and $S << 1$, is introduced in analyzing the leading-edge region. Physically, these two inequalities state that the shift of the mean-flow stagnation point from the leading edge, and the nose radius, are both small compared to the disturbance wavelength. Equivalently, the mean flow disturbance is weak when viewed on the scale of the wavelength. The solution can then be developed in a perturbation series of the form

$$h(\phi, \psi) = h_0 + \alpha_{eff}\sqrt{k}(h_1 + h_2 + h_3) + \sqrt{S}(h_4 + h_5 + h_6) + \ldots \qquad (5)$$

The parameter $\sqrt{S}$ is proportional to $\epsilon\sqrt{k}$, with the constant of proportionality dependent on the airfoil thickness distribution. We have chosen to use the parameter S in order to emphasize that the local effects due to airfoil thickness are determined by the nose radius of the airfoil. The expansion (5) shows that, in the local leading-edge region, the effects of camber and incidence angle can be combined into a single parameter. Furthermore, in this region the effects of camber and incidence angle and the effects of thickness can be analyzed separately and then superposed.

The solutions for h_0, h_1, h_2, h_3, h_4, h_5 and h_6 are developed utilizing Fourier transform methods. The mean-flow Mach number is assumed to be O(1), and hence wave propagation effects are important in the local leading-edge region. Since the airfoil appears semi-infinite, the Wiener-Hopf technique is applicable.

The function h_0 is the standard leading-edge field for the case of zero mean loading and zero thickness. We have separated the $O(\alpha_{eff}\sqrt{k})$ correction terms due to airfoil camber and incidence angle into three different functions representing different physical effects. The function h_1 represents the influence on the surface boundary condition of vorticity distortion by the mean-flow gradients due to camber and incidence angle, while h_2 is generated by volume sources related to the Ffowcs Williams and Hawkings quadrupoles. The function h_3 is generated by the interaction between the field h_0 and the local variations in mean-flow velocity and speed of sound contained in the operator L_1. Since these local variations have the same length scale as the acoustic wave, this interaction results in additional far-field sound radiation. Similarly, we also separate the $O(\sqrt{S})$ correction terms due to airfoil thickness into three different functions. The function h_4 represents the influence of airfoil thickness on the surface boundary condition, h_5 is generated by volume sources and h_6 results from the interaction of the field h_0 with the operator L_2.

In the outer region, all three operators L_0, L_1 and L_2 in Eq. (3) must be considered in the leading approximation. Thus, in this region the effects of airfoil camber and incidence angle and airfoil thickness must be considered simultaneously. However, the analysis is simplified by the fact that the gust wavelength is short compared to the scale on which the mean flow varies. The solution in the outer region has two components. The first is due to a distribution of "quadrupole" type volume sources. These sources generate local fluctuations, but make negligible contributions to the far-field sound. The second component is the ray-acoustic continuation of the asymptotic behavior of the local leading-edge solution. This is the true sound field. The ray field is most conveniently analyzed in polar (r, θ) coordinates with origin at the leading edge, and $\theta = 0$ corresponding to the airfoil upper surface. The mean-flow gradients produce significant phase distortions in the sound field which are captured quite naturally in our analysis. In order to obtain these phase distortions from an acoustic analogy model, quadrupole sources in addition to $\rho u_i u_j$ would have to be accounted for.

The ray-acoustic solution does not quite satisfy the condition of zero normal-velocity component on the airfoil surface, and local "transition" regions must be introduced near the airfoil upper and lower surfaces. The analysis in these transition regions involves rescaling of the ψ coordinate by the factor $\sqrt{k}$. This brings a second derivative into the local equation, allowing an additional boundary condition to be satisfied.

The ray-acoustic outer solution (suitably modified by transition region effects) is then scattered by the trailing edge. This process is analyzed by a local trailing-edge solution developed via the Wiener-Hopf technique. The $O(1)$ phase distortions discussed above significantly modify the scattered field produced by the trailing edge. The phase distortions present at the trailing edge are a cumulative influence over the whole airfoil chord. The phase distortion at the trailing edge due to airfoil camber and incidence angle is proportional to the airfoil lift coefficient, but calculation of the contribution due to airfoil thickness requires full knowledge of the thickness distribution. The asymptotic behavior of the local trailing-edge solution matches with an additional ray-acoustic solution for $O(1)$ values of θ, while the matching is with a diffraction-type field in the transition region behind the airfoil. Thus, the far-field sound consists of the direct ray field generated by the leading edge and the scattered field from the trailing edge. Additional details of the analysis can be found in the papers referenced above.

3 Parametric Study

In this section, we present results which illustrate the influence of airfoil shape and incidence angle on sound generated due to interaction with high-frequency vortical and entropic gusts. The sound field is a function of the gust characteristics as well as the airfoil parameters. In this paper we restrict attention to gusts which are two-dimensional and time harmonic.

Our primary interest is in the total sound power level. A suitable normalization for the acoustic power generated by the gust interaction is given by

$$NormalizedPower = \frac{Power/Span}{1/2\rho U_\infty^3 A^2 b} \tag{6}$$

where $A = |\mathbf{v}'|/U_\infty$ is the nondimensional upstream amplitude of the vortical gust. Since the gust amplitude is absorbed in the definition of Normalized Power, the only gust parameters that enter are the magnitude and orientation of the gust wavevector. We utilize the angle θ_g which the gust wavevector makes with respect to the mean velocity far upstream of the airfoil, and hold the streamwise component k of the wavenumber constant as θ_g is varied. This particular choice is made so that the frequency of the generated sound remains constant as θ_g is varied. For the case of entropy gusts, A is replaced by $|s'/2c_p|$ in the definition (6) of Normalized Power. The effects of airfoil loading and thickness are examined separately in this paper. We present results for zero-thickness, cambered airfoils and for symmetric Joukowski airfoils.

First consider the case of a zero-thickness, cambered airfoil. Earlier models of mean loading effects (Ffowcs Williams and Hawkings, 1969b; Goldstein et al., 1974; Mani, 1974) on gust interaction noise utilized the Ffowcs Williams and Hawkings acoustic analogy (Ffowcs Williams and Hawkings, 1969a), and modeled the mean flow by a row of two-dimensional point vortices superposed on a uniform flow. Hence, the only mean flow parameter which entered in these models was the vortex strength, or equivalently the airfoil lift coefficient. Our theory shows that, at high frequencies, the noise generation is concentrated in the local leading-edge region, with the trailing edge playing the secondary role of a scatterer of the leading-edge field. This suggests that the mean loading effects on the noise level should correlate on a local characteristic of the leading-edge region, rather than on a global parameter such as airfoil lift coefficient.

To examine this issue, in Fig. 2 we plot the noise level vs. lift coefficient for airfoils of four different shapes. The mean flow Mach number is M = 0.6, and the incident disturbance is a vorticity gust with gust angle θ_g = 45 deg. Curves 1, 2 and 3 are the results for airfoils of fixed shape at different incidence angles; curve 1 is a flat plate, curve 2 is a 5 percent camber circular arc, and curve 3 has a NACA 63 camber distribution. Curve 4 is the result obtained for a circular arc airfoil at zero incidence angle, but varying degrees of camber. It is evident from Fig. 2 that the airfoil lift coefficient does not correlate the influence of mean loading on gust interaction noise. The same data is plotted in Fig. 3, but here the abscissa is the effective leading edge incidence angle, α_{eff}. The deviations between the four curves in Fig. 3 are remarkably small, demonstrating that α_{eff} is an extremely useful correlating parameter for mean loading effects.

Returning to Fig. 2, the results for the different airfoils can be understood in terms of the relationship between α_{eff} and the lift coefficient. For example, consider curve 4, the circular arc airfoil with zero incidence angle but varying camber levels. It can easily be shown that, for a circular arc airfoil at zero incidence angle, there is no flow around the leading edge ($\alpha_{eff} = 0$). Hence, while the introduction of camber changes the airfoil lift, it does not modify the local flow in the leading edge region, and the far field sound is affected very little by changes in lift coefficient. The flat plate airfoil (curve 1) is at the other extreme, since all the lift is due to incidence angle effects, and for a given C_L this airfoil has the largest value of α_{eff}. The slight deviations from an exact correlation in Fig. 3 are produced by trailing-edge scattering. Our theory shows that scattering by the trailing edge becomes less important as the frequency increases, and that the correlation becomes exact as $kw = kM/\beta^2$ approaches infinity. However, we have found that this correlation is remarkably accurate even at modest values of kw. For example, the value of kw in Fig. 3 is 5.25, and Kerschen and Myers (1987) found deviations of less than 5 percent for a lower frequency $kw = 1.58$.

The above results show that scattering by the trailing edge has only a small influence on the total sound power level. (Trailing edge scattering has a much more significant influence on the directivity pattern, as illustrated in Kerschen and Myers (1987) and Tsai and Kerschen (1990).) Thus, in exploring the influence of gust and mean flow parameters on the total sound power level, we shall neglect the presence of the trailing edge. This allows us to define a normalized Leading Edge Power which absorbs the dependence on reduced frequency k. Next consider the dependence on Mach number. For small values of M, the analysis of Kerschen and Tsai (1989) showed that the leading-edge sound field is a pole of 3/2 order, intermediate between a monopole and a dipole. The fractional order pole arises because the airfoil chord appears semi-infinite when the trailing edge is neglected. The sound power level for a two-dimensional pole of 3/2 order scales as M^4. Hence, in defining a normalized Leading Edge Power it proves convenient to absorb an additional power of M. Thus we set

$$LeadingEdgePower = NormalizedPower \frac{k}{M}. \tag{7}$$

For the case of a zero-thickness flat plate for which $\alpha_{eff} = 0$, Leading Edge Power turns out to be independent of M for all subsonic Mach numbers. Hence, in the limit $\alpha_{eff} \rightarrow 0$ (and $S \rightarrow 0$) the Leading Edge Power is only a function of the gust wavevector orientation θ_g.

We first consider the influence of mean loading on the sound power generated by vorticity gusts at low Mach numbers. The velocity fluctuation associated with the vorticity gust is in the direction perpendicular to the gust wavevector. In Fig. 4, Leading Edge Power is plotted as a function of gust angle θ_g, for $M = 0.1$ and four values of the mean loading parameter $\alpha_{eff}\sqrt{k}$. For zero α_{eff}, noise is generated only by the airfoil surface blocking the normal component of gust velocity, and hence the curve has the simple shape $cos^2\theta_g$. When mean loading is introduced, the maximum of the curve increases in amplitude and its location shifts to positive values of θ_g. Results for the same values of α_{eff} at $M = 0.75$ are presented in Fig. 5. The results at $M = 0.75$ are qualitatively similar to those at $M = 0.1$, but the influence of mean loading is observed to be significantly stronger. It is well known that the mean lift coefficient for a given airfoil increases with Mach number, and within the framework of thin-airfoil theory this effect can be accounted for by introducing a Prandtl-Glauert factor $\beta = \sqrt{1 - M^2}$. In Fig. 6, we show that the inclusion of a Prandtl-Glauert factor in the mean loading parameter provides a reasonable collapse of the results for different Mach numbers.

Next consider the influence of mean loading on the sound power generated by entropy (or temperature) gusts. In Figs. 7 and 8, Leading Edge Power is plotted as a function of gust angle for $M = 0.1$ and $M = 0.75$, respectively, and four values of $\alpha_{eff}\sqrt{k}$. In contrast to the case of vorticity gusts, no sound is generated for the case of an entropy gust interacting with an unloaded leading edge ($\alpha_{eff} = 0$). Essentially, in this case there is no flow acceleration in the vicinity of the leading edge, and the entropy disturbances produce no velocity fluctuations. However, when α_{eff} is nonzero, the acceleration of the entropy disturbances generates Reynolds stresses and velocity fluctuations whose interaction with the leading edge generates sound. Here the maximum sound power occurs at $\theta_g = 0$, and the curve is symmetric with respect to θ_g. We can readily prove that the leading-edge power is proportional to α_{eff}^2. As in the case of vorticity gusts, Fig. 9 shows that the inclusion of a Prandtl-Glauert factor in the mean loading parameter provides a reasonable collapse of the results for different Mach numbers.

We now turn to the influence of airfoil thickness on sound generated by airfoil-gust interactions. First consider the case of vorticity gusts. In Fig. 10, Leading Edge Power is plotted as a function of gust angle for $M = 0.1$ and four values of the leading-edge Strouhal number S. As discussed above, for the case of zero thickness ($S = 0$) the power is proportional to $cos^2\theta_g$. The results for nonzero values of S have their maximum values at $\theta_g = 0$, and their shapes are similar to the case $S = 0$. A surprising feature of the low Mach number results presented in Fig. 10 is that moderate amounts of leading edge thickness actually decrease the sound radiation. However, Kerschen and Tsai (1989) showed that the minimum sound level occurs near $S = 0.3$; with further increases in the airfoil nose radius, the sound power level rises sharply.

The influence of airfoil thickness on sound generated by vorticity gusts at a higher Mach number $M = 0.75$ is presented in Fig. 11. The Leading Edge Power for $S = 0$ is seen to be identical to the result in Fig. 10, confirming that the zero-thickness case is independent of Mach number. For M = 0.75, the influence of leading-edge thickness is strongly dependent on the gust angle. For small gust angles, the sound power decreases as leading-edge thickness is first introduced. However, at larger gust angles the same values of leading-edge thickness increase the sound power. The curve for $S = 0.3$ has a different shape than that for $S = 0$, suggesting that a new source mechanism is becoming important.

Finally, consider the influence of airfoil thickness on sound generated by entropy gusts. Since an entropy gust generates sound only through interaction with a non-uniform mean flow, in this case the sound power level rises with increasing leading-edge thickness. Results for $M = 0.1$ are presented in Fig. 12. The most striking feature in Fig. 12 is the small amplitude of the sound power. In fact, Kerschen and Tsai (1989) showed that, in the low Mach number limit, the interaction of an entropy gust with leading-edge thickness is a two-dimensional dipole source. Hence, the dimensional sound power scales with U^5 in this limit, compared to U^4 for the vorticity gust interaction. Therefore, at low Mach numbers the interaction of an entropy gust with leading-edge thickness generates much less sound than the corresponding interaction for a vorticity gust. Sound power levels due to an entropy gust for the case $M = 0.75$ are presented in Fig. 13. In contrast to the behavior at low Mach numbers, the sound power levels in this case are comparable to those generated by vorticity gusts.

4 Summary

In this paper, we have examined the influence of airfoil shape and incidence angle on sound generated by interaction with high-frequency vortical and entropic gusts. We find that, in the high-frequency limit, the primary sound generation occurs in the local leading-edge region. Thus, it is the mean-flow properties in this local region which control the overall level of the radiated sound field. The airfoil camber and incidence angle influence the mean flow in the local leading-edge region only through a single parameter α_{eff}. We have shown that this parameter provides a remarkably accurate correlation for the influence of mean loading on total sound power level. The influence of airfoil thickness on the total sound power level enters through the nose radius of the airfoil in terms of the parameter $S = \omega r_n / U_\infty$.

For vorticity gusts, positive gust angles result in sound levels which increase with leading-edge loading, while for negative gust angles an increase in leading-edge loading may actually reduce the noise level. Turbomachinery viscous wake interactions typically involve positive gust angles. Except at high subsonic Mach numbers, moderate amounts of airfoil thickness also reduce the sound level. However, with further increases in the leading-edge radius, a rapid rise in sound power is found.

Entropy gusts generate sound only by interacting with mean-flow gradients, and hence are less important than vorticity gusts for small values of leading-edge loading or thickness. At low Mach numbers, the interaction of entropy gusts with leading-edge thickness is a much less efficient sound source than the corresponding interaction for leading-edge loading. At large values of $\alpha_{eff}\sqrt{k}$ (or S for O(1) Mach numbers), entropy gusts can generate sound levels that are of the same order or higher than those for vorticity gusts.

5 Acknowledgement

The work described herein was funded by NASA Lewis Research Center under grant NAG3-357. We gratefully acknowledge this support.

6 References

[1] Atassi, H.M., and Grzedzinski, J. 1989, "Unsteady Disturbances of Streaming Motions Around Bodies," *J. Fluid Mech.*, Vol. 209, pp. 385-403.

[2] Atassi, H.M., Subramaniam, S. and Scott, J.R. 1990, "Acoustic Radiation From a Lifting Airfoil in Nonuniform Subsonic Flows," AIAA Paper 90-3911.

[3] Ffowcs Williams, J. E. and Hawkings, D. L. 1969a, "Sound Generation by Turbulence and Surfaces in Arbitrary Motion," *Phil. Trans. Roy. Soc. A*, Vol. 264, pp. 321-342.

[4] Ffowcs Williams, J. E. and Hawkings, D. L. 1969b, "Theory Relating to the Noise of Rotating Machinery," *J. Sound Vib.*, Vol. 10, pp. 10-21.

[5] Goldstein, M. E. 1978, "Unsteady Vortical and Entropic Disturbances of Potential Flows Round Arbitrary Obstacles," *J. Fluid Mech.* Vol. 89, pp. 433-468.

[6] Goldstein, M. E., and Atassi, H. 1976, "A Complete Second Order Theory for the Unsteady Flow about an Airfoil due to a Periodic Gust," *J. Fluid Mech.*, Vol. 74, pp. 741-765.

[7] Goldstein, M. E., Rosenbaum, B. M., and Albers, L. U. 1974, "Sound Radiation from a High-Speed Axial-Flow Fan due to the Inlet Turbulence Quadrupole Interaction," NASA TN D-7667.

[8] Kerschen, E.J. and Myers, M.R. 1983, "Incidence Angle Effects on Convected Gust Interaction Noise," AIAA Paper 83-0765.

[9] Kerschen, E. J. and Myers, M. R. 1985, "Influence of Airfoil Mean Loading on Convected Gust Interaction Noise," *Proceedings of IUTAM Symposium on Aero and Hydro-Acoustics*, Lyon, Springer-Verlag.

[10] Kerschen, E. J. and Myers, M. R. 1986, "Perfect Gas Effects in Compressible Rapid Distortion Theory," *AIAA Journal*, Vol. 25, pp. 504-507.

[11] Kerschen, E. J. and Myers, M. R. 1987, "A Parametric Study of Mean Loading Effects on Airfoil Gust Interaction Noise," AIAA Paper 87-2677.

[12] Kerschen, E.J. and Tsai, C.T. 1989, "Influence of Airfoil Thickness on Convected-Gust Interaction Noise," AIAA Paper 89-1082.

[13] Mani, R., "Isolated Rotor Noise due to Inlet Distortion or Turbulence," NASA CR 2479, 1974.

[14] Myers, M. R. and Kerschen, E. J. 1984, "Effect of Airfoil Mean Loading on Convected Gust Interaction Noise," AIAA Paper 84-2324.

[15] Myers, M. R. and Kerschen, E. J. 1986, "Influence of Airfoil Camber on Convected Gust Interaction Noise," AIAA Paper 86-1873.

[16] Scott, J.R. and Atassi, H.M. 1989, "Numerical Solution of Periodic Vortical Flows About a Thin Airfoil," NASA TM-101998.

[17] Tsai, C.T. and Kerschen, E.J. 1990, "Influence of Airfoil Nose Radius on Sound Generated by Gust Interactions," AIAA Paper 90-3912.

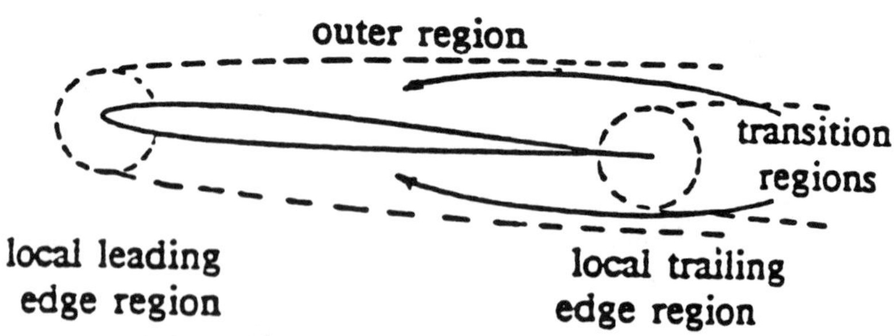

FIGURE 1. Asymptotic regions for the singular perturbation analysis of highfrequency airfoil-gust interactions.

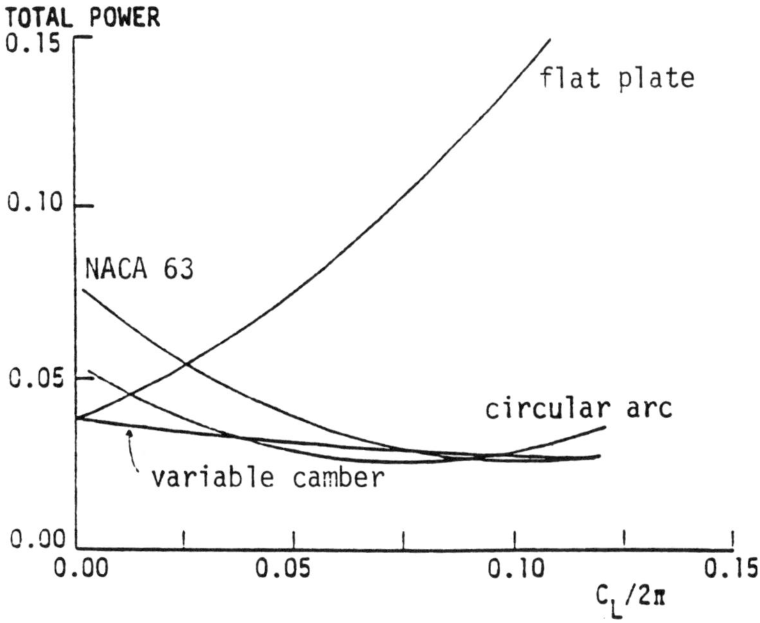

FIGURE 2. Total sound power vs. airfoil lift coefficient for four different airfoil shapes. Vorticity gust, $S = 0$, $M = 0.6, k = 7, \theta_g = 45$ deg.

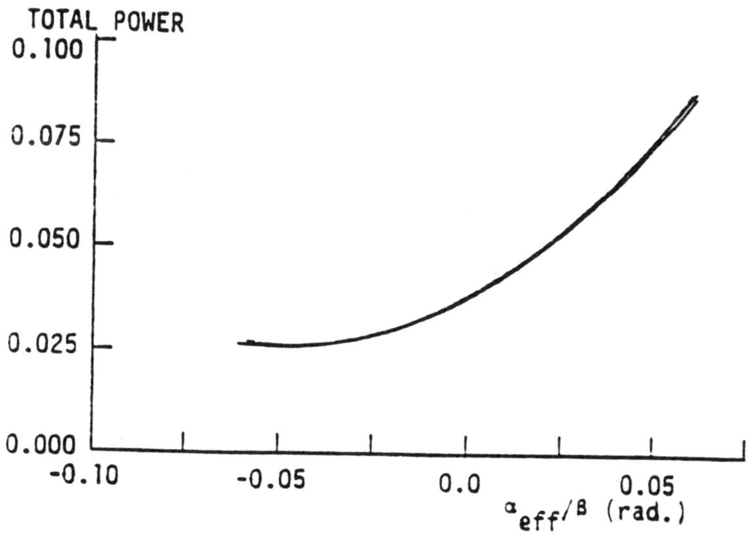

FIGURE 3. Total sound power vs. effective leading edge incidence angle for four different airfoil shapes. Vorticity gust, $S = 0$, $M = 0.6, k = 7, \theta_g = 45$ deg.

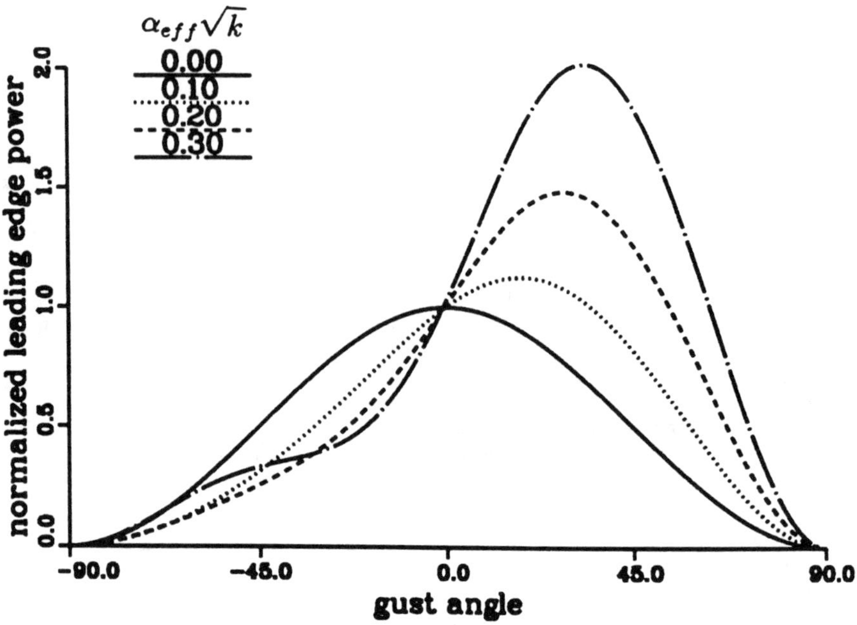

FIGURE 4. Leading edge power vs. gust angle, for four values of $\alpha_{eff}k^{1/2}$. Vorticity gust, $S = 0$, $M = 0.1$.

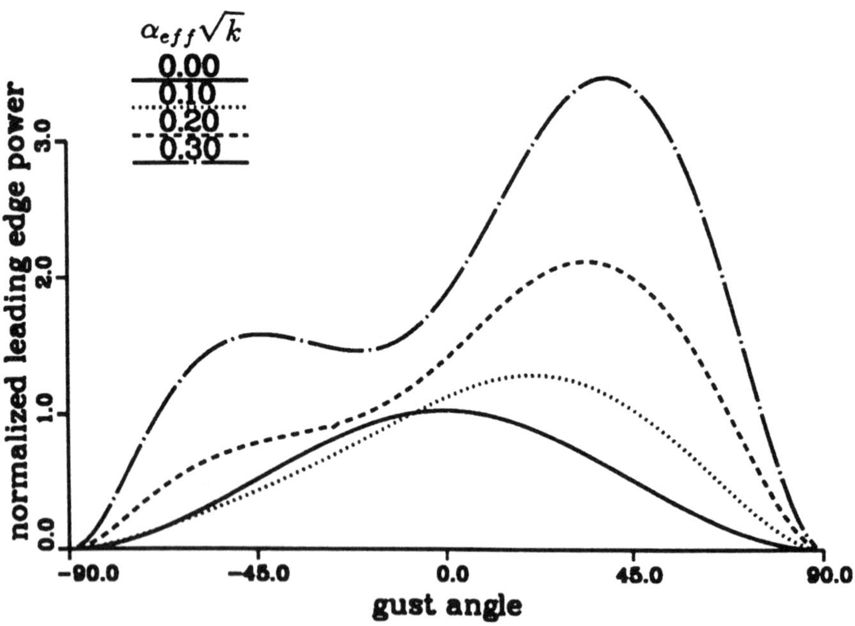

FIGURE 5. Leading edge power vs. gust angle, for four values of $\alpha_{eff}k^{1/2}$. Vorticity gust, $S = 0$, $M = 0.75$.

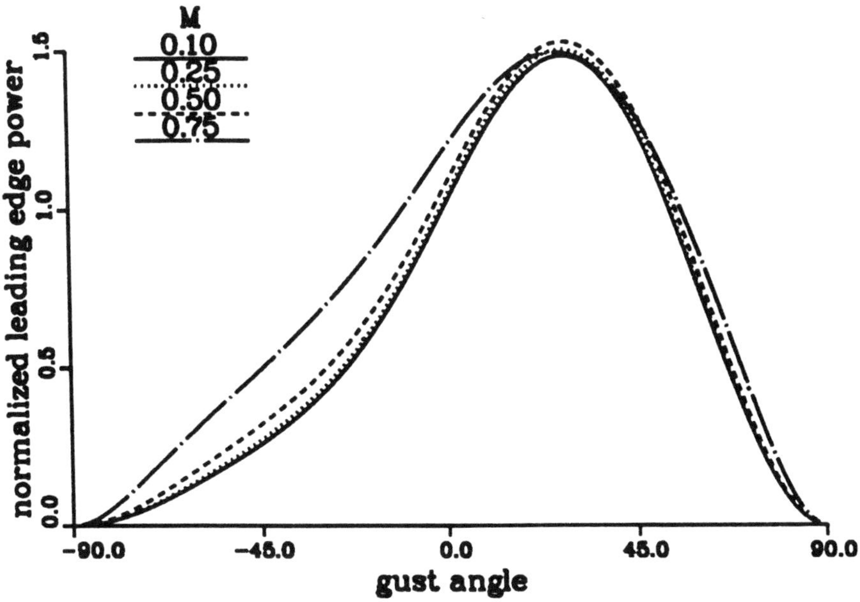

FIGURE 6. Approximate correlation of Mach number effects by use of a Prandtl-Glauert factor. Vorticity gust, $S = 0$, $\alpha_{eff}k^{1/2}/\beta = 0.2$.

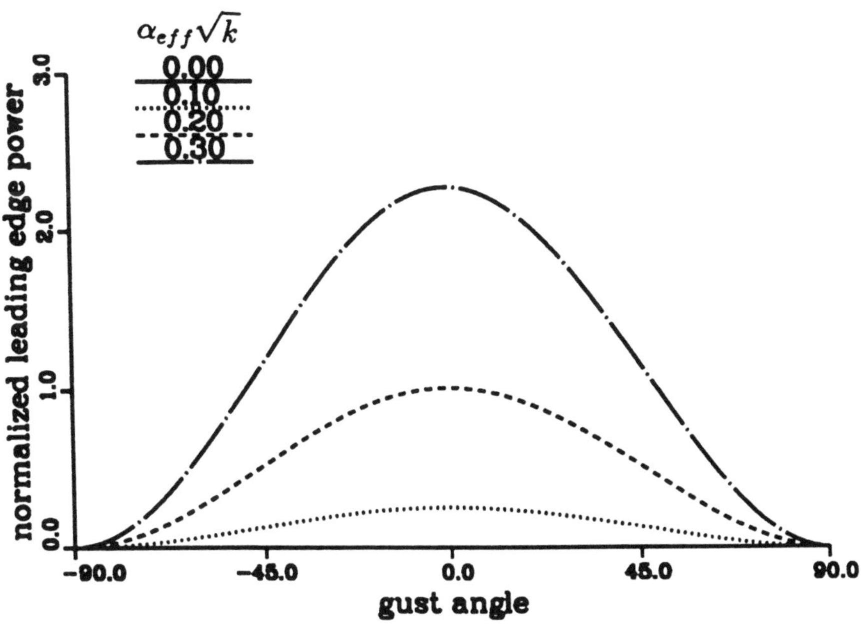

FIGURE 7. Leading edge power vs. gust angle, for four values of $\alpha_{eff}k^{1/2}$. Entropy gust, $S = 0$, $M = 0.1$.

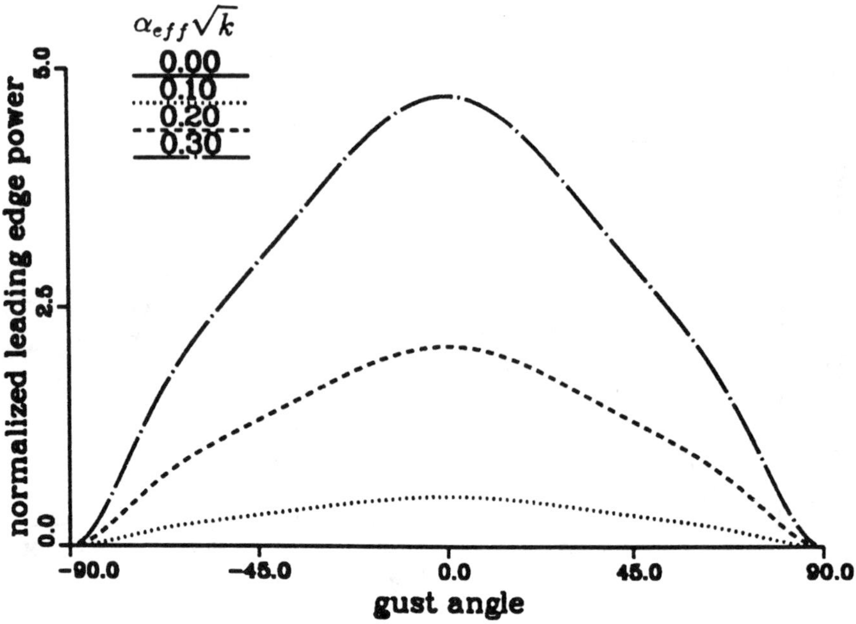

FIGURE 8. Leading edge power vs. gust angle, for four values of $\alpha_{eff}k^{1/2}$. Entropy gust, $S = 0$, $M = 0.75$.

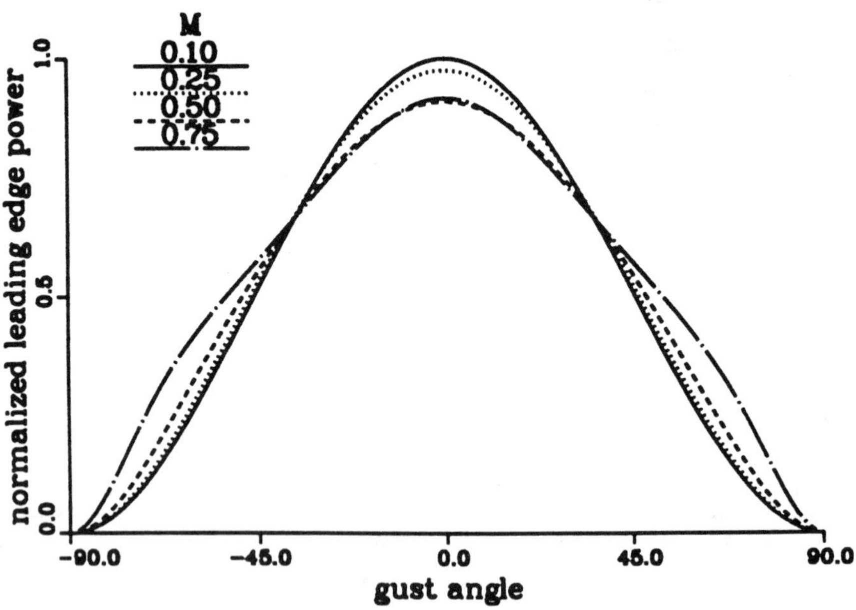

FIGURE 9. Approximate correlation of Mach number effects by use of a Prandtl-Glauert factor. Entropy gust, $S = 0$, $\alpha_{eff}k^{1/2}/\beta = 0.2$.

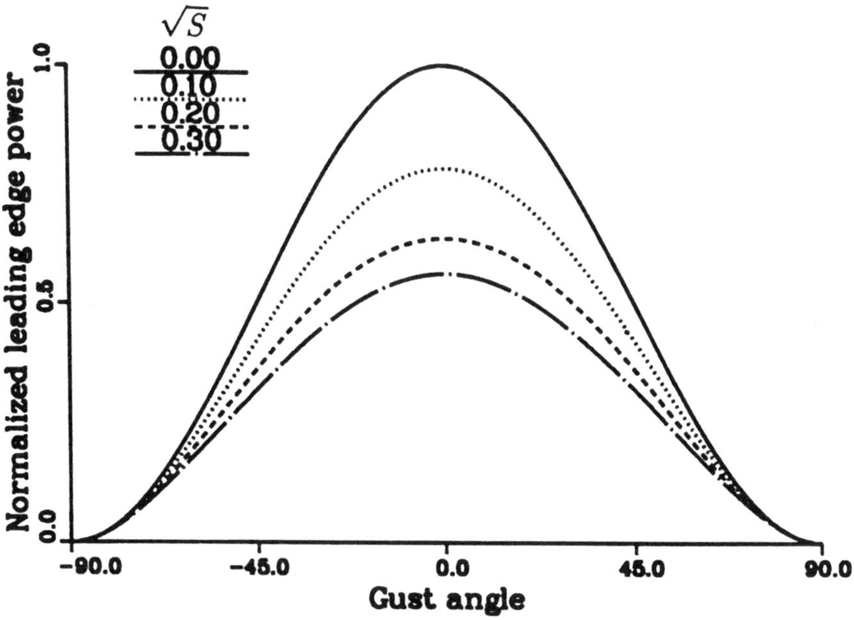

FIGURE 10. Leading edge power vs. gust angle, for four values of $S^{\frac{1}{2}}$. Vorticity gust, $\alpha_{eff} = 0$, $M = 0.1$.

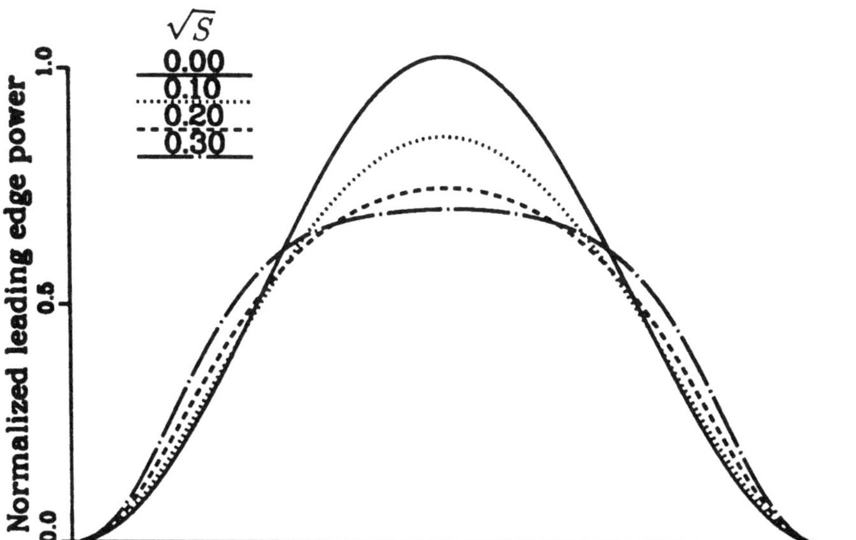

FIGURE 11. Leading edge power vs. gust angle, for four values of $S^{\frac{1}{2}}$. Vorticity gust, $\alpha_{eff} = 0$, $M = 0.75$.

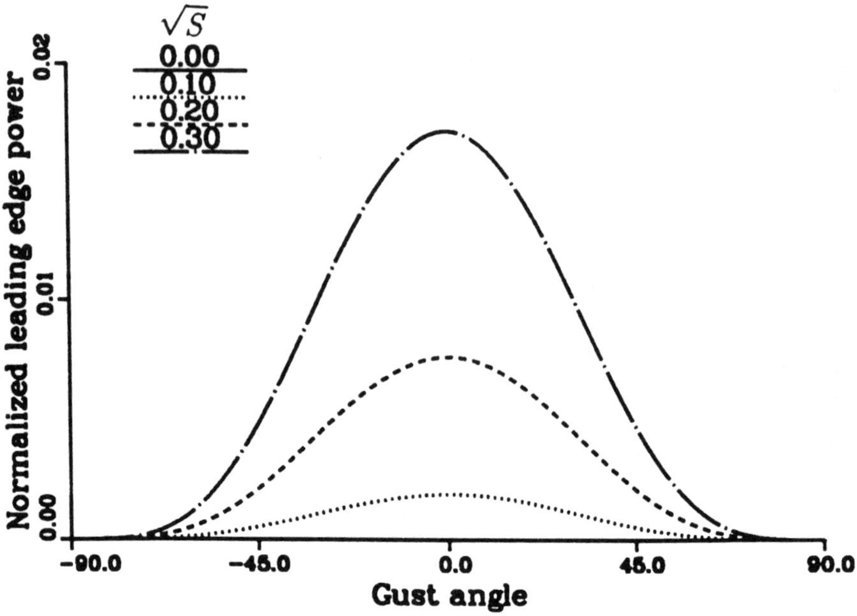

FIGURE 12. Leading edge power vs. gust angle, for four values of $S^{\frac{1}{2}}$. Entropy gust, $\alpha_{eff} = 0$, $M = 0.1$.

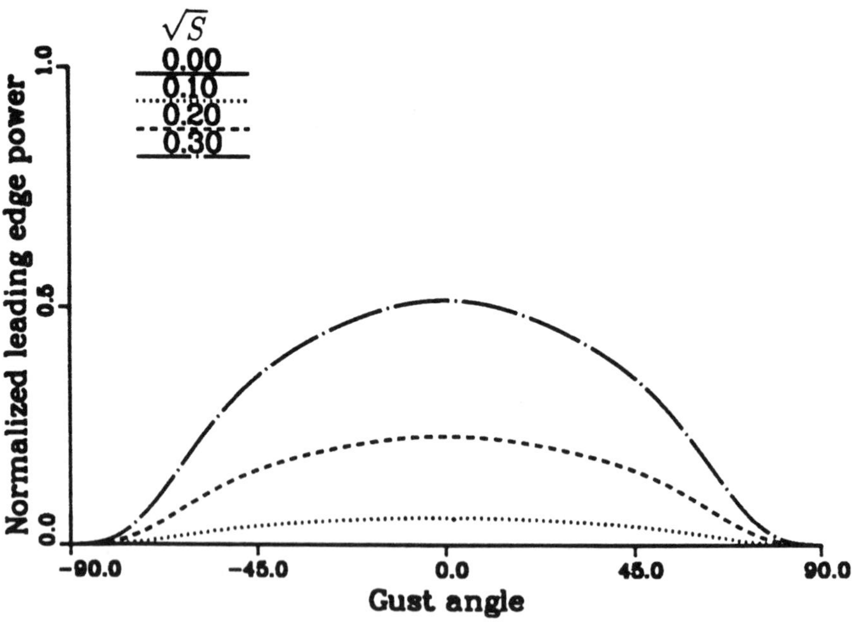

FIGURE 13. Leading edge power vs. gust angle, for four values of $S^{\frac{1}{2}}$. Entropy gust, $\alpha_{eff} = 0$, $M = 0.75$.

VII. Aeroelastic Instabilities and Control

A Comparison of Mistuning Effects on Flutter Between Stalled and Unstalled Cascades

Ichiro Fujimoto[1]
Hiroshi Nagai[2]
Hideo Tanaka[3]

Summary

A comparison is made between stalled and unstalled cascades paying attention to mistuning effects on flutter. With regard to the concept of mistuning, an attempt is made to utilize the fact that the actual cascade blades have fluctuation in natural frequency brought about through manufacturing process. Torsional flutter of mistuned cascades is considered both in a fully stalled condition where the flow around blades shows the leading edge separation and in an unstalled condition without flow separation. The flutter suppression effect by mistuning in blade natural frequency is confirmed for stalled cascades as well as for unstalled ones. Furthermore, emphasis is to be placed on the fact that a more remarkable flutter suppression effect is expected from slight mistuning in a stalled condition than in an unstalled one. Both in a stalled and in an unstalled condition, within range of normal blade manufacturing imperfections, the larger the standard deviation on blade natural frequency becomes, the more stable the cascade becomes, i.e., the cascade composed of blade groups having large fluctuation in natural frequency is more stable with regard to flutter than that whose blade natural frequencies are closely grouped. The blade group should therefore not be selected in such a way that the standard deviation decreases and the group itself consequently approaches to the tuned one. An irregularly distributed type of zigzag arrangement gives a fairly high flutter velocity, while an alternating arrangement, which was concluded to be optimum in an unstalled condition, does not necessarily give the satisfactory flutter suppression effect in a stalled condition. Special care must be taken to avoid a blade arrangement where the blades having exactly or nearly equal natural frequencies are successively or closely placed.

---

[1] Associate Professor,Takushoku University, Hachioji, Tokyo,Japan
[2] Lecturer,Sano Women's Junior College, Sano,Tochigi,Japan
[3] Professor,Tokai University, Hiratsuka,Kanagawa,Japan

Nomenclature

c = blade chord length
I = blade moment of inertia
K = blade torsional stiffness
k = $c\omega/U$ = reduced frequency
M = unsteady aerodynamic moment
N = number of blades in one cyclic group shown in Fig. 1
U = upstream flow velocity relative to blades
V_{cr} = critical flutter velocity
α = angular displacement
β = interblade phase angle
μ = mass ratio of blade
σ = standard deviation
τ = intergroup phase angle shown in Fig. 1
ω = flutter frequency
ω_i = natural frequency of i-th blade
ω_m = arithmetic mean of ω_i in a group

1. Introduction

There is an attempt to suppress cascade flutter of turbomachines by non-uniform design of cascade blades called mistuning or detuning. The concept of mistuning was treated by Whitehead [1] in 1964. Bendiksen [2] reviewed recent research on passive flutter control including mistuning.

In an unstalled condition, it has been found that flutter becomes hard to occur when mistuning is applied by changing only the elastic condition without changing any cascade geometry, the vibration with a uniform amplitude and a uniform interblade phase angle as seen in tuned cascades being suppressed[3][4]. By positively utilizing the above aspect from the viewpoint of flutter suppression, the effect of mistuning on unstalled cascade flutter has been analyzed, paying attention to the influences of a change in blade natural frequency, arrangement of blades, etc.. It has been numerically shown that the flutter velocity becomes highest when two kinds of blades having different natural frequencies are alternately arranged as far as the surveyed range is concerned[5][6].

As for stalled cascades, the effect of mistuning on flutter was surveyed only in the case of the alternating blade arrangement as a typical example[7]. The result shows that a more significant flutter suppression effect is expected from slight mistuning for stalled cascades than for unstalled ones. An example[8] of experimental researches is also reported for mistuned cascades in a stalled condition.

This paper extends the analyses developed in previous papers, paying attention to the fact that the actual cascade blades have fluctuation in natural frequency brought about through manufacturing process. Torsional flutter of mistuned cascades is considered both in a fully stalled condition where the flow around blades shows the leading edge separation and in an unstalled condition without flow separation. A comparison of mistuning effects is made between stalled and unstalled cascades mainly through the examination of following areas.:

(1)What effect does the blade arrangement have on flutter velocity when the cascade is composed of blades having arbitrarily different natural frequencies?

(2)What is the relation between the flutter velocity and the degree of mistuning in blade natural frequency?

2. Theoretical Foundation

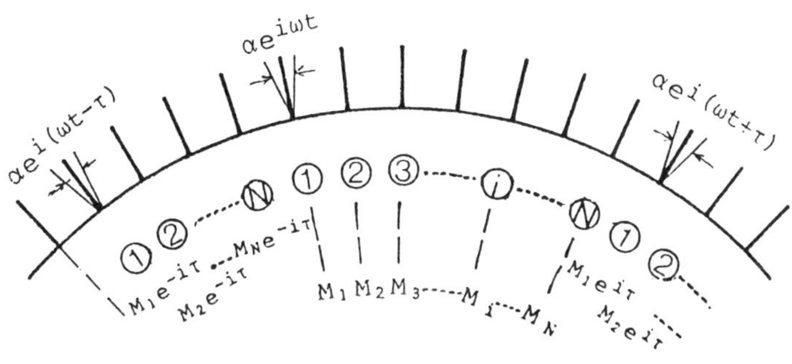

Fig. 1. Scheme of model cascade.

A model cascade is considered as shown in Fig.1, where a blade group consists of N different blades and is cyclically arranged by taking the cyclicity of flutter in annular cascades into consideration. For example, the i-th blade in a reference group and the corresponding i-th blade in its neighboring group vibrate with equal amplitudes and frequencies, and with a specified intergroup phase angle τ. The vibration of each blade is assumed to be in a one degree of freedom torsional mode. Only the elastic condition is changed to accomplish mistuning of the cascade without degrading any steady cascade specifications. In this paper, the equations of motion are considered as follows by simply changing the natural frequency, neglecting any friction term.

$$
\begin{array}{lllllll}
\alpha_1 M_1 + \alpha_2 M_N e^{-i\tau} & + \cdots\cdots & + \alpha_L M_{N+2-L} e^{-i\tau} & + \cdots\cdots & + \alpha_N M_2 e^{-i\tau} & = & I_1\ddot{\alpha}_1 + K_1\alpha_1 \\
\alpha_1 M_2 + \alpha_2 M_1 + \alpha_3 M_N e^{-i\tau} & & + \alpha_L M_{N+3-L} e^{-i\tau} & + & + \alpha_N M_3 e^{-i\tau} & = & I_2\ddot{\alpha}_2 + K_2\alpha_2 \\
\vdots \quad \vdots \quad \vdots & & \vdots & & \vdots & & \vdots \quad \vdots \\
\alpha_1 M_K + \alpha_2 M_{K-1} + & \cdots\cdots & + \alpha_L M_{N+K-L+1} e^{-i\tau} & + \cdots & + \alpha_N M_{K+1} e^{-i\tau} & = & I_K\ddot{\alpha}_K + K_K\alpha_K \\
\alpha_1 M_{L-1} + \alpha_2 M_{L-2} + \cdots & + \alpha_{L-1} M_1 & + \alpha_L M_N e^{-i\tau} & + \cdots\cdots & + \alpha_N M_L e^{-i\tau} & = & I_{L-1}\ddot{\alpha}_{L-1} + K_{L-1}\alpha_{L-1} \\
\alpha_1 M_L + \alpha_2 M_{L-1} + & \cdots\cdots & + \alpha_L M_1 + \alpha_{L+1} M_N e^{-i\tau} & + \cdots & + \alpha_N M_{L+1} e^{-i\tau} & = & I_L\ddot{\alpha}_L + K_L\alpha_L \\
\vdots \quad \vdots \quad \vdots & & \vdots & & \vdots & & \vdots \quad \vdots \\
\alpha_1 M_N + \alpha_2 M_{N-1} + & \cdots\cdots & + \alpha_L M_{N+1-L} & + \cdots\cdots & + \alpha_N M_1 & = & I_N\ddot{\alpha}_N + K_N\alpha_N
\end{array}
\tag{1}
$$

Here, I_i $(=mr^2)$, K_i $(=I_i\omega_i^2)$ and α_i $(=\alpha_i e^{i\omega t})$ denote the moment of inertia, torsional stiffness and angular displacement of vibration of the i-th blade respectively, and M_i represents the unsteady aerodynamic moment coefficient induced on the i-th blade by all 1st blades vibrating with the above mentioned intergroup phase angle τ. Equation (1) can be written as

$$[A_{ij}][\alpha_i] = 0 \qquad (i, j = 1, 2, \cdots, N) \tag{2}$$

where

$$A_{ij} = \begin{cases} M_{i-j} & (i > j) \\ M_{N+i-j+1}\, e^{-i\tau} & (i < j) \\ M_i + I_i(\omega^2 - \omega_i^2) & (i = j) \end{cases} \tag{3}$$

Here, ω denotes a blade vibration frequency and ω_i is the natural frequency of the i-th blade. Given the unsteady aerodynamic moment M_i , the eigenvalue ω/ω_1 and eigenvector $[\ \alpha_i\ ]$ are obtained from Eq.(2) as the flutter frequency and flutter mode respectively. The flutter velocity $U/c\omega_m$ is determined from Equation (4).

$$\frac{U}{c\omega_m} = \frac{1}{k}\frac{\omega}{\omega_1}\frac{\omega_1}{\omega_m} \tag{4}$$

where k $(=c\omega/U)$ is the reduced frequency. The calculation is proceeded with variation of the intergroup phase angle so as to give the minimum flutter velocity, which is defined as the critical flutter velocity. The term **flutter velocity** or V_{cr} used in this paper hereafter means such a critical flutter velocity.

Unsteady aerodynamic moment M_i due to the vibration of all 1st blades in a stalled condition is given through the calculation[6] by using the unsteady aerodynamic moment in the case of uniform vibration where all the blades vibrate with equal amplitudes and equal interblade phase angles. As for the latter unsteady moment, refer to reference [7] and [6] in the cases of stalled and unstalled cascades respectively.

3. Results and Discussion

The cascade conditions are specified such that the stagger angles of G = ± 30 deg, the solidity (chord/pitch ratio) of s = 1.0, the mass ratio of μ (= $2m/\rho\ c^2$) = 300, the torsional axis location of Pc = 0.5 (midchord position), and the attack angles of α = 30 deg for stalled cascades and of α = 4 deg for unstalled ones. The authors presume a calculation model where one group is composed of 8 blades or 20 blades in total and such blade groups are infinitely repeated.

3.1 Effects of Blade Arrangement

The critical value of flutter velocity V_{cr} (=$U/c\omega_m$) is expected to vary depending on how the blade arrangement is made up since the degree of mutual interaction among blades is influenced in aeroelastic sense by the blade arrangement. The simplified blade groups composed of only 8 blades are surveyed for the purpose of basic understanding of the effect of blade arrangement on flutter. The blade group composed of 20 blades is subsequently examined to give a more realistic condition.

3.1.1 Main Features of Stalled Cascades

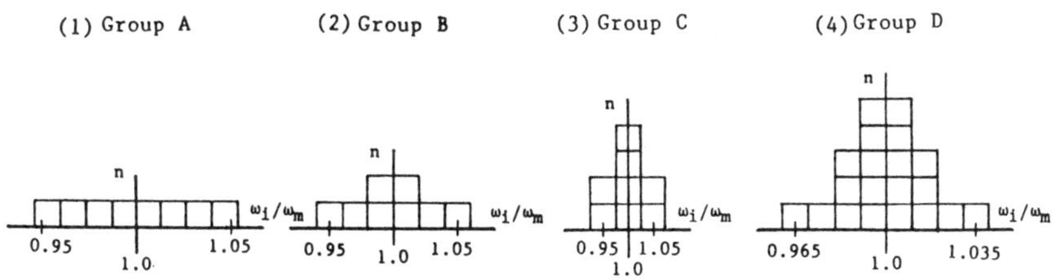

Fig. 2. Histogram of blade natural frequencies in groups of 8 and 20 blades.

Three kinds of distributions of natural frequencies in model blade groups A, B and C composed of 8 blades are first considered, as shown in Fig. 2(1)-2(3) in form of histogram. Six kinds of blade arrangements are examined for each group of blades. The flutter velocity V_{cr} and the relative flutter velocity normalized by V_{cr} of a tuned cascade are shown in Table 1, Table 2 and Table 3 together with the standard deviation σ on blade natural frequency.

Table 1 shows that the blade arrangements A-1 and A-2 give rather low flutter velocities, whereas A-5 and A-6 give rather high ones. The flutter velocity of either A-3 or A-4 lies in between these high and low flutter velocities.

Table 1. Effects of blade arrangement on flutter velocity for stalled cascades(group A, 8 blades).

Blade Arrangement for Stalled Cascade ($\alpha=30^\circ$, $G=30^\circ$)		Vcr $\left(\frac{Vcr}{(Vcr)_{Tuned}}\right)$	$\sigma \times 10^2$
A-1	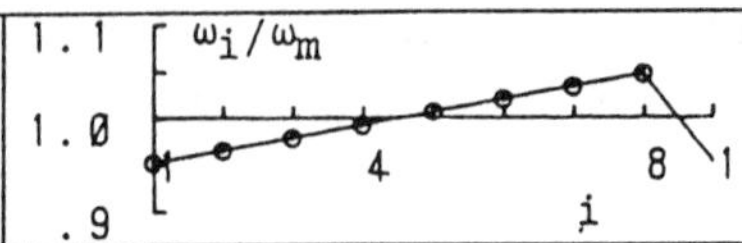	0.9536 (1.616)	3.117
A-2	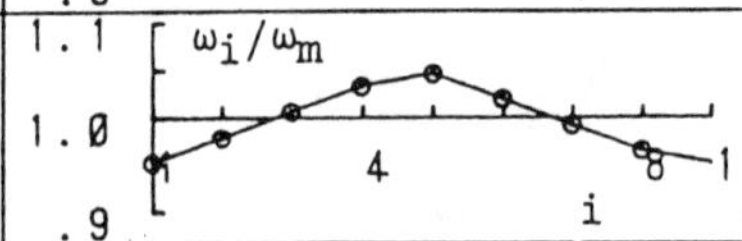	0.9492 (1.608)	3.117
A-3	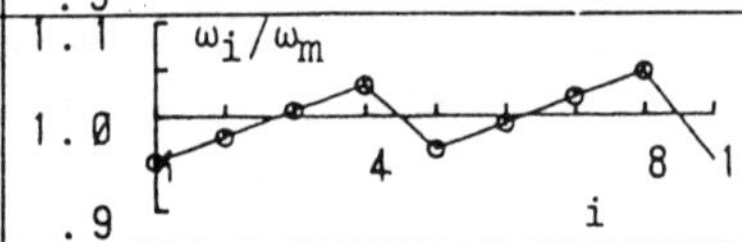	0.9604 (1.627)	3.117
A-4	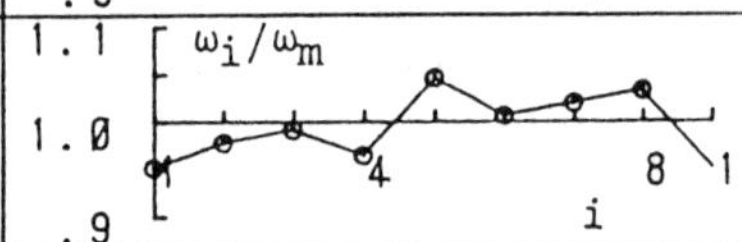	0.9604 (1.627)	3.117
A-5	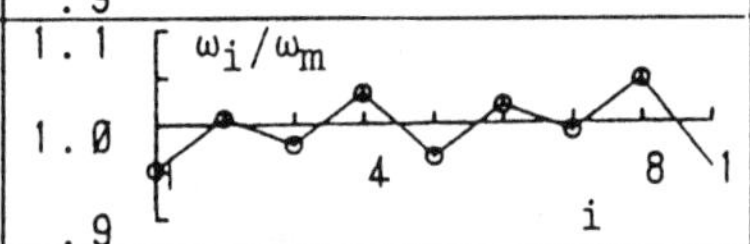	0.9636 (1.632)	3.117
A-6	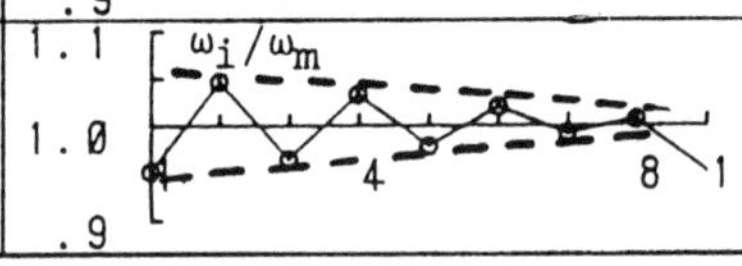	0.9636 (1.632)	3.117

The lowest flutter velocity is found in A-2, which is called the chevron arrangement, where the natural frequency increases and then decreases moderately in order of the blade number. The arrangement A-1, which is called the sawtooth arrangement, where the natural frequency increases in series, seems to have slightly more difficulty in aeroelastic coupling than A-2, since the former one has a considerable step-like difference in natural frequency between neighboring blades at the edge of the group. The arrangement A-1 therefore becomes slightly more stable on flutter than A-2.

The arrangement A-3, intended to have two sets of sawtooth patterns, gives a higher flutter velocity than A-1 does, since A-3 includes two step-like differences mentioned above, whereas A-1 includes one which tends to break aeroelastic coupling. The arrangement A-4 is designed to have low frequency blades in one half of the group and high frequency ones in the

Table 2. Effects of blade arrangement on flutter velocity for stalled cascades(group B, 8 blades).

Blade Arrangement for Stalled Cascade (α=30°, G=30°)		Vcr $\left(\frac{Vcr}{(Vcr)_{Tuned}}\right)$	σ x10²
B-1	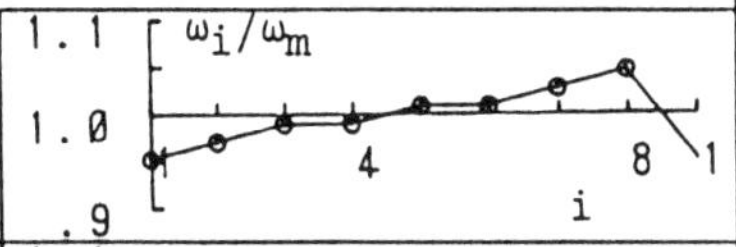	0.9508 (1.611)	2.857
B-2	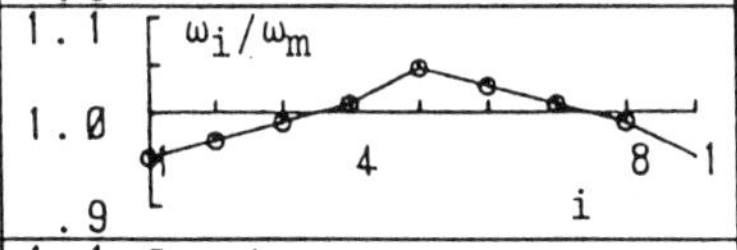	0.9554 (1.618)	2.857
B-3	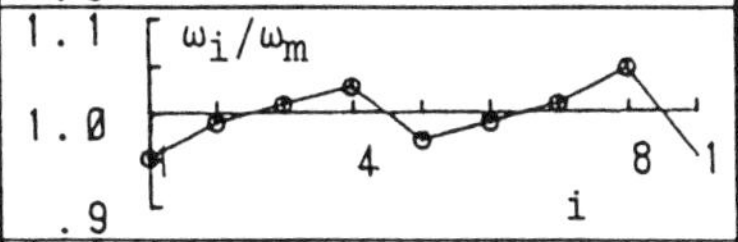	0.9643 (1.634)	2.857
B-4	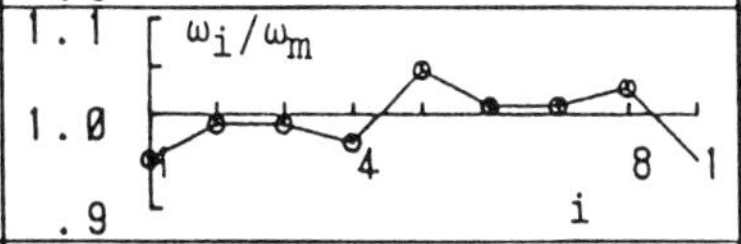	0.9493 (1.608)	2.857
B-5	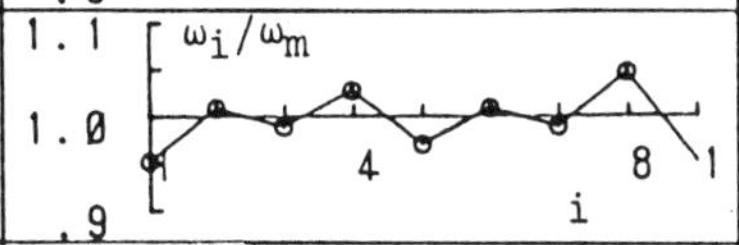	**0.9710 (1.645)**	2.857
B-6	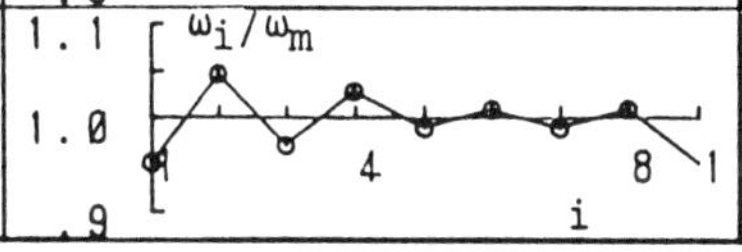	0.9643 (1.634)	2.857

other half. The step-like differences at the edges of the two halves of the arrangement also seem to increase the flutter velocity.

Fairly high flutter velocities are obtained by A-5 and A-6 which are called zigzag arrangements, where the high and low frequency blades are distributed by turns so as to keep much difference in natural frequency between neighboring blades. The arrangement A-5 is planned to be an irregularly distributed type of zigzag arrangement in order to break aeroelastic coupling among neighboring several blades, whereas A-6 is an orderly distributed zigzag type, i.e., envelopes of the frequency arrangement present monotonous curves, as indicated by dashed lines in Table 1. The arrangement A-5 looks slightly more stable than A-6. As a result, the irregular type zigzag arrangement of A-5 is found to be the most stable in these six kinds of blade arrangements.

Table 3. Effects of blade arrangement on flutter velocity for stalled cascades(group C, 8 blades).

<table>
<tr><th colspan="2">Blade Arrangement for Stalled Cascade (α = 30°, G = 30°)</th><th>Vcr $\left(\frac{Vcr}{(Vcr)_{Tuned}}\right)$</th><th>$\sigma \times 10^2$</th></tr>
<tr><td>C-1</td><td>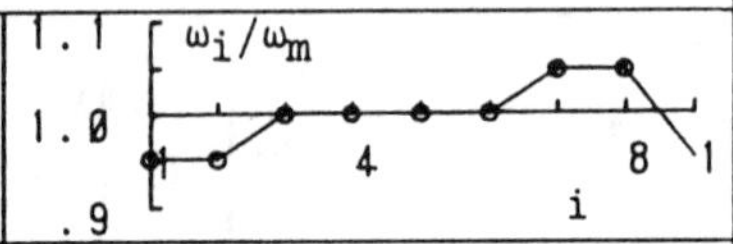
</td><td>0.8636 (1.463)</td><td>3.367</td></tr>
<tr><td>C-2</td><td>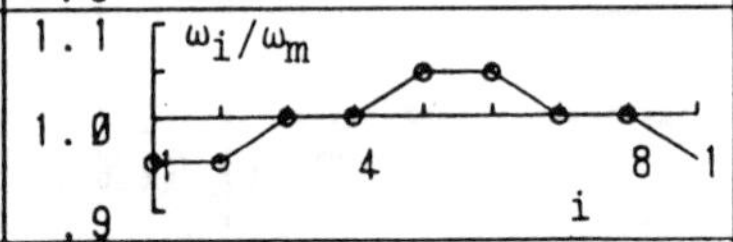
</td><td>0.8549 (1.448)</td><td>3.367</td></tr>
<tr><td>C-3</td><td>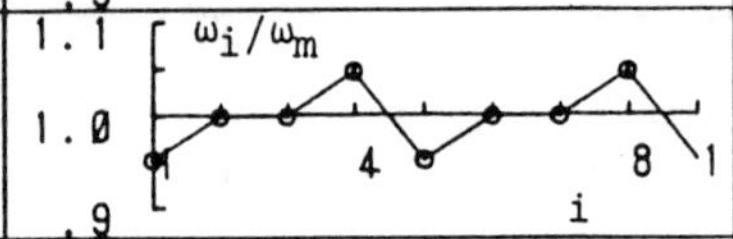
</td><td>0.8612 (1.459)</td><td>3.367</td></tr>
<tr><td>C-4</td><td>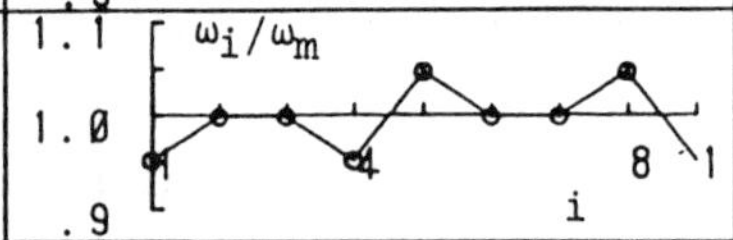
</td><td>0.8612 (1.459)</td><td>3.367</td></tr>
<tr><td>C-6</td><td>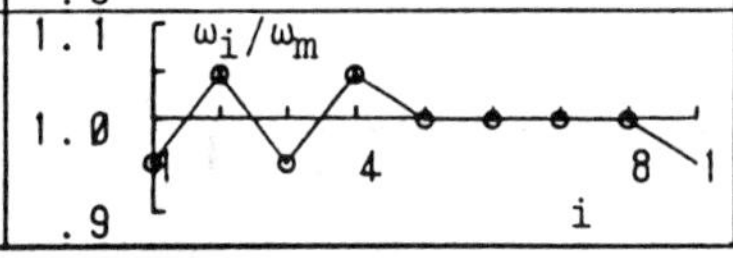
</td><td>0.8702 (1.474)</td><td>3.367</td></tr>
</table>

Table 2 is very similar in results to Table 1, showing that the most stable one is the irregular type zigzag arrangement of B-5. An inconsistency is observed, however, in the arrangement presenting lower flutter velocity. The arrangement B-4 gives the lowest flutter velocity and B-1 gives the next lowest. These considerably low flutter velocities can be expected since either arrangement includes two pairs of adjacent identical blades, which seem to stimulate aeroelastic coupling among the blades and therefore to reduce the flutter velocity.

Table 3 also looks similar in general characteristics to Table 1, showing that the zigzag arrangement of C-6 gives a fairly high flutter velocity and the chevron arrangement of C-2 gives the lowest one.

3.1.2 Comparison between Stalled and Unstalled Cascades

Table 4. Comparison of effects of blade arrangement on flutter velocity between stalled and unstalled cascades (group A, 8 blades, $\sigma = 3.117 \times 10^{-2}$).

Blade Arrangement ($\sigma = 3.117 \times 10^{-2}$)		Critical Flutter Velocity Upper: Vcr , Lower: Vcr/(Vcr)Tuned			
		Stalled $\alpha = 30^\circ$		Unstalled $\alpha = 4^\circ$	
		G = 30°	G = -30°	G = 30°	G = -30°
A-1	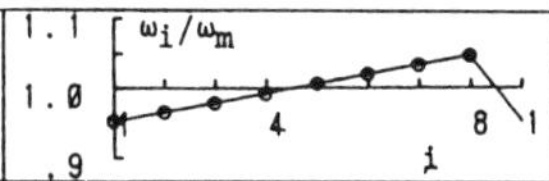	0.9536 (1.616)	1.737 (1.734)	2.607 (1.112)	2.249 (1.164)
A-2	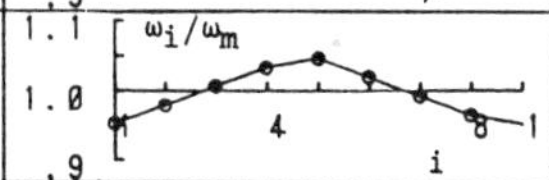	0.9492 (1.608)	1.733 (1.730)	2.573 (1.098)	2.181 (1.129)
A-3	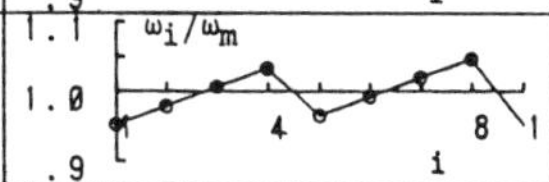	0.9604 (1.627)	1.747 (1.743)	2.620 (1.118)	2.251 (1.165)
A-4	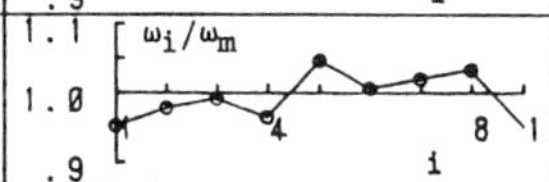	0.9604 (1.627)	1.742 (1.738)	2.619 (1.117)	2.266 (1.173)
A-5	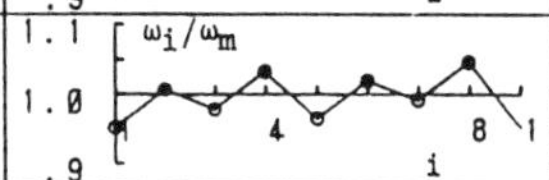	0.9636 (1.632)	1.749 (1.745)	2.631 (1.122)	2.299 (1.190)
A-6	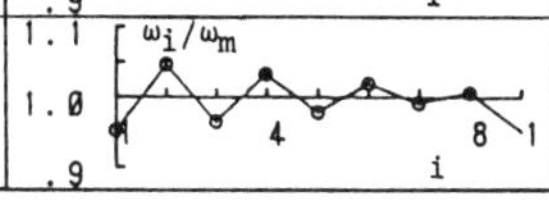	0.9636 (1.632)	1.747 (1.744)	2.623 (1.119)	2.296 (1.188)

Table 5. Comparison of alternating and other zigzag type blade arrangements.

Blade Arrangement for Compressor Cascade (G = 30°)		Vcr $\left(\frac{Vcr}{(Vcr)_{Tuned}}\right)$		$\sigma \times 10^2$
		Stalled $\alpha = 30^\circ$	Unstall-ed $\alpha = 4^\circ$	
α	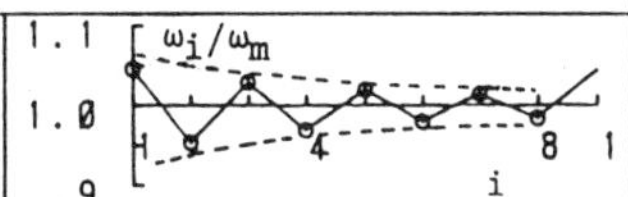	0.9662 (1.637)	2.612 (1.117)	3.017
β	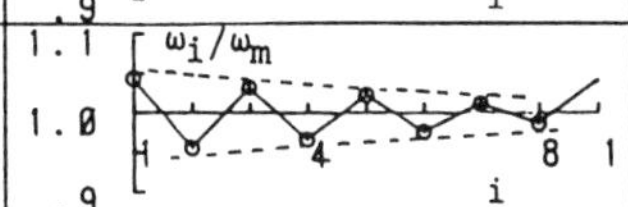	0.9031 (1.530)	2.623 (1.119)	3.015
γ	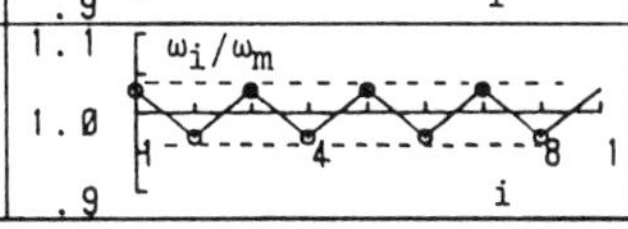	0.7312 (1.239)	2.647 (1.129)	3.000

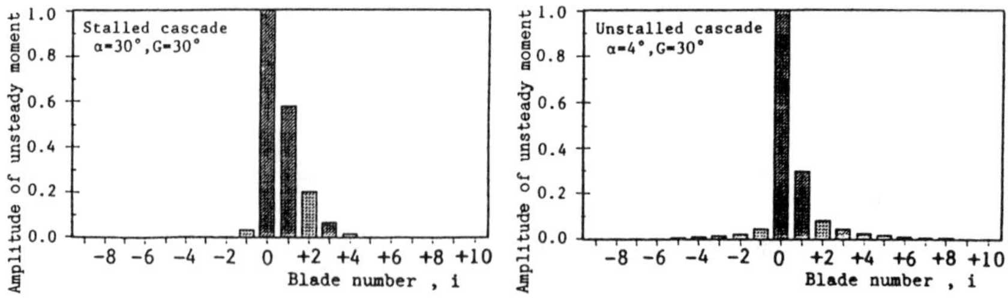

Fig. 3. Unsteady aerodynamic moment coefficients induced on blades due to center blade vibration.

As an example, the blade group A with the standard deviation of $\sigma = 3.117 \times 10^{-2}$ is examined to compare stalled and unstalled cascades both in the cases of the stagger angle G = 30 deg and G = −30 deg, as shown in Table 4. There is no difference between stalled and unstalled cascades concerning general features of the effects of the blade arrangements as mentioned in 3.1.1. The zigzag arrangements of A-5 and A-6, for example, also show fairly high flutter velocities. The irregular type zigzag arrangement of A-5, in particular, is found to be the most stable in any cascade condition. The relative flutter velocities, however, indicate a more significant flutter suppression effect for stalled cascades than for unstalled ones in this case of slight mistuning.

To survey further on the zigzag arrangement, flutter velocities of both cascades for various arrangements are shown in Table 5. The arithmetic mean natural frequency and the standard deviation σ are kept nearly constant during the calculation. The arrangement γ describes the alternating type, i.e., two kinds of blades having different natural frequencies are alternately arranged, as an extreme case where envelopes of the frequency arrangement become parallel lines. The envelopes for α have steeper slope than those for β. For unstalled cascades, the steeper the slope is, the lower the flutter velocity is, i.e., γ gives the highest flutter velocity. On the other hand for stalled ones, γ gives a considerably low flutter velocity.

The above unexpected results can be explained by considering the influence coefficient of unsteady aerodynamic moment, induced on each blade due to vibration of the center blade(No. 0 blade), as shown in Fig. 3 in terms of normalized amplitude. The influence coefficient of No. 2 blade is over 0.2 in a stalled condition while it is under 0.1 in an unstalled condition. This,

Table 6. Effects of blade arrangement on flutter velocity for stalled cascades (group D, 20 blades).

Blade Arrangement for Stalled Cascade (α = 30°, G = 30°)		Vcr $\left(\frac{Vcr}{(Vcr)_{Tuned}}\right)$	σ x10^2
D-1	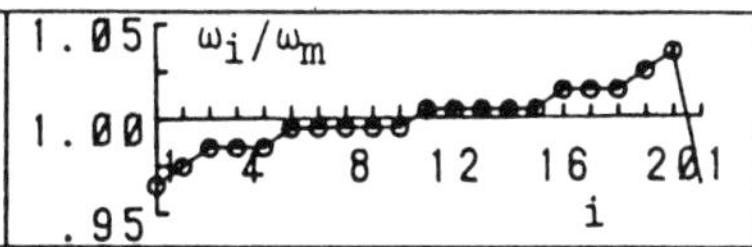	0.9344 (1.583)	1.628
D-2	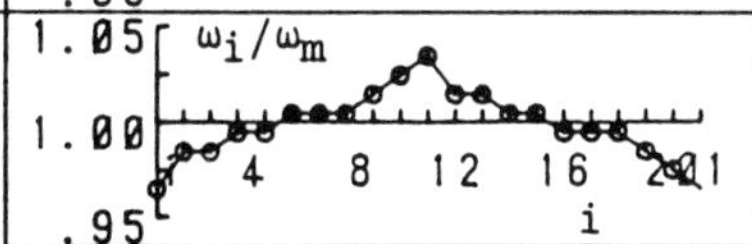	0.9206 (1.559)	1.628
D-3	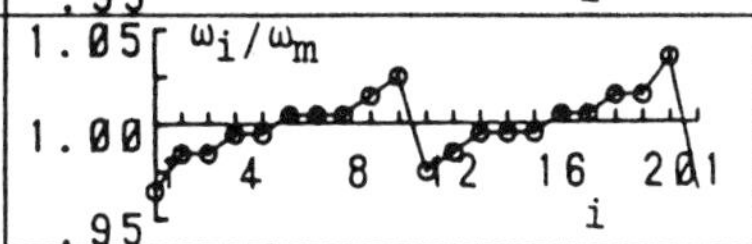	0.9408 (1.594)	1.628
D-4	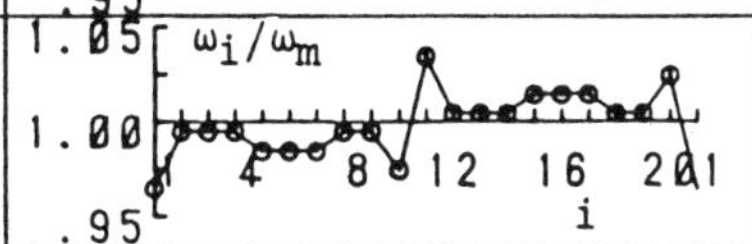	0.9290 (1.574)	1.628
D-5	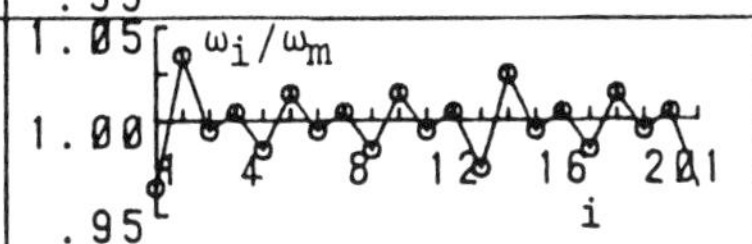	**0.9667 (1.637)**	1.628
D-6	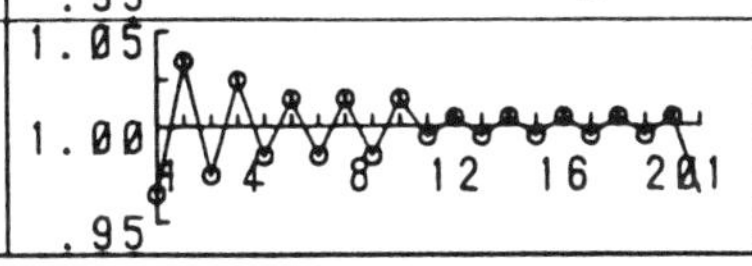	0.8507 (1.441)	1.628

in the case of stalled cascades compared with that of unstalled cascades, leads to the stimulation of aeroelastic coupling between the i-th blade and the $i+2$-th blade of the arrangement γ where every other blade is identical and therefore leads to quite a low flutter velocity. It seems insufficient for flutter suppression like in this case to break aeroelastic coupling only between adjacent two blades, i.e., i-th and $i+1$-th blade. It is thus found that the alternating arrangement, which was concluded to be optimum in an unstalled condition[6], does not necessarily give satisfactory flutter suppression effects, as indicated in a stalled condition.

Next, Table 6 gives the results for stalled cascades of a more realistic blade group D composed of 20 blades, ranging the fluctuation of $\pm$ 3.5 % in natural frequency specified as a normal manufacturing imperfections shown in Fig. 2(4). Corresponding results for unstalled cascades are supplied from

Table 7. Effects of blade arrangement on flutter velocity for unstalled cascades (group D, 20 blades, results from Ref. [6]).

Blade Arrangement for Unstalled Cascade ($\alpha = 4^\circ$, $G = 30^\circ$)		Vcr $\left(\frac{Vcr}{(Vcr)_{Tuned}}\right)$	$\sigma \times 10^2$
D-1	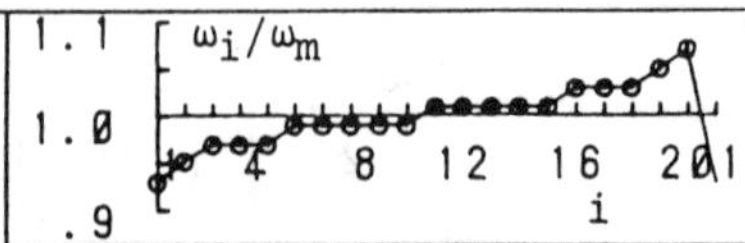	2.622 (1.119)	3.256
D-2	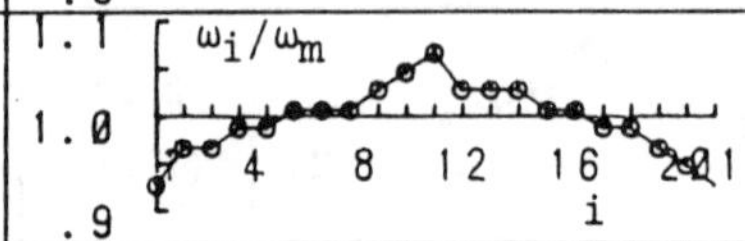	2.621 (1.119)	3.256
D-3	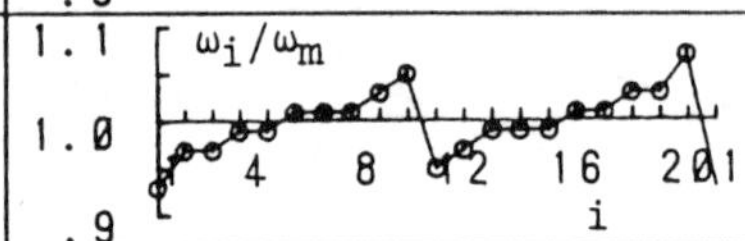	2.625 (1.120)	3.256
D-4	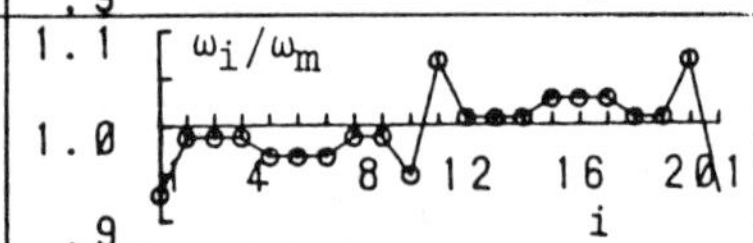	2.625 (1.120)	3.256
D-5	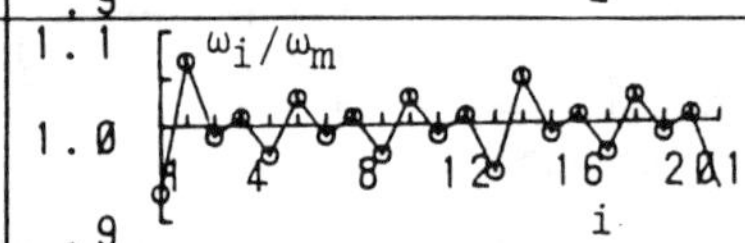	**2.632 (1.123)**	3.256
D-6	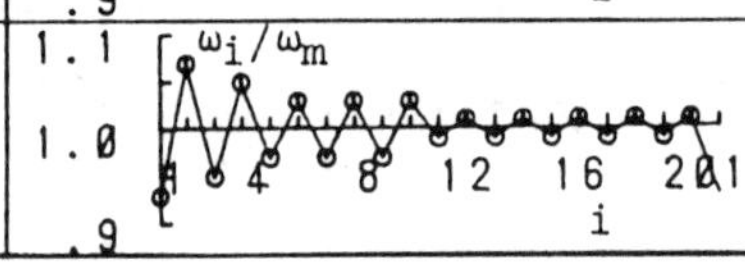	2.611 (1.114)	3.256

Ref. [6] as shown in Table 7. As can be expected from previous results, the irregular type zigzag arrangement of D-5 shows a substantially high flutter velocity in both cascades. However, the orderly type zigzag arrangement D-6 shows rather low flutter velocity, since it resembles the alternating arrangement γ, which possibly gives undesirable results, as mentioned above.

Generally speaking, an irregular type zigzag arrangement gives a fairly high flutter velocity, while an alternating arrangement, which was reported to be optimum in an unstalled condition, does not necessarily give the satisfactory flutter suppression effect in a stalled condition. Special care, in any case, must be taken to avoid a blade arrangement where the blades having exactly or nearly equal natural frequencies are successively or closely placed.

3.2. Effects of Fluctuation in Natural Frequency

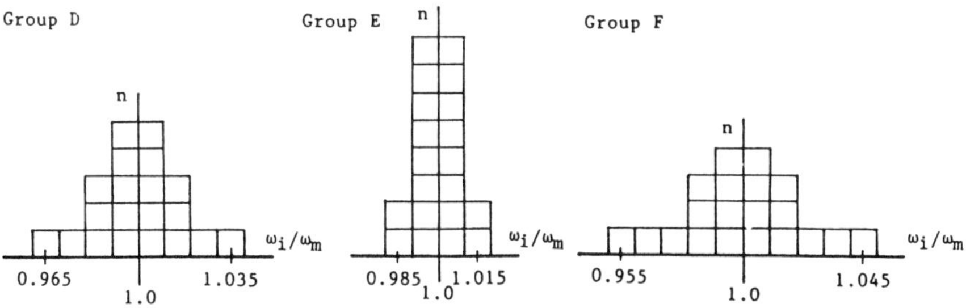

Fig. 4. Histogram of blade natural frequencies in groups of 20 blades.

Table 8. Effects of standard deviation on flutter velocity for stalled cascades (zigzag type blade arrangements).

Blade Arrangement for Stalled Cascade (α = 30°, G = 30°)		Vcr $\left(\frac{Vcr}{(Vcr)_{Tuned}}\right)$	σ x10²
D-5	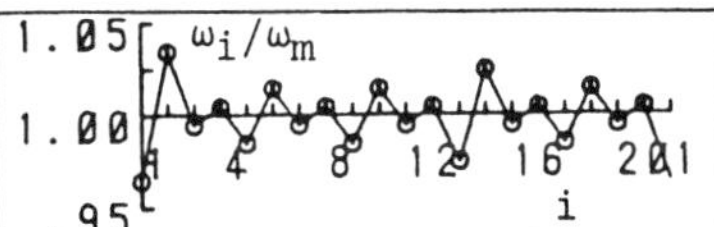	0.9667 (1.637)	1.628
E-5	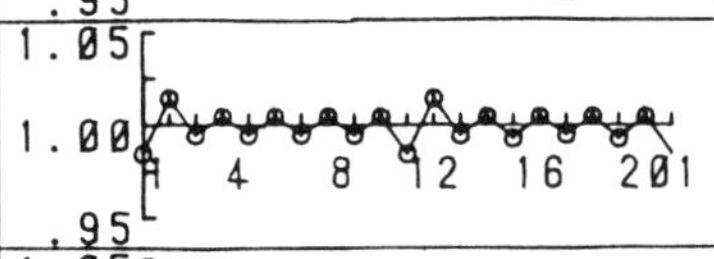	0.8836 (1.497)	0.826
F-5	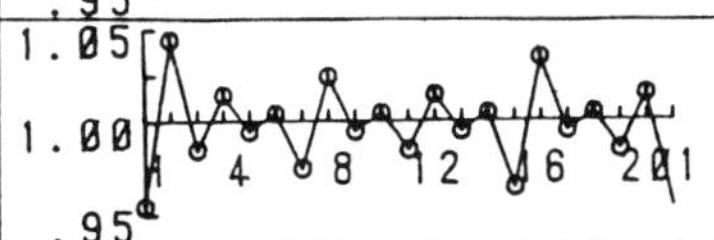	**0.9675 (1.639)**	2.264
Tuned Cascade	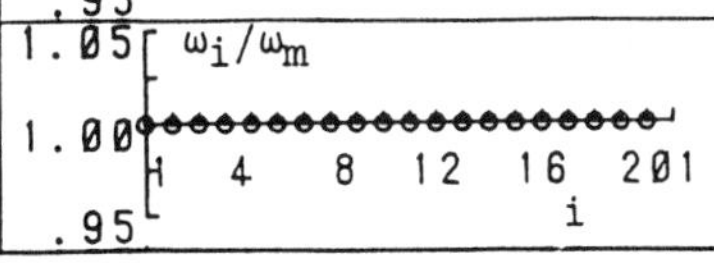	0.5903 (1.000)	0.000

Effects of variation in configuration of the natural frequency histogram on flutter are surveyed, i.e., the effects of fluctuation in natural frequency of a blade group. Three kinds of blade groups D, E and F are considered with regard to the fluctuation in natural frequency as shown in Fig. 4.

Table 9. Effects of standard deviation on flutter velocity for unstalled cascades (zigzag type blade arrangements, results from Ref. [6]).

Blade Arrangement for Unstalled Cascade ($\alpha=4^\circ$, $G=30^\circ$)		Vcr $\left(\frac{Vcr}{(Vcr)_{Tuned}}\right)$	$\sigma \times 10^2$
D-5	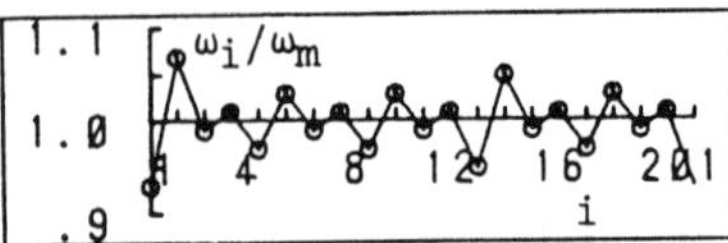	2.632 (1.123)	3.256
E-5	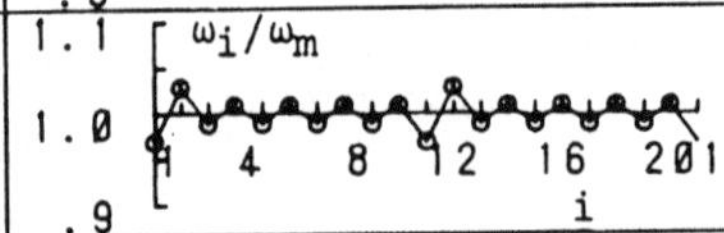	2.451 (1.046)	1.612
F-5	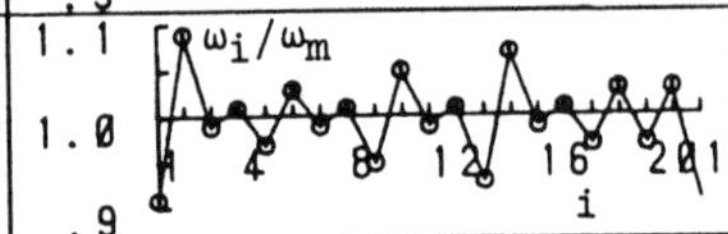	2.731 (1.166)	4.313
Tuned Cascade	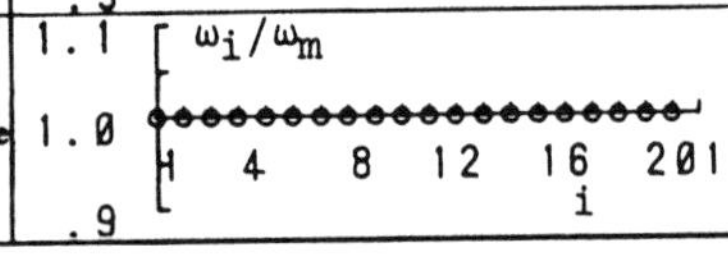	2.343 (1.000)	0.000

Group D is regarded as a reference group with normal manufacturing imperfections mentioned in the previous section. Group E is made up through the refinement to eliminate extremities from the reference group D. Group F corresponds to the case of a rougher manufacturing process. Thus the degree of fluctuation increases in order of E, D and F.

In the previous section, a preferable mistuning arrangement is found to be a so-called irregular type zigzag arrangement, from the viewpoint of flutter suppression. Three arrangements E-5, D-5 and F-5 are therefore examined in this section. The results are shown in Table 8 for stalled cascades together with the case of the tuned cascade. Corresponding results for unstalled cascades obtained from Ref.[6] are shown in Table 9. The standard deviation, σ, represents the degree of fluctuation in natural frequency. The lowest flutter velocity is seen in the case of the tuned arrangement as an extreme case of non-fluctuation in natural frequency, i.e., $\sigma=0$. It is revealed from both tables that within the range of normal blade manufacturing imperfections, the flutter velocity increases both for stalled and unstalled cascades, as the degree of fluctuation in blade natural frequency increases, i.e., as σ increases. The cascade composed of blade groups having large fluctuation in natural frequency is found to be more stable with regard to flutter than that whose blade natural frequencies are closely grouped, i.e., well-tuned. In other words, the more precise the blade manufacturing process becomes, the more unstable the cascade becomes. The blade group should therefore not be selected in such a way that the standard deviation decreases and consequently the group itself approaches to the tuned one.

3.3. Flutter Velocity and Flutter Mode

Previous papers[5][7] indicate that the flutter mode, giving the lowest flutter velocity, generally changes depending on the degree of mistuning. The modal change is surveyed with the change in standard deviation, taking the arrangement A-1 shown in Table 1 as an example.

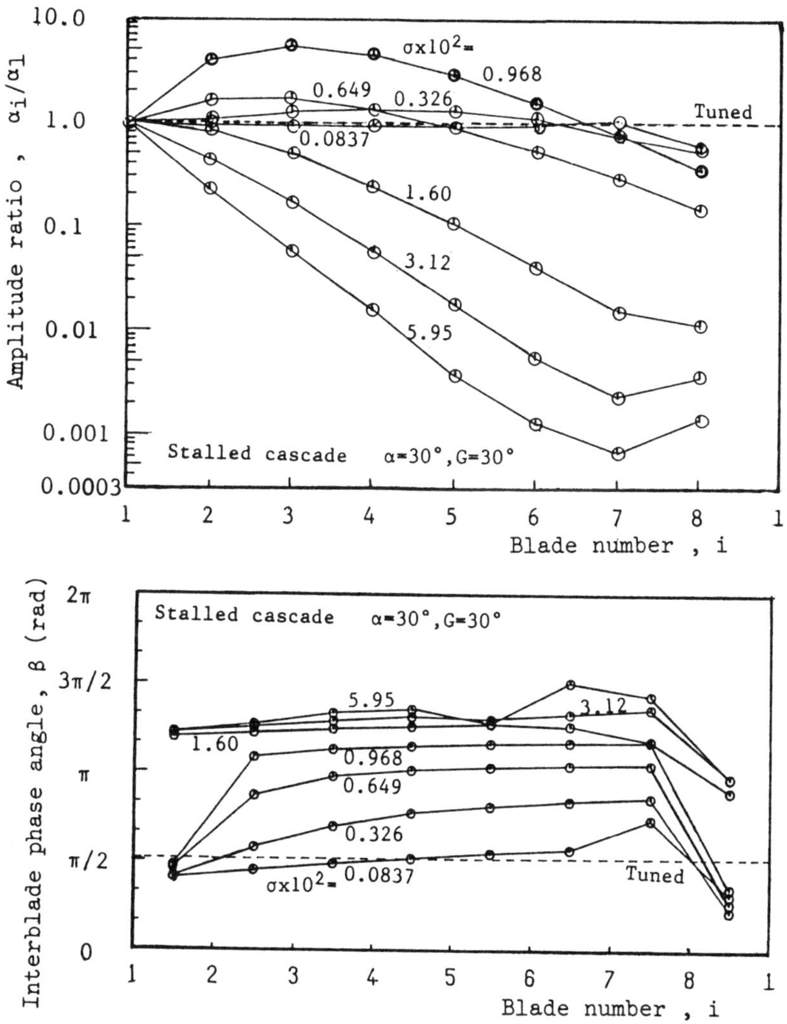

Fig. 5. Variation of flutter mode according to the degree of fluctuation in blade natural frequency for stalled cascades.

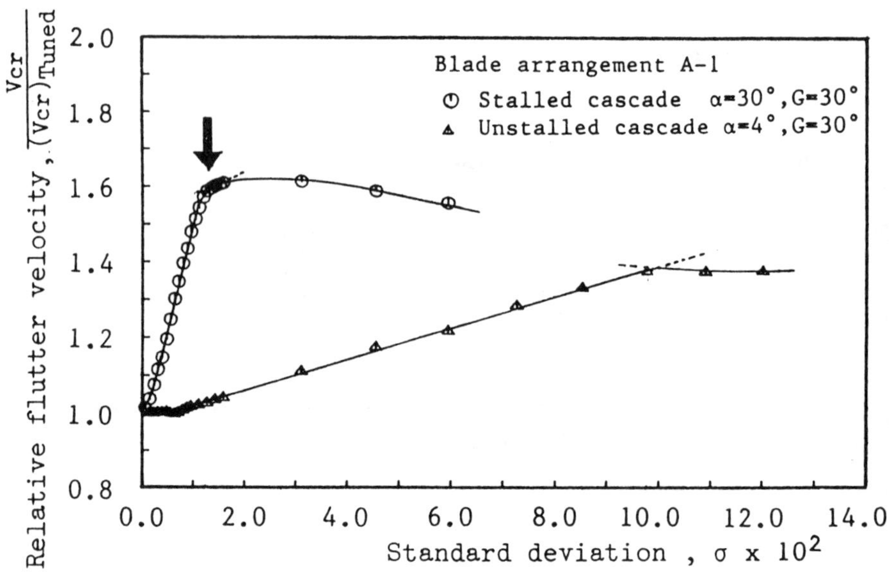

Fig. 6. Variation of flutter velocity for stalled and unstalled cascades.

For stalled cascades, Fig. 5 illustrates the effects of the standard deviation on blade vibration amplitude and interblade phase angle of A-1. The corresponding change in flutter velocity is shown in Fig. 6, in terms of the relative flutter velocity normalized by the flutter velocity of the tuned cascade, together with the case of unstalled cascades. It is evident from the figures that the flutter mode abruptly changes at a certain point between $\sigma = 0.968 \times 10^{-2}$ and $\sigma = 1.60 \times 10^{-2}$. Up to this point, marked with an arrow in Fig. 6, the flutter mode seems to be equivalent to one derived from the tuned flutter mode, although the envelope of either amplitude or phase angle varies and accordingly the flutter velocity increases rapidly with an increase in σ. A flutter suppression effect according to the degree of mistuning is noticeable in this region. On the other hand, beyond this point, flutter occurs in the mode where amplitudes of blades having relatively low natural frequencies are dominant, as compared with those having relatively high natural frequencies. In this region far from the tuned case, the envelope of phase angle remains almost unchanged and accordingly the curve of flutter velocity rise becomes rather flat, which implies no further progress in flutter suppression effect, with an increase in standard deviation.

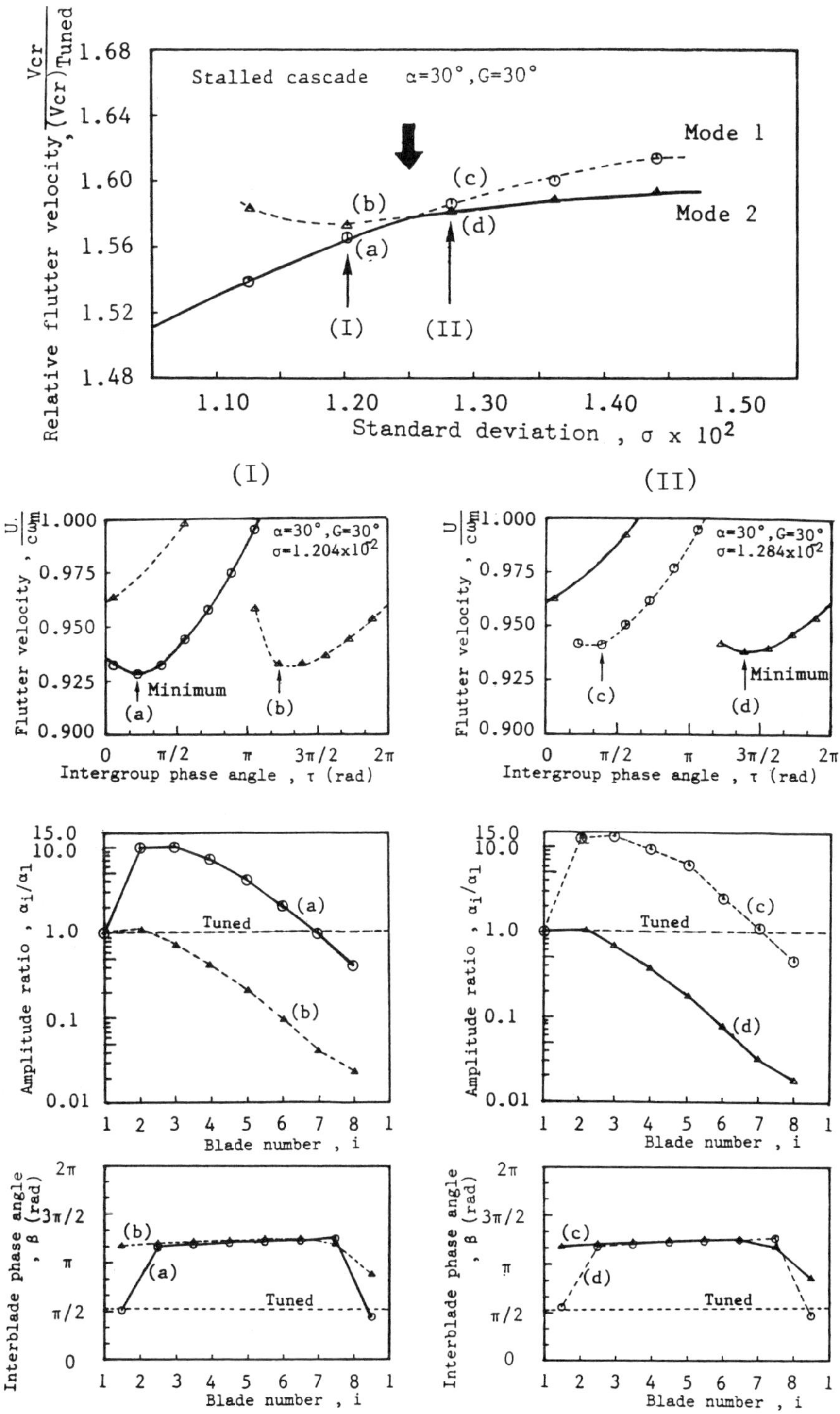

Fig. 7. Modal change for stalled cascades through critical point.

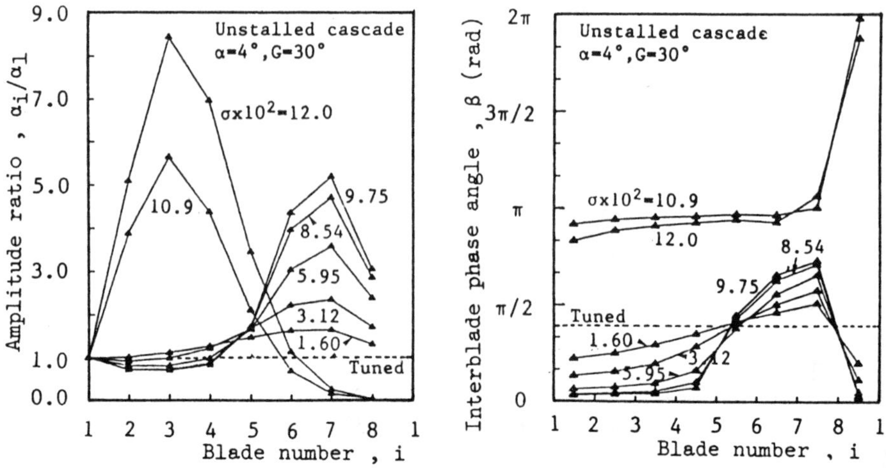

Fig. 8. Variation of flutter mode according to the degree of fluctuation in blade natural frequency for unstalled cascades.

The modal change around an arrow point in Fig. 6 is further illustrated in detail in Fig. 7. In the case of $\sigma = 1.204 \times 10^{-2}$ marked with (I), the curves of the flutter velocity give the minimum value at the point (a), where flutter occurs in Mode 1 which is equivalent to one derived from the tuned mode as mentioned above. The minimum flutter velocity at (a) in Mode 1 is lower than the next minimum found at (b) in Mode 2 . In the case of $\sigma = 1.284 \times 10^{-2}$ marked with (II), on the other hand, the minimum at (d) in Mode 2 becomes lower than the minimum at (c) in Mode 1. Therefore flutter occurs in Mode 2 at (d), where amplitudes of lower natural frequency blades are dominant, as explained above.

For unstalled cascades, no sudden modal change is seen up to the point between $\sigma = 9.75 \times 10^{-2}$ and $\sigma = 10.9 \times 10^{-2}$, as shown in Fig. 8. The standard deviation at that point is considerably large compared with the case of stalled cascades, as seen in Fig. 6. However, the flutter mode changes, in the same manner as the case of stalled cascades, from a derivative mode of the tuned one to the mode where amplitudes of blades having relatively low natural frequencies are dominant as well.

3.4. Overall View on Flutter Suppression

An overall view of the relationship between the relative flutter velocity, which corresponds to the critical flutter velocity of the mistuned cascade normalized by that of the tuned cascade, and the standard deviation is

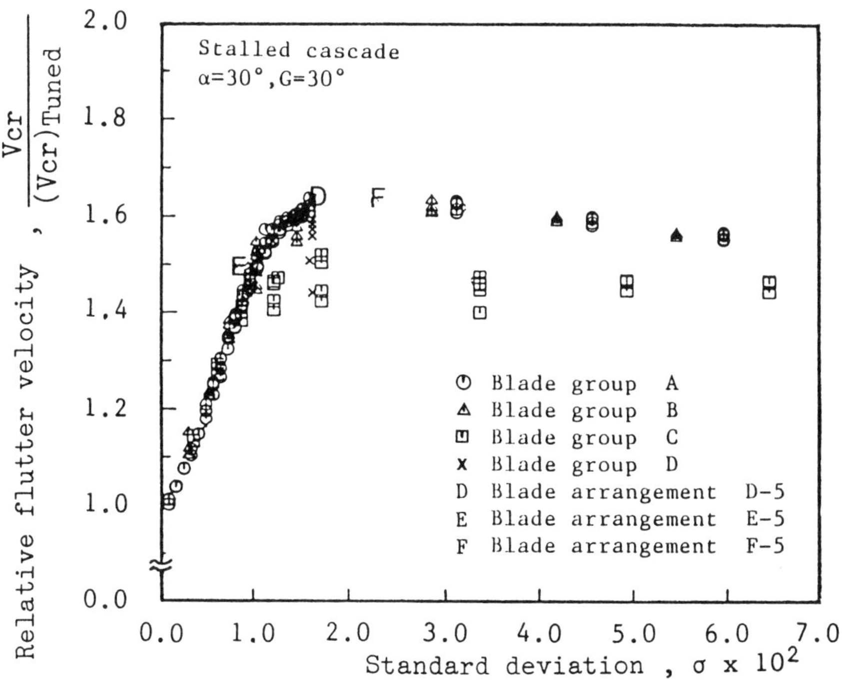

Fig. 9. Overall view of flutter velocity versus standard deviation for stalled cascades.

shown in Fig. 9, for stalled cascades. Figure 9 includes the results of the blade groups of A, B and C for various standard deviations together with the results of the group D and of arrangements E-5 and F-5. Also the results, obtained from Ref.[9], for unstalled cascades in similar mistuning conditions are shown in Fig. 10 to compare with the present results.

From both figures, the flutter suppression effect by mistuning in blade natural frequency is confirmed for stalled cascades as well as for unstalled ones. The flutter velocities for stalled cascades do not lie on a single curve in the region of relatively large standard deviations, whereas they lie on a straight line when σ becomes small enough, i.e., the flutter velocity is proportional to σ. Furthermore, emphasis is to be placed on the fact that a more significant flutter suppression effect is expected from slight mistuning in a stalled condition than in an unstalled one, which is obvious by comparison of the slopes of curves in both cases.

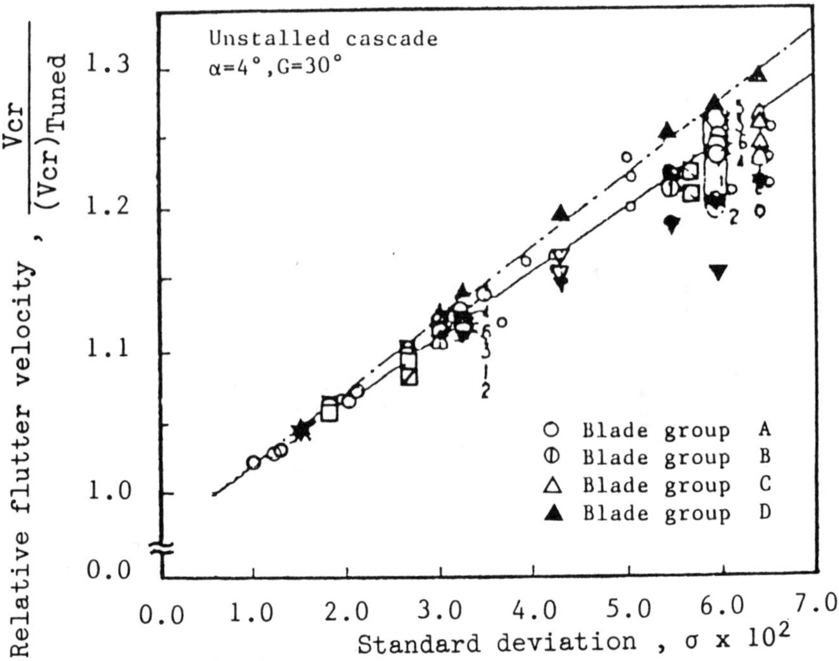

Fig. 10. Overall view of flutter velocity versus standard deviation for unstalled cascades (results from Ref.[9].).

4. Conclusions

A comparison of mistuning effects on flutter in torsional mode is made between stalled and unstalled cascades, paying attention to the fact that the cascade blades actually have fluctuation in natural frequency brought about through manufacturing process. The conclusions of the present study are summarized as follows:

(1) The flutter suppression effect by mistuning in blade natural frequency is confirmed for stalled cascades as well as for unstalled ones. Furthermore, emphasis is to be placed on the fact that a more significant flutter suppression effect is expected from slight mistuning in a stalled condition than in an unstalled one.

(2) Both in a stalled and an unstalled condition, within the range of normal blade manufacturing imperfections, the larger the standard deviation on blade natural frequency becomes, the more stable the cascade becomes. In other words, the cascade composed of blade groups having large fluctuation in natural frequency is more stable with regard to flutter than

that whose blade natural frequencies are closely grouped, i.e., well-tuned. The blade group should therefore not be selected in such a way that the standard deviation decreases and the group itself consequently approaches to the tuned one.

(3) As for the blade arrangement in a blade group, an irregularly distributed type of zigzag arrangement, which can be expected to break aeroelastic coupling among neighboring several blades, gives a fairly high flutter velocity both in stalled and unstalled conditions. However, an alternating arrangement as an extreme case of zigzag arrangements, which is planned to break aeroelastic coupling only between two adjacent blades and was concluded to be optimum in an unstalled condition, does not necessarily give the satisfactory flutter suppression effect in a stalled condition. Either for stalled or for unstalled cascades, special care must be taken to avoid a blade arrangement where the blades having exactly or nearly equal natural frequencies are successively or closely placed.

References

[1] Whitehead,D.S., (1964) "Torsional Flutter of Unstalled Cascade Blades at Zero Deflection," Great Britain A.R.C. R & M 3429, 1964.

[2] Bendiksen,O.O., (1988) "Recent Developments in Flutter Suppression Techniques for Turbomachinery Rotors", Journal of AIAA, Vol.4, No.2, 1988, pp. 164-171.

[3] Tanaka,H.,Hanamura,Y.,and Hattori, S.,(1975) "Flutter of NonUniform Cascade (1st Report, Theory)",(in Japanese) Transaction JSME, Vol.41, No.345, 1975, pp.1437-1446.

[4] Hanamura,Y.,Tanaka,H., (1977) "A Modification of Flutter Characteristics by Changing Elastic Nature of Neighbouring Blades in Cascade",Proc. of '77 Tokyo Joint Gas Turbine Congress, 1977, pp.418-425.

[5] Tanaka,H.,Hanamura,Y.,and Fujimoto,I., (1984) "Analyses of Cascade Flutter as a Multi-Eigenvalue Problem (on Cascade Composed of Two Groups of Blades Having Different Natural Frequencies)",Proc. of '83 Tokyo International Gas Turbine Congress, Vol.2, 1984, pp.651-658.

[6] Tanaka,H.,Fujimoto,I., (1987) "Flutter of Cascade Blades Composed of Blades Having Arbitrarily Different Natural Frequencies",Proc. of 4th International Symposium on Unsteady Aerodynamics and Aeroelasticity of Turbomachines and Propellers, (held in Aachen), 1987, pp.547-560.

[7] Fujimoto,I.,Nagai,H.,Tanaka,H.,Hirano,T.,and Ishii,S., (1989) "Mistuning Effects on Fully Stalled Cascade Flutter",Proc. of 5th International Symposium on Unsteady Aerodynamics and Aeroelasticity of Turbomachines and Propellers, (held in Beijing), 1989, pp.363-374.

[8] Fujimoto,I.,Tanaka,H.,Ishii,S.,and Hirano,T.,(1990) "An Example to Show Effects of Mistuning in Blade Natural Frequency on Turbomachine Cascade Flutter with Flow Separation",Motortechnische Zeitschrift 51.Jahrgang/Nr.6, Juni 1990, pp.276-279.

[9] Tanaka,H.,Fujimoto,I.,(1986) "Flutter of Non-uniform Cascade Having Arbitrary Distributions of Blade Natural Frequency.(2nd Report, General Survey on the Fluctuation in Natural Frequency and on the Blade Arrangement)",(in Japanese) Proc. of the Gas Turbine Society of Japan, (The Fall Conference held in Hiroshima), 1986, pp.103-108.

This article was processed by the author using the TEX macro package from Springer-Verlag.

Splitter Blades for Passive Turbomachine Flutter Control

Hsiao-Wei D. Chiang[1,2]
Sanford Fleeter[1]

ABSTRACT

Splitter blades as a passive flutter control technique is investigated by developing a mathematical model to predict the stability of an aerodynamically loaded splittered-rotor operating in an incompressible flow field. The splitter blades, positioned circumferentially in the flow passage between two principal blades, introduce aerodynamic and/or combined aerodynamic-structural detuning into the rotor. The two-dimensional oscillating cascade unsteady aerodynamics, including steady loading effects, are determined by developing a complete first-order unsteady aerodynamic analysis together with an unsteady aerodynamic influence coefficient technique. The torsion mode flutter of both uniformly spaced tuned rotors and detuned rotors are predicted by incorporating the unsteady aerodynamic influence coefficients into a single-degree-of-freedom aeroelastic model. This model is then utilized to demonstrate that incorporating splitters into unstable rotor configurations results in stable splittered-rotor configurations.

[1]School of Mechanical Engineering, Purdue University, Chaffee Hall, West Lafayette, IN 47907.
[2]Currently Aeromechanical Engineer, General Electric Aircraft Engines, One Neumann Way, Cincinnati , OH 45215.

NOMENCLATURE

$\overline{C}_p$	steady surface static pressure, $P/\frac{1}{2}\rho U_\infty^2$
C_p	unsteady pressure distribution, $P'/\rho U_\infty^2 \overline{\alpha}$
$[CM]^n$	motion-induced influence coefficient of airfoil n
DC_p	unsteady pressure difference, $\Delta P'/\rho U_\infty^2 \overline{\alpha}$
k	reduced frequency, $\omega b/U_\infty$
P'	unsteady pressure
$\vec{Q}$	complete flow field
$\vec{Q}_o$	steady mean flow
$\vec{Q}'$	unsteady flow
$\overline{U}_\infty$	far field uniform mean flow
a_o	mean incidence angle
$\overline{\alpha}$	amplitude of torsional oscillations
$\hat{\alpha}$	complex oscillatory displacement
b_o	interblade phase angle
b_d	detuned interblade phase angle
Φ	general velocity potential
Φ'	unsteady velocity potential
Φ'_C	circulatory unsteady velocity potential
Φ'_{NC}	noncirculatory unsteady velocity potential
G	steady circulation constant
Γ'	unsteady circulation constant
w_a	natural torsional frequency
w_o	reference frequency

INTRODUCTION

To develop compact lightweight advanced gas turbine engines requires the development of axial flow blade rows which can achieve higher pressure ratios without flow separation, accomplished by means of high solidity, low aspect ratio blade rows, with the number of airfoils and/or the airfoil chord increased. Unfortunately, these adversely affect aerodynamic efficiency, with the additional airfoils and end wall surface areas resulting in increased losses and the increased blockage reducing the flow capacity.

One approach to minimizing these performance penalties is the use of partial chord blades, termed splitters, between principal full chord blades. In particular, the splitters help to control trailing edge flow separation without the aerodynamic penalties associated with high solidity full chord blade rows, i.e., the splitters enable the aft rotor blade row to have increased flow turning, thereby increasing the work transfer. In addition, the splitters do not affect the blade row flow capacity as it is controlled by the entrance regions. with the splitters not affecting the entrance region solidity.

In addition to performance goals, advanced engines must also have improved reliability. Thus, it is necessary to address the effect of incorporating splitters on flow induced vibrations. In this regard, flutter is well known to have a negative impact on the development of advanced design axial flow rotors. Unfortunately, the incorporation of part-span shrouds to eliminate the flutter problem also results in a degradation of performance.

Two research techniques have been proposed to eliminate flutter from the operating envelope - structural detuning and aerodynamic detuning. Structural detuning, defined as designed blade-to-blade differences in the natural frequencies of a blade row, has been considered for passive aeroelastic control. Several studies of the effect of structural detuning on rotors operating in supersonic flow fields with subsonic axial components have shown that structural detuning has beneficial effects on flutter. However, structural detuning is not a universally accepted passive stability control mechanism because of the associated manufacturing, material, inventory, engine maintenance, control and cost problems.

Aerodynamic detuning is a relatively new concept for passive aeroelastic control. It is defined as designed blade-to-blade differences in the unsteady aerodynamic flow field of a blade row. Thus, aerodynamic detuning creates blade-to-blade differences in the unsteady aerodynamic forces and moments, thereby affecting the fundamental driving force. This results in the blades not responding in a classical traveling wave mode typical of a conventional uniformly spaced aerodynamically tuned rotor. Studies of rotors operating in both incompressible flow fields [1] and supersonic flow fields with subsonic axial components [2, 3] have shown that aerodynamic detuning is beneficial to flutter stability.

Splitter blades introduce combined aerodynamic-structural detuning into a rotor. The aerodynamic detuning results from the differences in the principal blade passage unsteady aerodynamics due to the splitters, with the structural detuning achieved through the higher natural frequencies of the splitter blades as compared to the full chord principal rotor blades. Hence, the incorporation of splitters into a rotor may not only result in improved aerodynamic performance, but also in increased flutter stability, i.e., the combined aerodynamic-structural detuning resulting from the splitter blades may take advantage of the enhanced flutter stability due to both aerodynamic and structural detuning while eliminating the difficulties associated with structural detuning.

For a supersonic axial flow with a subsonic axial component, splitters have been shown to be a viable passive control technique for flutter and forced response [4, 5]. However, these studies are based on classical linear theory, with the steady flow assumed to be uniform and the principal blades and splitters modeled as flat plates. Thus, the effects on the steady flow field of incorporating the splitters into the rotor as well as the effect of the resulting steady flow field on stability were not considered.

This paper is directed at investigating the effect on stability of incorporating splitter blades into a rotor. This involves the investigation of the viability of splitters as a passive torsion mode flutter control technique of an aerodynamically loaded rotor operating in an incompressible flow field. Thus, this research significantly extends the previous modeling and understanding of the fluid dynamics and aeroelasticity of splittered rotors. This is accomplished by developing a mathematical model and utilizing it to demonstrate the enhanced torsion mode stability associated with incorporating splitters into a loaded rotor.

A complete first order unsteady aerodynamic model is developed, i.e., the thin airfoil approximation is not utilized, to analyze the oscillating airfoil motion induced aerodynamics of both conventional uniformly spaced rotors without splitters and rotors incorporating variably spaced splitters between principal blades, including the effects of steady aerodynamic loading. An unsteady aerodynamic influence coefficient technique is then utilized, thereby enabling the stability of both conventional rotors and splittered rotors to be determined.

The analysis of the torsion mode stability characteristics of a rotor requires the prediction of the unsteady pressure resulting from the harmonic torsional motions of the cascade. The steady and unsteady aerodynamics acting on the typical two-dimensional airfoil sections of a blade row are determined by considering: (1) a single principal blade passage with periodic boundary conditions for the conventional tuned uniformly spaced rotor; and (2) two principal blade passages with two passage periodic boundary conditions for the aerodynamically detuned rotor. The flow field is assumed to be linearly comprised of a steady potential mean flow and an harmonic unsteady flow field. The steady and unsteady potential flow fields are individually described by Laplace equations, with both the steady and unsteady potentials further decomposed into circulatory and noncirculatory components. The steady flow field is independent of the unsteady flow. However, the unsteady flow is coupled to the steady flow field through the unsteady boundary conditions on the airfoil surfaces.

Mathematical Model - Tuned Cascade

Figure 1 presents a schematic representation of a thick, cambered airfoil cascade at finite mean incidence α_0 to the farfield uniform mean flow, executing torsion mode oscillations. The complete flow field is assumed to be comprised of a steady mean flow and an harmonic unsteady flow field, Equation 1. The unsteady flow field corresponds to the motion-induced unsteady flow field.

$$\vec{Q}(x,y,t) = \vec{Q}_0(x,y) + \vec{Q}'(x,y) \exp[ikt] \tag{1}$$

Steady Flow Field

For the steady flow of an incompressible, inviscid fluid, a velocity potential function can be defined. The complete flow field is then described by the Laplace equation.

$$\nabla^2\Phi_0(x,y) = 0 \tag{2}$$

where $\vec{Q}_0 (x,y) = \nabla\Phi_0(x,y)$.

Since the Laplace equation is linear, the velocity potential can be decomposed into components by the superposition principle. In particular, the steady potential is decomposed into noncirculatory and circulatory components.

$$\Phi_o(x,y) = \Phi_{NC}(x,y) + \Gamma\Phi_C(x,y) \tag{3}$$

where $\nabla^2\Phi_{NC} = 0$; $\nabla^2\Phi_C = 0$; and Γ is the unknown steady flow circulation constant.

To complete the steady flow mathematical model, farfield inlet, farfield exit, airfoil surface, wake dividing streamline, and cascade periodic boundary conditions must be specified.

The steady farfield inlet flow is uniform, with the mass flow rate specified by the farfield exit boundary conditions. Also, a zero normal velocity is specified on the airfoil surfaces.

The steady velocity potential is discontinuous along the airfoil wake dividing streamline. This discontinuity is satisfied by a continuous noncirculatory velocity potential, with the discontinuity in the circulatory velocity potential equal to the steady circulation, Γ. The Kutta condition is also applied, thereby enabling the steady circulation constant to be determined. It is satisfied by requiring the chordwise velocity components on the upper and lower airfoil surfaces to be equal in magnitude at the airfoil trailing edge. In addition, the cascade periodic steady velocity potential boundary conditions are satisfied by requiring both the normal and chordwise velocity components to be continuous between the upper and lower periodic boundaries.

Unsteady Flow Field

The unsteady potential component Φ' is also described by a Laplace equation. The solution is determined by decomposing this unsteady potential component into circulatory and noncirculatory components $\Phi'_C(x,y)$ and $\Phi'_{NC}(x,y)$ each of which is individually described by a Laplace equation.

$$\Phi' = \Phi'_{NC} + \Gamma'\Phi'_C \tag{4a}$$

$$\nabla^2\Phi'_C = 0; \quad \nabla^2\Phi'_{NC} = 0 \tag{4b}$$

where Γ' is the unsteady flow circulation constant.

The farfield inlet unsteady velocity potential boundary conditions are obtained by using a Fourier series to satisfy the periodicity condition at the far upstream. The farfield exit unsteady noncirculatory potential boundary condition is obtained in an analogous manner. Since the wake does not attenuate in the farfield, the unsteady circulatory potential farfield exit boundary condition is obtained by solving the Laplace equation at the farfield exit and satisfying the blade-to-blade periodicity condition [6]. Also, the airfoil surface boundary conditions specify that the normal velocity of the flow field must be equal to that of the airfoil.

$$\Phi'_{NC}|_{\text{far field inlet}} = -\left|\frac{S}{\beta_o}\right| \frac{\partial \Phi'_{NC}}{\partial n}|_{\text{farfield inlet}} \tag{5a}$$

$$\Phi'_{C}|_{\text{far field inlet}} = -\left|\frac{S}{\beta_o}\right| \frac{\partial \Phi'_{C}}{\partial n}|_{\text{farfield inlet}} \tag{5b}$$

$$\Phi'_{NC}|_{\text{far field exit}} = -\left|\frac{S}{\beta_o}\right| \frac{\partial \Phi'_{NC}}{\partial n}|_{\text{farfield exit}} \tag{5c}$$

$$\Phi'_{C}|_{\text{far field exit}} = -\Delta\Phi' e^{-ikx}\left[\frac{e^{ky}}{1-e^{(k-i\nu)S\cos\delta}} + \frac{e^{-ky}}{1-e^{-(k+i\nu)S\cos\delta}}\right] \tag{5d}$$

where $\nu = \dfrac{\beta_o + k\sin\delta/S}{S\cos\delta}$, $\Delta\Phi'$ is the unsteady velocity potential discontinuity at the farfield exit and β_o is the interblade phase angle.

$$\frac{\partial \Phi'_C}{\partial n}|_{\text{airfoil}} = 0 \tag{6a}$$

$$\frac{\partial \Phi'_{NC}}{\partial n}|_{\text{airfoil}} = W'(x,y) \tag{6b}$$

The normal velocity boundary condition on the oscillating airfoils, W'(x,y), is a function of both the position of the airfoil and the steady flow field. Thus, it is the boundary condition which couples the unsteady flow field to the steady aerodynamics. For an airfoil cascade executing harmonic torsion mode oscillations about an elastic axis location at x_o as measured from the leading edge, the normal velocity on the surfaces of the airfoil is

$$W'(x,y) = \bar{\alpha}\left\{\begin{array}{c}\dfrac{ik\left[(x-x_o) + y\partial f/\partial x\right] + U_o + V_o\partial f/\partial x}{\left[1+(\partial f/\partial x)^2\right]^{1/2}} + \\ \dfrac{\partial U_o/\partial y\left[(x-x_o)\,\partial f/\partial x + y\right]}{\left[1+(\partial f/\partial x)^2\right]^{1/2}} - \dfrac{\partial V_o/\partial y\left[(x-x_o) - y\partial f/\partial x\right]}{\left[1+(\partial f/\partial x)^2\right]^{1/2}}\end{array}\right\} \tag{7}$$

where $U_O = U_O(x,y)$ and $V_O = V_O(x,y)$ are the steady airfoil surface velocity components, f(x) denotes the airfoil profile, and $\overline{\alpha}$ is the amplitude of the torsional oscillations.

The unsteady velocity potential is discontinuous along the airfoil wake dividing streamline. This discontinuity is satisfied with a continuous noncirculatory and a discontinuous circulatory velocity potential. The unsteady circulatory velocity potential discontinuity is specified by requiring the pressure to be continuous across the wake and then utilizing the unsteady Bernoulli equation to relate the unsteady velocity potential and the pressure. The Kutta condition is also applied to the unsteady flow field, enabling the unsteady circulation constant, Γ', to be determined. It is satisfied by requiring no unsteady pressure difference across the airfoil chordline at the trailing edge. In addition, the cascade periodic unsteady potential boundary conditions are satisfied by requiring the normal and chordwise velocity components to be continuous in magnitude and to be discontinuous in phase, with the discontinuity equal to the interblade phase angle, β_O, between the upper and lower periodic boundaries.

The unsteady dependent variable of primary interest is the unsteady pressure P', from which the unsteady aerodynamic moment on the airfoil is calculated. It is determined from the solution for the steady flow field, the unsteady velocity potential, and the unsteady Bernoulli equation. Also, the unsteady airfoil surface boundary conditions were applied on the mean position of the airfoil. After transfer to the instantaneous airfoil position

$$P' = -\nabla\Phi_o \cdot \nabla\Phi' - ik\Phi' + \partial U_o/\partial y\,(-U_o x + V_o y) - \partial V_o/\partial y\,(V_o x + U_o y)$$

where the last two terms account for the transfer of the pressure value from the mean position of the airfoil to its instantaneous position.

The unsteady aerodynamic moment on the reference airfoil is calculated by integrating the unsteady surface pressure difference across the chordline.

$$M_R = \int \left[P'(x-x_o)dx + P'y\,dy\right] = \overline{\alpha}_R C_{\alpha\alpha} \tag{8}$$

where $\overline{\alpha}_R$ is the amplitude of oscillation of the reference airfoil and $C_{\alpha\alpha}$ is the conventional torsion mode unsteady aerodynamic moment coefficient.

Locally Analytical Solution

A boundary fitted computation grid generation technique is utilized for the numerical solution. A Poisson type grid solver is used to fit a C-type grid around a reference airfoil in the cascade. This method permits grid points to be specified along the entire boundary of the computational plane.

Laplace equations describe the complete flow field including the unknown velocity potentials. In the transformed (ξ,η) coordinate system, the Laplace equation takes on the following nonhomogeneous form.

$$\frac{\partial^2\overline{\Phi}}{\partial\xi^2} + \alpha\frac{\partial^2\overline{\Phi}}{\partial\eta^2} - 2\alpha\beta\frac{\partial\overline{\Phi}}{\partial\eta} - 2\gamma\frac{\partial\overline{\Phi}}{\partial\xi} = F(\xi,\eta \tag{9}$$

where the coefficients α, β, and γ are functions of the transformed coordinates ξ and η which are treated as constants in each individual grid element.

To obtain the analytical solution to the transformed Laplace equation, it is first rewritten as a homogeneous equation by defining a new dependent variable.

$$\frac{\partial^2\widehat{\phi}}{\partial\xi^2} + \alpha\frac{\partial^2\widehat{\phi}}{\partial\eta^2} - (\gamma^2 + \alpha\beta^2)\widehat{\phi} = 0 \tag{10}$$

where $\overline{\Phi} = \widehat{\phi}\ \exp\{\gamma\xi + \beta\eta\} - \dfrac{F(\gamma\xi + \beta\eta)}{2(\gamma^2 + \alpha\beta^2)}$

The following general solution is determined by separation of variables.

$$\widehat{\phi}(\xi,\eta) = [A_1\cos(\lambda\xi) + A_2\sin(\lambda\xi)][B_1\cos(\mu\eta) + B_2\sin(\mu\eta)] \tag{11}$$

where $\mu = [(\gamma^2 + \alpha\beta^2 + \lambda^2)/\alpha]^{1/2}$ and λ_1, A_1, A_2, B_1, and B_2 are constants to be determined from the boundary conditions.

Analytical solutions in individual computation grid elements are determined by applying proper boundary conditions to evaluate the unknown constants in this general velocity potential solution . The solution to the global problem is then determined through the applications of the global boundary conditions and the assembly of the locally analytical solutions.

Aerodynamic Model-Detuned Cascade

For the aerodynamically detuned rotor configuration of interest herein, i.e., variable circumferential spacing and chord length, an analogous cascade unsteady aerodynamic model is developed by considering two passages with two passage periodic boundary conditions.

In this model, the rotor will incorporate splitters (short chord airfoils) between each pair of full chord airfoils. As schematically depicted in Figure 2, the splitters are not required to have the same airfoil shape as the full chord airfoils, nor are they restricted to particular circumferential or axial positions between each pair of the full chord airfoils.

There are two distinct passages: (1) a reduced spacing, or increased solidity, passage; and (2) an increased spacing, or reduced solidity, passage There are also two distinct sets of airfoils, with the two reference airfoils denoted by R_0 and R_1. These individual sets of airfoils can be considered as cascades of uniformly spaced airfoils each with twice the spacing of the associated baseline aerodynamically tuned uniformly spaced cascade. The circumferential spacing between these two sets of airfoils, S_1 and S_2, is determined by specifying the level of aerodynamic detuning, ε.

$$S_{2,1} = (1 \pm \varepsilon)\, S \tag{12}$$

where S is the spacing of the baseline uniformly spaced cascade, and S_1 and S_2 denote the spacings of the detuned cascade.

An interblade phase angle for this aerodynamically detuned cascade configuration can be defined. In particular, each set of airfoils is individually assumed to be executing harmonic torsional oscillations with a constant aerodynamically detuned interblade phase angle, β_d, between adjacent airfoils of each set, Figure 2. Thus, this detuned cascade interblade phase angle is two times that for the corresponding baseline tuned cascade.

$$\beta_d = 2\beta_o \tag{13}$$

where β_o is the tuned baseline cascade interblade phase angle, defined between adjacent airfoils on the rotor.

For a rotor with N uniformly spaced blades, the values of β_o must satisfy the following condition.

$$\beta_o = \frac{2\pi r}{N}, \qquad r = 0, \pm 1, \pm 2, \ldots, \pm N - 1 \tag{14}$$

where ± refers to forward and backward traveling waves, respectively.

Steady Aerodynamics

The steady aerodynamic model developed for the baseline uniformly spaced cascade can be applied directly to the two reference passage model of the detuned cascade with the addition that the steady potential for the detuned cascade, Φ_{d0}

is decomposed into one circulatory $\Phi_{dNC}(x,y)$ and two circulatory components, one associated with each of the two reference airfoils, $\Phi_{CR_0}(x,y)$ and $\Phi_{CR_1}(x,y)$.

$$\Phi_{d0}(x,y) = \Phi_{dNC}(x,y) + \Gamma_{R_0}\Phi_{CR_0}(x,y) + \Gamma_{R_1}\Phi_{CR_1}(x,y) \tag{15}$$

where $\nabla^2\Phi_{dNC} = 0$; $\nabla^2\Phi_{CR_0} = 0$; $\nabla^2\Phi_{CR_1} = 0$; and Γ_{R_0} and Γ_{R_1} are the unknown steady flow circulation constants for the reference airfoils R_0 and R_1, respectively.

The circulatory component $\Phi_{CR_0}(x,y)$ is discontinuous along the wake dividing streamline of the reference airfoil R_0 and is continuous along the wake dividing streamline of the reference airfoil R_1, while $\Phi_{CR_0}(x,y)$ is discontinuous along the wake of R_1 and continuous along the wake of R_0. The steady potential discontinuity is satisfied by a continuous noncirculatory velocity potential, with the discontinuities in the circulatory components Φ_{CR_0} and Φ_{CR_1} equal to the steady circulation, Γ_{R_0} and Γ_{R_1}, respectively. Also, the steady circulation constants Γ_{R_0} and Γ_{R_1} are determined by simultaneously applying the Kutta condition to the two reference airfoils.

Unsteady Aerodynamics

The unsteady potential for the detuned cascade Φ'_d is also decomposed into one noncirculatory $\Phi'_{dNC}(x,y)$ and two circulatory components, one associated with each of the two reference airfoils.

$$\Phi'_d(x,y) = \Phi'_{dNC}(x,y) + \Gamma'_{R_0}\Phi'_{CR_0}(x,y) + \Gamma'_{R_1}\Phi'_{CR_1}(x,y) \tag{16}$$

where $\Phi'_{CR_0}(x,y)$, $\Phi'_{CR_1}(x,y)$, Γ'_{R_0}, and Γ'_{R_1} are defined analogous to the corresponding detuned cascade steady aerodynamic quantities.

Influence Coefficient Technique

The unsteady airfoil surface boundary conditions require that the cascaded airfoils oscillate with equal amplitudes. Also, the interblade phase angle between adjacent nonuniformly spaced airfoils must be specified. Neither of these requirements is appropriate for the aerodynamically detuned cascade. To overcome these limitations, an unsteady aerodynamic influence coefficient technique is utilized.

The unsteady aerodynamic moment acting on the two reference airfoils is expressed in terms of influence coefficients.

$$M_{R0,R1} = \left(\hat{\alpha}_{R0}[CM]^0_{R0,R1} + \hat{\alpha}_{R1}[CM]^1_{R0,R1}\right)e^{i\omega t} \tag{17}$$

where $\hat{\alpha}_{R0}$ and $\hat{\alpha}_{R1}$ are the unknown complex oscillatory displacements for the reference airfoils R_0 and R_1, respectively.

The unsteady aerodynamic moment influence coefficients $[CM]^0_{R0,R1}$ and $[CM]^1_{R0,R1}$ are determined by analyzing the two reference passages with the unsteady detuned cascade model developed above. $[CM]^0_{R0,R1}$ are determined by considering a unit amplitude motion of only the reference airfoil R_0, with the interblade phase angle β_d, as shown in Figure 2. The influence coefficients $[CM]^1_{R0,R1}$ are obtained in an analogous manner but with a unit amplitude motion of only the reference airfoil R_1.

AEROELASTIC MODEL

The equations describing the single-degree-of-freedom torsional motion of the two reference airfoils of the aerodynamically detuned cascade are developed by considering the typical airfoils. The elastic restoring forces are modeled by linear torsional springs at the elastic axis location, with the inertial properties of the airfoils represented by their mass moments of inertia about the elastic axis. The equations of motion, determined by Lagrange's technique, are

$$I_{\alpha R_0}\ddot{\alpha}_{R_0} + (1+2ig_{\alpha R_0})I_{\alpha R_0}\,\omega^2_{\alpha R_0}\,\alpha_{R_0} = M_{R_0}$$

$$I_{\alpha R_1}\ddot{\alpha}_{R_1} + (1+2ig_{\alpha R_1})I_{\alpha R_1}\,\omega^2_{\alpha R_1}\,\alpha_{R_1} = M_{R_1} \tag{18}$$

where $I_{\alpha R0,R1}$ are the mass moments of inertia about the elastic axis, $g_{\alpha R0}$ and $g_{\alpha R1}$ denote the structural damping coefficients for the reference airfoils, and the undamped natural frequencies are $\omega^2_{\alpha R0} = K_{\alpha R0}/I_{\alpha R0}$ and $\omega^2_{\alpha R1} = K_{\alpha R1}/I_{\alpha R1}$.

Considering harmonic time dependence of the reference airfoils and utilizing the total unsteady aerodynamic moments defined in Equation 18, the equations of motion are written in matrix form.

$$\begin{bmatrix} \mu_0 & [CM]^1_{R0} \\ [CM]^0_{R1} & \mu_1 \end{bmatrix} \begin{bmatrix} \bar{\alpha}_{R0} \\ \bar{\alpha}_{R1} \end{bmatrix} = 0 \tag{19}$$

where $\mu_0 = \mu_{R0} r^2_{\alpha R0} + [CM]^0_{R0} - (1+2ig_{\alpha R0})\, \mu_{R0}\, r^2_{\alpha R0}\, \gamma^2_{\alpha R0}\, \gamma$

$$\mu_1 = \mu_{R1} r^2_{\alpha R1} + [CM]^1_{R1} - (1+2ig_{\alpha R1})\, \mu_{R1}\, r^2_{\alpha R1}\, \gamma^2_{\alpha R1}\, \gamma$$

$$\mu_{R0,\,R1} = \frac{m_{R0,\,R1}}{\pi\rho b^2}; \qquad r^2_{\alpha R0,\,R1} = \frac{I_{\alpha R0,\,R1}}{m_{R0,\,R1}\, b^2}$$

$$\gamma^2_{\alpha R0,\,R1} = \frac{\omega_{\alpha R0,\,R1}}{\omega_o}; \qquad \gamma = \frac{\omega_o^2}{\omega^2}$$

and ω_0 = reference frequency.

Equation 19 represents a complex eigenvalue problem, with the eigenvalues being γ, the ratio of the reference frequency to the flutter frequency. The flutter stability of both conventional uniformly spaced full chord airfoil cascades and cascades incorporating splitters are specified by the real part of the eigenvalue, μ. When μ is negative, the amplitude of the harmonic motion of the airfoils will decay and the cascade is stable.

$$\frac{i\,\omega}{\omega_o} = \frac{i}{\sqrt{\gamma}} = \mu + i\nu = \text{Eigenvalue} \tag{20}$$

The required motion dependent unsteady aerodynamic loading is obtained by specifying the detuned interblade phase angle, β_d, with each permissible interblade phase angle utilized to find the most unstable mode. The permissible values for β_d being obtained from Equations 13 and 14 by setting N equal to the number of blades on the baseline uniformly spaced tuned rotor being analyzed. Since Equation 19 represents two equations, there will be two eigenvalues obtained for each value of β_d. These complex eigenvalues are related to the exponent ($i\omega$) which specifies the harmonic motion of the airfoils.

The effects of structural detuning are included through the frequency terms $\gamma_{\alpha R_o,\,R_1}$. These terms represent the ratios of the natural frequency to a specified reference frequency. Defining the reference frequency ω_0 as the torsion mode natural frequency, a structurally tuned cascade would have $\gamma_{\alpha R_o,\,R_1}$ equal to unity. For the case when the cascade is structurally detuned, the values of $\gamma_{\alpha R_o,\,R_1}$ are altered.

In this regard, splitters enable combined aerodynamic and structural detuning to be introduced into a rotor. In particular, the natural torsion mode frequency of an airfoil is a function of its chord length , thickness, and span as well as its material properties. As a result, the splitters may have a higher natural frequency than the full chord airfoils.

The vibrational characteristics of the full chord and splitter airfoils are determined by modeling each airfoil as a thin rectangular cross-section cantilevered slender beam. The torsion mode natural frequency, ω_α, is then specified.

$$\omega_\alpha = \sqrt{\frac{G}{2\pi^2\rho_m}\frac{T}{CL}} \tag{21}$$

where G is the material modulus of rigidity, L is the airfoil span, T is the airfoil thickness, C is the chord length and ρ_m is the material density.

Two types of splitters are of interest: (1) splitters with the same thickness-to-chord ratio as the full chord airfoils and (2) splitters with the same thickness as the full chord airfoils.

For the first type, the splitters and full chord airfoils have the same natural frequency when their material properties and spans are the same, with the splitters having a higher natural frequency when the material properties are different or the splitter span is smaller

than that of the full chord airfoils. Hence, these splitters incorporate either aerodynamic detuning or combined aerodynamic-structural detuning into a rotor design.

For the second case, the splitters have a higher natural frequency than the full chord airfoils as long as the two sets of airfoils have the same material properties and spans, with the possibility of the splitters and full chord airfoils having the same natural frequency if different materials are utilized or the splitters have increased span. Thus, these splitters can also introduce either aerodynamic detuning or combined aerodynamic-structural detuning into the rotor.

RESULTS

To take advantage of the enhanced flutter stability associated with combined aerodynamic-structural detuning and aerodynamic detuning alone while eliminating the difficulties associated with structural detuning, splitter blades are introduced into an unstable rotor design. In particular, alternate blades in a baseline unstable rotor are replaced by splitters, resulting in a twelve bladed rotor with six splitters and six full chord airfoils.

Baseline Rotor

The baseline rotor is uniformly spaced, characterized by a Gostelow cascade geometry [15] with a solidity of 1.67, a stagger angle of 40°, a mean flow incidence angle of 24°, and an airfoil mass ratio μ and radius of gyration r_α of 193.776 and 0.3957, respectively, values typical of the tip region of modern fan blades

Splittered-Rotor

Aerodynamic detuning is demonstrated by introducing splitters with the same thickness-to-chord ratio as the full chord airfoils, and thus, the same natural frequencies. The aerodynamic detuning results from both the splittered decreased chord and their circumferential position between adjacent full chord airfoils. Three circumferential splitter locations are considered: 40%, 50% and 60%.

Combined aerodynamic-structural detuning is accomplished with splitters having the same thickness but a different chord length than the full chord airfoils positioned such that the splitter trailing edge is aligned with that of the full chord airfoils.

Steady Aerodynamic Performance

Steady rotor performance is defined by the chordwise distributions of the airfoil surface steady static pressure coefficient. For the baseline and splittered-rotors, these are presented in Figure 3. As seen, the introduction of the splitters into the rotor has a noticeable effect on the steady loading distribution of the full chord airfoils, with this effect being a function of the splitter circumferential position.

Baseline Rotor Stability

For a midchord elastic axis location and a reduced frequency of 0.8, the torsion stability of the baseline rotor is shown in Figure 4. The 12 bladed baseline rotor is unstable in traveling wave modes characterized by an interblade phase angle of -30°.

Splittered-Rotor Stability

Aerodynamically detuning the unstable baseline rotor by the splitters affects stability, with the splittered-rotor stability a function of the splitter circumferential position, i.e., the aerodynamic detuning is a strong function of the splitter circumferential position, Figures 5 through 7. Positioning the splitters at 40% of the full chord airfoil passage spacing results in a stable splittered-rotor configuration. The splittered-rotor is neutrally stable when the splitters are at midpassage, becoming even more unstable than the baseline rotor with the splitters at 60% passage spacing. Also, the aerodynamic detuning results in a separation of the splitter and full chord airfoils, with the splitters always more stable.

Combined aerodynamic-structural detuning due to the splitters has a significant effect on rotor stability. In particular, the higher natural frequency of the splitters results in increased torsion mode stability for all of the detuned splittered-rotors as well as increased splitter and full chord airfoil separation, with the splitter circumferential position only affecting the stability margin. This stability enhancement is shown in Figure 8 for a midpassage circumferential splitter location.

SUMMARY AND CONCLUSIONS

A mathematical model has been developed and utilized to predict the effect of incorporating splitter blades on the torsion mode stability of a rotor operating in an incompressible flow field. The splitter blades, positioned circumferentially in the flow passage between two principal blades, introduce aerodynamic and/or combined aerodynamic-structural detuning into the rotor. The two-dimensional oscillating cascade unsteady aerodynamics, including steady loading effects, were determined by developing a complete first-order unsteady aerodynamic analysis together with an unsteady aerodynamic influence coefficient technique. The torsion mode flutter of both uniformly spaced tuned rotors and detuned rotors were then predicted by incorporating the unsteady aerodynamic influence coefficients into a single-degree-of-freedom aeroelastic model.

The viability of splitters as a passive torsion mode flutter control technique for an aerodynamically loaded rotor operating in an incompressible flow field was then considered, accomplished by applying this model to an unstable baseline twelve bladed rotors. This study demonstrated that aerodynamic detuning and combined aerodynamic-structural detuning due to incorporating splitters into unstable baseline rotors resulted in stable splittered-rotor configurations.

For splitters with only aerodynamic detuning, the aerodynamic damping and, thus, the splittered-rotor stability was found to be a function of the splitter circumferential position, being stable for splitters at 40% of the passage but even more unstable than the baseline rotor with splitters at 60% spacing. Combined aerodynamic-structural detuning due to the splitters had a significant effect on rotor stability, resulting in stability splittered-rotors for all splitter circumferential positions.

Thus, the improved aerodynamic performance potential of a rotor incorporating splitter blades also offers the potential of increased stability.

ACKNOWLEDGEMENTS

This research was sponsored, in part, by the NASA Lewis Research Center. Both the financial and technical support are most gratefully acknowledged.

REFERENCES

1. Chiang, H. and Fleeter, S., "Oscillating Aerodynamics and Flutter of an Aerodynamically Detuned Cascade In an Incompressible Flow, *AIAA Paper 89-0289,* January 1989.

2. Hoyniak, D. and Fleeter, S., "Aerodynamic Detuning Analysis of an Unstalled Supersonic Turbofan Cascade," *ASME Journal of Engineering for Gas Turbines and Power,* Vol. 108, No. 1, January 1986, pp. 60-67.

3. Fleeter, S. and Hoyniak, D., "Chordwise Spacing Aerodynamic Detuning for Unstalled Supersonic Flutter Stability Enhancement," *Journal of Sound and Vibration,* Vol. 115, No. 3, 1987, pp. 483-497.

4. Topp, D.A. and Flutter, S., "Splitter Blades as an Aeroelastic Detuning Mechanism for Unstalled Supersonic Flutter of Turbomachine Rotors," *ASME Journal of Turbomachinery,* Vol. 108, No. 2, October 1986, pp. 244-252.

5. Fleeter, S., Hoyniak, D., and Topp, D.A., "Control of Rotor Aerodynamically Forced Vibrations by Splitters," *AIAA Journal of Propulsion and Power,* September-October, 1988, pp. 445-451.

6. Verdon, J.M., "The Unsteady Flow in the Far Field of an Isolated Blade Row," *United Technologies Research Center Report R87-957333301*, April 1987.

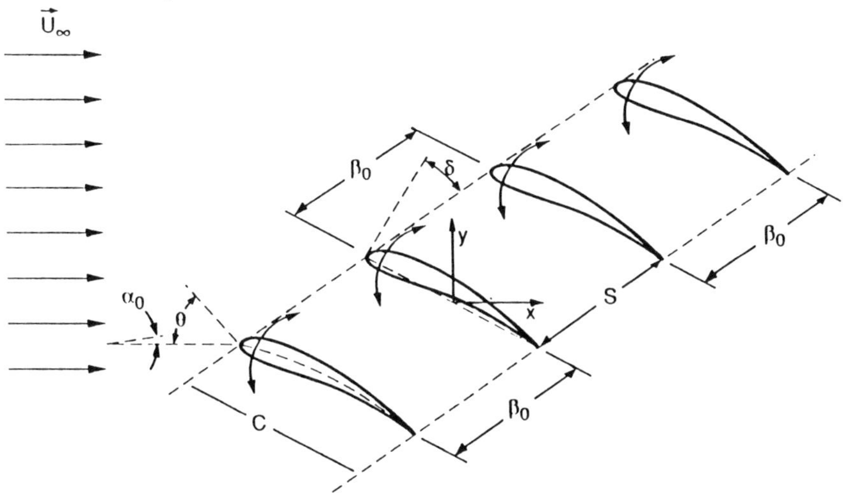

Figure 1. Cascade and flow geometry

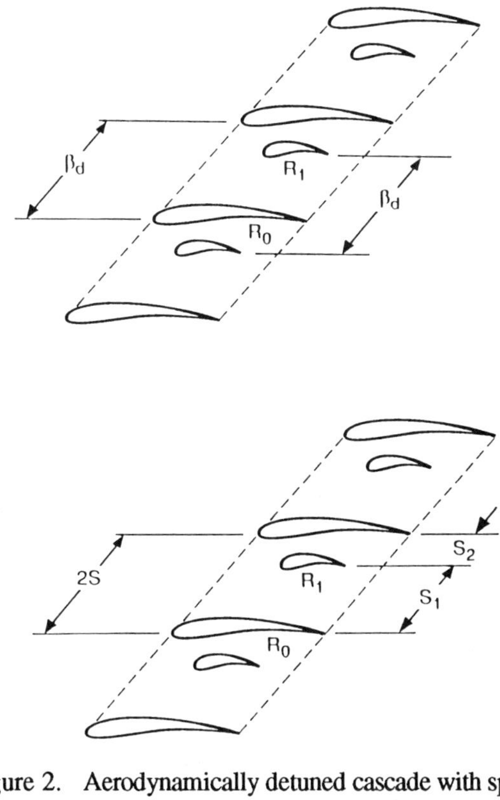

Figure 2. Aerodynamically detuned cascade with splitters

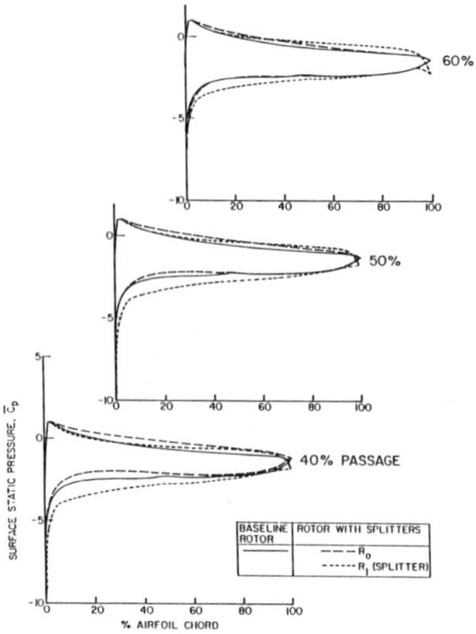

Figure 3. Steady aerodynamic performance

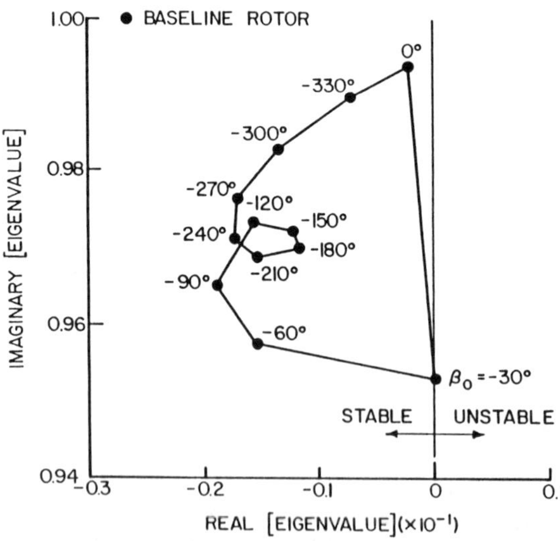

Figure 4. Baseline rotor torsional stability

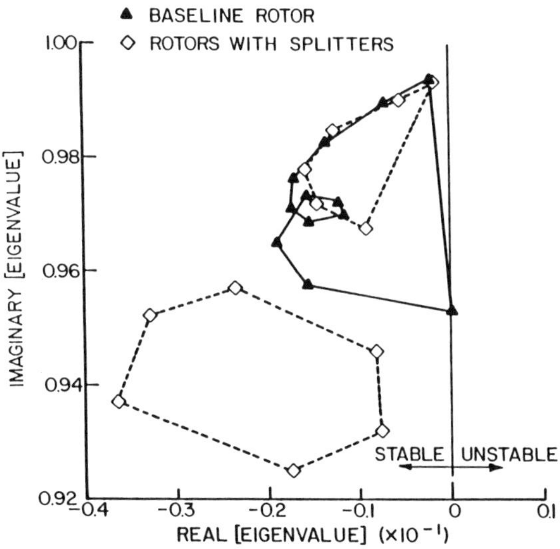

Figure 5. Stability of aerodynamically detuned splittered-rotor splitters at 40%passage

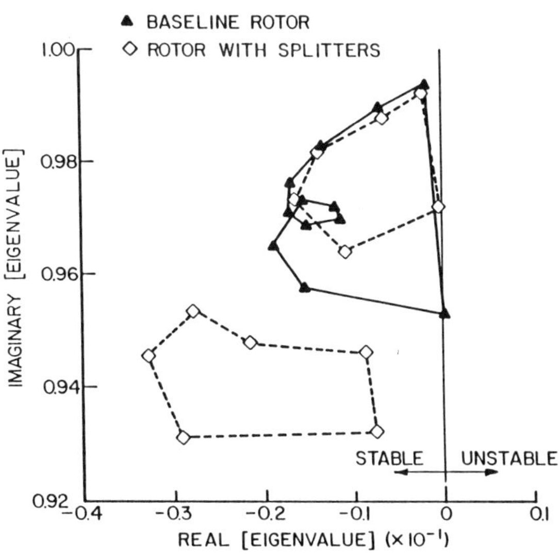

Figure 6. Stability of aerodynamically detuned splittered-rotor splitters at 50% passage

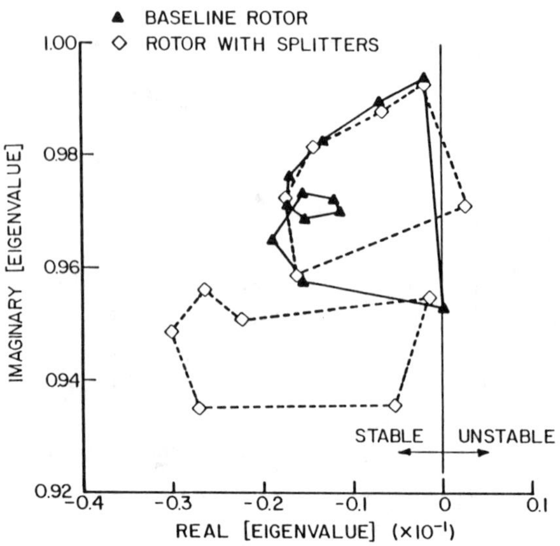

Figure 7. Stability of aerodynamically detuned splittered-rotor splitters at 60% passage

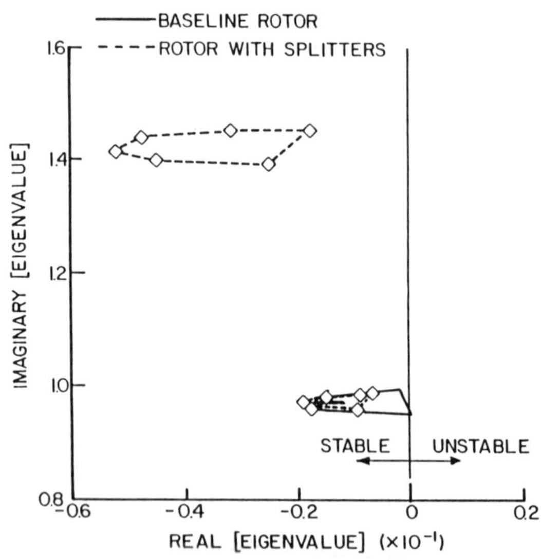

Figure 8. Stability of aerodynamic-structurally detuned splittered-rotor splitters at 50% passage

A Non-synchronous Vibration of Moving Blades Coupled with the Bleed Chamber Resonance in an Axial Compressor

N. Yamaguchi[1]
T. Sato[1]
S. Umemura[1]
T. Ohwaki[1]

ABSTRACT In a low-speed model test of an axial compressor for a gas turbine, a new type of non-synchronous vibration of moving blades, which resonate at apparently non-integral orders of the machine rpm, was discovered and its cause was identified. The moving blades, which were located immediately upstream of a bleed slit and chamber, were excited by a flow oscillation through the slit due to the resonance or near-resonance of the air column within the bleed chamber, which, in turn, was excited by the rotor revolution. The resonance condition proved to be the following simple formula:

$$mf_R = |f_B \pm f_S|$$

where f_R: rotor revolution frequency, f_B: blade first bending frequency, f_S: pressure fluctuation frequency of the bleed chamber resonance, and m: arbitrary integer.

1.1 Introduction

A large amount of interstage bleed air is indispensable for high-pressure-ratio axial compressors for gas turbines both for prevention of rotating stalls during the start-up and for providing cooling air for turbine blades and disks. It has been known since early times that the existence of bleed

[1]Takasago R&D Center, Mitsubishi Heavy Industries, Ltd., 2-1-1, Shinhama, Araicho, Takasago City, Japan.

ports and chambers for the purpose tends to excite blade vibrations and various countermeasures have been devised.

We happened to have observed a new kind of blade resonant vibration during a model test for development of a large-capacity axial compressor. We encountered the phenomenon during a low-speed model test for an investigation about the flow field around a bleed slit for bleeding a large quantity of air. The moving blades, which located just upstream of the bleed slit, showed severe resonant vibrations. The resonances appeared non-synchronously, i. e. at rotor r. p. m. s violating the condition for the blade resonance with integral orders of the rotor revolutions. The resonance depended much on the bleed system configuration and the quantity of bleed air.

Such kind of resonant vibrations have never been encountered in the past, and the present case is the first one as far as our experience is concerned. The conditions for the possible occurrence have not yet been fully understood.

The following content presents the experiment and analysis to explain the phenomena and the cause.

1.2 The model compressor

The test was originally intended for one of developmental tests for an axial compressor for a large capacity gas turbine. The compressor was designed to bleed a large amount of air for cooling as much as 20% of the main flow at an interstage location. So, just downstream of a moving-blade row, the outer casing wall was reduced in radius and the annulus area was correspondingly reduced. Bleed air was discharged into the bleed chamber through the circumferential slit, which was provided along the annulus reduction.

Flow field in such circumstances was being investigated on a low-speed single-stage compressor model provided with a bleed system similar to the designed one. During the time, the phenomena which will be described below appeared.

Table1.1 gives main figures for the experimental apparatus.

Figure1.1 gives a sectional view of the set-up. The tip diameter is 540mm, its hub-to-tip ratio is 0.626, and the maximum speed is 6000rpm. The flow and the static pressure of the compressor is shown in Fig.1.2, together with its operating point for the test.

The bleed system is shown in Fig.1.1. The bleed air leaves the tip of the moving blade, passes through a circumferential radial slit into an annular bleed chamber, then into four bleed pipes (flexible hoses here) arranged on the chamber external wall, and is finally sucked by an auxiliary fan. The radial slit has swirl-removal vanes arranged around the circumference. Figure1.3(a) and (b) show the situations.

TABLE 1.1. Model Compressor

Tip diameter	540mm
Hub-to-tip ratio	0.626
Stage and Arrangement	1, IGV + R + OGV
Moving blades	20 blades, 65mm chord length (at mean height)
Inlet guide vanes	32 vanes, chord length 55mm
Outlet guide vanes	24 vanes, chord length 79.4mm
Rotor speed	max. 6000rpm
Driving system	370kW electric motor + fluid coupling step-up gear

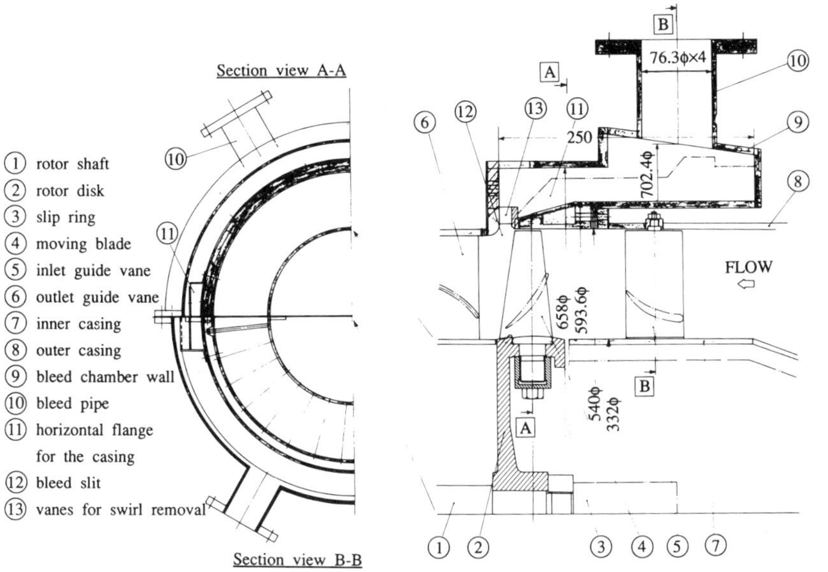

FIGURE 1.1. Sectional view of the model compressor

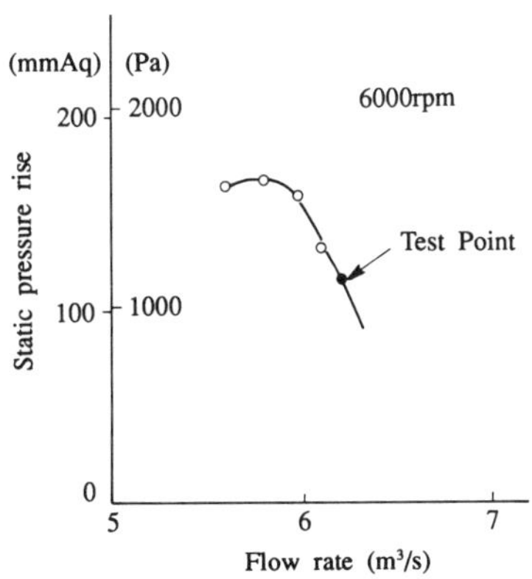

FIGURE 1.2. Overall performance of the model compressor

FIGURE 1.3. Model compressor (a) External view: Air comes from the suction ducting seen left hand side and is dumped into the discharge box seen behind. The flexible hoses are for the air bleeding. (b) The upper casing is removed (view from the suction side).

The moving blades, made of aluminum alloy, are bolted on to the rotor disk. Some of the blades were strain-gaged on the surface near the root. The output signals of the strain gauges were transmitted to outside through a slip ring. The blade natural frequency was 650 to 680Hz. The rotor r. p. m. was swept from 3500rpm to 6000rpm, whose integral orders of from 7 to 11 were to excite the blade natural vibration.

Miniature pressure transducfers (Kulite XCQ-093-50G) were mounted flush on the blade surfaces, and the signals were transmitted to outside also through the slip ring.

Pressure fluctuation within the bleed chamber was monitored with transducers mounted on the chamber wall.

1.3 Blade responses

The speed of the model compressor was swept from 3500rpm to 6000rpm slowly at a constant rate. During the time, the signals of the blade strains and pressures and the suction chamber pressures were recorded with a data recorder. For the respective test run the bleed air quantity was set at a constant one corresponding to a specified value at 6000rpm, which will be referred to as nominal bleed ratio in this paper.

Therefore, the bleed ratio, which is defined as the bleed quantity against the compressor main flow rate upstream, changes with the compressor speed. The following relation holds approximately, if n is the r. p. m. in

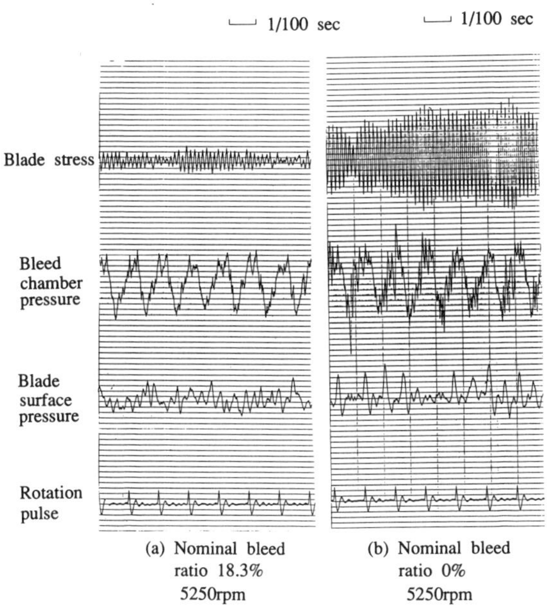

FIGURE 1.4. Comparison of the oscillograms for different bleed conditions

question:

$$\text{the actual bleed ratio} = (\text{the nominal bleed ratio}) \times \left(\frac{6000}{n}\right) \qquad (1.1)$$

The nominal bleed ratios were set at 0 (no bleed), approximately 9% and 18%, respectively. Hereafter "nominal bleed ratio" is abbreviated simply "bleed ratio."

Flow tests were conducted on several types of bleed configurations, in some of which very large responses of the blades occurred and the response curves were very different from the normally expected ones. For a typical example of the configurations shown in Fig.1.1, detailed measurements were conducted and the results are given below. Examples of the oscillograms for the measurements at 5250rpm are compared between the bleed ratios 0% and 18.3% in Fig.1.4. The differences between these are striking.

From the measured results, average maximum amplitudes (half amplitude, O-P) were read for each of the blade stress, the blade surface pressure and the chamber pressure, and are plotted in Fig.1.5 for the three bleed ratios.

For bleed ratio 18.3%, it is seen that the blade stress curve shows ordinary resonances with integral orders of the rotor speed, as expected, and its peak stress level is relatively low. On the other hand, however, for bleed ratios 9.1% and 0%, the situation is completely different. The amplitudes

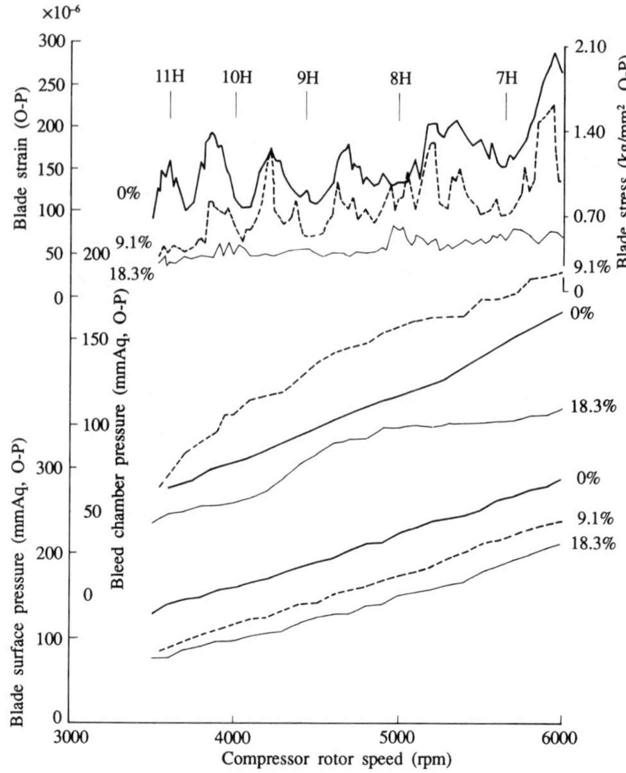

FIGURE 1.5. Oscillation amplitudes vs. rotor speed for various bleed conditions

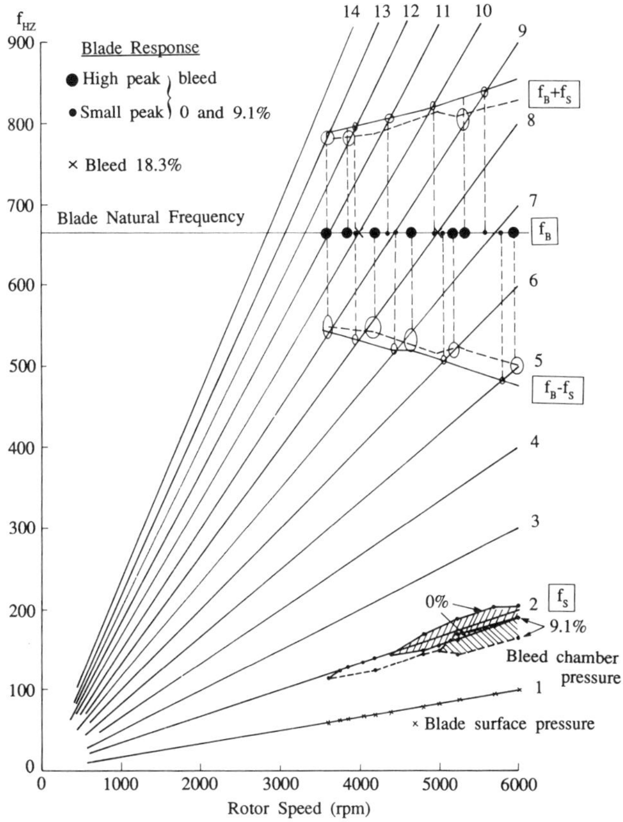

FIGURE 1.6. Influences of rotor speeds on the blade responses

are very large, and unexpectedly many peaks are seen to stick out in apparently random-fashion. In either condition, pressure fluctuations on the blade surface and the bleed chamber grow monotonically with the rotor speed. The smaller the bleed ratio, the larger the fluctuation level tends to be.

The stress peaks are plotted on the line for the blade natural frequency in the Campbell diagram, Fig.1.6. In the top part of the figure, symbol X means peak stress frequencies for bleed ratio 18.3%, which are seen to coincide with the integral orders of the rotor r. p. m. .

On the other hand, symbol • means peak stress frequencies of the blade for bleed ratios 9.1% and 0% in Fig.1.5. Large black circle means marked peaks commonly seen for 9.1% and 0%, while small black circle gives rather low peaks. Surprisingly, these black circles do not coincide with integral orders of the rotor r. p. m. .

The cause for the non-synchronous blade vibrations is the question.

1.4 Spectra of the oscillations

For further study on the mechanism of the excitation, spectra of the oscillations of the blade stress, the chamber pressure and the blade surface pressure were analyzed. The results are shown in Fig.1.7 (a), (b) and (c).

Figure1.7 (a) gives the situations for bleed ratio 18.3%. The spectra for the three quantities are given for several rotor speeds. The spectra of the chamber pressure do not have any predominant frequency but low-frequency turbulence. It corresponds well to the lower level of the blade vibration. Close inspection of the chamber pressure spectra shows a small peak at a frequency corresponding to 20 per-rev, which is corresponding to the blade passing frequency nZ.

Figure1.7 (c) shows the spectra for no bleed. The spectra of the bleed chamber pressure gives the following observation. Peaks are observed at frequencies corresponding to approximately 2 and 18 per-rev. The 2 per-rev frequency did not appear in Fig.1.7 (a). The 18 per-rev frequency differs from the 20 per-rev frequency in Fig.1.7 (a). Furthermore, a low frequency of roughly 10Hz is seen. The spectra of the blade surface pressure show 1 per-rev and its higher harmonics.

Figure1.7 (b) gives data for bleed ratio 9.1%. The chamber pressure also has approximately 2 and 18 per-rev frequencies, the latter being weaker than that for 0% bleed. Besides those, a third of the rotor r. p. m. is seen.

As can be seen from these spectra, when the non-synchronous blade vibration occurs, the chamber pressures have frequency components of approximately 2 per-rev and also 18 per-rev.

1.5 The chamber pressure fluctuations and the blade resonance

The peak frequencies obtained in the previous sections are inspected in the Campbell diagram, Fig.1.6. The predominant frequencies of the blade surface pressure are plotted in Fig.1.6 to give a straight line corresponding to 1 per-rev. This is a natural phenomenon.

The predominant frequencies of the bleed chamber pressure (f_S) for bleed ratios 0% and 9.1% tend to gather near the 2 per-rev line. The spectra have either two low peaks or some lump, which either shift or spread on both sides of the 2 per-rev line, depending on the rotor r. p. m. . Inspection of the spectra in Fig.1.7 (b) and (c) suggests that, of the two low peaks, the higher frequency one is caused by the rotor rotation and the lower frequency one may be close to the natural frequency of the chamber, being affected by the rotor revolution. These f_S's differ somewhat between bleed ratios 0% and 9.1%.

Now, let the chamber pressure frequency f_S be added to or extracted

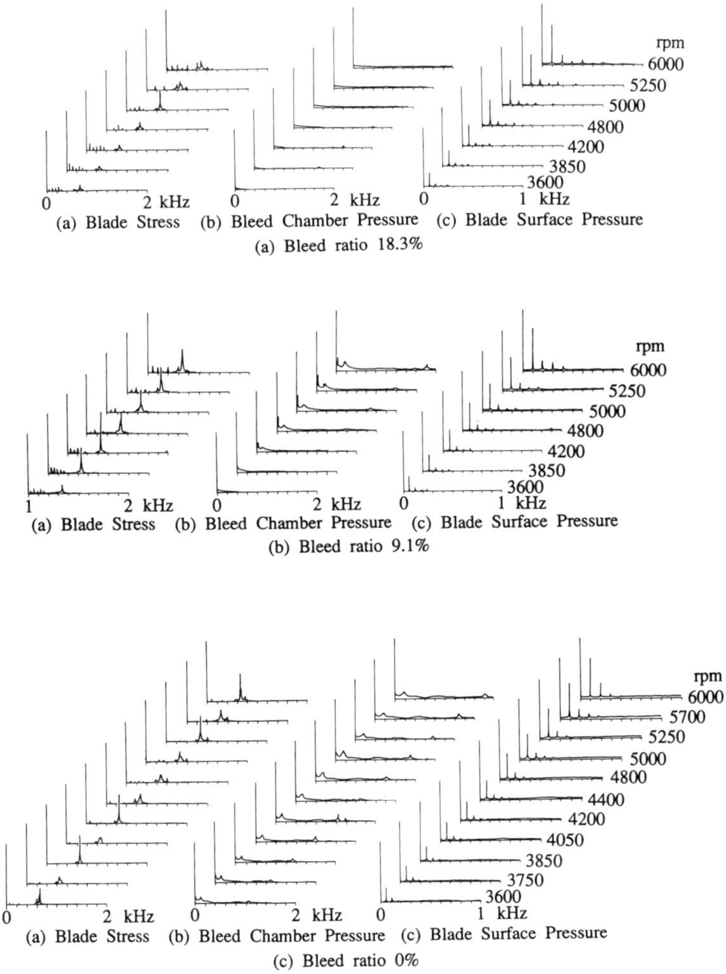

FIGURE 1.7. Spectral behavior of the oscillations for the three bleed conditions

from the blade natural frequency f_B;

$$f_o = f_B \pm f_S \tag{1.2}$$

f_o's are plotted both above and below the line for the blade natural frequency in Fig.1.6. As f_S, the lower frequency of the chamber pressure above described is adopted in eq.(1.2), for it may be closer to the natural frequency of the chamber air mass. The dotted line is for bleed ratio 9.1% and the solid line for 0%.

Then, at r. p. m. s where either of the two lines of f_o crosses or comes very near the lines for integral orders of the rotor r. p. s. , the resonances (the black circles) in question are found to occur.

Thus it may be said that the non-synchronous blade vibrations occur at the rotor r. p. m. s where f_o is coincident with integral orders of the rotor r. p. s. . The frequency f_S is the predominant frequency of the bleed chamber pressure, which seems to be relatively weak and random fashioned resonance of the air column, judging from the observed comparatively blunt peaks. The near-resonance frequency of the air column is considered to be near the 2 per-rev of the rotor r. p. s. within the speed range of the experiment, and be excited by the rotor revolution. The air movement is supposed to affect and excite the moving blades.

A simple calculation regarding the bleed chamber as a Helmholtz resonator estimates the natural frequency (f_S) to be roughly 140Hz. The frequency range which may easily be excited by the rotor revolution is the 2 per-rev in this test. Thus the pressure fluctuation appears to be affected by the 2 per-rev excitation. The observed condition proves it not to be a pure resonance, from the wide spread manner of the spectra. The natural frequency is considered to be nearer the lower-frequency one, which has been used for plotting $f_B \pm f_S$ in Fig.1.6. Non-stationary and random-fashioned resonance of the air mass may be excited at an open end at the bleed slit. Thus, air movement out of and into the slit by the resonance gives rise to distortion in the main flow, causing a non-stationary distorted flow field around the circumference, thus exciting the moving blades passing the field.

A sufficient amount of bleed air eliminates the near-resonance, and the blades cease to be excited.

1.6 Moving blade excitation in a fluctuating flow field

Let us examine what frequency excitation does affect the moving blades, which are passing through the field influenced by a periodic flow distortion having a frequency f_S.

The basic equations are as follows. Symbols are:

t : time (s),

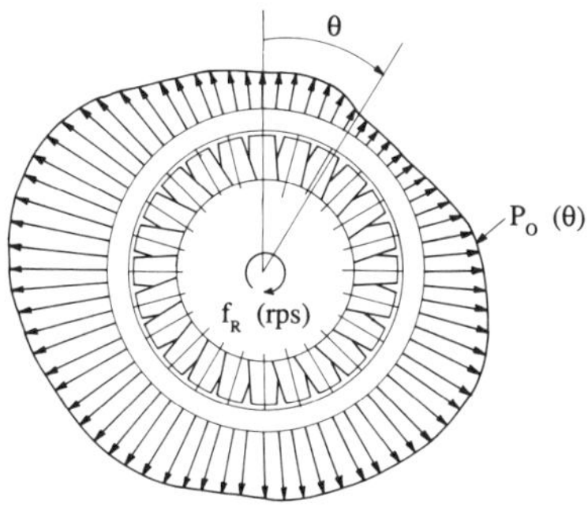

FIGURE 1.8. Flow fluid distortion around the compressor circumference

f_S : frequency of the flow field, in the static frame of reference (Hz),

f_R : rotor revolutions per second (Hz),

θ : angular coordinate (radian), and

$P(\theta, t)$: distortion flow field with respect to θ and t.

The flow field is assumed to vary in the following manner (Fig.1.8):

$$P(\theta, t) = P_O(\theta) \sin 2\pi f_S t \tag{1.3}$$

The distortion amplitude $P_O(\theta)$ is assumed to be able to be expanded into a Fourier series with respect to θ:

$$P_O(\theta) = \sum_{m=0}^{\infty} a_m \sin(m\theta + \psi_m) \tag{1.4}$$

Here,

m : the order of the series,

a_m : the content of the m-th component, and

ψ_m : the phase angle of the m-th component.

From eqs.(1.3) and (1.4)

$$P(\theta, t) = \sum_{m=0}^{\infty} a_m \sin(m\theta + \psi_m) \sin 2\pi f_S t \tag{1.5}$$

A moving blade is passing through the field with the following relation:

$$\theta = 2\pi f_R t \tag{1.6}$$

Substitution of eq.(1.6) into eq.(1.5) gives the following non-stationary flow field experienced by the moving blade:

$$P_S(t) = \sum_{m=0}^{\infty} a_m \sin(2\pi m f_R t + \psi_m) \sin 2\pi f_S t \tag{1.7}$$

The above equation is rearranged into the following:

$$P_S(t) = \frac{1}{2} \sum_{m=0}^{\infty} a_m [\cos\{2\pi(m f_R - f_S)t + \psi_m\} - \cos\{2\pi(m f_R + f_S)t + \psi_m\}] \tag{1.8}$$

Thus, the fluctuating frequency of the flow field in the relative frame of reference, namely the excitation frequency on the moving blade is given by

$$f_o = |m f_R \pm f_S| \tag{1.9}$$

Therefore, when

$$f_o = f_B \tag{1.10}$$

blade resonance occurs. Namely,

$$m f_R = |f_B \pm f_S| \tag{1.11}$$

gives the resonance condition of the moving blades.

The above result agrees well with our observation already described. Namely, owing to the unsteady flow field due to the bleed chamber resonance, the moving blades passing through the flow field suffer from excitation with frequencies $((m f_R \pm f_S))$. When the frequency coincides with the blade natural frequency f_B, blade resonant vibration occurs.

This is the reason why the moving blades fall into resonant vibration at apparently non-synchronous conditions.

1.7 Pressure fluctuation within the bleed chamber

The existence of approximately 18 per-rev frequency component within the bleed chamber can be explained also by a reasoning similar to the one described in the preceding section. Ordinarily, the blade number of 20 in this case would have given rise to the blade passing frequency, namely 20 per-rev frequency. In fact, for bleed ratio 18.3% where the non-synchronous vibration was not detected, the pressure fluctuation with 20 per-rev frequency was surely observed.

At a fixed point in the bleed chamber, the pressure variation in a sinusoidal manner of, for example, $\sin(2\pi f_R Z t)$ will be exerted by the moving blades passing near the point. If the flow field fluctuates with a frequency f_S and the passing blades, being influenced by the f_S phenomenon, change the amplitude of the above-said pressure effect, then the fixed point in the bleed chamber will receive the following reaction:

$$\sin(2\pi f_R Z t + \psi)\sin 2\pi f_S t$$
$$= \frac{1}{2}\cos\{2\pi(f_R Z - f_S)t + \psi\} - \frac{1}{2}\cos\{2\pi(f_R Z + f_S)t + \psi\} \quad (1.12)$$

Thus, in such circumstances, a phenomenon having frequencies $(f_R Z \pm f_S)$ is to be detected in the bleed chamber. If frequency f_S is assumed to be given approximately $2f_R$ as in the present case, the phenomenon has frequency of approximately $(Z \pm 2)f_R$. In the present case, approximately $20 \pm 2 = 18$ and 22 per-rev frequency is to be detected. The former was surely detected.

1.8 Conclusion

During developing a special type of interstage air bleed system by use of an experimental low-speed, single-stage axial compressor, we observed that the moving blades vibrated severely and non-synchronously, i. e. at r. p. m. s apparently independent of the blade resonance with integral orders of the rotor revolution. The moving blades were located just upstream of a circumferential bleed port.

Occurrence of such non-synchronous blade vibrations is very rare and has been limited only to the above one as far as our experience concerns. It seems to depend much on the bleed system configuration, and the conventional bleed systems employed practically have never had such phenomena. The conditions for the possible occurrence of the non-synchronous vibration are not yet fully understood.

Experimental and analytical study, however, has made clear the following cause and features of the phenomenon:

1. The blade vibration is excited by non-stationary distortion of the flow field due to periodic flow variation through the bleed port resulting from air-mass resonance or near-resonance within the bleed chamber.

2. The blade resonances appear non-synchronous. However, they are proved to occur at the rotor speeds satisfying the following condition:

   $$m f_R = f_B \pm f_S$$

   where f_R : r. p. s. of the rotor m : arbitrary integer f_B : blade natural frequency, and f_S : natural frequency of the air-mass in the bleed system.

3. The air-mass resonance or near-resonance is excited by a close encounter of the natural frequency of the chamber air-mass with some integer multiple of the rotor r. p. s. .

4. At the resonant condition of the air-mass, in addition to f_S, pressure oscillations with frequencies $f_R Z \pm f_S$ may be observed within the bleed chamber. Here, Z is the number of the blades. In ordinary conditions without the resonance, only a $f_R Z$ (and pherhaps its higher harmonics) is detected within the chamber.

5. A large quantity of bleed air may prevent the occurrence of the resonant vibration.

Aeroelasticity of Fan Blades with Complex Modes

Edmond Széchényi[1]

ABSTRACT Research into fan blade flutter generally considers structural modes that are pure in either bending or torsion. Reality is often rather different. Real blade modes generally have both bending and torsion components. Moreover, the effect of rotation on large twisted blades induces complex modes, that is modes where the bending and torsion components are not perfectly in-phase.
The aeroelastic stability of complex modes is studied experimentally in a straight cascade wind tunnel in both sub and supersonic flow on a typical fan configuration. Aeroelastic coefficients are measured in the presence of pure plunging and pure pitching as well as complex combinations of these two motions. It is shown that reliable coefficients for complex modes can be obtained by superposition of the coefficients for plunging and pitching. Using the validated principle of superposition, the effect of realistic blade modal parameters on cascade stability is studied.

1.1 Introduction

Fundamental research into fan blade flutter conducted in cascade tunnels generally considers structural modes that are pure in either bending or torsion (plunging or pitching) of two-dimensional models. Reality is often rather different. Real blade modes generally are combinations of bending and torsion. Moreover, in rotation gyroscopic forces induce slightly complex modes, that is modes where the motion of all the points of the blade is not exactly in-phase. This is particularly true of very large twisted blades. Such modes can be expressed as a combination of pure bending and pure torsion with a phase lag between these two components of motion.

Experimental research has generally been limited, for practical reasons, to pure pitching and plunging modes. Few experimental rigs allow well controlled complex motion of models. Moreover, the possibility of combinations is so vast that a systematic study would be a momentous undertaking. Given a suitable rig, simulation of a specific fan configuration

[1]O.N.E.R.A., B.P. 72, 92322 Chatillon, France

would of course be possible but design-stage stability predictions are then questionable due to the possible unreliability of predicted modes.

These difficulties can be overcome by assuming the superposability of component bending and torsional motions because this allows predictions for any shape of mode based solely on data obtained from pure plunging and pure pitching motions. This hypothesis is made in the present paper and is validated experimentally in a two-dimensional cascade tunnel at sub and supersonic flow speeds. The validity of this assumption opens the door for in-detail studies of the effect of modal parameters on cascade stability.

1.2 The experimental facility

Though this paper does not intend to go at length into a now well established experimental facility [SF80, SG80, Sz80] a brief description of the underlying experimental principles and procedures is necessary.

The cascade tunnel used is composed of 9 blades of which the 5 central ones can be considered to be subjected to true periodic flow. Thus measurements are restricted to a 5 blade cascade. This wind tunnel can be fitted with 3 different nozzles: Mach=1.6, Mach=1.2 and sub/transonic. Test blades are 120mm in length and have a chord of about 110mm.

Aeroelastic forces in a cascade are those induced on a blade by its own vibrating motion or by the vibrations of neighbouring blades. Aeroelastic measurements yield unsteady coefficients of lift and moment. These are either "direct", as in the case of forces acting on the vibrating blade itself, or "coupling" coefficients when the influence of the neighbouring blades is considered.

In the testing technique used, the assumption is made that the "direct" and "coupling" terms combine linearly so that a vectorial addition of these forces can be made. This is frequently referred to as the "influence coefficient" technique. Thus, the total coefficient for a blade in an infinite cascade is:

$$C_{total} = \sum_{n=-\infty}^{\infty} C_n e^{in\alpha} \tag{1}$$

where n is the number of the blade:

$= 0$ is the vibrating blade

> 0 are the upstream blades

< 0 are the downstream blades

C_n is the complex coefficient measured on blade n

α is the blade-to-blade phase angle which is not an experimental parameter but merely appears in the calculations.

This assumption considerably simplifies the experimental set-up. The practical consequence is that only one blade is made to vibrate and only one blade is instrumented. The relative position of these two blades is variable. In principle one should measure coupling effects as far as possible in a cascade; in practice measurements are carried out on 2 blades up-stream and 2 blades down-stream.

This influence coefficient technique has now been used successfully for well over a decade and is supported by a number of studies as for instance those of Hanamura et al. [HTY80] and of Schlaefli [Sch89].

The aeroelastic force measuring instrumentation consists of 30 miniature pressure transducers placed in a single chord of a blade. Lift and moment are obtained by integration of the pressures.

The blade vibrations required for aeroelastic measurements are provided by 4 linear hydraulic actuators placed 2 each on either side of the two-dimensional test blade. The 4 actuators are made to vibrate with relative amplitudes and phases such as to give the required plunging, pitching or complex motion.

1.3 The superposition of motions

1.3.1 The principle

Fan blade modes are only rarely pure in either bending or torsion and, in rotation, they are often slightly complex. However, the true motion of a chordline is always a linear combination of pure plunging and pure pitching, assuming of course that there is no chordwise deformation.

The principle of superposition assumes that the aeroelastic forces resulting from pure plunging and pitching can be combined linearly to give forces for any complex motion, just in the same way as the motions themselves are combined. This linear combination can be expressed as:

$$M_b + M_a = M_c$$

$$F_b + F_a = F_c$$

where

M is the aeroelastic moment about the mid-chord

F is the aeroelastic lift force

and the indexes signify -

b pitching motion about the mid-chord

a plunging motion

c a complex motion

These aeroelastic forces can be expressed in terms of non-dimensional aeroelastic coefficients as follows:

$$
\begin{aligned}
M_b &= \gamma\theta_0(m_b^{'}\sin\omega t + m_b^{''}\cos\omega t)\\
M_a &= \gamma\delta_0\mu(m_a^{'}\sin\omega t + m_a^{''}\cos\omega t)\\
M_c &= \gamma\delta_v\mu(m_c^{'}\sin\omega t + m_c^{''}\cos\omega t)\\
F_b &= \beta\theta_0/\mu(k_b^{'}\sin\omega t + k_b^{''}\cos\omega t)\\
F_a &= \beta\delta_0(k_a^{'}\sin\omega t + k_a^{''}\cos\omega t)\\
F_c &= \beta\delta_v(k_c^{'}\sin\omega t + k_c^{''}\cos\omega t)
\end{aligned}
$$

where

m_a and k_a are the complex aeroelastic moment and lift coefficients respectively, for pure plunging motion

m_b and k_b are the complex aeroelastic moment and lift coefficients respectively, for pure pitching motion

m_c and k_c are the complex aeroelastic moment and lift coefficients respectively, for any complex motion

θ_0 is the pitching amplitude angle

δ_0 is the plunging amplitude normalised by the half-chord

δ_v is the amplitude of a reference point v along the chord, normalised by the half-chord

δ_w is the amplitude of a second reference point w symmetrical to v relative to the mid-chord line

μ is a dimensionless quantity function of the positions of points v and w ($\mu = 1$ when v and w are at the leading and trailing edges respectively)

$\gamma = \frac{1}{2}c\beta$ and $\beta = -\frac{1}{2}\rho V^2 c$

$c =$ chord length

The motions are defined as:

Pure pitching $= \theta_0 \sin\omega t$

Pure plunging $= \delta_0 \sin(\omega t + \phi)$

Motion at $v = \delta_v \sin(\omega t + \lambda)$

Motion at $w = \delta_w \sin(\omega t + \psi + \lambda)$

and let $Q = \delta_w/\delta_v$

where

ϕ is the phase angle between the pitching and plunging components of a complex motion

λ is the phase angle between the pitching component and the motion at v

ψ is the phase angle between motions at v and w

The more general non-dimensional parameters describing the modal behaviour of a blade, θ_0/δ_0 and ϕ, can be expressed in terms of Q and as:

$$
\frac{\theta_0}{\delta_0} = \mu\left[\frac{(1 + Q^2 - 2Q\cos\psi)}{(1 + Q^2 + 2Q\cos\psi)}\right]^{\frac{1}{2}}
$$

and

$$\phi = \arctan \frac{(2Q \sin \psi)}{(1 - Q^2)}$$

In the case of a pure motion, stability of a blade is directly given by the sign of the imaginary aeroelastic coefficient. This is no longer the case for complex or composite motions. The work done by the aeroelastic forces, E, during one cycle must be calculated. A negative value of E shows instability:

A) In the case of aeroelastic coefficients on a single blade obtained with pure plunging and pitching motions, the superposition gives –

$$E = \int_0^{2\pi/\omega} [(M_b + M_a)\theta_0\omega \cos \omega t + (F_a + F_b)\delta_0 \frac{c}{2}\omega \cos(\omega t + \phi)]dt$$

$$E = \frac{1}{4}\gamma\delta_v^2 \Big[(1 + Q^2 - 2Q \cos \psi)\mu^2 m_b'' \\ +((1+Q^2)^2 - 4Q^2 \cos^2 \psi)^{\frac{1}{2}} (\mu^2 m_a' \sin \psi + \mu^2 m_a'' \cos \psi - k_b' \sin \psi + k_b'' \cos \psi) \\ +(1 + Q^2 + 2Q \cos \psi) k_a'' \Big] \tag{2}$$

B) In the case of coefficients on a single blade obtained with a complex motion (θ_0 and δ_0 are the amplitudes of the pitching and plunging components of the complex motion) –

$$E_c = \int_0^{2\pi/\omega} [M_c\theta_0\omega \cos(\omega t + \lambda) + F_c\delta_0 \frac{c}{2}\omega \cos(\omega t + \phi + \lambda)]dt$$

$$E_c = \frac{1}{4}\gamma\delta_v^2 \Big[2\mu^2 (m_c''(1 - Q \cos \psi) + m_c' Q \sin \psi) \\ +2(k_c''(1 + Q \cos \psi) - k_c' Q \sin \psi) \Big] \tag{3}$$

The stability of a cascade is of course not simply that of a single blade. It also depends on the aeroelastic blade-to-blade coupling. Thus, using the hypothesis described by equation 1, the total work done by the aeroelastic forces in a cascade can be easily shown to be:

A) When combining pure plunging and pitching motion, the superposition gives –

$$
\begin{aligned}
E = \gamma\pi\delta_0^2 \sum_n K_Q \Big[&\cos n\alpha\big(\mu^2 m''_{bn}(1+Q^2-2Q\cos\psi) \\
&+(m'_{an}\mu^2 - k_{bn}')2Q\sin\psi + (m''_{an}\mu^2 + k''_{bn})(1-Q^2) \\
&+k''_{an}(1+Q^2+2Q\cos\psi)\big) \\
&+\sin n\alpha\big(\mu^2 m'_{bn}(1+Q^2-2Q\cos\psi) \\
&+(m''_{an}\mu^2 - k''_{bn})2Q\sin\psi + (m'_{an}\mu^2 + k'_{bn})(1-Q^2) \\
&+k'_{an}(1+Q^2+2Q\cos\psi)\big)\Big]
\end{aligned}
\tag{4}
$$

where

$K_Q = (1+Q^2+2Q\cos\psi)^{-1}$
α is the blade-to-blade phase angle $= 2\pi m/N$
N is the number of rotor blades
m is the rotor-mode order: the number nodal diametral lines. (This mode order has negative values when the modes are regressive.)

B) In the case of a complex motion –

$$
\begin{aligned}
E_c = \gamma\pi\delta_0^2 \sum_n 2K_Q \Big[&\cos n\alpha\big(\mu^2(m''_{cn}(1-Q\cos\psi) + m'_{cn}Q\sin\psi) \\
&+k'_{cn}Q\sin\psi + k''_{cn}(1+Q\cos\psi)\big) \\
&+\sin n\alpha\big(\mu^2(m'_{cn}(1-Q\cos\psi) - m''_{cn}Q\sin\psi) \\
&+k''_{cn}Q\sin\psi + k'_{cn}(1+Q\cos\psi)\big)\Big]
\end{aligned}
\tag{5}
$$

The principle of superposition of motions can thus be verified by comparing combinations of pure plunging and pitching (equation 4) with results obtained directly from the corresponding complex motion (equation 5).

1.3.2 Experimental validation

The validation of the principle of superposition of motions was performed with experimental data obtained on a cascade simulating a standard modern fan. Various complex motions were tested with widely varying ratios of plunging to pitching and of their relative phase angle. In fact, phase angles larger than 40^0 were tested when in practice this parameter is unlikely to exceed 5^0. Correlatively, measurements were carried out with pure plunging and pitching. The substitution of the experimental aeroelastic coefficients into equations 2 and 3 allows a direct and immediate appraisal of the superposition when considering only the vibrating blade and no coupling terms.

Figure 1 shows typical results for both a subsonic and a supersonic case. The figure shows the work done, which has been normalised to give a parameter (η) directly proportional to the aerodynamic viscous damping coefficient, in terms of the parameter θ_0/δ_0 for specific values of ϕ.

A more complete validation is obtained when substituting the experimental coefficients in equations 4 and 5, thus taking into account the coupling terms. Figures 2 and 3 compare some direct measurements with those obtained with superposed motions. The damping (work done) parameter, η, is shown as a function of the rotor mode order m (see equations 4 and 5). Results for some very different values of the motion component parameters are shown.

The values obtained by the two completely independent measurements rarely give identical values but the trends are astonishingly close. These results certainly show the validity of the assumption within certain tolerances. This is particularly true in the case where only "direct" terms (no coupling) are considered (figure 1). Moreover, the differences in the results are not necessarily due to limitations in the validity: the testing is difficult and lengthy and many sources of error exist along the way, and in particular those due to the imperfect "periodicity" of the flow in the cascade when measuring coupling terms.

1.4 Complex modes and cascade stability

Figures 4 to 6 show examples of the influence of modal parameters on cascade stability. The damping (η) is given as a function of θ_0/δ_0. The figures cover values of this parameter going from near zero (pure plunging) to near infinity (pure pitching).Typical cases were chosen to illustrate the discussion.

Many parameters play a role: the pitching to plunging ratio (θ_0/δ_0), their relative phase angle (ϕ), the rotor mode (m), Mach number, pressure ratio, cascade geometry, etc... It is therefore not obvious that universal influences can be easily deduced.

However, the following trends are suggested:

— All the figures show that when the motion ratio (θ_0/δ_0) is smaller than about 0.1 or greater than 10, the damping is close to that obtained with pure plunging and pure pitching respectively. Between these limits damping can vary considerably and can become negative, even if it is positive for both pure motions.

— The damping systematically increases with the phase angle, ϕ, for all realistic values of this parameter, though the magnitude of this effect depends very much on flow conditions.

— The effect of the rotor mode order, m, can be very large. From figure 5 there is no obvious general trend. On the whole the positive modes (nodal lines revolving in the direction of rotation) seem to be more stable.
— Figure 6 shows the effect of angle of attack, in the case of subsonic and unstarted supersonic flow, and of pressure ratio in started supersonic flow. This figure illustrates the possibilities of this technique in studying a very wide range of fan conditions from a data base limited to pure motions.

1.5 Conclusions

Real fan blades generally have both bending and torsion components and the effect of rotation on large twisted blades induces complex modes. The experimental study of a wide range of real cases is extraordinarily complex and lengthy unless one adopts the principle of superposition of motions.

In the present work the validity of this principle was shown over a wide range of flow conditions and modal parameters. The principle was then used to gain some insight into the effects of blade modal parameters on the aeroelastic stability of a cascade having the geometry of a typical modern fan.
It is very clear that modal parameters greatly influence the stability limits of a fan. In any experimental simulation these must necessarily be accounted for correctly.

References

[Sch89] D. Schlaefli. Experimentelle untersuchung der instationaeren stroemung in oszillienden ringgittern. In *Thèse de Doctorat.* EPF Lausanne, 1989.

[SF80] E. Széchényi and R. Finas. Aeroelastic testing in a straight cascade wind tunnel. In *Proceedings of the 2nd International Symposium on Aeroelasticity in turbomachines.* Lausanne, 1980.

[SG80] E. Széchényi and J.P. Girault. A study of compressor blade stall-flutter in a straight cascade wind tunnel. In *Proceedings of the 2nd International Symposium on Aeroelasticity in turbomachines.* Lausanne, 1980.

[Sz80] E. Széchényi. Understanding fan blade flutter through linear cascade aeroelastic testing. In *Agardograph 298 vol 1, Ch.X*, 1980.

[HTY80] Y. Hanamura, H. Tanaka and K. Yamaguchi. A simplified method to measure unsteady forces acting on the vibrating blades in a cascade. In *Bulletin of the JSME, vol 23, No 180*, 1980.

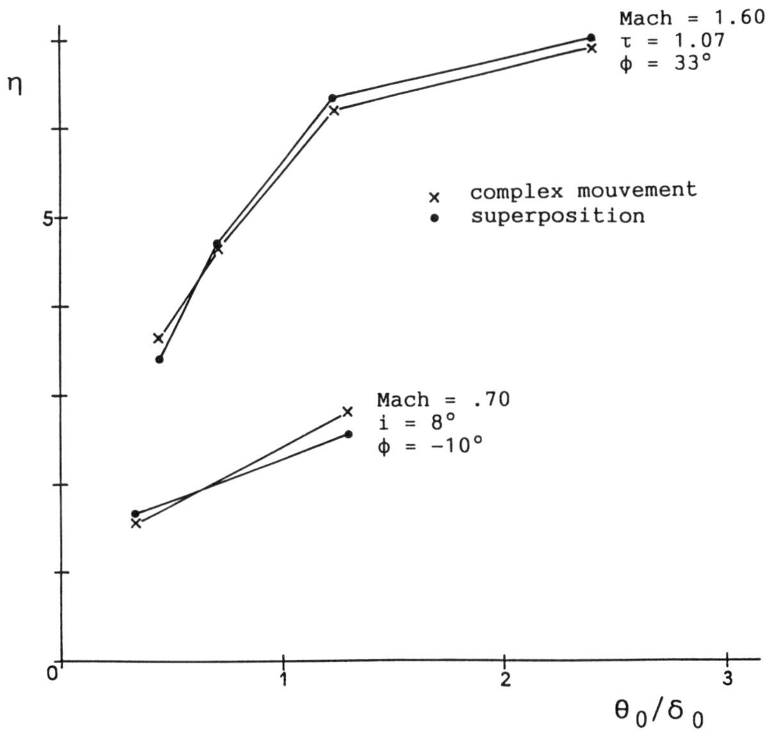

Figure 1. Validation of the principle of superposition of motions for "direct" aeroelastic terms only.

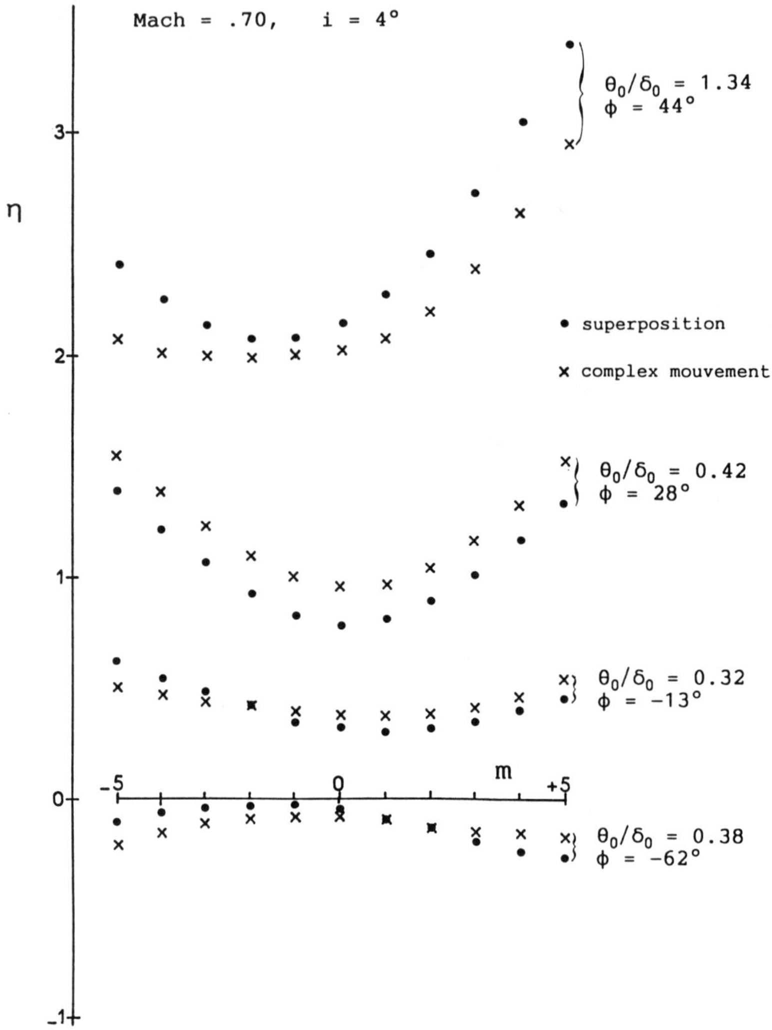

Figure 2. Validation of the principle of superposition of motions for a cascade - subsonic flow.

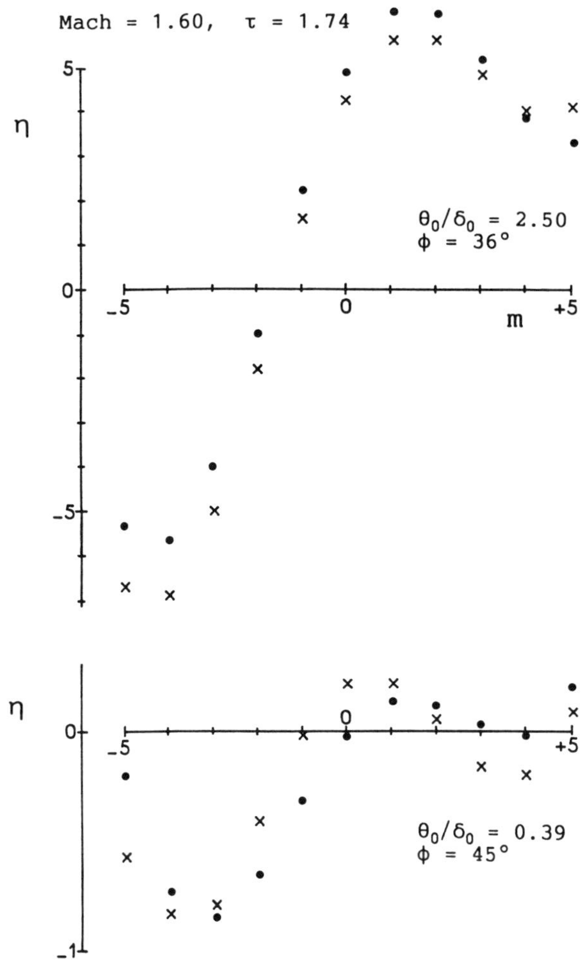

Figure 3. Validation of the principle of superposition of motions for a cascade - supersonic flow.

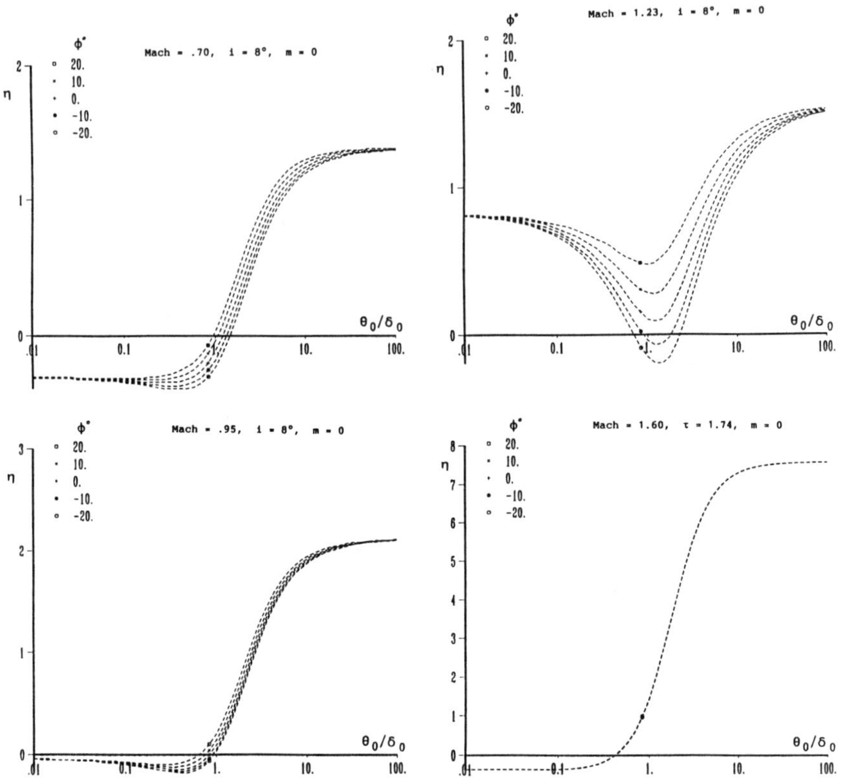

Figure 4. The effect of modal parameters on stability: the phase angle, ϕ.

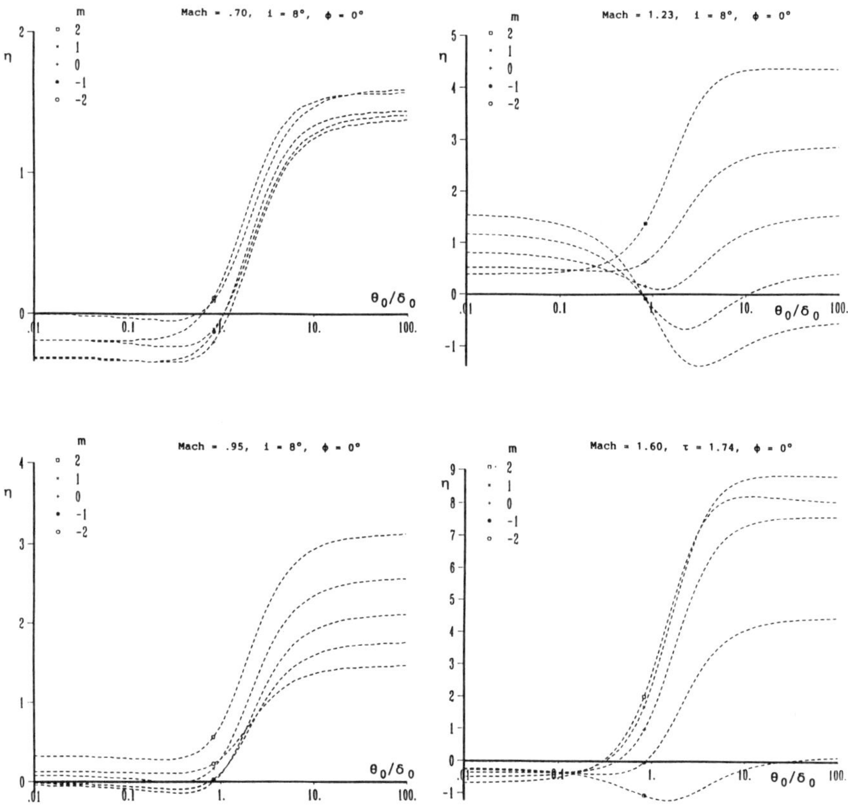

Figure 5. The effect of modal parameters on stability: the rotor mode order, m.

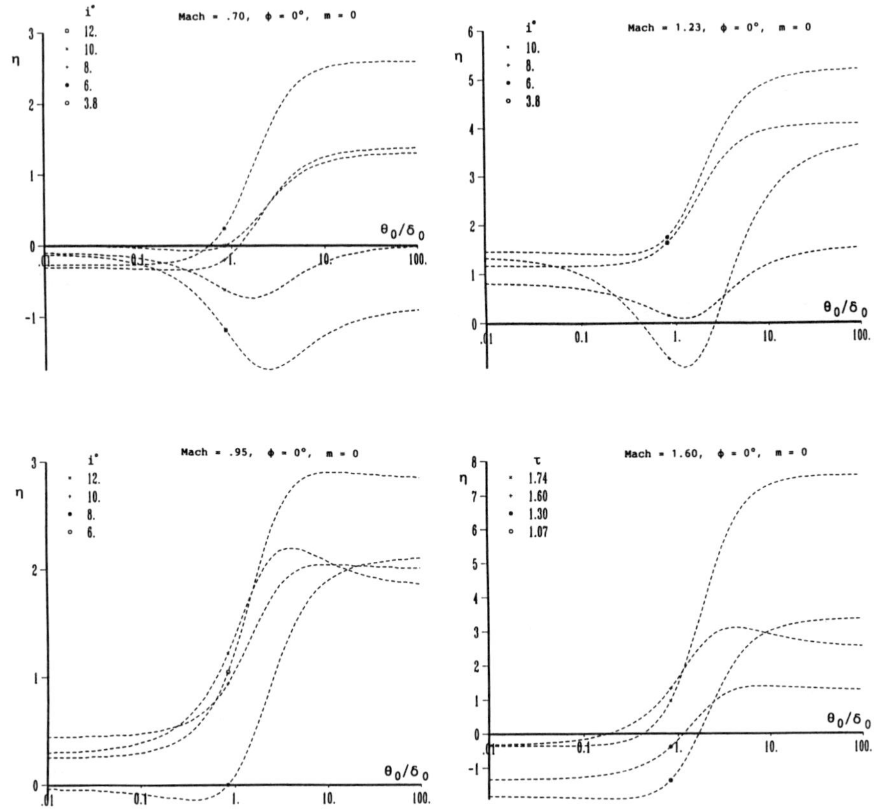

Figure 6. Examples of the effect of flow conditions and modal parameters on stability.

VIII. Standard Configurations for Unsteady Flows

Panel Discussion on Standard Configurations for Unsteady Flow Through Vibrating Axial-Flow Turbomachine-Cascades

T.H. Fransson[1]
J.M. Verdon[2]

Abstract

As part of the Symposium series "Unsteady Aerodynamics and Aeroelasticity of Turbomachines and Propellers" a set of cascade standard configurations, with emphasize on the unsteady aerodynamics, was established [Bölcs and Fransson, 1986]. These have been used extensively by several authors since the "Fourth Aeroelasticity Symposium" in 1987. As both experimental and numerical procedures presently undergo a rapid development it was at that time recognized that a continuous update of the standard configurations is a necessity.

The "Standard Configurations on Unsteady Flow Though Vibrating Axial-Flow Turbomachine Cascades" have been updated in the sense that three cascade geometries have been reduced in importance, whereas one has been extended to include experimental data at high incidence angles. The flat plate standard configuration has been extended to include two supersonic geometries that have been proven useful for comparisons in the past. Finally, a new analytical inviscid high subsonic/transonic compressor cascade geometry is proposed as "Standard Configuration 10", with the aim of detailed studies of shock wave influences on the unsteady load on vibrating blades. The Reynolds numbers have been included for the experimental test cases. In these proceedings of the symposium only the most important modifications concerning the standard configurations are given. The reader is referred to the updated report, in which also all results obtAined from different researchers are presented, for full details [Fransson and Verdon, 1991].

[1] Swiss Federal Institute of Technology, Lausanne, Switzerland. Presently at Royal Institute of Technology, Stockholm, Sweden.
[2] United Technologies Research Center, MS 129-20, 411 Silver Lane, East Hartford, CT 06108.

Nomenclature[3]

Symbol	Explanation	Dimension
$C(x)$	Camber distribution (eq. 2)	-
c	Blade chord	m
$\tilde{c}_l(t)$	Unsteady perturbation lift (= force normal to chord) coefficient vector per unit amplitude, positive in positive y-direction	-
$\tilde{c}_m(t)$	Unsteady perturbation moment coefficient vector per unit amplitude, positive in clockwise direction	-
$\tilde{c}_p(x)$	Real amplitude of unsteady pressure coefficient	-
$\tilde{c}_p(x,t)$	Unsteady perturbation pressure coefficient $\tilde{c}_p(x,t) = \frac{1}{h} \cdot \frac{\tilde{p}(x,t)}{\bar{p}_{w1} - \bar{p}_1} = \tilde{c}_p(x) \cdot e^{i(\omega t + \Phi_p)}$	-
$\bar{c}_p(x)$	Steady-state pressure coefficient $\bar{c}_p(x) = \frac{\bar{p}(x)}{\bar{p}_{w1} - \bar{p}_1}$	-
d	Maximum blade thickness ratio	-
f	Blade vibration frequency	Hz
h	Dimensionless (with chord) bending vibration amplitude	-
$H_\pm$	Thickness on upper/lower surfaces (Fig. 6)	-
H_C	Height of camber-line at midchord (eq. 2)	-
H_T	Maximum (full) blade thickness, normal to chord (eq. 1)	-
i	Complex notation = $(-1)^{0.5}$	
i	Incidence angle, from mean camberline at leading edge	o
k	Reduced frequency, $k = (c \cdot \omega) / (2 \cdot v_{ref})$	-
M	Mach Number	-
p	Pressure	mbar
R	Radius of camber line (eq. 2)	-
Re	Reynolds number = $(v_{ref} \cdot c) / \nu$	-
t	Time	sec
$T(x)$	Blade (full) thickness distribution	-
v	Velocity	m/sec
v_{ref}	Reference velocity $v_{ref}=v_1$ for compressor cascade $v_{ref}=v_2$ for turbine cascade	m/sec
x	Dimensionless (with chord) coordinate along chord	-
X	Dimensionless (with chord) blade surface coordinate (eq. 3)	-
y	Dimensionless (with chord) blade coordinate normal to chord	-
Y	Dimensionless (with chord) blade surface coordinate (eq. 3)	-

---

[3] The nomenclature is in agreement with the one previously used for the discussed standard configurations [Bölcs and Fransson, 1986, pp. 9-16], but extended for the tenth configuration.
Note: The nomenclature "eigen-damping", "eigen-influence" and "eigen-excitation" will be used here to indicate influence on a reference blade by the blade itself. "Self-excitation" is not used in order to separate the single blade-excitation from a cascade-excitation, which usually is called a "self-excitation".

α	Real amplitude of blade vibration pitching amplitude	rad
β	Flow angle, measured from axial, positive in direction of rotation	deg
γ	Chordal stagger angle, from axial	deg
$\Delta\tilde{c}_p(x)$	Real amplitude of unsteady perturbation pressure difference	-
$\Delta\tilde{c}_p(x,t)$	Unsteady perturbation pressure difference coefficient	-
δ	Blade bending vibration direction = $\tan^{-1}(h_y/h_x)$	deg
θ	Angle of mean camber line (eq. 3)	rad
Ξ	Aerodynamic damping coefficient	-
σ	Interblade phase angle, positive when blade "n+1" leads blade "n"	deg
τ	Dimensionless, with chord, pitch	-
ϕ	Phase angle. Positive when disturbance leads blade "n"	deg
ω	Circular frequency of the blades	rad/s
ν	Kinematic viscosity	m^2/s

Subscript

h	Bending motion
is	Isentropic values
p	Pressure
ref	Reference value
w	Stagnation values in relative frame of reference
1	Upstream flow conditions
2	Downstream flow conditions
1B	First bending mode
$-\infty$	Values at "infinity" upstream
α	Pitching motion or Position of pitch axis

Superscript

~	Time dependent perturbation values
-	Time-averaged values

Note:
A question as regards the definition of dimensionless pressures have been brought up during some discussions between different researchers working with the standard configurations. It must be remembered that all the standard configurations use the time-averaged dynamic pressure $(\bar{p}_{w1}-\bar{p}_1)$ as a means to obtain non-dimensional pressure coefficients, both for the steady-state and time-dependent aeroelastic test cases.

Introduction and Objectives

Vibration-related fatigue failures have been, and remain, a major concern for turbomachine manufacturers and operators worldwide. If such a failure appears it has major consequences and can become extremely expensive. It is not probable that blade vibration problems will disappear in the next few years while, although the quality of design methods increases, manufacturers constantly push the aerodynamics and materials used for blades into higher performances. Modern turbomachine blades are thus constantly at the high end of possibilities of the materials employed.

Evidence of a tremendous improvement in predicting the unsteady aerodynamic behavior of the high-speed flow through vibrating turbomachine-cascades, together with an increase in the number of well documented experimental data, can be found in the literature during recent years. The agreement between theoretical and experimental aeroelastic results ranges from extremely good to extremely bad, depending on such factors as geometry and flow conditions, but also on the experimental procedure and the numerical method. Generally it can be concluded that the low subsonic attached flow can be reasonably well measured and predicted, whereas the transonic flow conditions on realistic profile types show discrepancies between theory and experiments, and that virtually no data or prediction models are available for separated flow conditions. A large need exists thus for both well documented experimental test cases and for appropriate prediction models.

Within the framework of the symposia "Aeroelasticity in Turbomachines" a workshop, with the aim to define the state-of-the-art of prediction models and experimental test rigs for unstalled two-dimensional cascade flow, was performed with the conclusion that several successes could be found but that also some significant differences, both between theories and experiments and between different prediction models, had to be documented [Bölcs and Fransson, 1986].

At the 1987 Aeroelasticity symposium a task force[4] was thus created with the objectives to:

- follow up the literature on new experimental and theoretical procedures
- redefine the standard configurations to evolve together with the present need of aeroelastic information in transonic and high incidence flow regimes
- reduce, where possible, the number of aeroelastic sample cases for unstalled flow
- seek explanations for agreements and disagreements between prediction and experiment

---

4 S. Fleeter, E. Széchényi, J. M. Verdon, T. H. Fransson.

- present an updated version of standard configurations at the 1991 aeroelasticity meeting.

The present note constitutes a documentation in this context. Some of the standard configurations, as defined in the 1986 report [Bölcs and Fransson, 1986], have been reduced in importance[5]. Of the remaining configurations the number of aeroelastic cases has been modified, whereas one more configuration has been added in order for the sample cases to better reflect the fields in which basic research is needed today. Furthermore, some authors have recently asked for more information on time-dependent stalled and partially stalled flow. To this end the Reynolds numbers have been given, and the fifth standard configuration has been extended with a few highly viscous sample cases. It is foreseen that the standard configurations also in the future will be extended to stalled flow conditions as soon as detailed data on separated flow becomes available.

Standard Configurations

Previous standard configurations

The collection of two-dimensional standard configurations, for attached flow, compiled by Bölcs and Fransson [1986] consisted of 9 different cascade-geometries, with the objective to validate both experimental data and prediction models. During the project it was established that most of the data selected for purely subsonic flow gave a reasonable agreement with some prediction models, which validated both the experiments and the theories. The largest computational efforts thus went into Standard Configurations 1 (low subsonic flow compressor cascade), 4 (cambered transonic turbine cascade), 8 (flat plate) and 9 (double circular arc profiles with low camber), with some additional work put into configurations 5 (high subsonic compressor profile) and 7 (supersonic compressor). It was concluded from the project that certain geometries and flow conditions could be accurately predicted, whereas discrepancies existed for other sample cases.

Modifications of standard configurations

As the number of predictions on the configurations 2, 3 and 6 (see Bölcs and Fransson [1986]) were considerably smaller than on other cascade configurations, it is proposed that efforts for comparisons with low subsonic compressor blades and cambered transonic turbine blades should be reduced to configurations 1 and 4 (Figs. 1 and 2 below). Configuration 5 (Fig. 3) is today of larger interest than in the beginning of the project because of its systematic parameter-study from attached to

---

5 This corresponds to those with small participation or small industrial interest. (Standard configurations 2, 3 and 6 have been eliminated.)

stalled flow, although data are only available for one blade vibrating (here the question of superposition of influence coefficients in stalled flow may arise). Configuration 7, which treats supersonic inlet flow conditions for a compressor (Fig. 4), is also of present interest.

Flat plate cascades, double circular arc profiles and other analytically defined geometries are of large interest and are necessary in order to compare prediction models with each other and to draw physical conclusions from the results. Furthermore, modern compressor blades in the high subsonic and supersonic flow domains are often derivatives of such profiles, and there seems to be a renewed interest in cascades with supersonic leading edge locus. The flat plate and double circular arc geometries (configurations 8 and 9) are thus kept in a slightly redefined way (Figs. 5-6). Finally, a supplementary configuration is proposed, based on a modified NACA four digit series airfoil (Fig. 8)[6].

It is important to note that the configurations are presently still limited to two-dimensional flow conditions, with mostly attached flow. Although numerical results today are available for solving the Navier-Stokes equations with different (steady-state) turbulence model assumptions there are hardly no separated experimental cascade data available (see however the extension of "Standard Configuration 5" below). Furthermore, three-dimensional unsteady cascade effects are today not taken into full account experimentally. However, it is reasonable to expect that some data for separated and three-dimensional flow will appear in the not too distant future. All present sample cases are, furthermore, considered to be of uncoupled modes, although the coupling effects are of large importance for the stability of the blading. It is presently assumed that the coupled modes are obtained by superimposing the heaving and pitching motions.

The number of aeroelastic sample cases in the report are still out of necessity large. This can not be otherwise as the standard configurations should cover all velocity domains from low subsonic to supersonic velocities and both compressor and turbine geometries, and as the interblade phase angle is a parameter of major importance in turbomachine applications. Furthermore, it is well known that the overall time-dependent blade lift and moment coefficients may give reasonable agreement between different prediction models and with experimental data, although the unsteady blade surface pressure coefficient results may represent quite different trends and to a certain extent may indicate different physical interpretations. It is thus today even more important than at the outset of the workshop to represent, for different interblade phase angles and for different cascade and flow configurations, the pressure and suction surface time-dependent pressure coefficients separately.

---

6 Please note, however, that the original information and experimental results for all the other standard configurations still exist in the workshop data-base.

Because of space limitations, the reader is referred to the updated report on the standard configurations [Fransson and Verdon, 1991] for details about the configurations. All airfoil coordinates as well as the experimental data and numerical results presented, either in the first [Bölcs and Fransson, 1986] or the updated standard configuration report [Fransson and Verdon, 1991], exist on computer files and can be obtained[7] upon request.

First standard configuration (compressor cascade in low subsonic flow)

The first standard configuration, included by the courtesy of Dr. Frank Carta at the United Technologies Research Center [Carta 1982a,b; 1984], and its 15 recommended aeroelastic sample cases, gives an overview of different steady-state and time-dependent flow conditions at low Mach numbers (see Fig. 1, Table 1). Most prediction models give good agreement with the data, but some minor differences between the different models became apparent.

The most important conclusion from this standard configuration is that the detailed blade surface pressures, and thus the flutter limits, can be fairly accurately predicted for low incidence flows on this type of

Table 1: *First standard configuration: Experimental values for 15 recommended aeroelastic sample cases [Bölcs and Fransson, 1986, p. 60]*

	Time averaged				Time dependent parameters				
	M_1 (-)	i (°)	p_1/p_{w1} (-)	p_1/p_{w1} (-)	β_1 (°)	k (°)	α (-)	σ (°)	f (Hz)
1	0.18	2	0.9774	0.9818	62.0	0.122	0.5	-45	15.5
2	"	"	"	"	"	"	"	+45	"
3	0.17	6	0.9790	0.9852	62.5	0.122	0.5	-45	15.5
4	"	"	"	"	"	"	2.0	+45	"
5	"	"	"	"	"	"	"	-45	"
6	"	"	"	"	"	"	"	180	"
7	"	"	"	"	"	"	"	-135	"
8	"	"	"	"	"	"	"	-90	"
9	"	"	"	"	"	"	"	0	"
10	"	"	"	"	"	"	"	+90	"
11	"	"	"	"	"	"	"	+135	"
12	"	"	"	"	"	0.072	"	-90	9.2
13	"	"	"	"	"	0.151	"	"	19.2
14	"	"	"	"	"	0.301	"	"	38.4
15	"	"	"	"	"	0.603	"	"	76.8

7 From the first author of the present report.

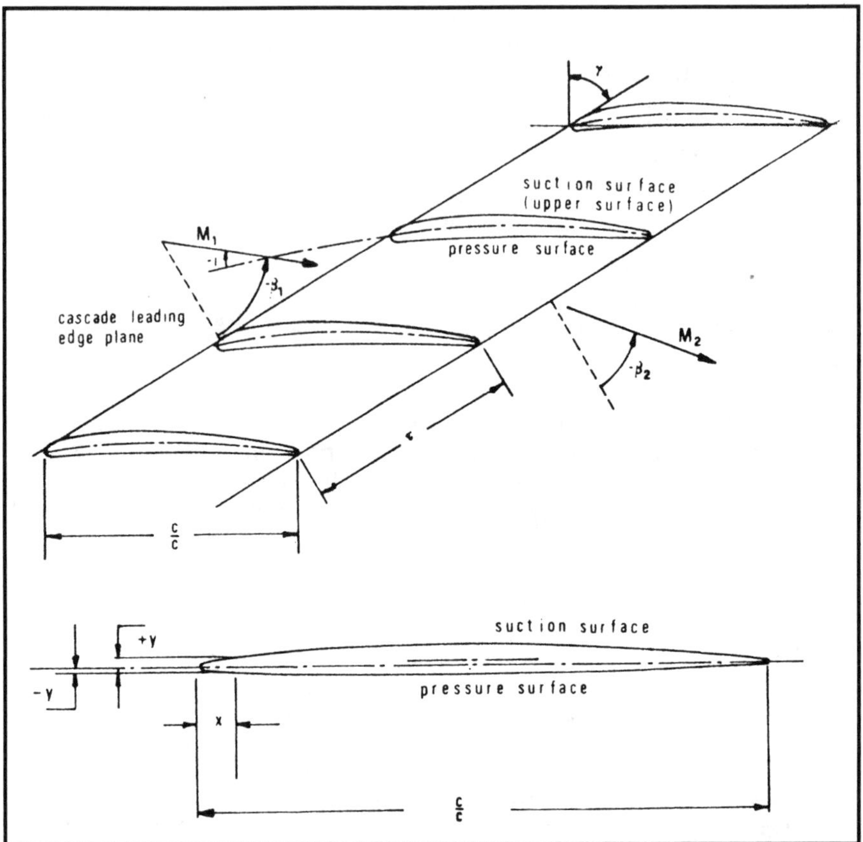

Maximum thickness at x=0.5	Vibration in pitch around (x_α, y_α)=(0.5,0.0115)	
Thickness / chord = 0.06	τ=0.75	i=variable (2°, 6°)
span=0.254 m	c=0.1524m	k=variable
γ=55°	camber=10°	
Working fluid: Air	α = 0.5^o, 2° (=0.0087, 0.0349 rad)	

Fig. 1: First standard configuration: Cascade geometry [Bölcs and Fransson, 1986, p. 58]

cascades. The reader is referred to section 7.1 in the original report on the standard configurations [Bölcs and Fransson, 1986] for details.

It is however of importance to note that although the comparisons of the results obtained are positive, the agreement between the data and the predictions, or between the different predictions, is not as good as would be wished from a theoretical point of view. It should especially be mentioned that, in several cases, a flat plate model gives as good agreement, or better, with the data as some prediction models that consider the blade thickness.

The aeroelastic sample cases originally defined have proven their value and are thus presently kept in the data-base. It is however important to point out that some uncertainties as regards to the exact value of the inlet flow angle still exist. It has been found by different researchers that a better agreement with the steady-state blade surface pressure distribution is obtained if the inlet flow angle is modified about 2° [Bölcs and Fransson, 1986, p. 61], or if a stream-tube variation is introduced.

Results on this standard configuration have recently been presented by Chiang and Fleeter [1989] and Huff [1989]. These authors have, as other researchers, found a good agreement with the experimental data.

Fourth standard configuration (cambered turbine cascade in transonic flow)

The fourth standard configuration, with data obtained at the Swiss Federal Institute of Technology [Bölcs et al, 1985], is shown in Fig. 2, with the aeroelastic sample cases given in Table 2. Please note that the blade coordinates are now given with more significant numbers than earlier [Fransson and Verdon, 1991] as it was pointed out that the coordinates showed some oscillatory behavior. The Reynolds number was not given previously and are included now as a basis for viscous calculations. Furthermore, experimental data exist for many more steady-state and time-dependent operating conditions (see for example Schläfli [1989]). These can be obtained upon request.

It is important to point out that the time-averaged dynamic pressure $(\bar{p}_{w1}-\bar{p}_1)$ is used, as for all experimental standard configurations in the present study, as the quantity with which the pressure coefficients are made dimensionless. This should be remembered when predictions are performed.

The results obtained so far show a good general agreement between the data and prediction models for shock-free flows, both for the time-dependent blade surface pressures and the unsteady forces. At transonic and supersonic outlet flow conditions major discrepancies are found, as presented by Bölcs and Fransson [1986]. It should however be pointed out that a re-evaluation of the original data has indicated that the experimental uncertainty for the inlet flow angle may be larger than originally thought, which may explain some differences in the experimental and computed steady-state pressure distributions, especially in the leading edge region. Most authors have thus introduced a stream-tube contraction (up towards 10%) into the calculations in order to compensate for leakage flow and boundary layer growth in the test facility. A better agreement with the steady-state blade surface pressure distributions is then generally obtained.

Since the first results were presented and analyzed [Bölcs and Fransson, 1986], further predictions have been performed on the cascade by Whitehead [1987], Servaty et al [1987], Gallus an Kau [1989], Kau and Gallus [1989], He [1989] and Carstens [1991a] with the same general tendencies as earlier for both subsonic and transonic flow conditions. It can probably be concluded that the experiments are good, but some aspects of the data can not be explained with present prediction models. Further numerical developments and experiments in transonic flow seem thus necessary for the future.

It has been pointed out [Carstens, 1991a] that the position of the time-averaged shock wave at the supersonic outlet flow velocity cases can be quite accurately predicted if a sufficiently fine mesh structure is used in the numerical calculations. The predicted shock strength did however not correspond to the measured one.

While the new profile definition (which corresponds to the one after which the profiles where fabricated) is more detailed than the one originally given, it has been found that also the new coordinates show some oscillatory behavior when a very fine mesh is used [Hoyniak, 1991; Carstens, 1991b]. This can give some spurious steady-state and unsteady results, and can be avoided by performing a smoothing of the coordinates [Carstens, 1991b].

Another problem that was discussed is the treatment of the blunt trailing edge. Most prediction models use, on this standard configuration, a modified airfoil shape towards the trailing edge in order to close in towards a sharp edge.

It can generally be stated that the turbine geometry defined, with its corresponding 8 aeroelastic sample cases, is still of importance for the understanding of flutter-phenomena and further developments of numerical prediction models. The original cases are thus kept in the database. Much more work has to be done to find explanations for the differences between the data and the predictions.

Table 2: *Fourth standard configuration: Experimental values for 8 recommended aeroelastic sample cases [Bölcs and Fransson, 1986, p. 103]*

	Time averaged					Time dependent parameters			
	M_1 (-)	β_1 (°)	Re_1 (-)	$M_{2,is}$ (-)	β_2 (°)	σ (°)	k (°)	$h^{(0)}$ (-)	δ (°)
1	0.19	-45	5.6×10^5	0.58	-71	-90	0.168	0.0038	60
2	0.26	"	6.6×10^5	0.76	"	"	0.128	"	"
3	0.28	"	8.2×10^5	0.90	"	"	0.107	"	"
4	0.29	"	8.5×10^5	1.02	"	"	0.095	0.0033	"
5	"	"	"	1.19	"	"	0.082	0.0038	"
6	0.28	"	8.2×10^5	0.90	"	180	0.107	0.0033	"
7	"	"	"	"	"	+90	"	"	"
8	"	"	"	"	"	0	"	"	"

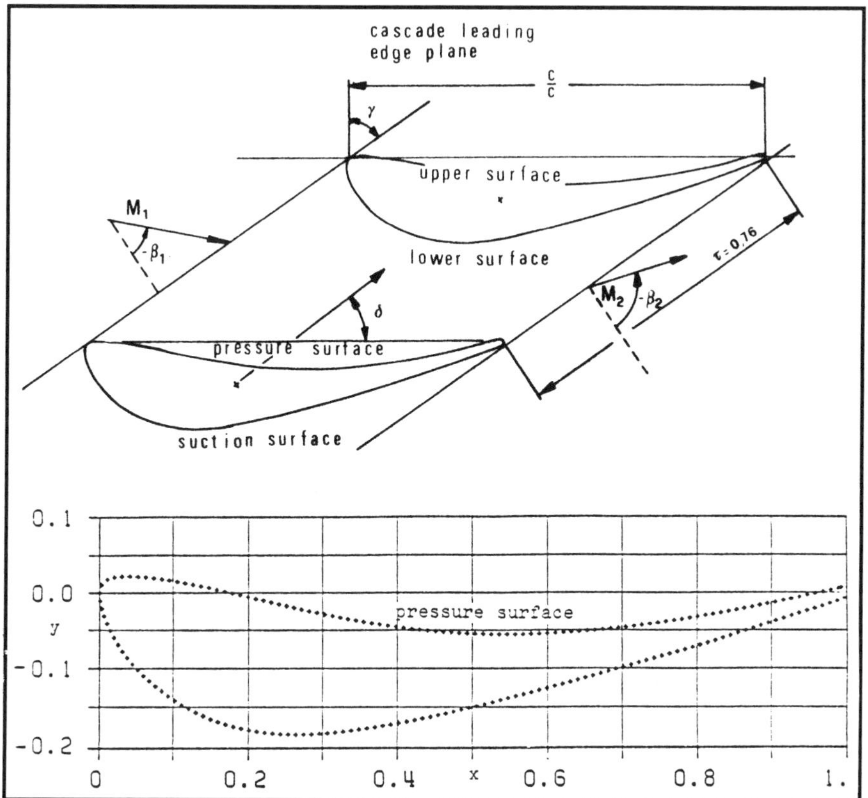

Span = 0.040m	Hub / tip = 0.8	c = 0.0744	$\tau =$ 0.67 (hub)
d = 0.17	Camber = 45°	$\gamma = 56.6^{o}$	0.76 (midspan)
$Re_1 = 5.6 - 8.2 \times 10^5$	M_{2is} = variable	Working fluid: Air	0.84 (tip)
f = variable	k = variable	σ = variable	
Vibration direction:	First bending mode $\delta = 60.4^{o}$		
Nominal flow conditions:	M_1=0.31; β_1=-44.1o;	M_2=0.90;	$\beta_2=-72.4^{o}$

Fig. 2: *Fourth standard configuration: Cascade geometry [Bölcs and Fransson, 1986, pp. 99-100].*

Fifth standard configuration (compressor cascade in high subsonic flow)

Results on the fifth standard configuration (Fig. 3, Table 3) are included by the courtesy of Dr. E. Széchényi at the "Office Nationale d'Etudes et de Recherche Aerospatiale" (ONERA) [Széchényi, 1984; Széchényi et al, 1981a,b]. The original experiments treated a large domain of incidence angles, ranging from attached to stalled flow conditions. In the 1986 report on the standard configurations only the small incidence angles ($i \leq 6^{o}$) cases were included, as at the time no models for prediction of the stalled cases were proposed. In the meantime a few viscous solvers have become available, and some researchers have asked for more information on the partially and fully stalled flow conditions. These updated cases are given in Table 3, and the experimental data are given in the full report [Fransson and Verdon, 1991]. In evaluating these data it should be kept in mind that the experimental results are obtained with **only one blade vibrating**. The data presented do thus not correspond to the time-dependent pressure coefficient in the travelling wave mode as for the other standard configurations, but instead to the eigen-influence of the reference blade on itself when all the other blades in the cascade are fixed.

Some researchers have pointed out that the original blade coordinates seem to have some "wiggles" in them when blown up for the numerical calculations. Unfortunately, it is presently not possible to give the profile definition with a better resolution. As with the fourth standard configuration a smoothing is necessary.

The question as how the pressure coefficients have been non-dimensionalized has been brought up. It has been pointed out that a

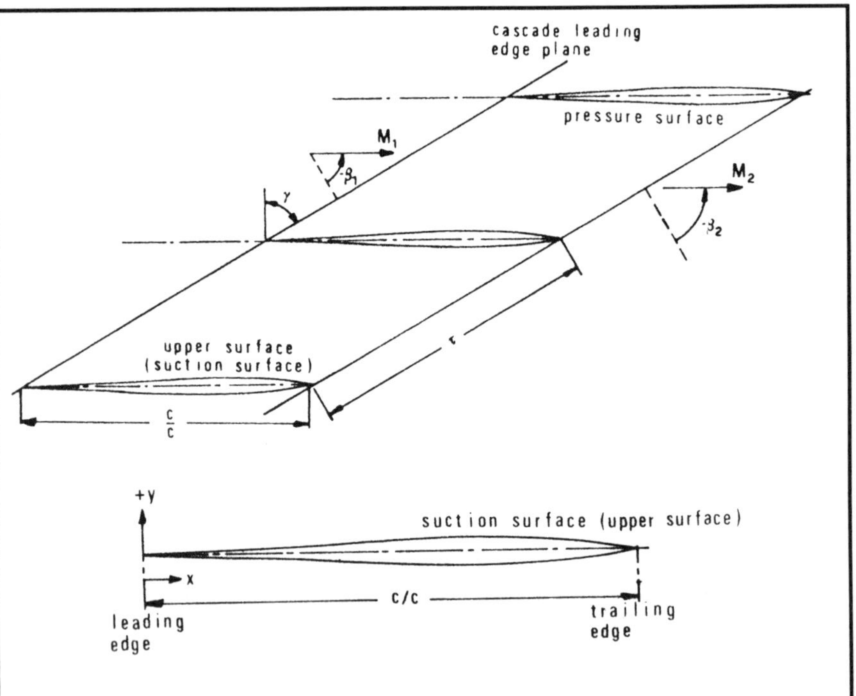

Working fluid: Air	Maximum blade thickness at x=0.67					
d = 0.027	c	=	0.090 m	span	=	0.120 m
τ = 0.95	camber	=	0°	γ	=	59.3^o
i = 2°->12°	M_1	=	0.5->1.0	α	=	0.00524 rad
f = 75->550 Hz	k	=	0.14->1.01	σ:	Only one blade vibrated	

Fig. 3: Fifth standard configuration: Cascade geometry [Bölcs and Fransson, 1986, p. 124].

better agreement between the data and the predictions was found if the unsteady data would have been scaled with $\{2(\bar{p}_{w1}-\bar{p}_1)\}$, i. e. twice the compressible dynamic pressure. A verification of the original results presented [Széchényi, 1991] has shown that the experimental data have all been non-dimensionalized with the measured upstream dynamic pressure $(\bar{p}_{w1}-\bar{p}_1)$, as is the case for all other experimental standard configurations in the present study.

Table 3: *Fifth standard configuration: Experimental values for 30 recommended aeroelastic sample cases [Bölcs and Fransson, 1986, p. 126].*

	Time averaged					Time dependent parameters				
	M_1 (-)	i (°)	p_{w1} (bar)	p_2/p_{w1} (-)	Re_1 (10^6)	α or h (° or -)	x_α/c (-)	f (Hz)	k (-)	
1	0.5	2	1.4	0.84	1.4	0.3	0.5	200	0.37	Attached
2	"	4	"	0.86	"	"	"	"	"	"
3	"	6	"	0.87	"	"	"	"	"	Part. separated
4	"	4	"	0.86	"	"	"	75	0.14	Attached
5	"	"	"	"	"	"	"	125	0.22	"
6	"	"	"	"	"	"	"	300	0.54	"
7	"	"	"	"	"	"	"	550	1.02	"
8	"	6	"	0.87	"	"	"	75	0.14	Part. separated
9	"	"	"	"	"	"	"	125	0.22	"
10	"	"	"	"	"	"	"	300	0.56	"
11	"	"	"	"	"	"	"	550	1.02	"
12	"	8	"	0.87	"	"	"	200	0.37	"
13	"	10	"	0.88	"	"	"	"	"	Fully separated
14	"	12	"	0.89	"	"	"	"	"	"
15	"	10	"	0.88	"	"	"	75	0.14	"
16	"	"	"	"	"	"	"	125	0.22	"
17	"	"	"	"	"	"	"	300	0.56	"
18	"	"	"	"	"	"	"	550	1.01	"
19	0.6	"	"	0.84	1.6	"	"	200	0.31	"
20	0.7	"	"	0.80	1.7	"	"	"	0.27	"
21	0.8	"	"	0.77	1.8	"	"	"	0.24	"
22	"	"	2.0	"	"	"	"	"	0.23	"
23	0.9	"	"	0.72	2.0	"	"	"	0.21	"
24	1.0	"	"	0.69	2.1	"	"	"	0.19	"
25	0.5	6	1.4	0.88	1.4	"	0.0	125	0.22	Part. separated
26	"	"	"	0.87	"	"	0.92	"	"	"
27	"	10	"	0.87	"	"	0.0	"	"	Fully separated
28	"	"	"	0.88	"	"	0.92	"	"	"
29	"	6	"	0.87	"	0.001	-	200	0.37	Part. separated
30	"	10	"	0.88	"	"	-	"	"	Fully separated

- Sample cases 12-30 are new.
- Only the center blade is oscillated. The others are fixed.
- y_α=0. for aeroelastic sample cases 1-28 above.
- Cases 29-30 are normal-to-chord heaving oscillations.
- Cases 21 and 22 have different inlet stagnation pressure.

Again, as for Standard Configurations 1 and 4, an inconsistence exists between the inlet and outlet flow conditions. Some researchers have compensated this by modifying the inlet flow angle and some by an introduction of a stream tube contraction ratio. Others have left the original values given, and show therefore a less good correlation with the steady-state blade surface pressure distribution. How much these differences in the set-up of the theoretical steady-state flow conditions make out for the time-dependent flow is unclear, but less differences are usually found, between the different prediction models, for the unsteady flow than for the time-averaged.

Please note that cases 21 and 22 below are for the same flow conditions, with the only difference being the steady-state stagnation pressure. This can give some indications about stagnation pressure influence and the experimental accuracy and should be considered while analyzing the results.

Sidén [1991a,b] presents results from a Navier-Stokes solver on incidence angles 2°, 4° and 6° on this standard configuration. He shows a large unsteady pressure in the leading edge region where the experiments indicate a larger value than predictions with a linearized potential model. The viscous solver gives however a considerable overshoot compared to the data. Still, these results indicate the importance of considering viscous effects also at fairly low incidence angles on compressor blades with sharp leading edges. Széchényi [1991] points out that the viscous separation bubble can be well trendwise predicted by a recently developed coupled inviscid/boundary layer code at ONERA. These results will be published next year (some are included in the full report [Fransson and Verdon, 1991]).

Seventh standard configuration (compressor cascade in supersonic flow)

The seventh standard configuration (Fig. 4 and Table 4) was tested at Detroit Diesel Allison, and included in the workshop report by courtesy of the sponsoring agent, D. R. Boldman at NASA Lewis Research Center [Boldman, 1983; Riffel and Rothrock, 1980]. A question as regards to the validity of the profile coordinates did arise at the 1991 Aeroelasticity meeting. These have been verified [Boldman, 1991], and it is confirmed that the original data were correct. The agreement between the data and the predictions performed is not satisfactory for any of the 12 aeroelastic sample cases. It is probable that some of the discrepancies come from the viscous effects in the experiment, some from experimental accuracies and some from the prediction models. However, not enough predictions have been performed to analyze either the data or predictions for supersonic cascades with thickness. It will thus be extremely interesting and important to see if future prediction models will give a significantly better agreement with the data.

It should be pointed out that, as with the other experimental Standard Configurations, the pressure coefficients are scaled with the steady-state upstream dynamic pressure ($\bar{p}_{w1}-\bar{p}_1$). Furthermore, the Reynolds

Table 4: *Seventh standard configuration: Experimental values for 12 recommended aeroelastic sample cases [Bölcs and Fransson, 1986, p. 156].*

Aeroelastic Test Case	Time averaged parameters					
	M_1 (-)	β_1 (°)	p_{w2}/p_{w1} (-)	p_2/p_{w1} (-)	M_2 (-)	β_2 (°)
1-6	1.315	-64.0	0.958	1.04	1.25	-62.8
7-12	"	"	0.957	1.45	0.99	-63.6

a: Time-averaged values

	Time dependent parameters							
	$f_{nominal}$ (Hz)	f_{actual} (Hz)	k (-)	$\sigma_{nominal}$ (°)	$\sigma(-1)$ (°)	$\sigma(0)$ (°)	$\sigma(+1)$ (°)	$\sigma(+2)$ (°)
1	725	725	0.44	180	-173	-187	-178	-180
2	"	714	0.43	-90	-91	-65	-81	-80
3	"	-	0.44	-45	-	-	-	-
4	"	724	"	0	+4	+6	-8	0
5	"	745	0.45	+45	+48	+53	+54	+50
6	"	724	0.44	+90	+91	+91	+84	+90
7	"	"	"	180	-172	-168	-163	-170
8	"	725	"	-45	-66	-69	-61	-66
9	"	724	"	0	-4	+8	-5	0
10	"	"	"	+45	+66	+59	+60	+60
11	"	725	"	+90	+92	+74	+86	+85
12	"	"	"	+120	+96	+115	+98	+100

	Time dependent parameters (continuation)					
	$\alpha_{nominal}$ (rad)	$\alpha(-2)/\alpha(0)$ (-)	$\alpha(-1)/\alpha(0)$ (-)	$\alpha(0)$ (rad)	$\alpha(+1)/\alpha(0)$ (-)	$\alpha(+2)/\alpha(0)$ (-)
1	.00349	0.76	0.52	.00122	1.27	0.0
2	"	0.87	0.20	"	0.67	"
3	"	-	-	-	-	-
4	"	0.34	0.61	.00379	0.75	"
5	"	0.71	0.78	.00206	0.32	"
6	"	0.56	0.59	.00187	0.97	"
7	"	1.21	0.85	.00131	0.60	"
8	"	0.54	0.44	.00178	0.34	"
9	"	0.64	0.96	.00202	1.11	"
10	"	0.36	0.65	.00386	0.50	"
11	"	0.66	0.96	.00227	0.73	"
12	"	0.35	0.40	.00129	0.25	"

b: Time-dependent values.

number is now given. This lies in the range of 1.1-1.6 •10^6 [Riffel and Rothrock, 1980, p. 11] for the performed tests.

Gerolymos et al [1990] have presented calculations on this geometry. A stream sheet contraction of 0.85 was introduced for these computations in order to get a reasonable agreement with the steady-state outflow data. Trendwise agreement can be found in the unsteady pressures on the blade surfaces and, as is often the case, the experimental and numerical stability limits of the cascade agree fairly well.

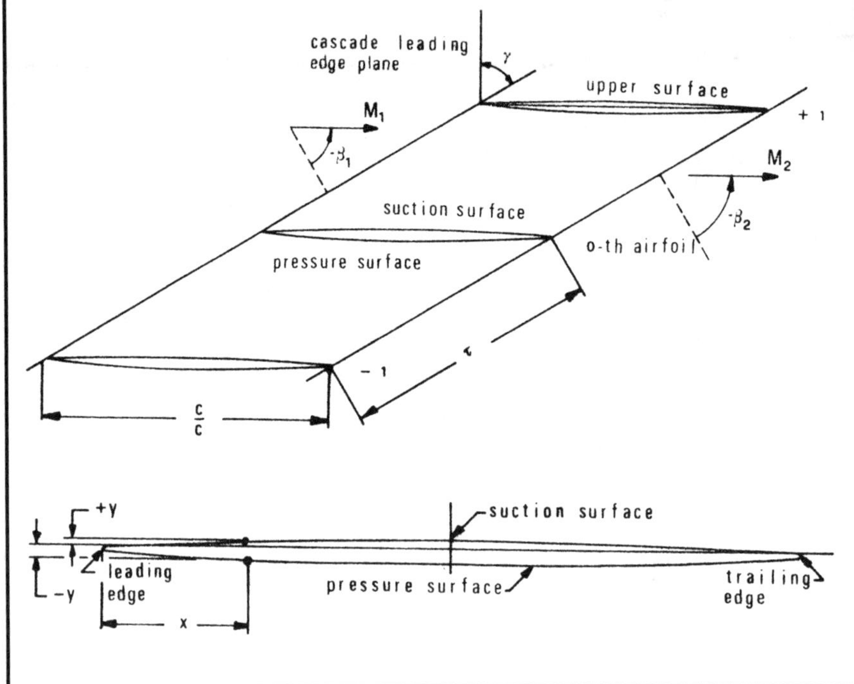

Working fluid: Air	Vibration in pitch mode around $(x_\alpha, y_\alpha)=(0.5, 0.0)$	
d = 0.034	c = 0.0762 m	span = 0.0762 m
τ = 0.855	camber = -1.3^o	γ = 61.55^o
β_1 = 64.0^o	M_1 = 1.315	α = $0.06\text{-}0.2^o$
f = 710->730 Hz	k = variable	Re1 = 1.1-1.6 x 10^6

***Fig. 4:** Seventh standard configuration: Cascade geometry [Bölcs and Fransson, 1986, p. 154].*

Eighth standard configuration (flat plate cascades at different flow conditions)

As flat plate configurations are the most important to compare against as regards to the numerical completeness and accuracy of non-analytical prediction models, it is important that such configurations cover as wide a scope as possible, especially as regards to Mach number variations. The number of aeroelastic sample cases will thus forcibly be large, but this is not a major problem as the analytical flat plate solutions used as baseline

Table 5: *Eighth standard configuration: 54 recommended aeroelastic sample cases [Bölcs and Fransson, 1986, p. 168; Verdon and McCune, 1975].*

Aeroel. Sample Case	Time-Averaged Conditions M_1 (-)	Normal shock?	γ (^{o})	τ (-)	Time-Dep. k (-)	σ (^{o})
1	Incomp.	-	60	0.75	1.0	90
2	"	-	45	"	"	"
3	"	-	30	"	"	"
4	"	-	0	"	"	"
5	0.5	-	60	"	"	"
6	0.6	-	"	"	"	"
7	0.7	-	"	"	"	"
8	0.8	-	"	"	"	"
9	0.9	-	"	"	"	"
10	0.95	-	"	"	"	"
11	0.8	-	45	"	"	"
12	"	-	30	"	"	"
13	"	-	0	"	"	"
14	"	-	60	0.5	"	"
15	"	-	"	1.0	"	"
16	1.1	-	"	0.75	"	"
17	"	at LE	"	"	"	"
18	"	at TE	"	"	"	"
19	1.2	-	"	"	"	"
20	"	at LE	"	"	"	"
21	"	at TE	"	"	"	"
22	1.3	-	"	"	"	"
23	"	at LE	"	"	"	"
24	"	at TE	"	"	"	"
25	1.4	-	"	"	"	"
26	"	at LE	"	"	"	"
27	"	at TE	"	"	"	"
28	1.5	-	"	"	"	"
29	"	at LE	"	"	"	"
30	"	at TE	"	"	"	"
31	1.3	-	"	0.5	"	"
32	"	-	"	1.0	"	"
33	"	at LE	"	0.5	"	"
34	"	"	"	1.0	"	"
35	"	"	45	0.75	"	"
36	Incomp.	-	60	0.75	0.5	"
37	0.5	-	"	"	"	"
38	0.7	-	"	"	"	"
39	0.8	-	"	"	"	"
40	1.3	-	"	"	"	"
41	1.4	-	"	"	"	"
42	1.5	-	"	"	"	"

Table 5: *Continued*

Aeroel. Sample Case	Time-Averaged Conditions M_1 (-)	Normal shock?	γ (^{o})	τ (-)	Time-Dep. k (-)	σ (^{o})
43 ="A"	1.345	-	59.5	0.7886	0.5	0
44	"	-	"	"	"	60
45	"	-	"	"	"	120
46	"	-	"	"	"	180
47	"	-	"	"	"	240
48	"	-	"	"	"	300
49 ="B"	1.281	-	63.4	0.6711	0.5	0
50	"	-	"	"	"	60
51	"	-	"	"	"	120
52	"	-	"	"	"	180
53	"	-	"	"	"	240
54	"	-	"	"	"	300

comparisons are usually fairly fast, and researchers who may want to compare a numerical result against an analysis will only choose the domain of current interest and perform a few calculations.

The 35 original aeroelastic sample cases will be kept (Fig. 5a and Table 5). The time-dependent cases 36 to 42, corresponding to some of the earlier cases but with a reduced frequency of k=0.5 instead of k=1.0, have been added. This lower reduced frequency (k=0.5) was recommended as the reduced frequency of k=1.0 showed some differences (which may eventually be attributed to numerical effects) between the different prediction models. To these aeroelastic test cases are added two supplementary steady-state conditions (Figs. 5b, c). These, denoted as aeroelastic test cases 43-54 and suggested by Dr. Verdon, are the well known "Cascades A and B" as defined by Verdon [1973] and Verdon and McCune [1975]. These cascade configurations were originally carefully selected to highlight wave reflections in supersonic flow and have been used frequently in the past as comparison for a number of analyses. They have provided useful insight into the physical phenomena of unsteady supersonic flow on vibrating flat plates and interesting, often positive, results from the comparisons. It is believed that they also in the future should be of special interest as baseline comparisons for numerical models for supersonic flow, also for supersonic leading edge locus configurations.

Further information and interesting results on "Cascades A and B" have been presented by, among others, Gerolymos et al [1990], Whitehead [1990], Fleeter and Hoyniak [1989], Topp and Fleeter [1986], Verdon [1977a,b]. More results on flat plates have been presented by several authors the last few years, among them Giles and Haimes [1991], Namba and Toshimitsu [1990], Huff [1989], Huff and Reddy [1989], Bakhle et al [1989], Verdon [1989a], and Shroeder and Fleeter [1989], to mention just a few.

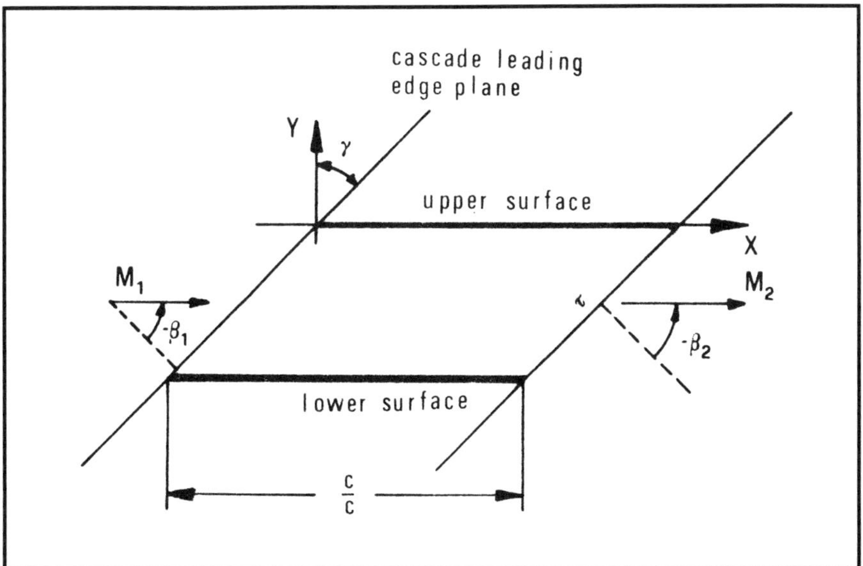

d = 0.0	c = 0.1 m	σ = variable
τ = 0.5->1.0	camber = 0°	γ = 0->63.4^o
i = 0°	M_1 = 0->1.5	α = 2.0^o
Vibration in pitch, (x_α, y_α)=(0.5,0.0)		k = 0.5, 1.0

Fig. 5a: *Eighth standard configuration: General cascade geometry [Bölcs and Fransson, 1986, p. 168].*

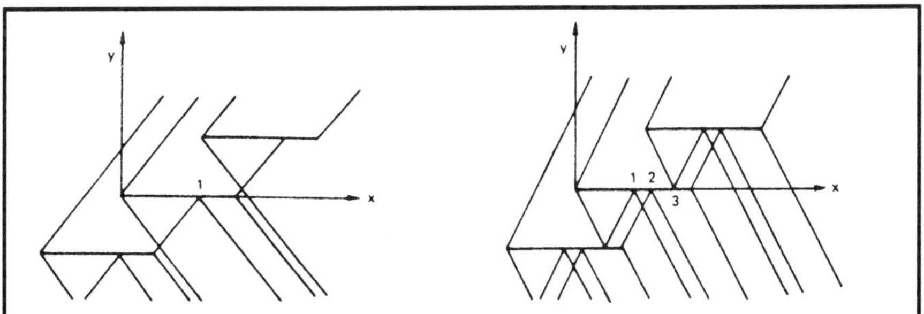

b: Verdon "Cascade A" c: Verdon "Cascade B"

Fig. 5b,c: *Eighth standard configuration: Geometry of cascades "A" and "B" from Verdon and McCune [1975, p.2]. Included are also the leading and trailing edge wave configurations for aeroelastic sample cases 43-54.*

Ninth standard configuration (double circular arc cascades at different flow conditions)

The ninth standard configuration is a continuation of the eighth, but includes both thickness and camber effects. The geometry of the blades was proposed by Dr. J. M. Verdon from United Technologies Research Center (Fig. 6). At the present time it is proposed to keep the originally defined 21 aeroelastic sample cases for further studies (Table 6). However, it has been noted by the calculations done up to now that the high reduced frequency (k=1.0) gave some unexpected numerical problems which should not be the main purpose of the investigation. The same configurations are thus proposed, as sample cases 22-42, at the lower reduced frequency of k=0.5.

Among others, Whitehead [1990], Li et al [1990], Huff [1989] and Verdon [1989a] have presented numerical results on unsteady flow through vibrating DCA cascades, and Buffum and Fleeter [1989a,b, 1988] and Giordano and Fleeter [1990] have presented experimental data.

The conclusions to be drawn are that promising results are obtained on this standard configuration, but that some unexplained effects still exist. The agreement between different prediction models is usually not as good as one would wish when the inlet flow Mach numbers are high or the blade thickness non-negligable.

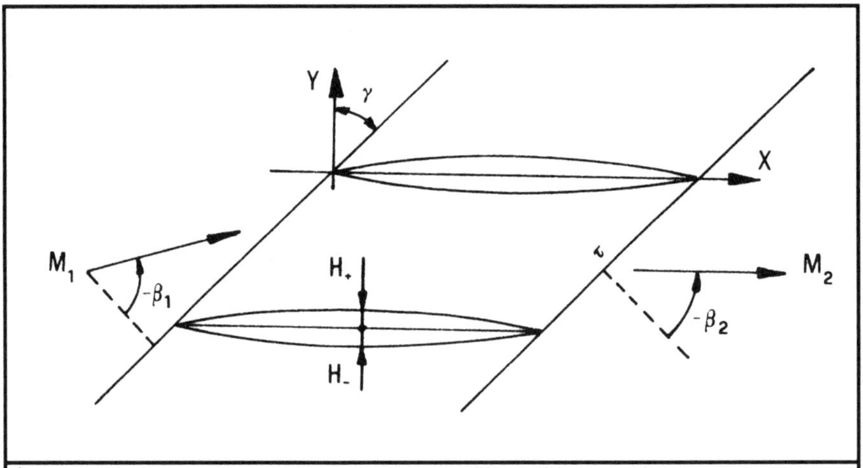

Symmetric/flat-bottomed circular arc profiles with equation:

$$y_\pm(x) = \mathrm{sgn}\,(H_\pm)\cdot\left\{|H_\pm| - R_\pm + \left[R_\pm^2 - (x-0.5)^2\right]^{0.5}\right\} \text{ if } H_\pm \neq 0$$

$$y_\pm(x) = 0 \quad \text{if } H_\pm = 0$$

$$R = \frac{\{H^2 + 0.5^2\}}{2|H|}$$

sgn (H) = ± 1 for H > 0 / H < 0;	$(\ )_+$ = upper surface $(\ )_-$ = lower surface	
Maximum thickness at x = 0.5	d = 0.01 - 0.1	c = 0.1 m
camber = 0° (for symmetric profiles)	τ = 0.75	γ = 45°, 60°
M_1 = variable (0.0 -> 1.5)	i = 0° (for M_1 < 1.)	α = 2.0°
Pitching (x_α, y_α) = (0.5, camberline)	k = 1.0	σ = 90°

Fig. 6: *Ninth standard configuration: Cascade geometry [Bölcs and Fransson, 1986, p. 180].*

Table 6: *Ninth standard configuration: 42 recommended aeroelastic sample cases [Bölcs and Fransson, 1986, p. 181].*

	Time-Averaged Conditions						Time-Dependent	
	M_1	γ (°)	β_1(°)	H_+ (-)	H_- (-)	d (-)	k (-)	σ (°)
1	0.0	60	-60	0.01	0.01	0.02	1.0	90
2	"	"	"	0.02	0.02	0.04	"	"
3	"	"	"	0.03	0.03	0.06	"	"
4	"	"	"	0.05	0.05	0.10	"	"
5	0.5	"	"	0.01	0.01	0.02	"	"
6	0.7	"	"	0.005	0.005	0.01	"	"
7	"	"	"	0.01	0.01	0.02	"	"
8	"	"	"	0.015	0.015	0.03	"	"
9	"	"	"	0.02	0.02	0.04	"	"
10	0.8	"	"	0.01	0.01	0.02	"	"
11	1.3	"	"	"	"	"	"	"
12	1.4	"	"	"	"	"	"	"
13	1.5	"	"	"	"	"	"	"
14	0.0	45	-45	"	"	"	"	"
15	0.5	"	"	"	"	"	"	"
16	0.7	"	"	"	"	"	"	"
17	0.8	"	"	"	"	"	"	"
18	0.5	"	-45	0.05	0.0	0.05	"	"
19	0.7	"	"	"	"	"	"	"
20	0.8	"	"	"	"	"	"	"
21	0.9	"	"	"	"	"	"	"
22	0.0	60	-60	0.01	0.01	0.02	0.5	"
23	"	"	"	0.02	0.02	0.04	"	"
24	"	"	"	0.03	0.03	0.06	"	"
25	"	"	"	0.05	0.05	0.10	"	"
26	0.5	"	"	0.01	0.01	0.02	"	"
27	0.7	"	"	0.005	0.005	0.01	"	"
28	"	"	"	0.01	0.01	0.02	"	"
29	"	"	"	0.015	0.015	0.03	"	"
30	"	"	"	0.02	0.02	0.04	"	"
31	0.8	"	"	0.01	0.01	0.02	"	"
32	1.3	"	"	"	"	"	"	"
33	1.4	"	"	"	"	"	"	"
34	1.5	"	"	"	"	"	"	"
35	0.0	45	-45	"	"	"	"	"
36	0.5	"	"	"	"	"	"	"
37	0.7	"	"	"	"	"	"	"
38	0.8	"	"	"	"	"	"	"
39	0.5	"	-45	0.05	0.0	0.05	"	"
40	0.7	"	"	"	"	"	"	"
41	0.8	"	"	"	"	"	"	"
42	0.9	"	"	"	"	"	"	"

Tenth standard configuration (Modified cambered NACA 0006 cascade at subsonic and transonic flow conditions)

The tenth Standard Configuration, included by proposal of Dr. J. M. Verdon at the United Technologies Research Center [1987], is a two-dimensional compressor cascade of modified NACA 0006 profiles (Fig. 7) that operates at subsonic inlet and exit conditions. The geometry is given by Verdon [1987] and is repeated here for convenience.

The cascade has a stagger angle, γ, of 45° and a gap/chord ratio, τ, of unity. The blades are constructed by superimposing the thickness distribution of a modified NACA four digit series airfoil on a circular arc camber line. The thickness distribution is given by:

$$T(x) = H_T \cdot \{2.969\, x^{0.5} - 1.26\, x - 3.516\, x^2 + 2.843\, x^3 - 1.036\, x^4\}, \quad 0 \le x \le 1 \qquad (1)$$

where H_T is the nominal blade thickness. The coefficient of the x^4 in eq. (1) differs from that used in the standard NACA airfoil definition (i.e., -1.015) so that the example blades will close in as wedge-shaped trailing edges.

The camber distribution is given by:

$$C(x) = H_C - R + \sqrt{\{R^2 - [x - 0.5]^2\}}, \quad 0 \le x \le 1 \qquad (2)$$

where H_C (>0) is the height of the camber-line at midchord and

$$R = \frac{\{H_C^2 + 0.25\}}{2\, H_C}$$

is the radius of the circular arc camber line. The surface coordinates of the reference blade are therefore given by:

$$[X,Y]_{\pm} = [x \mp 0.5\, T(x) \sin(\theta), C(x) \pm 0.5\, T(x) \cos(\theta)], \quad 0 \le x \le 1 \qquad (3)$$

where the signs + and - refer to the upper (suction) and lower (pressure) surfaces, respectively, and $\theta = \tan^{-1}(dC/dx)$. For the present example we set H_T=0.06 and H_C=0.05 to study the unsteady aerodynamic response of a vibrating cascade of cambered NACA 0006 airfoils.

We consider two different steady-state inlet operating conditions. In the aeroelastic cases 1-16 the inlet Mach number, M_1, and flow angle, β_1, are 0.7 and 55°, respectively, for cases 17-32, M_1=0.8 and β_1=-58° (see Table 7). The flow through the cascade is assumed to satisfy a Kutta condition at blade trailing edges and, therefore, only inlet flow information must be specified. For M_1=0.7 and β_1=-55°, the mean or steady flow through the cascade is entirely subsonic; for M_1=0.8 and

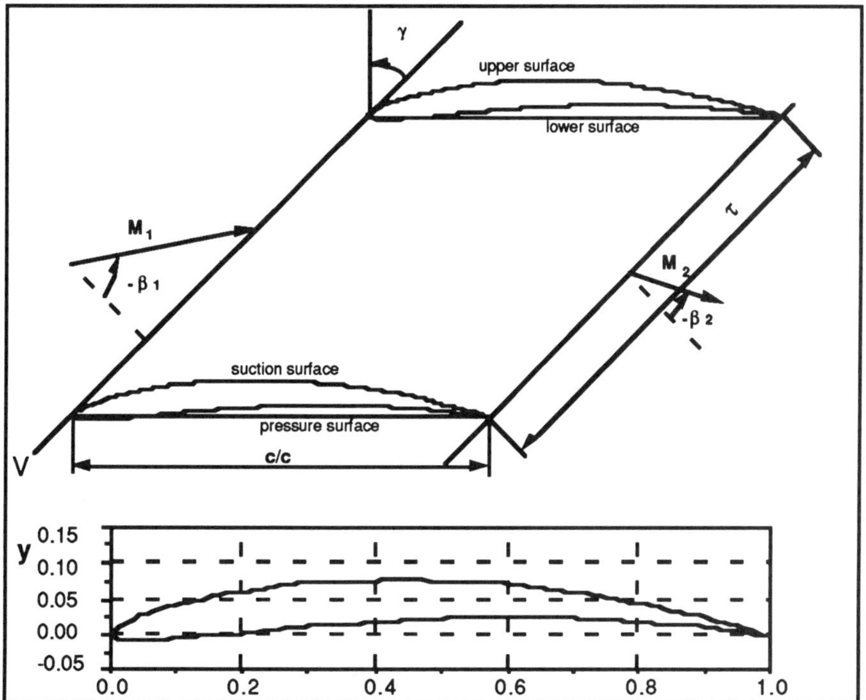

Stagger angle: $\gamma=45^{o}$	Gap/chord: $\tau=1.0$
Pitch axis: $(x_{\alpha},y_{\alpha})=(0.5,0.05)$	Camber: $\approx 22.8^{o}$
Reduced frequency: k = variable	Interblade phase angle: σ = variable

Fig. 7: Tenth standard configuration: Modified cambered NACA 0006 cascade.

$\beta_1=-58^{o}$ it is transonic with a normal shock occurring in each blade passage.

As aeroelastic test cases we consider single-degree-of-freedom blade heaving, normal to chord, and pitching motions at four different frequencies, k=0.25, 0.50, 0.75 and 1.0, and at interblade phase angles lying in the range $-\pi\leq\sigma\leq\pi$. The amplitude of the heaving motion, h, is 0.01; that of the pitching motion, α, is 2°. The blade pitching axis lies at midchord, i.e. $(x_{\alpha},y_{\alpha})=(0.5,0.05)$.

We are interested in the following aerodynamic response information for each of the two inlet operating conditions given, at reduced frequencies of k=0.25, 0.50, 0.75 and 1.0 (see Table 7 for details):

1: The time-averaged blade surface pressure coefficient, $\bar{c}_p(x)$ and Mach number.

Table 7: *Tenth standard configuration: 32 recommended aeroelastic sample cases (the pitch axis is located at (x_α, y_α)=(0.5,0.05)).*

Test case	Time averaged M_1 (-)	β_1 (°)	Time dependent α (°)	h (-)	σ (°)	k (-)
1	0.7	-55	2	---	0	0.25
2	"	"	"	---	90	"
3	"	"	"	---	0	0.50
4	"	"	"	---	90	"
5	"	"	"	---	0	0.75
6	"	"	"	---	90	"
7	"	"	"	---	0	1.0
8	"	"	"	---	90	"
9	"	"	---	0.01	0	0.25
10	"	"	---	"	90	"
11	"	"	---	"	0	0.50
12	"	"	---	"	90	"
13	"	"	---	"	0	0.75
14	"	"	---	"	90	"
15	"	"	---	"	0	1.0
16	"	"	---	"	90	"

Test case	Time averaged M_1 (-)	β_1 (°)	Time dependent α (°)	h (-)	σ (°)	k (-)
17	0.8	-58	2	---	0	0.25
18	"	"	"	---	90	"
19	"	"	"	---	0	0.50
20	"	"	"	---	90	"
21	"	"	"	---	0	0.75
22	"	"	"	---	90	"
23	"	"	"	---	0	1.0
24	"	"	"	---	90	"
25	"	"	---	0.01	0	0.25
26	"	"	---	"	90	"
27	"	"	---	"	0	0.50
28	"	"	---	"	90	"
29	"	"	---	"	0	0.75
30	"	"	---	"	90	"
31	"	"	---	"	0	1.0
32	"	"	---	"	90	"

2: a: Amplitude, $\tilde{c}_p(x)$, and phase lead angle, $\Phi_p(x)$, of the unsteady blade surface pressure coefficient for heaving and pitching motions at $\sigma=0^o$ and $\sigma=90^o$.

b: Amplitude, $\Delta\tilde{c}_p(x)$, and phase lead angle, $\Phi_{\Delta p}(x)$, of the unsteady blade surface pressure difference coefficient for heaving and pitching motions at $\sigma=0^o$ and $\sigma=90^o$.

3: a: The amplitude, $\tilde{c}_l$, and phase lead angle, Φ_l, of the unsteady lift coefficient per unit amplitude vs interblade phase angle for the heaving motions at $-\pi\leq\sigma\leq\pi$.

b: The amplitude, $\tilde{c}_m$, and phase lead angle, Φ_m, of the unsteady moment coefficient per unit amplitude vs interblade phase angle for the pitching motions at $-\pi\leq\sigma\leq\pi$.

4: The aeroelastic damping coefficient, Ξ, vs interblade phase angle for the heaving and pitching motions at $-\pi\leq\sigma\leq\pi$.

Usab and Verdon [1990], Verdon [1989b, 1987] and Whitehead [1990] have already presented results on this cascade. Results from other prediction models were also recently presented [Huff, 1991; Hall, 1991]. Some very promising results have been obtained (these will be included in the complete updated report presently in preparation [Fransson and Verdon, 1991]).

When comparing these results with each other it must be considered that no analytical results exist for this standard configuration. Only the mutual agreement between several similar theoretical methods can thus indicate the accuracy of the models.

Conclusions

A large amount of work has been put into the field of unsteady flow through vibrating blade rows in the period since 1987, both on the experimental and the theoretical side. The standard configurations for unsteady flow through vibrating axial flow turbomachine cascades, as originally defined, have been used by many researchers with the aim to compare prediction models against the data but also to analyze results for parametric studies. Unfortunately, due to the high computing costs involved in parametric studies with numerical time-dependent models, it is generally not possible to perform such investigations in order to better understand the physical reasons for aeroelastic effects on a special blade configuration.

Possibilities to predict aeroelastic phenomena have largely increased the last few years, to include aspects such as three-dimensionality and/or viscosity. In some cases these possibilities have overtaken the detailed experimental data available. As an example can be mentioned the prediction of time-dependent viscous flow (in most cases today still with steady-state turbulence models) on oscillating cascades, and the enormous difficulties to accurately measure enough details to be able to judge the quality of the viscous numerical predictions. As computers become more and more powerful, the computing cost will reduce simultaneously as the cost of an experimental investigation will increase. Comparisons between different predictions will thus probably increase in importance, at the expense of comparisons with experimental data. Comparisons between theoretical models, and with analytical solutions, are of large interest and necessity, as only through such efforts can the accuracy between the different models be assessed and the theoretical assumptions validated in detail. Only experimental test cases can, on the other hand, give information about the validity of all assumptions being made in the predictions. It is thus in the near future necessary to accurately measure, in detail, three-dimensional, transonic and viscous flow effects in both linear and annular cascades, and at realistic reduced frequencies. This is an extremely difficult and expensive task and all contributions in this regard will be warmly welcomed. In this procedure it must however not be forgotten that two-dimensional aeroelastic problems are far from understood, and that basic comparisons, such as against different flat plate analyses, are of extreme importance to establish the fundamental validity of any model.

Although great improvements are presently made in the field of numerical computations of unsteady flow, a general theory that can treat the subsonic through supersonic flow domain, on compressors, fans and turbines, including viscous effects is still very far away.

All the configurations discussed in the present study (and a large majority of all investigations performed on unsteady flow) consider rigid profiles, both in span- and chordwise directions. It is regularly shown by several authors that the strip theory assumption is of limited usefulness and that three-dimensional spanwise effects can be of large importance. Furthermore, as it in the future will be important to also consider higher modes than pure bending and pitching, coupled and chord-wise non-rigid blade vibrations at high frequencies is a large challenge.

Prediction of aeroelastic phenomena of the new generation of swept fan blades will be necessary. The use of composite blades will open up new aeroelastic challenges and will probably couple the unsteady aerodynamics and the structural mechanics parts in turbomachinery closer together than what is presently the case.

From the above it can be concluded that the data collected as the "Standard Configurations" will be of use also in the next future, while new experiments must be initiated for the validation of the next generation of prediction models. For these the exact treatment of unsteady shock waves and unsteady local separations seem to be the present most challenging theoretical parts before going into details about global separation.

Due to the success of the present comparative study, it has regularly been suggested that similar standard configurations should be defined also for gust response. This idea was specifically put forward during the panel discussion on the "Standard Configurations on Unsteady Flow Through Vibrating Axial-Flow Turbomachine Cascades" at the Aeroelasticity meeting in Notre Dame on September 15-19, 1991, through a suggestion that the same geometries as used for the cascades in the present study should be employed [Atassi, 1991]. This would help to avoid many problems, such as the previously mentioned difficulties in experimental blade coordinates, which by now to a large extent are known on the present standard configurations.

References

Atassi, H.; 1991
"Communication during the Panel Discussion", *Sixth International Symposium on Unsteady Aerodynamics, Aeroacoustics and Aeroelasticity of Turbomachines and Propellers, September 15-19, University of Notre Dame, USA, 1991.*
Bakhle, M. A.; Keith, T. G. K.; Kaza, K. R. V.; 1989
"Application of a Full-Potential Solver to Bending-Torsion Flutter in Cascades", *AIAA Paper 89-1386-CP, 1989.*
Bölcs, A; Fransson, T. H.; 1986
"Aeroelasticity in Turbomachines: Comparison of Theoretical and Experimental Cascade Results", *Communication du Laboratoire de thermique appliquée et de turbomachines, EPF-Lausanne, Switzerland, No. 13, 1986.*
Boldman, D. R.; 1991
"Private Communication"
Boldman, D. R.; 1983
"Private Communication"
Buffum, D. H.; Fleeter, S.; 1989a
"The Aerodynamics of an Oscillating Cascade in a Compressible Flow Field", *ASME Paper 89-GT-271, 1989.*
Buffum, D. H.; Fleeter, S.; 1989b
"Experimental Investigation of Transonic Oscillating Cascade Aerodynamics", *AIAA paper 89-0321, AIAA 27th Aerospace Sciences Meeting, January 9-12, Reno, Nevada, 1989.*
Buffum, D. H.; Fleeter, S.; 1988
"Investigation of Oscillating Cascade Aerodynamics by an Experimental Influence Coefficient Technique." *AIAA paper 88-2815, AIAA 24th Joint Propulsion Conference, July 11-13, Boston, Massachusetts, 1988.*
Carstens, V.; 1991a
"Calculation of the Unsteady Transonic 2D Cascade Flow for Aeroelastic Applications", *International Forum on Aeroelasticity and Structural Dynamics, Aachen, Germany, June 3-6, 1991.*

Carstens, V.; 1991b
"Communication during the Panel Discussion", *Sixth International Symposium on Unsteady Aerodynamics, Aeroacoustics and Aeroelasticity of Turbomachines and Propellers, September 15-19, University of Notre Dame, USA, 1991.*

Carta F. O.; 1984
"Private Communication"

Carta F. O.; 1982a
"Unsteady gapwise periodicity of oscillating cascaded airfoils", *ASME paper 82-GT-286, 1982.*

Carta F. O.; 1982b
"An experimental investigation of gapwise periodicity and unsteady aerodynamic response in an oscillating cascade (3 Volumes)", *NASA-CR 3513, 1982.*

Chiang, H.; Fleeter, S.; 1989
"Oscillating Incompressible Aerodynamics of a Loaded Airfoil Cascade", *AIAA Journal, Vol. 27, No. 4, pp. 446-454, 1989.*

Fleeter, S.; Hoyniak, D.; 1989
"Aeroelastic Detuning for Stability Enhancement of Unstalled Supersonic Flutter", *International Journal of Turbo and Jet-Engines, No. 6; pp. 17-26, 1989.*

Fransson, T. H.; Verdon, J. M.; 1991
"Updated Report on Standard Configuration for Unsteady Flow Through Vibrating Axial-Flow Turbomachine-Cascades", *Report Royal Institute of Technology, Stockholm, Sweden, to appear 1991.*

Gallus, H. E.; Kau, H.-P.; 1989
"Blade Loads Due to Unsteady Flow in Turbomachinery Cascades", *Proceedings of the Fifth International Symposium on "Unsteady Aerodynamics and Aeroelasticity of Turbomachines and Propellers", pp. 53-68, Beijing, China, September 18-21, 1989.*

Gerolymos, G. A.; Blin, E.; Quinio, H.; 1990
"Comparison of Inviscid Computations with Theory and Experiment in Vibrating Transonic Compressor Cascades", *ASME Paper, 1990.*

Giles, M.; Haimes, R.; 1991
"Validation of a Numerical Method for Unsteady Flow Calculations",*ASME Paper 91-GT-271, 1991.*

Giordano, D. D.; Fleeter, S.; 1990
"Investigation of Oscillating Airfoil Shock Phenomena", *AIAA Paper 90-0695, 1990.*

Hall, K. C.; 1991
"Communication during the Panel Discussion", *Sixth International Symposium on Unsteady Aerodynamics, Aeroacoustics and Aeroelasticity of Turbomachines and Propellers, September 15-19, University of Notre Dame, USA, 1991.*

He, L.; 1989
"An Euler Solution for Unsteady Flows Around Oscillating Blades." *ASME Paper 89-GT-279, 1989.*

Hoyniak, D.; 1991
"Communication during the Panel Discussion", *Sixth International Symposium on Unsteady Aerodynamics, Aeroacoustics and*

Aeroelasticity of Turbomachines and Propellers, September 15-19, University of Notre Dame, USA, 1991.

Huff, D. L.; 1991
"Communication during the Panel Discussion", *Sixth International Symposium on Unsteady Aerodynamics, Aeroacoustics and Aeroelasticity of Turbomachines and Propellers, September 15-19, University of Notre Dame, USA, 1991.*

Huff, D. L.; Reddy, T. S. R.; 1989
"Numerical Analysis of Supersonic Flow Through Oscillating Cascade Sections by Using a Deforming Grid", *AIAA paper 89-2805, AIAA/ASME/SAE/ASEE 25th Joint Propulsion Conference, Monterey, California, July 10-13, 1989.*

Huff, D. L.; 1989
"Numerical Analysis of Flow Through Oscillating Cascade Sections", *AAA paper 89-0437, AIAA 27th Aerospace Sciences Meeting, January 9-12, Reno, Nevada, 1989.*

Kau, H. P.; Gallus, H. E.; 1989
"Numerical Investigation of Unsteady Flow in Oscillating Turbine and Compressor Cascades", *AGARD Conference Proceedings No. 468, "Unsteady Aerodynamic Phenomena in Turbomachines", pp. 2.1-2.11, Luxembourg, August 28-30, 1989.*

Li, C.-C.; Messiter, A. F.; van Leer, B.; 1990
"Unsteady Transonic Cascade Flow with In-Passage Shock Wave", *AIAA Journal, Vol. 28, No. 6, pp. 1135-1138, 1990.*

Namba, M.; Toshimitsu, K.; 1990
"Effects of Chordwise Displacement and Nonrigid Section Deformation on Unsteady Aerodynamic Response of Subsonic and Supersonic Oscillating Cascades", *ASME Paper 90-GT-246, 1990.*

Platzer, M. F.; Carta, F. O.; Editors; 1987
"AGARD Manual on Aeroelasticity in Axial-Flow Turbomachines. Volume 1: Unsteady Turbomachinery Aerodynamics", *AGARD-AG-298, Vol. 1, 1987.*

Platzer, M. F.; Carta, F. O.; Editors; 1988
"AGARD Manual on Aeroelasticity in Axial-Flow Turbomachines. Volume 2: Structural Dynamics and Aeroelasticity", *AGARD-AG-298, Vol. 2, 1988.*

Riffel R. E.; Rohtrock M. D.; 1980
"Experimental determination of unsteady blade element aerodynamics in torsion mode cascade, final report", *NASA CR-159831, Vol. 1, 1980.*

Schläfli, D.; 1989
"Experimentelle Untersuchung der instationären Strömung in Oszillierenden Ringgittern" *PH. D. Thesis No. 800, EPF-Lausanne, 1989.*

Servaty, S. Gallus, H. E.; Editors; 1987
"Unsteady Aerodynamics and Aeroelasticity of Turbomachines and Propellers", *Proceedings of the Fourth International Symposium, Aachen, West Germany, September 6-10, 1987.*

Sidén, L. D. G.; 1991a
"Numerical Simulation of Unsteady Viscous Compressible Flows Applied to Blade Flutter Analysis", *ASME Paper 91-GT-203, 1991.*

Sidén, L. D. G.; 1991b
"Numerical Simulation of Viscous Compressible Flows Applied to Turbomachinery Blade Flutter", *Ph. D. Thesis, Chalmers University of Technology, Gothenbourg, Sweden, 1991.*
Széchényi E.; 1991
"Communication during the Panel Discussion", *Sixth International Symposium on Unsteady Aerodynamics, Aeroacoustics and Aeroelasticity of Turbomachines and Propellers, September 15-19, University of Notre Dame, USA, 1991.*
Széchényi E.; 1984
"Private Communication"
Széchényi E.; Finas R.; 1981a
"Aeroelastic testing in a straight cascade wind tunnel", *Communication de L'ITA/ EPFL, No. 10, pp. 143-149,1981.*
Széchényi E.; Girault J.-Ph.; 1981b
"A study of compressor blade stall flutter in a straight cascade wind-tunnel", *Communication de L'ITA/ EPFL, No. 10, pp. 163-169,1981.*
Topp, D. A.; Fleeter, S.; 1986
"Splitter Blades as an Aeroelastic Detuning Mechanism for Unstalled Supersonic Flutter of Turbomachine Rotors." *ASME Paper 86-GT-99, 1986.*
Usab, W. J.; Verdon, J. M.; 1990
"Advances in the Numerical Analysis of Linearized Unsteady Cascade Flows", *ASME Paper 90-GT-11, 1990.*
Verdon, J. M.; 1990
"Linearized Unsteady Aerodynamics for Turbomachinery Aeroelastic Applications", *AIAA Paper 90-2355, 1990.*
Verdon, J. M.; 1989a
"The Unsteady Flow in the Far Field of an Isolated Blade Row", *Journal of Fluids and Structures, No. 3, pp. 123-149, 1989.*
Verdon, J. M.; 1989b
"The Unsteady Aerodynamic Response to Arbitrary Modes of Blade Motion", *Journal of Fluids and Structures, No. 3, pp. 255-274, 1989.*
Verdon, J. M.; 1987
"Linearized Unsteady Aerodynamic Theory"", *AGARD Manual on Aeroelasticity in Axial-Flow Turbomachines, Vol. 1, Eds. M. F. Platzer and F. O. Carta, AGARD-AG-298, chapter 2, March 1987.*
Verdon J. M.; 1977a
"Further developments in the aerodynamic analysis of unsteady supersonic cascades (Part I of 2)", *ASME paper 77-GT-44, 1977.*
Verdon J. M.; 1977b
"Further developments in the aerodynamic analysis of unsteady supersonic cascades (Part II of 2)", *ASME paper 77-GT-45, 1977.*
Verdon J. M.; McCune J. E.; 1975
"Unsteady supersonic cascade in subsonic axial flow", *AIAA journal, Vol. 13, No. 2, pp. 193-201, 1975.*
Verdon J. M.; 1973
"The unsteady aerodynamics of a finite supersonic cascade with subsonic axial flow.", *J. of applied mechanics, Vol. 40, No. 3, pp. 667-671, 1973.*

Whitehead, D. S.; 1990
"A Finite-Element Solution of Unsteady Two-Dimensional Flow in Cascades", *International Journal fro Numerical Methods in Fluids, Vol. 10, pp. 13-34, 1990.*
Whitehead, D. S.; 1987
"Flutter of Turbine Blades", *Proceedings of the Fourth Symposium on Unsteady Aerodynamics and Aeroelasticity of Turbomachines and Propellers, pp. 437-452, Aachen, West Germany, September 6-10, 1987.*